Creatures of the Dark

The Nocturnal Prosimians

Creatures of the Dark

The Nocturnal Prosimians

Edited by

L. Alterman
Clark College
Dubuque, Iowa

Gerald A. Doyle
University of the Witwatersrand
Johannesburg, South Africa

and

M. Kay Izard
Duke University
Durham, North Carolina

Plenum Press • New York and London

Library of Congress Cataloging-in-Publication Data

Creatures of the dark : the nocturnal prosimians / edited by L. Alterman, Gerald A. Doyle, and M. Kay Izard.
p. cm.
"Proceedings of an International Conference on Creatures of the Dark: the Nocturnal Prosimians, held June 9-12, 1993, in Durham, North Carolina"--T.p. verso.
Includes bibliographical references and index.
ISBN 0-306-45183-2
1. Prosimians--Congresses. 2. Nocturnal animals--Congresses. I. Alterman, L. II. Doyle, Gerald A. (Gerald Anthony), 1931- . III. Izard, M. Kay. IV. International Conference on Creatures of the Dark: the Nocturnal Prosimians (1993 : Durham, N.C.)
QL737.P9C74 1995
599.8'1--dc20 95-43741
CIP

Proceedings of an International Conference on Creatures of the Dark: The Nocturnal Prosimians, held June 9–12, 1993, in Durham, North Carolina

ISBN 0-306-45183-2

A Division of Plenum Publishing Corporation
233 Spring Street, New York, N. Y. 10013

10 9 8 7 6 5 4 3 2 1

Printed in the United States of America

PREFACE

The papers in this volume are representative of those presented at a conference entitled "Creatures of the Dark: The Nocturnal Prosimians," held at Duke University, June 9-12, 1993. The purpose of the conference, attended by more than 100 scientists, was to assemble, for the first time ever, scholars from diverse fields with a common interest in the nocturnal prosimian primates. The history of the precursors of this meeting are outlined in the Historical Perspective by Doyle (this volume). Most of the invited papers are presented here in modified form, as are several papers originally presented as posters. Two papers are included that were not presented, due to scheduling conficts and health considerations. Some papers, delivered from the podium, are not included in this volume.

Interactions among conference participants resulted in many revisions to the contributions, as did the comments of reviewers and the editors. Several papers which contained new data or new interpretations of familiar phenomena met with constructive criticism, resulting in modification of the original papers. We thank all of the contributors for their patience and cooperation, and commend the numerous reviewers who generously donated their time and expertise.

We greatly appreciate funding from the National Institutes of Health, the Wenner-Gren Foundation for Anthropological Research, the Duke University Center for International Studies, the Duke University Primate Center, and Drs. Charles Putman of Duke University and Malcolm Gillis, currently of Rice Universtiy.

L. Alterman
G.A. Doyle
M.K. Izard

HOW PRIMATES INVENTED THE RAINFOREST AND VICE VERSA

Robert W. Sussman

Department of Anthropology
Washington University
St. Louis, MO 63130

INTRODUCTION

Fruiting trees in tropical forests depend for their existence on the animals that eat their fruit. Seeds falling below the parent plant are less likely to survive than those dispersed at a distance (Janzen, 1970; Clark and Clark, 1984; Howe et al., 1985). Primates, along with birds and bats, are the most important dispersers of seeds in the tropics (Howe, 1980, 1989; Terborgh, 1986; Fleming et al., 1987; Stiles, 1989). Of course, this was not always the case, and the evolutionary history of tropical rainforests and primates are intricately related. In fact, one might say they helped create one another through a long process of diffuse coevolution (Herrera, 1984).

EVOLUTION OF THE ANGIOSPERMS

Angiosperms first appeared approximately 120 million years ago (mya) during the Early Cretaceous Period. At this time, large gymnosperm forests consisting mainly of non-flowering conifers, cycads, and ferns were common throughout the earth (Tiffney, 1981; Crane, 1987; Upchurch and Wolf, 1987). Mammals and birds had just begun to appear. The herbivore community was dominated by huge dinosaurs that were ecological generalists (Bakker, 1978). The first angiosperms were small shrubs and fast growing, colonizing plants. They occurred mainly in unstable aquatic environments such as along river edges, or as small understory plants beneath the large dominant gymnosperms. The flowers of these small plants were pollinated by primitive unspecialized insects, and the seeds were very small and dispersed by the wind (Doyle and Hickey, 1976; Tiffney, 1984; Crepet and Friis, 1987; Friis and Crepet, 1987; Wing and Tiffney, 1987a).

Early angiosperms were able to establish themselves in disturbed habitats or in areas affected by fire because they possessed a number of life-history traits that allowed them to reproduce rapidly. These so-called "r selected" traits

Creatures of the Dark: The Nocturnal Prosimians
Edited by L. Alterman *et al.*, Plenum Press, New York, 1995

included small stature, rapid growth and development, and an ability to spread more quickly than conifers and other nonflowering plants (Doyle and Hickey, 1976; Niklas et al., 1980; Tiffney, 1984; Crane, 1987). These traits actually may have led to the initial evolution, spread, and success of early angiosperms because they were better adapted to handle the assaults of Cretaceous dinosaurs. On the other hand, herbivorous dinosaurs might have reinforced the r selected traits of early angiosperms by selecting against delayed reproduction and large, edible seeds, and by maintaining relatively open vegetation. Thus, dinosaurs likely played a key role in creating the environment for the mid-Cretaceous radiation of small shrubs and weedy varieties of flowering plants (Bakker, 1978, 1986; Coe et al., 1987; Wing and Tiffney, 1987a).

Just as browsing by large dinosaurs helped maintain the r selected forms of early angiosperms, the angiosperms reinforced the dominance of these large herbivores. This relationship probably remained stable for 40-50 million years (Coe et al., 1987; Wing and Tiffney, 1987a and b). It was only after the extinction of the large dinosaurs in the late Cretaceous that primitive angiosperms and some small, yet unknown, primate ancestors together helped create the first rainforests.

Undoubtedly, the extinction of dinosaurs dramatically altered the landscape in which flowering plants were evolving. By the late Cretaceous most of the archaic gymnosperms had also disappeared due either to the same event that led to dinosaur extinctions or because of direct competition with angiosperms (Tiffney, 1981; Wing and Tiffney, 1987a). With the demise of the dinosaurs many of the pressures that maintained the r selected features of the angiosperms also disappeared. The massive disturbances created by giant herbivores no longer occurred and flowering plants were able to grow more densely (Coe et al., 1987; Wing and Tiffney, 1987a). Denser plant communities would decrease the amount of light and energy reaching the ground, thus favoring the success of larger seeds with greater stored reserves. With reduced mortality, the necessity for rapid reproduction would decrease and the ability of individual plants to grow longer and larger would be enhanced. This, too, would allow plants to invest more resources into individual seeds. Generally, large trees tend to have large seeds and fruit (Silvertown, 1981; Tiffney, 1984). These factors taken together could have resulted in the rapid evolution of angiosperms with extended growth, large stature, and larger seeds (Tiffney 1984; Wing and Tiffney, 1987 a and b).

This is the historical and ecological context in which the ancestors of primates evolved. Cretaceous and most Paleocene mammals were small, nocturnal insectivores (Crompton and Jenkins, 1979; Collinson and Hooker, 1987). Those few that were inclined to feed on plant material could not, like the large herbivorous dinosaurs, eat the entire plant (Tiffney, 1984). Few birds were probably adapted to a diet of foliage. In fact, only two species out of the thousands alive today feed primarily on plants (Cracraft, 1973; Rasmussen, pers. comm.). Both mammals and birds would have been restricted to the high energy parts of plants such as flowers, seeds, and seed coverings. Though dinosaurs likely dispersed small seeds, the angiosperm seeds of the Cretaceous were insignificant food items to the giant dinosaurs, much like grass seeds are to cows

(Wing and Tiffney, 1987a). However, these and the larger, more nutritious seeds beginning to appear in the Paleocene would have been more attractive to small mammalian omnivores and herbivores (Tiffney, 1984; Wing and Tiffney, 1987b). Obviously, it would not be advantageous for a plant's seeds to be eaten by seed predators. Yet, since larger seeds could not be dispersed by the wind, it was beneficial to have animals disperse these heavier seeds. Angiosperms solved this dilemma by providing an attractive reward to potential dispersers: the hard, often toxic seeds were covered with a nutritious fruity cover (van der Pijl, 1982; Tiffney, 1984).

Plants with berries appeared by the late Cretaceous and early Paleocene, and fleshy fruits became relatively common. Although still rare, larger seeds and fruit began to appear. Some large flowering trees of stable, climax forests were present (Niklas et al., 1980; Tiffney, 1981, 1984; Upchurch and Wolfe, 1987; Wing and Tiffney, 1987a and b). This was the beginning of a change from the dominance of abiotic (such as wind) seed dispersal to an increase in the importance of animal or biotic dispersal agents (Tiffney, 1984). The diversity of angiosperms continued to increase across the Cretaceous-Paleocene boundary and into the Paleocene, coinciding with a major radiation of mammals and birds (Colbert, 1969; Lillegraven et al., 1979; Tiffney, 1984; Olson, 1985; Friis et al., 1987; Wing and Tiffney, 1987a and b).

EVOLUTION OF PRIMATES

During the Paleocene, a number of small, nocturnal mammals began to take advantage of the new niches created by the diversifying flowering plants. Probably first attracted to the insects that fed upon the leaves and reproductive parts of the plants, some of these small mammals were more or less tree dwelling and began to feed on the plant parts themselves (Kay and Cartmill, 1977; Szalay and Dagosto, 1980; Collinson and Hooker, 1987; Beard, 1990; Kay et al., 1990). The newly radiating angiosperms offered an array of previously unexploited resources including flowers, fruits, floral and leaf buds, gums, nectars, and also the insects that fed upon these nutritious items.

Among the newly radiating mammals were the plesiadapoids, ranging in size from small and mouse-like to as large as a domestic cat. Plesiadapoids shared a number of physical traits with modern primates. These traits include features of the skull, such as the petrosal bulla in some forms (but see MacPhee et al., 1983; and Kay et al., 1990), and of the postcrania, that imply some ability to climb about in the trees (Szalay and Dagosto, 1980; Beard, 1991). A number of plesiadapoids also had teeth that were similar to those of primates and were adapted for an herbivorous diet (eg. Simpson, 1935; Szalay, 1968, 1969; Gingerich, 1986; but see Wible and Covert, 1987; Martin, 1990).

Throughout the Paleocene, the number of mammals with adaptations for plant-feeding continued to increase (Collinson and Hooker, 1987). This is exemplified in the fossil record of the plesiadapoids. Like most early mammals, the first plesiadapoids were insectivorous. Adaptations for eating plants did not appear in this group until the mid-Paleocene, 6 million years later. By the late Paleocene, however, many genera of plesiadapoids were eating plant material and a wide range of plant-eating adaptations had evolved (Kay and Cartmill, 1977). The Paleocene saw the beginning of a new interdependence between angiosperms and mammalian seed dispersers and of a long-term obligatory relationship between these lifeforms. However, modern rainforests and modern primates had yet to appear on the scene.

EVOLUTION OF RAINFORESTS

Although there was a diverse array of large and small angiosperm seeds present in the Paleocene and the range of seeds was almost as great as that of modern floras, large seeds were still not abundant. Thus, there were still few seeds available for dispersal by mammals and birds. Adaptations for herbivory and arboreality continued to increase throughout this period but true mammalian herbivores were still few in number (Tiffney, 1984; Collinson and Hooker, 1987; Wing and Tiffney, 1987a and b). The coevolutionary spiral between angiosperms and their dispersers was beginning though, and, as a result, the rate of speciation in these forms increased (Tiffney, 1981; Wing and Tiffney, 1987a). However, this relationship did not reach fruition until the Eocene Epoch, approximately 55 mya.

If it is true that a major impetus for the diversification of certain mammals during the Paleocene was the concurrent radiation of angiosperms and the ability of flowering plants to occupy new adaptive zones, then we would expect certain groups to evolve more efficient means of exploiting these resources. At the same time, angiosperms would evolve better means of protecting themselves from predation and more efficient means of exploiting animals as dispersal agents. These coevolutionary interactions appear to have reached a threshold at the Paleocene-Eocene boundary (Hickey, 1981; Tiffney, 1981, 1984; Wing and Tiffney, 1987b; Upchurch and Wolfe, 1987).

By the beginning of the Eocene, angiosperms evolved a range of seed sizes as great as that found in modern forests and, for the first time, very large seeds appeared in large numbers. This dramatic increase in large seeds reflects the importance of animal dispersal. It also implies that angiosperms had an increased ability to form large, closed canopy forests. Indeed, large angiosperm trunks did not become common until the Eocene and trees producing large seeds dominated the forests by the mid-Eocene (Tiffney, 1984; Upchurch and Wolfe, 1987).

In addition to structural changes, a major rearrangement of taxonomic composition of the angiosperm community occurred across the entire Northern Hemisphere (Wolfe, 1975; Tiffney, 1981). The Paleocene-Eocene boundary saw the disappearance of many archaic taxa and the rise of a diverse array of modern families and genera of flowering plants. The result was the first angiosperm-dominated, closed-canopy tropical rainforest of modern aspect. Thus, modern evergreen tropical rainforests appeared and became widespread during the Eocene (Niklas et al., 1980; Tiffney, 1981, 1984; Upchurch and Wolfe, 1987).

Paralleling this modernization of the world's flora were major changes in the fauna. The fossil record indicates a major radiation of birds with the number of modern families increasing greatly in the Eocene, and fruit-eating birds making their first appearance at this time (Tiffney, 1984; Olson,

1985). Among the mammals, rodent, lagomorph, artiodactyl, and perissodactyl radiations began in the Eocene (Tiffney, 1984). Bats and bat-dispersed fruits also appeared early in this era (Tiffney, 1984). Along with these major changes is the appearance and radiation of our first primitive primate ancestors, the adapids and omomyids (for reviews, see Fleagle, 1988; Conroy, 1990; Martin, 1993). The evolution of these Eocene primates, as well as that of fruit bats and vegetarian birds, is directly related to the evolution of improved means of exploiting flowering plants. Furthermore, the extinction of the plesiadapoids at the end of the Paleocene is likely to be related to the rapid radiation of primates, bats, and rodents and their improved ability to exploit insects, fruits and flowers (Sussman and Raven, 1978; Maas et al., 1988).

We do not know which mammalian lineage, if any of those known from the fossil record, gave rise to primates during the late Paleocene (Covert, 1986; Wible and Covert, 1987)or earlier (Martin 1993). In contrast to Paleocene plant-eating mammals like the plesiadapoids, Eocene primates possessed many modern primate traits. These included divergent toes and thumbs with flattened nails producing effective grasping organs, and convergent, forward-facing eyes indicative of stereoscopic vision (Cartmill, 1992). Grasping hands and feet allowed the earliest primates far greater access to fruits and flowers, as well as to plant-visiting insects, and greater efficiency in locomoting and foraging in the small terminal branches of bushes and trees than plesiadapoid-like mammals (Cartmill, 1972, 1974, 1992; Sussman, 1991). Although birds and bats could reach the terminal branches of large rainforest trees by flying, primates needed their grasping appendages to obtain the same advantage. In fact, primates are one of the few taxonomic groups of non-flying animals to exploit regularly the terminal branch niche of the tropical forest.

Because their eyes are convergent and face forward like cats and some predatory birds, Cartmill (1992) believes that the earliest primates were visually-oriented predators. However, some animals that have these adaptations, such as sloths, koalas, and some phalangerid marsupials are strictly vegetarian. Many birds that do not have convergent eyes are highly insectivorous. Furthermore, most galagos, lorises and lemurs, are mainly frugivorous and use hearing as the main sense when they do hunt for prey (see Sussman, 1991). In fact, some fruit bats possess a complex visual system similar to that of primates (Allman, 1982; Pettigrew, 1986, 1989), which most likely is the result of convergent adaptations (Theile et al., 1991; Martin, 1993).

Both the earliest primates and fruit bats were active at night and were feeding on and manipulating items of small size, at very close range, under low-light conditions. This requires acute powers of discrimination and precise coordination. Tree squirrels, which do not have convergent eyes and have claws, cut large fruits and nuts from small branches and then move to large horizontal supports to feed (Garber and Sussman, 1984). Primates, in contrast, usually consume items where they are acquired. Most primates are able to venture out on very small supports, maintain a firm grasp with the hindlimbs, and detach food items with their mouth or forelimbs. Possibly, the visual adaptations of Eocene primates were related to the greater visual sensitivity needed to exploit small food items available on the newly diversifying angiosperms and the necessity of maneuvering on fine terminal branches in order to reach these resources (Rasmussen, 1990; Sussman, 1991; Cartmill, 1992).

The establishment of biological interactions between angiosperms and their dispersers is almost certainly reflected in the rapid appearance of modern families of plants and animals in the Eocene. As stated by Wing and Tiffney (1987a: p. 200):

> .... diversification of frugivorous mammals (and presumably birds) in the early Paleocene to early Eocene, and the increasing abundance of large angiosperm diaspores in the late Paleocene, indicate that the development of modern types of tetrapod/angiosperm dispersal interactions was concentrated in a 10 My interval from the latest Cretaceous to the earliest Eocene. The evolution of larger seeds in many lines of angiosperms, in conjunction with the coevolution of a mechanism allowing their dispersal, were probably two of the key developments that permitted angiosperms to form closed-canopy vegetation of the type found in many parts of the world today.

The evolution of primates must be seen in this context.

CONCLUSIONS

The evolution of Eocene primates parallels that of other herbivorous mammals, fruit-eating birds, and modern angiosperm rainforests. These organisms were linked in a tight coevolutionary relationship. At present, forest-dwelling primates along with frugivorous bats and birds are the most important seed dispersers in the tropics. Most likely, the ability of these animals to feed on a variety of items in the fine terminal branches of flowering plants was the most critical impetus for the major adaptive shift seen in Eocene primates. The first known primates, the Eocene adapids and omomyids, appear to be a more efficient version of the omnivorous and plant-eating Paleocene mammals, and were a product of a long coevolutionary interaction with the angiosperms.

ACKNOWLEDGEMENTS

I would like to thank Lon Alterman, Gerry Doyle, and Kay Izard for inviting me to participate in this volume and in the symposium in which these papers were first delivered. I also appreciate the support of the Duke University Primate Facility in this venture. A number of people provided comments during the formulation of this paper and/or on the manuscript. These include Lon Alterman, Matt Cartmill, Jim Cheverud, Glenn Conroy, Alwyn Gentry, Charles Hildebolt, Rob Hoffman, David Lorence, Jane Phillips-Conroy, Bob Martin, Tab Rasmussen, Peter Raven, Rich Smith, Ian Tattersall, and one anonymous reviewer. I appreciate their input.

REFERENCES

Allman, J.M., 1982, Reconstructing the evolution of the brain in primates through the use of comparative neurophysiological and neuroanatomical data, *in*: "Primate Brain Evolution," E. Armstrong, D. Falk, eds., Plenum, New York.

Bakker, R.T., 1978, Dinosaur feeding behavior and the evolution of flowering plants, *Nature*. 274:661-663.

Bakker, R.T., 1986, How dinosaurs invented flowers, *Nat. Hist.* 11:30-38.

Beard, K.C., 1990, Gliding behavior and palaeoecology of the alleged primate family Paromomyidae (Mammalia, Dermoptera). Nature. 345: 340-341.

Beard, K.C., 1991, Vertical postures and climbing in the morphotype of Primatomorpha: Implications for locomotor evolution in primate history, *in*: Origine(s) de la Bipédie chez les Hominidés (Cahiers de Paléoanthropologie), Editions du CNRS, Paris.

Cartmill, M., 1972, Arboreal adaptations and the origin of the order Primates, *in*: "The Functional and Evolutionary Biology of Primates," R. Tuttle, ed., Aldine, Chicago.

Cartmill, M., 1974, Rethinking primate origins, *Science*. 184:436-443.

Cartmill, M., 1992, New views on primate origins, *Evolutionary Anthropology* 1:105-111.

Clark, D.B. and Clark, D.A., 1984, Spacing mechanisms of a tropical rain forest, evaluation of the Janzen-Connell model, *Amer. Natur.* 124:769-788.

Coe, M.J., Dilcher, D.L., Farlow, J.O., Jarzgen, D.M., and Russell, D.A., 1987, Dinosaurs and land plants, *in*: "The Origins of Aniosperms and Their Biological Consequences," E.M. Friis, W.G. Chaloner, P.R. Crane, eds., Cambridge University Press, Cambridge.

Colbert, E.H., 1969, "Evolution of the Vertebrates a History of the Backboned Animals Through Time," Wiley, New York.

Collinson, M.E. and Hooker, J.J., 1987, Vegetational and mammalian faunal changes in the Early Tertiary of southern England, *in*: "The Origins of Angiosperms and Their Biological Consequences," E.M. Friis, W.G. Chaloner, P.R. Crane, eds., Cambridge University Press, Cambridge.

Conroy, G.C., 1990, "Primate Evolution," Norton, New York.

Covert, H.H., 1986, Biology of early Cenozoic primates. *in*: "Comparative Primate Biology Vol: 1: Systematics, Evolution, and Anatomy," D.R. Swindler, J. Erwin, eds., Alan R. Liss, New York. Crepet, W.L. and Friis, E.M., 1987, The evolution of insect pollination in angiosperms. *in*: "The Origins of Angiosperms and Their Biological Consequences," E.M. Friis, W.G. Chaloner, P.R. Crane, eds., Cambridge University Press, Cambridge.

Cracraft, J., 1973, Continental drift, paleoclimatology, and the evolution and biogeography of birds. *J. Zool. London* 169:455-545.

Crane, P.R., 1987, Vegetational consequences of angiosperm diversification, *in*: "The Origins of Angiosperms and their Biological Consequences," E.M. Friis, W.G. Chaloner, P.R. Crane, eds., Cambridge University Press, Cambridge.

Crompton, A.W. and Jenkins, F.A., 1979, Origin of mammals, *in*: "Mesozoic Mammals," Lillegraven, J.A., Kielan-Jaworowska, Z., Clemens, W.A. eds., University of California Press, Berkeley.

Doyle, J.A. and L.J. Hickey, 1976, Pollen and leaves from the mid-Cretaceous Potomac Group and their bearing on early angiosperm evolution, in: "Origin and Early Evolution of Angiosperms," C.B. Beck ed., Columbia Univ. Press, New York.
Fleagle, J.G., 1988, "Primate Adaptation and Evolution," Academic Press, New York.
Fleming, T.H., Breitwisch, R.L., and Whitesides, G.W., 1987, Patterns of tropical vertebrate frugivore diversity, Ann. Rev. Ecol. System 18:91-109.
Friis, E.M., Chaloner, W.G., and Crane, P.R., 1987, Introduction to the angiosperms, in: "The Origins of Angiosperms and Their Biological Consequences," E. M. Friis, W.G. Chaloner, P.R. Crane, eds., Cambridge University Press, Cambridge.
Friis, E.M. and Crepet, W.L., 1987, Time of appearance of floral features, in: "The Origins of Angiosperms and Their Biological Consequences," E.M. Friis, W.G. Chaloner, P.R. Crane, eds., Cambridge University Press, Cambridge.
Garber, P.A. and Sussman, R.W., 1984, Ecological distinctions between sympatric species of Saguinus and Sciurus, Amer. Jour. of Physical Anthropology 65:135-146.
Gingerich, P.D., 1986, Plesiadapis and the delineation of the Order Primates, in: "Major Topics in Primate and Human Evolution," B.A. Wood, L.B. Martin, P. Andrews, eds., Cambridge University Press, Cambridge.
Howe, H.F., 1980, Monkey dispersal and waste of a neotropical fruit, Ecology 61:944-959.
Howe, H.F., 1989, Scatter- and clump-dispersal and seedling demography: hypothesis and implications, Oecologia 79:417-426.
Howe, H.F., Schupp, E.W., and Westley, L.C., 1985, Early consequences of seed dispersal for neotropical tree (Virola surinamensis), Ecology 66:781-791.
Herrera, C.M., 1984, Determinants of plant-animal coevolution: the case of mutualistic dispersal of seeds by vertebrates, OIKOS 44:132-144.
Hickey, L.J., 1981, Land plant evidence compatible with gradual, not catastrophic, change at the end of the Cretaceous, Nature 292:529-531.
Janzen, D.H., 1970, Herbivores and the number of tree species in tropical forests, Amer. Natur. 104:501-528.
Kay, R.F. and Cartmill, M., 1977, Cranial morphology and adaptations of Palaecthon nacimienti and other Parmomyidae (Plesiadapoidea, ? Primates), with a description of a new genus and species, J. of Hum. Evol. 6:19-53.
Kay, R.F., Thorington, R.W. and Houde P., 1990, Eocene plesiadapiform shows affinities with flying lemurs not primates, Nature 345:342-344.
Lillegraven, J.A., Kielan-Jaworowska, Z. and Clemens, W.A. (eds.), 1979, "Mesozoic Mammals," Univ. California Press, Berkeley.
Maas, M.C., Krause, D.W., and Strait, S.G., 1988, Decline and extinction of plesiadapiforms in North America: Displacement or replacement. Paleobiology, 14:410-431.
MacPhee, R.D.E., Cartmill, M., and Gingerich, P.D., 1983, New Palaeogene primate basicrania and the definition of the order Primates. Nature 301:509-511.
Martin, R.D., 1990, "Primate Origins and Evolution: A Phylogenetic Reconstruction," Princeton University Press, Princeton.
Martin, R.D., 1993, Primate origins: plugging the gaps. Nature 363:223-234.

Niklas, K.J., Tiffney, B.H., and Knoll, A.H., 1980, Apparent changes in the diversity of fossil plants: a preliminary assessment, *in*: "Evolutionary Biology Vol. 12," M.K. Hecht, W.C. Steere, B. Wallace, eds., Plenum, New York.

Olson, S.L., 1985, The fossil record of birds, *in*: "Avion Biology, Vol. 3," D. Farner, J. King and J.K. Parkes, eds., academic Press, New York.

Pettigrew, J.D., 1986, Flying primates? Megabats have the advanced pathway from eye to midbrain, *Science* 231:1304-1306.

Pettigrew, J.D., 1989, Phylogenetic relationships between microbats, megabats and primates (Mammalia: Chiroptera and Primates), *Phil. Trans. Soc. London* 325:489-559.

Pijl, L. van der, 1982, "Principles of Dispersal in Higher Plants," Springer-Verlag, Berlin.

Rasmussen, D.T., 1990, Primate origins: lessons from a neotropical marsupial, *Amer. J. Primatol.* 22:263-277.

Silvertown, J.W., 1981, Seed size, life span, and germination date as coadapted feature of plant life history, *Am Naturalist 118:860-864*.

Simpson, G.G., 1935, The Tiffany fauna, Upper Paleocene. 2. Structure and relationships of Plesiadapis, *Am. Mus. Novit.* 816:1-30.

Sussman, R.W., 1991, Primate origins and the evolution of angiosperms. *Am. J. Primatol.* 23:209-223.

Sussman, R.W. and Raven P.H., 1978, Pollination by lemurs and marsupials: an archaic coevolutionary system, *Science* 200:731-736.

Stiles, E.W., 1989, Fruits, seeds, and dispersal agents, *in*: "Plant-Animal Interactions," W.G. Abrahamson, ed., Mcgraw-Hill, New York.

Szalay, F.S., 1968, The beginnings of primates, *Evolution* 22:19-36.

Szalay, F.S., 1969, Mixodectidae, Microsyopidae, and the insectivore- primate transition. *Bull. Am. Mus. Nat. Hist.* 140:193-330.

Szalay, F.S. and Dagosto, M., 1980, Locomotor adaptations as reflected on the humerous of Paleogene primates, *Folia Primatol.* 34:1-45.

Terborgh, J., 1986, Community aspects of frugivory in tropical forests, *in*: "Fruguvores and Seed Dispersal," A. Estrada, T.H. Fleming, eds. Dr. W. Junk, Dordrecht.

Thiele, A., Vogelsang, M., Hoffmann, K.-P., 1991, Pattern of retinonectal projection in the megachiropteran bat Rousettus aegyptiacus, *J. Comp. Neurol.* 314:671-683.

Tiffney, B.H., 1981, Diversity and major events in the evolution of land plants, *in*: "Paleobotany, Paleoecology, and Evolution, Vol. 2," K.J. Niklas, ed., Praeger, New York.

Tiffney, B.H., 1984, Seed size, dispersal syndromes, and the rise of the angiosperms: evidence and hypothesis, *Ann. Missouri Bot. Gard.* 71:551-576.

Upchurch Jr., G.R. and Wolfe J.A., 1987, Mid-Cretaceous to Early Tertiary vegetation and climate: evidence from fossil leaves and woods, *in*: "The Origins of Angiosperms and Their Biological Consequences," E.M. Friis, W.G. Chaloner and P.R. Crane, eds., Cambridge University Press, Cambridge.

Wible, J.R. and Covert, H.H., 1987, Primates: Cladistic diagnosis and relationships, *J. Hum. Evol.* 16:1-22.

Wing, S.L. and Tiffney, B.H., 1987a, The reciprocal interaction of angiosperm evolution and tetrapod herbivory, *Review of Paleobotany and Palynology* 50:179-210.

Wing, S.L. and Tiffney, B.H., 1987b, Interactions of angiosperms and herbivorous tetrapods through time, *in*: "The Origins of Angiosperms and Their Biological Consequences," E.M. Friis, W.G. Chaloner and P.R. Crane, eds., Cambridge University Press, Cambridge.

Wolfe, J.A., 1975, Some aspects of plant geography of the northern hemisphere during the Late Cretaceous and Tertiary, *Ann. Missouri Bot. Gard.* 62:264-279.

"VISUAL PREDATION," HABITAT STRUCTURE, AND THE ANCESTRAL PRIMATE NICHE

Robin Huw Crompton

Department of Human Anatomy and Cell Biology
The University of Liverpool
PO Box 147
Liverpool L69 3BX UK

INTRODUCTION

Many of the characters of the basic suite of primate adaptations have at one time or the other been ascribed to arboreality. Frederick Wood Jones (1916) proposed that tree climbing lead to selection for a grasping forelimb, and Grafton Elliott Smith (1924) that grasping feet and hands were an adaptation for agility in a complex three-dimensional environment. Smith argued that these developments lead to replacement of the sense of smell by the senses of vision and touch, and consequent reduction of the snout. Collins (1921) and Le Gros Clarke (1959) noted a link between frontally-directed, convergent orbits and stereoscopic vision, which, Le Gros Clarke suggested, was advantageous for estimation of distances in arboreal leaping. But Cartmill (1972; 1974a & b; 1992) has challenged this "arboreal hypothesis" (Howells, 1947), indicating that the primate facility in arboreal movement *per se* could be matched, or even exceeded, by animals such as squirrels, with clawed, convergent digits, enhanced sense of smell, and laterally-facing orbits. He showed that clawed cheiridia actually offer locomotor advantages on large diameter vertical supports over clawless, grasping cheiridia (Cartmill, 1972a), and noted that "Despite their laterally directed eyes (and presumed lack of stereoscopy) squirrels of several genera may leap 13 to 17 body lengths from tree to tree, which compares favorably with the 20 body lengths reported for the saltatory lemuroid *Propithecus verreauxi*" (Cartmill, 1974b).

Cartmill (1974b) argues that the existence of such counter-examples invalidates the arboreal hypothesis: "If the primate evolutionary trends have not been characteristic of other lineages of arboreal mammals, we may conclude that there is something wrong with the arboreal theory in its received form, and any explanation of the primate trends must involve a more detailed description of the habitus of the ancestral primate" (Cartmill, 1972). Admitting that, despite the squirrels' abilities for arboreal life *per se*, squirrels cannot support themselves on fine branches for the extended periods possible in small mammals with grasping feet and hands, so that "grasping feet and prehensile tails permit their possessors to exercise more caution in crossing narrow gaps between supports" (Cartmill 1974a), he asks "why do not prosimians, cebids and cattharrines have hands and feet like a marmoset's, combining the best features of both adaptive extremes?" He gives three alternative answers.

The first possibility is that the typical primate traits are primitive retentions from the ancestral mammal. This he rejects, in the absence of adequate data from fossils, primarily on the basis of the comparative anatomy of sensory organs and of the hands and feet. The second is what Cartmill (1974a) calls the "Vertical Clinging and Leaping Theory" of Napier & Walker (1967). Their paper defined a new category of locomotor behavior, typified by clinging to vertical supports, and a strongly saltatory locomotion, which is common amongst the Malagasy lemurs, galagos and tarsiers. Among these primates Napier and Walker found a set of characteristics of the musculoskeletal system in common, which could be functionally related to the common behavioral traits. The authors appeared to suggest that this behavior was part of the basic adaptive suite of early primates: in "...all the postcranial bones reputedly assigned to Eocene prosimians...all the skeletal characters point to these animals being Vertical Clingers and Leapers" (Napier & Walker, 1967, quoted by Cartmill, 1974a). They suggested that the locomotor behavior arose as a means of predator avoidance. Now, Napier, Cartmill notes, had elsewhere suggested (Napier, 1967) that during the Eocene, certain primates began to specialize on eating fruits and leaves, which occur at the ends of fine branches in the peripheral canopy. However, "the limitations imposed by the vertical clinging and leaping habit upon feeding behaviour...are quite marked; the larger forms, especially, are at a disadvantage when feeding in a small branch milieu" (Napier & Walker, 1967, quoted by Cartmill, 1974a). Cartmill therefore observes, among other objections, that the latter statement implies that these prosimians are better adapted to clinging on large vertical supports, which "would be an excellent reason for retaining...clawed, squirrel-like hands and feet" (Cartmill, 1974a). The Vertical Clinging and Leaping (VCL) adaptation, he implies, cannot therefore be used to explain the *loss* of claws and the development of grasping hands and feet.

Cartmill (1972; 1974 a & b; 1992) thus presents a third alternative: the "visual predation hypothesis". He observed that almost all non-primates that have forward-directed eyes, such as owls, hawks and cats, are "visually directed predators" (Cartmill, 1972). While, despite their arboreality, squirrels show few if any adaptive similarities to primates, cats, although terrestrial, share with primates a high degree of orbital frontality and convergence, and their brains show close similarities to those of primates. Some of these, such as the common existence of "binocular depth cells" in the striate cortex, are specifically related to stereoscopic vision. Thus, Cartmill suggested, "optic convergence and stereoscopy in primates have the same adaptive significance they have in cats" (Cartmill, 1974b). Parallels to the basic primate traits in the locomotor system, Cartmill suggests, are to be found in small marsupials such as *Cercartetus* and *Burramys*, and in chameleons, where "grasping extremities represent a predatory adaptation, permitting prolonged and stealthy locomotion on slender terminal branches in pursuit of insects" (Cartmill, 1974b). Amongst primates, he reminds us, the greatest degree of orbital convergence is to be found not in the galagos, but in lorises, which do not leap, but "move with silent deliberation." Similarly, morphology in other predatory prosimians, didelphids and chameleons facilitates similar visually guided predation." (Cartmill, 1974a). The earliest primates, Cartmill (1972) suggests, were "shrub-layer insectivores", like *Cercartetus*, and the living primate representatives of this *habitus* include *Loris*, *Nycticebus*, *Phaner*, *Microcebus*, *Galago* and *Tarsius*. Such primates "carefully stalk and strike suddenly with their prehensile hands" (Cartmill, 1972). The early primate evolutionary trends are represented in the extreme by the lorisines, which are "more highly specialized than any other living strepsirhine for the mode of life whose adoption lead to the differentiation of the order Primates from the other placentals" (Cartmill, 1972). Cartmill, however, favors the much less specialized *Microcebus* as a model for the *habitus* of ancestral primates, where "a shift in feeding habits, from nosing out forest-floor insects to spying and snatching insects among terminal branches, appears to account for the specializations distinctive of all Neogene primates" (Cartmill, 1972).

While a major competing hypothesis, that of Szalay (1972), which suggested that primates were derived from an arboreal, scansorial, tree-shrew-like archontan ancestor by an early adaptive shift towards herbivory, is now generally rejected (Martin, 1990; Cartmill, 1992), Cartmill's "visual predation hypothesis" has had a signal influence on the development of thinking on primate origins. It is still generally accepted as the most likely explanation of the basic suite of primate characteristics. However, in the twenty-two years since Cartmill's first (1972) paper on the "visual predation hypothesis", it has become clear that comparative evidence on the pattern of occurrence of orbital frontality, which Cartmill employed in support of his hypothesis, is inconsistent and inconclusive. The anatomy of squirrels, it appears, does not support Cartmill's (1972) argument against the traditional depth-perception view of orbital frontality quite as strongly as he originally suggested. Squirrels lack truly laterally-facing orbits, but simply have much *less* frontality and *less* visual field overlap than do primates (40°-60° [Cartmill, 1974b; Allmann, 1977] versus 120-140° overlap in *Galago*). Histological evidence from both the lateral geniculate body and visual cortex suggests binocular integration and even stereoscopic vision may indeed occur (Allman, 1977). Equally, of course, it is now clear that orbital frontality is *not* a prerequisite for "life in the trees". Birds such as the woodcock, *Scolopax rusticola*, with eyes more caudally than rostrally placed, so that only monocular cues are available, can still fly rapidly in forest environments (Walker, pers. comm.) although woodcocks are largely restricted to open forest, where branches are not densely packed, and will avoid undergrowth (Bruun, 1972).

If the relationship between orbital frontality and depth perception in an arboreal environment is confused, that between orbital frontality and visual predation is confounded. Some visually-guided predators have *laterally* directed eyes, as Cartmill (1992) himself pointed out, such as mongooses, treeshrews and robins (both European robins, *Erithecus rubecula*, and American robins, *Turdus migratorius*). There are also cases where orbital frontality is found in a non-predator: one such is the frugivorous oilbird, *Steatornis caripensis* (Pettigrew and Konishi, pers. comm. to Allman, 1977), and some degree of optical convergence is also found in fruitbats (Cartmill, 1992).

Cartmill found, in his reassessment of the "arboreal" theory, that the existence of "counter-examples" indicated that it was not arboreality *per se* that lead to the basic suite of primate characters. In the same way, the above, and other, exceptions to the generalizations embodied in the "visual predation hypothesis" appear to suggest that neither was it "spying and snatching insects among terminal branches" *per se*. Probably, there is a substantial amount of truth in both hypotheses, but what is required is, indeed, a "more detailed description of the *habitus* of the ancestral primate" (Cartmill, 1972). I hope that the following reconsideration of Cartmill's hypothesis, in the light of aspects of prosimian biology pertinent to our understanding of the ancestral primate and ancestral simian niche, may contribute a little to that goal. The aspects on which I shall concentrate are: the niche and evolution of tarsiers; the leaping adaptation; and the relationship between leaping locomotion, habitat structure and diet. First, however, it is necessary to review some aspects of the inter-relationship between stereoscopic vision, arboreality, and primate dietary and locomotor adaptations.

STEREOPSIS, ARBOREALITY AND VISUAL PREDATION

Since the time when Le Gros Clarke (1959) wrote about the relationship between orbital frontality, depth perception and arboreality, we have learned that the perception of depth does not absolutely depend on binocular stereoscopic vision (*stereopsis*). Parallax, perspective and shading can all act as monocular cues to depth and provide some three-dimensionality of vision to people chronically blind in one eye. In fact, recent work suggests that graded depth perception can be achieved when the target is visible only in one eye (Liu *et al.*, 1994); monocular depth cues can thus not only give qualitative depth cues but at least some quantitative depth information. It is common knowledge, however, that monocular depth perception is still not the equal of binocular stereopsis. As Kerr & Forrester (1976) remind us, few people could manage to thread a needle using monocular vision. An analogous comparison can be made between the display created by standard 3-D graphics workstations, which provide monocular three-dimensional information by calculating true perspective and shading, and the much greater illusion of solidity provided by virtual-reality headsets. The latter, which present the left and right eyes respectively with the left and right components of a stereopair, provide the visual cortex with stereoscopic cues with which to create a "real" 3-D visual world.

Stereopsis does not normally depend on the presence of the monocular depth cues (Allman & McGuinness, 1988). Depth perception is achieved in the V3 and V5 (middle temporal, MT) areas of the primary visual cortex, which receive projections from layers IVa, IVb, IVβ and IVcα of region V1 of the primary visual cortex (Hubel & Wiesel, 1977; Allman 1977; Allman & McGuinness, 1988). One system in particular, the "magno-IVb-thick-stripe-MT" system has been linked to stereoscopic depth perception. It is sensitive to motion, but less informative about stationary targets, and color blind (Walker, pers. comm.) Matching of luminance-defined edges in both eyes has been regarded as the basis of binocular depth perception (Julesz, 1971), and it has been argued that the neural mechanism for binocular stereoscopic reconstruction is based on estimation of the "Characteristic Disparity" between the output of two overlapping visual fields (Wagner & Frost, 1993). Wagner and Frost regard Characteristic Disparity as analogous to "Characteristic Delay", the neural counterpart of "Interaural Time Difference", which owls use to reconstruct depth information from binaural sound data. Full binocular stereopsis, as opposed to monocular depth perception, does, then, seem to involve comparison of neural disparity between two extensively overlapping fields of vision, and does, therefore, require orbital frontality. However, excessive approximation of the two eyes will *at some point* actually impair the accuracy of reconstruction of the stereo image by excessively reducing the angle between the two axes of vision.

Cartmill (1972; 1974a & b; 1992) remarks: "Primate-like approximation of the orbits increases visual field overlap, but decreases parallax, reducing the distance over which visual field disparities can provide distance clues. In a leaping animal, selection should act against the extreme orbital approximation seen in tarsiers and higher primates. This expectation is borne out by a comparison of lorises with galagos; the slow moving *Loris* and *Nycticebus* have more convergent and closely approximated orbits than the saltatory galagos, where wider interorbital space allows stereoscopic ranging over greater distances." (Cartmill, 1974b). Thus, he argues that optic convergence "must have evolved in animals that needed a wide field of stereoscopic vision *at close range*" (Cartmill, 1992, italics added). Allman (1977) came to a similar conclusion, recalling Le Gros Clarke's (1924) comment that tarsiers seize prey with the hands before killing them, and, indeed, providing an illustration of a tarsier using such a predatory snatch as Cartmill regards as the key to primate orbital convergence.

Cartmill (1972; 1974a & b; 1992) seems however to have missed work by Julesz (1971) which suggested that the role of stereoscopic vision is not so much depth perception *per se*, which can be achieved using monocular vision, as the detection of "crypsis" (objects, such as prey, camouflaged against a background). This would be a major enhancement to the value of orbital frontality for a visually guided predator. However, hawks and eagles probably have less orbital frontality than do most primates, and can detect prey from a great distance. Their eyes, however, are still eminently adapted to the detection of cryptic prey, and must certainly be able to measure substantial distances accurately, as a falcon has little possibility for altering the length of its dive once committed to the "stoop". Accurate stereoscopic reconstruction is *not,* therefore, incompatible with accurate perception over a considerable distance. However, owls, which have greater orbital frontation, when hunting in woodland, tend to fly near ground level, indicating that the detection of cryptic prey at great distances may not be possible in nocturnal species.

While Cartmill (1992) himself acknowledged that the non-universality of orbital frontality in visually guided predators is a major flaw in his argument, he welcomed as a "patch" for this flaw an observation of Allman & McGuinness (1988). These authors pointed out that a particular visual advantage of orbital frontality accrues to *nocturnal* predators (eg. cats, owls and tarsiers). In vertebrate eyes, images projected onto the retina by the center of the lens are well focused, but those passing obliquely through the peripheral lens are relatively blurred. Predators with laterally facing orbits thus cannot clearly see objects which are immediately ahead of their muzzle, *unless* the iris is "stopped down". While "stopping down" the iris is possible for diurnal predators, such as hawks and eagles, it is not an option for nocturnal predators, which thus tend to have eyes which are rotated forwards so that the central part of the lens can be employed. Such "stopping down" also, of course, increases the depth of field.

A high degree of orbital frontality, and hence of binocular vision, may thus have evolved in *nocturnal* predators, such as felids and owls, as a means of prey detection, providing both enhanced scotopic acuity and stereoscopic perception of cryptic targets. But, once again, comparative data does not universally support this otherwise attractive idea. Charles-Dominique (1977) describes the prey capture of *Galagoides* demidovii* and other Makoukou galagines thus: "The bushbabies move around very rapidly by a mixture of running and leaping. The resulting commotion in the foliage often provokes insects to flee, and they are then exposed to auditory or visual localization." In his experiments, he found that *Galagoides demidovii* and *Euoticus* were consistently capable of localizing and following mobile insects through a (plywood) barrier. These findings would suggest that auditory cues may be as important as vision for prey capture in these galagines. Further, he states that "In captivity, a cricket can escape detection in litter on a cage-floor as long as it remains immobile...but a bushbaby will spot it as soon as there is the slightest movement..." Charles-Dominique's observations, albeit anecdotal, suggest that galagos, at least are actually *not* well-equipped for visual detection of cryptic insects, being insensitive to immobile prey, although very sensitive to moving prey. We have noted that at least one stereoscopic perception system, the "magno-4b-thick-stripe-MT" system, is also insensitive to immobile targets. These observations, together, tend to negate a hypothesis that the visual detection of cryptic prey *alone* lead to the development of orbital convergence and binocular stereopsis in ancestral primates.

Cartmill (1972; 1974a & b) argued that not only stereoscopic vision, but also manual dexterity, are adaptations to "visually directed predation" upon insects by ancestral primates, among fine branches in the shrub layer. While the pygmy possum, *Cercartetus,* and chameleons, which Cartmill (1974b) gave as examples of non-primates convergent with primates in locomotor adaptations, do hunt in this zone; cats, hawks and owls, Cartmill's parallels for convergence in the visual apparatus, do not. Cats and owls at least both take primarily terrestrial prey, as does the Bornean tarsier (Niemitz, 1984a & b; Crompton, 989; Crompton & Andau, 1986). However, some viverrids *do* sometimes hunt in the "fine branch niche", although, like squirrels, lacking grasping extremities, their locomotor capabilities in fine branches are limited. Such viverrids all have frontally directed

---

*I do not personally subscribe to the generic separation of *Otolemur* and *Galagoides* from *Galago*, as I believe it inappropriate to split genera without very good reason. However, at the request of the editors, I have agreed to conform to this division in this paper, for the sake of consistency with other contributors.

orbits,and all are nocturnal. They include the African palm civet *Nandinia binotata* in West Africa (Charles-Dominique, 1977; Cartmill 1992); the large-spotted genet *Genetta tigrina* in Southern Africa; and in South-East Asia, the banded linsang, *Prionodon lisang*, and the binturong, *Arctictis binturong* (pers. observ.). These, however, are not specialist predators, but mixed animalivores/frugivores. ("Insectivores" amongst both the Carnivora and Primates nearly all take small vertebrates as well as insects, so "animalivores" is a more accurate term.) For Cartmill, however, the forward-facing orbits of these viverrids have little to do with their current adaptations; their frontally-facing orbits were "probably inherited from visually predatory ancestors" (Cartmill, 1992). This might be true. However, oilbirds, which were cited by Allman (1977) as an exception to the universality of visual predation among animals with frontally-directed eyes, have a high degree of orbital frontality, and are apparently both totally nocturnal and totally frugivorous (BBC: The Natural World, April 1994). In the total darkness of caves, Trinidadian oilbirds use audible echo-location clicks to aid perception in extreme low light conditions, as do some fruitbats. Like fruitbats, too, oilbirds are dependent on food resources located primarily in the fine branches of the peripheral canopy of trees. Are the frontally directed eyes of oilbirds also to be attributed to a visually predatory ancestor?

Cartmill (1974b) compared the adaptation of the last common ancestor of the primates to that of "*Tarsius, Microcebus, Loris, Arctocebus* and smaller galagines". All of these are small, and all are nocturnal. As Martin (1990) indicates, nocturnality is the norm in mammals under 5 kg. All are under Kay's upper threshold of 1 kg for habitual insectivory (Kay, 1984), and all subsist at least partially on insects. As Martin (1990) remarks, all share (at a low frequency of occurrence) a locomotor performance called "cantilevering" (Walker, 1967; 1974) which is closely associated with insect-catching behavior (Charles-Dominique, 1977; Crompton, 1980; Crompton & Andau, 1986; Oxnard *et al.*, 1990) and which is usually accompanied by a stylized two-handed grab of the prey. However, while "cautious movement" ("stealth") is certainly important for the foraging strategy of the lorisines, rapid, active motion ("speed") is characteristic of the foraging of the galagines (Walker, 1969), as it is for mouse lemurs and tarsiers. The connection, proposed by Cartmill (1974b) between grasping hands and "cautious, well-controlled movements in pursuit of prey on slender supports" can thus only be made for the lorisines. Further, while vision is indeed the primary close-range means of prey detection in *Tarsius*, we have seen that in galagos, audition may be equally significant. Yet again, as Cartmill (1974a) himself observes in a footnote, in lorisines, olfactory cues may be of particular importance.

Cartmill (1974a & b) noted that facility of extended arboreal motion in squirrels and tree shrews is largely confined to large branches and tree trunks. However, an assumption that clawed digits only permit motion on large trunks, and should always be lost in species which exploit the fine branch niche, is no longer tenable. Cartmill (1992) cites Garber (1980) on *Saguinus oedipus*, which, despite their possession of secondary claw-like nails, "chiefly feed on insects among the twigs and vines of the forest understorey" (Cartmill, 1992) while the same "claws" offer them the advantage of access to large tree trunks for gummivory. Cartmill (1992) does not make it clear whether, on the basis of Garber's data he therefore rejects his earlier view that grasping hands permit "cautious, well controlled movements in pursuit of prey on slender supports". To do so, under the "visual predation hypothesis" would leave the close-range predatory act alone responsible for both grasping hands and frontally directed orbits. However, any limitations of *S. oedipus*' "clawed" hands for motion in the fine branch zone also apply, presumably, in the visually-directed manual snatch of prey. Thus, on the basis of this argument, loss of claws in ancestral primates cannot be attributed to *either* "cautious motion in pursuit of prey on slender supports" or the predatory act itself.

However, it would be a mistake to exaggerate the locomotor capabilities of clawed primates. We know already, from the viverrid case, that claws do not prevent hunting in the periphery of trees. The question is one of resolution in our definition of "fine branches". Garber's (1980) paper did not subdivide supports under 2.5 cm diameter, and a 2.5 cm support can hardly be called a "fine branch", and certainly not a "twig". *S. oedipus* do not show anything like the speed, facility or variety of locomotion of *Microcebus* or *Galagoides demidovii* when moving on branches and twigs less than 1 cm, and even more less than 0.5 cm, in diameter (pers. observ. on captive *S. oedipus* and *Galagoides demidovii*, and on captive and free-ranging *Microcebus*). The situation can only be resolved by further detailed field studies of these species with more resolution in estimation of support diameters.

The possession of claw-like structures (keeling) occurs together with sap and gum feeding on large tree trunks in at least two prosimian lineages. These are Galaginae (eg. *Euoticus* [Charles-Dominique, 1977] and *Otolemur crassicaudatus* [pers. observ.]) and Cheirogaleidae (*Phaner*, Tattersall, 1982). In *Daubentonia*, which feeds on grubs and insects on tree trunks and large branches (Tattersall, 1982), fully developed secondary "claws" are found. We can conclude that a strong and very specific adaptive link exists between the two characters, large branch/tree trunk feeding; and development of keeled nails or full secondary claws, and that access to food resources available on large substrates is more important to some species than any disadvantage to their locomotion in the fine branch niche.

Size, diet, nocturnality, and the commonality of a rather infrequent prey capture method (cantilevering) thus link the group of animals which Cartmill (1972; 1974a & b; 1992) uses as a model for the common primate ancestor. However, while stereopsis and acuity of night vision may well have played a major role in the dietary adaptation of early primates, "visual predation" in this group of animals is so diverse with respect to foraging strategy, locomotion and habitat that it does not provide a satisfactory basis on which *alone* to account for the familiar suite of primate adaptations.

Charles-Dominique & Martin (1970) and Martin (1972) reflect the distinctions in foraging strategies and locomotion I have noted, by excluding the slow-climbing gracile lorisines from their model for the common primate ancestor. Their living representatives of this model are limited to *Microcebus* and *G. demidovii*. More recently, however, Martin (1990) has, like Cartmill, retained *Tarsius* in the list, and proposes, as a primate ancestor, "a small-bodied, fully arboreal creature with grasping feet and a pattern of hindlimb domination, specifically adapted for active locomotion in the fine-branch niche". Indeed, Martin (1990) believes that his "image of the ancestral primate fits quite closely with that of a small-bodied, visually oriented insect predator advanced by Cartmill". The two models have in common body size and the "fine branch niche". But Martin's (1990) model is not closely similar to Cartmill's. "Cautious" movement in the "fine-branch niche" has become "active locomotion" in the same niche; grasping hands have become grasping feet, and instead of an hypothesis in which predation is central, Martin's model is one in which a role is also played by fast movement in a complex, three-dimensional environment.

The majority of cheirogaleines and galagines are mixed insectivores/frugivores/gummivores, rather than specialist animalivores (Martin, 1990), rather like the viverrids we have mentioned above. Moreover, the larger galagos and dwarf lemurs, at least, are *opportunistic* animalivores: animal prey is taken when possible, but foraging strategy is governed rather by the distribution of fruit and gums (Crompton, 1980 *et seq.* and *cf.* Rasmussen, 1990). The smallest living prosimians, *G. demidovii* and *Microcebus*, however, take a higher proportion of animal food than do dwarf lemurs and larger galagos (Charles-Dominique, 1977; Tattersall, 1987; Martin, 1990), and are more restricted to the "fine branch niche" (Oxnard *et al.*, 1990). The "fine branch niche" is usually the arboreal zone in which flowers and fruit are most available, but these flowers and fruits also attract insects. In this zone, made up of the peripheral canopy of trees and of undergrowth, discontinuities are many, and so, fairly obviously, flying insects are likely to be more abundant than flightless insects. Flying insects tend to have a relatively random dispersion pattern. Therefore, a randomly-oriented arrangement of closely packed small supports offers a predator the best access, in the maximum volume of space, in which to find them.

Sussman (this volume) presents a model in which primate visual and prehensive abilities evolved in the Cretaceous, in parallel with the appearance of trees bearing small colored fruit. This model is to some extent a more sophisticated revival of Polyak's (1957) interpretation of the origins of color vision, and represents a modification of earlier work (Sussman & Raven, 1978; Sussman, 1991) in which (according to Cartmill, 1992) Sussman had suggested that the basic suite of primate adaptations arose as an adaptation to a diet of flowers and nectar. "These adaptations would have allowed the Eocene prosimians far greater access to fruits and flowers as well as to many plant-visiting insects, making them much more efficient at locomoting and foraging in the small terminal branches..." (Sussman & Raven 1978, quoted in Cartmill, 1992). Cartmill (1992) objects that the dentitions of early primates are not those of dedicated nectar-eaters, but this criticism does not apply to the more generalized diet discussed in the phrase he himself quotes (above). Sussman's later paper (1991) suggests a role for suspension in the origin of grasping extremities, in allowing safe access to fruits in the terminal branches. As Cartmill (1992) admits, comparative evidence from the marsupial *Caluromys* (Rasmussen, 1990) suggests that grasping *feet* may assist in movement on fine branches in search of fruit, while insects are taken opportunistically, being located visually and grabbed with the *hands*. Cartmill (1992) objects to Sussman's link of grasping extremities to frugivory on the grounds that the grasping hallux of ancestral marsupials predates the diversification of fruit-bearing angiosperms. This objection would not, of course, apply if the grasping hallux evolved in response to leaf and bud eating. Cartmill therefore returns to close-range predation: "It makes sense that having a wide field of sharp stereoscopic vision directly in front of it would help a visually predatory animal be sure of hitting its prey accurately on the first strike. However, such specializations are hardly needed when the prey is a banana." (Cartmill, 1992). The humor tends to obscure the fact that the fruit taken by small-bodied nocturnal prosimians are not usually large fruit like bananas, but small, *and often very inconspicuous.*

If monocular vision can supply information on depth, and arboreality *per se* does not require extensive overlap of the two visual fields, can there be any other role for orbital frontality than the provision of short-range stereoscopic vision for accurate manual strikes on prey immediately in front of the predator, and/or the detection of cryptic prey? Sussman's model points to one possibility: the detection of small fruit in the terminal branches. This is of particular interest if we adopt the Charles-Dominique & Martin (1970) model of the primate ancestor, a small-bodied insectivore/frugivore like *Galagoides demidovii* or *M. murinus* rather than a specialist animalivore like *T. bancanus*. The requirements of a nocturnal predator for visual detection of crypsis are surely not very different from the visual requirements of a nocturnal frugivore or nocturnal leaper in the fine branch niche. In all three cases, a small target (prey, fruit or branch) must be recognized as a discrete entity at a certain distance from a highly complex background. This small target is often obscured by objects in front of it, and shadowing effects (ignoring color, as we are only concerned with nocturnal primates) act to break up the outline. Locomotor requirements are thus as well served as are dietary requirements, in the fine-branch niche, by the development of stereopsis and scotopic acuity.

It is clear from all the above that an understanding of the relationship between locomotion and habitat structure is an essential component of reconstruction of the *habitus* of the common primate ancestor. It was Napier (1963) who first saw the link between habitat structure and locomotion, discussing in particular the different pattern of continuity or discontinuity typical of the various forest strata, and relating this to the locomotor behavior appropriate for movement within them. Brachiation and leaping, he indicated, are two ways in which primates cross discontinuities in the supports available to them.

Tarsiers, galagos and mouse lemurs, as well as the larger Malagasy lemurs and indriids, all use leaping as a means of crossing discontinuities, but the larger species also tend to show some propensity for suspensory behavior (Napier & Walker, 1967). VCL locomotion, they argued, offers further benefits for predator avoidance, and the universality of VCL-related features in the limb anatomy of Eocene primates suggested to these authors that VCL may represent the ancestral locomotor condition for primates.

However, studies by Stern & Oxnard (1973) and Oxnard *et al.* (1990) among others have indicated that the VCL category contains at least three distinct morphological and behavioral groups: an indriid group, linked to *Hapalemur* and *Lemur catta*, which latter species also have links with a group containing *Varecia, Eulemur* and the Cheirogaleinae together with the galagos, (with the exception of *Galago alleni* and *G. moholi*); and a group containing the latter two galagos together with the tarsiers. The lemur-cheirogaleine-galagine group can be regarded as representing animals morphologically unspecialized for leaping, while the indriid group and the tarsier-galago group are both morphologically specialized, but in different ways: by elongation of the femur in the indriid group, and by elongation of the foot in the galago-tarsier group.

Further, Martin (1972) has suggested that the ancestral primate was unlikely to have been a specialized or committed leaper like a tarsier or an indri. Rather, it was simply hindlimb-dominated, with a locomotor pattern like that of *Microcebus* and *Galagoides demidovii*. In these species, no preference for support orientation is exhibited, and movement is more or less equally by leaping, walking and climbing (Crompton, 1980; Crompton *et al.*, 1987). This form of locomotion is typical of many small, primitive mammals (see Preuschoft *et al.*, this volume). In the same way that, according to Gebo (1986), Napier & Walker indicated that VCL features may have been preadaptive for both quadrupedalism and brachiation, Rollinson & Martin (1981, quoted in Gebo, 1986) then envisioned this small, nocturnal, generalized ancestral primate evolving into a medium-sized, diurnal, quadrupedal simian with a preference for horizontal supports. Gebo (1986; 1988) found that all living strepsirhines and the Eocene Adapidae share derived conditions of the foot, which rule out the latter as anthropoid ancestors. This implies, in contradiction to Rollinson & Martin (1981) that a common mouse lemur-galago ancestor may already have become too derived to represent a good locomotor model for the anthropoid ancestor. In contrast, Gebo argued that living anthropoids, tarsiers and omomyids share features of the foot, some of which can be regarded as representing the primitive primate condition, and some which represent the ancestral simian condition, so that a common haplorhine-omomyid ancestor might be a better model for some aspects of the ancestral primate, as well as the ancestral simian. However, Gebo's study does not address the type of locomotor behavior such ancestors exhibited, or the habitat in which they lived.

Thus, in reconstruction of the ancestral primate niche, a contrast between the relationships between locomotion, habitat and diet in leaping specialists, such as the tarsier, and the relationships obtaining in less specialized leapers such as the mouse lemur, seems to be at the heart of the question. In the second half of this paper, I present data from field and experimental studies which I believe is highly pertinent to this contrast.

TARSIER EVOLUTION: TERRESTRIALITY OR PHYLETIC DWARFISM?

Recent discoveries of fossils which appear to have close relationships to simians from the Eocene, and even the Palaeocene, have forced a reappraisal of the phylogenetic relationships of Eocene primates and extant taxa (Martin, 1993). The apparent absence of some haplorhine features in the omomyids, together with much earlier dates for possible tarsioids, has in particular encouraged fresh interest in the possibility of close tarsier-simian links. In this context, Martin (1990) drew attention to work by Jouffroy *et al.* (1984) on the lower limb of *Tarsius*. They found that the "extreme" length of *both* the fore and hind limbs in relation to the spine distinguish *Tarsius* from all other leaping prosimians. Further, they indicated that the extreme shortness of the cuboid in *Tarsius* leads to foot postures which are parasagittal, like those of *terrestrial* mammals, not laterally deviated, like those of other, arboreally adapted, leaping prosimians. Along with this feature (the significance of which has been disputed by Gebo [1986; 1988]) they argued that the character of bony tibiofibular fusion also links tarsiers to terrestrially adapted mammals, such as rabbits and elephant shrews. Martin, (1990) therefore suggested that the common simian-tarsier ancestor may well have been terrestrial, or rather, to some degree adapted to exploitation of the ground (Martin, pers. comm.) Further, he argued that the VCL locomotor complex must have evolved independently at least three times, in lemurs, bushbabies, and tarsiers. The last common ancestor of these three would not have been a committed leaper, but would have merely displayed generalized, hind-limb dominated locomotion, as an inhabitant of the "fine-branch niche".

Tarsiers, alone among nocturnal prosimians, lack a tapetum, presumably making up for the loss of light reflected back onto the retina by the sheer size of the eye and the number of rod cells which the retina can thus contain. The tarsier retina is also unusual amongst nocturnal prosimians in possessing a fovea. If tarsiers are indeed closely related to the basal simian, they may well be secondarily nocturnal. Further, as Martin (1990) pointed out, tarsiers are remarkably small in comparison to other haplorhines. Haplorhines have a modal body weight of around 5 kg, which is also a general mammalian threshold for diurnality. The modal body weight of nocturnal prosimians, however, is 500 g. However, the the range of body weights in tarsiers is even lower, 50-140 g. The logic of the argument seems to lead to the conclusion that tarsiers are not only secondarily nocturnal, but also secondarily small, in short, as Martin (1990) suggests, phyletic dwarfs.

The possible role of phyletic dwarfism in the evolution of tarsiers was first suggested by Crompton (1989), on the basis of locomotor, dietary, and biogeographic evidence. Firstly, tarsier locomotion appears to be a more "extreme" expression of the behavioral differences between the small-bodied *Galago moholi*, and the large bodied *O. crassicaudatus*, at the season when these differences are most clearly expressed. In the dry season, when its secondary diet of gums is less available, *G. moholi* utilizes the near-ground sapling trunk zone, and its locomotion displays high frequencies of leaping between these vertical supports, and leaping down to the ground to catch insects. In *T. bancanus*, however, this behavior is typical all year round. The behavioral differences between the two galagos were more extreme when the locomotion of juveniles of each species was compared (Crompton, 1983). I argued (Crompton, 1984) that size-related differences between the two galago species in dietary energetics had brought about contrasts in their foraging strategy and hence, ultimately, their current distinctions in locomotor behavior and limb anatomy. This further lead me to the conclusion (Crompton, 1989) that size-related heterochronic changes after a speciation event might have been the mechanism for locomotor differentiation of the two galagos. Thus, by extension, *Tarsius*' behavior could be seen as the product of similar processes.

Secondly, the four then-recognized species of living tarsier: *T. bancanus*, *T. syrichta*, *T. spectrum* and *T. pumilus*, I argued, appear, on the basis of our rather limited current knowledge, to form a size gradient. The newly recognised *T. dianae* (Niemitz *et al.*, 1991) also appears to fit on this gradient, between *T. syrichta* and *T. spectrum*. Within this gradient, I showed that contrasts in social behavior, habitat use and locomotion between the two best known species, the largest: *T. bancanus* and the second smallest: *T. spectrum*, are consistent with the hypothesis that *T. bancanus* is relatively stenotopic, and K-selected, while *T. spectrum* is relatively eurytopic, and r-selected. (*T. bancanus* is solitary, while *T. spectrum* is pair living; *T. bancanus* is confined to a narrower vertical range than *T. spectrum* and *T. bancanus'* locomotion is much more dominated by leaping than is the locomotion of *T. spectrum*.) R-selection and eurytopy generally correspond to a depauperate (species-poor), relatively unstable environment, while K-selection and stenotopy are usually associated with stable, diverse environments (MacArthur & Wilson, 1967; Pianka, 1970). The rather dry lowland forest of Sulawesi, the habitat of *T. spectrum*, is indeed depauperate and unstable compared to the lowland evergreen rainforest of Borneo, the habitat of *T. bancanus* (Groves, 1976). The montane forest habitat of *T. pumilus* might be expected to be even more depauperate.

All things being equal, a larger animal should have a larger home range. *T. bancanus* (Crompton & Andau, 1987) does indeed have a much larger home range size than that of either *T. spectrum* (MacKinnon & MacKinnon, 1980) or *T. dianae* (Tremble *et al.*, 1993). This would also be expected from a contrast in diet: *T. bancanus'* diet comprises only large invertebrates, while a great range in size is reported for the invertebrates taken by *T. spectrum* (MacKinnon & MacKinnon, 1980). Home range sizes are also positively correlated with "Average Daily Metabolic Needs" (Harvey & Clutton-Brock, 1981), so we might expect *T. bancanus* to have relatively large total energy requirements. Jablonski & Crompton (1994) have argued that *T. bancanus'* diet of large invertebrates is itself a response to the high cost of locating and catching isolated insect prey, and maximizes the relative reward for a successful hunt. Large invertebrates are not abundant in tropical rainforest, so that a large home range, and an efficient means of traversing it, are in turn required in order to locate adequate prey. Niemitz's (1984a) suggestion that *T. bancanus* (primarily because of the costs of thermoregulation) operates on very tight energy balances supports the apparent existence of a requirement for energetic efficiency. He further suggests (Niemitz *et al.*, 1984) that as a response to this challenge, torpor, with lowered body temperatures, occurs during the day, and the intrascapular brown fat body is used on waking each evening as an quickly-available source of energy. However, McNab & Wright (1987) found no evidence of torpor in the next largest tarsier, *T. syrichta*, and Crompton & Andau (1987) note that considerable heat will be produced during locomotion, which may offset thermoregulation costs.

Although total energy costs should be relatively high in *T. bancanus*, basal metabolic energy costs, and particularly the costs of thermoregulation, should be higher for the smaller tarsiers, and a behavioral/dietary response favouring a faster food turnover would be expected. *T. bancanus* appears to use caecal bacteria to digest chitin, and has long gut passage times for an animalivore (Jablonski & Crompton, 1994). One would predict that the smaller tarsiers would not utilize such low-quality foodstuffs, which are perhaps forced on *T. bancanus* by the rarity of edible fruits in the dipterocarp forests of Borneo. At present, we lack the data on diet, locomotion and home range in the smaller tarsiers which would enable us fully to analyze intrageneric energy relationships, and the consequences of smaller body size. However, the argument that tarsiers have undergone phyletic dwarfism seems well supported by the existence of size-related distinctions among the living species, although some of these might possibly represent the consequences of an *increase* rather than a *decrease* in size.

LEAPING, INSECTIVORY AND VISUAL PREDATION

Leaping locomotion is an inherently expensive means of travel, since it is non-cyclic, and no possibility exists for conservation of energy between jumps. Small bodied leapers like *Tarsius, Microcebus, Galagoides demidovii* and *Galago moholi* face the problem of the absolute shortness of their limbs, which reduces the amount of time during which they can apply force to the ground before take-off (Bennet-Clark, 1975; 1977; Emerson, 1985). They thus have to exert very high forces over the short period available (Demes & Günther, 1989). Bennet-Clark has suggested that internal elastic energy stores must exist in order for small leapers like *Galago moholi* to make leaps of the distances recorded. However, data on kangaroo rats (Biewener *et al.*, 1981) indicates that there is little possibility for storage of energy in the tendons of saltatory animals of this size, and our recent work (unpublished research by Alexander & Crompton) has confirmed this for *G. moholi*.

Moreover, using kinematic analysis, we have shown (Crompton *et al.* 1993; Sellers & Crompton, 1994) that of six prosimians studied, all of which include a substantial element of leaping in their locomotor repertoire: *M. murinus*, *Mirza coquereli*, *Cheirogaleus medius*, *O. garnettii*, *L. catta*, and *G. moholi*, only the last species takes off at the energy-efficient angle of 45° at all jump distances (and see Treff, 1970). *T. bancanus* also appears to use a 45° takeoff angle (Peters & Preuschoft, 1984). While the longest, or, energetically, the cheapest jumps can be performed at a takeoff angle of 45°, lower angles result in much higher kinetic energy costs. The curve for kinetic energy cost versus angle of takeoff is relatively flat near 45°, so that small differences in angle do not greatly increase the kinetic energy cost, but the relationship between the metabolic and kinetic energy cost of jumping (Walton & Anderson, 1988; Fedak *et al.*, 1982) is such that the metabolic energy involved in a jump might be 20 times greater than the kinetic cost, and the "mechanical efficiency" some 5%. An increase of kinetic output of 1 joule (J), which would result from a takeoff angle 15° flatter (or steeper) than 45°, would thus require a minimum 20 J increase in metabolic input. For *T. bancanus*, (the only species for which we have suitable data) which makes a minimum 1,417 jumps a night, averaging 1.12 meters (Crompton & Andau, 1986), the increase in energy demand arising from a 15° flatter or steeper trajectory would be some 28 kilojoules. This represents some 15% of the average daily energy *intake* of 190 kilojoules, and nearly half of the *available* energy of 63 kJ as estimated by Niemitz (1984a). How could the considerable increases in energy demand needed for flat trajectories be tolerated by kinematically unspecialized leapers such as cheirogaleines?

Leaping, optimized for distance, is not a fast means of locomotion. It typically occurs at speeds around 4 m/s, compared with ground speeds of 12-15m/s which can be achieved by cursorial quadrupeds (Günther *et al.*, 1991). But for *G. moholi* leaping remains of far greater importance than quadrupedalism, making up 672 versus 173 meters in every kilometer of travel. Hopping contributes a further 55 meters (Crompton, 1980, 1984) to each kilometer. It seems likely that such a high frequency of leaping is a consequence of an inertial conformation of the body segments of *G. moholi* which appears to limit the performance increases that are possible in quadrupedal locomotion (Smith, 1987). We have shown (Crompton *et al.*, 1993) that *G. moholi*'s actual maximum performance is less than half that predicted by regressions (Lindstedt *et al.*, 1991) of $^{V}0_2$ measurements (Taylor *et al.*, 1982) against body size for 20 mammalian species. Where speed is important, quadrupedalism is not an option for *G. moholi* (and by extension for *Tarsius*).

Further, leaping offers a means of crossing discontinuities (Napier, 1963) and hence, an escape from detours (Crompton, 1980; 1984; Crompton & Andau 1986). Where detours would involve vertical climbing, energy costs and musculoskeletal loads incurred would be very high, (Preuschoft *et al.*, 1992) since body movements are performed directly against the gravity vector. Indeed, in both *T. bancanus* and *G. moholi*, height change is accomplished by jumping, while climbing occurs most often in the context of "detailed foraging", or active searching for food (Crompton, 1980, 1984; Crompton & Andau, 1986). Climbing, in *G. moholi*, is statistically associated with fine supports arranged at low angles. Such supports are found in the "fine branch niche", but lie in the periphery of trees, rather than in undergrowth. We have noted that this zone is at least seasonally rich in flowers, fruit and flying insects.

The habitual performance of high trajectory leaps, however, requires an habitat open enough for safe ballistic leaping. We have shown (Crompton *et al.* 1993) that ballistic, energetically efficient leaps are also slow. Low angle, inefficient leaps, on the other hand, are fast. Flat trajectories also permit passage though closed, tangled habitats (Charles-Dominique, 1977). Ballistic leaps would inevitably be easier for predators (usually raptors) to predict, and are thus the dangerous option, particularly in the open habitat where they can occur: the low sapling-trunk zone immediately above the ground. Thus, for leapers like

M. murinus, M. coquereli and *C. medius*, inefficient, energy expensive leaps may be tolerable because they allow a safe but fast means of moving in dense habitats.

There may also be mechanical reasons why leapers such as *Microcebus* do not adopt energy-efficient trajectories, but tolerate the higher musculoskeletal forces involved in low, fast jump trajectories. Body size alone does not account for the differences observed in takeoff trajectories: the smallest and the largest species in our study, *M. murinus* and *L. catta*, both leap at less than the energetically optimal angle. Theoretical analysis of segment inertial properties (Peters & Preuschoft, 1984; Preuschoft, 1989) indicates that not all small jumpers (and probably not all large jumpers) are the same. Mechanically, the closer body segments are kept to the centre of gravity (which is usually in the pelvic region), the less the overall torque tending to rotate the body towards the pull of gravity. Since most jumps require the body to be moved against gravity, an extended trunk flattened against a horizontal or low-angled support may be subject to excessive torques, and jumping may require substantial additional input of energy. *G. moholi* and *Tarsius* not only avoid these problems by crouching into a mechanically efficient ball-shaped posture, which minimizes the body moment of inertia in the sagittal plane (Peters & Preuschoft, 1984); but use their long, thin tails in a downwards "flick" on takeoff which further counteracts these torques (Preuschoft 1989; Günther *et al.*, 1991). But *O. crassicaudatus, O. garnettii, Mirza coquereli* and *M. murinus* all leap with extended body postures (Crompton, 1980; Oxnard *et al.*, 1990) so that body posture (together with relatively lower hindlimb mass in all these species) is likely to lead to body and segment moments of inertia too high for tail motion to counteract. Thus, one can speculate that they may not be able to align their bodies to the optimal angle of 45°, without the expense of so much metabolic energy in relation to the kinetic energy of the jump that the mechanical efficiency drops to unacceptable levels (possibly as little as 1%). Further, *Tarsius, G. moholi*, and the indriid leapers, leaping in long, parabolic leaps, are able to use motion of certain body segments (the tail for *G. moholi* and *Tarsius*; the arms for the indriids, Demes & Günther, 1989) to bring about in-flight rotation of the body, so that the hindlimbs make the first contact with the terminal support, and can be used to absorb the much higher forces involved in landing (Peters & Preuschoft, 1984). Unspecialized leapers such as *Microcebus* and *O. crassicaudatus* land forelimb-first, whatever the length of leaps (Crompton, 1980; Oxnard *et al.*, 1990), and some at least appear to use branch elasticity to absorb shock (Crompton *et al.*, 1993). The biomechanics, and thus ultimately the morphology, of jumping prosimians is not, therefore, simply a matter of whether they are large or small, but also whether they are specialists or not. Only by optimizing the inertial configuration of the body do prosimians appear to be able to jump at the energetically optimal angle. The evolutionary cost of these behavioral and morphological adaptations may be abandonment of quadrupedalism as a viable alternative locomotor mode, and, almost inevitably, further specialization towards jumping locomotion.

HABITAT STRUCTURE AND VISUAL PREDATION

Thus, the energetically inefficient leaping of many unspecialized prosimians appears to offer a safe means of exploitation of dense arboreal habitats. Leaping, in these species, is likely to perform the biological roles of predator avoidance and speed of movement in a zone possessing many small discontinuities, in situations where time pressure operates, and climbing is therefore not possible.

For the specialist leapers we have so far discussed, *G. moholi* and *T. bancanus*, leaping allows energetically efficient horizontal and vertical movement in a zone comprised of few, but large discontinuities, where vertical supports predominate, and hence where vertical climbing is the only alternative for height change. However, in contradiction to Crompton's (1984) earlier view, parabolic leaps are not in themselves the safest means of movement. They allow paths to be predicted easily by predators, and if the distance, flexibility, and position of the target support is poorly estimated, may easily result in injury (as witnessed by the large numbers of healed fractures observed by Niemitz [1977] in wild-caught *T. bancanus*). Why, then, was this specialization originally adopted?

In both species, leaping is not only a means of locomotion, but also a means of predation. As reported by Fogden (1974) and all subsequent field studies (Niemitz, 1984b; Crompton & Andau, 1986; Crompton, 1989; Jablonski & Crompton, 1994; *contra* Roberts & Cunningham, 1986) in *T. bancanus* ground-dwelling prey are visually located from "perches" clinging to vertical tree-trunks a meter or so above ground level, allowing a substantial area of ground to be examined. Then, a leap to the ground is made, the prey seized manually and transferred to the mouth, and a leap made back up to a perch before the prey is masticated and swallowed. In *T. bancanus*, predatory behavior at or near ground level is the chief means of prey capture at all seasons. Tarsiers are, as Niemitz (1984a) points out, rather like owls, and indeed forest owls also pounce on their prey from a distance of a meter or so.

In *G. moholi* however, this behavior is primarily seen in the dry season, when ground prey are the main source of food (Crompton, 1984), and audition may be as important as vision in prey location. The *T. bancanus* "perching" height of around one meter corresponds approximately to the mean leap length (as reported by Niemitz [1984b] and Crompton & Andau [1986]). Thus, visual detection of cryptic prey, and estimation of distance, the major roles, according to Julesz (1971), of primate stereopsis, often occur at the same mean distance as that between the supports used in non-predatory locomotion.

"Visually directed predation" would be an accurate term for the behavior we have just described in *Tarsius*. Stereopsis and acute scotopic vision are obviously relevant to the location and capture of prey hidden in the leaf litter. However, this kind of dietary adaptation is firmly associated with exploitation of a niche made up of the ground and the first 1.5 m above it, containing primarily vertical sapling trunks of around 4-6 cm in size.

Cartmill (1974a & b; 1992) claimed that the range of accurate stereoscopic vision in primates is limited by the crowding together of the eyes on the front of the face, and therefore, that orbital frontality develops because "having a wide field of sharp stereoscopic vision directly in front of it would help a visually predatory animal be sure of hitting its prey accurately on the first strike" (Cartmill, 1992, and see Allman 1977). Accurate, visually-directed strikes by the hands are not, however, the key element of tarsier predation, at least in *T. bancanus*, despite the impression Allman (1977) gained from Le Gros Clarke's (1924) observations of captive animals. The elongated fingers of *Tarsius* are clearly primarily adapted (Niemitz, 1984b) for support for support on relatively large, 4-5 cm. supports, not fine branches. Further, my own detailed observations of free-ranging tarsiers indicate that tarsiers use these elongated fingers to search about within the leaf litter for hidden prey and to trap and hold down prey while transferring them to the mouth. Thus, the locomotor element, as well as the close-range stereopsis element, of the "visual predation hypothesis" is not well supported by the *Tarsius* case. A close-range manual strike *is* typical of the galagos, lorises and mouse lemurs, but here may be directed by the ears as much as by the eyes.

"Visually directed predation", in *T. bancanus*, is not usually a close-range phenomenon, and there seems little to distinguish the perception of prey from the perception of the target support during locomotion. "Motion" of the target, which is needed to stimulate at least some stereoscopic perception systems, is of course relative to the primary receptors, and can be achieved by motion of the eyes, or in *Tarsius*, of the head, as well as of the target itself.

While close-range manual strikes are indeed an element of the predatory behavior of prosimians such as *Microcebus* and *G. demidovii*, "visually directed predation" is not a satisfactory term with which to describe their niche. Charles-Dominique's (1977) work indicates that *Galagoides demidovii* is apparently quite poor at the visual detection of immobile, cryptic prey, and audition seems at least as important as vision in short range tracking of prey position. In the "fine branch niche" occupied by *Microcebus* and *G. demidovii*, located in the peripheral canopy and undergrowth, not only are fruit and flowers available, but also the insects attracted to them. A large number of supports gives access to a large volume of space for access to flying insects, caught opportunistically using binocular or binaural distance cues. Stereopsis, in this zone, cannot readily be ascribed to the need to detect cryptic, immobile insects, since they are not the typical prey. However, the same density of supports which favours the use of flat leaping trajectories *would* favour fine-grained nocturnal stereopsis. It provides a visually complex, confusing background against which to distinguish a variety of mobile and immobile targets, both dietary items (fruit, as well as insects) and locomotor substrates.

Hands, rather than feet, are used for grasping the landing support by the leapers in this zone (although they are still hindlimb-dominated with respect to propulsion), and, as we have noted, a forelimb-powered locomotor mode, climbing, also occurs here, in the context of "detailed foraging". Thus, in this zone, the anatomical adaptations involved in Cartmill's (1972, 1974a & b) "visual predation hypothesis" can be explained on locomotor as well as dietary grounds. Both climbing and flat trajectory leaps favour grasping hands, and the complexity and density of supports favours stereoscopy - as indeed, safe leaping does in the open zone occupied by *Tarsius*.

While *Tarsius*' adaptation is now well understood, much less is known about the reasons for adoption of a very similar specialization for parabolic leaping by the indriids. Their habitat in a way is the inverse of tarsier habitat: well separated vertical trunks, but with the resource to be exploited above, in the canopy, not below, on the ground. Although subsisting on an energetically poorer diet than tarsiers, energy-efficient leaping at the much greater body size of the indriid leapers may more readily be aided by elastic energy storage in tendons, and the richochetal leaping used by most indriids suggests strongly that this does occur. As Napier & Walker (1967) and Stern & Oxnard (1973) indicate, forelimb elongation is a characteristic of many the larger prosimians, and it is taken to an extreme in the indriids, where suspensory postures and even armswinging locomotion (Warren & Crompton, 1994) are found. Elongation of the forelimbs not only serves suspensory feeding and locomotion in the terminal branches, but permits support on large diameter supports (see Preuschoft *et al.*, 1992). Cartmill's (1974a) objection to the so-called VCL hypothesis is thus invalid.

CONCLUSIONS

"Visually directed predation" is not a satisfactory description of the likely dietary behavior of a generalized primate ancestor similar to *Microcebus*. A consideration of locomotor mechanics and locomotor ecology, however, helps to clarify the reasons why such a generalized ancestral primate would develop grasping hands and stereoscopic vision. Since long-distance leaps are expensive for such species, because of their inappropriate body conformation, locomotion favors occupation of forest habitats with greater support density. Predator avoidance, however, is another major biological need served by occupation of the same zone. Access to *each* of the elements of the probable diet is *equally* well served by the likely adaptive features of a generalized, small-bodied ancestral primate. In combination, they allow rapid, safe movement, as well as slow detailed foraging, in the "fine branch niche". Visual isolation and three-dimensional location of diverse targets (not only insects, but equally small branches and fruit) using monochromatic, scotopic vision, in this densely-packed, complexly-shaded forest zone, would have been well served by the twin benefits of orbital frontality: stereopsis and enhanced scotopic acuity. Since scotopic acuity is a major benefit of orbital frontality, to the extent to which orbital frontality is an element of the adaptive suite of ancestral primates, nocturnality is also implied as a basic primate adaptation, whether it be an ancestral or derived character.

T. bancanus is, in its dietary behavior, a "visually guided predator". However, the highly specialized locomotion and adaptations of this species are distinctively different from those of *Microcebus* and most galagines, and do not resemble the model put forward by Cartmill (1972 *et seq.*) *T. bancanus*' adaptations have little to do with the fine-branch niche, and prey detection and the predatory pounce, are not, here, a close-range phenomenon. Further, *T. bancanus* cannot readily serve as a living model for the niche of early primates. Some of its smaller congeners might prove a better bet, but, on the basis of current biogeography, *T. bancanus* is probably closest to the ancestral tarsier, and the smaller species are probably phyletic dwarfs. Acceptance of *Tarsius* as a model for the ancestral primate has other difficulties. For example, if we accept Charles-Dominique's (1977) data, we would have to accept that capabilities for visual detection of crypsis once acquired, in a *Tarsius*-like ancestor could later be lost, in galagos, in favor of audition!

Nor can *Tarsius* readily act as a model for the niche of the ancestral simian. While available evidence does suggest strongly that phyletic dwarfism has occurred in the lineage leading to modern *Tarsius* from the common haplorhine ancestor, and *Tarsius'* anatomy is indeed optimized for efficient *exploitation* of the ground, these anatomical adaptations are probably unrelated to terrestrial *locomotion*, and unlikely to have been present in a common ancestor. They can more parsimoniously be explained as derived specializations for distance- and energy-optimized ballistic leaping, of which anatomical features of the ankle joints favoring purely parasagittal limb motion (as reported by Jouffroy *et al.*, 1984) are almost certainly a component. Unless we could imagine a nocturnal tarsier-like animal, feeding in the same way, and presumably, in the same stratum, as the living Bornean tarsier, but less specialized with respect to its postcranial anatomy, giving rise to both an ancestral haplorhine and an ancestral strepsirhine stock, the "visual predator" model needs considerable modification. Could the omomyids possibly fill the bill? At present, we can do no more than speculate.

However, it is an inescapable fact that orbital frontality, in cats and in owls is associated with predatory behavior. For cats, and in the lorises, with their highly convergent eyes, the pounce is indeed a close-range phenomenon, as indicated by Cartmill (1972; 1974a & b; 1992). For owls, however, as well as tarsiers, it occurs from some distance. The two methods of prey capture are not closely similar. Rather than being associated with the "fine branch niche", amongst the prosimians, *specialized* predatory behavior is associated rather with occupation of the structurally different, open sapling trunk and ground zone, where a specialized postcranial anatomy is also favoured. Postcranial adaptations for speed are also present in cats, of course, where a good case for a "visual predation hypothesis" exists. The case of *Loris*, together with that of *Tarsius*, suggests rather that in primates the specific benefits of orbital frontation for specialized visual predation were exploited secondarily. Orbital frontality is more likely to have first appeared as a consequence of the more general benefit that accrues, for a small-bodied primate similar to *Microcebus*, in the fine-branch niche. This is provision of scotopic acuity and depth perception for the location of diverse targets, fruit and branches as well as insects, in a complexly-shaded environment. I thus tend to concur with Rasmussen (1990) in regarding insectivory as more probably opportunistic than specialized in the common primate ancestor. Evidence presented here suggests the importance of the grasping hand may not only be related to climbing, and manipulation of food. Equally important is its value for safe landing in unspecialized prosimian leapers which use low-angle takeoffs optimized for speed and safety rather than energetic efficiency.

While Cartmill's "visual predator hypothesis" has considerable validity, I conclude that only a multifactorial explanation of the locomotor, visual and dietary adaptations appropriate to a small bodied generalized primate in the fine-branch niche is likely to be totally satisfactory.

ACKNOWLEDGEMENTS

I was stimulated to write this paper by the many helpful discussions engendered by the friendly atmosphere at the "Creatures of the Dark" meeting. I wish to thank L. Alterman and M.K. Izard for the invitation to the symposium, and Lon and family for their generous hospitality. My thanks to R.D. Martin, H. Preuschoft, and R.W. Sussman for helpful suggestions, and particular thanks to A. Walker for drawing my attention to Julesz's work, and enlightening me on stereopsis. I am also indebted to M.M. Günther, W.I. Sellers and R.McN. Alexander for cooperations and discussions which have contributed some of the ideas presented here. None of the above, of course, necessarily endorse any of my conclusions. My research has been sponsored by The Science and Engineering Research Council, The Medical Research Council, The Erna and Victor Hasselblad Foundation, The Royal Society, Tektronix UK Ltd., The WorldWide Fund for Nature (Hong Kong) and the L.S.B. Leakey Fund.

REFERENCES

Allman, J. (1977). Evolution of the Visual System in Primates. In J.M. Sprague and A.N. Epstein (Eds.), Progress in Psychobiology and Physiological Psychology 7. (pp. 1-53.) New York: Academic Press.

Allman, J. & McGuinness, E. (1988) Visual Cortex in Primates. In H.D. Steklis & J. Erwin (Eds.,) *Comparative Primate Biology 4*: Neurosciences (pp. 279-326). New York: Alan Liss.

BBC, The Natural World (1994, April) *Vampires, Devilbirds and Spirits.*

Bennet-Clark, H.C. (1975) The energetics of the jump of the locust, *Schistocerca gregaria. Journal of Experimental Biology, 47,* 59-76.

Bennet-Clark, H.C. (1977) Scale effects in Jumping Animals. In T.J. Pedley (Ed.): *Scale Effects in Animal Locomotion* (pp. 185-201.) London: Academic Press.

Biewener, A. Alexander, R.McN. & Heglund, N.C. (1981). Elastic strain energy in the hopping of kangaroo rats. *Journal of Zoology (London), 195,* 369-383.

Bruun, B. (1972). *The Hamlyn Guide to Birds of Britain and Europe.* London: Hamlyn.

Charles-Dominique, P. (1984). *Ecology and Behavior of Nocturnal Primates.* London: Duckworth.

Cartmill, M. (1972). Arboreal adaptations and the origin of the order Primates. In R.H. Tuttle (Ed.) *Functional and Evolutionary Biology of Primates* (pp. 97-122). Chicago: Aldine-Atherton.

Cartmill, M. (1974a). Pads and Claws in Arboreal Locomotion. In F.A. Jenkins, Jr. (Ed.) *Primate Locomotion* (pp. 45-83). New York: Academic Press.

Cartmill, M. (1974b). Rethinking Primate Origins. *Science, 184,* 436-443.

Cartmill, M. (1992). New Views of Primate Origins. *Evolutionary Anthropology 1 (3),* 105-111.

Charles-Dominique, P. (1977). *Ecology and Behavior of Nocturnal Primates.* London: Duckworth.

Charles-Dominique, P. & Martin, R.D. (1970). Evolution of lorises and lemurs. *Nature, 227,* 257-60.

Collins, E.T. (1921). Changes in the visual organs correlated with the adoption of arboreal life and with the assumption of the erect posture. *Transactions of the Opthalmology Society,United Kingdom, 41,* 10-90.

Crompton, R.H. (1980). Galago *locomotion.* Doctoral dissertation, Harvard University, Cambridge, (United States of America). Ann Arbor: University Microfilms International.

Crompton, R.H. (1983). Age Differences in Locomotion of 2 Subtropical Galaginae. *Primates, 4(2),* 241-259.

Crompton, R.H. (1984). Habitat Structure, Foraging and Locomotion in Two Species of *Galago.* In P. Rodman & J. Cant (Eds.) *Adaptations for Foraging in Non-human Primates* (pp.74-111). New York: Columbia University Press.

Crompton, R.H. (1989). Mechanisms for Speciation in *Galago* and *Tarsius. Human Evolution, 4 (2),* 105-116.

Crompton, R.H. & Andau, P.M. (1986). Locomotion and Habitat Utilization in Free Ranging *Tarsius bancanus*: a preliminary report. *Primates, 27 (3),* 337-355.

Crompton, R.H. & Andau, P.M. (1987). Ranging, Activity Rhythms and Sociality in Free-Ranging *Tarsius bancanus*: a preliminary report. *International Journal of Primatology 8, (1)*, 43-71.

Crompton, R.H. Oxnard, C.E. & Liebermann, S.S. (1987). Morphometrics and Niche Metrics in Prosimian Evolution 1: An Initial Approach to Measuring Locomotion, Habitat and Diet. *American Journal of Physical Anthropology, 73,* 149-177.

Crompton, R.H., Sellers, W.I. & Günther, M.M. (1993). Energetic efficiency and ecology as selective factors in the saltatory adaptation of prosimian primates. *Proceedings of the Royal Society, Series B, 254,* 41-45.

Demes, B. & Günther M. M. (1989). Biomechanics and Allometric Scaling in Primate Locomotion and Morphology. *Folia Primatologica, 53,* 125-141.

Emerson, S.B. (1985). Jumping and Leaping. In M. Hildebrand, D.M. Bramble, K.F. Liem & D.B. Wake (Eds.) *Functional Vertebrate Morphology.* London: Belknap.

Fedak, M.A., Heglund, N.C. & Taylor, C.R. (1982). Energetics and Mechanics of Terrestrial Locomotion II. *Journal of Experimental Biology, 79,* 23-40.

Fogden, M.P.L. (1974). A preliminary study of the western tarsier, *Tarsius bancanus* Horsfield. In R.D. Martin, G.A. Doyle & A. C. Walker (Eds.) *Prosimian Biology* pp.151-165. Pittsburgh: University of Pittsburgh Press.

Garber, P. (1980). Locomotor behavior and feeding ecology of the Panamanian tamarin (*Saguinus oedipus geoffroyi* (Callithricidae, Primates). *International Journal of Primatology, 1,* 185-201.

Gebo, D.L. (1986). Anthropoid origins - the foot evidence. *Journal of Human Evolution, 15* (6), 421-430.

Gebo, D.L. (1988). Foot Morphology and Locomotor Adaptation in Eocene Primates. *Folia Primatologica, 50 (1),* 3-41.

Groves, C.P. (1976). The origin of the mammalian fauna of Sulawesi (Celebes). *Zeitschrift für Saügetierekunde, 41 (4)*, 201-206.

Günther, M.M., Ishida, H., Kamakura, H. & Nakano,Y. (1991). The jump as a fast mode of locomotion in arboreal and terrestrial biotopes. *Zeitschrift für Morphologie und Anthropologie, 78 (3)* 341-372.

Harvey, P.H. & Clutton-Brock, T.H. (1981). Primate Home Range Size and Metabolic Needs. *Behavioral Ecology and Sociobiology, 8,* 151-155.

Howells, W.W. (1947). *Mankind so far.* Garden City, New York: Doubleday.

Hubel, D.H. & Wiesel, T.N. (1977). Functional architecture of macaque monkey visual cortex. *Proceedings of the Royal Society, Series B, 198,* 1-59.

Jablonski, N.G. & Crompton, R.H. (1994) Feeding Behavior, Mastication and Toothwear in the Western Tarsier, *Tarsius bancanus. International Journal of Primatology, 15 (1),* 1-31.

Jouffroy, F.-K., Berge, C. & Niemitz, C. (1984). Comparative study of the lower extremity in the genus *Tarsius.* In C. Niemitz, (Ed.) *Biology of Tarsiers* (pp. 168-190). Stuttgart: Gustav Fischer.

Julesz, B. (1971). *Foundations of Cyclopean Perception.* Chicago: University of Chicago Press.

Kay, R.F. (1984). On the use of anatomical features to infer foraging behavior in extinct primates. In P. Rodman & J. Cant (Eds.) *Adaptations for Foraging in Non-human Primates* (pp. 22-53). New York: Columbia University Press.

Kerr, A.W.S. & Forrester J.M. (1976). Visual and Auditory Systems. In R. Passmore and J.S. Robson (Eds.) *A Companion to Medical Studies 1* (pp 26.1-26.32) . Oxford: Blackwell.

Le Gros Clarke, W.E. (1924) Notes on the living Tarsier. *Proceedings of the Zoological Society, London, Part 1,* 217.

Le Gros Clarke, W.E. (1959). *Antecedents of Man.* Edinburgh: Edinburgh University Press.

Lindstedt, S.L., Hokanson, J.F., Wells, D.J. Swain, S.D., Hoppeler, H. & Navarro, V. (1991). Running energetics in the pronghorn antelope. *Nature, 353,* 748-750.

Liu, L., Stevenson, S.B. & Schor C.M. (1994). Quantitative stereoscopic depth without binocular correspondence. *Nature, 367,* 66-69.

MacArthur, R.H. & Wilson, E.O. (1967). *The Theory of Island Biogeography.* Princeton, New Jersey: Princeton University Press.

MacKinnon, J. & MacKinnon, K. (1980). The behavior of wild spectral tarsiers. *International Journal of Primatology, 1,* 361-379.

Martin, R.D. (1990). *Primate Origins and Evolution.* London: Chapman and Hall.

Martin, R.D. (1972). Adaptive radiation and behavior of the Malagasy lemurs. *Philosophical Transactions of the Royal Society, 264,* 295-352.

Martin, R.D. (1993). Primate Origins: plugging the gaps. *Nature*, 363: 223-234.

McNab, B.K. & Wright P.C. (1987). Temperature regulation and oxygen consumption in the Phillippine Tarsier, *Tarsius syrichta. Physiological Zoology*, 60 (5), 596-600.

Napier, J.R. (1963). Brachiation and Brachiators. *Symposia of the Zoological Society of London, 10*, 183-195.

Napier, J.R. (1967). Evolutionary aspects of primate locomotion. *American Journal of Physical Anthropology*, 19, 337-389.

Napier, J.R. & Walker, A.C. (1967). Vertical Clinging and Leaping - a newly recognized category of locomotor behavior of Primates. *Folia Primatologica*, 6, 204-219.

Niemitz, C. (1977). Zur Funktionsmorphologie und Biometrie der Gattung *Tarsius*, Storr, 1780. Herleitung von Evolutionensmechanismen bei einen Primaten. *Courier Forschungsinstitut Senckenberg, 25*, 1-161.

Niemitz, C. (1984a). Synecological and feeding behavior of *Tarsius*. In C. Niemitz (Ed.) *Biology of Tarsiers* (pp. 59-75). Stuttgart: Gustav Fischer.

Niemitz, C. (1984b) Locomotion and posture of *Tarsius bancanus*. In C. Niemitz (Ed.) *Biology of Tarsiers* (pp. 191-225). Stuttgart: Gustav Fischer.

Niemitz, C., Klauer, G. & Eins, S. (1984). The interscapular brown fat body in *Tarsius bancanus*, with comparisons to *Tupaia* and man. In C. Niemitz (Ed.) *Biology of Tarsiers (*pp.258-273). Stuttgart: Gustav Fischer.

Niemitz, C. Nietsch, A., Warter, S., & Rumpler, Y. (1991). *Tarsius dianae:* A New Primate Species from Central Sulawesi (Indonesia). *Folia Primatologica*, 56 (2) 105-116.

Oxnard, C.E. Crompton, R.H. & Liebermann, S.S. (1990) *Animal Lifestyles and Anatomies*. Seattle: Washington University Press.

Peters, A. & Preuschoft, H. (1984) External biomechanics of leaping in *Tarsius* and its morphological and kinematic consequences. In C. Niemitz, (Ed.) *Biology of Tarsiers* (pp. 227-255). Stuttgart: Gustav Fischer.

Pianka, E.R. (1970). On r- and K-selection. *American Naturalist, 104*, 592-597.

Polyak, S.D. (1957). *The Vertebrate Visual System*. Chicago: University of Chicago Press.

Preuschoft, H. (1989). Quantitative Approaches to Primate Morphology. *Folia Primatologica, 53*, 82-100.

Preuschoft, H. Witte, H. & Demes, B. (1992). Biomechanical Factors that Influence Overall Body Shape of Large Apes and Humans. In S. Matano, R.H. Tuttle, H. Ishida & M. Goodman (Eds.) *Topics in Primatology,Volume 3: Evolutionary Biology, Reproductive Endocrinology and Virology* (pp. 259-289). Tokyo: University of Tokyo Press.

Rasmussen, D.T. (1990). Primate Origins: Lessons from a neotropical marsupial. *American Journal of Primatology*, 22, 263-277.

Roberts, M. & Cunningham, B. (1986). Space and Substrate Use in Captive Western Tarsiers, *Tarsius bancanus. International Journal of Primatology*, 7 *(2)*, 113-130.

Rollinson, J. & Martin, R.D. (1981). Comparative aspects of primate locomotion with special reference to arboreal cercopithecines. *Symposia of the Zoological Society of London, 48*, 377-427.

Sellers, W.I & Crompton, R.H. (1994 *in press)*. A System for 2- and 3-D Kinematic and Kinetic Analysis of Locomotion, and its Application to Analysis of the Energetic Efficiency of Jumping in Prosimians. *Zeitschrift für Morphologie und Anthropologie*.

Smith, G.E. (1924). *The Evolution of Man*. London: Oxford University Press.

Smith, R.M. (1987). *Biomechanics of the locomotion of* Galago senegalensis. Doctoral dissertation, University of Arizona, Tucson. Ann Arbor: University Microfilms Inc.

Stern, J.T. & Oxnard, C.E. (1973). Primate locomotion: some links with evolution and morphology. *Bibliotheca Primatologia, 4*, 10-93. Basel: Karger.

Sussman, R.W. (1991). Primate Origins and the Evolution of Angiosperms. *American Journal of Primatology, 23*, 209-223.

Sussman, R.W. & Raven P.H. (1978). Pollination by lemurs and marsupials: An archaic coevolutionary system. *Science, 200*, 731-736.

Szalay, F.S. (1972). Palaeobiology of the earliest primates. In R.H. Tuttle (Ed.) *The Functional and Evolutionary Biology of Primates*. (pp. 3-35). Chicago: Aldine-Atherton.

Taylor, C.R., Heglund, N.C., & Maloiy, G.M.O. (1982). Energetics and mechanics of terrestrial locomotion I. Metabolic energy consumption as a function of speed and body size in birds and mammals. *Journal of Experimental Biology*, 971-21.

Tattersall, I. (1982). *The Primates of Madagascar.* New York: Columbia University Press.

Treff, H.A. (1970). Der Absprungwinkel beim schrägen Sprung des Galago (*Galago senegalensis*) *Zeitschrift für Vergleichende Physiologie, 67,* 120-132.

Tremble, M. Muskita, Y. & Supriatna, J. (1993). Field observations of *Tarsius dianae* at Lore Lindu National Park, Central Sulawesi, Indonesia. *Tropical Biodiversity, 1 (2),* 67-76.

Wagner, H. & Frost, B. (1993). Disparity-sensitive cells in the owl have a characteristic disparity. *Nature, 364,* 796-798.

Walton M. & Anderson, B.D. (1988). The aerobic cost of saltatory locomotion in the Fowlers's toad (*Bufo woodhousei fowleri*) *Journal of Experimental Biology, 136,* 273-288.

Walker, A. (1967). Locomotor adaptations in recent and fossil Madagascan lemurs. Doctoral dissertation, University of London.

Walker, A. (1969). The locomotion of the lorises, with special reference to the potto. *East African Wildlife Journal, 7,* 1-5.

Walker, A. (1974). Locomotor adaptations in Past and Present Prosimian Primates. In F.A. Jenkins Jr. (Ed.) *Primate Locomotion* (pp. 349-381). New York: Academic Press.

Warren, R.D. and Crompton, R.H. (1994) A Comparative Study of Locomotion in Two Species of Nocturnal Lemur (*Avahi occidentalis* and *Lepilemur edwardsi*). Paper presented at the Swansea Meeting of the Society for Experimental Biology.

Wood Jones, F. (1916) *Arboreal Man.* London: Arnold.

PATTERN AND PROCESS IN STREPSIRHINE PHYLOGENY

J.C. Masters,[1] R.J. Rayner[2] and I. Tattersall[3]

[1]Department of Anthropology
State University of New York
Stony Brook, New York 11794-4364, USA

[2]Bernard Price Institute for Palaeontological Research
University of the Witwatersrand
Private Bag 3, P.O. Wits
2050 South Africa

[3]Department of Anthropology
American Museum of Natural History
Central Park West at 79th Street
New York, New York 10024-5192, USA

INTRODUCTION

Although Darwin's ideas concerning the factors and forces responsible for the generation of diversity varied somewhat from one edition of *The Origin of Species* to the next, by the time he came to write the sixth edition his thoughts had crystallized (Darwin, 1872, p. 129):

> "Although isolation is of great importance in the production of new species, on the whole I am inclined to believe that *largeness of area* is still more important, especially for the production of species which shall prove capable of enduring for a long period, and of spreading widely. Throughout a great and open area, not only will there be a better chance of favourable variations, from the large number of individuals of the same species there supported, but the conditions of life are much more complex from the large number of already existing species; and if some of these many species become modified and improved, others will have to be improved in a corresponding degree, or they will be exterminated. Each new form, also, as soon as it has been much improved, will be able to spread over the open and continuous area, and will thus come into competition with many other forms." (our emphasis)

According to recent taxonomies, the Afro-Asian lorisids are comprised of 11 extant galago species (Nash *et al.*, 1989) and five lorises (Napier and Napier, 1985), totalling 16 species. [It is certainly possible that there are more of these cryptic species waiting to be discovered, but not in numbers that would affect this analysis.] By contrast, a conservative estimate of the number of extant strepsirhine species on Madagascar is 32 (data accumulated from Petter and Charles-Dominique, 1979; Jenkins, 1987; Crovella and Rumpler, 1992; Crovella *et al.*, 1993), and this figure is likely to increase with more intensive genetic and behavioral studies. If the subfossil taxa (minimally estimated at 12 species by Jenkins, 1987) are included, the Malagasy radiation can be seen to have produced at least 44 recent strepsirhine species - almost three times the number currently alive on the continental landmass.

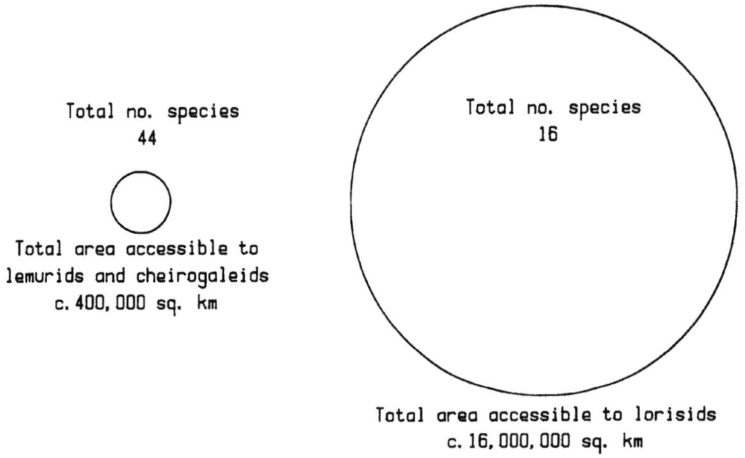

Figure 1. Comparison of approximate areas available to strepsirhines on Afro-Asia and Madagascar, together with numbers of species actually present.

We estimated the areas occupied by these animals, both on Madagascar and on the Afro-Asian mainland, from the geographic distributions presented by Petter and Petter-Rousseaux (1979). From this estimate we were able to calculate a rough figure of the area broadly accessible to strepsirhine primates on the two landmasses. These comparative areas are represented graphically in Figure 1. The area which supports a mere 16 lorisid species was estimated to be greater than 16,000,000 km^2, while the Malagasy species were distributed over an area approximating 400,000 km^2 - i.e., one 40th of the mainland area.

The strepsirhine speciation pattern thus runs directly counter to Darwin's prediction - an intriguing problem for any biologist interested in evolutionary processes. Darwin's position was an entirely logical one: if natural selection and competition are indeed the primary motive forces in the production of species, then the more opportunity they have to act, the more effective they should be. This apparent conundrum led us to explore a range of alternative explanations that have been employed elsewhere to account for discrepancies in species richness between clades, and to assess their appropriateness as explanations for the strepsirhine radiations.

A second, even more intriguing, problem in strepsirhine evolution concerns the way in which these animals came to occupy their present, somewhat fragmented distribution. In the second part of this paper, we investigate the likelihood of the various zoogeographic hypotheses that have been advanced, in the light of current knowledge of the paleohistory of Gondwana.

EVOLUTIONARY EXPLANATIONS FOR DISCREPANCIES IN SPECIES RICHNESS AMONG CLADES

Speciation sits at the interface between micro- and macroevolutionary theories, and both kinds of processes need to be considered in attempts to understand its causation. We have identified four classes of explanation that could potentially be important in understanding the strepsirhine radiations: (1) classical selection models, (2) models arising out of hierarchy theory, (3) biogeographic hypotheses, and (4) a null hypothesis proposing different rates of extinction.

Classical Selection Models

Natural Selection and Competition. These are both microevolutionary processes, and are inextricably linked, since competition forms the major driving force behind Darwin's "struggle for life." Hence, we shall consider these two processes together. We have already alluded to the fact that the strepsirhine pattern does not fit Darwin's predictions for the direct involvement of these forces in generating species diversity. More recently, however, the role of natural selection and competition in speciation has come to be viewed rather differently. In the neo-Darwinian construction, it is not the *presence* of these forces, but their *absence* which is viewed as being fundamental to evolutionary radiation. Huxley (1942, pp. 323-4), in formalizing the Modern Synthesis, wrote:

> "Decreased selection-pressure permits radiation. This is true not only for species or subspecies but for entire groups. In the former case the result is higher variability, in the latter more extensive evolutionary divergence and radiation... The principle can be generalized in relation to competitor-pressure as well as predator-pressure."

Further (p. 383):

> "...divergence is normally slow, but occasionally, as on oceanic islands and other places where the intensity of selection is relaxed, it may be much more rapid and more extensive than usual."

Contemporary authors share this view. For example, Jablonski (1989, p. 364) has written:

"Rebound intervals provide settings of unbridled radiation in which innovations can be captured and new adaptive zones occupied in relative freedom from pre-emptive competition that typifies clade interactions during background times."

We have termed this pre-emptive competition "the ghost of competition yet to come" (Masters and Rayner, 1993). This form of the argument is particularly relevant to explanations regarding the evolution of the strepsirhines, since it has been used directly to explain the explosive radiation of the Malagasy forms by several authors. A good example is the following description by Hill (1972, p. 148):

"Having once gained sanctuary in Madagascar, the ancestral lemurs underwent rapid evolutionary deployment unhindered by simian competition or predator hazard. They rapidly exploited every potential ecological situation, and niches were extremely varied from rain-forest, montane forest, savannah, rocky outcrops, marshes, swamps (with their reed beds) and even, according to some workers, an aquatic environment... The result was a wealth of speciation and subspeciation unheard of elsewhere..."

Or, as Gibbons (1992, p. 308) put it recently,

"...lemurs give anthropologists a glimpse of an alternate world that might have evolved if apes and monkeys had never come onto the scene. Since monkeys never made it to Madagascar, they weren't able to crowd out lemurs, which flourished there as a result."

While it may well be true that the existence of free "ecological space" could allow for the survival of a new form once it had evolved, we believe that to hold that space responsible for the actual origin of evolutionary novelties describes an extremely unlikely scenario for speciation. Most species are highly resistant to major evolutionary change (Gould, 1982), and avoid the destructive effects of natural selection whenever possible. Many specific adaptations, especially those relating to successful feeding and reproduction, are closely fitted to the environment in which speciation occurred. A habitat shift would incur strong selective penalties against any organism which strayed from the environment to which it was adapted. As a result, organisms tend to remain within their preferred habitats, and specific distributions tend to be largely predictable in terms of environmental characteristics. If, however, the environment changes in response to climatic fluctuations, fossil evidence indicates that organisms will tend to seek out conditions similar to those they inhabited previously (Masters and Rayner, 1993). Speciation will only occur in trap situations from which migration is prevented, either by a hostile surrounding environment or topographic barriers.

A vacant ecological "niche" to which an organism was not yet adapted (and Lewontin, 1983, has argued that vacant niches have no existence in reality) would constitute no inducement for that organism to move out of its preferred habitat, in the absence of strong competitive or selective pressures forcing it to do so. And yet the crux of this argument is that radiation occurs in the absence of any such pressures. We believe that this is an unlikely explanation for the discrepancy in diversity evinced by the two geographic strepsirhine groups.

Sexual Selection. Lande (1981; 1982) proposed a model of rapid speciation which involved the joint evolution of female mating preferences and the secondary sexual characters of males. West-Eberhard (1983), too, has argued that "runaway" sexual selection may be responsible for rapid rates of speciation in some groups. Sexual selection, a microevolutionary theory, comprises two processes: intrasexual selection (also referred to as "male-male competition", since in most cases it is the males that compete), and intersexual selection (often termed "mate"/"female choice"). Mate choice is based either on mating territories which have been won by intrasexual selection, or on epigamic characters which are assumed to be heritable.

The invocation of either intra- or intersexual selection generally implies the existence of some recognizable morphological correlate of increased reproductive success in the sex under selection. Intrasexual selection may be reflected in sexual dimorphism in body size, or in the development of defensive or offensive weapons as secondary sexual characters. Intersexual selection generally implies the elaboration of calls, mating behaviors, or adornments either of the pelage or of the mating territory.

In nocturnal or crepuscular animals, long-range signalling is accomplished almost entirely through non-visual modalities. This limits considerably the avenues through which sexual selection might operate. Studies of body size dimorphism by Kappeler (1990) and by Jenkins and Albrecht (1991) showed the majority of prosimians to be essentially monomorphic. Where some degree of sexual dimorphism was evident, there was no obvious correlation with variance in male reproductive success.

Dixson (1987; 1989) has suggested that sexual selection might take place through a form of "female choice" of particular penile structures or adornments (spines, elongated bacula). Once again, however, no consistent pattern emerged that might support the contention that the higher rate of speciation on Madagascar can be explained in terms of the greater intervention of sexual selection. In fact the most elaborate penile structures tended to be found among the Afro-Asian species, where speciation has apparently occurred far less often (see also Dixson, this volume).

Models Arising out of Hierarchy Theory

The main tenets of hierarchy theory, as expounded by Vrba (1984), Vrba and Eldredge (1984), Eldredge (1985, 1986), and Vrba and Gould (1986) are as follows:

(i) that the patterns and processes operating at higher levels of biological organization (species and clades) have a dynamic of their own that is, if not irreducible to the often temporary allelic frequency changes of microevolution, at least related to them in a nonadditive manner (Vrba, 1984);

(ii) the same general evolutionary processes involving the sorting and selection of variation are likely to operate at all levels of organization, not just at the level of the gene and the organism (Vrba and Gould, 1986); and

(iii) any particular focal level will be affected by events at higher and lower levels through downward and upward causation (Vrba and Eldredge, 1984).

There are two potentially important hypotheses in this class:

Species Selection. This hypothesis, formulated by Stanley (1975) and developed by Vrba (1984), proposes that there are characters emergent at the species level (e.g., characteristic population size, degree of gene exchange between populations, and size and shape of the species distribution - see Vrba and Eldredge, 1984) that predispose certain species to undergo speciation more often. This is thus a higher level analogue of an elevated birth rate at the organismal level.

We found nothing in the literature on population structures to suggest that the ranges of the Malagasy species were inherently more likely to fragment than those of the Afro-Asian species. From information currently available, lorisids show no greater predisposition towards interpopulation migration and gene exchange than do lemurids or cheirogaleids, although more extensive field studies could prove otherwise. Certainly the slow locomotor mode adopted by lorisines would argue *against* high levels of migration between populations - not in favor of them.

General difficulties associated with demonstrating species selection (Vrba, 1984) led us to set it aside for the time being, in favor of more testable models.

The Effect Hypothesis. This model, proposed by Vrba (1980), focuses much lower down on the hierarchy than does species selection. It suggests that the characteristics of organisms (or even of their components) may incidentally create higher level trends and differentials in clade diversity, by sorting upwardly through the hierarchy. Vrba's example was breadth of resource use, and she reasoned as follows: it is the organisms, not the species, that interact with their environments and determine the species' ecological requirements. Organisms with a narrow ecological tolerance (stenotopes) will experience environmental changes as significant habitat deterioration long before their generalist (eurytopic) counterparts. This means that stenotopic individuals will encounter natural selection pressures more often and more intensely than eurytopes, causing them to undergo genetic change and speciation at a higher rate. A clade of stenotopes will soon become highly speciose (or extinct, if the species are unable to respond genetically), while a eurytopic clade will appear relatively depauperate. The first response of many biologists might be to suggest that the speciose clade possesses some advantageous trait that enables it to invade a number of niches successfully, whereas in reality its speciose nature is an incidental effect of a set of rather ill-adaptive traits possessed by individuals lower in the hierarchy.

The problem with applying this narrow version of the model to strepsirhines is that, by and large, they are great generalists, on Afro-Asia and Madagascar alike. Tattersall (1992) investigated the habitat ranges of several Malagasy genera, including *Propithecus*, *Lemur*, and *Eulemur*, and found all of them to occupy a wide variety of habitat types, indicating broad resource tolerance. It is true that there are more stenotopic groups on Madagascar than on the mainland, viz. the hapalemurs and the lepilemurs, but for the effect hypothesis to be the major factor in explaining the discrepancy in species abundance, it would be necessary to show that these groups account for a significantly greater part of the diversification than do the other, more generalist groups.

The genus *Hapalemur* contains only three species. *Lepilemur* is a little more promising, as Petter and Charles-Dominique (1979) have estimated that there may be as many as nine cryptic species within this genus on the basis of long range vocalizations. Lepilemurs are also extremely narrow in their dietary habits. However, even if there were as many as nine species within this group, they would still only constitute between 20 and 25% of the Malagasy radiation. Thus it appears to us that, while the effect hypothesis may account for the disproportionate radiation of certain taxa within the Malagasy clade, it does not explain the overall discrepancy between the Malagasy and mainland groups.

Of course, resource use is not the only domain of the effect hypothesis. Is it possible, perhaps, that the Malagasy forms are more susceptible to chromosomal rearrangement than are the Afro-Asian forms, and thus speciate more often? The studies of chromosomal evolution in strepsirhines by Rumpler and Dutrillaux (1990) do not support such an hypothesis. We thus turned to the third class of explanation.

Biogeographic Hypotheses

Whereas the previous two classes of explanation place their major emphasis on intrinsic biological factors and processes, the biogeographic models accord a major causative role in speciation to where the organisms occur on the earth's surface. Martin (1972) proposed an interpretation of the Malagasy radiation based on MacArthur and Wilson's (1967) theory of island biogeography. This was suggested by the complex nature of Madagascar's topography, which enabled him to divide the geographic distributions of the lemurs into distinct climatic and vegetational zones demarcated by major physical barriers. He considered speciation to be partially the result of prolonged geographic isolation in these zones.

This phase of Martin's model has much in common with Cracraft's (1985) more general geomorphological complexity hypothesis, which states that speciation rate is controlled primarily by large scale changes in lithospheric complexity. Cracraft reasoned that, since most speciation involves the geographic splitting of species ranges, it is the complexity of the habitat in which species occur rather than characteristics of the species themselves, that is primarily responsible for speciation.

Madagascar indeed has a complex geomorphology. Approximately one third of its surface today is taken up by an uninhabitable central plateau, which serves as a highly efficient barrier to migration. [However there are indications that it might have supported migration routes in the past (Tattersall, 1982).] The most important aspect of geomorphological complexity is the manner in which it interacts with changes in climate: more complex landscapes undergo more vicariance events than do more uniform ones. In addition, the island is organized longitudinally to cover a maximum range of latitudes, ensuring a wide range of climatic regimes. When global changes cause climatic shifts on Madagascar, species become trapped in isolated refuges. When these relict habitats deteriorate further, the trapped populations come under severe selection, and speciate or go extinct. By contrast, when climates change and habitats shift on the continental landmass species can generally track their preferred environmental conditions by moving with them. This is the model we believe to be most effective in explaining the speciation differential on the two landmasses.

Martin's hypothesis, however, went further than this. While geographic isolation formed the first essential component of his model, competition between breeding units was invoked as the second phase necessary for adaptive radiation. Occasional migration between neighboring vegetational zones would have resulted in competition between closely related species, and pressure for increased specialization for distinct ecological niches. No reason was given as to why such a situation did not prevail in some parts of the Afro-Asian prosimian distributions, where the absence of geographic barriers would have made migration events far more frequent. We have already indicated what we believe to be weaknesses in the competitive explanation for speciation, as well as the problems involved in identifying lemurs as specialists (above). Hence we believe Cracraft's model to be a sufficient explanation for the difference in species abundance between the strepsirhine clades.

The Null Hypothesis: Differential Rates of Extinction

There is one further possible explanation for this observation, and this is that speciation rates have been similar on the two landmasses, but extinction has occurred more frequently on Africa and Asia. The 16 lorisid species are therefore the few remaining twigs of a once highly speciose bush, as species-rich as the Malagasy lemurs. The admittedly sparse fossil record (reviewed in Masters, 1988) offers no support for such an hypothesis. Further, studies of both blood enzymes and highly repeated DNA sequences indicate that the lorisids form a close-knit genetic group of relatively recent descent, while the lemurs show very much higher levels of genetic diversity (Masters *et al.*, 1994; Crovello *et al.*, 1994). Because of the implications of these findings for theories of strepsirhine phylogenetic history, we attempt in the next section to relate them to recent information regarding the paleohistory of the two landmasses.

STREPSIRHINE PHYLOGENY AND GONDWANAN PALEOHISTORY

The primary argument in strepsirhine phylogeny concerns the nature of the relationship between the Malagasy and mainland forms. The argument centers around structural and behavioral similarities that are to be found in the Malagasy cheirogaleids and the small-bodied African galagos. Adherents of the separate lineage view (originally proposed by Charles-Dominique and Martin, 1970) consider these shared characteristics to be primitive retentions, or parallel elaborations thereof. The alternative view, proposed by

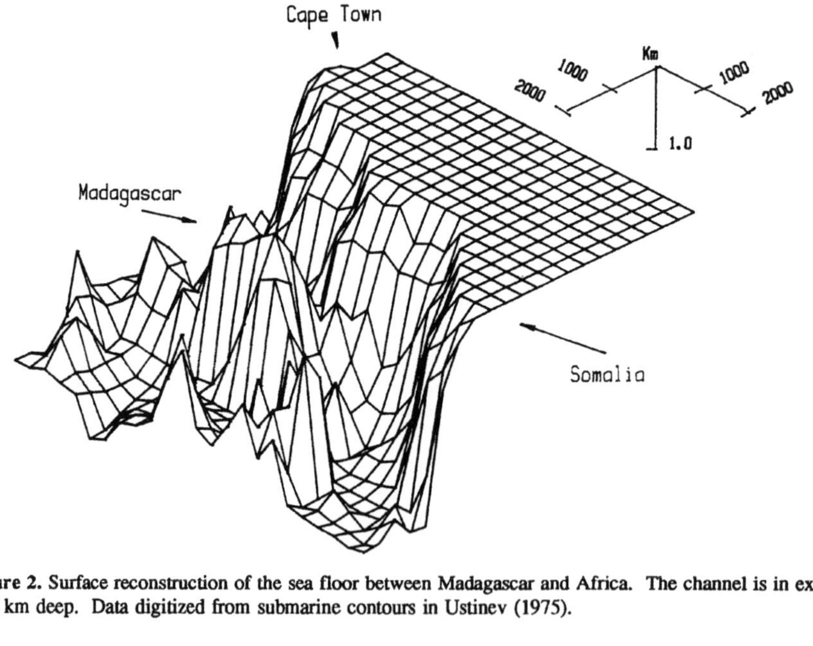

Figure 2. Surface reconstruction of the sea floor between Madagascar and Africa. The channel is in excess of 2 km deep. Data digitized from submarine contours in Ustinev (1975).

Szalay and Katz (1973), holds that these shared features are not primitive primate, but advanced strepsirhine traits. Thus the crux of the argument is: "Are the mainland galagos and lorises descended from a cheirogaleid or are the cheirogaleids evolved from galagos that found their way to Madagascar?" (Fleagle, 1988; p. 100).

The paleohistories of Africa and Madagascar bear strongly on this argument. The present day Mozambique channel which separates the landmasses is approximately 500 km wide from East to West at its narrowest point, and more than twice that distance from North to South. In addition, the channel is extremely deep - in excess of 2 km at its deepest point (Figure 2) - indicating that even a significant drop in sea level (e.g., 300 m) would make little difference to Madagascar's isolation.

Of course, Africa and Madagascar were once joined as part of the supercontinent of Gondwana. Marine geophysical data from the Somali basin (Rabinowitz *et al.*, 1983; Coffin and Rabinowitz, 1987) have produced a model of sea floor spreading which comprises a North-East to South-West spreading ridge and two North-South transform faults (Figure 3). Madagascar therefore moved almost directly southward from the coast of Somalia to its present position. This is supported both by paleomagnetic data from Karoo sediments in Madagascar, and by the latitudinal distribution of identical Permian fossil floras from Africa and Madagascar (Rayner, 1992).

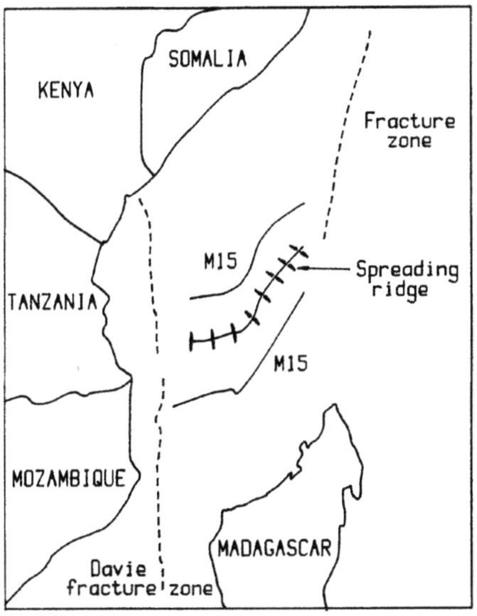

Figure 3. Diagrammatic representation of sea floor spreading between Madagascar and Africa. The position of M15, either side of the spreading ridge, and the site of the transform faults have been estimated from Rabinowitz *et al.* (1983).

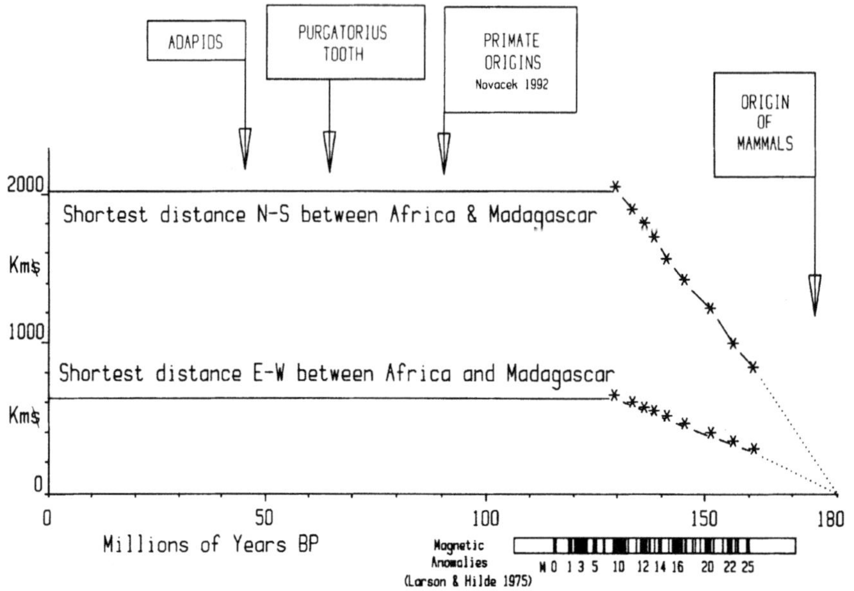

Figure 4. Graph showing the shortest East-West and North-South distances between Africa and Madagascar over time. Significant events in the evolution of primates, and the dates of the magnetic anomalies (taken from Larson and Hilde, 1975) have been added.

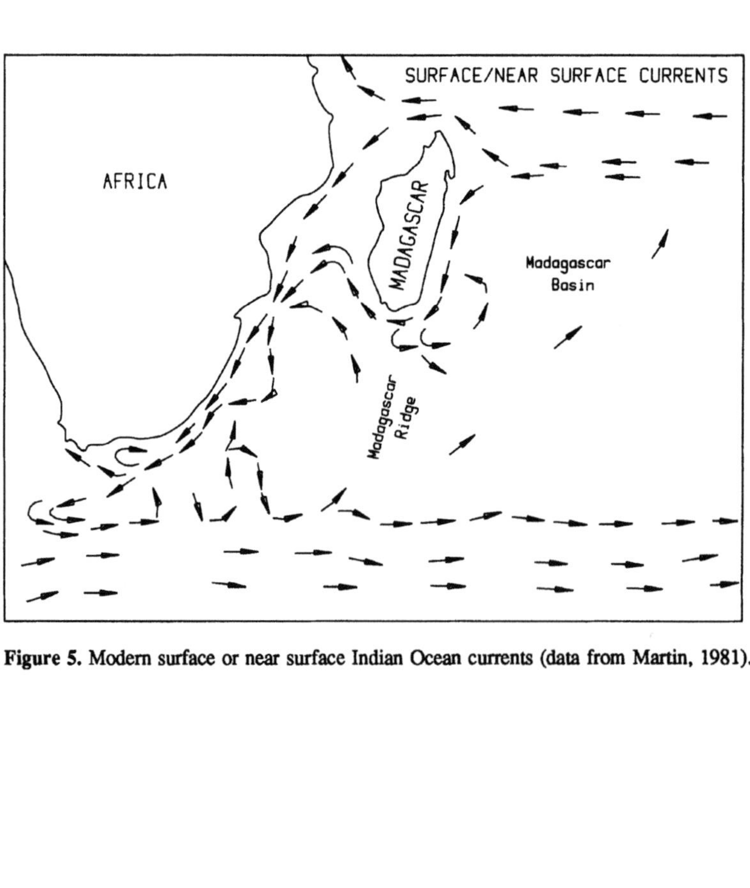

Figure 5. Modern surface or near surface Indian Ocean currents (data from Martin, 1981).

(a)

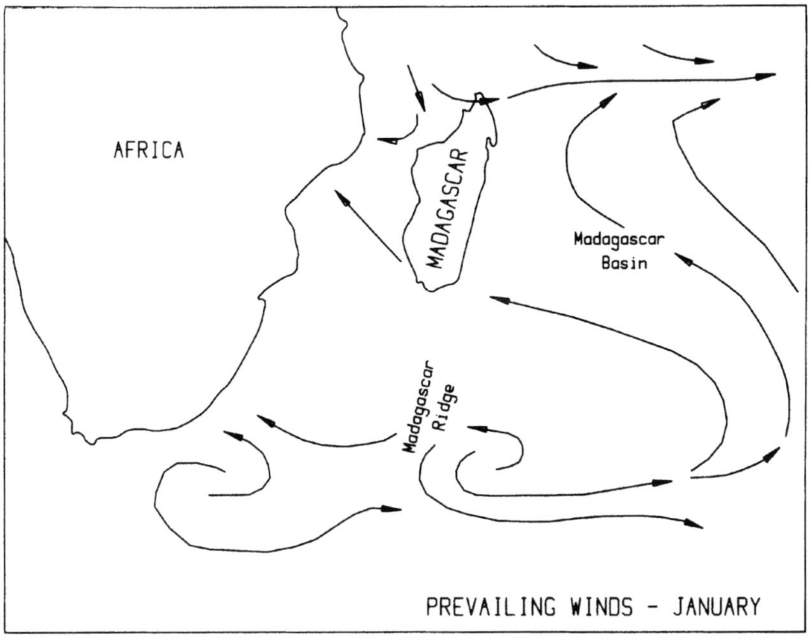

(b)

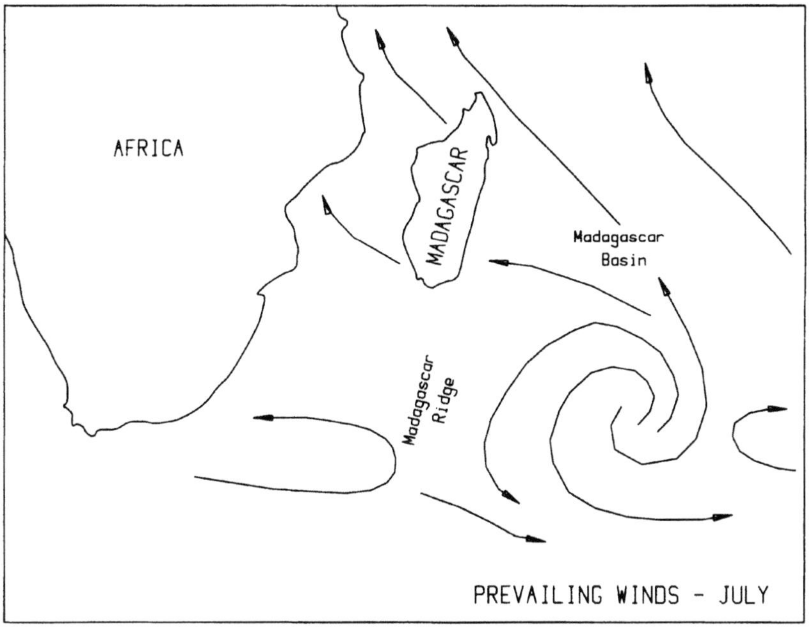

Figure 6. Prevailing winds between Africa and Madagascar in (a) summer (January), and (b) winter (July) (data from Burroughs, 1977).

The geophysical model indicates that the break-up began during the time of the magnetic anomaly M25, which has been dated at around 160 million years ago (Mya), and that Madagascar reached its present position relative to Africa in M10 times, approximately 120 Mya (Larson and Hilde, 1975). Figure 4 is a graphical representation of this process, and shows the distance between Africa and Madagascar from 160 Mya to the present. Superimposed on the graph are major events in the primate fossil record: the origin of mammals (between 190-200 Mya); the estimated origin of primates (90 Mya, according to Novacek, 1992); the oldest putative primate fossil (*Purgatorius triceratops* of 65Mya; Van Valen and Sloan, 1965); and the earliest adapid fossil, *Europolemur*, from the early Eocene of Germany (50 Mya; von Koenigswald, 1979).

A host of questions immediately arise from this information. The first of these is, if the cheirogaleid-like ancestral strepsirhine rafted across from Africa to Madagascar in the Late Paleocene-Early Eocene, long after Madagascar had reached its present position, why did such a dispersal event occur only once in primate history? Why is it that the monkeys could not have made such a crossing? The high degree of faunal and floral endemicity on Madagascar suggests that dispersal events are rather rare. In order to test the likelihood of such a migration further, we examined data on the surface water currents (Figure 5) and the prevailing winds (Figure 6) in this area of the Indian Ocean. In both cases, the dominant flow direction is from Madagascar to Africa, not the other way around. Paleo-oceanographic studies (e.g., Barraclough-Fell, 1967) indicate that the current directions have not changed substantially since the middle Cretaceous (c.100 Mya). On these data alone, then, it appears to us that, if any migrations have occurred, it is far more likely that animals crossed from Madagascar to Africa than the other way. Such a model supports the Szalay-Katz hypothesis of strepsirhine relationships, and is in turn supported by the relative amounts of genetic diversity in the lemur and loris clades (see above).

If we accept the Szalay-Katz model, however, one problem still remains: How did ancestral strepsirhines get to Madagascar in the first place? It seems to us that primate historians should not yet discard the notion that Primates may be a much older group than is indicated by the rather poor fossil record. Tattersall (1982) and Martin (1990; 1993) have both intimated that the ancestors of the modern strepsirhines may have been alive and diversifying by the later part of the Cretaceous; and one reading of our paleobiogeographic information suggests that this diversification could have taken place even earlier. If only because of the relative implausibility of alternative scenarios, we cannot yet reject the possibility that the primate colonization of Madagascar occurred when the island was either still attached to Africa, or when the distance between them was negligible, around 150 Mya.

Acknowledgments

We thank L. Alterman, K. Izard and G. Doyle for inviting us to participate in the "Creatures of the Dark" conference at Duke University, and the staff of the Duke University Primate Center for making that participation so enjoyable. We are grateful to K. Reed and O. Pearson for their insightful comments on an earlier version of the manuscript, and to M. Moniz for his assistance in its production. We acknowledge the continued financial support of the Foundation for Research Development, Pretoria, and the University of the Witwatersrand, Johannesburg.

REFERENCES

Barraclough-Fell, H. (1967). Cretaceous and Tertiary surface currents of the oceans, *Oceanograpy and Marine Biology: An Annual Review, 5,* 317-341.

Burroughs, R.H., III. (1977). Structural and sedimentological evolution of the Somali Basin: Paleo-oceanographic interpretations. PhD Thesis, Woods Hole Oceanographic Institute and MIT, MA.

Charles-Dominique, P., & Martin, R.D. (1970). Evolution of lorises and lemurs. *Nature, 227,* 257-260.

Coffin, M.F., & Rabinowitz, P.D. (1987). Reconstruction of Madagascar and Africa: Evidence from the Davie Fracture Zone and western Somalia Basin. *Journal of Geophysical Research, 92,* 9385-9406.

Cracraft, J. (1985). Biological diversification and its causes. *Annals of the Missouri Botanical Garden, 72,* 794-822.

Crovella, S., & Rumpler, Y. (1992). Confirmation of the specific status of *Hapalemur aureus* (Primates, Strepsirhini) by restriction genomic DNA banding patterns. *Human Evolution, 7,* 63-67.

Crovella, S., Montagnon, D., & Rumpler, Y. (1993). Highly repeated DNA analysis and systematics of the Lemuridae, a family of Malagasy prosimians. *Primates, 34,* 61-69.

Crovella, S., Masters, J.C., & Rumpler, Y. (1994). Highly repeated DNA sequences as phylogenetic markers among the Galaginae. *American Journal of Primatology, 32,* 177-185.

Darwin, C. (1872). *The origin of species,* 6th ed. London: John Murray.

Dixson, A.F. (1987). Observations on the evolution of the genitalia and copulatory behaviour in male primates, *Journal of Zoology, 213,* 423-443.

Dixson, A.F. (1989). Sexual selection, genital morphology, and copulatory behavior in male galagos, *International Journal of Primatology, 10,* 47-55.

Eldredge, N. (1985). *Unfinished synthesis. Biological hierarchies and modern evolutionary thought.* New York: Oxford University Press.

Eldredge, N. (1986). Information, economics, and evolution. *Annual Review of Ecology and Systematics, 17,* 351-369.

Fleagle, J.G. (1988). *Primate adaptation and evolution.* San Diego: Academic Press.

Gibbons, A. (1992). Anthropologists bet on their latest data in Las Vegas. *Science, 256,* 308-309.

Gould, S.J. (1982). Free to be extinct. *Natural History, 91(8),* 12-16.

Hill, W.C.O. (1972). *Evolutionary biology of the Primates.* London: Academic Press.

Huxley, J. (1942). *Evolution. The modern synthesis,* 1st ed. London: George Allen & Unwin.

Jablonski, D. (1989). The biology of mass extinction: A palaeontological view. *Philosophical Transactions of the Royal Society of London, Series B, 325,* 357-368.

Jenkins, P.D. (1987). *Catalogue of Primates in the British Museum (Natural History) and elsewhere in the British Isles. Part IV: Suborder Strepsirrhini, including the subfossil Madagascan lemurs and family Tarsiidae.* London: British Museum (Natural History).

Jenkins, P.D., & Albrecht, G.H. (1991). Sexual dimorphism and sex ratios in Madagascan prosimians. *American Journal of Primatology, 24,* 1-14.

Larson, R.L., & Hilde, T.W.C. (1975). A revised time scale of magnetic reversals for the Early Cretaceous and Late Jurassic. *Journal of Geophysical Research, 80,* 2586-2594.

Kappeler, P.M. (1990). The evolution of sexual size dimorphism in prosimian primates. *American Journal of Primatology, 21,* 201-214.

Lande, R. (1981). Models of speciation by sexual selection on polygenic traits. *Proceedings of the National Academy of Sciences, USA, 78,* 3721-3725.

Lande, R. (1982). Rapid origin of sexual isolation and character divergence in a cline. *Evolution, 36,* 213-223.

Lewontin, R.C. (1983). Gene, organism and environment. In D. S. Bendall (Ed.), *Evolution from molecules to men* (pp.273-285). Cambridge: Cambridge University Press.

MacArthur, R.H., & Wilson, E.O. (1967). *The theory of island biogeography.* Princeton: Princeton University Press.

Martin, A.K. (1981). Evolution of the Agulhas current and its palaeo-ecological implications. *South African Journal of Science, 77,* 547-554.

Martin, R.D. (1972). Adaptive radiation and behaviour of the Malagasy lemurs. *Philosophical Transactions of the Royal Society of London, Series B, 264,* 295-352.

Martin, R.D. (1990). *Primate origins and evolution. A phylogenetic reconstruction.* London: Chapman and Hall.

Martin, R.D. (1993). Primate origins: plugging the gaps. *Nature, 363,* 223-234.

Masters, J. (1988). Speciation in the greater galagos (Prosimii: Galaginae): Review and synthesis. *Biological Journal of the Linnean Society, 34,* 149-174.

Masters, J.C., & Rayner, R.J. (1993). Competition and macroevolution: The ghost of competition yet to come? *Biological Journal of the Linnean Society, 49,* 87-98.

Masters, J.C., Rayner, R.J., Ludewick, H., Zimmermann, E., Molez-Verriere, N., Vincent, F., & Nash, L.T. (1994). Phylogenetic relationships among the Galaginae as indicated by erythrocytic allozymes. *Primates, 35,* 177-190.

Napier, J.R., & Napier, P.H. (1985). *The natural history of the Primates.* Cambridge, MIT Press.

Nash, L.T., Bearder, S.K., & Olson, T.R. (1989). Synopsis of *Galago* species characteristics. *International Journal of Primatology, 10,* 57-80.

Novacek, M. (1992). Mammalian phylogeny: shaking the tree. *Nature, 356,* 121-125.

Petter, J.-J., & Charles-Dominique, P. (1979). Vocal communication in prosimians. In G. A. Doyle & R. D. Martin (Eds.), *The study of prosimian behavior* (pp.247-305). New York: Academic Press.

Petter, J.-J., & Petter-Rousseaux, A. (1979). Classification of the prosimians. In G. A. Doyle & R. D. Martin (Eds.), *The study of prosimian behavior* (pp.1-44). New York: Academic Press.

Rabinowitz, P.D., Coffin, M.F., & Falvey, D. (1983). The separation of Madagascar and Africa. *Science, 220,* 67-69.

Rayner, R.J. (1992). Botanical evidence for the palaeoposition of Madagascar. *Journal of African Earth Sciences, 14,* 411-415.

Rumpler, Y., & Dutrillaux, B. (1990). Chromosomal evolution and speciation in primates. *Cell Biology Reviews, 23,* 1-112.

Stanley, S.M. (1975). A theory of evolution above the species level. *Proceedings of the National Academy of Sciences, USA, 72,* 646-650.

Szalay, F.S., & Katz, C.C. (1973). Phylogeny of lemurs, galagos and lorises. *Folia primatologica, 19,* 88-103.

Tattersall, I. (1982). *The primates of Madagascar.* New York: Columbia University Press.

Tattersall, I. (1992). Systematic versus ecological diversity: the example of the Malagasy primates. In N. Eldredge (Ed.), *Systematics, ecology and the biodiversity crisis* (pp.25-39). New York: Columbia University Press.

Ustinev, G.B. (Ed.) (1975). *Geological-geophysical atlas of the Indian Ocean.* Moscow: Academy of Sciences of the USSR - Main Administration of Geodesy and Cartography under the Council of Ministers of the USSR.

Van Valen, L., & Sloan, R.E. (1965). The earliest primates. *Science, 150,* 743-745.

Vrba, E.S. (1980). Evolution, species and fossils: how does life evolve? *South African Journal of Science, 76,* 61-84.

Vrba, E.S. (1984). What is species selection? *Systematic Zoology, 33,* 318-328.

Vrba, E.S., & Eldredge, N. (1984). Individuals, hierarchies and processes: towards a more complete evolutionary theory. *Paleobiology, 10,* 146-171.

Vrba, E.S., & Gould, S.J. (1986). The hierarchical expansion of sorting and selection: Sorting and selection cannot be equated. *Paleobiology, 12,* 217-228.

Von Koenigswald, W. (1979). Ein Lemurenrest aus dem eozanen Olschiefer der Grube Messel bei Darmstadt, *Palaeontologische Zeitschrift, 53,* 63-76.

West-Eberhard, M. J. (1983). Sexual selection, social competition and speciation. *Quarterly Review of Biology, 58,* 155-183.

PREDATION, POLLINATION AND TORPOR IN TWO NOCTURNAL PROSIMIANS: *CHEIROGALEUS MAJOR* AND *MICROCEBUS RUFUS* IN THE RAIN FOREST OF MADAGASCAR

Patricia C. Wright[1] and Louise B. Martin[2]

[1]Department of Anthropology
[2]Doctoral Program in Anthropological Sciences
SUNY at Stony Brook
Stony Brook, NY 11794-4364

INTRODUCTION

Cheirogaleus is the only known genus of primate with extended periods of torpor (Petter, 1978). Previous field studies of the two smallest rain forest species of nocturnal primates, *Cheirogaleus major*, the greater dwarf lemur and *Microcebus rufus*, the rufous mouse lemur, were based on anecdotal observations (Petter *et al.*, 1977) or short-term studies (Harcourt, 1987; Martin, 1972; Hladik, 1979; Ganzhorn, 1989). Both species eat fruit, flowers, and insects, are seasonal breeders, have two to three offspring and protect their offspring in leaf nests (Petter *et al.*, 1977). These data do not in themselves explain why in sympatric species, one species has an extended inactive period, but the other does not. A goal of this study was to shed light on some of the factors which might have contributed to the evolution of this behavior.

Descriptions of torpor in *C. medius*, the fat-tailed dwarf lemur, a smaller, dry forest species (250 g), suggest that dormancy in the dry season is associated with a seasonal shortage of food. *C. medius* become torpid as early as March, and individuals reemerge from hollow tree trunks at the beginning of the wet season in November (Hladik *et al.*, 1980; Petter, 1978). From one to five individuals were found together in torpor (Petter, 1978). During the rainy season *C. medius* stores fat under the skin and in the tail, increasing their body weight by 75g (about 40%; Hladik *et al.*, 1980). Petter (1978) suggested that *C. medius* can lose up to 100g during torpor.

The rain forest species, *C. major* (350-450g), has previously been reported to have a shorter period of dormancy (Petter *et al.*, 1977; Tattersall, 1982). From July through October they hide in the leaf litter at the base of large trees (Petter *et al.*, 1977). *C. major* has also been observed to store fat in its tail throughout the rainy season. Ganzhorn (1989) suggested that *C. major* feeds on a wide variety of foods during its active months. Why do Cheirogaleids go into torpor in Madagascar? Could it be predation pressure, competition for food, or what?

METHODS

Study Site

This study took place at the Ranomafana National Park (RNP) Biological Research Station in southeastern Madagascar (Fig. 1), 21°15'S; 47°26'E in the Province of Fianarantsoa. The rain forest is on steep slopes of 900-1000m in elevation, in an area classified as montane rain forest. It receives an average annual rainfall of 2500mm (Wright, 1992). Most rain occurs in December through March, with September and October receiving the least (Fig. 2). Temperatures in June-September are low, with minimums of 7°C. Maximum daytime temperatures occur in October-May with highs of 37°C. The vegetation is characterized by a 20-30m canopy forest and is transitional between lowland rain forest and high elevation sclerophyllous forests. Twelve species of primates occur in RNP, four are nocturnal (Table 1).

Nocturnal Primate Censuses

This research began on September 3 and continued until May 1 of 1993. Inventories of nocturnal primates sighted along rain forest trails between 1930 and 2400 were taken in September, October, and February. Group size, height in trees, activity, and location were recorded. These censuses were used to infer torpor patterns of *C. major* and to determine group size and population densities of both *M. rufus* and *C. major.*

Trapping

Four live traps were placed every 50m, for 500m, along two trails 1 km apart in rain forest with a closed canopy of 25-30m. To catch both *C. major* and *M. rufus* at each trap site, two small (22.2cm x 6.6cm x 6.6cm) traps and two larger (39.6cm x 12cm x 12cm) traps were placed on horizontal branches between 0.5m to 2.5m from the ground. Trap lines were baited and tied open for 1 wk before they were set in order to habituate the animals to finding the traps. These traps were set for 15 nights in September and 15 nights in December. In October, 20 traps were set at 3-7m high where *C. major* had previously been sighted. For the first two weeks, traps were baited with cheese, honey, peanut butter, small pieces of pineapple, or banana. Because the preferred bait was banana, all traps were subsequently baited with banana only. Since only one *C. major* was trapped low, only the more elevated traps were in use five days a week until April 15.

Morphometric data including body weights, body and testes measurements, tail circumference and reproductive condition were collected on both species. Anal temperature was recorded with a digital thermometer. *M. rufus* were marked using colored bird bands placed on the forearm. *C. major* were fitted with Telonics configuration CHP-1H transmitters on cable tie collars. Individual *C. major* were located before dusk using a hand-held, three element Yagi antenna and followed beginning at dusk for 4-6 hrs over five night when possible. Scan samples were taken every five minutes on activity. *Ad libitum* observations on nest building were recorded. Nightly path lengths were paced and plotted on a map of the study site to determine home range size.

Plant Phenology

Phenology data on 100 randomly selected trees of 25 species were taken bimonthly over six annual cycles (Meyers & Wright, 1993). In addition, 40 fruit traps were set 25m apart and collected twice monthly according to the method in Terborgh (1983) in order to determine seasonal variation in fruit productivity.

RESULTS

Botanical Censuses

Results from botanical plots and selected phenology tree samples taken over a 6 yr period in Ranomafana National Park have shown that fruit is most abundant from December until March and scarce from April until September (Overdorff, 1991; Meyers & Wright, 1993). Fruit trap data have further demonstrated that fruit abundance is low for nearly 6 months a year (Fig. 3).

Nocturnal Census

Results of the nocturnal censuses showed that *M. rufus* was active during both winter and summer months (Fig. 4). In contrast, *C. major* was not sighted before the spring equinox. The first *C. major* sighted was September 26, two nights after the equinox. Numbers of *C. major* sighted during October increased to one every 2hr, and did not decrease until March. In January through March juveniles as well as adults were seen. In April, three *C. major* were seen, but by May there were none active. The mean height of *C. major* sightings was 12.2m (SD = 4.4, range 7-20m), while the mean height of *M. rufus* sightings was 6.6m (SD = 3.1m, range 2-12). Both *M. rufus* and *C. major* were seen alone most of the time. *C. major* was not seen in groups, and *M. rufus* was seen in pairs in 1.6% of sightings. *C. major* and *M. rufus* were never seen in association during the censuses.

Trapping Results

Sex Ratios, Weights and Reproductive Condition

During the month of September, 18 male and two female *M. rufus* individuals were trapped. The mean body weight was 43g (SD = 6.3, range 34-54g). None of the females were pregnant based on palpation. There was no obvious difference in body weights between males and females, although a small sample size of females precluded statistical analysis. The sex ratio of *M. rufus* captured in December was nearly equal (4 males, 6 females).

Males had a mean body weight of 49g (SD = 4.8, range 45-55g), a slight increase from the September weights. All females in December were pregnant. Females weighed a mean of 62g (SD = 9.3, range 50-75g). On December 18 a 75g female gave birth to twins each weighing 5.5 g.

Only seven (2 males, and 5 females) *C. major* were captured, and none were recaptured. Captures occurred only once in the traps set 1 m high, and six times in traps at 7 m. Females were palpated and none was pregnant. Four of the seven individuals were captured early in October when they presumably first emerged from torpor, one was captured in December and the final two were juveniles captured in March and April. Weights of adults ranged from 235- 470g (mean = 356g).

Measurements

There was no significant sexual dimorphism in the measurements taken from *C. major* or *M. rufus*. Body temperatures did fluctuate in both species, but the greatest differences were seen in *C. major* (Fig. 5). Body temperatures of 34.3°C were recorded in October, after emergence from torpor but increased to 38.5°C in December. The lowest body temperature was recorded in April at 32.4°C. These data suggest that *C. major* may lower body temperature, as has been documented for captive *C. medius* (Russell, 1974).

Tail circumferences increased from September to January in both *M. rufus* and *C. major*, indicating that fat was being deposited. Further evidence for fat deposition was obtained on the dissection of a *C. major* female killed by a ring-tailed mongoose (*Galidea elegans)* in March. Her body weight was 470g, the largest weight recorded. Fat deposits, removed and weighed, were 32% of body weight (150g).

Table 1. RNP primate species

Nocturnal	Diurnal
Avahi laniger	Propithecus diadema edwardsi
Microcebus rufus	Hapalemur griseus
Cheirogaleus major	Eulemur rubriventer
Lepilemur sp.	Eulemur fulvus rufus
Daubentonia madagascariensis	Varecia variegata
	Hapalemur aureus
	Hapalemur simus

Table 2. Plants Species Eaten by *C. major*

Month	Vernacular Name	Latin Name	Family Name	Plant Type	Plant Part
October	Kimbaletaka	*Mammea sp.*	Clusiaceae	Tree	fruit
October	Voara	*Ficus sp,*	Moraceae	Tree	fruit
October	Maniny		Araliaceae	Tree	flower
October	Tavolo	*Ravensara sp.*	Lauraceae	Tree	fruit/flower
December	Vahimberana	*Strongylodon sp.*	Fabaceae	Vine	flower
January	Fandramanana	*Aphloia theaeformis*	Flacourtiaceae	Tree	fruit
February	Vahiambaniakondro	*Plectaneia sp.*	Apocynaceae	Vine	young leaves
March	Goavy	*Psidium cattleyanum*	Myrtaceae	Tree	fruit
March	Bararata	*Gaertnera sp,*	Rubiaceae	Tree	flower
March	Hazotoho	*Psychotria sp.*	Rubiaceae	Tree	fruit/flower
March	Tongoalahy	*Bakarella grisea*	Loranthaceae	Epiphyte	unripe fruit
March			Orchidae	Epiphyte	flower

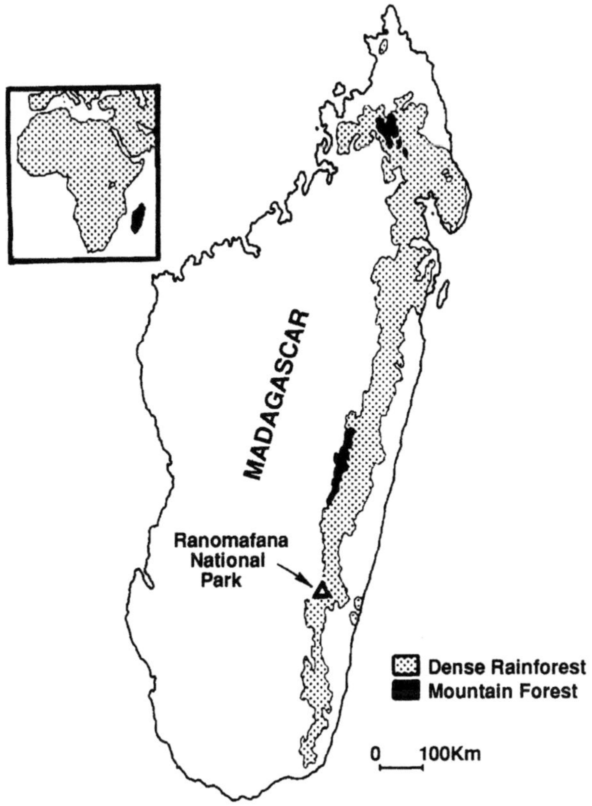

Figure 1.

Behavioral Observations of *Cheirogaleus major*

Activity - During the night, *C. major* spent 21% of 5 min-samples feeding, and 51% resting. A quarter of the time was spent traveling and moving from branch to branch. It is possible that this activity includes time spent foragingfor insects, but it was difficult to observe this behavior.

Diet - The diet of *C. major* was primarily fruit and flower nectar (Fig. 6). Fruits eaten had fleshy, sweet arils (Rubiaceae, Moraceae, Clusiaceae) (Table 2).

Flower nectar was an important part of the diet during November and December, when the common canopy vine *Strongylodon sp.* (Fabaceae) was covered with many flowers and no leaves. These large vines spread over 100m or more from 6m high to the top of the canopy. *Strongylodon sp.* has a large, 80mm long, bright yellow flower with a reservoir at the base of the flower which contains nectar. *C. major* was observed in the early evening walking quadrupedally along a 10-15cm diameter vine. The animal parted the petals of the flower with its hands, and licked the nectar. The duration of licking from each blossom was 2-7 min, after which the animal proceeded to an adjacent blossom. Some blossoms were visited with no licking, and none were eaten. The animal left the vine after 25-35min. There was never more than one *C. major* seen in a vine. Although *M. rufus* was seen licking blossoms of *Strongylodon* 12 times during December, they were never seen feeding in the same vine as a *C. major*. Young vine leaves were also eaten occasionally. Because of the difficulty of observation high in the canopy at night, insect foraging is probably underestimated. Animals caught a flying insect with both hands on three occasions. Fecal samples (n = 7) were examined for insect parts, but no chitin was seen.

Nest Building Behavior - *C. major* and *M. rufus* females build nests before giving birth (Martin, 1972). Four instances of nest building by *C. major* were observed in December and January. Nests were located from 6-12m high. On December 4, female A put a new green leaf, *Harungana madagascariensis* (Clusiaceae) into an existing nest made of dry leaves. This nest was first discovered by researchers on October 26, but was never found to be occupied by any species previously. Female A was not observed to use the nest again after this night.

On December 17, female B constructed a nest by biting off at the petioles 18 large *Dombeya sp.* (Sterculaceae) leaves and carrying them in her mouth 5m to a nest site located 6m high wedged between two trees *Tambourissa sp.* (Monimiaceae) and *Dichaetanthera sp.* (Melastomataceae). On December 23, the same female carried four *Dombeya sp.* (Sterculaceae) leaves by mouth to a second nest site more than 12m high located 25m from her previous nest. Finally on January 18, this female was observed biting off a leaf and carrying it to a third nest site located 50m from the second site and 75m from the first nest site.

In January 1991, three adult sized *C. major* emerged from a 20m high leaf nest, 0.3m in diameter. Unlike an aye-aye nest, this nest was not near the tree trunk or in a substantial tree fork.

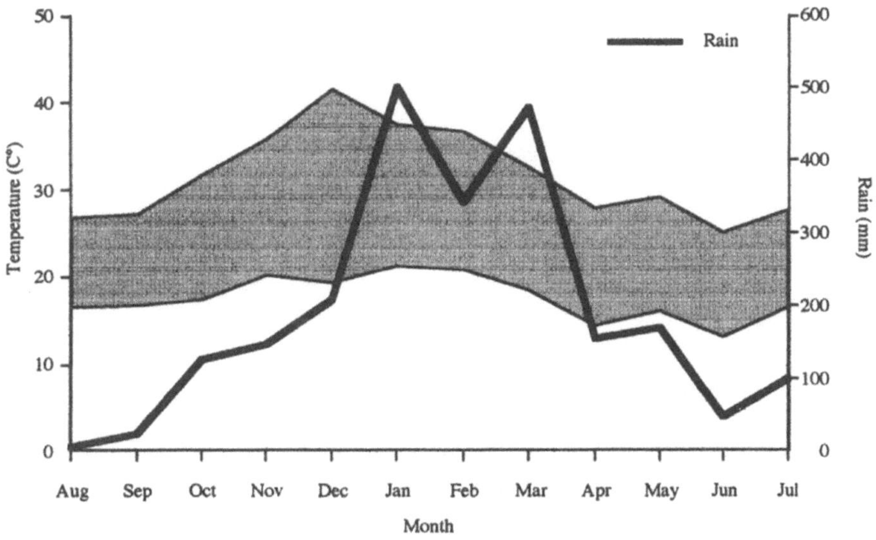

Figure 2. Rainfall and temperature at Ranomafana National Park.

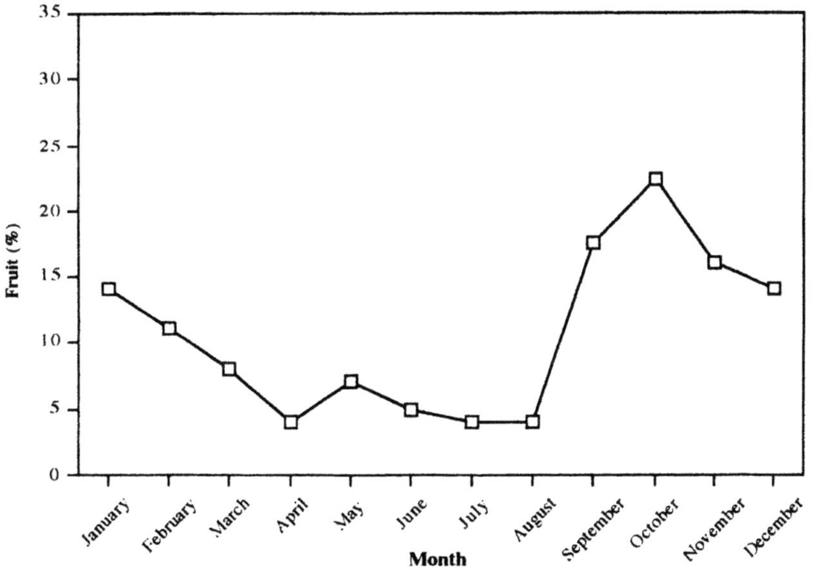

Figure 3. Fruiting seasonality in Ranomafana (Madagascar). Percentages of traps containing fruit each month. Data from Overdorff (1991).

M. rufus constructed leaf nests in January. A female *M. rufus* constructed a nest of 28 fleshy, large leaves 2m high. The female clipped each petiole and carried each leaf by mouth from a tree 2m from the nest. All *M. rufus* nests observed (n = 8) were close to the ground (1-3m high).

Home Range and Population Densities - Radio tracking data suggest that adult female *C. major* have territories which encompass 4ha. Males and juveniles were occasionally seen in these home ranges. Data from trapping or nocturnal censuses in this study underestimate the actual population densities of *C. major*. Home range data were not obtained for *M. rufus*, but trap/retrap data suggest a territory size of about 1ha.

Predation Pressure

Torpor may be a good strategy against predation pressures. We examined the incidence of predation on the two primate species to test this hypothesis.

No instances of predation on *C. major* were seen from September until February. A nest constructed by a radio collared female on January 18, had two small *C. major* near it on February 15. A boa constrictor, *Sanzinia madagascariensis* had been sighted 25m from this nest 10 days previous. On February 15, a boa constrictor 1 m 40cm ate a radio collared *C. major* female below this nest.

On March 14, a guide observed a ring-tailed mongoose, take a live *C. major* by the head from a hole 27cm deep x 23cm wide 1m from the base of a tree with a diameter at breast height of 25cm. On examination, after the mongoose dropped the dead female, the left eye was gone and the inside corner of the right eye was punctured. The right ear was torn at the base. There was a slash on the inside of the left thigh, and one under the left elbow.

Slow and fat *C. major* have been killed by *Galidea elegans* in two instances during the months of March and April of the previous year (A. Merenlender & C. Hemingway, pers obs.). *Galidea* have not been observed to prey upon *C. major* in other months. These data suggest that torpor does not seem to be a good predator defense.

DISCUSSION

Torpor

Many animals including bats, bears, and lizards go into torpor to avoid seasonal scarcity of food (Schmidt-Neilson, 1983). On the island of Madagascar, it has been documented that some species of rain forest tenrecs go into torpor for 4-6 months from May until October

(Eisenberg and Gould, 1970). However, the Malagasy rain forest rodents which are frugivorous and all weigh less than 225g (*Nesomys rufus* and *N. audeberti),* do not hibernate. Much of their diet is seeds, which are available longer than fruits (Ryan *et al.* 1993).

The only genus of primate with extended torpor is *Cheirogaleus* (Petter *et al.*, 1977) Both the rain forest (356g) and smaller dry forest species (25 g) increase body fat and become dormant in winter months. Sympatric species of primates with higher and lower body masses do not go into extended torpor. This includes the 45g insectivorous mouse lemur (*Microcebus rufus*) and the 900g bamboo eating *Hapalemur griseus* and leaf eating *Avahi laniger*. If the major components of the diet of *C. major* are fruits and flowers, then the evolution of torpidity may be associated with fruit scarcity. Indeed, fruits and flowers are less abundant in May through September than in the months October through June (Fig. 7). It also should be noted that there are few frugivorous primates in this rain forest community. *Eulemur rubriventer* and *E. fulvus rufus*, frugivorous primate species, supplement their diet with leaves during the winter season (Overdorff, 1991). *Varecia variegata*, a 3.5kg prosimian consuming over 80% fruits, does not hibernate (Morland, 1990; White, 1991). However, during the winter months at RNP the distance traveled by *V. v. variegata* in one day is extensive (2-3km, White *et al.*, in press). It is known that frugivorous primates often extend their range and increase daily path length in periods of fruit scarcity (Terborgh, 1983; Wright, 1986; Janson, 1988) Because of the higher energetic needs of smaller animals, this tactic of ranging more widely to find fruit may not be available to *Cheirogaleus*.

Is it possible that predation pressure is a factor in selecting for torpor patterns? There is growing evidence that predation on lemurs is an important selective factor (Sauther, 1989; Goodman *et al.*, 1993). In a dry forest habitat during July and August, one pair of owls had eaten close to 400 *M. murinus* (Goodman *et al.*, 1991). Also, at Beza Mahafaly the diurnal hawk *(Polyboroides radiatus*) was observed preying on *Propithecus verreauxi* and *Lemur catta* (Goodman *et al.*, 1993). Since these predation events all occurred during June through September, perhaps torpor may be a good defense against raptor predation for small lemurs.

However, data from this study show that at RNP, hawk, mongoose and boa predation can increase during the season when *C. major* and *M. rufus* are heavier in weight and lethargic. The incident of a mongoose digging up a hibernating C. major also suggests that torpor is not a good tactic to avoid predation pressures.

Why is it that in other rain forests torpor has not evolved in mammals? For example, in South American rain forests, small (60-270g), nocturnal, frugivorous, marsupials do not hibernate (Charles-Dominique, 1983; Charles-Dominique *et al.*, 1991; Emmons, 1990). The only nocturnal South American primate, *Aotus*, weighs 1000g and is active year-round (Wright, 1989, 1994). The smaller bodied primates, the tamarins, are diurnal, highly frugivorous, and are known to lose weight in the season of fruit scarcity (Goldizen *et al.*, 1988). Direct observations of *S. fuscicollis* showed that the abundance of fruits of species eaten by the tamarins was less during these months than at other times of year.

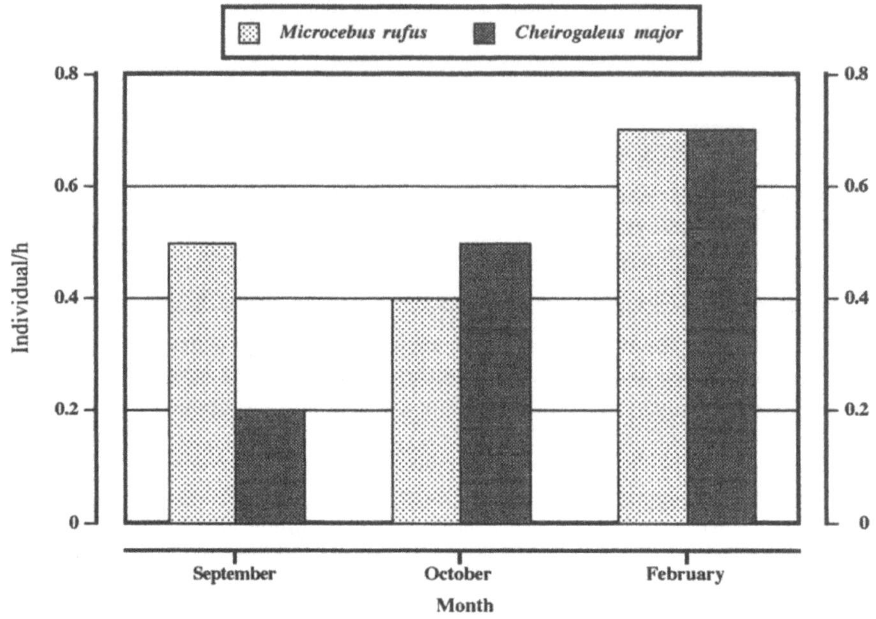

Figure 4. Nocturnal Census Results. Number of individuals sighted per hour at night at Ranomafana.

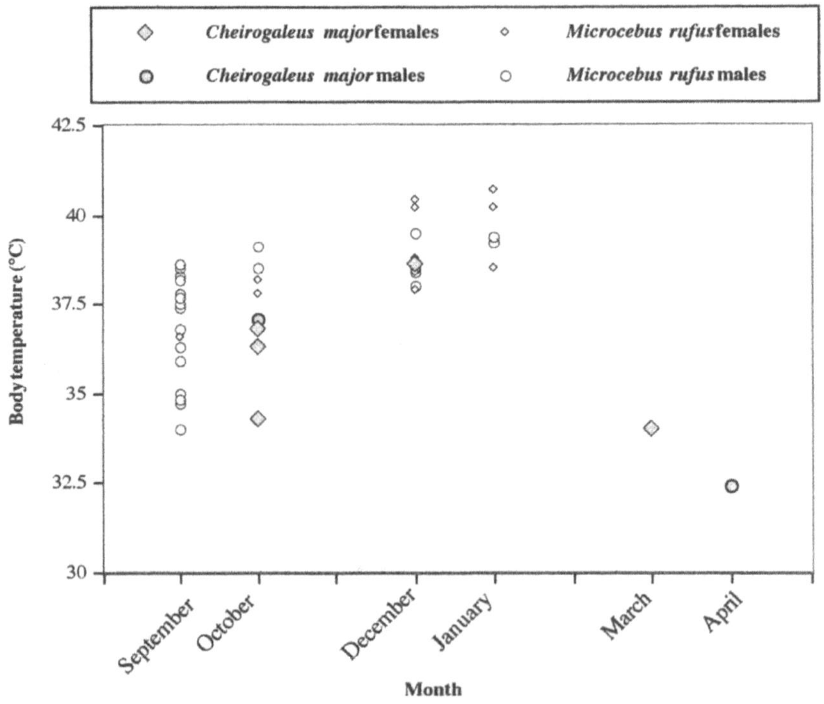

Figure 5. Body Temperature Variations by Month in *Cheirogaleus major* and *Microcebus rufus* at RNP.

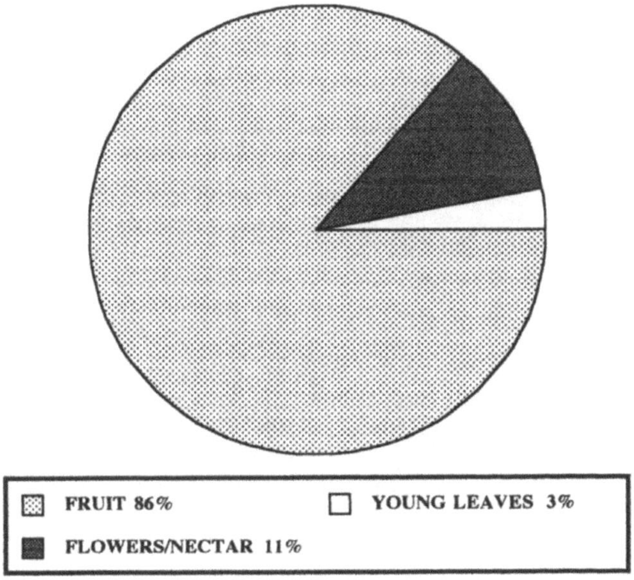

Figure 6. Percentages of *Cheirogaleus major* Diet. Percentage of feeding time on different food types calculated from five minute samples. N=158.

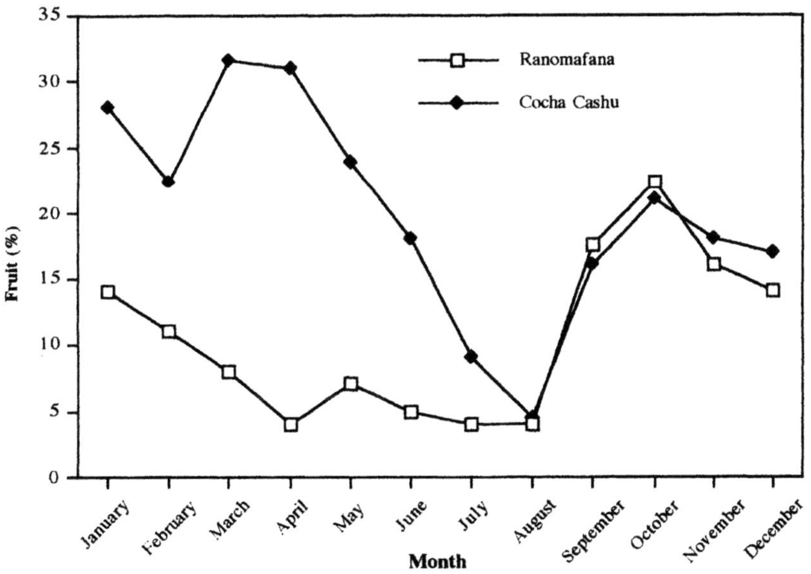

Figure 7. Fruiting seasonality in Ranomafana (Madagascar) and Cocha Cashu (Peru). Percentages of traps containing fruit each month. Data from Wright (1985) and Overdorff (1991).

Data from three annual cycles suggest that fruit availability at Cocha Cashu, Peru drops substantially through the months May through September (Terborgh, 1983; Wright, 1985; Janson, 1988). The mainstay of tamarin diet during these months of fruit scarcity is nectar (Terborgh & Stern, 1988; Goldizen *et al.*, 1988). Nectar feeding has been documented for both neotropical primates (Janson *et al.*, 1981; Terborgh & Stern, 1987; Garber, 1984; Sussman & Kinzey, 1984; & Wright, 1989; 1994) and Malagasy primates (Sussman & Raven, 1978; Sussman & Tattersall, 1976; Hladik *et al.* 1980; Sussman, 1991; Overdorff, 1992). The important difference between nectar-feeding in South American primates, and the nectar feeding of the rain forest Malagasy primates including *C. major* and *M. rufus*, is the seasonal timing. The Malagasy rain forest nectar is available when fruits are abundant, and not when fruits are scarce.

Phenology and fruit trap data from RNP show that overall fruit abundance is less in the rain forest of Madagascar and the period of fruit scarcity lasts longer than in the Amazon rain forest (Fig. 7; Wright, in press). Data from the rain forest of Manu National Park in Peru confirms that the period of fruit scarcity is limited to 3 mos (Terborgh, 1983; Janson, 1984; 1988); accordingly the Madagascar rain forest has fewer species of frugivorous birds or bats than any of the continental rain forests (Eisenberg, 1981; Langrand, 1990).

CONCLUSION

This study of radio collared *C. major* augments previous research that suggests that torpor evolved in *Cheirogaleus* as a defense against periods of food scarcity (Petter, 1978; Hladik *et al.,* 1980, Ganzhorn, 1988). Due to the extended period of fruit/nectar scarcity (twice as long as seen in other rain forests), and the lack of "keystone foods" (Terborgh, 1986) like nectar or gum to survive the period of minimum fruit production in Madagascar, these Malagasy primates minimize their energy expenditures by exhibiting a six month torpor. On the other hand, *M. rufus*, with a body mass ten times smaller than *C. major,* is able to feed on a few small-sized fruits and insects and remain active.

ACKNOWLEDGMENTS

The government of Madagascar, the Ministries of Water and Forests and of Universities, gave us permission for this research. The assistance of Association Nationale pour la Gestion des Aires Proteges (ANGAP) and the RNP staff is acknowledged. Many thanks to V. Razafimahatratra who provided insect sampling data, and G. Rakotonarina for data on phenology of trees. Without the expert field assistance of R. Randriamapionina, J.M. and P. Raliva this study would not have been possible. Thanks to S. Atsalis and L. Harste for stimulating conversations and support. S. Zack provided advice on banding methods. E. Raholimavo assisted with data collection. J. Cadle, C. Hemingway, J.E. Rafidison, A. Merenlender, E. Rajeriarison, E. Balko all contributed information on predation of lemurs. The Douroucouli Foundation and the John D. and Catherine T. MacArthur Foundation provided funds for this research. J. Jernvall gave advice and technical assistance with graphics. We thank K. Izard and L. Alterman for the opportunity to present these data at the Creatures of the Dark conference and for organizing the conference.

REFERENCES

Bearder, S.K. (1987). Lorises, bushbabies, and tarsiers: Diverse societies in solitary foragers. In B. B. Smuts, D. L. Cheney, R. M. Seyfarth, R.E. Wrangham, & T.T. Struhsaker (Eds.), *Primate Societies*, Chicago and London: The Oxford University Press, pp. 11-24.

Bearder, S.K., & Martin, R. D. (1980). The social organization of a nocturnal primate revealed by radiotracking. In C.J. Amlaner Jr., & D. W. Macdonald (Eds.), *A Handbook On Biotelemetry and Radio Tracking*, (pp. 633-648) Oxford: Pergamon Press.

Charles-Dominique, P. (1977). *Ecology and Behavior of Nocturnal Primates*, New York: Columbia University Press.

Charles-Dominique, P. (1983). Ecology and social adaptations in didelphid marsupials: comparison with eutherians of similar ecology. In J. F. Eisenberg, & D. G. Kleiman (Eds.), *Advances in the Study of Mammalian Behavior* (pp. 395-442). Lawrence, Kansas: American Society of Mammalogists.

Charles-Dominique, P., Atramentowicz, M., Charles-Dominique, M., Gerard, H., Hladik, C. M., & Prevost, M. F. (1991). Les mamiferes frugivores arboricoles nocturnes d'une foret Guyanaise: interrelations plantes-animaux. *Revue d'ecologie (La terre et la vie) 35*: 341-435.

Eisenberg, J. F. (1981). *The Mammalian Radiations.* Chicago: The University of Chicago Press.

Eisenberg, J. F., & Gould, E. (1970). The tenrecs; a study in mammalian behavior and evolution. *Smithsonian Contributions to Zoology* 27:v-138.

Emmons, L. (1990). *Neotropical Rainforest Mammals A Field Guide*, Chicago: University of Chicago Press.

Garber, P. (1984). Proposed nutritional importance of plant exudates in the diet of the Panamanian tamarin *(Saguinus oedipus goeffreyi), International Journal of Primatology* 1:185-201.

Ganzhorn, J.U. (1989). Niche separation of seven lemur species in the eastern rainforest of Madagascar. *Oecologia* 79:279-286.

Gautier-Hion, A. (1980). Seasonal variations of diet related to species and sex in a community of *Cercopithecus* monkeys. *Journal of Animal Ecology* 49: 237-269.

Goldizen, A. W., Terborgh, J., Cornejo, D. T., Porras, D. T., & Evans, R. (1988). Seasonal food shortage, weight loss, and the timing of births in saddle-back tamarins *(Saguinus fuscicollis) Journal of Animal Ecology* 57:893-901.

Goodman, S. M., Creighton, G. K., & Raxworthy, C. (1991). The food habits of the Madagascar long-eared owl (*Asio madagascariensis*) in south eastern Madagascar. *Bonn. Cool. Beitr.* pp.21-26.

Goodman, S. M., O'Connor, S., & Langrand, O. (1993). A review of predation on lemurs: implications for the evolution of social behavior in small, nocturnal primates. In P. M. Kappeler, & J. Ganzhorn (Eds.), *Lemur Social Systems and Their Ecological Basis*, (pp. 51-66). New York: Plenum Press.

Harcourt, C. S. (1987). Brief trap/retrap study of the brown mouse-lemur (*Microcebus rufus*). *Folia Primatologia* 49: 209-211.

Hladik, C. M. (1979). Diet and ecology of prosimians. In G. A. Doyle & R. D. Martin (Eds.), *The Study of Prosimian Behavior* (pp. 307-357). New York: New York Academic Press.

Hladik, C. M., Charles-Dominique, P., & Petter, J. (1980). Feeding strategies of five nocturnal prosimians in the dry forest of the west coast of Madagascar. In P. Charles-Dominique, H. M. Cooper, A., Hladik, C. M. Hladik, E. Pages, G. F. Pariente, A. Petter-Rousseaux, & A. Schilling (Eds.), *Nocturnal Malagasy Primates*, (pp. 41-73). New York: New York Academic Press.

Janson, C. H. (1984). Female choice and mating system of the brown capuchin monkey *Cebus apella (Primates: Cebidae)*. *Zeitschrift fur Tierpsychologie* 65:177-200.

Janson, C. H. (1988). Food competition in brown capuchin monkeys (*Cebus apella*): Quantitative effects of group size and tree productivity. *Behaviour* 105:53-76.

Janson, C. H., Terborgh, J. W., & Emmons, L.H. (1981). Non-flying mammals as pollinating agents in the Amazonian rainforest. *Reproductive Botany Biotropica* 14: 1-6.

Langrand, O. (1990). *A Field Guide to the Birds of Madagascar*, New Haven, Conn.: Yale University Press.

Martin, R. D. (1972). A preliminary field study of the lesser mouse lemur (*Microcebus murinus J. F. Miller 1777*). *Zeitschrift fur Tierpsychologie, Supplement* 9: 43-89.

McNab, B. K. (1983). Food habits, energetics, and population biology in small mammals. *American Naturalist* 116:106-124.

Meyers, D. M., & Wright, P. C. (1993). Resource tracking: food availibility and *Propithecus* seasonal reproduction. In J. U. Ganzhorn & P. K. Kappeler (Eds.), *Lemur Social Systems and Their Ecological Basis*. (pp. 179-192). New York, Plenum Press.

Morland, H. S. (1990). Parental behavior and infant development in ruffed lemurs (*Varecia variegata*) in a northeast Madagascar rain forest. *American Journal of Primatology*. 20: 253-265.

Overdorff, D. J. (1991). *Ecological Correlates To Social Structure In Two Prosimian Primates In Madagascar: Eulemur fulvus rufus* and *Eulemur rubriventer* Ph.D. Duke University.

Overdorff, D. J. (1992). Differential patterns in flower feeding by *Eulemur fulvus rufus* and *Eulemur rubriventer* in Madagascar. *American Journal of Primatology* 28: 191-203.

Petter, J.-J., Albignac, R., & Rumpler, Y. (1977). Mammiferes lemuriens (Primates prosimiens). *Faune de Madagascar* No. 44. ORSTOM-CNRS, Paris.

Petter, J.-J. (1978). Ecological and physiological adaptations of five sympatric nocturnal lemurs to seasonal variations in food production. In D. J. Chivers & J. Herbert (Eds.), *Recent Advances in Primatology Vol. 1.* (pp. 211-23). London, Academic Press.

Russell, J. (1974). Body temperature fluctuations in *Cheirogaleus medius.* In R. D. Martin, G. A. Doyle & A. C. Walker (Eds.), *Prosimian Biology* (pp.) , London: Duckworth.

Ryan, J. M., Creighton G. K. & Emmons L. H. (1993). Activity patterns of two species of *Nesomys* (Muridae: Nesomyinae) in a Madagascar rain forest, *Journal of Tropical Ecology* 9:101-107.

Sauther, M. L. (1989). Antipredator behavior in troops of free-ranging *Lemur catta* at Beza Mahafaly Special Reserve Madagascar. *International Journal of Primatology* 10:595-606.

Schmidt-Neilson, (1983). *Animal Physiology. Adaptation and Environment* (ed 3). Cambridge University Press, Cambridge.

Sussman, R. W. (1991). Primate origins and the evolution of angiosperms. In *American Journal of Primatology* 23: 209-223.

Sussman, R. W., & Kinzey W. G. (1984). The ecological role of the Callitrichidae: a review. *American Journal of Physcial Anthropology* 64:419-49.

Sussman, R. W., & Raven, P.H. (1978). Pollination by lemurs and marsupials: an archaic coevolutionary system. *Science* 200:731-736.

Sussman, R. W., & Tattersall, I. (1976). Cycles of activity, group composition, and diet of *Lemur mongoz mongoz Linnaeus, 1876* in Madagascar. *Folia Primatologica* 26:270-283.

Tattersall, I. (1982). *The Primates of Madagascar.* Columbia Univ. Press, New York, N.Y.

Terborgh, J. (1983). *Five New World Primates: A Study in Comparative Ecology.* Princeton University Press, Princeton, New Jersey.

Terborgh, J. (1986). Keystone plant resources in the tropical rain forest. In M. Soule (Ed.), *Conservation Biology: The Science of Scarcity and Diversity* (pp.330-344), Sunderland Massachusetts, Sinauer Associates, Inc.

Terborgh, J., & Stern, M. (1987). The surreptitious life of the saddle backed tamarin. *American Scientist* 75:260-2.

White, F. J., (1991). Social organization, feeding ecology and reproductive strategy of ruffed lemurs, *Varecia variegata*. In A. Ehara, T. Kimura, O. Takenaka, M. Iwamoto (Eds.) *Primatology Today: Proceedings of the XIII Congress of the International Primatological Society, Nagoya Kyoto*, (pp. 81-84). Amsterdam: Elsevier Science Publishers.

White, F. J., Overdorff, D. J., Balko, E. A., & Wright, P. C. (in press). Distribution of *Varecia variegata* in Ranomafana National Park, Madagascar. *International Journal of Primatology*.

Wright, P. C. (1985). Costs and benefits of nocturnality to the night monkey (*Aotus*). Ph.D. thesis, City University of New York, New York.

Wright, P. C. (1986b). Ecological correlates of monogamy in *Aotus* and *Callicebus*. In J. G. Else & P. C. Lee (Eds.) *Primate Ecology and Conservation* (pp. 159-167), New York: Cambridge University Press.

Wright, P. C. (1989). The nocturnal monkey niche in the New World. *Journal of Human Evolution* 18:635-658.

Wright, P.C. (1992). Primate ecology, rain forest conservation and economic development: building a national park in Madagascar. *Evolutionary Anthropology* 1: 25-33.

Wright, P.C. (1994). Night watch on the Amazon. *Natural History* 103 (5): 44-51.

Wright, P.C. (in press). Behavioral and ecological comparisons of Neotropical and Malagasy primates. In W. G. Kinzey (Ed.), *New World Primates: Ecology Evolution and Behavior*, New York, SUNY Press.

DETERMINANTS OF NIGHTTIME ACTIVITY IN "DIURNAL" LEMURID PRIMATES

Deborah J. Overdorff[1] and Michele A. Rasmussen[2]

[1]Department of Anthropology
University of Texas- Austin
Austin, TX 78712

[2]Department of Biological Anthropology and Anatomy
Duke University
3705-B Erwin Rd The Wheeler Bldg.
Durham, NC 27705

INTRODUCTION

Traditionally primates have been described as displaying either nocturnal (active only at night) or diurnal (active only during daylight hours) activity cycles. However, researchers investigating the behavior and ecology of Malagasy prosimian primates discovered that many species in the genus *Eulemur* traditionally labeled as diurnal also were active at night (*E. f. rufus:* Sussman, 1974; *E. mongoz*: Sussman and Tattersall, 1976; Tattersall and Sussman, 1975; *E. f. mayottensis*: Tattersall, 1977; Tattersall, 1979; *E. rubriventer*: Overdorff, 1988; *E. macaco*: Colquhoun, 1993; *E. coronatus* and *E. f. sanfordi* : Freed, pers. comm.). While *E. mongoz* appears to seasonally switch from a diurnal activity cycle to a completely nocturnal activity pattern, other species such as *E. fulvus* and *E. rubriventer* have been observed to be active irregularly during both the day and night. These observations led Tattersall (1988) and Fleagle (1988) to propose and define the word "cathemeral" as a possible label for this unusual activity pattern. However, it has not yet been determined whether activity during the day and night is a seasonal phenomenon or a year-round pattern typical for Malagasy lemurids. There is evidence that some New World primates may be active day and night. For example, howling monkeys move and feed at night, although rarely (Dahl and Hemingway, 1988). In addition, Wright (1985, 1989) observed that the traditionally nocturnal night monkey (*Aotus trivirgatus*) was active during the day in Paraguay presumably due to the rarity of large diurnal raptors.

To date, three main hypotheses have been offered as explanations for this unusual activity pattern in Malagasy primates. First, a combination of both day and night activity may represent a transitional stage between nocturnality and diurnality. As large diurnal mammals went extinct in Madagascar following the arrival of colonists two thousand years ago (Dewar, 1984), nocturnal primates took advantage of the newly available diurnal niches. Therefore, lemurs observed today are "evolving" the behavior and morphology of diurnal primates (Martin, 1972, Tattersall, 1982). However, this explanation is problematic given the absence of a tapetum lucidum in *Eulemur* species, which is not consistent with the idea that all lemurs were primitively nocturnal (Wolin and Massopust, 1970; Jacobs and Deegan, 1993). Tattersall's (1982) alternative explanation is that irregular nighttime activity may be ancestral to all "diurnal" lemurid primates, or at least represents a stable behavioral trait. Finally, Engqvist and Richard (1991) present a more ecologically based hypothesis. Lemurs may switch to a diet comprised of fibrous food items (e.g. leaves, flowers) when fruit is scarce or unavailable. Because these primates are typically small in body size and possess relatively simple digestive tracts, high-fiber foods must be consumed in large amounts over an extended time period, resulting in activity across a twenty-four hour cycle interspersed with rest periods.

Of the approaches listed above, Engqvist and Richard's hypothesis is the most testable. In this paper, we test their hypothesis using two types of data: 1) field data on the activity and behavior of two species of *Eulemur* in southeastern Madagascar, and 2) food passage rate tests conducted on five lemur species at the Duke University Primate Center (DUPC). In addition, we examine information on *Eulemur* dentition and gut morphology gathered from the literature. Field data are used to test the prediction that the degree of nightly activity correlates with the percentage of time spent feeding on leaves or flowers. In addition, we investigate other possible behavioral and ecological correlates to nightly activity such as climate, moonphase, food availability, percentage of time spent feeding, distance traveled the previous day and night, and overall degree of activity. We also use experimental data from passage rate tests conducted at the DUPC combined with functional analyses of masticatory and digestive efficiency from the literature to determine if lemurs are morphologically adapted to process fibrous foods. If not, as Engqvist and Richard hypothesize, then we predict that in these species 1) dentition will display little shearing crest development, 2) measures of masticatory efficiency will be low, 3) gut morphology will be generalized and inefficient at breaking down/digesting dietary fiber, and 4) food passage rates will be rapid in species that move both during the day and night relative to strictly diurnal species of equal body size.

METHODS

Field Study

Study Site and Subjects. The study site was in the Ranomafana National Park (RNP), a large (40,000 ha^2) southeastern stretch of rainforest in Madagascar located between 47 degrees 6'-47° 42' E longitude and 21° 0'-21° 24'S latitude. RNP is a continuous forest ranging from montane cloud forest (1500 m) to lowland rainforest (500 m). The main study site, known as Vatoharanana, was approximately 5 km south of the Duke research station and consisted of a high montane rainforest (1125 m) with an annual rainfall of 2300 mm during the study period. This region supports ten other sympatric primate species in addition to the two study species.

Two groups of *E. rubriventer* (RB, group size = 3) and one group of *E. fulvus rufus* (RL, group size = 9) were followed consistently throughout the study. On each observation day, a male and female focal animal were followed from 0600 to 1800 hr; different animals were followed on consecutive days. Focal samples were balanced between all adult individuals in each study group and each study group was sampled eight days per month from August 1988 through August 1989 with the help of a field assistant. No data are available on *E. f. rufus* for April, however, as they migrated from the study site. A second study group of *E. f. rufus* (group size = 7) was observed throughout the study on an opportunistic basis (n = 33 observation days). In addition, some all night follows (from 1800-0600 hr; RL: n = 8, RB: n = 3) and partial night follows (6 hr duration from 1800-2400 hr; RL: n = 11, RB: n = 3) were conducted on one group of each species. Moonphase was noted during all night samples. The nights with no moonlight are noted on Table 2.

Phenology, climate, diet, and activity. Phenological sampling was conducted twice a month to document food availability. These methods are described in detail elsewhere (Overdorff, 1992). In addition, the rainfall and maximum and minimum temperature were recorded daily.

A combination of focal animal and all occurrence sampling was used to collect data on feeding (Altmann, 1974). Observers recorded five types of data for each feeding bout initiated by the focal animal: duration, size of resource, number of individuals feeding with the focal animal (subgroup size), number and type of food removed by the focal animal, and location of the resource. Using these data, the following aspects of each lemur species' diet were quantified daily and could be used to test predictions regarding nightly activity: total amount of time an individual fed, and percentage of time spent feeding on ripe fruit, unripe fruit, mature and new leaves, flowers, and other food items. The methods used to calculate each of these variables are described in Overdorff (1993a).

Additionally, each focal animal's activity was noted at five minute intervals using scan samples during all day, all night, and partial night follows. Behaviors were designated as active (feed, travel, move, groom, play, scan) or non-active (rest). These data were used to calculate the percentage of time a focal animal spent active or resting during the day or night. The total number of scans for active versus rest behaviors based on partial and all night follows of groups were summed in each half-hour interval from 1800 to 0600 hr to determine the percentage of time spent active.

Because visibility was impaired during night samples, observers recorded whether any group member was feeding by noting the beginning and end time of a feeding bout (if possible), and the type of food being eaten. The food resource was labeled as described in the next section.

Ranging. Trails were mapped in 50 m intervals using a 50 m measuring tape and a compass (Overdorff, 1993b). All trees greater than 5 cm diameter at breast height used by either lemur species for food or resting sites were mapped onto a two-dimensional map of the study site using Pathfinder 1.5 Program designed especially for this project. Each tree visited was labeled with a numbered tree tag and mapped onto the study site map by measuring distance from previously mapped landmarks and taking a compass bearing.

These data were used to calculate daily path lengths and to estimate the distance the study group traveled at night. Daily path lengths (DPL) were calculated for every sample in which a focal animal was followed for at least 12 hr. Days were longest in December but animals were not followed any longer each day than in any other month. DPL was reconstructed using Pathfinder 1.5 by connecting trees and/or trail markers in the order they were visited or crossed each day. Pathfinder assumes paths traveled between trees were linear. DPL estimates for males and females were similar because both of the focal animals fed in the same trees at the same time.

Nightly path lengths (NPL) were reconstructed in two ways. First, NPL estimates were calculated using Pathfinder by connecting the point where animals were left the previous evening (no earlier than 1800 hr) to the point were they were found the following morning (no later than 0600 hr) on consecutive all day follows. Obviously, it is possible that groups will move more than once at night so that this estimate is the minimal possible distance groups traveled at night. We also recognize that animals do not move in straight lines from one point to the next. Our assumption here is that the distance traveled is a relative measure of nighttime activity. To verify this assumption, we used the same technique to estimate DPL using the data described in the above section. DPL estimatations underestimated actual DPL by less than a third and were correlated to actual measures in both lemur species (RL: r_S=.72; RB: r_S=.75, $p < .0001$). Second, direct measures of NPL were calculated using the same techniques to estimate DPL when animals were followed at night. When animals were followed during partial night samples, the distance traveled was added to the estimated distance for the remainder of the night.

Statistics. Data were analyzed on a daily basis except when data were compared to seasonal food availability. Data then were averaged in semi-monthly intervals ($n = 24$) to coincide with phenological sampling. The Mann-Whitney U-test was used to determine differences between species and between study groups of the same species. NPL estimates were compared to direct NPL using a Mann-Whitney test to determine if estimates may overestimate or underestimate NPL. Because data were ordinal and non-linear, Spearman's rank correlation coefficient was used to test for correlations between NPL when animals moved each night with the following variables: DPL, percentage of time spent feeding on different food types described above, and climate variables (rainfall and temperature).

Food Passage Rate Tests

Measures of food passage rates were made by one of us (M. A. R.) over five consecutive experimental trials for each of five lemur species during a five day period. The thirteen study animals representing five lemur species were held in adjacent outdoor runs at the DUPC. Each species was represented by at least two similarly sized animals (Table 1). Reports of nighttime activity have been made for *E. mongoz*, *E. fulvus*, and *E. rubriventer* in Madagascar. *Lemur catta* is similar to these lemurs in body size and overall morphology, but to date has been described as a diurnal primate in the literature. *Hapalemur griseus* was included for comparison because as a bamboo-specialist it is highly folivorous. *H. griseus* is also diurnal (but see Tattersall, 1988 and

Tien *et al.*, 1988). Before the onset of the experiments, the enclosures of the study animals were completely cleaned, and all old fecal material was removed. Prior to the daily provisioning of their usual diet of fresh fruit, vegetables, and Purina Monkey Chow™, the study animals were offered equal numbers of fresh banana pieces. In trials 2 through 4 cucumber slices were substituted for banana for *Hapalemur griseus* as this species preferred cucumber over banana. Soft plastic markers were concealed in the banana and cucumber slices (Milton, 1981; Ganzhorn, 1986). These markers were color-coded according to the date and time they were administered to facilitate the recording of passage times. The study animals were watched to ensure that the plastic markers were ingested, and the time of consumption was duly recorded. The study animals were then checked at 30 to 60-min intervals and all fecal matter was collected, with the date and time of its appearance recorded. All collected fecal material was placed in Ziploc© freezer bags and frozen for analysis once the trials were completed.

For all three *Eulemur* species and *L. catta* the markers were clearly visible in feces upon collection. To ensure that markers had not been passed earlier, all collected fecal material was analyzed upon completion of the feeding trials. This was achieved by partially thawing the frozen samples and grinding them with a pestle and mortar. The only markers found using this method were in the fecal material of *H. griseus*. These were extracted and checked for color, with their date and time of appearance subsequently recorded. Passage rates were estimated based on the time of first appearance (TFA) of markers. If, for example, markers were not present 3 hrs after feeding but apparent 3.5 hrs after feeding, TFA was estimated to span 185-210 min The mean time of this upper and lower limit was used as the estimate of passage rate, yielding a value of 200 minutes (rounding up to the next five-min interval). Species' food passage rates were compared using the Mann-Whitney U-test.

Table 1. Study subjects for food passage rate trials at the Duke University Primate Center.

Species	n	Age and Sex	Body Weight (g)[1]
Eulemur mongoz	1	Adult male	1570
	1	Adult female	1450
	1	Subadult female	1110
E. f. sanfordi	2	Adult male	2015
	1	Adult females	2205, 2635
E. rubriventer	1	Adult male	1800
	1	Subadult male	1450
Lemur catta	2	Adult males	2750, 2305
Hapalemur griseus	1	Adult male	1260
	2	Adult females	930, 860

1-Weights taken from DUPC records

RESULTS

Field Results- NPL

Table 2 summarizes the total number of estimated and all-night NPL collected for each lemur species. There was no significant difference between NPL generated from partial and all night follows and estimated distances for either lemur species (Mann-Whitney U-test RL: $Z = 1.40$, n.s.; RB: $Z = 1.43$, n.s.). Two patterns emerged when *E. f. rufus* and *E. rubriventer* were compared. First, *E. f. rufus* did not move as often at night as *E. rubriventer* (RL: $n = 11$, RB: $n = 3$, $G = 4.85$, d.f. $= 1$, $p < .05$). Second, both lemur species traveled less at night than during the day although when *E. f. rufus* traveled at night, they traveled further than *E. rubriventer* (Table 3).

Table 2. Summary of all nightly path lengths (estimates, all night, and partial night follows for *E. f. rufus* and *E. rubriventer*).

E. f. rufus					*E. rubriventer*				
	Date	**Distance**	**Note**	**Moon**		**Date**	**Distance**	**Note**	**Moon**
1988	09/01-09/02	0	N		1988	09/07-09/08	188		
	09/02-09/03	0	N			09/08-09/09	486		
	09/03-09/04	521	*N			09/13-09/14	253		None
	09/07-09/08	310				09/14-09/15	249		None
	09/08-09/09	0	N			09/15-09/16	114		None
	09/20-09/21	290				09/20-09/21	74		
	09/22-09/23	0	N			09/21-09/22	320		
	09/28-09/29	139	*N			09/27-09/28	254		
	09/29-09/30	0	N			10/04-10/05	337		
	10/11-10/12	182		None		10/05-10/06	313		
	10/12-10/13	179		None		10/06-10/07	135		
	10/13-10/14	480		None		10/26-10/27	173		
	10/14-10/15	0	N			11/09-11/10	200		None
	10/18-10/19	186				11/10-11/11	205	*N	None
	10/19-10/20	207				11/11-11/12	143		
	11/01-11/02	330				12/06-12/07	241		
	11/02-11/03	531	*N			12/07-12/08	96		
	11/03-11/04	187				12/21-12/22	317		
	11/14-11/15	115				12/28-12/29	289	*N	
	11/15-11/16	243				12/29-12/30	70		
	11/16-11/17	446			1989	01/12-01/13	335		
	11/17-11/18	413				02/01-02/02	43		
	11/18-11/19	350				02/02-02/03	350		
	11/30-12/01	584	*N			02/07-02/08	204		
	12/01-12/02	435				04/06-04/07	229		
	12/02-12/03	142		None		04/07-04/08	0	N	
	12/13-12/14	203				04/11-04/12	48		
	12/14-12/15	0	N			04/12-04/13	87		
	12/15-12/16	407				04/13-04/14	409		
1989	01/05-01/06	176				03/02-03/03	177		

Table 2 (con). Summary of all nightly path lengths (estimates, all night, and partial night follows for *E. f. rufus* and *E. rubriventer*)

E. f. rufus Date	Distance	Note	Moon	*E. rubriventer* Date	Distance	Note	Moon
01/23-01/24	0	N		03/03-03/04	204		
01/24-01/25	557			03/04-03/05	195		
02/21-02/22	454			05/03-05/04	341	*N	
02/22-02/23	379			05/04-05/05	321		
02/23-02/24	150	*N		05/09-05/10	269		
02/25-02/26	125			06/05-06/06	0	N	
03/13-03/14	162			06/06-06/07	199		
03/14-03/15	501			06/07-06/08	105		
03/15-03/16	251			06/08-06/09	94		
03/16-03/17	266	*N		06/20-06/21	212		None
03/24-03/25	200	*N	None	06/21-06/22	292		None
03/25-03/26	130		None	07/13-07/14	0	N	
05/24-05/25	155		None	07/14-07/15	72		
05/25-05/26	276			07/15-07/16	172		
05/26-05/27	133			07/25-07/26	182		
06/02-06/02	396			08/01-08/02	61		
06/13-06/14	192						
06/14-06/15	174	*N					
06/15-06/16	313						
06/28-06/29	180						
06/29-06/30	169						
07/18-07/19	0	N	None				
07/19-07/20	142		None				
07/20-07/21	0	N	None				
08/02-08/03	0	N					
08/03-08/04	518						
All samples:				**All samples:**			
AVERAGE	230			**AVERAGE**	197		
STDEV	171			**STDEV**	114		
N =	56			N =	46		

*N = all samples where animals were followed throughout the Night, N = partial night follows
In the Moon category, "None" indicates new moon.

Table 3. Comparison of daily and nightly travel patterns between *E. f. rufus* and *E. rubriventer.*

	Daily Path Length			Nightly Path Length[1]				
	Mean	SD	n	Mean	SD	n	Z	p
E. f. rufus	961.00	435.17	126	286.37	106.32	45	7.96	.0001
E. rubriventer	444.0	208.34	120	211.34	142.33	41	6.42	.0001

[1]Mann Whitney U test, difference between *E. f. rufus* and *E. rubriventer* NPL: $Z = 2.13$, $p < .03$.

The relationship of NPL to food seasonality and to daily travel patterns was examined for each species. NPL did not vary monthly for either lemur species (RL: $H = 8.62$, RB: $H = 10.65$, d.f. = 12, n.s.) and did not fluctuate with food availability (Table 4). In addition, the amount traveled daily or during the previous night did not seem to influence how far *E. rubriventer* traveled at night. In contrast, *E. f. rufus* traveled more at night following days when DPL was longer. Additional dietary, activity, and climatic variables were examined in relationship to NPL. *E. f. rufus* tended to travel more at night after feeding on unripe fruit during the previous day while *E. rubriventer* traveled more at night when they fed less daily. Finally, it was noted that moonphase did not influence NPL. Both lemur species traveled at night when the moon was full, half-full or absent (Table 2).

Table 4. Correlation of NPL with diet, activity, and climatic variables in *E. f. rufus* and *E. rubriventer.*

	E. f. rufus	*E. rubriventer*
DIET/ACTIVITY		
Fruit-ripe	-0.07	-0.07
Fruit-unripe	**0.30	0.21
Leaves-mature	-0.06	-0.04
New leaves	0.27	0.15
Flowers	0.01	0.07
Other foods	0.15	0.09
Total time feeding	-0.04	**-0.36
% Resting	-0.08	0.12
CLIMATE		
Rainfall	0.14	-0.10
Max. Temperature	0.09	0.24
Min. Temperature	*0.28	0.13
TRAVEL		
DPL	***0.39	-0.08
NPL	-0.12	-0.03

* $p < .05$, ** $p < .02$, *** $p < .01$

All night samples yielded some general information about type of food eaten and activity pattern. The type of food eaten at night during all night follows did not differ from the kinds of foods that either lemur species ate during the day. Both lemur species often used the same trees visited during the day. *E. f. rufus* and *E. rubriventer* fed three to four times at night at approximately two hour intervals. *E. rubriventer* also rested for longer periods of time than *E. f. rufus* at night (Fig. 1).

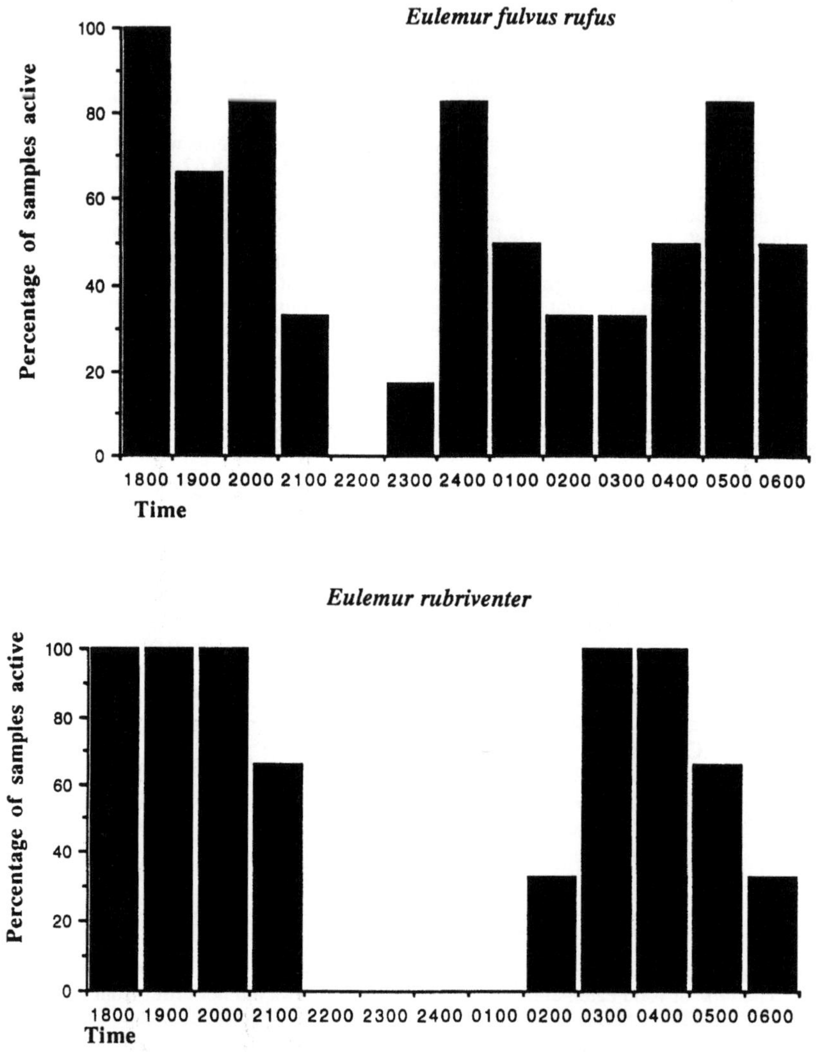

Fig. 1 Nightly activity cycle for E. f. rufus and E. rubriventer. Bars represent the percentage of samples during which each species was active during all night follows.

Table 5. Passage rates for five lemur species based on time of first appearance (TFA) of markers.

Trial / Species	1	2	3	4	5	Mean, ±SD (min)
E. mongoz [1]	205	150	135	80	85	131, 51.4
E. fulvus	100	85	155	70	60	94, 37.3
L. catta [2]	115	*335	140	90	170	170, 96.9
E. rubriventer [3]	180	*270	155	200	175	196, 44.4
H. griseus	3050	1605	2420	----	----	2358, 725

Underlined values indicate exact passage rate time (observer saw defecation of marker).
1- No significant difference between *E. mongoz* and *E. fulvus* ; U = 7.5, Z = 1.04.
2- No significant difference between *E. mongoz/E. fulvus* and *L. catta* ; U = 13.0, Z = 1.47
3- *E. rubriventer* has longer food passage rate time compared to *E. mongoz, E. fulvus*, and *L. catta* ; U = 10.5, A = 2.36, p < .02.

* see text

B. Food Passage Rate Tests

Table 5 shows the times of first appearance (TFA) of the plastic markers in the feces of the five lemur species used in the experimental feeding trials. Overall, there was no statistical difference in TFA among *L. catta, E. fulvus*, and *E. mongoz*. However, *E. rubriventer* had a significantly longer TFA compared to these three species The 270 and 335 min TFA recorded for *E. rubriventer* and *L. catta* respectively in trial 2 are probably overestimates for these species, and can best be accounted for by the fact that these animals were fed their normal diet more than two hours after the provisioning of the experimental banana. This lag between experimental feeding time and normal feeding time only occurred during trial 2, and only involved these two species. If the delay in provisioning of these lemurs' normal diet is corrected for, the passage rate time for trial 2 is approximately 180 min for *E. rubriventer* and 195 min for *L. catta* , which is more consistent with the other results. If trial 2 results for these two species are excluded from the analysis, *E. rubriventer's* passage rate times are still significantly longer than those of *L. catta, E. fulvus* and *E. mongoz*.

H. griseus clearly has a significantly longer passage rate than the other lemurs used in this study. Experimental food was only provided for four consecutive days when it became apparent that these animals were taking longer than 24 hours to pass the markers. On the last day of the feeding trials markers from trial 4 had not appeared; therefore, there are no passage rates given for trials 4 and 5. Of the 13 animals used in the feeding trials the three *H. griseus* were the only subjects that left several pieces of experimental food uneaten, and only four markers were recovered from fecal samples out of the 30 that were administered over the course of the experiment. The 3050 min passage time recorded for trial 1 seems very high when compared to the TFA of trials 2 and 3, and may represent the uppermost limits of food passage time in this species rather than the minimum passage time. However, no markers were retrieved from earlier fecal samples, so this rate remains as the TFA for trial 1 in *H. griseus*.

DISCUSSION

Results from this study diverged from what we predicted based on Engqvist and Richard's (1991) observations. Slightly different dietary variables were correlated with nightly movement in *E. f. rufus* and *E. rubriventer* in Madagascar. *E. f. rufus* moved more at night when they incorporated more unripe fruit in their daily diets. Fruit, particularly unripe fruit, can pose similar digestibility problems as leaves and flowers by varying in fiber content, degree of toughness, and presence of secondary compounds such as phenolics that are difficult to digest (Barton, 1989; Ganzhorn, 1988; 1989, Waterman *et al.*, 1983; Lucas, 1989; Rogers *et al.*, 1990). In addition, fecal samples that were collected in the field to determine how fruits and seeds were processed indicated that half to three quarters of the unripe fruit rind eaten by *E. f. rufus* passed through the digestive tract unprocessed. *E. f. rufus* may have to feed more often during both the day and night to meet individual nutritional requirements if they are 1) feeding on unripe fruit that is not broken down in small pieces and 2) fruit passes through their gut quickly. Therefore, we suggest that these lemurs may travel and forage at night in addition to their daytime movements when they incorporate difficult to digest foods in their diet.

In contrast, *E. rubriventer* traveled more at night when they spent less time feeding daily. In general, *E. rubriventer* fed less often on fibrous foods than *E. f. rufus*: they ate mostly ripe fruit and few leaves (Overdorff 1993b). In addition, in all fecal samples collected in the field, only seeds appeared and fruit rind was never detected. Therefore, the activity patterns of *E. rubriventer* may be influenced more by the amount of time spent feeding daily than dietary content. If they are feeding on ripe fruits which can be masticated more easily and are a higher quality food item relative to unripe fruit and leaves, they may not have to feed as often as *E. f. rufus*. Experimental tests of food passage rates showed that *E. rubriventer* can process foods more slowly than *E. f. rufus* which also may account for the differences in activity cycle. During the night, *E. f. rufus* alternated active and nonactive behaviors every two to three hours while *E. rubriventer* engaged in longer bouts of activity and rest (4-5 hours). This same pattern was observed during the day as well (Overdorff, 1991).

Results from the food passage rate tests conducted at DUPC indicate that *Eulemur* species can pass a relatively easily digestible food like banana through the gut very rapidly. Fast passage rates indirectly support the hypothesis that these prosimians may be challenged by seasonal increases in folivory (Milton, 1981; 1984); they must consume large amounts of leafy material presumably over a longer activity cycle as they lack the ability to hold fibrous foods in the gut for extended periods. However, it is important to realize that the potential for quick passage times observed in this study may reflect a trade-off between maximizing the absorption of nutrients found in fibrous foods and minimizing exposure to the potentially harmful toxins and chemical compounds found in leaves. In addition, given that these lemurs are consistently frugivorous throughout the year, passage rates that are too long would hinder the digestion of easily accessed sugars found in fruits.

Despite the potential for rapid passage rates shown here, evidence from previous studies of *Eulemur* dental and digestive morphology and efficiency do not support the idea that these species are strongly limited by an inability to handle dietary fiber. Several workers have found an association between primate diets high in structural carbohydrates (including chitin from insect exoskeletons, and cellulose, hemicellulose and lignin from plant cell walls) and molar morphology with pronounced shearing crest development (Hill, 1953; Kay, 1975; Kay and Hylander, 1978; Kay *et al.*, 1978; Kay and Sheine, 1979; Sheine, 1979; Maier, 1984; Anthony and Kay, 1993). Foods with a high dietary fiber or chitin content are better digested if an animal's masticatory apparatus can efficiently grind leaf or insect material prior to swallowing (Kay, 1975; Sheine and Kay, 1977; Sheine, 1979; Kay and Sheine, 1979). Kay and Hylander (1978) analyzed molar shape in 22 primates including 13 Malagasy prosimians, taking into account such functional measures as shearing blade length, cusp height, crushing basin size and length of contact between molars. Based on these measures *Eulemur fulvus* and *E. rubriventer* fell in with the folivorous class of primates, along with other leaf-eating Malagasy prosimians such as *Hapalemur, Lepilemur, Avahi, Indri,* and *Propithecus*. These findings were supported in Sheine's (1979) comparison of four prosimian species. Sheine determined coefficients of masticatory effectiveness for *L. catta, E. fulvus, Varecia variegata,* and *Otolemur crassicaudatus* by measuring their respective abilities to reduce the size of carrot pieces. She found that the coefficient of masticatory effectiveness for *E. fulvus* was significantly greater than those of the frugivorous *Otolemur* and *Varecia*. Given that *Lemur catta* displays a diurnal activity cycle in the wild (Sussman, 1972; Sauther, 1992), it is interesting to note Sheine's finding that this species has the highest coefficient of masticatory effectiveness of the four prosimians she studied. Kay *et al.*(1978) also found that despite its largely frugivorous diet (Sussman, 1974), *L. catta*'s molar dimensions are not significantly different from those of the more folivorous *E. f. rufus*. Although *E. mongoz* has not been included in any of these comparisons, Maier's (1984: 313) description of this species' dental morphology led him to conclude that "the dental morphology of *E. fulvus*...and *E. mongoz* is essentially the same". However, without detailed measurements of *E. mongoz*'s molar shearing crests, it may be premature to assume that this species shares the masticatory effectiveness of other *Eulemur* species based on gross molar morphology.

Relatively few detailed comparative studies have been made of the digestive anatomy of Malagasy prosimian primates. General information can be found in Chivers and Hladik (1980, 1984) who quantitatively describe dietary and gut specialization indices for a number of primate and other mammalian species in order to show the consistent relationship between diet composition and digestive adaptations (including stomach, caecum and colon capacity, and total surface area available for absorption). The most comprehensive description of lemur gut morphology can be found in Hill (1953). Based largely on Hill's findings, Tattersall (1982) described the gross digestive anatomy of Malagasy prosimians, comparing the shape, length and structure of the esophagus, stomach, duodenum, small intestine, caecum and colon among members of the Lemuridae. Only Sheine and Kay (1977), Sheine (1979), and Klein (1991) have directly compared digestive efficiency among prosimian species. In a series of experimental feeding trials, Sheine (1979) obtained cellulose digestibility measures for *L. catta*, *E. fulvus*, *V. variegata*, and *O. crassicaudatus*. She found that *L. catta* and *E. fulvus* digested 29.9% and 21.2% of ingested cellulose respectively, compared to only 5.9% for *V. variegata* and 0.8% for *O. crassicaudatus*. However, *L. catta* and *E. fulvus* were only able to maintain these proportions of cellulose digestion if food particle size was small; when Sheine increased the particle size of the experimental food, digestibility measures for the four study species did not differ significantly from zero. Combining these results with the measures of masticatory effectiveness described above, Sheine concludes that there is a high correlation between molar shearing crest development and cellulose digestion capabilities in these four prosimians. As would be expected, field studies of *E. fulvus* and *L. catta* have shown that the proportion of leaves in their diet in some regions (Sauther, 1992; Sussman, 1972) is considerably higher than that for both *O. crassicaudatus* which eats mostly fruit and gum (Bearder, 1987) and the frugivorous *V. variegata* (Morland, 1991; White, 1991). In another study, Klein (1991) measured and compared neutral-detergent fiber digestibility in *H. griseus* and *E. fulvus* and found that the smaller bamboo-specialist *Hapalemur* could digest 76.6% of dietary fiber while *E. fulvus* could only digest 58.9%. While lower than that of *H. griseus*, the digestibility value for *E. fulvus* compares favorably with the fiber-digesting abilities of other mammals. Klein (1991) reports values of 34.0% for the much larger chimpanzee, *Pan troglodytes*, and 25.1% and 50.0% respectively for koalas (*Phascolarctos cinereus*) and common brushtail possums (*Trichosurus vulpecula*) which are both *Eucalyptus* leaf specialists.

Curiously, *L. catta* , which is reported to be a diurnal species in the wild, has a higher measure of masticatory effectiveness than *Eulemur* species and can therefore digest cellulose effectively provided food pieces are small. In addition, food passage rates in ring-tailed lemurs are not significantly different from those of *E. mongoz* and *E. fulvus*. Given Engqvist and Richard's hypothesis, we might ask why the morphologically similar *L. catta* does not display the same activity patterns as *Eulemur* species. Specific qualities of their diet may make this possible. Sauther (1992) observed that in some seasons, *L. catta* feed primarily on young leaves and leaf buds which are easy to break down when processed (Barton, 1989; Rogers *et al.* 1990). In other seasons, they rely on small, ripe, soft bodied fruits which also are relatively easy to process. Therefore, the combination of a slightly higher degree of masticatory effectiveness than *Eulemur* species, and consumption of foods that are easier to process may allow this species to remain strictly diurnal.

We might add that without more detailed studies of year-round activity cycles in lemurs it is difficult to establish whether nighttime activity in "diurnal" lemurs is only a seasonal phenomenon. Engqvist and Richard (1991) cite a number of workers who have reported seasonal changes in activity during the day and night in *E. f. mayottensis*, *E. f. rufus*, *E. mongoz* and *E. rubriventer*, but the majority of these authors base their findings on studies that spanned only a few consecutive months of the year (Tattersall and Sussman, 1975; Sussman and Tattersall, 1976; Tattersall, 1976; 1977; 1979; Harrington, 1978; Overdorff, 1988). To our knowledge, no detailed studies to date have examined the fluctuations in nighttime activity of free-ranging Malagasy lemurid primates over an entire 12-month period. Until such studies are conducted, any attribution of nightly activity to seasonal variations in the environment must be tentative.

Therefore, field data, food passage rate tests on species of *Eulemur*, and morphological information gathered from the literature lead us to three conclusions regarding dietary and morphological correlates with day and night activity. First, based on observations of *E. f. rufus* and *E. rubriventer* in the wild, diet has some influence on nightly activity but not exactly as Engqvist and Richard (1991) predict. Second, food passage rates have the potential to be extremely short. Comparable studies on similar sized primates in other primate radiations must be conducted however, to determine whether these results are an artifact of small body size. Third, their dentition and gut morphology indicate that the lemur species examined here are able to process and effectively break down leafy food items. Although lacking large body size and digestive specializations for folivory, *Eulemur* species can subsist on a leafy diet through a combination of well-developed molar shearing crests and adequate cellulose breakdown processes in the digestive tract.

Properties characterizing food resources like unripe fruit other than dietary fiber may pose challenges to these relatively generalized prosimians. Given the results found here, we might expect that these properties are not readily countered by the ability to orally reduce food particle size. Instead, such foods as unripened fruit may require processing abilities in the gut which lemurids lack. Such limitations of their digestive tracts combined with swift passage rates may lead some *Eulemur* species to adopt a day and night activity cycle so that individual daily nutritional needs are met. Alternatively, the results obtained in this study for *E. rubriventer* indicate that this species regulates its nighttime activity based on its feeding rates of the previous day. This suggests that some lemurs are taking advantage of a flexible 24-hour activity cycle to compensate for periods during the day when less food is ingested for environmental and/or behavioral reasons. We suggest that examination of the properties of unripe fruit, comparative studies of lemurid gut morphology, and investigation of the year-round activity patterns in Malagasy *Eulemur* species should provide further insights into this unique behavior as it occurs in primates.

ACKNOWLEDGMENTS

D. J. O. would like to acknowledge M.R. Rakotoandriana and M.G. Rakotoanrivo of the Department of Water and Forests of Madagascar; M. Berthe Rakotsamimanana and M. Benjamin Andriamihaga of the Ministry of Higher Education for permission to work in Madagascar, and the Missouri Botanical Garden for plant identifications. B. Fox, C. van Schaik, I. Tattersall, and F. Ankel-Simons provided helpful comments concerning activity cycles in primates. Three anonymous reviewers provided useful comments concerning the organization of this manuscript. Special thanks go to M. E. Winslett for his patience and field assistance and his contribution of the Pathfinder program. D.J.O.'s field research was funded by National Science Foundation grant BNS-8819559, Wenner-Gren Foundation, Douroucouli Foundation, Duke University Travel Award, and National Sigma Xi Grand in Aid. M. A. R. also thanks the staff of the DUPC, K. Glander and B. Coffman for their assistance.

REFERENCES

Altmann, J. (1974). Observational Study of behavior: Sampling Methods. *Behaviour, 49*, 227-265.

Anthony, M.R.L. & Kay, R.F. (1993). Alouattine primates: Reflections on the comparative method. *American Journal of Science, 293A*, 356-382.

Barton, R.A. (1989). *Foraging Strategies, Diet and Competition in Olive Baboons* . Ph. D., University of St. Andrews.

Bearder, S.K. (1987). Lorises, bushbabies, and tarsiers: diverse societies in Solitary Foragers. In B.B. Smuts, D.L. Cheney, R.M. Seyfarth, R.W. Wrangham & T.T. Struhsaker (Eds.), *Primate Societies* , (pp. 11-24). Chicago:University of Chicago Press.

Chivers, D.J. & Hladik, C.M. (1980). Morphology of the gastrointestinal tract in primates: comparisons with other mammals in relation to diet. *Journal of Morphology, 166*, 337-386.

Chivers, D.J. & Hladik, C.M. (1984). Diet and gut morphology in primates. In D.J. Chivers (Ed.), *Food Acquistion and Processing in Primates* , (pp. 213-230). New York:Plenum Press.

Colquhoun, I.C. (1993). The socioecology of *Eulemur macaco*: a preliminary report. In P.M. Kappeler & J.U. Ganzhorn (Eds.), *Lemur Social Systems and Their Ecological Basis* (pp. 11-23). New York: Plenum Press.

Dahl, J.F. & Hemingway, C.A. (1988). An unusual activity pattern for the mantled howler monkey of Belize. *American Journal of Physical Anthropology, 75*, 201.

Dewar, R.E. (1984). Extinctions in Madagascar: The loss of the subfossil fauna. In P.S. Martin & R.G. Klein (Eds.), *Quarternary Extinctions* (pp. 574-593). Phoenix:University of Arizona Press.

Engqvist, A. & Richard, A. (1991). Diet as a possible determinant of cathemeral activity patterns in primates. *Folia Primatologica, 57,* 169-172.

Fleagle, J.G. (1988). *Primate Adaptation and Evolution.* San Diego:Academic Press, Inc.

Ganzhorn, J.U. (1986). Feeding behavior of *Lemur catta* and *Lemur fulvus. International Journal of Primatology, 7,* 17-30.

Ganzhorn, J.U. (1988). Food partitioning among Malagasy primates. *Oecologia, 75,* 436-450.

Ganzhorn, J.U. (1989). Primate species separation in relation to secondary compounds. *Journal of Human Evolution, 4,* 125-132.

Harrington, J.E. (1978). Diurnal behavior of *Lemur mongoz* at Ampijoroa, Madgascar. *Folia Primatologica, 29,* 291-302.

Hill, W.C.O. (1953). *Primates: Comparative Anatomy and Taxonomy I. Strepsirhini.* Edinburgh:The University Press.

Jacobs, G.H. & Deegan, J.F. (1993). Photopigments underlying color vision in Ringtailed lemurs (*Lemur catta*) and brown lemurs (*Eulemur fulvus*). *American Journal of Primatology, 30,* 243-256.

Kay, R.F. (1975). The functional adaptations of primate molar teeth. *American Journal of Physical Anthropology, 43,* 195-216.

Kay, R.F. & Hylander, W.L. (1978). The dental structure of mammalian folivores with special reference to primates and Phalangeroidea (Marsupialia). In G.G. Montgomery (Eds.), *Arboreal Folivores* , (pp. 173-187). Washington D.C.:Smithsonian Institute Press.

Kay, R.F. & Sheine, W.S. (1979). On the relationship between chitin particle size and digestibility in the primate *Galago senegalensis. American Journal of Physical Anthropology, 50,* 301-308.

Kay, R.F., Sussman, R.W. & Tattersall, I. (1978). Dietary and dental variations in the genus *Lemur,* with comments concerning general dietary-dental correlations among Malagasy strepsirhines. *American Journal of Physical Anthropology, 49,* 119-128.

Klein, M. (1991). *Digestibility of Detary Fiber and Passage Rates in Brown Lemurs and Gentle Lemurs* . M. S., Duke University.

Lucas, P.W. (1989). A new theory relating seed processing by primates to their relative tooth sizes. In L.H. Schmitt, L. Freedman & N.W. Bruce (Eds.), *Proceedings of the Australasian Society for Human Biology, No. 2, The Growing Scope of Human Biology* , (pp. 37-48). Center for Human Biology.

Maier, W. (1984). Tooth morphology and dietary specialization. In D.J. Chivers (Ed.), *Food Acquistition and Processing in Primates* , (pp. 303-330). New York:Plenum Press.

Martin, R.D. (1972). Adaptive radiation and behaviour of the Malagasy lemurs. *Philosophical Transcripts of the Royal Society, London, B264,* 295-352.

Milton, K. (1981). Food choice and digestive strategies of two sympatric primate species. *The American Naturalist, 117,* 496-505.

Milton, K. (1984). The role of food-processing factors in primate food choice. In P. Rodman & J. Cant (Eds.), *Adaptations for Foraging in nonhuman priamtes* , (pp. 249-279). New York:Columbia University Press.

Morland, H.S. (1991). *Social Organization and Ecology of the Black and White Ruffed Lemurs (Varecia variegata) in Lowland Rain Forest, Nosy Mangabe, Madagascar* . Ph.D., Yale University.

Overdorff, D.J. (1988). A preliminary report on the activity cycle and diet of the red-bellied lemur (*Eulemur rubriventer*) in Madagascar. *American Journal of Primatology,* 143-153.

Overdorff, D.J. (1991). *Ecological Correlates to Social Structure in Two Prosimian Primates in Madagascar: The Rufous Lemur (Eulemur fulvus rufus) and The Red-bellied Lemur (Eulemur rubriventer)* . Ph. D., Duke University.

Overdorff, D.J. (1992). Differential patterns in flower feeding by *Eulemur fulvus rufus* and *Eulemur rubriventer* in Madagascar. *American Journal of Primatology, 28,* 191-203.

Overdorff, D.J. (1993a). Similarites, differences, and seasonal patterns in the diets of *Eulemur rubriventer* and *Eulemur fuvlus rufus* in the Ranomafana National Park, Madagascar. *International Journal of Primatology, 14,* 721-753.

Overdorff, D.J. (1993b). Seasonal and reproductive correlates with ranging patterns in two prosimian primates in Madagascar. In P.M. Kappeler & J.U. Ganzhorn (Eds.), *Lemur Social Systems and Their Ecology* , (pp. 167-178), New York:Plenum.

Rogers, M., Maisels, F., Williamson, E.A., Fernandez, M. & Tutin, C.E.G. (1990). Gorilla diet in the Lope Reserve, Gabon: A nutritional analysis. *Oecologia, 84*, 326-339.

Sauther, M.L. (1992). *The Effect of Reproductive State, Social Rank, and Group Size on Resource Use Among Free-ranging Ringtailed Lemurs (Lemur catta) of Madagascar* . Ph. D., Washington University.

Sheine, W.S. (1979). *The Effect of Variations in Molar Morphology on Masticatory Effectiveness and Digestion of Cellulose in Prosimian Primates* . Ph.D., Duke University.

Sheine, W.S. & Kay, R.F. (1977). An analysis of chewed food particle size and its relationship to molar structure in the primates *Cheirogaleus medius* and *Galago senegalensis* and the insectivoran *Tupaia glis*. *American Journal of Physical Anthropology, 47*, 15-20.

Sussman, R.W. (1972). *An Ecological Study of Two Madagascar Primates: Lemur fulvus rufus Audebert and Lemur catta Linnaeus* . Ph. D., Duke University.

Sussman, R.W. (1974). Ecological distinctions in sympatric species of *Lemur*. In R.D. Martin, G. A. Doyle & A.C. Walker (Eds.), *Prosimian Biology* , (pp. 75-108). London:Duckworth.

Sussman, R.W. & Tattersall, I. (1976). Cycles of activity, group composition, and diet of *Lemur mongoz mongoz* Linnaeus 1766 in Madagascar. *Folia Primatologica, 26*, 270-283.

Tattersall, I. (1976). Group structure and activity rhythm in *Lemur mongoz* (Primates, Lemuriformes) on Anjouan and Moheli Islands, Comoro Archipelago. *Anthropological Papers of the American Museum of Natural History, 53*, 367-380.

Tattersall, I. (1977). Ecology and behavior of *Lemur fulvus mayottensis*. *Anthropology Papers of the American Museum of Natural History, 54*, 421-482.

Tattersall, I. (1979). Patterns of activity in the Mayotte lemur, *Lemur fulvus mayottensis*. *Journal of Mammology, 60*, 314-323.

Tattersall, I. (1982). *The Primates of Madagascar.* New York:Columbia University Press.

Tattersall, I. (1988), Cathemeral activity in primates: a definition. *Folia Primatologica, 49*, 200-202.

Tattersall, I. & Sussman, R.W. (1975). Observations on the ecology and behaviour of the mongoose lemur *Lemur mongoz mongoz* Linnaeus (Primates, Lemuriformes), at Ampijoroa, Madagascar. *Anthropological Papers of the American Museum of Natural History, 52*.

Tien, D., Rakotoarivelo, R. & Rasolonjatovo, W. (1988). Particularités de l'activité jounalière chez un jeune *Hapalemur griseus* (Primates: Lemuridae) isolé dès la naissance. *Mammalia, 52*, 453-464.

Waterman, P.G., Choo, G.M., Vedder, A.L. & Watts, D. (1983). Digestability, digestion-inhibitors and nutrients of herbaceous foliage and green stems from an African montane flora and comparison with other tropical flora. *Oecologia, 60*, 244-249.

White, F.J. (1991). Social Organization, feeding ecology, and reproductive strategy of ruffed lemurs, *Varecia variegata*. In A. Ehara, T. Kimura, O. Takenaka & M. Iwamoto (Eds.), *Primatology Today* , (pp. 81-84). Amsterdam:Elsevier Science Publishers.

Wolin, L.R. & Massopust, L.C. (1970). Morphology of the primate retina. In C. R. Noback & W. Montagna (Eds.), *The Primate Brain* (pp. 1-27). New York: Appleton-Century Crofts Educational Division/Meridith Co.

Wright, P.C. (1985). *Costs and Benefits of Nocturnality for Aotus (the night monkey)* . Ph. D., City University of New York.

Wright, P.C. (1989). The nocturnal primate niche in the New World. *Journal of Human Evolution, 18*, 635-658.

LIFE HISTORY VARIATION AMONG NOCTURNAL PROSIMIANS

Peter M. Kappeler

AG Verhaltensforschung/Ökologie
Deutsches Primatenzentrum
Kellnerweg 4
37077 Göttingen, Germany

INTRODUCTION

Principal life history traits determine the size of an organism at birth, how fast and how long it will grow, at what age and size it will mature, the number, size and sex of its offspring, as well as its reproductive and total life span. These traits are connected by numerous trade-offs, involving differential allocation of energy to growth, maintenance and reproduction. Because interactions among life history traits result in phenotypic adaptations that determine individual fitness through their direct effect on survival and reproduction, life histories are the key to understanding the action of natural selection (Gadgil and Bossert, 1970; Stearns, 1976, 1992).

Many different ways of combining life history traits and of resolving trade-offs have resulted in a fascinating diversity of life history strategies. Among primates, for examples, gorilla females give birth to a single offspring for the first time when they are about 10 years old and have additional offspring every 4-5 years thereafter. Mouse lemur females, on the other hand, reach sexual maturity within their first year of life, and produce one or two litters of 2-3 young every subsequent year. The other primates fall somewhere in between the two extremes of this slow-fast continuum (Harvey and Clutton-Brock, 1985; Harvey *et al.*, 1987).

Causes of Life History Variation

Two types of explanations have been proposed to account for the diversity of mammalian life histories. One is based on the observation that, across species, life histories tend to be highly correlated with body size, brain size and metabolic rate (Stearns, 1976, 1983; Western, 1979; McNab, 1980, 1986; Western and Ssemakula, 1982; Hennemann, 1983), which in turn are highly correlated among themselves (Martin, 1981; McNab, 1988; Harvey and Krebs, 1990). It has therefore been suggested that only one of these key variables is the target of selection, and that life history variables are constrained by morphological and physiological scaling principles to covary with them (Millar, 1981, 1984; Western and Ssemakula, 1982; Millar and Zammuto, 1983; Stearns, 1983).

An alternative view is that different life history strategies, which reflect optimal fecundity schedules, are the result of selection acting on a range of possible combinations of life history variables that are adapted to different mortality schedules and only broadly constrained by scaling principles (Gadgil and Bossert, 1970; Pianka, 1970; Pianka and Parker, 1975; Boyce, 1979, 1988; Michod, 1979; Harvey and Zammuto, 1985; Sutherland *et al.*, 1986; Harvey and

Creatures of the Dark: The Nocturnal Prosimians
Edited by L. Alterman *et al.*, Plenum Press, New York, 1995

Read, 1988; Harvey *et al.*, 1989; Read and Harvey, 1989; Promislow and Harvey, 1990; Harvey *et al.*, 1991).

Recent comparisons across mammalian orders supported the second explanation by demonstrating that the slow-fast continuum persists after controlling for the effects of body mass (Read and Harvey, 1989), and that variation in life history traits is not correlated with the proposed underlying mechanisms of allometric scaling, i.e., brain size, brain growth and metabolic rate, once the effects of body size are removed (Read and Harvey, 1989; Promislow and Harvey, 1990). In contrast, several studies of life history variation among prosimian primates supported the first explanation (but see Allman *et al.*, 1993) by suggesting that differences in basal metabolic rate, in particular, may be responsible for the characteristic range of variation in prosimian reproductive and developmental rates (Harvey *et al.*, 1987; Rasmussen, 1986; Rasmussen and Izard, 1988; Martin, 1984; Martin and MacLarnon, 1985; Young *et al.*, 1990).

The objective of this paper is to address this contradiction and to re-examine potential causes of interspecific differences in pre- and postnatal maternal reproductive investment among nocturnal prosimians. These variables summarize variation in gestation length, litter size, litter mass and infant growth rates, and therefore capture the most salient life history differences in this group (Charles-Dominique, 1977; Izard *et al.*, 1985; Izard and Rasmussen, 1985; Izard, 1987; Izard and Nash, 1988; Izard *et al.*, 1988; Rasmussen and Izard, 1988). Unlike previous studies, I controlled for the confounding effects of both body size and common ancestry before examining patterns and causes of life history variation.

Predictions

Life history variation has been linked to differences in body size, brain size, metabolic rate or age-specific mortality schedules. Body size is the most obvious predictor of many life history traits because it parallels their slow-fast continuum. Although body size defines the boundary conditions for life history variation, there is considerable size-independent variation (Read and Harvey, 1989). Among nocturnal prosimians, for example, the same-sized *Loris tardigradus* and *Cheirogaleus medius* produce average litters of one and three, respectively (Foerg, 1982; Izard and Rasmussen, 1985). Body size is, therefore, expected to covary with most life history traits, but we should expect significant residual variation.

Brain size has been invoked as an important determinant of life history variation because neural tissue grows slowly and therefore limits developmental rates (Sacher and Staffeldt, 1974), possibly because neonatal brain size and brain growth may be limited by maternal metabolic turnover (Martin, 1981). However, brain and body size are closely correlated, and correlations between brain size and some life history traits may therefore simply reflect their common dependence on body size (Pagel and Harvey, 1989; Harvey and Krebs, 1990). Thus, brain size and life history traits should covary independently of body size if brain size alone contributes to life history variation (cf. Allman *et al.*, 1993).

An intuitively compelling argument has been made for the direct dependence of life history traits on metabolic rate. According to this hypothesis, reproductive rates are as high as metabolic rates permit (Millar, 1977; McNab, 1980, 1986). Because of the well-known relationship between body size and metabolic rate (McNab, 1988), metabolic rate should be positively correlated with life history traits independent of their joint association with body size, if basal metabolic rate causes life history variation.

Finally, fecundity schedules are expected to be related to mortality schedules because fecundity must balance mortality in stable populations. Life history strategies of species in age-structured populations should therefore be adapted to age-specific schedules of mortality (Cole, 1954; Williams, 1966; Michod, 1979; Partridge and Harvey, 1988; Harvey *et al.*, 1989; Promislow and Harvey, 1990). Because causes of mortality are primarily extrinsic, i.e., resulting from environmental factors, it is more likely that mortality determines fecundity than vice versa. For the same reason, and in contrast to the other explanations, this hypothesis offers an opportunity to link life history variation to ecology.

A conclusive test of these evolutionary hypotheses is difficult for several reasons. Firstly, tests of their predictions have to be based on the comparative method because most primate life history traits elude experimental manipulation (Harvey and Pagel, 1991; Stearns, 1992). Although the comparative method cannot establish causality, it is, however, powerful enough to rule out certain hypotheses. Secondly, patterns of interspecific variation in life history variables may be confounded by phylogenetic effects because closely-related species may share certain combinations of traits through common ancestry (Harvey and Pagel, 1991; Stearns, 1992). Speciose taxa may, therefore, inflate sample size and produce spurious relationships. Finally, the conclusions drawn from the subsequent analyses should be considered preliminary until information on all crucial variables is available for the majority of prosimian species.

MATERIALS AND METHODS

I collected information on the following life history variables from the literature (Petter-Rousseaux, 1962; Doyle *et al.*, 1971; Charles-Dominique, 1977; Dixson and van Horn, 1977; Doyle, 1979; van Horn and Eaton, 1979; Charles-Dominique *et al.*,1980; Eaglen and Simons, 1980; MacKinnon and MacKinnon, 1980; Foerg, 1982; Nash, 1983; Niemitz, 1984; Izard and Rasmussen, 1985; Izard *et al.*, 1985; Harcourt, 1986; Izard and Simons, 1986; Wright *et al.*, 1986; Bearder, 1987; Izard, 1987; McNab and Wright, 1987; Izard and Nash, 1988; Izard *et al.*,1988; Rasmussen and Izard, 1988; Haring and Wright, 1989; Zimmermann, 1989; Kappeler, 1991; Niemitz *et al.*, 1991) and from the records of the Duke University Primate Center (DUPC):

Adult Female Mass (AFM): mean body mass (g) of sexually mature, non-pregnant females.
Gestation Length (GL): mean number of days (d) between copulation and birth.
Neonatal Mass (NM): mean body mass (g) of an individual infant on the day of birth.
Litter Size (LS): mean number of young per litter.
Litter Mass (LM): mean body mass (g) of an average litter (NM*LS).
Relative Litter Mass (RLM): litter mass as a proportion of maternal mass (LM/AFM).
Growth Rate (GR): mean growth rate (g/d) of an individual infant during early postnatal development. These estimates are based on growth curves fitted to mixed cross-sectional, longitudinal data sets, using a non-parametric regression technique (Cleveland,1979; Moses *et al.*,1992). For details see Kappeler (1992).
Litter Growth Rate (LGR): average postnatal growth (g/d) of an average litter per day (LS*GR).
Fetal Growth Rate (FGR): average prenatal growth (g/d) of an average litter per day (LM/GL).
Basal Metabolic Rate (BMR): average absolute basal metabolic rate (i.e., oxygen consumption rate) of a resting, fasting adult in its thermoneutral zone (ml O_2/h). These data were extracted from the literature (Hildwein and Goffart, 1975; Müller,1979, 1983; Müller and Jaksche, 1980; McCormick, 1981; Dobler, 1982; Müller *et al.*, 1985). It should be noted that most of these values were derived from only a few individuals of both sexes which did not contribute to the other life history data.
Brain Mass (BM): average adult brain mass (g). These data were also obtained from the literature (Bauchot and Stephan, 1966; Stephan and Bauchot, 1965; Armstrong, 1985).

I log-transformed prenatal growth rates (FGR), relative litter mass (RLM) and postnatal growth rates (LGR) and used them as dependent variables in the subsequent interspecific comparisons. I also log-transformed AFM, BM and BMR and calculated Pearson´s product-moment correlation coefficients among these six variables. Furthermore, I used partial correlation analyses to examine the relationship between pairs of variables independent of body size.

I calculated relative values for each dependent variable by measuring the residual variation around a best-fit line generated by least-squares regression of the respective life history trait on maternal body mass. I compared the average residual variation among the Lorisiformes, Lemuriformes and Tarsiiformes by one-way ANOVA. Significant ANOVAs were followed by post-hoc multiple comparisons, using the Tukey-Kramer method for unequal sample sizes. All statistical procedures followed Sokal and Rohlf (1981).

Taxonomic Dependence

Because previous studies indicate that taxonomy may explain a large proportion of the variation in primate life history traits (Harvey and Clutton-Brock,1985), I performed several nested ANOVAs with unequal sample size and Satterthwaite's approximation to examine the distribution of interspecific variation in reproductive traits across taxonomic levels. Because many genera in the present data set are represented by a single species, I compared variation among species within families, within families among infraorders, and among lemurs, lorises and tarsiers.

Because two characters may be associated across species due to common descent, rather than independent evolution, species' values in quantitative comparative studies are not statistically independent (Harvey and Pagel, 1991). Felsenstein (1985) pointed out that differences between species represent evolutionary independent events that can be used in comparative studies to examine whether evolutionary change has been correlated between two or more traits. It is also possible to estimate values for ancestral nodes from known species values by calculating standardized linear contrasts, which represent weighted sums of all lower taxa values that converge at a particular node (Grafen, 1989; Harvey and Purvis, 1991). I employed Grafen's method without specifying branch lengths to calculate phylogenetically independent contrasts, using software written by A. Purvis (see Pagel, 1992).

To control for the effects of phylogeny and body size simultaneously, I first calculated contrasts between body mass and each life history variable separately. After giving all body mass contrasts a positive sign, I forced a linear regression through the origin and measured the residual variation in life history contrasts for each trait. Finally, I plotted pairs of size-corrected residuals against each other and forced a regression through the origin to describe their relationship (Grafen, 1989; 1992; Harvey and Keymer, 1991; Harvey and Pagel, 1991; Harvey and Purvis, 1991; Lynch, 1991; Garland *et al.*, 1992; Pagel, 1992).

This new comparative method requires detailed information about phylogenetic relationships among species and higher taxa. As the true bifurcating phylogeny of prosimians is largely unknown, I based these tests on a phylogenetic hypothesis generated by studies of chromosomal evolution (Rumpler and Dutrillaux, 1986; Ishak *et al.*,1988; Rumpler *et al.*, 1990; Ishak *et al.*, 1992; Jung *et al.*, 1992; Crovella *et al.*, 1993). These relationships among nocturnal species are shown in Figure 1. Analyses based on alternative phylogenetic hypothesis yielded qualitatively identical results and are not presented here.

RESULTS

The data used in this study are summarized in Table 1. Most variables are highly correlated with one another (Table 2). Body mass is significantly correlated with all other variables and explains between 35% and 66% of the interspecific variation in these three reproductive variables. Thus, large nocturnal prosimians have higher pre- and postnatal litter growth rates than smaller species, but relative litter mass decreases with body size. The relationship between basal metabolic rate and life history variables is similar to that between body mass and life history traits (Table 2). Interestingly, brain size is only correlated with relative litter mass, but not with pre- and postnatal developmental rates (Table 2).

Pre- and postnatal litter growth rates are positively correlated with each other (Figure 2a). A significant positive correlation between these two variables also exists after controlling for the effects of body mass ($r = 0.71$, $P<0.01$), although the relative position of individual species along the continuum has changed (Figure 2b).

The major radiations of nocturnal prosimians differ in average life history traits after differences in body size among lemurs, lorises, galagos and tarsiers are controlled for. Comparisons of relative life history variables reveal that lemurs have significantly higher average relative prenatal litter growth rates than lorises, whereas those of galagos and tarsiers are intermediate and similar to each other (Figure 3).

Malagasy species also have the highest average relative postnatal litter growth rates among nocturnal prosimians, whereas tarsier females invest relatively little in postnatal growth on a daily basis (Figure 4). However, the average prenatal litter mass in relation to maternal size is significantly larger in tarsiers than in any of the other prosimian groups (Figure 5).

Table 1. Selected life history characters for nocturnal prosimians[1].

Genus	Species	AFM	BMR	BM	FGR	RLM	LGR
Otolemur	crassicaudatus	1242	421.0	10.3	0.54	0.059	9.74
Otolemur	garnettii	1028			0.42	0.052	5.60
Galagoides	demidovii	69	63.2	3.4	0.11	0.174	
Galagoides	zanzibaricus	136			0.13	0.110	0.52
Galagoides	alleni	260			0.18	0.092	
Galago	elegantulus	300		5.1			
Galago	moholi	155	120.1		0.17	0.134	2.11
Galago	senegalensis	266	198.0	4.8	0.15	0.080	1.96
Nycticebus	coucang	1195	286.8	12.5	0.27	0.044	4.42
Nycticebus	pygmaeus	376			0.17	0.081	2.06
Loris	tardigradus	193	130.5	6.9	0.06	0.052	0.78
Perodicticus	potto	989	356.0	14.0	0.20	0.033	3.19
Arctocebus	calabarensis	298	129.8	7.2	0.24	0.107	2.01
Lepilemur	ruficaudatus	845		7.6			
Lepilemur	mustelinus	594			0.27	0.067	
Cheirogaleus	medius	282	267.9	3.1	0.43	0.094	4.86
Cheirogaleus	major	443		7.1	0.78	0.123	
Mirza	coquereli	302			0.25	0.074	3.70
Microcebus	murinus	109	99.2	1.8	0.26	0.148	1.92
Microcebus	rufus	49					
Allocebus	trichotis	85					
Avahi	laniger	1316		11.5			
Daubentonia	madagascariensis	2573		45.2		0.053	
Tarsius	bancanus	117			0.16	0.239	0.44
Tarsius	dianae	110					
Tarsius	spectrum	113		4.7	0.19	0.256	
Tarsius	syrichta	117	76.7	3.6	0.14	0.222	0.48

[1]See Methods for definition.

Table 2. Correlations between life history variables[1].

	BM	BMR	FGR	LGR	RLM
AFM	0.87***	0.93***	0.59**	0.80***	-0.81***
BM		0.67*	0.16	0.38	-0.69***
BMR			-0.62*	0.86**	-0.83**
FGR				-0.81***	-0.18
LGR					-0.64**

[1]Pearson's product moment correlation coefficient. See Table 1 for sample sizes and text for definitions.
*P<0.05, **P<0.01, ***P<0.001.

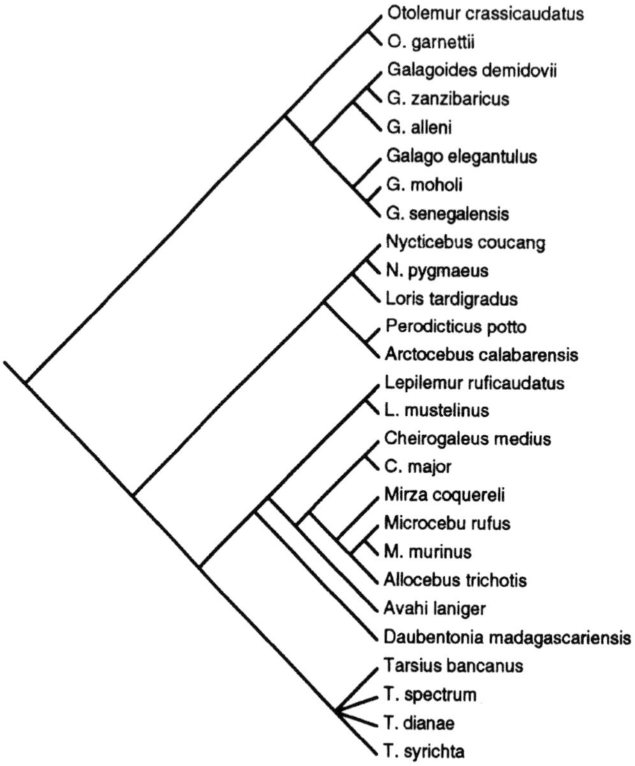

Figure 1. Hypothetical phylogenetic relationships among nocturnal prosimians based on patterns of chromosomal evolution.

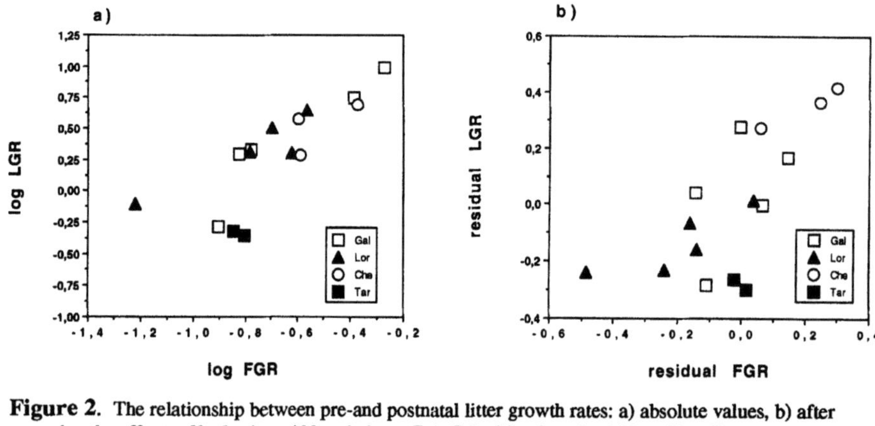

Figure 2. The relationship between pre-and postnatal litter growth rates: a) absolute values, b) after removing the effects of body size. Abbreviations: Gal: Galagidae, Lor: Lorisidae, Che: Cheirogaleidae, Tar: Tarsidae.

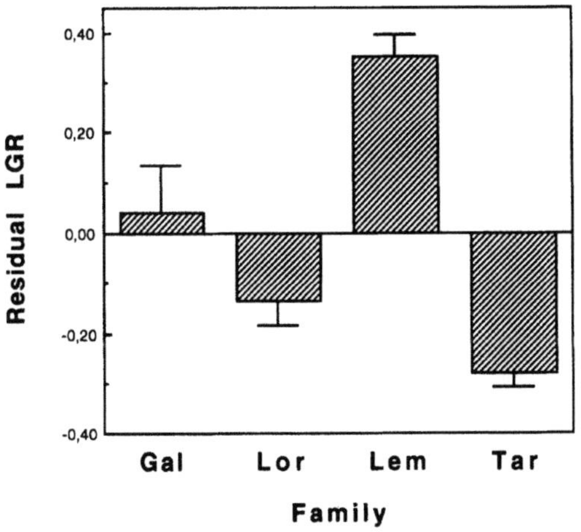

Figure 3. Average prenatal litter growth rates after controlling for the effects of body size. Mean (+-SE) shown for Galagidae (Gal), Lorisidae (Lor), Lemuriformes (Lem) and Tarsidae (Tar). The overall heterogeneity ($F=6.00$, $P<0.01$) is due to a significant difference between Lem and Lor ($P<0.05$).

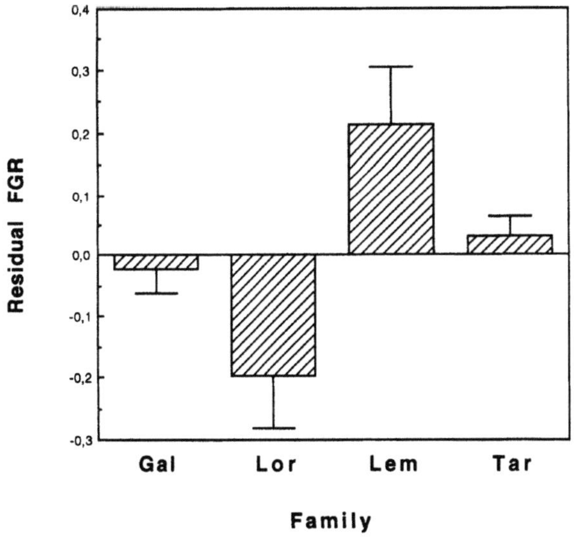

Figure 4. Average postnatal litter growth rates after controlling for the effects of body size. Legend as in Figure 3. The overall heterogeneity ($F=9.68$, $P<0.01$) is due to significant differences between Lem and Lor ($P<0.01$) and Lem and Tar ($P<0.01$).

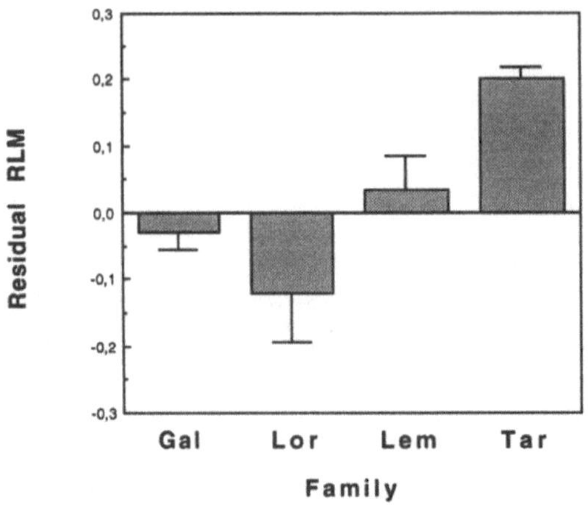

Figure 5. Average relative litter mass after controlling for the effects of body size. Legend as in Figure 3. The overall heterogeneity ($F=5.59$, $P<0.01$) is due to significant differences between Lor and Tar ($P<0.01$) and Gal and Tar ($P<0.05$).

I found that average prenatal litter growth rates do not vary significantly across taxonomic levels, and that the vast majority of variance in this variable was located within families (Table 3). Thus, even closely-related species vary considerably in prenatal maternal investment, suggesting that this trait is little constrained by common ancestry. Mean postnatal growth rates, on the other hand, varied among infraorders, but not among families (Table 3). Thus, there is more variation among higher taxa, indicating that variation at lower taxonomic levels is phylogenetically constrainted. I found a similar effect for relative litter mass, which varied significantly among infraorders, but not among families (Table 3).

I found little evidence for an independent role of brain size and metabolic rate in determining life history variation among nocturnal prosimians. Brain size was not associated with fetal growth rate rate or with relative litter mass, but I found a significant positive correlation with postnatal litter growth (Figure 6). Basal metabolic rate was negatively associated with fetal growth rate and relative litter mass and uncorrelated with litter growth rate (Figure 7). The independent contrasts of life history variables, on the other hand, were positively correlated with each other (Figure 8), indicating that certain suites of traits have co-evolved.

Unfortunately, tests of the relationship between life history traits and age-specific schedules of mortality are currently not possible because the relevant demographic data are unavailable.

Table 3. Taxonomic variation in life history traits[1].

Taxonomic level	**FGR**	**LGR**	**RLM**
Among Infraorders	1.7%	36.7%*	49.4%***
Among Families Within Infraorders	14.5%	9.3%	16.4%
Among Genera Within Families	83.8%	54.0%	34.2%

[1]Percent variation according to nested ANOVA. For FGR: $F_{2,2}=1.53$, NS; for LGR: $F_{2,1}=4,05$, $P<0.05$, $F_{1,11}=0.14$, NS; for RLM: $F_{2,3}=11.57$, $P<0.001$, $F_{3,15}=2.42$, NS.

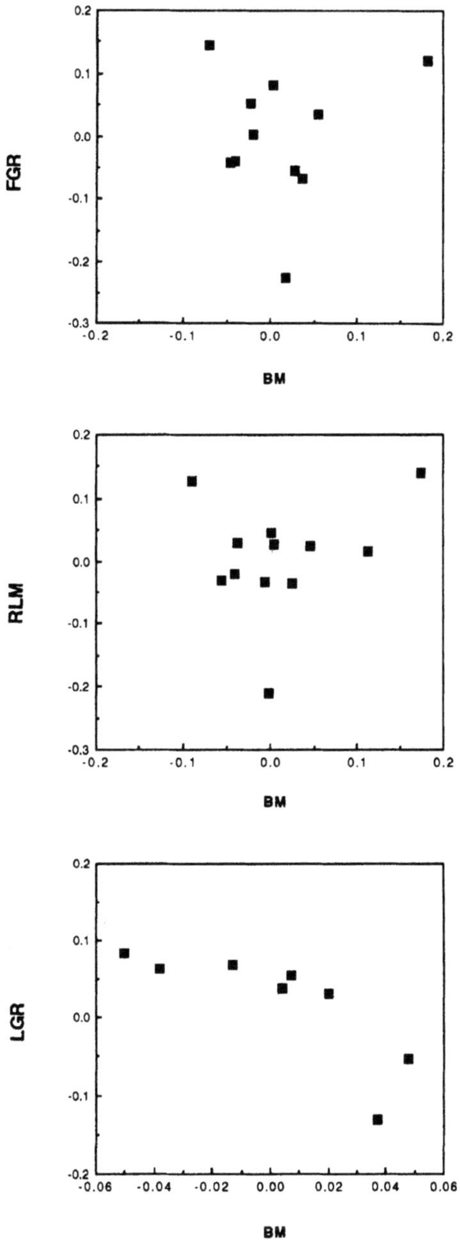

Figure 6. Phylogenetically independent contrasts between brain mass and reproductive traits after removing the effects of body size. For BM and FGR: b=1.67, r=0.11, NS (regression through the origin); for BM and RLM: b=0.28, r=0.22, NS; for BM and LGR: b=-1.66, r=0.72, $P<0.05$.

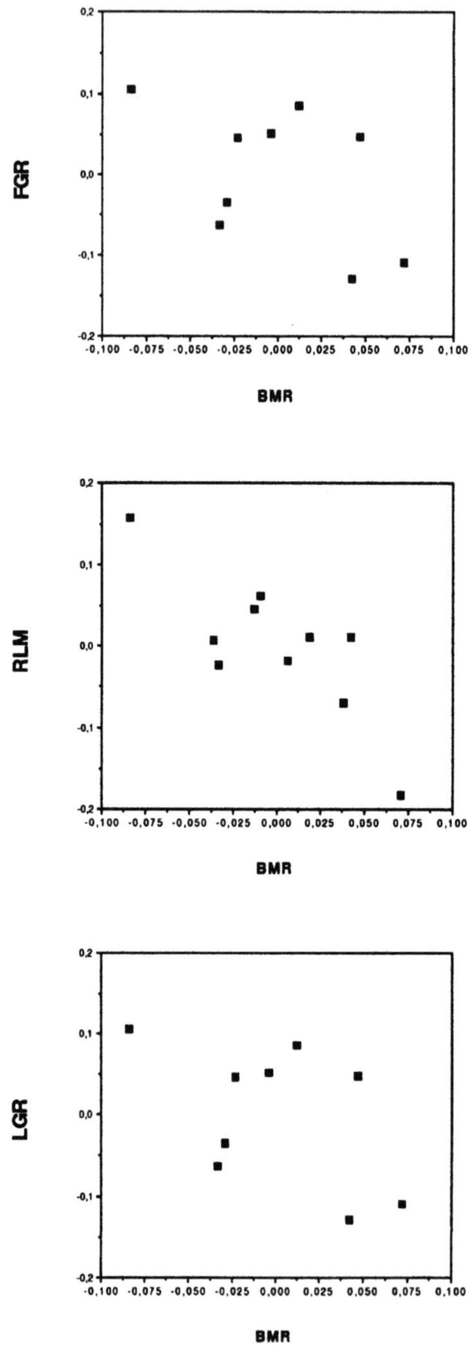

Figure 7. Phylogenetically independent contrasts between basal metabolic rate and reproductive traits after removing the effects of body size. For BMR and FGR: b=-1.61, r=0.70, P<0.05 (regression through the origin); for BMR and RLM: b=-1.58, r=0.81, P<0.01; for BMR and LGR: b=--0.91, r=0.51, NS.

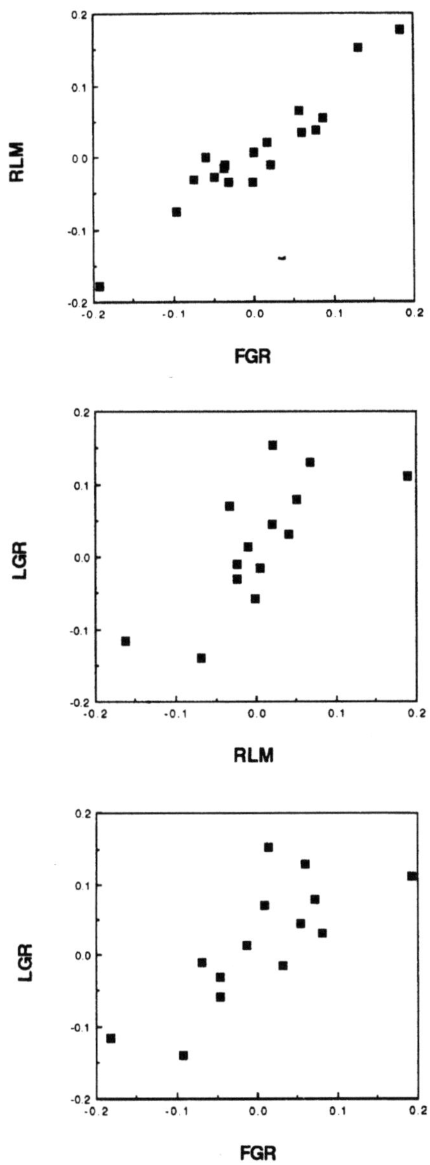

Figure 8. Phylogenetically independent contrasts among reproductive traits after removing the effects of body size. For FGR and RLM: b=0.85, r=0.95, P<0.001 (regression through the origin); for RLM and LGR: b=0.84. r=0.71, P<0.01; for FGR and LGR: b=0.75, r=0.75, P<0.01.

DISCUSSION

Determinants of Life History Variation

These analyses demonstrate that there is striking variation in life history traits related to maternal reproductive investment among nocturnal prosimians, and that part of this variation is related to differences in body size and common ancestry, but that differences in brain size and basal metabolic rate are not associated with this variation in the predicted way. The negative correlation between basal metabolic rate and two reproductive traits was an unexpected result. It indicates that size-independent evolutionary increases in BMR were accompanied by a decrease in fetal growth rates and a reduction in relative litter mass, whereas changes in BMR had no effect on postnatal litter growth. Size-independent evolutionary increase in relative brain mass, on the other hand, was associated with a decrease in postnatal growth, but was not associated with evolutionary changes in FGR and RLM.

Pre- and postnatal maternal investment rates, however, are positively correlated independently of body size and phylogenetic effects. Because variation in these reproductive traits can be arranged along a slow-fast continuum, that is independent of body size and other constraints, selection for particular combinations of traits, rather than passive scaling principles, is most likely responsible for the observed diversity of developmental and reproductive rates. This conclusion differs from previous studies, whose results may have been affected by smaller sample size (Rasmussen, 1986; Rasmussen and Izard, 1988), by not controlling for the relationship between body mass and metabolic rate (Young *et al.*, 1990), or by not controlling for phylogenetic dependence (Martin and MacLarnon, 1988). Furthermore, this conclusion is supported by recent findings that relative brain mass and basal metabolic rate are not correlated with relative female age at first reproduction and maximum life-span of strepsirhine primates (Allman *et al.*, 1993).

The hypothesis that slow and fast life histories are associated with different age-specific mortality schedules could not be tested. However, the strong association between mortality schedules and life history variation found in other mammalian groups should encourage long-term demographic studies of nocturnal prosimians to generate information on age-specific fecundity and mortality needed to test this hypothesis. In contrast to explanations invoking scaling principles, which accounted for only part of the observed variation, mortality schedules could provide a framework for a comprehensive understanding of the adaptive basis of prosimian life history variation.

Phylogenetic Conservatism

Life history variation among nocturnal prosimians is characterized by intruiging phylogenetic conservatism, i.e., much variation among higher taxa but little variation within them. Such similarities in life history variables among closely-related species could be due to functional constraints associated with a certain morphological "Bauplan", genetic drift or adaptive responses to fundamentally different selection pressures (Partridge and Sibly, 1991; Stearns, 1992).

Because the four major groups of nocturnal prosimians display significant variability in relative life history traits, the evolution of different life history strategies is not simply the result of differences in body size. Because prosimians are known for their low basal metabolic rates (Müller, 1985; Kurland and Pearson, 1986; Ross, 1992a), which are up to 60% below the expected values for mammals of their size (McNab, 1988), BMR has always been used to explain taxonomic variation in life history. For example, differences in basal metabolic rates were linked to differences in prenatal maternal investment rates between strepsirhine and haplorhine primates (Martin and MacLarnon, 1988), and to differences in gestation length, lactation length and postnatal growth rate between two species of galagos and lorises (Rasmussen and Izard, 1988). However, virtually all nocturnal species are hypometabolic (Ross, 1992a), and, as demonstrated by this study, BMR is not positively correlated with reproductive variables after controlling for body size. Thus, differences in BMR among lemurs, galagos, lorises and tarsiers are unrelated to the observed pattern of life history variation.

Differences in energy allocation and/or in changes in energy production during reproduction may provide a proximate explanation for the variation in pre- and postnatal growth rates among taxonomic groups (Nicoll and Thompson, 1987; Richard and Nicoll, 1987), although this is not very likely (Read and Harvey, 1989). Nevertheless, measurement of active daily metabolic rate, maximal metabolic rate and additional data on milk yield and milk energy content (see Oftedal and Tilden, this volume) would provide valuable insights into the energetics of prosimian reproduction and may also illuminate the potential proximate causes of phylogenetic conservatism.

Ultimately, the striking differences in life history strategies among the major prosimian radiations may be based on different mortality schedules. Although this hypothesis could not be evaluated in the present study, it will be considered in this discussion because alternative explanations fail and because pre- and postnatal growth rates among mammals in general vary according to its prediction (Pontier *et al.*, 1993). Age-specific mortality itself has intrinsic and extrinsic causes (Promislow and Harvey, 1990). Because intrinsic mortality is primarily related to the energetic costs of reproduction, high rates of adult mortality are unlikely to select for fast life history strategies because their high energetic costs contribute to the intrinsic causes of adult mortality. Thus, if a causal relationship between mortality and fecundity exists, it must be primarily driven by the extrinsic causes of mortality.

Predation is probably the most important cause of extrinsic mortality for nocturnal prosimians (Goodman *et al.*, 1993). If life history strategies vary as a function of mortality schedules, species with different life history strategies should therefore face predation risks of different intensity. This differential risk is independent of the number and density of predators because similar-sized nocturnal prosimians with fast and slow life histories occur sympatrically throughout much of Africa (Charles-Dominique, 1977; Bearder, 1987). It is, therefore, possible that the risk of predation can be modified by behavior, and that species with slow life histories, such as lorises and tarsiers, reduce their risk of predation through cryptic behavior (van Schaik and van Hooff, 1983; van Schaik, 1983; Terborgh and Janson, 1986) and possibly chemical defense mechanisms (MacKinnon and MacKinnon, 1980; Alterman, this volume).

The consequences of environmental seasonality may constitute a second important determinant of extrinsic mortality. The few months of ecologically favorable conditions have probably resulted in strong directional selection for the fast life history of fat-tailed dwarf lemurs (*Cheirogaleus medius*), for example (Hladik, 1980; Hladik *et al.*, 1980). The existing evidence in support of a predictable relationship between various life history variables and environmental factors, such as habitat type, diet and climatic variability, is mixed, however (Nash,1983; Harvey and Clutton-Brock,1985; Ross,1991,1992b). Information on seasonal variation in resource availability and climate in other prosimians is, therefore, needed to examine whether these are generally important sources of extrinsic mortality.

Life History Speed and Maternal Behavior

Nocturnal prosimians not only display striking variability in life history strategies, but also in several aspects of maternal behavior and infant care pattern. More specifically, they differ in the presence or absence of maternal nests and in the duration of mother-infant association during the night. Most newborn galagos and cheirogaleids, as well as *Daubentonia* infants, are left in nests or tree holes during their first days of life. The same shelters are used also by adults throughout the year. Lorises, tarsiers, lepilemurs and *Avahi*, on the other hand, have no such nests. They either carry their young from birth or park them in the vegetation during their activity periods.

The existence of maternal nests may be a consequence of the tendency of these species to produce litters. Lorises and tarsiers, in contrast, may have eliminated the need for maternal nests by producing relatively large and precocial infants that can be parked immediately after birth. Thus, the least developed nocturnal prosimian neonates are initially left in nests, whereas those with relatively long prenatal development can be parked soon after birth. There are two exceptions, however. First, *Loris* infants are born almost naked and are carried by the mother during the first month of life before they are parked for the first time (Rasmussen, 1986), suggesting that the presence of maternal nests may be more closely associated with litters than with the developmental state of the infants *per se*. Second, *Daubentonia* deposit their single offspring in a nest. A lack of detailed information on the developmental state of aye-aye neonates and maternal behavior in this species make this case difficult to interpret. Information on the gestation length and neonatal mass of *Avahi* and *Phaner*, which appear to carry their single young from birth, may help to clarify the relationship between nesting/parking and infant development among nocturnal prosimians.

The Relative Speed of Prosimian Life Histories

Nocturnal prosimians provide an evolutionary link between higher primates and other mammals (Martin, 1990). Moreover, because prosimians combine both ancestral and derived primate life history traits, it is interesting to contrast their life history strategies with those of other primates and mammals.

Among primates, nocturnal prosimians as a group are closer to the fast end of the continuum of life history speeds than are the anthropoids, whereas primates as a group, compared to other orders of mammals are close to the slow end. Although some prosimians have relatively shorter gestation periods than anthropoids, the two sub-orders do not differ significantly from each other in relative gestation length (Harvey and Clutton-Brock, 1985; Martin and MacLarnon, 1988). Compared to other mammals, primates have relatively long gestation periods, which are similar to those of the artiodactyls and bats, and longer than those of most carnivores, insectivores and rodents (Kihlstrom, 1972; Martin and MacLarnon, 1985;1988).

Prosimian neonatal mass is almost three times below that for anthropoids of the same body size (Leutenegger, 1973; Harvey *et al.*, 1987), but prosimians tend to have larger litter sizes (Chapman *et al.*, 1990). However, the difference in prosimian and anthropoid neonatal mass is not due to differential efficiencies of the epitheliochorial and hemochorial placentae (Martin and MacLarnon, 1988). In general, primate neonates, like ungulates and cetaceans, are relatively precocial, i.e., they have a well-developed hair coat and eyes and ears open at birth (Martin and MacLarnon, 1985).

Anthropoid mothers allocate more resources to prenatal growth per unit time than prosimian females (Martin and MacLarnon,1988; Ross, 1988; Young *et al.*, 1990), and New and Old World monkeys, do not differ in this respect from each other (Ross, 1991). Together with carnivores, however, primates in general have the lowest rates of maternal investment in individual fetal growth, compared to artiodactyls, insectivores and rodents (Gittleman and Oftedal,1987; Martin and MacLarnon,1988). This difference is further magnified, once prenatal growth rates of the entire litter are considered (Martin and MacLarnon, 1988).

Finally, some prosimians and callitrichids have the highest relative postnatal growth rates among primates (Kirkwood, 1985; Ross, 1991). Compared to other mammals, however, prosimian growth rates are only moderate and clearly slower than those of seals, ungulates, cetaceans and lagomorphs (Case, 1978). Thus, although nocturnal prosimians exhibit several ancestral mammalian reproductive traits, such as producing litters, nest building and infant parking, their overall life history strategies are clearly more primate-like and reflect an evolutionary shift towards producing precocial young.

SUMMARY

The nocturnal prosimians exhibit a fascinating diversity of life history strategies. The objective of this paper was to examine patterns and potential causes of this variation. Comparative analyses revealed that Malagasy species have relatively higher pre- and postnatal litter growth rates than true lorises and that lemurs also have higher postnatal litter growth rates than tarsiers. In relation to maternal mass, tarsiers produce a significantly larger litter mass than lorises and bushbabies. Relative litter mass and fetal and postnatal litter growth rates are positively correlated with body mass. Moreover, relative litter mass and postnatal litter growth do not vary among species independent of common ancestry. Phylogenetically independent contrasts, which control for both phylogenetic and allometric effects, showed that these reproductive traits are (1) not consistently associated with evolutionary changes in brain size, (2) not associated with evolutionary changes in basal metabolic rate, as expected, and (3) highly and positively correlated among themselves. The resulting slow-fast continuum in reproductive traits is, therefore, most likely the result of selection acting on a range of life history traits, which is possibly adapted to different mortality schedules, and only broadly constrained by scaling principles.

ACKNOWLEDGMENTS

I greatly appreciate the cooperation and assistance of R. Ange, B. Coffman, J. Drake, D. Haring, B. Hess, K. Izard, A. Katz, M. Ravosa, F. Stewart and K. Weissenseel in compiling the information presented in this paper. Thanks to C. van Schaik for discussion and comments, to G. Doyle and an anonymous reviewer for comments on this manuscript, and to A. Purvis for kindly providing a copy of C.A.I.C.. This is publication number 594 of the DUPC.

REFERENCES

Allman, J., McLaughlin, T., & Hakeem, A. (1993). Brain weight and life span in primate species. *Procedings of the Natliona Academy of Science USA 90*, 118-122.

Armstrong, E. (1985). Relative brain size in monkeys and prosimians. *American Journal of Physical Anthropology 66*, 263-273.

Bauchot, R., & Stephan, H. (1966). Données nouvelles sur l'éncephalisation des insectivores et des prosimiens. *Mammalia 30*, 160-196.

Bearder, S. K. (1987). Lorises, bushbabies, and tarsiers: diverse societies in solitary foragers. In B. B. Smuts, D. L. Cheney, R. M. Seyfarth, R. W. Wrangham & T. T. Struhsaker (Eds.), *Primate societies* (pp.11-24). Chicago: University of Chicago Press.

Boyce, M. S. (1979). Seasonality and patterns of natural selection for life histories. *American Naturalist 114*, 569-583.

Boyce, M. S. (1988). Evolution of life histories: theory and patterns from mammals. In M. S. Boyce (Ed.), *Evolution of life histories of mammals* (pp.3-30). New Haven: Yale University Press.

Case, T. J. (1978). On the evolution and adaptive significance of post-natal growth rates in terrestrial vertebrates. *Quarterly Review of Biology 53*, 243-282.

Chapman, C. A., Walker, S., & Lefebvre, L. (1990). Reproductive strategies of primates: the influence of body size and diet on litter size. *Primates 31*, 1-13.

Charles-Dominique, P. (1977). *Ecology and behaviour of nocturnal primates*. New York: Columbia University Press.

Charles-Dominique, P., Cooper, H. M., Hladik, A., Hladik, C. M., Pages, E., Pariente, G. F., Petter-Rousseaux, A., Petter, J.-J., & Schilling, A. (Eds.) (1980). *Nocturnal Malagasy Primates: Ecology, Physiology and Behavior*. New York: Academic Press.

Cleveland, W. S. (1979). Robust locally weighted regression and smoothing scatterplots. *Journal of the American Statistical Association 74*, 829-836.

Cole, L. C. (1954). The population consequences of life history phenomena. *Quarterly Review of Biology 29*, 103-137.

Crovella, S., Montagnon, D., & Rumpler, Y. (1993). Highly repeated DNA analysis and systematics of the Lemuridae, a family of Malagasy prosimians. *Primates 34*, 61-69.

Dixson, A. F., & van Horn, R. (1977). Comparative studies of morphology and reproduction in two subspecies of the greater bushbaby *Galago crassicaudatus crassicaudatus* and *Galago crassicaudatus argentatus*. *Journal of Zoology, London 183*, 517-526.

Dobler, H. J. (1982). Temperaturregulation und Sauerstoffverbrauch beim Senegal- und Zwerggalago (*Galago senegalensis, Galago demidovii*). *Bonner Zoologische Beiträge 33*, 33-59.

Doyle, G. A., Andersson, A., & Bearder, S. K. (1971). Reproduction in the lesser bushbaby (*Galago senegalensis moholi*) under semi-natural conditions. *Folia Primatologica 14*, 15-22.

Doyle, G.A. (1979). Development of behavior in prosimians with special reference to the lesser bushbaby, *Galago senegalensis moholi*. In G.A. Doyle & R.D. Martin (Eds.),*The Study of Prosimian Behavior* (pp.157-206). New York: Academic Press.

Eaglen, R. H., & Simons, E. L. (1980). Notes on the breeding biology of thick-tailed and silvery galagos in captivity. *Journal of Mammalogy 61*, 534-537.

Felsenstein, J. (1985). Phylogenies and the comparative method. *American Naturalist 125*, 1-25.

Foerg, R. (1982). Reproduction in *Cheirogaleus medius*. *Folia Primatologica 39*, 49-62.

Gadgil, M., & Bossert, W. H. (1970). Life historical consequences of natural selection. *American Naturalist 104*, 1-24.

Garland, T., Harvey, P. H., & Ives, A. R. (1992). Procedures for the analysis of comparative data using phylogenetically independent contrasts. *Systematic Biology 41*, 18-32.

Gittleman, J. L., & Oftedal, O. T. (1987). Comparative growth and lactation energetics in carnivores. *Symposia of the Zoological Society, London 57*, 41-77.

Goodman, S. M., O'Connor, S., & Langrand, O. (1993). A review of predation on lemurs: implications for the evolution of social behavior in small, nocturnal primates. In P. M. Kappeler & J. U. Ganzhorn (Eds.), *Lemur social systems and their ecological basis* (pp.51-66). New York: Plenum Press.

Grafen, A. (1989). The phylogenetic regression. *Philosophical Transactions of the Royal Society, London B 326*, 119-157.
Grafen, A. (1992). The uniqueness of the phylogenetic regression. *Journal of theoretical Biology 156*, 405-423.
Harcourt, C. S. (1986). *Galago zanzibaricus*: birth seasonality, litter size and perinatal behaviour of females. *Journal of Zoology, London 210*, 451-457.
Haring, D. M., & Wright, P. C. (1989). Hand-raising a Phillipine tarsier, *Tarsius syrichta*. *Zoo Biology 8*, 265-274.
Harvey, P. H., & Clutton-Brock, T. H. (1985). Life history variation in primates. *Evolution 39*, 559-581.
Harvey, P. H., & Keymer, A. E. (1991). Comparing life histories using phylogenies. *Philosophical Transactions of the Royal Society, London B 332*, 31-39.
Harvey, P. H., & Krebs, J. R. (1990). Comparing brains. *Science 249*, 140-146.
Harvey, P. H., Martin, R. D., & Clutton-Brock, T. H. (1987). Life histories in comparative perspective. In B. B. Smuts, D. L. Cheney, R. M. Seyfarth, R. W. Wrangham & T. T. Struhsaker (Eds.), *Primate societies* (pp.181-196). Chicago: University of Chicago Press.
Harvey, P. H., & Pagel, M. D. (1991). *The comparative method in evolutionary biology*. Oxford: Oxford University Press.
Harvey, P. H., Pagel, M. D., & Rees, J. (1991). Mammalian metabolism and life histories. *American Naturalist 37*, 556-566.
Harvey, P. H., Promislow, D. E. L., & Read, A. F. (1989). Causes and correlates of life history differences among mammals. In V. Standen & R. A. Foley (Eds.), *Comparative socioecology* (pp.305-318). Oxford: Blackwell.
Harvey, P. H., & Purvis, A. (1991). Comparative methods for explaining adaptations. *Nature 351*, 619-624.
Harvey, P. H., & Read, A. F. (1988). How and why do mammalian life histories vary? In M. S. Boyce (Ed.), *Evolution of life histories of mammals* (pp.213-232). New Haven: Yale University Press.
Harvey, P. H., Read, A. F., & Promislow, D. E. L. (1989). Life history variation in placental mammals: unifying the data with theory. In P. H. Harvey & L. Partridge (Eds.), *Oxford surveys in evolutionary biology* (pp.13-31). Oxford: Oxford University Press.
Harvey, P. H., & Zammuto, R. M. (1985). Patterns of mortality and age at first reproduction in natural populations of mammals. *Nat.*ure *315*, 319-320.
Hennemann, W. W. (1983). Relationship among body mass, metabolic rate, and the intrinsic rate of natural increase in mammals. *Oecologia 56*, 104-108.
Hildwein, G., & Goffart, M. (1975). Standard metabolism and thermoregulation in a prosimian. *Perodicticus potto*. *Comparative Biochemistry and Physiology 50A*, 201-213.
Hladik, A. (1980). The dry forest of the west coast of Madagascar: climate, phenology, and food available for prosimians. In P. Charles-Dominique, H. M. Cooper, A. Hladik, C. M. Hladik, E. Pages, G. F. Pariente, A. Petter-Rousseaux, J.-J. Petter & A. Schilling (Eds.), *Nocturnal malagasy primates: ecology, physiology, and behavior* (pp.3-40). New York: Academic Press.
Hladik, C. M., Charles-Dominique, P., & Petter, J.-J. (1980). Feeding strategies of five nocturnal prosimians in the dry forest of the west coast of Madagascar. In P. Charles-Dominique, H. M. Cooper, A. Hladik, C. M. Hladik, E. Pages, G. F. Pariente, A. Petter-Rousseaux, J.-J. Petter & A. Schilling (Eds.), *Nocturnal malagasy primates: ecology, physiology, and behavior* (pp.41-73). New York: Academic Press.
Ishak, B., Warter, S., Dutrillaux, B., & Rumpler, Y. (1988). Phylogenetic relation between Lepilemuridae and other lemuriform families. *American Journal of Primatology 15*, 275-280.
Ishak, B., Warter, S., Dutrillaux, B., & Rumpler, Y. (1992). Chromosomal rearrangements and speciation of sportive lemurs (*Lepilemur* species). *Folia Primatologica 58*, 121-130.
Izard, M. K. (1987). Lactation length in three species of *Galago*. *American Journal of Primatology 13*, 73-76.
Izard, M. K., & Nash, L. T. (1988). Contrasting reproductive parameters in *Galago senegalensis braccatus* and*Galago senegalensis moholi*.*International Journal of Primatology 9*, 519-527.
Izard, M. K., & Rasmussen, D. T. (1985). Reproduction in the slender loris, *Loris tardigradus malabaricus*. *American Journal of Primatology 8*, 153-166.
Izard, M. K., & Simons, E. L. (1986). Infant survival and litter size in primigravid and multigravid Galagos.*Journal of Medical Primatology 15*, 27-35.
Izard, M. K., Weisenseel, K. A., & Ange, R. L. (1988). Reproduction in the slow loris (*Nycticebus coucang*). *American Journal of Primatology 16*, 331-339.
Izard, M. K., Wright, P. C., & Simons, E. L. (1985). Gestation length in *Tarsius bancanus*. *American Journal of Primatology 9*, 327-331.
Jung, K. Y., Crovella, S., & Rumpler, Y. (1992). Phylogenetic relationships among lemuriform species determined from restriction genomic DNA banding patterns. *Folia Primatologica 58*, 224-229.
Kappeler, P. M. (1991). Patterns of sexual dimorphism in body weight among prosimian primates. *Folia Primatologica 57*, 132-146.
Kappeler, P. M. (1992). Female dominance in Malagasy primates. Ph.D. Thesis, Duke University, Durham, NC.
Kihlstrom, J. E. (1972). Period of gestation and body weight in some placental mammals. *Comparative Biochemistry and Physiology 43A*, 673-679.
Kirkwood, J. K., (1985). Patterns of growth in primates. *Journal of Zoology, London 205*, 123-136.

Kurland, J. A., & Pearson, J. D. (1986). Ecological significance of hypometabolism in nonhuman primates: allometry, adaptation, and deviant diets. *American J.ournal of Physical Anthropology 71*, 445-457.
Leutenegger, W. (1973). Maternal-fetal weight relationships in primates. *Folia Primatologica 20*, 280-293.
Lynch, M. (1991). Methods for the analysis of comparative data in evolutionary biology. *Evolution 45*, 1065-1080.
McCormick, S. A. (1981). Oxygen consumption and torpor in the fat-tailed dwarf lemur (*Cheirogaleus medius*): rethinking prosimian metabolism. *Comparative Biochemistry and Physiology A68*, 605-610.
McNab, B. K. (1980). Food habits, energetics, and the population biology of mammals. *American Naturalist 116*, 106-124.
McNab, B. K. (1986a). The influence of food habits on the energetics of eutherian mammals. *Ecological Monographs 56*, 1-19.
McNab, B. K. (1986b) Food habits, energetics, and the reproduction of marsupials. *Journal of Zoology, London 208*, 595-614.
McNab, B. K. (1988). Complications inherent in scaling the basal rate of metabolism in mammals. *Quarterly Review of Biology 63*, 25-54.
McNab, B. K., & Wright, P. C. (1987). Temperature regulation and oxygen consumption in the Phillipine tarsier,*Tarsius syrichta*. *Physiological Zoology 60*, 596-600.
MacKinnon, J., & MacKinnon, K. (1980). The behavior of wild spectral tarsiers. *International Journal of Primatology 1*, 361-379.
Martin, R. D. (1981). Relative brain size and basal metabolic rate in terrestrial vertebrates. *Nature 293*, 56-60.
Martin, R. D. (1984). Scaling effects and adaptive strategies in mammalian lactation. *Symposia of the Zoological Society, London 51*, 87-117.
Martin, R. D. (1990). *Primate origins and evolution*. London: Chapman and Hall.
Martin, R. D., & MacLarnon, A. M. (1985). Gestation period, neonatal size, and maternal investment in placental mammals. *Nature 313*, 220-223.
Martin, R. D., & MacLarnon, A. M. (1988). Comparative quantitative studies of growth and reproduction. *Symposia of the Zoological Society, London 60*, 39-80.
Michod, R. E. (1979). Evolution of life histories in response to age-specific mortality factors. *American Naturalist 113*, 531-550.
Millar, J. S. (1977). Adaptive features of mammalian reproduction.*Evolution* 31:370-386.
Millar, J. S. (1981). Pre-partum reproductive characteristics of eutherian mammals.*Evolution 35*, 1149-1163.
Millar, J. S. (1984). The role of design constraints in the evolution of mammalian reproductive rates. *Acta Zoologica Fennica 171*, 133-136.
Millar, J. S., & Zammuto, R. M. (1983). Life histories of mammals: an analysis of life tables. *Ecology 64*, 631-635.
Moses, L. E., Gale, L. C., & Altmann, J. (1992). Methods for the analysis of unbalanced, longitudinal growth data. *American Journal of Primatology 28*, 49-59.
Müller, E. F. (1979). Energy metabolism, thermoregulation and water budget in the slow loris (*Nycticebus coucang*). *Comparative Biochemistry and Physiology 64A*, 109-119.
Müller, E. F. (1983). Thermoregulation and energy budget of prosimians. *Bonner Zoologische Beiträge 34*, 29-71.
Müller, E. F. (1985). Basal metabolic rates in primates: the possible role of phylogenetic and ecological factors. *Comparative Biochemistry and Physiology A81*, 707-711.
Müller, E. F., & Jaksche, H. (1980). Thermoregulation, oxygen consumption, heart rate and evaporative water loss in the thick-tailed bushbaby (*Galago crassicaudatus*). *Zeitschrift für Säugetierkunde 45*, 269-278.
Müller, E. F., Nieschalk, U., & Meier, B. (1985). Thermoregulation in the slender loris (*Loris tardigradus*). *Folia Primatologica 44*, 216-226.
Nash, L. T. (1983). Reproductive patterns in galagos (*Galago zanzibaricus* and *Galago garnetti*) in relation to climatic variability. *American Journal of Primatology 5*, 181-196.
Nicoll, M. E., & Thompson, S. D. (1987). Basal metabolic rates and energetics of reproduction in therian mammals: marsupials and placentals compared. *Symposia of the Zoological Society, London 57*, 7-28.
Niemitz, C. (1984). *Biology of tarsiers*. New York; Fischer.
Niemitz, C., Nietsch, A., Warter, S., & Rumpler, Y. (1991). *Tarsius dianae*: a new primate species from Central Sulawesi (Indonesia). *Folia Primatologica 56*, 105-116.
Pagel, M. D. (1992). A method for the analysis of comparative data. *Journal of theoretical Biology 156*, 431-442.
Pagel, M. D., & Harvey, P. H. (1989). Taxonomic differences in the scaling of brain on body weight among mammals. *Science 244*, 1589-1593.
Partridge, L., & Harvey, P. H. (1988). The ecological context of life history evolution. *Science 241*, 1449-1455.
Partridge, L., & Sibly, R. (1991). Constraints in the evolution of life histories. *Philosophical Transactions of the Royal Society, London B 332*, 3-13.
Petter-Rousseaux, A. (1962). Recherche sur la biologie de la réproduction des primates inferieurs. *Mammalia 26 Suppl.1*, 1-88.
Pianka, E. R. (1970). On r- and K-selection. *American Naturalist 104*, 592-596.
Pianka, E. L., & Parker, W. S. (1975). Age-specific reproductive tactics. *American Naturalist 109*, 453-464.

Pontier, D., Gaillard, J. M., & Allainé, D. (1993). Maternal investment per offspring and demographic tactics in placental mammals. *Oikos 66*, 424-430.
Promislow, D. E. L., & Harvey, P. H. (1990). Living fast and dying young: A comparative analysis of life-history variation among mammals. *Journal of Zoology, London 220*, 417-437.
Rasmussen, D. T. (1986). Life history and behavior of slow lorises and slender lorises: implications for the Lorisine-Galagine divergence. Ph.D. Thesis, Duke University, Durham, NC.
Rasmussen, D. T., & Izard, M. K. (1988). Scaling of growth and life history traits relative to body size, brain size, and metabolic rate in lorises and galagos (Lorisidae, Primates). *American J.ournal of Physical Anthropology 75*, 357-367.
Read, A. F., & Harvey, P. H. (1989). Life history differences among the eutherian radiations. *Journal of Zoology, London 219*, 329-353.
Richard, A. F., & Nicoll, M.E. (1987). Female social dominance and basal metabolism in a Malagasy primate, *Propithecus verreauxi. American Journal of Primatology 12*, 309-314.
Ross, C. (1988). The intrinsic rate of natural increase and reproductive effort in primates. *Journal of Zoology, London 214*, 199-219.
Ross, C. (1991). Life history patterns of New World monkeys. *International Journal of Primatology 12*, 481-502.
Ross, C. (1992a). Basal matabolic rate, body weight and diet in primates: an evaluation of the evidence. *Folia Primatologica 58*, 7-23.
Ross, C. (1992b). Environmental correlates of the intrinsic rate of natural increase in primates. *Oecologia 90*, 383-390.
Rumpler, Y., & Dutrillaux, B. (1986). Evolution chromosomique des prosimiens. *Mammalia 50*, 82-107.
Rumpler, Y., Warter, S., Rabarivola, C., Petter, J. J., & Dutrillaux, B. (1990). Chromosomal evolution in Malagasy lemurs: XII. Chromosomal banding study of *Avahi laniger occidentalis* (Syn: *Lichanotus laniger occidentalis*) and cytogenetic data in favour of its classification in a species apart - *Avahi occidentalis. American Journal of Primatology 21*, 307-316.
Sacher, G. A., & Staffeldt, E. F. (1974). Relation of gestation time to brain weight for placental mammals: implications for the theory of vertebrate growth. *American Naturalist 108*, 593-616.
Sokal, R. R., & Rohlf, F. J. (1981). *Biometry*. New York: Freeman.
Stephan, H., & Bauchot, R. (1965). Hirn-Körpergewichtsbeziehungen bei den Halbaffen (Prosimii). *Acta Zoologica XLVI*, 209-231.
Stearns, S. C. (1976). Life-history tactics: a review of the ideas. *Quarterly Review of Biology 51*, 3-47.
Stearns, S. C. (1983). The influence of size and phylogeny on patterns of covariation among life-history traits in the mammals. *Oikos 41*, 173-187.
Stearns, S. C. (1992). *The evolution of life histories*. Oxford: Oxford University Press.
Sutherland, W. J., Grafen, A., & Harvey, P. H. (1986). Life history correlations and demography. *Nature 320*, 88.
Terborgh, J., & Janson, C. H. (1986). The socioecology of primate groups. *Annual Reviews in Ecology and Systematics 17*, 111-135.
van Horn, R. N., & Eaton, G.G. (1979). Reproductive physiology and behavior in prosimians. In G.A. Doyle & Martin, R. D. (Eds.), *The study of prosimian behavior* (pp.79-122). New York: Academic Press.
van Schaik, C. P. (1983). Why are diurnal primates living in groups? *Behaviour 87*, 120-144.
van Schaik, C. P., & van Hooff, J. A. R. A. M. (1983). On the ultimate causes of primate social systems. *Behaviour 85*, 91-117.
Western, D. (1979). Size, life history and ecology in mammals. *African Journal of Ecology 17*, 185-204.
Western, D., & Ssemakula, J. (1982). Life history patterns in birds and mammals and their evolutionary significance. *Oecologia 54*, 281-290.
Williams, G. C. (1966). *Adaptation and natural selection*. Princeton: Princeton University Press.
Wright, P. C., Izard, M. K., & Simons, E. L. (1986). Reproductive cycles in *Tarsius bancanus. American Journal of Primatology* 11, 207-215.
Young, A. L., Richard, A. F. & Aiello, L. C. (1990). Female dominance and maternal investment in strepsirhine primates. *American Naturalist 135*, 473-488.
Zimmermann, E. (1989). Aspects of reproduction and behavioral and vocal development in Senegal bushbabies (*Galago senegalensis*). *International Journal of Primatology 10*, 1-16.

SEXUAL SELECTION AND THE EVOLUTION OF COPULATORY BEHAVIOR IN NOCTURNAL PROSIMIANS

A. F. Dixson

MRC Group on Development & Integration of Behaviour
Sub Department of Animal Behaviour
University of Cambridge, Madingley
Cambridge CB3 8AA

INTRODUCTION

The simplistic view of nocturnal prosimians as "solitary" primates leading impoverished social lives is now known to be incorrect. Although many prosimians are relatively non-gregarious, they occupy varied and complex social networks which are maintained by means of olfactory and vocal signals, as well as by direct encounters during nocturnal activities (Charles-Dominique, 1977; Schilling, 1979; Bearder, 1986). Bearder has defined five types of social organization among the nocturnal primates, all of which involve varying degrees of home range overlap and range size differentials between the sexes (Figure 1). Mating systems may differ from social systems, however, as has been amply demonstrated by fieldwork on diurnal anthropoid species. Examples include the existence of monogamy and polyandry in some callitrichids (Sussman and Garber, 1987; Goldizen, 1987), and multi-male influxes into one male units in *Erythrocebus patas* and in various *Cercopithecus* species (Cords, 1986; 1988). Since, in some nocturnal prosimians, the home ranges of a number of females may overlap that of a single male (e.g., in *Perodicticus potto*: Type 2 system in Figure 1), it is tempting to speculate that the male may monopolize sexual access to these females, producing something akin to a polygynous "one-male group". However, this review will present evidence that sperm competition (Parker, 1970) and sexual selection by female choice (Eberhard, 1985) have influenced the evolution of the genitalia and copulatory behavior in males of many nocturnal prosimian species. It will be argued that many of these animals have "dispersed" mating systems in which females are able to mate with a number of males during the peri-ovulatory phase of the ovarian cycle.

Creatures of the Dark: The Nocturnal Prosimians
Edited by L. Alterman *et al.*, Plenum Press, New York, 1995

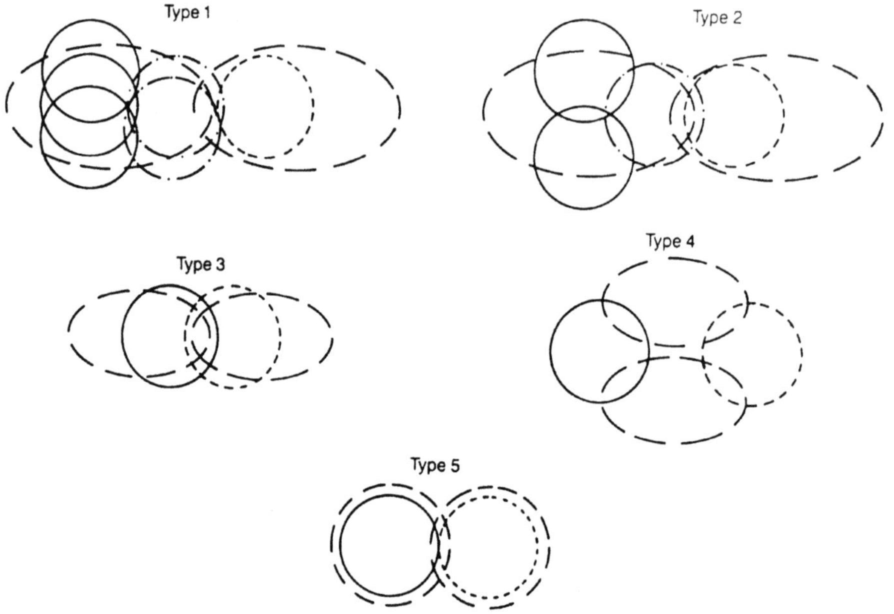

Figure 1. Diagrammatic representation of patterns of home range overlap in nocturnal primates (after Bearder, 1986). Circles = female ranges; ovals = male ranges (except in Type 5 where both sexes are represented by circles). Type 1. e.g. *Galago senegalensis; G. alleni;* Type 2. e.g. *Perodicticus potto*; Type 3. e.g. *Tarsius bancanus*; Type 4. e.g. *Microcebus murinus*: Type 5. e.g. *Eulemur mongoz; Aotus* spp.

Relative Testes Size and Sperm Competition in Prosimians

Among the anthropoid primates, species with multimale/multifemale mating systems tend to have heavier testes, relative to body weight, than is the case for pair-living (monogamous) forms or those which live in one-male (polygynous) groups (Short, 1979; Harcourt *et al.*, 1981). Sexual selection has led to the evolution of larger testes in multimale/multifemale mating systems because females may mate with a number of partners and sperm competition (Parker, 1970) favors males which can supply larger numbers of spermatozoa per ejaculate. Unfortunately, few accurate data have been collected on testicular weights in prosimian species, but some workers have published information on the lengths and widths of testes, and I have collected additional data on several species. Using this information, it was possible to calculate testicular volumes for 13 prosimians (nine nocturnal and four diurnal species) as well as for the owl monkey (*Aotus*) which is the only nocturnal anthropoid. A log-log plot of mean testis volume vs. bodyweight was constructed using these data (Figure 2). The principal axis line in Figure 2 (Slope = 0.718) derives from a previous study of 39 primate species which also included various monkeys, the apes and man (Dixson, 1987a). All of the prosimians represented in Figure 2 except one *(Avahi laniger)* have large testes relative to their body weights; the values fall above the principal axis line. This applies to the Galaginae (e.g., *Otolemur garnettii; G. senegalensis)*, Lorisinae *(P. potto*; *Nycticebus pygmaeus)*, and Cheirogaleinae *(Microcebus murinus; Cheirogaleus major)* as well as to various diurnal lemurs which live in multimale/multifemale groups (e.g., *Lemur catta; Eulemur fulvus*). These data indicate that sperm competition probably occurs in various nocturnal prosimians, as well as in some diurnal lemurs, in the same way as in anthropoids (e.g., macaques and chimpanzees) which have multimale/multifemale mating systems.

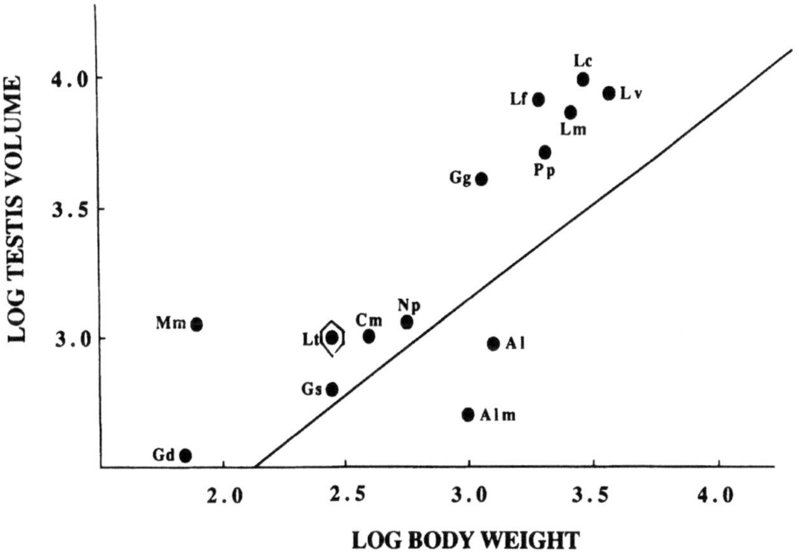

Figure 2. Logarithmic plot of mean testis volume vs. body weight in 14 species of nocturnal primates. Np. *Nycticebus pygmaeus;* Lt. *Loris tardigradus;* Pp. *Perodicticus potto;* Gs. *Galago senegalensis;* Gg. *Otolemur garnettii:* Gd. *Galagoides demidoff* Mm. *Microcebus murinus;* Cm. *Cheirogaleus major*; Al. *Avahi laniger*; Lc. *Lemur catta*; Lf. *Eulemur fulvus*; Lm. *Eulemur macaco*; Lv. *Varecia variegata*; Alm. *Aotus lemurinus.* In the case of *L. tardigradus*, mean testis volume was calculated by dividing testicular weight by 1.05 (the specific gravity of the tissue). Data from Ramakrishna and Prasad (1967) on *L. tardigradus* cover all 12 months of the year. These monthly values all fall within the diamond-shaped symbol on the graph.

Some complicating factors must be acknowledged when interpreting the data in Figure 2. Thus, it is surprising that *Varecia variegata* which is said to be pair-living in the wild (White, 1991), has large testes in relation to body weight, similar to those of lemur species which live in multimale/multifemale groups. In this regard it may be significant that social organization is variable in *V. variegata*, so that Pereira, *et al.* (1988) suggest that ruffed lemurs have a fission-fusion society like that of chimpanzees. Sperm competition and large testes would be expected under these conditions. *Avahi laniger* has small testes in relation to body weight (Figure 2) which is consistent with the observation that these animals occur in pairs or small groups of two to four individuals in the wild (Pollock, 1979). The social organization of these animals, is poorly known, however, (Bearder, 1986) so that it would be premature to conclude that they are monogamous. The data in Figure 2 refer to a single specimen of *Avahi*, measured by Petter-Rousseaux (1964), who commented that it was "towards the end of the active period" of spermatogenesis. This point raises an additional problem, namely that many prosimians are strictly seasonal breeders, in which the testes increase markedly in size during the annual mating season. Given that such mating seasons can be very short (e.g., < two weeks in *L. catta*: Jolly, 1966), selection may have led to the evolution of large testes because males must produce sufficient spermatozoa within a relatively short period of time, rather than because of sperm competition. However, it is unlikely that the very large testes which occur in various prosimians can be explained solely on this basis. For example, the potto is not a seasonal breeder (at least in Gabon: Charles Dominique, 1977), yet it has large testes. Likewise fieldwork conducted by Ramakrishna and Prasad (1967) near Bangalore on the slender loris (*Loris tardigradus*) showed that spermatogenesis continued throughout the year. While the testes were largest in animals collected between October and December, monthly mean values for testes size vs. body weight (n = 138 animals) all lie well above the principal axis line in Figure 2 and do not indicate that the effect depends upon a seasonal cycle. It is possible that an additive effect may occur in some prosimians, and that the testes are especially large in forms where both seasonal mating and sperm competition occur.

The large testes of many nocturnal prosimians thus probably reflect the occurrence of sexual selection via sperm competition. The "dispersed" mating systems of nocturnal forms, such as galagos, pottos, angwantibos, lorises and mouse lemurs probably result in situations in which females can mate with a number of males, and this in turn favors males with large testes capable of providing maximal numbers of motile spermatozoa in the ejaculate (Moller, 1988; 1989). It is likely that similar relationships between testes size and body weight may exist in some other nocturnal prosimian species. The aye aye *(Daubentonia madagascariensis)* is a prime example. Hill (1953) cited Kollmann to the effect that the aye aye has very "voluminous" testes, but no quantitative measurements are available in the literature. Since several centers now maintain captive colonies of these fascinating animals, it should be a simple matter to measure the testes and calculate testicular volumes. Likewise, for some other genera (e.g., *Arctocebus)* even simple measurements of testis size are lacking and could easily be obtained from captive animals.

Copulatory Patterns: Effects of Sexual Selection and Natural Selection

In the majority of primate species, males ejaculate during a single, relatively brief, intromission with pelvic thrusting (Dixson, 1983; Dewsbury and Pierce, 1989). However, males of some species employ multiple brief intromissions (MI) with thrusting to attain ejaculation, or a single prolonged intromission (PI) copulatory pattern occurs in which the male commonly maintains insertion during the post-ejaculatory interval. These more complex MI and PI patterns occur only in certain primates which have multimale/multifemale or dispersed mating systems (Dixson, 1987a; 1991a). Thus, the PI copulatory pattern has been recorded in a number of nocturnal prosimians (e.g., various galagos, the angwantibo, slow loris and aye aye) which have dispersed mating systems as well as in various diurnal monkeys (e.g., spider monkeys, woolly monkeys and stumptail macaques) which live in multimale/multifemale groups. Data on intromission durations in species which exhibit PI copulatory patterns are summarized in Table 1. It has been proposed that both sexual selection and natural selection have played a part in the evolution of this coital specialization (Dixson, 1991a). Since all species which exhibit the PI pattern have either a dispersed or multimale/multifemale mating system, it can be argued that prolonged copulations are adaptive in some circumstances where fertile females mate with more than a single male. Lengthy intromissions, such as the 260 min recorded for *O. garnettii* by Eaton, *et al.* (1973), may act as part of a "mate guarding" strategy; the male maintaining insertion for as long as possible in order to prevent access by other partners. Prolonged intromissions might also facilitate sperm transport, the glans penis serving to keep the ejaculate in close contact with the female's os cervix. In some galagos, repeated bouts of pelvic thrusting during intromission may indicate the occurrence of multiple ejaculations, but no studies have been reported which have examined this possibility. However, if a PI pattern increases male reproductive success, why is this form of copulation not more widespread among primates, and particularly among many monkeys with multimale/multifemale mating systems (e.g., *Saimiri, Miopithecus, Macaca or Papio spp)* Figure 3 provides a possible answer to this question. All primate species which exhibit the PI pattern are either small-bodied, cryptic nocturnal prosimians (all of which are arboreal in habit) or larger bodied arboreal anthropoids. Terrestrial monkeys (e.g., most macaques and baboons) and small-bodied arboreal types (such as *Saimiri)* may lack the PI pattern because predation, or some other negative selection pressure, has prevented its evolution. The anthropoids which exhibit PI

Table 1. Maximum intromission durations, adult male body weights and mating systems in primates which exhibit the prolonged single intromission (PI) copulatory pattern.

Species	Body weights (g)	Intromission durations (min.)	Mating System	Source
Nocturnal Prosimians				
1. *Otolemur garnettii*	1142	260*	D	Eaton *et al.*, 1973
2. *Galago senegalensis*	237	12	D	Dixson, 1989
3. *Galagoides demidoff*	63	60	D	Charles-Dominique, 1977
4. *Microcebus murinus*	80	>3	D	Martin, 1973
5. *Arctocebus calabarensis*	259	4	D	Dixson, 1989
6. *Loris tardigradus*	285	17	D	Dixson, 1989
7. *Nycticebus coucang*	1110	7	D	Dixson, 1989
8. *Daubentonia madagascariensis*	2800	5	D	Winn, pers. comm.
Diurnal Anthropoids				
9. *Brachyteles arachnoides*	15000	8.3	MM	Milton, 1985
10. *Ateles belzebuth*	8462	>10	MM	Klein, 1971
11. *A. paniscus*	6600	>10	MM	Roosmalen and Klein, 1988
12. *A. fusiceps*	8890	25	MM	Eisenberg, 1976
13. *Lagothrix lagotricha*	6670	14	MM	Nishimura, 1988
14. *Macaca arctoides*	10,050	>3	MM	Goldfoot *et al.*, 1975
15. *Pongo pygmaeus*	73,388	46**	D	Nadler, 1977

* In this case intromission was maintained for 260 min. after ejaculation had occurred.

** Median duration of copulation was 14 min. D = Dispersed; MM = Multimale/multifemale copulatory patterns include large arboreal monkeys with prehensile tails (e.g., *Brachyteles:* > 15 kg; *Ateles*: > 7 kg), and the orang-utan which, as a skillful quadrumanous climber weighing over 70 kg., has little to fear from predators other than man. The only semi-terrestrial monkey which utilizes the PI pattern is *Macaca arctoides* and its intromission duration is the shortest listed in Table 1. In the case of many nocturnal prosimians, their cryptic habits may have allowed them to utilize PI copulatory patterns despite small body size and the possible risks of predation by carnivores, snakes or owls. Interestingly, the slow moving lorisines copulate hanging upside down beneath branches (e.g., the potto: Charles Dominique, 1974; slow loris: Zimmerman *et al.*, 1979; slender loris: Izard and Rasmussen, 1985). In this position the mating pair may be less visible to predators and also the animals might release their grip and thus fall to the ground if attacked This form of defensive behavior does occur in pottos, although it has not been observed during mating (Charles-Dominique, 1977).

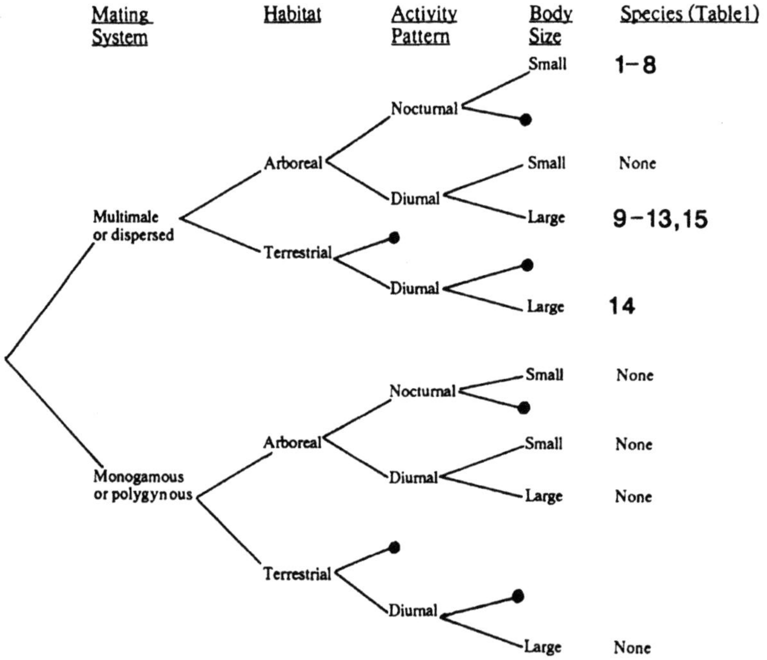

Figure 3. Occurrence of PI copulatory patterns in primates: interactions between the mating system and ecological factors (from Dixson, 1991a). —● indicates that no forms exist in a given category, e.g. there are no nocturnal terrestrial primates. Numbers refer to the species listed in Table 1.

The Evolution of Complex Penile Morphologies

Hill (1972) commented that "it is very remarkable, considering that the organs have the same rather limited functions to perform, how varied the male genitalia of primates are in their morphology - a circumstance which lends support to their value in assessing relationships". The prosimians are certainly notable for the complexity of their penile morphologies; in many species the glans penis is unusual in shape, adorned with keratinized spicules ("penile spines") and contains a prominent os penis or baculum (Figure 4). The functions of such male genitalia may be rather less "limited" than envisaged by Hill and by other students of primate systematics. There is now ample evidence that sexual selection has influenced the evolution of intromittent organs throughout the animal kingdom. Eberhard (1985) has proposed that in those species where females mate with a number of males, the penis may serve as an "internal courtship device", encouraging the female to "make use" of the male's gametes by transporting or storing them and thus increasing the likelihood of successful fertilization. Small adaptive features of penile morphology or copulatory behavior might thus become subject to rapid "sexual selection by female choice" in the same manner as some masculine secondary sexual characters (e.g., the plumage of peacocks and argus pheasants: Darwin, 1871). It is important to emphasize that Eberhard's use of the term "female choice" in no way implies any conscious decision making by the female during copulation and does not relate to the question of precopulatory mate choice. Rather, this intriguing hypothesis concerns those anatomical and physiological qualities of the female genitalia which influence and "test" the males' genital morphology and/or coital pattern with consequent effects upon reproductive success. Eberhard's hypothesis has been tested, and supported, in studies of penile evolution in primates (Dixson, 1987a, 1987b, 1989, Dixson and Mundy, 1994; Verrel, 1992). I shall concentrate here upon two features of penile morphology in nocturnal prosimians; the penile bone or baculum and the keratinized penile spines.

The form and functions of the baculum - *Tarsius* is unusual because it lacks a baculum but it is present in other nocturnal prosimians and is remarkably elongated in many cases (e.g., *Otolemur, Perodicticus, Loris, Nycticebus, Microcebus, Cheirogaleus* and *Daubentonia:* Dixson, 1987b). It is also common for the distal pole of the baculum to project beyond the tip of the glans penis with the urethral opening on its perineal surface (e.g., see *Arctocebus calabarensis*: Figure 4F). There is a correlation between increased baculum length and the occurrence of PI copulatory patterns among Old World primates (Dixson, 1987b). Nocturnal prosimians which exhibit prolonged intromissions (> 3 minutes), such as *O. garnettii*, *G. demidoff* and *L. tardigradus*, have much longer bacula in relation to their body weights than diurnal prosimians (e.g., *L. catta*) or anthropoids (e.g., *Saimiri, Callithrix, Cercopithecus* and *Papio* spp) which utilize brief intromission copulatory patterns (Figure 5). For some nocturnal prosimian species, data are available on baculum lengths and body weights but we lack detailed information about their copulatory behavior. Examples include *Euoticus elegantulus* and *P. potto;* data for both species fall close to the principal axis line for primates which exhibit PI copulatory patterns (Figure 5). It can be predicted with some confidence that a PI pattern occurs in these species.

The baculum may fulfil several functions which assist male fertility and hence promote reproductive success. First, the bone may strengthen the penis during prolonged copulations and protect the urethra from compression as the male ejaculates. A similar function has been proposed for the baculum in canids (Ewer, 1973), in which a "genital lock" occurs between the sexes during copulation. Second, the projecting tip of the baculum may contact the os cervix of the female during copulation, thus facilitating sperm transport into the cervix. Finally, the baculum might stimulate the female as the male makes intromitted pelvic thrusts. Such stimulation might be important in two ways; to facilitate the female's immobile, sexually receptive posture and to activate neuroendocrine responses which assist sperm transport and capacitation. I shall return to these possible effects of penile stimulation below, when discussing the role of the penile spines in nocturnal prosimians.

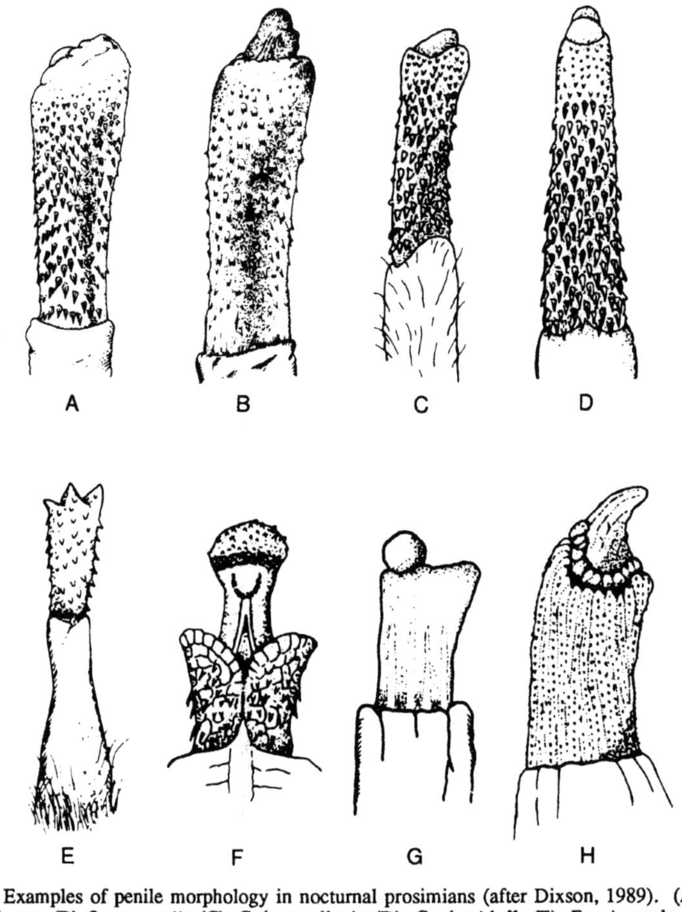

Figure 4. Examples of penile morphology in nocturnal prosimians (after Dixson, 1989). (A) *Otolemur crassicaudatus*; (B) *O. garnettii*; (C) *Galago alleni;* (D) *G. demidoff*; (E) *Euoticus elegantulus*; (F) *Arctocebus calabarensis;* (G) *Nycticebus coucang*; (H) *Tarsius bancanus*. Right lateral views of penes except D (dorsal) and F (perineal).

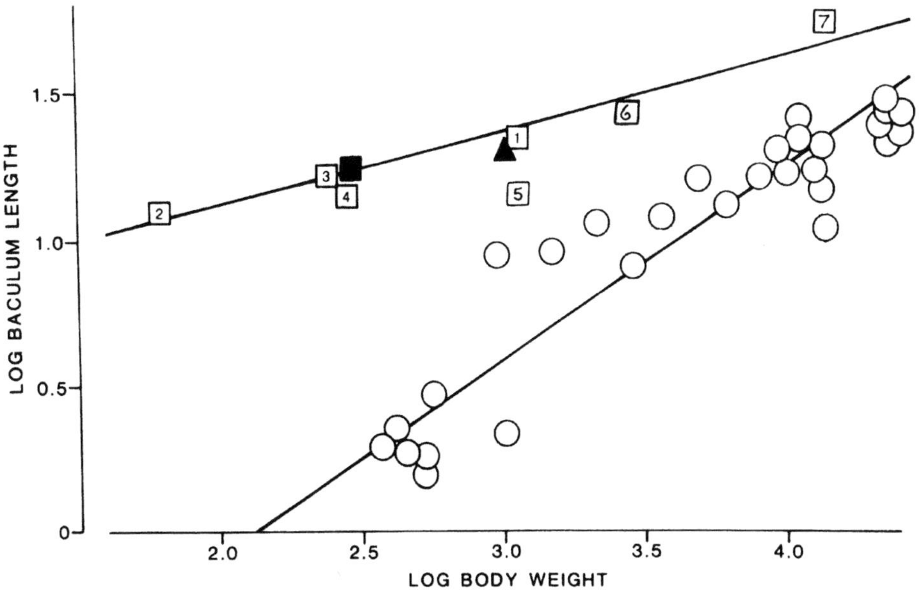

Figure 5. Baculum length in relation to patterns of copulatory behavior in primates. Open squares denote species with long intromission times and/or postejaculatory maintenance of intromission. Open circles denote species in which intromission is usually brief and is terminated promptly after ejaculation. The long intromission species are numbered as follows: (1) *Otolemur garnettii;* (2) *G. demidoff;* (3) *G. senegalensis*; (4) *L. tardigradus*; (5) *Nycticebus concang*: (6) *Daubentonia madagascariensis*; (7) *Macaca arctoides.* Two additional species are plotted for which only anatomical data are available: *Perodicticus potto* (Δ); *Euoticus elegantulus.*( ). Species which exhibit brief intromissions (open circles) are not numbered individually but they include various diurnal prosimians (e.g. *Lemur catta*) and anthropoids (e.g. *Saimiri sciureus, Cebus apella,* Macaca mulatta). Full details are available in a previous report (Dixson, 1987b).

The form and functions of penile spines - Like the baculum, the occurrence of the keratinized "penile spines" in many prosimians, monkeys and apes represents a primitive character of the order Primates. Similar structures occur on the glans penis in some other mammals, such as rodents, where they have been studied as an aid to determining taxonomic relationships (Hooper and Musser, 1964), or to analyze their functions in sexual behavior (Beach and Levinson, 1950; Phoenix *et al.*, 1976). In primates, as in rodents, the spines exhibit marked interspecific and intergeneric variations in size and shape. They may be classified into three types:

1. Simple spines: small, single-pointed structures of moderate length (e.g., Common marmoset: Figure 6A).
2. Robust spines: single-pointed but much enlarged structures, often thickened at the base (e.g., *G. demidoff* : Figure 6B).
3. Complex spines: multi-pointed, robust spines (e.g., *O. garnettii:* Figure 6C; *M. murinus:* Figure 6D).

Type 1, simple spines, are the most widely distributed among primates. Type 2, robust spines, are found in a few anthropoids (e.g., *Macaca arctoides*: *Ateles belzebuth*) but more commonly in prosimians (e.g., *E. elegantulus: O. crassicaudatus, G. demidoff; A. laniger; P. furcifer).* Complex, multi-pointed structures are restricted to certain prosimians (e.g., *O. garnettii; M. murinus; Eulemur fulvus).* It has been reported that some members of the subfamily *Lorisinae* bear low hexagonal plates upon the glans penis rather than pointed structures (e.g., in *Loris, Nycticebus and Perodicticus:* Hill, 1958). However, specimens of *L. tardigradus* and *N. coucang* which I examined by scanning electron microscopy bore conical penile spines. I have been unable to verify whether spines occur in *P. potto* (they were absent in two specimens examined), but the morphology of the penile spines in other *Lorisinae* (including *Arctocebus:* Figure 4F) conforms to the pattern for other primates.

In terms of evolution, Type 1 spines probably represent the primitive condition. Complexity has been achieved either by increasing the size and thickness of individual spines or by fusion of two or more spines to produce multiple points. This process can occur within a single species on different regions of the glans penis. Thus in *O. garnettii* or *E. fulvus,* Type 1 spines occur on the distal glans, whilst those in the mid or basal region are Type 3, complex structures. This observation raises the possibility that, within a single individual, the various spine morphotypes may fulfil different functions.

Several functions have been proposed for penile spines in rodents and some of these may be applicable to nocturnal prosimians. The spines are typically hollow at the base and each contains a papilla of dermal tissue. The spines overlie dermal tactile receptors in the glans penis so that, as a male mounts, their deflection during pelvic thrusting may enhance the tactile feedback required to achieve intromission and/or ejaculation (Beach and Levinson, 1950; Hart, 1978). This suggestion is supported by experiments on the common marmoset *(Callithrix jacchus)* which showed that removal of Type 1 spines leads to an increase in the intromission latency and in the duration of intravaginal pelvic thrusting (Dixson, 1991b). In prosimians where more than one spine morphotype occurs on the glans, simple spines are located distally where they are ideally positioned to enhance sensory feedback during the pre-intromission phase of pelvic thrusting.

A second possibility is that penile spines impart tactile stimulation to the female, as well as to the male, during copulation. Such stimulation might be important for "orgasmic" or neuroendocrine responses in females. However, there is little evidence for the occurrence of such phenomena in female nocturnal prosimians. Orgasmic responses have been described in females of various anthropoid species (Allen and Lemmon, 1981; Dixson, 1990) but not in prosimians; nor is there any proof that female orgasm hastens sperm transport or affects fertility (in women: Masters and Johnson, 1966; Bancroft, 1989). However, given the large size and rasping action of Type 2 and Type 3 penile spines (Figure 6) it seems likely that female prosimians might "perceive" them during copulation. No studies of prosimians have examined whether copulatory stimuli might affect the fine-tuning of the preovulatory luteinizing hormone "surge". Whilst primates are known to be spontaneous ovulators, it is not beyond the bounds of possibility that the PI copulatory patterns of various nocturnal prosimians, and their impressive penile spines, may influence the female's pituitary-gonadal axis.

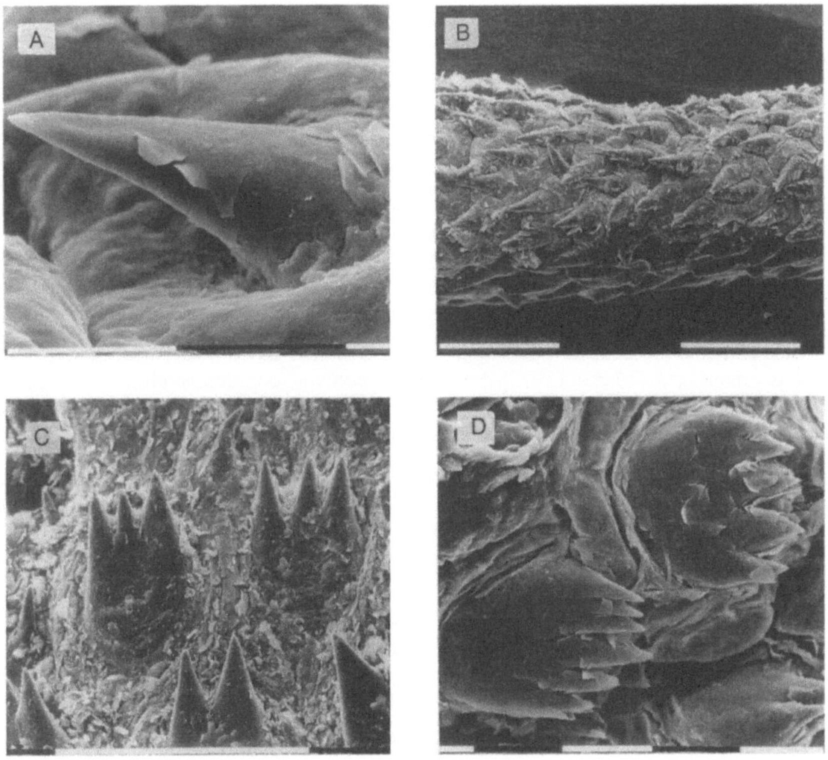

Figure 6. Scanning electron micrographs of penile spines in various primates. (A) Type 1, simple spine of *Callithrix jacchus* Scale = 0.1 mm. (B) Type 2, robust spines on the glans penis of *Galago demidoff*. Scale = 1.0 mm. (C) Type 3, Complex spines of *Otolemur garnettii*. Scale - 1.0 mm. (D) Complex spines of *Microcebus murinus*. Scale = 0.1 mm.

Coital stimulation may also have some effect upon the female's immobile, sexually receptive posture. In the rat, for example, penile stimulation during pelvic thrusting enhances the display of lordosis by the receptive female. This applies to stimulation as the male mounts and thrusts against the female's perineal area (Pfaff, 1980), and to stimulation received during intromission (Rodriquez-Sierra, *et al.*, 1975). Female prosimians exhibit limited periods of sexual receptivity and, in some cases, adopt a lordotic posture during copulation (Dixson, 1983). It is possible, therefore, that "sexual selection by female choice" (Eberhard, 1985) has encouraged the evolution of complex penile morphologies which stimulate the female and induce her to remain immobile. This might be especially important in those nocturnal prosimians which exhibit PI copulatory patterns.

Anatomists have occasionally suggested that the enlarged penile spines found in some prosimians might act as "grappling spurs" during copulation (Le Gros Clark, 1959; Hill, 1972). Indeed, during mating tests conducted under laboratory conditions, it is often difficult to separate copulating *O. garnettii;* the penis is firmly lodged in the vagina (pers. obs.). Charles-Dominique (1977), described a similar phenomenon in *G. demidoff.* Thus, although Dewsbury and Pierce (1989) deny the existence of a copulatory "lock" in primates, I believe that some of the nocturnal prosimian species listed in Table 1 qualify as "locking" species and that their enlarged penile spines are important in this context. A possible parallel occurs in the Australian hopping mouse *(Notomys alexis)* in which the male's elongated glans penis is studded with robust, Type 2 spines (Breed, 1981). A true genital lock occurs during copulation in the hopping mouse, and this is probably facilitated by the penile spines (Dewsbury and Hodges, 1987). The elongated glans and large penile spines of various nocturnal prosimians (e.g., in *galagos)* indicate the existence of a similar mechanism. In *Daubentonia,* the specialized arrangement of erectile tissue and the occurrence of recurved penile spines, indicates that some form of genital lock may occur during copulation in this species. Detailed observations of mating in captive aye ayes are required to resolve this question.

A final possible function for penile spines relates to the occurrence of "copulatory plugs" in various prosimians. Martin (1973) described a "conical" copulatory plug in *Microcebus murinus* and Manley (1967) recorded that "a massive and solid copulatory plug is deposited in the case of the angwantibo and slender loris". *G. senegalensis* does not often produce a copulatory plug, however, so that plugs are not a universal characteristic of prosimians, including those species in which large penile spines are present. Various functions have been suggested for copulatory plugs in mammals. In the rat, the male's plug serves to minimize loss of spermatozoa from the female tract and hence to encourage sperm transport (Blandau, 1945; Matthews and Adler, 1977). That plugs might act as physical barriers to subsequent matings has also been suggested (Voss, 1979) but experimental work on rodents has failed to support this "chastity belt" hypothesis (Dewsbury, 1988).

Mechanisms to dislodge the copulatory plugs of rival males have been described in rodents and some parallels may exist among primates. In the rat, the male's multiple intromission copulatory pattern, coupled with reflexive "cupping" responses of the glans penis during intromissions, can dislodge a plug deposited by a previous male (Wallach and Hart, 1983). Spines on the glans penis are also thought to play some role in this process (Sachs, 1982; Sachs and Meisel, 1988). It might be tempting, therefore, to argue that the very large penile spines of some nocturnal prosimian species might serve to remove, or at least to break up, plugs from previous matings. However, none of the nocturnal prosimians is known to utilize a multiple intromission copulatory pattern and some species which possess large spines do not produce a copulatory plug.

Intermale Competition and Copulatory Behaviour

Few direct observations of copulatory behavior in nocturnal prosimians have been made under field conditions. Despite the difficulties involved, observers have witnessed a number of males congregating around a single estrous female in *O. crassicaudatus* (Clark, 1985), *G. senegalensis* (Bearder, 1986) and *N. coucang* (Barrett, 1985). In *Daubentonia,* several males have been found sleeping close to the nest of an estrous female during the daylight hours (Sterling, 1992). The extent to which females of these, or of other, prosimian species mate with multiple partners during the peri-ovulatory period is not known. However, because the home range of a male potto, demidoff's galago or mouse lemur may overlap a number of female ranges (Figure 1) this is no reason to suppose that the females copulate only with this male. Given the nocturnal environment of these primates and the specializations of their copulatory behavior and genital morphology, it is unlikely that all females confine copulations to a single partner. Indirect evidence of intermale competition is plentiful. First, in various nocturnal species, adult males practice mutual avoidance, or they are actively aggressive towards each other. Home range overlap between such males is often minimal (e.g., in *G. senegalensis*: Bearder and Doyle, 1976; *M. murinus*: Martin, 1973; *Lepilemur mustelinus:* Charles- Dominique and Hladik, 1971 and *P. potto:* Charles-Dominique, 1977). His studies of five sympatric nocturnal prosimians in Gabon led Charles-Dominique to the conclusion that "there is active competition between males giving rise to selection for the most vigorous among them. The smallest and weakest males are hence excluded to various degrees from contact with the female home ranges". Charles-Dominique examined adults and immature individuals of *G. demidoff, G. alleni, E. elegantulus* and *A. calabarensis* for signs of healed wounds. Damaged ears, tails, digits, eyes and body scars were present in 13-23% of adults examined and were two to four times more common in males than in females. Immature animals, by contrast, showed no signs of involvement in intraspecific fights.

There is also ample evidence that males of various nocturnal prosimian species adopt a variety of reproductive strategies which result, presumably, from intermale competition for ranging areas and for access to females. Studies of mouse lemurs, for instance, have shown that home ranges are smaller (0.2-0.5 1 hectares) as are ratios of males:females (1:4) in S. Madagascar where the forests are drier and ecological conditions are more challenging. In N.W. Madagascar, home ranges are larger and more males are present in populations which inhabit deciduous forest (Martin, 1973; Pages-Feuillade, 1988). The more dominant or "central" males in the population nuclei exclude other males and are presumed to experience a reproductive advantage (Martin, 1973). In an extensive series of experiments on captive *M. Murinus,* Perret (1992) has shown that dominant males are aggressive towards subordinates, perform most copulations and exhibit higher plasma testosterone levels. Subordinates also exhibit a greater binding capacity for testosterone by testosterone binding globulin, indicating that "free" testosterone levels may be lower than in dominant males. Perret (1992) has also shown that chemical cues in the urine of dominant males cause significant decreases in circulating testosterone in subordinates. Hence the scent marking behavior used by so many nocturnal prosimian species may play a role in inter-male competition and in the physiological suppression of lower ranking males. The urine of a castrated, dominant mouse lemur is still capable of producing such pheromonal effects upon subordinates (Schilling and Perret, 1987) and current evidence indicates that the adrenal glands of dominant males are important in secreting the active substances excreted in the urine (Perret, 1992).

The intrasexual competition which occurs in *M. murinus* has parallels in some other nocturnal prosimians. In *G. demidoff* for instance, adult males belong to four social categories; Central A, Central B, peripheral and "vagabond" males have been described (Charles-Dominique, 1972). These males vary in body weight; Central A males are heaviest (75 g) and occupy the largest home ranges (1.8 hectares) allowing greatest access to female ranging areas. Central B males are much lighter (56 g) but they are apparently tolerated within the central area of female ranges by Central A males. Whilst it has been presumed that such Central B males have "virtually no contact with females" (Charles-Dominque, 1977) it seems possible that they might engage in "sneak" copulations and so contribute to the gene pool. Peripheral males are larger (61 g) and occupy home ranges adjacent to the central female area. It is believed that such peripheral positions are temporary, so that at least some males eventually succeed in obtaining a more central ranging area. Likewise "vagabond" males, which are younger post-pubertal males of nomadic habit, probably represent a transitory phase in social development. Vagabonds tend to migrate from their natal areas whereas females remain in the region where they were born. Vagabonds have been described in some other species, e.g. in the potto where Charles-Dominique (1977) found that they tend to weigh less and to have slightly smaller testes than "sedentary" territorial males. In *G. senegalensis,* bigger "A" males (226 g), smaller "B" males (211 g) and migrating vagabonds have also been identified (Bearder, 1986). "A" males have greater contact with females and chase "B" males from their ranges. Bearder (1986) pointed out that "the distribution of male ranges is strongly influenced by the distribution of females, and male range size bears no obvious relationship to environmental resources". He also expressed the view that "B" males "attempt to find and mate with estrous females that are temporarily unguarded by the resident "A" males".

Some aspects of intermale competition and mating behavior in the dispersed mating systems of various nocturnal prosimians are comparable to those which occur in diurnal anthropoid primates. In the mandrill *(Mandrillus sphinx)* for instance, studies of semi free-ranging animals in Gabon have shown that two classes of adult males occur; group associated "fatted" males and peripheral or solitary "non-fatted" males. "Fatted" males (so-called because of their large rumps and flanks) have larger testes, higher plasma testosterone levels and brighter secondary sexual colouration than non-fatted individuals (Wickings and Dixson, 1992). Females mate with a number of males during the annual mating season, but DNA fingerprinting studies have demonstrated that the most dominant fatted males in the group sire the offspring (Wickings, *et al.*, 1993). In the orang-utan, which is unusual among anthropoids in having a dispersed (non-gregarious) mating system, massive territorial adult males have the greatest mating success. Younger adults, which lack full development of secondary sexual characteristics (cheek flanges, throat sac etc.) live a more mobile existence and mate opportunistically with females (Mackinnon, 1974; Galdikas, 1985). In captivity, the presence of a fully developed male can delay development of the secondary sexual characteristics of others (Kingsley, 1982). Despite this, young male orang-utans which lack cheek flanges are fertile by 6-7 years of age (Dixson, *et al.*, 1982), and are perhaps equivalent to the vagabonds or B males of certain nocturnal prosimians. Indeed these examples serve to show that differences between the mating systems of diurnal and nocturnal primates may be less absolute than previously thought. Although they are relatively non-gregarious, the galagos, pottos, lorises and cheirogaleids exhibit intermale competition for mating opportunities, whilst the effects of sexual selection via sperm competition and female choice parallel those which occur in multimale/multifemale groups of many diurnal primates.

Female Sexuality and Female "Choice"

There are several ways in which female nocturnal prosimians might exert some influence or "choice" in sexual contexts, such that the occurrence of copulation is not solely the outcome of male initiative and intermale competition. First, the female might be selectively receptive to certain males, while refusing others. There is ample evidence that receptivity in female prosimians is limited in duration, under rigid hormonal control, and coincides with the peri-ovulatory period (Table 2). Females are sexually unreceptive at other times and refuse males' mounting attempts if they occur. Receptive females may also show sexual preferences for particular males. Female greater galagos, which were ovariectomised and treated with oestradiol, copulated with "familiar" males during pair-tests while refusing males they had not encountered previously. Male galagos inspected and attempted copulations with ovariectomised females, even before hormonal treatment, indicating that the ovaries are not essential for female sexual attractiveness under these conditions (Dixson, 1978). It is interesting that under natural conditions males of various nocturnal prosimian species "visit" and begin to "court" neighboring females even when the latter are not sexually receptive. Charles-Dominique (1977), for instance, conducted ingenious experiments in the rainforests of N.E. Gabon whereby he placed liane "bridges" bearing electrical recording devices within the home ranges of pottos. He was thus able to map the nocturnal "visiting" behavior of a male potto with some precision (Figure 7). This male visited a particular female regularly at 2-4 day intervals, seeking proximity to her but not necessarily establishing physical contact until she entered estrus. Such behaviour may be crucial in establishing social relationships between the sexes as well as allowing males to monitor the females' reproductive status.

A second possible avenue of female choice concerns sexual invitational behavior or proceptivity (Beach, 1976). Few studies of the nocturnal prosimians have produced any evidence that females actively solicit copulations during estrus. Lipschitz (1988) observed that estradiol-treated, ovariectomised lesser galagos were more proceptive towards male partners than untreated females were. Female proceptivity was expressed by approaching and presenting to males. Eaton *et al.*, (1973) saw no such proceptive behavior in female greater galagos, but Lipschitz (1988) used much larger laboratory cages for his pair tests and provision of sufficient space can be crucial in allowing female primates to express their proceptivity (Wallen, 1982). Sexual presentation displays have also been recorded for captive Bornean tarsiers *(Tarsius bancanus:* Wright *et al.*, 1986b). Perhaps proceptive behaviour also occurs in females of other nocturnal prosimian species, but the necessary fine-grained studies of sexual behavior have yet to be reported, particularly for free-ranging animals. One possibility is that as females approach estrus they might begin to range more

Table 2. Length of the ovarian cycle and duration of sexual receptivity in various prosimians.

Species	Average length of ovarian cycle	Duration of sexual receptivity	Source
Tarsius bancanus	24 days	1 - 3 days	Wright *et al*, 1986a; 1986b
Otolemur garnettii	44 days	5.8 days	Eaton *et al.*, 1973
Galago senegalensis	31.7 days	1 - 3 days	Manley, 1966; Doyle *et al.*, 1967
Perodicticus potto	37 - 39 days	?	Ioannou, 1966; Manley, 1966
Arctocebus calabarensis	36 - 45 days	?	Manley, 1966
Loris tardigradus	29 - 40 days	2 days	Izard and Rasmussen, 1985
Nycticebus coucang	42.3 days	2 - 5 days	Manley 1966; Zimmerman, 1988
Cheirogaleus medius	19.7 days	2 - 3 days	Foerg, 1982
Microcebus murinus	38 days	2 - 4 hours	Perret 1986, Lebec, 1984
Varecia variegata	30 - 36 days	1 day	Shideler *et al.*, 1983
Lemur catta	39 days	21.66 hours	Van Horn and Resko, 1977
Lemur coronatus	34 days	1 day	Kappeler, 1987
Propithecus verreauxi	?	12 - 36 hours	Richard, 1976

widely and so attract the attention of males more frequently. Increases in locomotor activity (wheel running) as well as sexual behavior have long been known to occur during estrus in rats. In discussing this, Young (1961) stated that "it is probable, if our knowledge of other species were more complete, that an increase in general activity, if not in running activity, would be found to be associated with the mating period". Among the primates, increased locomotor activity during periods of heightened sexual activity has been noted in *Brachyteles*

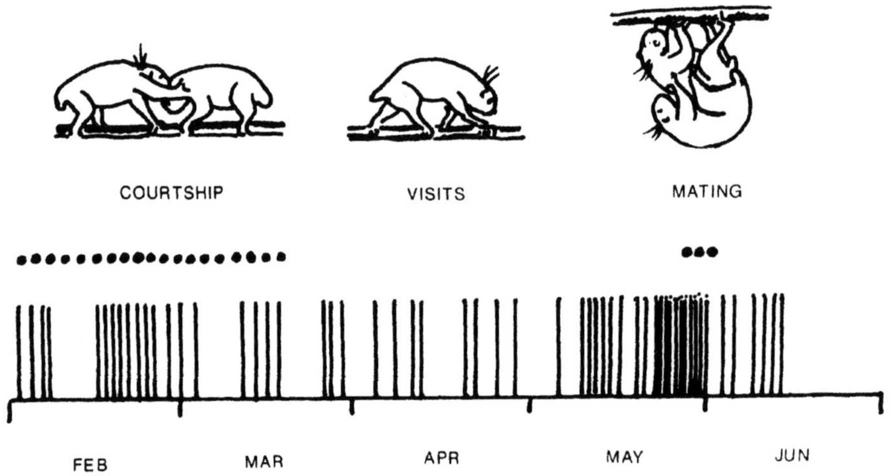

Figure 7. "Visits" made by a male potto to a female during the period March-May 1967. The male made use of a liane bearing an electrical contact which allowed each visit to be logged accurately. Data are from field studies by Charles Dominique (1977) and have been redrawn and modified from his work.

arachnoides (Milton, 1985) and chacma baboons (Saayman, 1970). Treatment of ovariectomised rhesus monkeys with estradiol cypionate increases their activity but this effect does not occur if castrated males are treated with the estrogen (Shinoda *et al.*, 1991). It is not known whether female nocturnal prosimians exhibit activity changes during ovarian cycles and whether they actively approach males when in estrus. Active approaches by females may be rendered less necessary, however, if they are highly attractive sexually to prospective mates. This third aspect of female sexuality, sexual attractiveness, is also relevant to discussions of female choice. Among the prosimians there is abundant evidence that scent-marking displays and chemical communication play a crucial role in social and sexual contexts. (Schilling, 1979; Perret, 1992; Schilling and Evans, 1994). Females frequently mark substrates with urine, genital and cutaneous secretions when in oestrus and males inspect such marks, or the females' genitalia prior to copulation. Female mouse lemurs are attractive to males for approximately 10 days before they become receptive, yet receptivity lasts for only 2-4 hrs. In this species, the females' urine has a pheromonal action and stimulates the production of plasma testosterone in males. Female slender lorises scent mark frequently in the period preceding copulation, leaving a urine trail which the male follows and overmarks with his own urine (Izard and Rasmussen, 1985). Greater galagos employ no less than six behavioral patterns in relation to olfactory communication; urine-washing, rhythmic micturition, chest-rubbing, anogenital rubbing, face rubbing and foot rubbing (Eaton *et al.*, 1973; Andrew and Klopman, 1976; Bearder and Doyle, 1976; Clark, 1982). Urine-washing is particularly well adapted for the distribution of chemical cues within a three-dimensional, arboreal home range (Charles-Dominique, 1977). Homologous displays occur in some other galagos (e.g., *G. senegalensis; G. demidoff, G. alleni*) as well as in the slender loris. In these and other nocturnal prosimians, it is entirely possible that olfactory cues serve to attract a number of males to the vicinity of a female as she approaches estrus. This would facilitate intermale competition and also female choice, whether at the precopulatory level or via multiple matings with resulting sperm competition. This argument for a broadcast function of olfactory cues and female attractiveness in some nocturnal prosimians is analogous to the function proposed for female sexual swellings in promoting intermale competition, including sperm competition, in the multimale groups of various Old World monkeys and chimpanzees (Clutton Brock and Harvey, 1976; Harvey and May, 1989).

One final aspect of female "choice" which I wish to discuss is the timing of estrus among females within the same population. Absolute synchrony of the timing of receptivity and ovulation among females might be expected to diminish the likelihood that a single male could exert a monopoly over them. A single male would be unable to visit, court and copulate with all partners under such conditions. However, there is evidence that, although a broad strategy of estrus synchony exists in some prosimians, there are appreciable gaps between females in its exact timing. This is the case, for example, in *L. catta* (Pereira, 1991) a species which has an annual mating season of less than two weeks during which individual females are in estrus for less than 24 hours (Jolly, 1966; Evans and Goy, 1968). I hope that future field studies of nocturnal prosimians, such as galagos, lorises, pottos and cheirogaleids will examine the dynamics of oestrous synchrony in relation to intermale competition and copulatory behaviour. In *O. garnettii* for example the annual mating season occurs during August on the Islands of Pemba and Zanzibar, and in eastern Tanzania (Haddow and Ellice, 1964; Masters *et al.*, 1988). Nash and Harcourt (1986) who studied these galagos in a Kenyan coastal forest, found that male home ranges averaged 17 hectares and overlapped the smaller (12 hectare) ranges of several females. Since these galagos can travel rapidly, covering up to 1.6 km. per night, it is likely that they can visit all parts of the home range between dusk and dawn. Laboratory studies of *O. garnettii* have shown that females have an ovarian cycle which averages 44 days but that vaginal cornification lasts for 12.4 days, on average, and females are sexually receptive for 5.8 days. Tremendous variability occurs between captive females, however, so that periods of receptivity can vary between 2-10 days (Eaton *et al.*, 1973). Sexual receptivity is under rigid hormonal control in

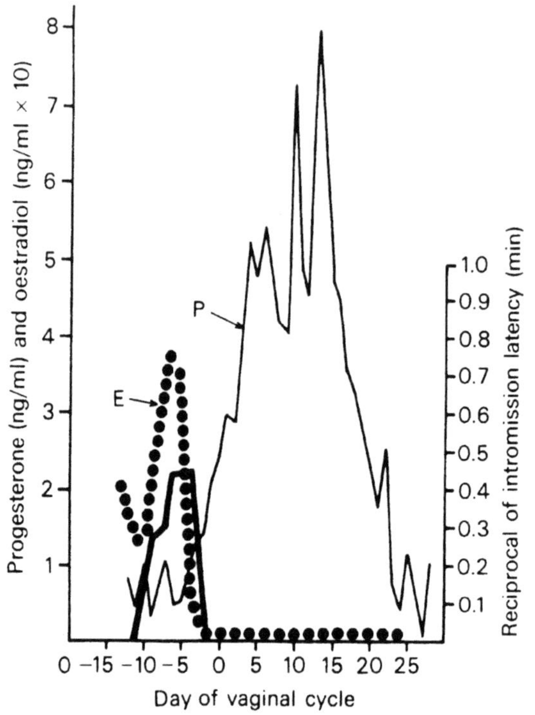

Figure 8. Changes in circulating hormone levels and sexual receptivity during the ovarian cycle of the greater galago. E = estradiol; P = progesterone. The reciprocal of intromission latency provides a useful index of sexual receptivity in this species (From Dixson, 1983; redrawn and modified from Eaton *et al.*, 1973).

O. garnettii (Figure 8) as is almost certainly the case in all prosimians. A vaginal closure membrane occurs in this species, as in some other prosimians (e.g., *Microcebus*), so that the vagina is only patent during the period of vaginal cornification. However, the receptive period is long enough to envisage the occurrence of multiple copulations with a number of partners under natural conditions. In this regard it is most interesting to note that copulation itself may shorten the period of vaginal cornification and/or sexual receptivity in *O. garnettii.* Eaton *et al.*, (1973) pair-tested *O. garnettii* to observe behavioral changes during the ovarian cycle (Figure 8), but they separated the animals once intromission occurred in order to prevent ejaculations and unwanted pregnancies. Under these conditions, the duration of vaginal cornification ranged from 7-24 days ($\bar{X}$ ± S.D.: 12.4 ± 3.7 days). Yet in a subsequent study where partners were caged together constantly and copulations were undisturbed (Dixson and Van Horn, 1977), the duration of vaginal cornification was reduced (R = 5-12 days; Mean ± S.D. = 7.0 ± 2.04 days). That coital stimuli may also serve to shorten the receptive period is also supported by studies of a diurnal prosimian, *Lemur catta.* Laboratory studies in which ejaculation was prevented by terminating pair tests showed that female ringtails remained receptive for 22 hours on average (Van Horn and Resko, 1977), whereas tests with completed copulations produced a much shorter estimate (10 hrs or less: Evans and Goy, 1968). In his field study of *L. catta* at Berenty in Madagascar, Koyama (1988) found that females remained receptive for just 4 hours but that they mated with a number of males in succession during this period. The high degree of coital stimulation received (in one case a female mated with five males and received 27 ejaculations) may have contributed to a reduction in the duration of sexual receptivity. It is possible that similar phenomena might occur in other prosimians, exemplifying Eberhard's (1985) "sexual selection by female choice" at the copulatory level.

SUMMARY AND CONCLUSIONS

The thesis advanced in this chapter is that the mating systems of various nocturnal prosimian species are more complex than currently realized and that sexual selection has played a crucial role in the evolution of copulatory behaviour and genital structure in these primates. While field work indicates that the sexes may associate in pairs in a few nocturnal prosimians (e.g., in *Avahi* or *Tarsius)* it is far more usual for adults to inhabit individual, overlapping home ranges. Range overlap among adult males tends to be restricted, however, (e.g., in *P. potto, O. garnettii)* whereas the ranges of females, which are often smaller, overlap extensively with those of males. Clearly a variety of arrangements has evolved and Bearder (1986) has described five types of social organization amongst nocturnal primates. Nor are the animals "solitary" in the sense of being socially isolated; a rich repertoire of olfactory cues (affecting the vomeronosal system as well as the main olfactory system: Izard, 1990) and vocalizations links them together in a social network. Although most species forage alone at night, some form sleeping associations during the day time (e.g, *M. murinus; Galago demidoff,* etc.). Nor should we conclude that sexual contacts are restricted among nocturnal prosimians and that females mate exclusively with a single male whose home-range coincides with their own. Intermale competition is intense, and in certain species males can be classified into distinct types which differ in body weight and social status (e.g., the Central A, Central B, peripheral and vagabond males of *G. demidoff).* More successful males acquire central ranges overlapping those of the breeding females (e.g., in *M. murinus* Martin, 1973), while others occupy more peripheral regions' and younger "vagabond" males migrate away from the natal area in search of suitable niches within the social framework (e.g, in various lorisids: Charles Dominique, 1977; Clark, 1985). Intermale competition can produce profound effects upon testicular function of subordinates in captive *M. murinus* (Perret, 1992), and these effects, mediated in part by chemical cues, doubtless have their parallels in the natural environment of nocturnal prosimians. One may ask how successful the more central males are in siring offspring, and whether peripheral males or vagabonds succeed in copulating with females in estrus and in securing fertilizations? There are currently no answers to these questions and the field is ripe for detailed studies of sexual behavior combined with DNA fingerprinting analyses of paternity in free-ranging populations. Application of molecular genetics to studies of mating systems in anthropoid primates is producing useful information (Martin *et al.* 1992; De Ruiter and Inoue, 1993) and it should be possible to extend these approaches to the nocturnal prosimians which can be trapped (and retrapped) for collection of the necessary blood or hair samples.

Although intermale competition prior to copulation among nocturnal prosimians may limit access to females during the peri-ovulatory period, there is evidence that females mate with a number of partners. Much of this evidence is indirect but it is nonetheless compelling. A number of males have been observed to congregate near an oestrus female in various galago species, in lorises and the aye aye. Further, this review has marshalled evidence that sperm competition occurs in various nocturnal primates and that sexual selection by "female choice" (Eberhard, 1985) has influenced the evolution of complex penile morphologies and copulatory patterns in some species. Such specializations can only occur in mating systems where females mate with a number of males rather than with a single individual. The specialized copulatory patterns of some nocturnal prosimians (e.g., lorises angwantibos, galagos, aye ayes, etc.), which involve prolonged intromissions, are associated with complex penile morphologies (such as an elongated baculum and enlarged penile spines). The possible adaptive advantages conferred by these specializations have been discussed at length and I will not repeat them here. Suffice to say that prolonged copulations and highly specialized penile morphologies almost certainly evolved because they confer a reproductive advantage, by excluding other males from mating and by enhancing sperm uptake and transport or capacitation within the female reproductive tract (Dixson, 1987a; 1991a). There is even some evidence that copulatory stimuli influence the duration of sexual receptivity in galagos and ringtailed lemurs. The neuroendocrine mechanisms which govern sexual behavior in female prosimians remain poorly studied. Prosimians are the only primates which exhibit a rigid control of receptivity by ovarian hormones; they alone exhibit "estrus" in the true sense of that term. Studies of neuroendocrinology and sexual behaviour in female prosimians should provide some valuable comparisons with work on anthropoid species (Dixson, 1990).

Sexual selection, by intra- and intersexual competition, probably occurs in most nocturnal prosimians. In this respect, their mating systems have more in common with the multimale/multifemale systems of certain anthropoids (e.g., Macaques, baboons and chimpanzees), than with monogamous forms (e.g., gibbons or *Callicebus* monkeys). Since nocturnal prosimians are less gregarious socially, the term "dispersed" has been proposed to describe their mating systems (Dixson, 1987a). The "dispersed" mating systems of galagos, lorises, cheirogaleids, aye ayes and others doubtless contain some complex variations, but a central theme probably exists, just as is the case for the various social organizations of nocturnal prosimians (Bearder, 1986). I hope that future fieldwork, combined with laboratory research on molecular genetics, reproductive physiology and behaviour, will produce some exciting advances. Despite the rarity of many prosimians, especially the Malagasy forms, some nocturnal species (e.g., many galagos) still flourish in the wild and may be bred in ample numbers in captivity. Field and laboratory studies are feasible given the will and the necessary funding for research.

REFERENCES

Allen, M. L. and Lemmon, W. B., 1981, Orgasm in female primates. *Am. J. Primatol.* 1: 15-34.

Andrew, R. J. and Klopman, R. B., 1976, Urine-washing: comparative notes, in: R. D. Martin, G. A. Doyle and A. C. Walker, eds., Prosimian Behaviour, Duckworth, London.

Bancroft, J. H., 1989, Human Sexuality and its problems. (2nd Ed.) Churchill Livingstone, Edinburgh.

Barrett, E. B. M., 1985, The ecology of some nocturnal arboreal mammals in the rainforest of peninsula Malaysia. Ph.D. Thesis, University of Cambridge.

Beach, F. A., 1976, Sexual attractivity, proceptivity and receptivity in female mammals. *Horm. Behav. 7:* 105-138.

Beach, F. A. and Levinson, G., 1950, Effects of androgen on the glans penis and mating behaviour of castrated rats. *J. Exp. Zool.* 114: 159-171.

Bearder, S. K., 1986 Lorises, bushbabies and Tarsiers: diverse societies of solitary foragers, in, Primate Societies, B. B. Smuts, D. L. Cheney, R. M. Seyfarth, R. W. Wrangham and T. T. Struhsaker eds., Chicago University Press, Chicago.

Bearder, S. K. and Doyle, G. A., 1974, Ecology of bushbabies, *Galago senegalensis* and *Galago crassicaudatus* with some notes on their behaviour in the field, in: Prosimian Biology, R. D. Martin, G. A. Doyle and A. C. Walker, eds., Duckworth: London.

Bearder, S. K., and Doyle, G. A., 1976, Ecology of bushbabies, *Galago senegalensis* and Galago crassicandatus, with some notes on their behaviour in the field, in: Prosimian Behaviour, R. D. Martin, G. A. Doyle and A. C. Walker, eds., Duckworth, London.

Blandau, R. J., 1945, On factors involved in sperm transport through the cervix uteri of the albino rat. *Am. J. Anat.* 73: 253-272.

Breed, W. G., 1981, Unusual anatomy of the male reproductive tract in *Notomys alexis (Muridae). J. Mammal.* 62: 373-375.

Charles-Dominique, P., 1972, Ecologie et vie sociale de *Galago demidovii* (Fischer, 1808, Prosimii). Z. *Tierpsychol. Suppl.* 9: 7-41.

Charles-Dominique, P., 1974, Vie sociale de *Perodicticus potto* (Primates, Lorisides). Etude de terrain en foret équatoriale de l'ouest African au Gabon. Mammalia 38: 355-379.

Charles-Dominique, P. 1977, Ecology and Behaviour of Nocturnal Primates, Duckworth, London

Charles-Dominique, P. and Hladik, C. M., 1971, Le lépilémur du sud de Madagascar: écologie, abmentation et vie sociale. Terre et Vie 25: 3-66.

Clark, A. B., 1982, Scent marks as social signals in *Galago crassicaudatus*. I. Sex and reproductive status as factors in signals and responses. *J. Chem. Ecol.* 8: 1133-1151.

Clark, A. B., 1985, Sociality in a nocturnal "solitary" prosimian: *Galago crassicaudatus* Int. *J. Primatol.* 6: 581-600.

Clark, W. Le Gros., 1959, The Antecedents of Man, Edinburgh Univ. Press, Edinburgh.

Clutton-Brock, T. H. and Harvey, P. H., 1976, Evolutionary rules and primate societies, in: Growing Points in Ethology , P. P. G. Bateson and R. A. Hinde, eds., Cambridge University Press, London.

Cords, M., 1986, Forest guenons and patas monkeys: male-male competition in one-male groups, in: Primate Societies, B. B. Smuts, D. L. Cheney, R. M. Seyfarth, R. W. Wrangham and T. T. Struhsaker eds., University of Chicago Press, Chicago.

Cords, M., 1988, Mating systems of forest guenons: a preliminary review, in: A Primate Radiation: Evolutionary Biology of the African Guenons, A. Gautier-Hion, F. Boubiere, J. P. Gauter and J. Kingdom, eds., Cambridge University Press, Cambridge.

Darwin, C., 1871, The Descent of Man and Selection in Relation to Sex. John Murray, London.

De Ruiter, J. R. and Inoue, M., 1993, Paternity, male social rank and sexual behaviour. *Primates* 34: 553-555.

Dewsbury, D. A., 1988, A test of the role of copulatory plugs in sperm competition in deer mice *(Peromyscus maniculatus).*

Dewsbury, D. A. and Hodges A. W., 1987, Copulatory behavior and related phenomena in spiny mice *(Acomys cahirinus)* and hopping mice *(Notomys alexis), J. Mammal.* 69: 49-57.

Dewsbury, D. A., and Pierce, J. D., 1989, Copulatory patterns of primates as viewed in broad mammalian perspective *Am. J. Primatol.* 17: 51-72.

Dixson, A. F., 1978, Effects of ovariectomy and estradiol replacement therapy upon the sexual and aggressive behavior of the greater galago *(Galago crassicaudatus crassicaudatus). Horm. Behav.* 10: 61-70.

Dixson, A. F., 1983, The hormonal control of sexual behaviour in primates. *Oxford Rev. Repro. Biol.* 5: 131-219.

Dixson, A. F., 1987a, Observations on the evolution of the genitalia and copulatory behaviour in male primates, *J. Zool. Lond.* 213: 423-443.

Dixson, A. F., 1987b, Baculum length and copulatory behavior in primates. *Amer. J. Primatol.* 13: 51-60.

Dixson, A. F., 1989, Effects of sexual selection upon the genitalia and copulatory behaviour in male galagos. *Int. J. Primatol.* 10: 47-55.

Dixson, A. F., 1990, The neuroendocrine control of sexual behaviour in female primates. *Ann. Revs. Sex. Res.* 1: 197-226.

Dixson, A. F., 1991a, Sexual selection, natural selection and copulatory patterns in male primates. *Folia primatol.* 57: 96-101.

Dixson, A. F., 1991b, Penile spines affect copulatory behaviour in a primate. *(Callithrix jacchus). Physiol. Behav.* 49: 557-562.

Dixson, A. F., Knight, J., Moore, H. D. M. and Carman, M., 1982, Observations on sexual development in male orang-utans *(Pongo pygmaeus). Int. Z. Yrbk.* 22: 222-227.

Dixson, A. F. and Mundy, N. 1994. Sexual behavior, sexual swelling and penile evolution in chimpanzees. *(Pan troglodytes). Arch. Sex. Behavior.* 23: 267-280.

Dixson, A. F. and Van Horn, R. N., 1977, Comparative studies of morphology and reproduction in two subspecies of the greater bushbaby, *Galago crassicaudatus crassicaudatus* and G. C. *argentatus. J. Zool. Lond.* 183, 517-526.

Doyle, G. A. Pelletier, A. and Bekker, T., 1967, Courtship mating and parturition in the lesser bushbaby *(Galago senegalensis moholi)* under semi-natural conditions. *Folia Primatol.* 7: 169-196.

Eaton, G. G., Slob, A. K. and Resko, J. A., 1973, Cycles of mating behaviour, oestogen and progesterone in the thick-tailed bushbaby *Galago crassicaudatus crassicaudatus.* under laboratory conditions. *Anim. Behav.* 21: 309-315.

Eberhard, W. G., 1985, Sexual Selection and Animal Genitalia, Harvard University Press, Harvard.

Eisenberg, J. F., Communication mechanisms and social integration in the black spider monkey, *Ateles fusciceps robustus* and related species. Smithson, Contr. Zool. 231: 1-108.

Evans, C. S. and Goy, R. W., 1968, Social behaviour and reproductive cycles in captive ring-tailed lemurs *(Lemur catta). J. Zool. Lond.* 156: 181-197.

Ewer, R. F., 1973, The Carnivores, Cornell Univ. Press, New York.

Foerg, R., 1982, Reproduction in *Cheirogaleus medius, Folia primatol.* 39: 49-62.

Galdikas, B. M. F. 1985. Adult male sociality and reproductive tactics among orang-utans at Tajung, Puting. *Folia primatol.* 45: 9-24.

Goldfoot, D.A., Slob, A. K., Scheffler, G., Robinson, J. A., Wiegand, S. and Cords, J., 1975, Multiple ejaculations during prolonged sexual tests and lack of resultant serum testosterone increases in male stumptail macaques *(M. arctoides). Arch. Sex. Behav.* 4: 547-560.

Goldizen, A. W., 1987, Facultative polyandry and the role of infant carrying in wild saddleback tamarins *(Saguinus fuscicollis), Behav. Ecol. Sociobiol.* 20: 99-109.

Haddow, A. J. and Ellice, J. M., 1964, Studies on bushbabies *(Galago* spp.) with special reference to the epidemiology of yellow fever. *Trans. Soc. Trop. Med. Hyg.* 58: 521-538.

Harcourt, A. H., Harvey, P. H., Larson, S. G. and Short, R. V., 1981, Testis weight, body weight and breeding system in primates, *Natural Lond.* 293: 55-57.

Hart, B. L., 1978, Hormones, Spinal reflexes and sexual behaviour, in: Biological Determinants of Sexual Behaviour, J. B. Hutchison, ed., John Wiley and Sons, Chichester.

Harvey, P. H. and May, R. M., 1989, Out for the sperm count. *Nature Lond.* 337: 508-509.

Hill, W. C. O., 1953, Primates: Comparative Anatomy and Taxonomy Vol. I, Strepsirhini, Edinburgh University Press, Edinburgh.

Hill., W. C. O., 1958, External genitalia, in: Primatologia. Handbook of Primatology, Vol. 3, H. Hofer, A. H. Schultz and D. Starck, eds., S. Karger, Basel.

Hill, W. C. O., 1972, Evolutionary Biology of the Primates, Academic Press, London.

Hooper, E. and Musser, G., 1964, The glans penis in neotropical cricetines (Muridae) with comments on classification of muroid rodents. Misc. Publ. Mus. Zool. Univ. Mich. 123: 1-57.

Ioannou, J. M., 1966, The oestrous cycle of the potto, *J. Reprod. Fertil.* 11: 455-457.

Izard, M. K., 1990, Social influences on the reproductive success and reproductive endocrinology of prosimian primates, in: F. Bercovitch and T. Ziegler, eds. Socioendocrinology of Primate Reproduction, Wiley-Liss, New York.

Izard, M. K. and Rasmussen, D. T., 1985, Reproduction in the slender loris *(Loris tardigradus malabaricus). Am. J. Primatol.* 8: 153-165.

Jolly, A., 1966, Lemur Behavior, Chicago University Press, Chicago.

Kappeler, P.M., Reproduction in the crowned lemur *(Lemur coronatus)* in captivity, *Am. J. Primatol.* 12: 497-503.

Kingsley, S., 1982, Causes of non-breeding and the development of the secondary sexual characteristics in the male orang-utan: a hormonal study, in: L. E. M. de Boer, ed., The Orang-Utan: it's Biology and Conservation, Dr. Junk, The Hague.

Klein, L. L., 1971, Observations on copulation and seasonal reproduction in two species of spider monkeys, *Ateles belzebuth* and *A. geoffroyi.* Folia primatol. 15: 233-248.

Koyama, N., 1988, Mating behavior of ring-tailed lemurs *(Lemur catta)* at Berenty, Madagascar. Primates, 29: 163-175.

Lebec, A., 1984, Relation entre le comportement agressif due microcébe mâle, les autres comportements et la physiologie sexuelle. Role de quelques facteurs déterminants. Thése Zéme Cycle, Universié Paris VI.

Lipschitz, D. W., 1988, Endocrine factors determining female *Galago senegalensis moholi.* mating behaviour, M.Sc. Thesis, University of Witwatersrand.

Mackinnon, J. R., 1974, Behaviour and ecology of wild orang-utans *(Pongo pygmaeus) Anim. Behav.* 22: 3-74.

Manley, G. H., 1966, Reproduction in lorisoid primates, Symp. Zool. Soc. Lond. 15: 493-509.

Manley, G. H., 1967, Gestation periods in the Lorisidae, *Int. Z. Ybk.* 1: 80-81.

Martin, R. D., 1973, A review of the behavior and ecology of the lesser mouse lemur *(Microcebus murinus* J. F. Miller 1777), in: Michael and J. H. Crook, eds., Ecology and Behaviour of Primates, R. P. Academic Press, London.

Martin, R. D., Dixson, A. F. and Wickings, E. J., 1992, Paternity in Primates, Genetic Tests and Theories. Karger, Basel.

Masters, W. H. and Johnson, V. E., 1966, Human Sexual Response, Churchill, London.

Masters. J. C., Lumsden, W. H. R. and Young, D. A., 1988, Reproductive and dietary parameters in wild greater galago populations. *Int. J. Primatol.*

Matthews, M. K. and Adler, N. T., 1977, Facilitative and inhibitory influences of reproductive behavior on sperm transport in rats. *J. comp. physiol. psychol.* 91: 727-741.

Milton, K. M., 1985, Mating patterns of woolly spider monkeys, *Brachyteles arachnoides:* implications for female choice. *Behav. Ecol. Sociobiol.* 17: 53-59.

Moller, A. P., 1988, Ejaculate quality, testes size and sperm competition in primates, *J. Human Evol.* 17: 479-488.

Moller, A. P. 1989, Ejaculate quality, testes size and sperm production in mammals, *Funct. Ecol.* 3: 91-96.

Nadler, R. D., 1977, Sexual behavior of captive orang-utans. *Arch. Sex. Behav.* 6: 457-475.

Nash, L. T. and Harcourt, C. S., 1986, Social organization of galagos in Kenyan coastal forests: II. *Galago garnettii Am. J. Primatol.*10: 357-369.

Nishimura, A., 1988, Mating behavior of woolly monkeys, *Lagothrix lagotricha,* at La Macarena, Colombia. Field Stud. N. W. Monkeys 1: 19-27.

Pagés-Feuillade, E., 1988, Modalités de l'occupation de l'espace et relations inter-individuelles chez un prosimian malgache, *Microcebus murinus. Folia Primatol.* 50: 204-220.

Parker, G. A., 1970, Sperm competition and its evolutionary consequences in the insects, *Biol. Rev.* 45: 525-567.

Pereira, M. E., 1991, Asynchrony within estrous synchrony among ringtailed lemurs. (Primates: Lemuridae). *Physiol. Behav.* 49: 47-52.

Pereira, M. E., Sieligson, M. A. and Macedonia, J. M., 1988, The behavioral repertoire of the black and white ruffed lemur, *Varecia variegata variegata. Folia Primatol.* 51: 1-32.

Perret, M., 1986, Social influences on oestrous cycle length and plasma progesterone concentrations in the female lesser mouse lemur *(Microcebus murinus). J. Reprod. Fertil.* 77: 303-311.

Perret, M. 1992. Environmental and social determinants of sexual function in the male mouse lemur *(Microcebus murinus). Folia Primatol.* 59: 1-25.

Petter-Rousseaux, A., 1964, Reproductive Physiology and behaviour of the Lemuroidea, in: Evolutionary and Genetic Biology of Primates Vol. 2, in: J. Buettner-Janusch ed., Academic Press, New York.

Phoenix, C. H., Copenhaver, K. H. and Brenner, R. M. 1976, Scanning electron microscopy of penile papillae in intact and castrated rats. *Horm. Behav.* 7: 217-227.

Pfaff, D. W., 1980, Estrogens and Brain Function, Springer, New York.

Pollock, J. I., 1979, Spatial distribution and ranging behavior in lemurs, in: G. A. Doyle and R. D. Martin, eds., The Study of Prosimian Behavior, Academic Press, New York.

Ramakrishna, P. A. and Prasad, M. R. N., 1967, Changes in the male reproductive organs of *Loris tardigradus lydekkerianus (Cabrera). Folia Primatol.* 5: 176-189.

Richard, A., 1976, Patterns of mating behaviour in *propithecus verreaxi verreauxi*, in: A. Doyle and A. C. Walker eds., Prosimian Behaviour, R. D. Martin, G. Duckworth, London.

Rodriquez-Sierra, J. F., Crowley, W. R. and Komisaruk, B. R., 1975, Vaginal stimulation in rats induces prolonged lordosis responsiveness and sexual receptivity, *J. Comp. Physiol. Psychol.* 89: 79-85.

Roosmalen, M. G. M. Van and Klein, L. L., 1988, The spider monkeys, genus *Ateles*, in Ecology and Behavior of Neotropical Primates, Vol. 2, R. A. Mittermeier, A. B. Rylands, A. Coimbra-Filho and A. B. Fonseca, eds., World Wildlife Fund, Washington D. C.

Saayman, G. S., 1970, The menstrual cycle and sexual behaviour in a troop of free-ranging chacma baboons *(Papio ursinus). Folia primatol.* 12: 81-110.

Sachs, B. D., 1982, Role of the rat's striated penile muscles in penile reflexes, copulation and induction of pregnancy. *J. Reprod. Fertil.* 66: 433-443.

Sachs, B. D. and Meisel, R. L., 1988, The physiology of male sexual behavior, in: E. Knobil and J. Neill eds., The Physiology of Reproduction, Vol. 2, Raven Press, New York.

Schilling, A. 1979, Olfactory communication in Prosimians, in: The Study of Prosimian Behavior, G. A. Doyle and R. D. Martin, eds., Academic Press, New York.

Schilling, A. and Evans, C., 1994, Vomerolfaction capacity in lemuroids, in: L. Alterman, G. A. Doyle and M. K. Izard (eds.). Creatures of the Dark: the Nocturnal Prosimians. Plenum, New York.

Schilling, A. and Perret, M., 1987, Chemical signals and reproductive capacity in a male prosimian primate *(Microcebus murinus). Chem. Senses* 12: 143-158.

Shideler, S. E. Lindburg, D. G. and Lasley, B. L., 1983, Estrogen-behavior correlates in the reproductive physiology and behavior of the ruffed lemur *(Lemur variegatus), Horm. Behav.* 17: 249-263.

Shinoda, A., Kraemer, G. W. and Goy, R. W., 1991, Effects of estrogen on activity in rhesus monkeys, in: A. Ehara, T. Kimura, O. Takenaka and M., Iwamoto, eds., Primatology Today, Elsevier, Amsterdam.

Short, R. V., 1979, Sexual selection and its component parts, somatic and genital selection, as illustrated by man and the great apes. *Adv. Study Behav.* 9: 131-158.

Sterling, E. J., 1992, Feeding behavior of the aye-aye *(Daubentonia madagascariensis)* on Nosy Mangabe, Madagascar. Paper read at the XIVth Congress of the International Primatological Society, Strasbourg, August 1992.

Sussman, R. W. and Garber, P. A., 1987, A new interpretation of the social organization and the mating system of the Callitrichidae, *Int. J. Primatol.* 8: 73-92.

Van Horn R. and Resko, J. A., 1977, The reproductive cycle of the ring-tailed lemur *(Lemur catta):* sex steriod levels and sexual receptivity under controlled photosperiods. *Endocrinology* 101: 1579-1586.

Verrell, P. A., 1992, Primate penile morphologies and social systems: further evidence for an association. *Folia primatol.* 59: 114-120.

Voss, R., 1979, Male accessory glands and the evolution of copulatory plugs in rodents, *Occ. Pap. Mus. Zool. Univ. Mich.* 689: 1-17.

Wallach, S. J. R. and Hart, B. L., 1983, The role of the striated penile muscles of the male rat in seminal plug dislodgement and deposition. *Physiol. Behav.* 31: 815-821.

Wallen, K., 1982, Influence of female hormonal state upon rhesus monkey sexual behavior varies with space for social interactions. *Science,* 217: 375-377.

White, F. J., 1991, Social organization, feeding ecology and reproductive strategy of ruffed lemurs, *Varecia variegata.* in: A. Ehara, T. Kimura, O. Takenaka and M. Iwamoto, eds., Primatology Today, Elsevier, Amsterdam.

Wickings, E. J. and Dixson, A. F., 1992, Testicular function, secondary sexual development and social status in male mandrills *(Mandrillus sphinx). Physiol. Behav.* 52: 909-916.

Wickings, E. J., Bossi, T. and Dixson, A. F., 1993, Reproductive success in the mandrill *(Mandrillus sphinx);* correlations of male dominance and mating success with paternity as determined by DNA fingerprinting. *J. Zool. Lond.* 231: 563-574

Wright, P. C. Izard, M. K. and Simons, E. L., 1986a, Reproductive cycles in *Tarsius bancanus, Am. J. Primatol.* 11: 207-215.

Wright, P. C. Toyama, T. M. and Simons, E. I., 1986b, Courtship and copulation in *Tarsius bancanus. Folia primatol.* 46: 142-148.

Young, W. C., 1961, The hormones and mating behavior, in: Sex and Internal Secretions, Vol. 2, W. C. Young, ed., Williams and Wilkins, Baltimore.

Zimmerman, E., 1988, Reproduction, physical growth and behavioral development in slow loris *(Nycticebus coucang, Lorisidae). Hum. Evol.* 4: 171-179.

Zimmerman, E., Zimmerman, P. and Zimmerman, A., 1979, Sociale communikation bei plumploris *(Nycticebus coucang). Z. Des Kolner Zoo* 22: 25-36.

THE BIOENERGETICS OF REPRODUCTION IN PROSIMIAN PRIMATES: IS IT RELATED TO FEMALE DOMINANCE?

Christopher D. Tilden[1] and Olav T. Oftedal[2]

[1]Department of Biological Anthropology and Anatomy
Duke University
3705-B Erwin Road
Durham, NC 27705

[2]Department of Zoological Research
National Zoological Park
Smithsonian Institution
Washington, DC 20008

INTRODUCTION

Primates have been characterized as having relatively low reproductive rates, as indicated by extended periods of gestation, lactation, and infant dependency (Sacher and Staffeldt, 1974; Martin, 1975; Eisenberg, 1981; Ankel-Simons, 1983; Harvey and Clutton-Brock, 1985; Rasmussen, 1986; Rasmussen and Izard, 1988). Nonetheless, among primates there have evolved diverse patterns of reproduction and parental care. For example, in some species care is provided solely by the mother, while in other species other members of the family are primary care-givers. Infants may be carried nearly continuously; or they may be left in a nest, in tree holes, or on branches for extended periods of time. Some species may exhibit communal care and nursing. Rearing patterns are especially diverse among prosimian primates. Patterns vary from those of the social, diurnal lemurs of Madagascar in which females of most species carry their dependent young, and in some species with the assistance of family members (Klopfer and Klopfer, 1970; Klopfer, 1974; Wright, 1990) to those of the small, solitary, nocturnal bushbabies and lorises of Asia and Africa in which females provide nearly all care, and frequently must park their infants to forage for food (Doyle *et al.*, 1969; Rosenson, 1971, 1972; Harcourt, 1984; Rasmussen, 1986).

Differences in parental care may be associated with different rates of energy expenditure, but the magnitude of reproductive energy expenditure in primates has received little attention. In other mammals, the energetic demands placed upon females may be greatly magnified during reproduction (Hanwell and Peaker, 1977; Oftedal, 1985; Gittleman, 1988; Gittleman and Thompson, 1988; Thompson, 1992). Given the paucity of data on primates, any suggested linkages between reproductive energetics and life histories have been speculative. Here we examine the assertion that high energetic costs

Creatures of the Dark: The Nocturnal Prosimians
Edited by L. Alterman *et al.*, Plenum Press, New York, 1995

associated with reproduction among some of the lemuriform primates of Madagascar were responsible for the evolution of female dominance among these Malagasy primates. According to this hypothesis, female dominance is a means of achieving priority access to limited food resources that are required to support both gestation (Jolly, 1984; Young *et al.*, 1990) and lactation (Eisenberg *et al.*, 1972; Jolly, 1984; Richard and Nicoll, 1987). The purported high energy requirements of reproduction among lemuriforms are particularly significant, it is argued (Richard and Nicoll, 1987; Young *et al.*, 1990), because prosimians are characterized by low rates of metabolism (Müller, 1979; Müller and Jaksche, 1980; Daniels, 1984; McNab and Wright, 1987; Ross, 1992) and thus presumably have constrained energy budgets.

Unfortunately, this attractive hypothesis has little direct experimental support. Researchers have been forced to rely on indirect estimates of energy expenditure such as rates of pre- and postnatal mass gain and the relative mass and developmental state of neonates (Eisenberg *et al.*, 1972; Jolly, 1984; Young *et al.*, 1990; Kappeler, 1992). In the course of studies of body composition and lactation performance on captive prosimians at the Duke University Primate Center (DUPC), we obtained direct data on maternal energy transfer both before and after birth (Tilden, 1993; Tilden and Oftedal, unpublished data). These data have allowed us to refine comparisons of reproductive energetics among lemuriform and lorisiform prosimians, and thus to reevaluate the hypothesized relationship between female social dominance and reproductive burden.

METHODS AND TERMINOLOGY

Maternal energy investment in offspring during gestation was examined by chemical analysis of newborn infants of two species of lemurs (the brown lemur *Eulemur fulvus* and the black lemur *E. macaco*) and two species of bushbabies (the thick-tailed bushbaby *Otolemur crassicaudatus* and Garnett's bushbaby *Otolemur garnettii*). Only infants that died within the first 24 hours after birth and that were of apparently normal weight and condition were used. The entire body was ground, dried and assayed for water content by oven drying to constant weight, ash by incineration in a muffle furnace, fat by overnight extraction with petroleum ether in Soxhlet extractors, and protein by the Kjeldahl method using copper sulfate as catalyst and 0.1N hydrochloric acid for titration of ammonia (Association of Official Analytical Chemists, 1990). Energy content (kcal) was calculated as the sum of protein (=nitrogen x 6.25) and fat energy content using energy values for protein and fat in Brouwer (1965). The details of this study are described by Tilden (1993) and will be published elsewhere; only the calculated energy concentrations are used in this paper. Species estimates of energy investment relative to maternal size and gestation length employed life history data collected at the DUPC (Table 1).

Maternal energy output during lactation was calculated based on milk energy concentration and milk yield measurements. In comparative studies, there has been an attempt to collect milk samples during mid-lactation, a period during which milk composition is relatively stable (Oftedal, 1984b). In the absence of definitive data on the shape of the lactation curve, we followed convention by assuming that mid-lactation

Table 1. Selected life-history data for energetic comparisons.

Species	FW (kg)	GL* (days)	LS*	NW (g)	GR1 (g/day)	GR2 (g/day)
Eulemur fulvus	2.30	120	1.2	67	0.56	4.1
Eulemur macaco	2.83	127	1.4	57	0.45	6.9
Otolemur crassicaudatus	1.07	136	1.6	45	0.33	7.6
Otolemur garnettii	1.09	130	1.1	33	0.25	6.0

FW = maternal mass; GL = gestation length; LS = average litter size; NW = neonatal mass; GR1 = prenatal growth rate; GR2 = postnatal growth rate. All data were collected at the DUPC. Data are original except that denoted by an asterisk (*), which were taken from Kappeler (1992).

occurs prior to substantial intake of solid foods by the young, i.e. from about 30-90 days postpartum in the species studied. Mid-lactation milk samples were collected manually after the administration of oxytocin, and were assayed for water content by oven drying (Association of Official Analytical Chemists, 1990), for protein and non-protein nitrogen (NPN) by digestion with sulfuric acid and hydrogen peroxide followed by direct Nesslerization (Belek and Jenness, 1962), for fat by the Roese-Gottlieb method of ether extraction (Association of Official Analytical Chemists, 1990), and for total sugar by the phenol-sulfuric acid procedure (Dubois *et al.*, 1956; Marier and Boulet, 1959). Using energetic factors of Perrin (1958), we calculated milk energy concentration as:

Gross energy (kcal/100g) = 9.11(Fat %) +5.86(Protein%) + 3.95(Sugar %) + 7.4(NPN%).

Milk yield was measured by isotope dilution over fifteen day intervals after administering deuterium oxide to infants at about 30 and 45 days postpartum (see Oftedal, 1984b and Oftedal and Iverson, 1987 for a review of isotope dilution methods). Details of these studies are given by Tilden (1993) and will be published elsewhere; only milk energy concentrations and estimated milk energy outputs at mid-lactation are reported here. Data were also collected on lactation length, defined as the period during which milk can be manually expressed from the mammary glands.

In comparisons among species, erroneous conclusions may be drawn if allometric effects are not taken into account. Both birth mass and milk output vary with body size, but not in direct proportion to maternal mass. For example, across a broad array of mammals, milk energy output at peak lactation scales to maternal metabolic mass ($kg^{0.75}$) rather than maternal mass, although there appear to be grade differences among different mammalian groups (Oftedal, 1984a, 1984b). Since rates of metabolism and energy expenditure tend to scale to metabolic mass (maternal $mass^{0.75}$, or MS), and given the widespread use of 0.75 as an allometric correction factor (Blaxter, 1989), we have expressed energy costs relative to MS for purposes of interspecific comparisons. Young *et al.* (1990) and Kappeler (1992) concluded that a lower exponent best represents the allometric scaling of metabolic rate among prosimians. However, the prosimian data are limited in scope and uneven in quality, and the introduction of a different exponent for metabolic mass would complicate broader taxonomic comparisons. Further study is needed to clarify the allometric relationships of energetic processes in prosimians.

Reproductive energetics cannot be accurately discussed without the use of specific terminology. We will use the following terms:

Gross energy refers to total combustible energy (as measured by bomb calorimetry) in a food, tissue or product (such as milk).

Metabolizable energy refers to the gross energy in food minus energy lost as excreta (urine, feces, methane) when food is metabolized. Whereas gross energy is an inherent property of the material in question, the metabolizable energy of a food may vary according to species and the physiological state of the consumer.

Basal metabolic rate refers to the energy expenditure of an adult, non-reproductive animal that is within thermoneutrality, is at rest during the quiescent period of its daily cycle and is not digesting or absorbing nutrients.

Resting metabolic rate is similar to basal metabolic rate but it not as restrictive as it may be measured during the active period of the daily cycle in animals that are not demonstrably postabsorptive or within thermoneutrality.

Maintenance energy requirement refers to the amount of energy required to sustain energy expenditures of an active but nonreproductive and non-growing animal without any loss or gain of stored energy in the body.

Fetal, neonatal and milk energy refer to gross energy (kcal/g) in the fetus, neonate or milk, but do not include the metabolic costs to the mother of producing these "products."

Efficiency of fetal deposition or milk secretion refers to the fraction of the metabolizable energy consumed by a reproductive animal above and beyond maintenance energy requirements that is actually deposited as fetal energy or secreted as milk energy. Efficiencies may be quite low both because of metabolic costs and because energy is required for the synthesis of other tissues such as the placenta, uterus, and mammary glands.

ENERGY TRANSFER DURING GESTATION

Primates are characterized by long gestations and relatively well-developed neonates (Leutenegger, 1973; Sacher and Staffeldt, 1974; Martin and MacLarnon, 1985, 1988). While gestation length is absolutely longer in anthropoid than in prosimian primates, this distinction disappears when body mass is taken into account. In fact, relative to adult body mass, gestation length is significantly longer among lorisiform prosimians than among lemuriforms or anthropoids (Young *et al.*, 1990; Kappeler, 1992). Lemuriformes have the shortest relative gestation lengths among primates.

Relative to maternal mass, no distinction in litter mass exists between Lemuriformes and Lorisiformes, but anthropoid primates produce heavier "litters" than prosimian females (Leutenegger, 1973; Wright, 1990; Young *et al.*, 1990). Anthropoids also tend to produce more fetal tissue (mass) per day, although some litter-producing prosimians do produce as much fetal mass per day as some anthropoids (Young *et al.*, 1990). Although anthropoids gain weight more rapidly in utero, it has been argued that lemurs face higher reproductive stress because of prosimians' presumably low metabolic rates and constrained energy budgets. Because lemuriform and lorisiform prosimians give birth to litters with similar relative mass at birth and lemurs do so in less time, daily fetal weight gain ("prenatal maternal investment rate" of Young and Kappeler) is relatively higher in lemurs (Young *et al.*, 1990; Kappeler, 1992). Young *et al.* (1990) contend that more rapid prenatal development in lemuriforms than lorisiforms represents a more energy-intensive strategy during gestation that is linked to, and behaviorally facilitated through, the pattern of female feeding priority and social dominance exhibited by some lemurs.

What is overlooked in this analysis is that mass is an indirect estimate of, and not the equivalent of, energy. Due to differences in the levels of water, protein, and fat in the young, the energy concentration of neonatal tissues is known to vary among mammalian taxa. During fetal development, there is a progressive decline in body water as the amount of extracellular fluid declines (Spray and Widdowson, 1950; Salmon-Legagneur, 1967; Adolph, 1970; Adolph and Heggeness, 1971) while the concentration of fat generally increases from conception to parturition (Spray and Widdowson, 1950; Salmon-Legagneur, 1967). Therefore, at birth altricial species with short gestations typically have a relatively low fat fraction and a high proportion of lean body mass (Spray and Widdowson, 1950; Salmon-Legagneur, 1967; Adolph, 1970; Adolph and Heggeness, 1971). Because lean body mass is predominantly water, these neonates tend to have low energy concentration at birth. However, species may vary in the time at which perinatal fat deposition occurs and in the rate of deposition. Because the energy concentration per gram of fat (9.3 kcal/gram) is very high (Brouwer, 1965), those species that deposit more fat prenatally are characterized by high energy content relative to body mass.

In this study, although *Eulemur* species had greater litter masses than *Otolemur* spp. they also had significantly lower neonatal energy concentration (0.77 kcal/g) than *Otolemur* (1.27 kcal/g). As a result, litter energy as a proportion of maternal metabolic size was significantly greater in bushbaby than lemur "litters" (Table 2). Despite longer gestations, *Otolemur* also had significantly higher average fetal energy deposition rates (Table 2). We caution that these average energy deposition rates do not reflect the fact that most fetal growth (and especially fat deposition) occurs in the last third of gestation. Nonetheless, our data on energy deposition indicate that litter mass and mass gain are inadequate measures of prenatal maternal investment. Actual energy deposition rates are higher in at least some lorisiform primates than among some lemuriforms. Thus the conclusion of Young *et al.* (1990) that prenatal maternal investment rates are greater among lemuriform than lorisiform prosimians appears to be incorrect.

Table 2. Neonatal energy composition and fetal energy deposition.

	Ave. Litter Wt. (g)	Energy* (kcal/g)	Ave. Litter Energy (kcal)	Ave. Litter Energy/MS (kcal/MS)	Energy Deposition/MS (kcal/MS/day)
Eulemur					
E. fulvus	80	0.77	62	33	0.28
E. macaco	80	0.77	62	28	0.22
Otolemur					
O. crassicaudatus	72	1.27	96	87	0.64
O. garnettii	36	1.27	48	43	0.33

*Energy concentration based on pooled generic data for *Eulemur* (*E. fulvus*, N=2; *E. macaco*, N=1) and *Otolemur* (*O. crassicaudatus*, N=2; *O. garnettii*, N=1). Life history data from Table 1.

ENERGY TRANSFER DURING LACTATION

While gestation entails quantifiable incremental energetic costs for female mammals, lactation is the most energetically demanding phase of reproduction for these females (Salmon-Legagneur, 1967; Oftedal, 1984a, 1984b, 1985; Gittleman and Oftedal, 1987; Gittleman, 1988; Gittleman and Thompson, 1988; Thompson, 1992). An energetic bottleneck occurs during lactation that must be supported through a marked increase in food intake, mobilization of stored tissue energy, or both. Milk production may entail a several-fold increase in maternal food intake relative to consumption during non-reproductive states (Sadleir, 1984; Loveridge, 1986: Sauther and Nash, 1987, Tilden, unpublished data).

Few quantitative comparisons have been made of postnatal maternal reproductive investment among primates, and most have focused on indirect estimates such as mass gain (Kappeler, 1992). The small size and developmental status of prosimian neonates, by comparison to anthropoid young, has also been used to infer rapid rates of growth after birth (Eisenberg, 1981; Jolly, 1984), which would place increased demands on the mother. However, such comparisons are complicated by the effects of body size. A suite of life-history variables including litter size, developmental status, and growth rates (as measured through a variety of methods, including time to double birth weight) are all affected by allometry as well as phylogeny, and interactions among these variables makes their analysis complex (Leutenegger, 1973; Clutton-Brock and Harvey, 1983; Martin, 1984; Oftedal, 1984b, 1986; Harvey and Clutton-Brock, 1985; Martin and MacLarnon, 1985, 1988). The expression "rapid growth" can have many interpretations, but usually refers to mass gain of an infant relative to its size (not the size of the mother). Yet a rapidly growing offspring may place a small demand on the mother if the infant is small. If the intent is to use growth as an estimate of demand on the mother, it is important that growth of the entire litter be considered in relation to the size and energy requirements of the mother. Kappeler (1992) found that, relative to maternal body size, postnatal rates of litter growth did not differ significantly among lemuriform and lorisiform primates, and thus suggested that "postnatal maternal investment" was similar in the two groups. However, he conceded that weight gain, as an indirect measure of investment, might not fairly represent maternal energy transfer because of potential differences in milk composition and production required to produce a similar result: infant mass accumulation. The composition and energy content of mass gain may differ among taxa, clouding simple comparisons based on mass.

Among primates, maternal energy transfer has been measured by isotope methods in two anthropoid genera: humans and baboons (Buss and Voss, 1971; Coward *et al.*, 1979; Roberts *et al.*, 1985; Schoeller and Fjeld, 1988). These anthropoid primates are large compared to all prosimians (and most other anthropoids) and have very low milk energy transfer rates in comparison to other mammals; relative to maternal size, daily rates of milk energy transfer in both baboons and human are considerably less than in non-primate mammals (Oftedal, 1984b). Since so few data are available, it is not certain if this pattern of reproductive output is characteristic of all primates. However, reproductive and developmental rates are reduced in primates compared to other mammals (Martin, 1975; Eisenberg, 1981; Ankel-Simons, 1983; Harvey and Clutton-Brock, 1985; Rasmussen and Izard, 1988). The fact that primate milks are typically dilute (Oftedal, 1984b) also seems consistent with the perception of low rates of energy transfer. However, some mammals with dilute milks, such as equids, compensate by producing large volumes of milk (Oftedal *et al.*, 1983; Oftedal and Jenness, 1988).

Our analyses of prosimian milks indicate that species within a genus may produce very homogeneous milks, but that substantial differences may be found in milk energy concentration among genera (Figure 1). For generic comparisons, data were pooled (using individual female means as data points) and analysis of variance was employed, using unplanned pairwise comparison of means (Sokal and Rohlf, 1981). *Eulemur* milk was remarkably dilute, with a gross energy concentration of about 0.5 kcal/g, which was significantly less than that of the closely related lemuriform, *Varecia* (0.8 kcal/g). The energy concentrations of milk among the lorisiforms *Nycticebus* and *Otolemur* were more than twice that of *Eulemur* and were significantly greater than those of both *Eulemur* and *Varecia.*

Measurements of milk energy output further suggest that lactation is more energy-intensive in at least some lorisiform primates than in their Malagasy cousins. Milk energy output at mid-lactation was significantly greater among both bushbabies than among the lemurs (Figure 2). When genera were compared, milk energy output was much higher among *Otolemur* than *Eulemur*, both absolutely (despite the smaller female size of *Otolemur* species) and when corrected for maternal metabolic mass ($p<.01$).

Comparisons of peak energy output provide an indication of the intensity of lactation, but not the total energetic investment over the entire lactation period. Total energy output in milk is a function of the duration of lactation, the shape of the lactation curve describing milk yield, and changes in milk energy concentration with the stage of lactation. Although few quantitative data are available for these parameters, we developed crude, "ballpark" estimates for comparative purposes. We assumed that average daily milk yield over the course of lactation was 70% of yield at mid-lactation (by analogy to ungulates for which average daily yield is about 60-80% of peak yield; Oftedal, 1985), that milk energy concentration remained constant at mid-lactation levels (although this is unlikely), and that lactation length was 90 days for bushbabies and 160 days for lemurs (see Table 3). Although all of these assumptions are subject to challenge,

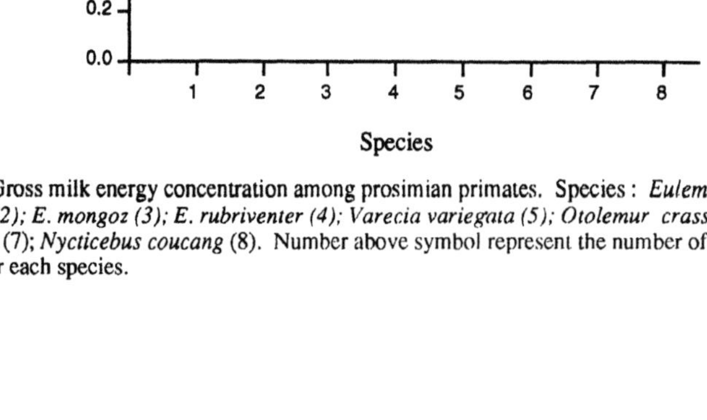

Figure 1. Gross milk energy concentration among prosimian primates. Species : *Eulemur fulvus* (1); *E. macaco (2); E. mongoz (3); E. rubriventer (4); Varecia variegata (5); Otolemur crassicaudatus* (6); *O. garnettii* (7); *Nycticebus coucang* (8). Number above symbol represent the number of milk samples analyzed for each species.

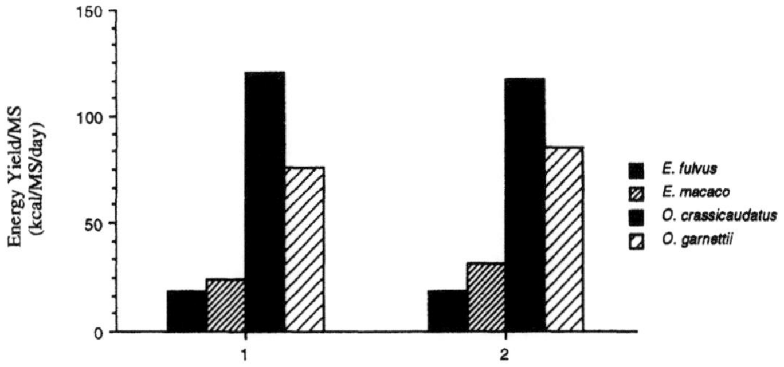

Figure 2. Milk energy output over two study periods. Period 1 is approximately 30-45 days postpartum, and Period 2, 45-60 days after birth. Energy output is not significantly different between lemurs, but differences between bushbaby species are significant. When pooled, the two genera are significantly different.

this exercise suggests total energy outputs of 5100-7500 kcal/kg$^{0.75}$ among bushbabies and only 2100-3100 kcal/kg$^{0.75}$ among lemur species. Thus, despite the shorter lactation in bushbabies than in lemurs, the estimated total milk energy transfer of bushbabies is about twice that of lemurs relative to maternal metabolic size.

Demonstration of higher peak and total milk energy outputs among bushbabies than among lemurs, despite similar litter growth rates (Kappeler, 1992), shows that litter growth rate does not account for species differences in the composition of gain and/or the efficiency of conversion of milk energy to tissue energy. Our results further call into question the female dominance hypothesis that assumes higher energetic burdens of reproduction in lemuriform than lorisiform primates, although the study of two genera admittedly limits any discussion of broad phylogenetic patterns.

ENERGY REQUIREMENTS OF REPRODUCTION

Data on energy output in reproductive products such as fetal tissues and milk can be used to estimate the energy requirements associated with reproduction if the efficiencies with which these products are formed are known (Oftedal, 1985). Unfortunately, there have been few definitive studies of the energetic efficiencies of mammalian reproduction, and most of these have been with taxa other than primates (Prentice and Prentice, 1988). Studies with domestic cattle and sheep demonstrate that efficiencies are not constant, but may vary according to maternal diet, mobilization of stored nutrients by themother, and the composition of the final product (Agricultural Research Council, 1980). As a first estimate of the reproductive costs of prosimians, we have extrapolated efficiency data

Table 3. Lactation Length of Individual Females.

Individual	Dates[1]	Length
Eulemur macaco		
Daphne	3/15/89-9/25/89	194
Latona	3/18/89-9/25/89	191
Eulemur fulvus		
Bozy	5/18/89-9/25/89	130
Nicole	4/7/89-9/19/89	165
Otolemur crassicaudatus		
Kiboko	10/19/89-1/19/90	92
Otolemur garnettii		
Obsidian	12/22/89-3/26/90	104
Calamine	5/22/89-8/30/89	100

[1]Dates refer to the period during which milk could be manually expressed from the mammary gland.

from domestic animals, but caution thate these values may mask important phylogenetic differences associated with dietary, physiological and reproductive adaptations.

In the factorial method, energetic costs of specific reproductive processes are estimated separately and are then compared to estimated maintenance requirements (Oftedal, 1985). This may be particularly valuable in comparing reproductive costs in taxa with different metabolic rates and presumably different maintenance requirements. For present purposes, maintenance requirements were estimated as two times basal or resting metabolic rates using literature values for metabolic rates (Müller and Jaksche, 1980; Daniels, 1984) and assuming a respiratory quotient of 0.78 in converting oxygen consumption to energy equivalents (Maynard et al., 1979; Tilden, 1993). Maintenance requirements are, very roughly, about 200%-250% of BMR in small primates and other small mammals (Power, 1991; Thompson, 1992).

In gestating cattle and sheep, 11% of the increase in metabolizable energy intake above maintenance is deposited in the fetus (Agricultural Research Council, 1980); of the remainder, some is incorporated into growth of associated tissues (placenta, uterus, fluids) and the mammary glands, and some goes to support increased metabolic costs. We estimated the energetic cost of gestation by dividing litter energy content at birth by 0.11. Of course, some of this investment may be recovered by the mother after birth when the supportive tissues degenerate. Estimated energy requirements for gestation were 256-301 kcal/kg$^{0.75}$ and 392-790 kcal/kg$^{0.75}$ in lemurs and bushbabies, respectively (Table 4). Over the course of gestation these values represent only a small increase over maintenance requirements. The increase is about 2.5-3% in lemurs and about 3-6.5% in bushbabies. The aggregate costs of gestation are equivalent to 3-4 and 4-9 days of maintenance energy requirements in lemurs and bushbabies, respectively.

Milk secretion is considerably more expensive. The efficiency of milk secretion has been estimated as 60-70% in domestic ruminants (Agricultural Research Council, 1980), but is thought to be as high as 75-80% in humans (Prentice and Prentice, 1988). The latter estimates are imprecise due to the difficulties in measuring changes in maternal body composition during pregnancy, and are probably too high. We assumed an efficiency of 70%. From this, we calculated that daily energy requirements during mid-lactation increased by 33-48% in lemurs and by 129-192% in bushbabies in comparison to maintenance requirements (Table 4). We infer that energy requirements for lactation may be grossly equivalent to about 37-54 days of maintenance energy requirements in lemurs and 81-121 days in bushbabies.

Table 4. Energy requirements of gestation and maintenance.

	Species			
	(1)	(2)	(3)	(4)
Litter energy (kcal/MS)	33	28	87	43
Gest. energy reqs.[1] (kcal/MS)	301	256	790	393
Maintenance req.[2] (kcal/MS/day)	80	83	88	89
Gestation/maintenance[3] (%)	3.1	2.4	6.6	3.4
Daily milk energy (kcal/MS/day)	18.7	27.9	118.9	80.4
Lact. energy reqs.[4] (kcal/MS/day)	26.7	39.9	169.6	114.9
Lactation/maintenance[5] (%)	33	48	192	129

Figures are all based on average litter size. MS = maternal metabolic mass. Species designations are: *Eulemur fulvus* (1), *E. macaco* (2), *Otolemur crassicaudatus* (3), and *O. garnettii* (4).

[1]Total energy requirements of gestation (litter energy/0.11, see text for discussion).

[2]Daily requirements inferred from resting metabolic rates of *E. fulvus* (0.32ml O_2/g/h; Daniels, 1984) and *O. crassicaudatus* (0.434ml O_2/g/h; Müller and Jaksche, 1980).

[3]Daily energy requirements of gestation as a percentage of daily maintenance.

[4]Daily milk energy output divided by 0.70 (see text).

[5]Daily requirements of lactation as a percentage of daily maintenance.

Although crude, the estimates illustrate that lactation is markedly more expensive than gestation, and that the reproductive requirements of *Eulemur* species appear to be less than, not more than, the requirements of *Otolemur* species. Relative to maintenance requirements, the combined requirements of gestation and lactation in bushbabies are apparently comparable to some non-primate mammals, but these requirements appear extremely low in lemurs. In this respect, energy requirements for reproduction in the lemuriform primates may be comparable to the maintenance equivalents for reproduction in anthropoid primates (Tilden, 1993).

CONCLUSION

Consistent differences were found between species of *Otolemur* and *Eulemur*, all of which indicated a difference in reproductive intensity. By comparison to bushbabies, lemurs had: 1) neonates that were lower in gross energy concentration, 2) lower rates of fetal energy deposition relative to maternal metabolic size, 3) milks of lower energy concentration, 4) lower peak milk energy yields relative to maternal metabolic size, and 5) lower estimated energy requirements for reproduction, expressed relative to maintenance requirements. It appears that at least some lemuriforms have relatively low energetic costs of reproduction rather than the higher reproductive burden that has been postulated based on an analysis of prenatal growth rates, the relatively altricial state of neonates and the prevalence of female dominance among many species of lemurs (Jolly, 1984; Young *et al.*, 1990; Kappeler 1992).

We suggest, however, that modification of this long standing hypothesis allows an alternative explanation of the prevalence of female dominance in lemuriforms that is consistent with the present findings. Among species of the Carnivora, postnatal litter growth rates and predicted milk energy outputs appear to be lower among herbivorous species, presumably because of the difficulties in acquiring and processing sufficient energy from such foods to sustain high rates of milk output (Gittleman and Oftedal, 1987; Gittleman, 1988). It is possible that the low reproductive outputs of lemurs are the consequence of similar energetic constraints, whether due to seasonality of food supplies, low digestibility of foods, or extreme intra-group competition for food in large groups. Successful reproduction is possible only if females have adequate nutrient and energy intakes. Female dominance (or at least feeding priority) could have evolved in a context where priority access to foods would assure adequate intakes for reproductive females. In this view, reproductive energy output of lemur females remains low in spite of female dominance. We are not aware of any direct evidence of unique nutritional constraints on lemurs that do not apply to primate taxa in which female dominance is rare. Nevertheless, compared to bushbabies lemurs do possess markedly seasonal schedules of reproduction and growth, seemingly corresponding to seasonal climatic patterns in Madagascar (Pereira, 1993). This suggests that resource availability may indeed impose unique constraints on reproductive lemur females. It is possible that lemurs not characterized by female dominance, facing the same environmental stress as females in those species with female dominance, have adopted different strategies for the acquisition of nutrients and energy (Pereira and McGlynn, submitted), or have adopted strategies for energy conservation. Many lemurs exhibit traits such as hypometabolism, seasonal torpor, sun bathing and group huddling that have been interpreted as energy-conserving adaptations (Jolly, 1966; Charles-Dominique et al., 1980; Daniels, 1984; Müller, E. F. , 1985).

We acknowledge that the study of reproductive performance in just two genera does not permit generalizations about broader phylogenetic patterns. However, our results hint at a much broader diversity in reproductive rates among primates than is generally recognized, and suggest that the generalization that all primates are characterized by reduced reproductive and developmental rates should be critically reexamined. We predict that the diverse mating patterns, parental care and rearing systems, and types of social organization among the primates are associated with equally diverse patterns of energy transfer and reproductive energy requirements. We hope that our results will stimulate further study on this topic; a broader database is needed to address more thoroughly many of the issues raised herein.

ACKNOWLEDGEMENTS

We would like to thank Elwyn Simons and the staff of the Duke University Primate Center for providing access to the resources of the facility and for technical support. Kay Izard, in particular, deserves special thanks. Technical support was also provided by Michael Jakubasz in the Nutrition Laboratory, Department of Zoological Research, National Zoological Park. As members of C.D.T.'s doctoral committee, Elwyn Simons, Pat Wright, Carel van Schaik, Ken Glander, and Richard Kay made numerous and varied contributions to this project. Friderun Ankel-Simons provided insight and support, and Mike Power made valuable comments on the manuscript. Financial support was provided to C.D.T. by the Duke University Graduate School, the Friends of the National Zoo, the Chicago Zoological Society, and by a Grant-in-Aid of Research from Sigma Xi, the Scientific Research Society.

REFERENCES

Adolph, E. F. (1970). Physiological stages in the development of mammals. *Growth, 34*, 113-124.

Adolph, E. F., & Heggeness, F. W. (1971). Age changes in body water and fat in fetal and infant mammals. *Growth, 35*, 55-63.

Agricultural Research Council. (1980). *The Nutrient Requirements of Ruminant Livestock.* Farnham Royal, England: Commonwealth Agricultural Bureaux.

Ankel-Simons, F. (1983). *A Survey of Living Primates and Their Anatomy.* New York: MacMillan Publishing Co.

Association of Official Analytical Chemists. (1990). *Official Methods of Analysis. Fifteenth edition.* Arlington, Virginia: Association of Official Analytical Chemists.

Belek, J., & Jenness, R. (1962). Dephosphorization of casein by heat treatment. I. In caseinate solutions. *Journal of Dairy Science, 45*, 12-19.

Blaxter, K. L. (1989). *Energy Metabolism in Animals and Man.* Cambridge: Cambridge University Press.

Brouwer, E. (1965). Report of subcommittee on constants and factors. In K. L. Blaxter (Ed.), *Energy metabolism: proceedings of 3rd symposium* (pp. 441-443). London: Academic Press.

Buss, D. H., & Voss, W. R. (1971). Evaluation of four methods for estimating the milk yield of baboons. *Journal of Nutrition, 101*, 901-910.

Charles-Dominique, P., Cooper, H. M., Hladik, A., Hladik, C. M., Pages, E., Pariente, G. F., Petter-Rousseaux, A., Petter, J. J. & Schilling, A. (1980). *Nocturnal Malagasy Prosimians.* New York: Academic Press.

Clutton-Brock, T. H., and Harvey, P. H. (1983). The functional significance of variation in body size among mammals. In J. F. Eisenberg & D. G. Kleiman (Eds.), *Recent advances in the study of mammalian behavior* (pp. 632-663). Shippenburg, PA: American Society of Mammologists.

Coward, W. A., Sawyer, M. B., Whitehead, R. G., Prentice, A.M. & Evan, J. (1979). New method for measuring milk intakes in breast-fed babies. *Lancet, 1979(2)*, 13-14.

Daniels, H. L. (1984). Oxygen consumption in *Lemur fulvus*: deviation from the ideal model. *Journal of Mammalogy, 65(4)*, 584-592.

Doyle, G.A., Andersson, A. & Bearder, S. K. (1969). Maternal behavior in the lesser bushbaby (*Galago senegalensis moholi*) under semi-natural conditions. *Folia Primatologica, 11*, 215-238.
Dubois, M., Gilles, K. A., Hamilton, J. K., Rebers, P.A. & Smith, F. (1956). Colorimetric method for determination of sugars and related substances. *Analytical Chemistry, 28(3)*, 350-356.
Eisenberg, J. F. (1981). *The Mammalian Radiations*. Chicago: University of Chicago Press.
Eisenberg, J. F., Muckenhirn, N. A. & Rudran, R. (1972). The relationship between ecology and social structure. *Science, 176*, 863-874.
Gittleman, J. L. (1988). Behavioral energetics of lactation in a herbivorous carnivore, the red panda (*Ailurus fulgens*). *Ethology, 7*, 13-24.
Gittleman, J. L., & Oftedal, O. T. (1987). Comparative growth and lactational energetics in carnivores. In A. Loudon & P. A. Racey (Eds.), *Reproductive energetics in mammals* (pp. 41-77). Oxford: Oxford University Press.
Gittleman, J. L., & Thompson, S. D. (1989). Energy allocation in mammalian reproduction. *American Zoology, 28*, 863-875.
Hanwell, A., & Peaker, M. (1977). Physiological effects of lactation on the mother. *Symposium of the Zoological Society of London, 41*, 297-312.
Harcourt, C. S. (1984). The behaviour and ecology of galagos in Kenyan coastal forest. Ph.D. thesis, Cambridge University, Cambridge, England.
Harvey, P. H., & Clutton-Brock, T. H. (1985). Life history variation in primates. *Evolution, 39*, 559-581.
Jolly, A. (1966). *Lemur Behavior*. Chicago: Chicago University Press.
Jolly, A. (1984). The puzzle of female feeding priority. In M. E. Small (Ed.), *Female Primates: Studies by Women Primatologists* (pp. 197-215). New York: Alan Liss.
Kappeler, P. M. (1992). Female dominance in Malagasy prosimians. Ph.D. thesis, Duke University, Durham, NC.
Klopfer, P. H. (1974). Mother-young relations in lemurs. In R. D. Martin, G. A. Doyle & A. C. Walker (Eds.), *Prosimian Biology* (pp. 273-292). Pittsburgh: University of Pittsburgh Press.
Klopfer, P. H., & Klopfer, M. S. (1970). Patterns of maternal care in three species of *Lemur*. I. Normative description. *Zeitschrift für Tierpsychologie, 27*, 984-996.
Leutenegger, W. (1973). Maternal-fetal weight relationships in primates. *Folia Primatologica, 20*, 280-293.
Loveridge, G. G. (1986). Bodyweight changes and energy intake in cats during gestation and lactation. *Animal Technology, 37*: 7-15.
Marier, J. R., & Boulet, M. (1959). Direct analysis of lactose in milk and serum. *Journal of Dairy Science, 4*, 1390-1391.
Martin, R. D. (1975). The bearing of reproductive behavior and ontogeny on strepsirhine phylogeny. In W. P. Luckett and F. S. Szalay (Eds.), *Phylogeny of the Primates: a multidisciplinary approach* (pp. 265-297). New York: Plenum Press.
Martin, R. D. (1984). Scaling effects and adaptive strategies in mammalian lactation. *Symposium of the Zoological Society of London, 51*, 87-117.
Martin, R. D., & MacLarnon, A. M. (1985). Gestation period, neonatal size and maternal investment in placental mammals. *Nature, London, 313 (17)*, 220-223.
Martin, R. D., & MacLarnon, A. M. (1988). Comparative quantitative studies of growth and reproduction. *Symposium of the Zoological Society of London, 60*, 39-80.
Maynard, L. A., Loosli, J. K., Hintz, H. F. & Warner, R. G. (1979). *Animal Nutrition. Seventh edition.* New York: McGraw-Hill Book Co.
McNab, B. K., & Wright, P. C. (1987). Temperature regulation and oxygen consumption in the Phillipine tarsier *Tarsius syrichta*. *Physiological Zoology, 60*, 596-600.

Müller, E. F. (1979). Energy metabolism, thermoregulation, and water budget in the slow loris (*Nycticebus coucang*, Boddaert 1785). *Comparative Biochemistry and Physiology, 64A*, 109-119.
Müller, E. F. (1985). Basal metabolic rates in primates: the possible role of phylogenetic and ecological factors. *Comparative Biochemistry and Physiology, 81A*, 707-711.
Müller, E. F., & Jaksche, H. (1980). Thermoregulation, oxygen consumption, heart rate and evaporative water loss in the thick-tailed bushbaby (*Galago crassicaudatus* Geoffroy, 1812). *Zeitschrift für Säugetierkunde, 45*, 269-278.
Oftedal, O. T. (1984a). Body size and reproductive strategy as correlates of milk energy output in lactating mammals. *Acta Zoologica Fennica, 17*, 183-186.
Oftedal, O. T. (1984b). Milk composition, milk yield and energy output at peak lactation: a comparative review. *Symposium of the Zoological Society of London, 51*, 33-85.
Oftedal, O. T. (1985). Pregnancy and lactation. In R. J. Hudson & R. G. White (Eds.), *Bioenergetics of Wild Herbivores* (pp. 216-238). Boca Raton, FL: CRC Press.
Oftedal, O. T. (1986). Growth rate and milk composition: a critical appraisal. In L. J. Filer & S. J. Fomon (Eds.), *The breastfed infant: a model for performance. Report on the 91st Ross conference on pediatric research* (pp. 50-58). Columbus, OH: Ross Laboratories.
Oftedal, O. T., Hintz, H. F. & Schryver, H. F. (1983). Lactation in the horse: milk composition and intake by foals. *Journal of Nutrition, 113*, 2196-2206.
Oftedal, O. T., and Iverson, S. J. (1987). Hydrogen isotope methodology for measurement of milk intake and energetics of growth in suckling young. In A. C. Huntley, D. P. Costa, G. A. J. Worthy & M. A. Castellini (Eds.), *Approaches to Marine Mammal Energetics* (pp. 67-96). Lawrence, KS: Allen Press.
Oftedal, O. T., and Jenness, R. (1988). Interspecies variation in milk composition among horses, zebras, and asses (Perissodactyla, Equidae). *Journal of Dairy Research, 55*, 57-66.
Pereira, M. D. (1993). Seasonal adjustment of growth rate and adult body weight in ringtailed lemurs. In P. M. Kappeler & J. U. Ganzhorn (Eds.), *Lemur Social Systems and Their Ecological Basis* (pp. 205-221). New York: Plenum Press.
Pereira, M. E., Kaufman, R., Kappeler, P. M. & Overdorff, D. J. (1990). Female dominance does not characterize all of the Lemuridae. *Folia Primatologica, 55*, 96-103.
Perrin, D. R. (1958). The calorific value of milk of different species. *Journal of Dairy Research, 25*, 215-220.
Power, M. L. (1991). Digestive function, energy intake and the response to dietary gum in captive callitrichids. Ph.D. thesis, University of California at Berkeley, Berkely, CA.
Prentice, A. M., & Prentice, A. (1988). Energy costs of lactation. *Annual Review of Nutrition, 8*, 63-79.
Rasmussen, D. T. (1986). Life history and behavior of slow lorises and slender lorises. Ph.D. thesis, Duke University. Durham, NC.
Rasmussen, D. T., & Izard, M. K. (1988). Scaling of growth and life history traits relative to body size, brain size, and metabolic rate in lorises and galagos (Lorisidae, Primates). *American Journal of Physical Anthropology, 75*, 357-367.
Richard, A. F., & Nicoll, M. E. (1987). Female social dominance and basal metabolism in a Malagasy primate, *Propithecus verreauxi*. *American Journal of Physical Anthropology, 12*, 309-314.
Roberts, S. B., Cole, T. J. & Coward, W. A. (1985). Lactational performance in relation to energy intake in the baboon. *American Journal of Clinical Nutrition, 41*, 1270-1276.
Rosenson, L. M. (1971). Social behavior of the greater bushbaby. Ph.D. thesis, Duke University, Durham, NC.

Rosenson, L. M. (1972). Observations of the maternal behaviour of two captive greater bushbabies (*Galago crassicaudatus argentatus*). *Animal Behavior, 20*, 677-688.

Ross, C. (1992). Basal metabolic rate, body weight and diet in primates: an evaluation of the evidence. *Folia Primatologica, 58*, 7-23.

Sacher, G. A., & Staffeldt, E. F. (1974). Relation of gestation time to brain weight for placental mammals: implications for the theory of vertebrate growth. *American Naturalist, 108(963)*, 593-615.

Sadleir, R. M. F. S. (1984). Ecological consequences of lactation. *Acta Zoologica Fennica, 171*, 179-182.

Salmon-Legagneur, E. (1967). Prenatal development in the pig and some other multiparous animals. In G. A. Lodge and G. E. Lamming (Eds.), *Growth and Development of Mammals* (pp. 158-191). London: Butterworths.

Sauther, M. L., & Nash, L. T. (1987). Effect of reproductive state and body size on food consumption in captive *Galago senegalensis braccatus*. *American Journal of Physical Anthropology, 73*, 81-88 .

Schoeller, D. A., & Fjeld, C. R. (1986). Measurement of human milk consumption by deuterium oxide kinetics. In L. J. Filer & S. J. Fomon (Eds.), *The breastfed infant: a model for performance. Report on the 91st Ross conference on pediatric research* (pp. 13-22). Columbus, OH: Ross Laboratories.

Sokal, R. R., & Rohlf, F. J. (1981). *Biometry*. New York: W. H. Freeman and Co.

Spray, C. M., & Widdowson, E. M. (1950). The effect of growth and development on the composition of mammals. *British Journal of Nutrition, 4*, 332-353.

Thompson, S. D. (1992). Gestation and lactation in small mammals: basal metabolic rate and the limits of energy use. In T. E. Tomasi & T. H. Horton (Eds.), *Mammalian energetics: interdisciplinary views of metabolism and reproduction* (pp. 213-259). Ithaca, NY: Comstrock Publishing Associates (Cornell University Press).

Tilden, C. D. (1993). The reproductive energetics of prosimian primates. Ph.D. thesis, Duke University, Durham, NC.

Wright, P. C. (1990). Patterns of paternal care in primates. *International Journal of Primatology, 11(2)*, 89-102.

Young, A. L., Richard, A. F. and Aiello, L. C. (1990). Female dominance and maternal investment in strepsirhine primates. *American Naturalist, 135*, 473-488 .

HISTORY, ANATOMY, SUBFOSSIL RECORD AND MANAGEMENT OF *DAUBENTONIA MADAGASCARIENSIS*

Elwyn L. Simons

Duke University Primate Center and
Department of Biological Anthropology and Anatomy
Duke University
Durham, North Carolina

HISTORY OF STUDY

Although discovered by Pierre Sonnerat in 1780 the scientific study of the aye aye has been relatively sporadic, especially during the first half of this century. In the 70 years previous to a series of quite recent importations and studies little research in the field or in captivity was conducted, with the exception of the research and field work of Petter *et al.* (1977).

The aye-aye is ranked in its own family, Daubentoniidae (Gray, 1870). It is the only primate species that is considered endangered at the specific, generic and familial level. Its ranking as being so critically endangered came about in part, because the species is nocturnal, ranges widely, and inhabits remote areas. For these reasons aye-aye were believed to be rarer than they actually are. Killing is widespread. Unlike other Malagasy animals that are considered "fady" (taboo or bad luck), when seen in villages aye-aye are usually hunted down and killed in order to take away the bad luck that they are believed to carry. This species is , unfortunately, attracted to villages because it eats coconuts, mangoes and sugar cane that are planted by the villagers. Being killed because of a fady is unique, among lemurs, to the aye-aye. During the 1960's and 70's the aye-aye was often reported as being extremely rare, ocurring only in disapearing coastal forests, nearing extinction and about to disapear. As a conservation measure, several individuals of this species were placed in a reserve on the island of Nosy Mangabe in 1966. In the 1980's and 90's Ganzhorn, Meyers, and I, along with others working in association with the Duke University Primate Center (DUPC) or independently have discovered that aye-aye are widely distributed and survive in many areas. Most recently, in 1991, G. Rakotoarasoa and I located a large population of aye-aye in the Manasamody hills about 40 to 50 km southwest of Ansohihy near the west coast of Madagascar.

When Sonnerat reached Paris in 1782, an aye-aye he brought back was studied by the Comte de Buffon, who regarded it as a kind of squirrel, but one which, perhaps, had a relationship to *Tarsius*, an animal that Buffon also believed to be a rodent (Buffon, 1789).

In 1789, Gmelin assigned the animal to the squirrel genus*Sciurus*, as*Sciurus madagascariensis* (Gmelin 1789). A few years later, Geoffroy transferred the animal to its own genus *Daubentonia* (Geoffroy 1795). This name was given in honor of the 18th century French naturalist and contemporary of Buffon, Louis Jean-Marie Daubenton. Some of the other generic names applied to this animal early on, now considered synonymous with *Daubentonia*, are aye-aye, *Cheiromys*, and *Chiromys*. The enormous, ever-growing incisors and large bushy tail of Sonnerat's original specimen contributed to the aye-aye's being mistaken for a squirrel, but the naturalist Schreber (1774), after studying the hands and feet, understood that it was related to lemurs.

It was not until about eighty years after Sonnerat's findings that several additional specimens of Daubentonia arrived in Europe between 1859 and 1865. These new materials made a second phase of analysis possible. The new finds were the subject of two monographs (Peters, 1866; Owen, 1866). At the same time behavioral information was reported in a letter from H. Sandwich to Owen and by two French students, Lienard and Soumagne, (see references in Owen 1866). The French surgeon, A. Vinson, had living aye-aye on the island of Reunion, where the first captive births took place (Vinson, 1855).

Jones (1986) summarized information available from records kept at the world's zoological gardens. Between August, 1862 and September, 1932, seven different aye-aye were held at the London Zoo. Between November, 1880 and March, 1932 thirteen others were at the Jardin des Plantes in Paris. Between April, 1887 and September, 1937 twelve additional aye-aye were held in various Zoological Gardens in the Netherlands, Belgium, Ireland and Germany. In spite of these many exportations, the species is not known to have arrived alive in North America before the importation of two male aye-aye to the DUPC on December 30, 1987. In 1884, thirty five individuals were captured for display in European zoos, but of these only one arrived alive in the Jardin des Plantes. It seems that prior to 1988, no zoo or other institution ever held, or caged, male and female aye-aye together in captivity for any length of time. Both sexes caged together could only have occurred at the Jardin des Plantes between September 1927 and September 1929, and perhaps also at Berlin for a period of time in 1913. This explains why captive breeding was not reported until 1992. It was not because aye-aye are difficult to breed that none were born in captivity during the last hundred years.

As mysterious as the anatomy and affinities of this animal seemed at first are the reasons for the name, aye-aye, which has been associated with it from Sonnerat on. Alternate variants of the spelling are: hai hai (pronounced 'high') or hay hay (pronounced 'hay'). One persistent theory is that the species emits a call which resembles the sound "aye aye". Those who know the animal best do not believe that this is the case. This sound has not been heard at the DUPC nor has it been recorded by an observer in the wild. Nevertheless, the Malagasy people have named many lemurs onomatopoeically from the sounds they make, which probably accounts for the idea.

A second theory, originally proposed by Sonnerat, is that he coined this name for the animal from the cries of alarm made by people shouting "aiee!- aiee!" when they saw the strange animal. On several occasions I have heard Malagasy villagers exclaim "aye-aye" or "aie-yi" when frightened or in pain. Several similar expletives occur in other languages, for instance in German (ei-ei-ei). The name, aye-aye, for this animal must be of long-standing usage. It is known all over the island of Madagascar and even in remote areas where the animal does not occur. It seems most unlikely that Sonnerat really coined this name.

A third theory is that in some dialects "heh heh", which could be linguistically related to "hey hey", means 'I don't know". It is supposed by those who offer this meaning that "heh heh" is a reply produced by villagers that have no name for the animal or who do not wish to speak it name because it is "fady".

ANATOMY

Much of the early anatomical work on *Daubentonia* centered around its highly aberrant dental anatomy and the odd structure of its middle finger. Other distinctions noted include the curious shape of its skull and the consequent klinorhynchy it exhibits (see Cartmill, 1974). Cranial and mandibular proportions are greatly affected by the enormous incisors and short, muscular neck.

Attempts at association of this species with other animals were confounded by the construction and arrangement of the adult teeth, which bear almost no relationship to those of any other primate species. The enormous incisors are ever-growing and their growth centers are situated far back in the adult skull and mandibles. Ever-growing teeth are most unusual in mammals and have originated perhaps not more than half a dozen times among Mammalia and never among Primates. The cheek teeth, even at the time of eruption, bear no clear-cut surface patterns that could relate the animal to other lemurs. In addition, from the time of Peters (1866), the identification of the type and number of the milk teeth has been interpreted wrongly (Ankel-Simons in press). Among mammals, the deciduous dentition sometimes reflects dental patterns dental patterns seen in ancestral forms. X-rays of a newborn aye aye show that the mandibular milk incisors and canines are implanted in a procumbent manner, similar to lemuroid toothcombs. This indicates that the aye aye dentition is derived from an ancestral form with a lemuroid toothcomb.

Other anatomical peculiarities of the aye-aye are its extremely large, mobile ears; the inguinally located mammae and the long flexible fingers, especially the thin third digit which is the most wire-like of all the digits and swivels freely from its basal joint.

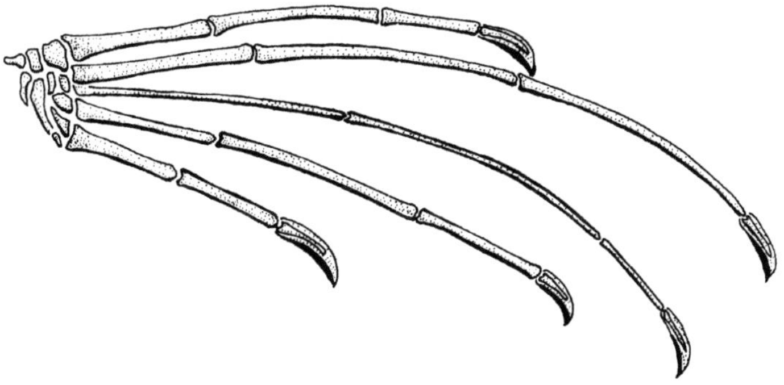

Figure 1: Handskeleton of *Daubentonia madagascariensis*. From Ankel-Simons 1983.

Nails on all digits other than on the great toes are narrowed from side to side and consequently are claw like. When foraging, the animal taps tree branch surfaces with the third digit while holding the ears tilted forward. If something is detected, the bark is torn open by the large anterior incisors and the larva extracted with the wire-like digit, as reported by Owen (1866) and Petter (1977). Drinking of liquids is also achieved by using the third finger as a pump. The digit is rapidly drawn back and forth between the mouth and liquids such as coconut milk or the liquid contents of oranges or raw eggs.

Recently Erickson has studied aye aye foraging behavior in detail (Erikson, 1991). He has termed this manner of searching "percussive foraging". It appears that the gently tapping action of the third finger produces different sound values over cavities inside wood, or else that the digit itself is so sensitive that it picks up variations in vibration above cavities. Erickson found that cavities without grubs were opened almost as frequently as were those which contained them, suggesting that searching is mostly for cavities only, not for grub. This type of feeding has been suggested to be an adaptive equivalent to woodpecker feeding habits (Cartmill, 1974), there being no woodpeckers in Madagascar. Since we now know that aye-aye also frequently eat fruits, seeds, nectar, nuts and fungi, this extends their range of food items greatly beyond the feeding adaptations of most woodpeckers. In the eating of ramy nuts (*Canarium*) and coconuts, aye-aye use both their gnawing abilities and the third digit.

Since the aye-aye's unusual physical adaptation all appear to be related to its manner of feeding, more needed to be known about its feeding habits in the wild. New data just became available from the dissertation of Sterling (1993). She shows, that even though the aye aye are highly selective in their feeding habits, foods other than grubs — such as ramy seeds and ravinala nectar — are important dietary components.

In part following Cartmill (1974) , as Sterling (1993) points out, a tradition has developed that the aye aye's morphological specialization's arose and are used for a woodpecker-like feeding dependence on woodboring grubs. Sterling states that among others Hill (1953) and Stevens (1988) make the case. The predominance of wood feeding larvae in the he aye aye's diet has been questioned by others, including Petter (1977) and Iwano and Iwakawa (1988). These authors have stated that the seed and nut feeding of the aye aye makes them analogous to the squirrels for which they were originally mistaken by Gmelin (1789). Like woodpeckers, squirrels also do not occur in Madagascar. Coconuts now make an important element in the diet of the aye-aye (Petter, 1977). Other adaptations of the aye aye resemble squirrels besides the ability to gnaw into coconuts and seeds. These include their construction of nests resembling those of squirrels as well as their large, bushy tails, used as a wrap-around "blanket" during sleeping.

SUBFOSSILS

The living aye-aye is the largest nocturnal primate and is several times larger than a strictly insectivorous mammal should be. Now that Sterling's (1993) ecological study has shown the aye aye to be basically more omnivorous than insectivorous, its large size is easier to understand. The subfossil remains of aye-aye are intriguing because they demonstrate that in the recent past a much larger species existed. Distribution of the subfossils in sites restricted to the mainly "xeric forests" of the southern and southwestern coasts poses an even greater mystery as the what this huge aye-aye could have fed on.

This large species, named by Lamberton (1934) comes from three sites: 1) Tsirave (locality of the type specimen), south of the Mangoky river in south central Madagascar; 2) Lamboharana on the Bay of Assassins, southwest coast; and 3) Anavoha (or Beloha) on the southern coast. The material from Lamboharana was collected in 1901 by Grandidier and consisted of three perforated incisors, two of which have recently been rediscovered and reported on by MacPhee and Raholimavo (1988). Perhaps more interesting are the partial postcranial skeletons found at Tsirave and Anavoha (see Fig. 2) which, although lacking any dental or cranial remains, are similar in size and complete enough to allow for a range of probable size estimates. Lamberton (1934) described, figured and presented tables of comparative measurements for *Daubentonia* (= *Chiromys*)*robusta* (=*robustus*) that well demonstrated the subfossil species to be considerably larger than the living aye aye. Albrecht *et al.* (1990) reported that femoral length of *D. robusta* is 19% greater then that of *D. madagascariensis*. Similarly in *D. robusta* the diameter of the femoral head is 59% greater, the length of the humerus 35% greater and the diameter of the humeral head 64% greater. From this they concluded that the extinct species weighed at least as much as the living one. Albrecht *et al.* (1990) calculated regressions on the average mid-shaft circumference of humerus and femur in *D. robusta* which gave an estimated body weight for the subfossil aye-aye of 13.5 kg. Regressions calculated from femoral head surface area in *D. robusta* produced an estimate of 6.7 kg. These latter estimates suggest that the animal weighed two and a half to five times as much as the average living aye-aye.

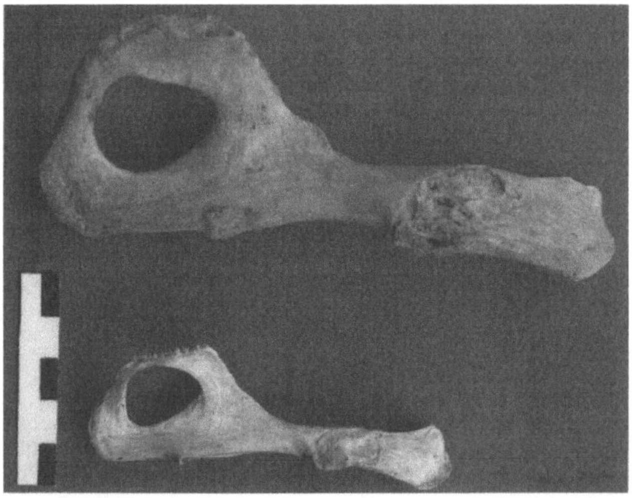

Figure 2: Comparison of the innominate of *Daubentonia robusta* (above), and *Daubentonia madagascariensis* (below). Scale in cm. L.R. Godfrey photo.

CAPTIVE MANAGEMENT

Because the aye-aye is so highly endangered, recent incentive has been taken to breed this species in captivity, so as to create a reserve population that may, as necessary, contribute to the restocking or reintroduction of wild populations. At present two zoological gardens in Madagascar, Tsimbazaza and Ivoloina, together hold three pairs of aye aye. At the DUPC, there are nine individuals and at the Jersey Wildlife Preservation Trust (JWPT) there are a further seven individuals. Two others recently arrived at Vincennes Zoo, Paris. Captive born offspring are to be exchanged eventually between these and such other institutions as may eventually contribute to captive conservation of *D. madagascariensis*.

In temperate climates it has not, so far, proven practical to house aye-aye outdoors where they would be difficult to monitor and to protect from winter temperatures. Aye-aye are extremely active when awake and need large indoor rooms with dimension of at least 15x15x15 feet. A complex network of branches and vines should fill such rooms and if possible aye-aye should be provided with artificial food puzzles or natural logs where they can forage for food. Haring*et al.* (1994) have recently reviewed captive managment of the aye-aye in detail.

Although in the wild, female and male aye aye do not range together as pairs they can be kept together in captivity on a long-term basis and they will breed and tolerate young without mishap.

A gestation length of 158 days has been reported by the JWPT, while DUPC records suggest 170 or 172 days. Males can be removed for a period of time after the birth of an infant, so as to avoid possible conflict. However, there is no evidence to date that males try to interfere in any way with newborn infants. The female protects the newborn in next boxes and the female may fill such boxes if supplied with durable (non wilting leaves) such as bamboo or juniper branches before parturition. When kept as pairs, aye-aye should have two or more nest boxes, each approximatley 60 cm per side, that are securely mounted well above the floor. Nest boxes may be open in front except for a 15 cm lip at the bottom to protect infants from falling out. After birth, infants are uncoordinated and unable to climb or hold on. If the mother wishes to relocate the infant it is carefully carried in the mouth. For proteciton agianst injury from falls, room floors should be covered with wood chips, "Dri-Dek"--vinyl matting--or any similar substance.

Branches forming the substrate for living in the aye aye rooms should be attached to walls with steel eyebolts or lag bolts. Branches and nestboxes should be made of unpainted and untreated wood because *Daubentonia* chew into wood, especially branches, in order to keep their incisors sharpened. This chewing and shredding of bark and branches caused by the need to sharpen teeth is important to their well-being. Tomes (1894) reports a case of an aye aye in London which was given no opportunity to sharpen its incisors. These teeth continued growing unabated.

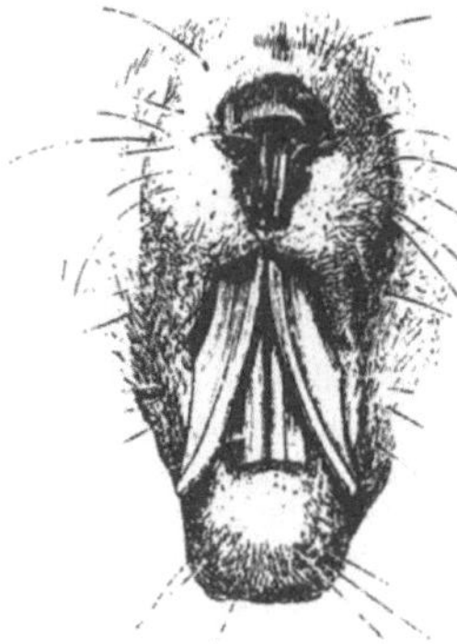

Figure 3 : Front dentition of an aye aye that was not provided with the means to use it's upper incisors sufficiently: they have grown very long. From Tomes, 1894.

When sharpening or grinding down teeth aye aye typically grip a hard branch with the upper teeth and rotate the lowers in a circular motion against the hard wood--thus sharpening the edges of the lower incisors. Upper incisors appear to be sharpened against the lowers and by other grinding motions. In one case at DUPC when a female was found to have somewhat overgrown incisors, a metal file was mounted on a branch and this animal used it to grind down the excess growth. Damaged and ripped up branches in aye aye cages are a necessary product of their normal behavior and are not a result of poor maintenance.

Aye aye in captivity have been successfully kept at temperatures around 20°-25°C. Warmed air, circulated indoors in winter, may lead to dry, itchy skin and increased level of scratching. This may be corrected by misting, spraying from a hose or by other means. Most aye-aye live in regions of high humidity.

Room lights may be controlled with automatic timers set for a reversed light cycle adjusted twice a week so as to maintain a photoperiod corresponding to Madagascar seasons (around 16 S). Rooms may be lit with red-light fixtures that provide twilight-like illumination necessary for behavioral observation. The bright lights can be set to go off at noon, allowing for room cleaning in the morning and behavioral study in the afternoon and evening.

DIET

It appears that aye-aye derive much of their water needs from food, but they should be supplied continuously with fresh water in stainless steel drinking bowls. Stoppered, hanging water bottles may be damaged by exploratory gnawing.

Food items accepted include mealworms, waxworms and a variety of fruits and vegetables that are relatively high in sugar and fat content. Food usually preferred includes raw (washed) eggs, grubs, pieces of coconut, avocado, mango (seed especially), sugar cane, peanuts, corn on the cob, various nuts, cucumbers, sectioned oranges, banana, and pieces of melons and apples.

Newly captured aye-aye do not always recognize grubs in dishes but will readily extract them from slits or holes drilled into bamboo canes This searching resembles wild foraging For initial feeding after capture, mangoes or gruel made of honey, sweetened condensed milk mixed with high protein baby cereal and other additives such as cream cheese and raw eggs are usually accepted. Aye-aye prefer to carry away from the feeding tray food items one at a time to a preferred location for eating. Consequently, larger pieces, such as whole eggs or oranges, are better sized foodstuffs than small pieces.

CAPTIVE WELL BEING

More than most other primates, aye aye are often solitary and, as stated above, male and female territories do not appear to completely overlap (Sterling 1993). It does not seem to stress animals to establish male-female pairs, although females sometimes are hostile at first. Such hostility may be circumvented by introducing the male first to the cage, room or enclosure, or by temporarily housing the female in a cage within the cage housing the male. Mother and offspring together or pairs are most natural, but there is evidence that young may continue to nurse overlong when caged together longterm with the mother, there being no opportunity for the offspring to establish spatial, that is, territorial independence. Little evidence has been gathered to date about the advisability of same sex pairing. Judging from what little is known from the captive behavior of most other lemurs, male-male pairing might well be without disturbance but female-female pairing of unrelated animals is most likely to be immediately stressful or sooner or later lead to fighting.

D. madagascariensis has a convoluted brain that is relatively large for a prosimian of its size and weight and this species is highly inquisitive. These animals need things to do. As has already been pointed out, to maintain their captive well being, cages must be large and well stocked with "cage furniture". Individual aye-aye often move great distances overnight in the wild and when caged, and not feeding, actively range and circle through their rooms. Eventually, with radio collars and properly heated sleeping boxes and feeding stations, it may be possible to house this species out of doors so as to further increase the space available to them. Even so, the risk of loss from predators or winter conditions should be taken seriously since each captive animal is almost irreplaceable. In the west of Madagascar, aye aye are sometimes seen out and about as early as around 4:00 p.m. If this is a typical behavior it might not be difficult to habituate them to be fed at that time in outdoor circumstances, so that technicians could evaluate their status during daylight.

In order to enrich life in captivity, food foraging puzzles have proven valuable. Food such as corncobs can be suspended from "bungee" cords. On the arrival of aye aye at the DUPC in 1987, the two animals would not take grubs from a dish. I drilled holes in sections of bamboo, put grubs inside and they readily fed. Sections of bamboo stems that have been drilled with many mealworms inserted can be attached to branches or hung from cords so that aye aye may hunt from them by percussive foraging. Principally younger aye aye will play or swing on ropes and vines. Circular pathways established by means of inter-attached branches can be placed at several levels of a room, with some branches or vines leading from one to another and to nest boxes and feeding stations. Wire mazes containing food items that have to be extracted might also prove entertaining.

As mentioned, sectioned natural logs from the forest floor, chosen for having galleries containing beetle larvae, can be provided and will give much interest to animals searching for grubs. The addition of small trees with flexible branches supply substrates for jumping and climbing as well as material for nest building. Each person concerned with the care of this truly inquisitive species can use their own ingenuity to help invent other diversions or puzzles.

ACKNOWLEDGMENTS

I thank L. R. Godfrey for providing data on probable ranges in weight of *Daubentonia robusta* and and for supplying a photograph of the fossil. Thanks also go to F.A. Ankel-Simons for helpful suggestions and for reviewing the manuscript. Funds for determining the range of the western aye-aye were provided by the DUPC and through a gift from Verna C. Simons. This is DUPC Publication Number 569.

REFERENCES

Albrecht, G.H., Jenkins, P.D. and Godfrey, L.R. (1990). Ecogeographic size variation among the living and subfossil Prosimians of Madagascar. *American Journal of Primatology. 22*: 1-50.

Ankel-Simons, F.A. (1983). A Survey of Living Primates and their Anatomy. MacMillian Publishing Company., New York and London, 311 pp.

Ankel-Simons, F.A. (1995). A Historical Note about the Deciduous Dentition of the Aye aye, *Daubentonia madagascariensis*. In preparation.

Buffon, C. G.L. (1789). Historie Naturelle, Générale et Particulière, Supplementary Ed. Vol. 7, Imprimerie Royale, Paris.

Cartmill, C.A. (1974). *Daubentonia, Dactylopsila*, woodpeckers and klinorhynchy. In: "Prosimian Biology", R. D. Martin, G.A. Doyle and A.C. Walker, Eds. Duckworth, London.

Erickson, C.J. (1991). Percussive foraging in the aye aye, *Daubentoniamadagascariensis* . *Animal Behavior . 41*: 793-801.

Gmelin, J.F. (1789) Systema Naturae per Regina Tria. Natura, secundum classes. ordines, genera, species cum charcteribus et differentiis". Editio Decima Tertia T.1: 152.

Gray, J.E. (1870). Catalogue of monkeys, lemurs and fruit eating bats in the collection of the British Museum, London. *Proceedings of the Zoological Society . II*, 828-831.

Haring, D.M., Hess, W.R. Coffman, B.S. and T.M. Owens (1994). Natural history and capitve management of the aye-aye (*Daubentoniamadagascariensis*) at the Duke University Primate Center (DUPC). International Zoo Yearbook . 33, 16pp.

Hill, W.C.O. (1953) Primates: Comparative Anatomy and Taxonomy, I-Strepsirhini. Edinburgh U. Press, Edinburg.

Iwano, T. and Iwakawa, C. (1988). Feeding behavior of the aye-aye *Daubentonia madagascariensis* on nuts of ramy *Canariummadagascariensis. Folio Primatology. 50:* 135-142.

Jones, M.L. (1986). Sucesses and failures of captive breeding. In: Primates: The Road to Self-Sustaining Populations. K. Benirschke, Ed. Springer-Verlag, New York.

Lamberton, C. (1934). Contribution a la connaissance de la faune subfossile de Madagascar. Note IX: *Chiromys robustus* sp. nov. Lamb. *Mémoires du Academie Malagache 17*: 40-46.

MacPhee, R.D.E. and Raholimavo, E.M. (1988). Modified subfossil aye-aye incisors from southwestern Madagascar: Species allocation and paleoecological significance. *Folia Primatology. 51*: 126-142.

Owen, R. (1866). On the aye aye. Trans. Zoology Society. London. 5: 33-101.

Peters, W., (1866). Über die Säugethiergattung *Chiromys* (Aye aye). *Abhandlungen der Königlich preußischen Akadademie der Wissenschaften, Berlin*, 79-100.

Petter, J.J., Albinac, R. and Rumpler, Y.(1977). eds. Faune de Madagascar V. 44: Mammifères Lemuriens (Primates Prosimiens). Orstom, Paris.

Petter, J.J., (1977). The aye aye. In: Primate Conservation. Prince Rainier III and G.H. Bourne, Eds., Academic Press, New York.

Schreber, J.C.D. (1774-1855). Die Säugethiere in Abbildungen nach der Natur mit Beschreibungen. Theil 1-17 suppl. Bd. 1-5, Erlangen.

Sonnerat, M. (1782). "Voyage aux Indes Orientales et a la Chine". Faune, Paris.

Sterling, E.J. (1993). Behavioral Ecology of the Aye-aye on Nosy Mangabe, Madagascar". Doctoral Dissertation, Yale University, 239pp.

Stevens, C.E. (1988). Comparative Physiology of the Vertebrate Digestive System. Cambridge U. Press, Cambridge.

St. Hilaire, E.G. (1795). Extrait d'un Mémoire sur un nouveau genre de quadrupedes de l'ordre des Ronguers (Glires L.) Decade Philos Litt Pol. 4 (28): 195, Paris.

Tomes, C.S. (1894). A Manual of Dental Anatomy, Human and Comparative. J & A Churchill, 11 Burlington Street, London.

Vinson, A. (1855). Description de l' Aye - Aye apporté vivant au Museum d'histoire naturelle de l'île de la Réunion. Paris, Comptes Ren. Academy Science 2 Sem 638-640.

PHYLOGENETIC RELATIONS AMONG PROSIMII WITH SPECIAL REFERENCE TO LEMURIFORMES AND MALAGASY NOCTURNALS

Bernard Dutrillaux[1] and Yves Rumpler[2]

[1]Institut Curie, URA N° 118 CNRS, Structure et Mutagenèse chromosomiques, Section Biologie, 26 rue d'Ulm, 75231 Paris cedex 05, France
[2]Université Louis Pasteur, Faculté de Médecine, Institut d'Embryologie, 67085 Strasbourg cedex, France

INTRODUCTION

Numerous chromosome banding studies on Primates have been done since 1973. These studies allowed us to establish phylogenetic relationships. It was possible to reconstruct ancestral karyotypes according to the parsimony principle: chromosomes shared by several species, particularly when they were distantly related, were considered to be ancestral.

When the work was extended to other mammalian orders such as Lagomorpha, Rodentia (Viegas-Péquignot *et al.*, 1986) and Carnivora (Couturier and Dutrillaux, 1986), fairly similar ancestral karyotypes could be reconstructed. These results suggested that only a few chromosomal changes occurred from the ancestral eutherian mammals to the emergence of the orders in question. Thus, the ancestral karyotype of Primates does not strongly differ from those of other orders. When suborders of Primates were compared, their ancestral karyotypes also appeared quite similar, and some species among Anthropoidea and Prosimii retained only slightly modified karyotypes. These numerous similarities among orders of Mammals, and within suborders of Primates, support the validity of the comparisons, barring multiple convergences.

ANCESTRAL KARYOTYPE

The presumed ancestral karyotype of all Primates, PRI, was reconstructed from the crossed comparison of about 100 species of Primates (Dutrillaux *et al.*, 1986) and approximately the same number of non-Primate eutherian mammals. This karyotype

(Figure 1) is composed of about 58 chromosomes, all acrocentric except the X chromosome. Most chromosomes, from *1* to *17*, are consistently observed in a number of living species, suggesting their ancestral status. The identification of the smaller chromosomes (*18* to *27*), is tentative, and they may differ numerically and morphologically from the ancestral karyotype, PRI. Their variation, therefore, need not be taken into consideration for phylogenetic reconstruction. Starting from PRI and comparing it to the living species in question, a cladogram positioning more than 100 species of primates was proposed (Dutrillaux *et al.*, 1986; Rumpler and Dutrillaux, 1990). The part of this cladogram limited to Prosimians is shown in Figure 2, and a number of branchings and ambiguities that have not been resolved by cytogenetics will be discussed.

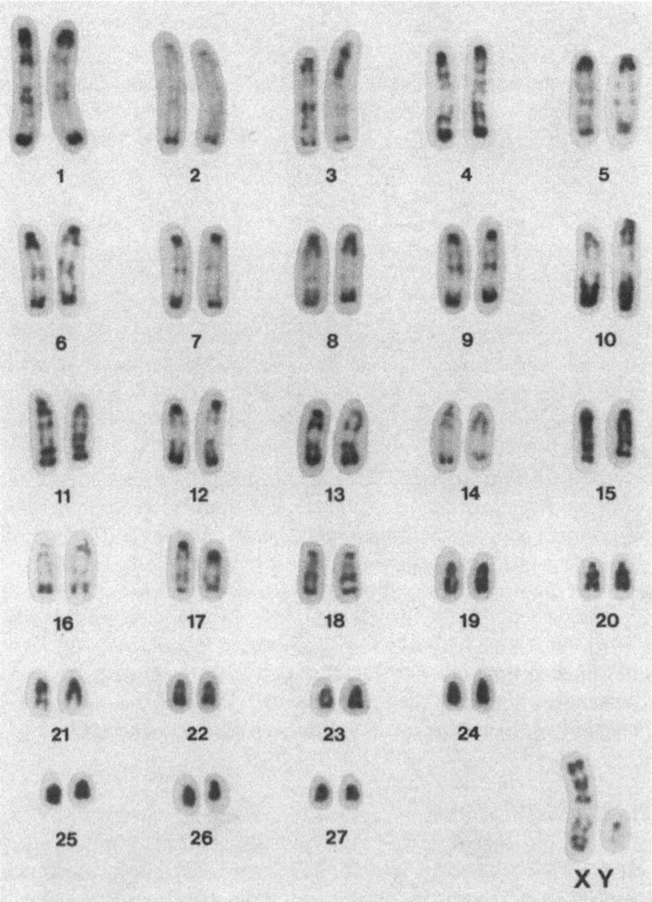

Figure 1. Reconstructed hypothetical karyotype of the common ancestor for all Primates.

THE DIVERGENCE BETWEEN SIMIANS AND PROSIMIANS

It was assumed that all primates share a common trunk separating them from the other Eutherian mammals. Chromosome rearrangements which might have occurred during this common chromosomal evolution are expected to characterize the karyotype of ancestral primates. Unfortunately, the ancestral karyotype of the Primates is similar to that which could be reconstructed for Carnivora, Rodentia and Lagomorpha (Dutrillaux *et al.*, 1986). At present a common trunk for all primates cannot be proposed by cytogenetics, and other approaches are needed. For this reason, we directly branched Prosimians and Simians from the ancestral Primate karyotype (Figure 2).

COMMON TRUNK FOR ALL PROSIMIANS

It is reasonable to postulate two chromosomal trunks. One is common to Anthropoids, the other common to Prosimians. The exitence of an ancestral karyotype for the Anthropoids is suggested by three derived chromosomes, not observed in non prosimian karyotypes, These are likely to have occurred at the moment of their emergence. The situation is not so clear for Prosimians, as no derivative chromosome appears to be common to this group. The only cytogenetic argument in favour of a common trunk is the absence of a derivative chromosome, PRI *3*, observed both in Anthropoids and in non-primate mammals. This chromosome is presumed to have undergone fission, giving rise to two small acrocentric chromosomes.

However this rearrangement appears differently in the three groups, Lorisiformes, Lemuriformes and Daubentonia. Two hypotheses can be proposed. 1) A fission gave rise to a shorter chromosome *3* which subsequently underwent additional fissions on the branch leading to LEM and to LOR, and was translocated on the branch leading to DMA. According to this hypothesis, a short common trunk would exist for Lorisiformes and Lemuriformes, characterized by a single observable chromosomal rearrangement. 2) Alternatively three different and independent rearrangements of PRI *3* occurred on the branches leading to the LEM, LOR and DMA, and no common trunk exists. The difficulty in finding changes common to all Prosimians suggests, at the very least, that a divergence occurred quite early to give the Lemuriformes (LEM and DMA) and the Lorisiformes (Figure 2).

THE DIVERGENCE BETWEEN LORISIFORMES AND LEMURIFORMES

The divergence between Lorisiformes and Lemuriformes is well documented. In particular, the Lorisformes have accumulated about 10 chromosome changes that are not observed in the Lemuriformes. For this reason, the ancestral karyotype of the Lorisformes is very different from the ancestral Lemuriforme karyotype. Among Lorisformes, an accumulation of six pericentric inversions clearly distinguishes Galagidae from Lorisidae. Each Lorisid species studied has a well-defined karyotype, yet ancestral karyotype at the branching points of each species remains unclear. The same situation exists for the Galagidae. The uncertainty of the branching for both Galagidae and Lorisidae may result from the occurrence of non-dichotomic (network) evolution prior to a separation of the taxon (Figure 2). Among all the Lorisformes studied, *Otelemur crassicaudatus* is the species that has retained the most ancestral karyotype

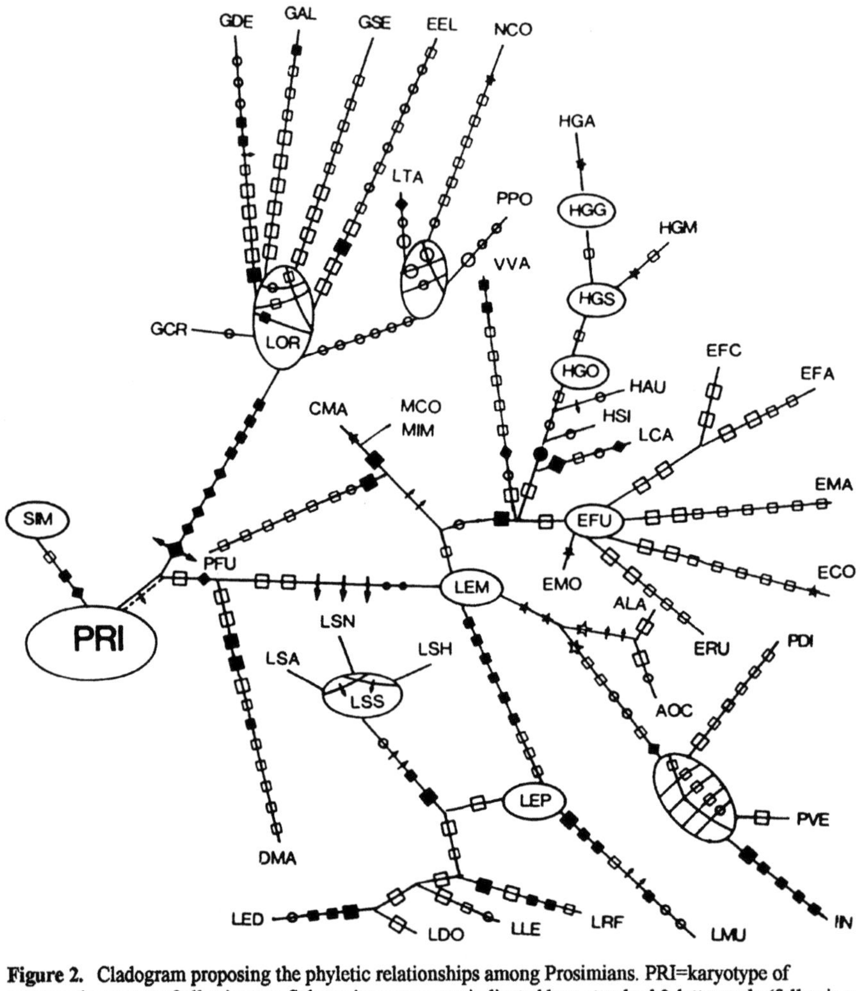

Figure 2. Cladogram proposing the phyletic relationships among Prosimians. PRI=karyotype of presumed ancestor of all primates. Subspecies names are indicated by a standard 3-letter code (following Rumpler & Dutrillaux, 1990).
□ Robertsonian translocation; ■ other translocation; O pericentric inversion; ● paracentric inversion; ↑ fission; ☆ shift; ★ gain of heterochromatin; ♦ unspecified rearrangement. Symbols of greater size correspond to chromosomal rearrangements able to lead to an immediate reproductive barrier.

Subspecies names are indicated by 3-letter codes:

ALA	*Avahi laniger*	HGO	*Hapalemur griseus occidentalis*
AOC	*Avahi occidentalis*	HGS	*Hapalemur griseus ssp.*
ATR	*Allocebus trichotis*	HSI	*Hapalemur simus*
CMA	*Cheirogaleus major*	IIN	*Indri indri*
CME	*Cheirogaleus medius*	LCA	*Lemur catta*
DMA	*Daubentonia madagascariensis*	LDO	*Lepilemur dorsalis*
ECO	*Eulemur coronatus*	LED	*Lepilemur edwardsi*
EEL	*Eulemur elegantulus*	LLE	*Lepilemur leucopus*
EFA	*Eulemur fulvus albocollaris*	LMU	*Lepilemur mustelinus*
EFC	*Eulemur fulvus collaris*	LRF	*Lepilemur ruficaudatus*
EFF	*Eulemur fulvus fulvus*	LSS	*Lepilemur septentrionalis*
EMA	*Eulemur macaco*	LTA	*Loris tardigradus*
EMO	*Eulemur mongoz*	MCO	*Mirza coquereli*
ERU	*Eulemur rubriventer*	MIM	*Microcebus murinus*
GAL	*Galago alleni*	NCO	*Nycticebus coucang*
GCR	*Galago crassicaudatus*	PDI	*Propithecus diadema*
GDE	*Galago demidovii*	PFU	*Phaner furcifer*
GSE	*Galago senegalensis*	PIN	*Pachylemur insignis*
HAU	*Hapalemur aureus*	PPO	*Perodicticus potto*
HGA	*Hapalemur griseus alaotrensis*	PVE	*Propithecus verreauxi*
HGG	*Hapalemur griseus griseus*	VVA	*Varecia variegata*
HGM	*Hapalemur griseus meridionalis*		

Other codes correspond to reconstructed ancestral karyotypes for

LEM	Lemuriformes
LEP	Lepilemuridae
LOR	Lorisiformes
PRI	Primates
SIM	Anthropoidea.

(Figure 3). It differs from that of the postulated common ancestor of all Lorisiformes by a single chromosomal rearrangement. However, this does not mean that this species is representative of ancestral Lorisiformes unless other approaches also confirm ancestral characteristics.

The evolution of Lemuriformes is more complicated. A short common trunk is strongly suggested by the presence of two derivative chromosomes characterizing this group: one derived from the translocation of chromosome PRI *1* to a small acrocentric. The other is derived from the translocation of two small acrocentrics, which gave rise to a characteristic chromosome of the lemurs (chromosome *22* of *Eulemur fulvus*) (Figure 4).

EVOLUTION OF THE LEMURIFORMES

Emergence of *Daubentonia* and the monophyletic origin of the Malagasy settlement

From the proposed common trunk, two branches emerge. One leads to LEM, the ancestor of all Lemuriformes, and bears seven rearrangements. The other, leading to *Daubentonia*, has accumulated at least twelve rearrangements. These observations suggest the occurrence of a long evolution common to all Lemuriformes, except *Daubentonia*, which branched off first (Rumpler *et al.*, 1988). This early divergence of *Daubentonia* explains why this species may be classified among Lemuriformes or in a group apart, Daubentoniformes. On the other hand, the existence of this common trunk argues in favor of a monophyletic prosimian settlement of Madagascar.

Evolution of the other lemurs

All of the other lemurs diverged from a common point that is assumed to correspond to their last common ancestor, LEM (Figure 2). A trifurcation exists from this point.

1. One branch leads to Lepilemuridae, for which a long common trunk is postulated followed by a bifurcation. These nocturnal species can be unambiguously positioned by the cytogenetic approach, including true species (*Lepilemur mustelinus, L. leucopus, L. edwardsi, L. ruficaudatus, L. dorsalis, L. septentrionalis*) and subspecies (*L. s. septentrionalis, L. s. sahafarensis, L. s. ankaranensis, L. s. andrafiamenensis*). The inferred chromosomal rearrangements that took place during the evolution of the sportive lemur are consistent with the chronological sequence of rearrangements that would be expected to produce a reproductive barrier compatible with the geographic distribution of species in this genus (Ishak *et al.*, 1988).

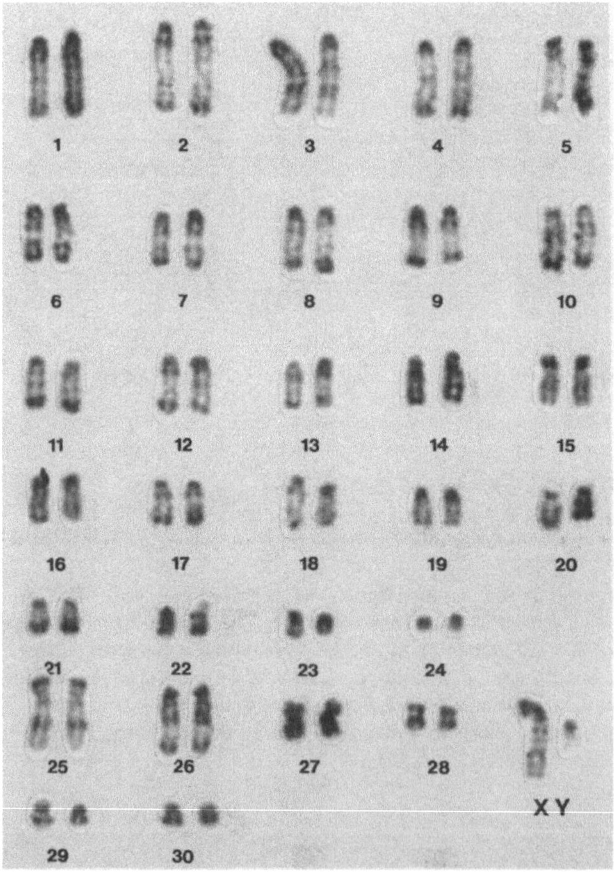

Figure 3. Karyotype of *Otolemur crassicaudatus.*

2. A second branch leads to Indriidae. Their common trunk is very short. The two nocturnal species, *Avahi laniger* and *A. occidentalis*, were first isolated. A long common trunk exists for the diurnal species *Propithecus* and *Indri*. Cytogenetic studies do not allow us to propose a dichotomic evolution for this group, and no unequivocal scheme can yet be proposed. Since *P. verreauxi* and *Indri* share four chromosome rearrangements, one could conclude that their ancestors have undergone a common evolution. However, *Indri* has accumulated five complex chromosomal changes that may parallel the acquisition of a characteristic phenotype. This suggests that the genus *Propithecus* could be ancestral to that of *Indri*.

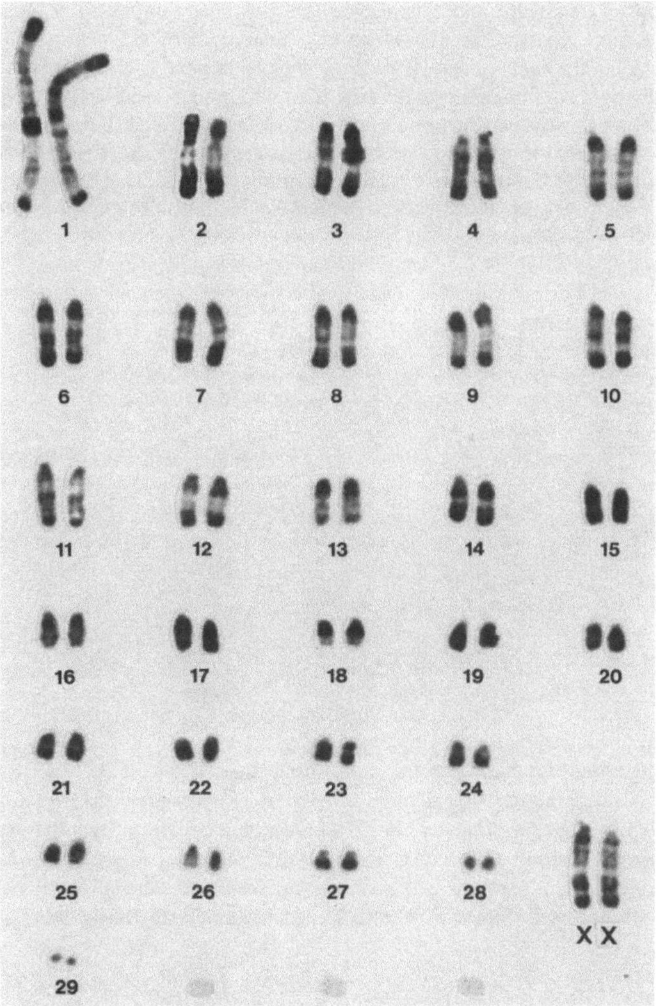

Figure 4. Karyotype of *Eulemur fulvus*.

3. The last branch is very short and bears only one rearrangement which is common to the two remaining families: Cheirogaleidae and Lemuridae.

a) Branching of the Cheirogaleidae. Cheirogaleid species diverged quickly. Whereas the karyotype of *Phaner* underwent multiple rearrangements, that of *Microcebus, Mirza, Allocebus* and *Cheirogaleus* remained similar to the ancestral karyotype. No structural rearrangement separated the four genera. However, juxtacentromeric heterochromatin of *Cheirogaleus* has staining properties different from that of the other genera. Fortunately, the distinction among Cheirogaleidae genera can also be observed by studies of highly repeated DNA (HR-DNA). These studies allowed the reconstruction of the cladogram shown in Figure 5.

These studies revealed the existence of common electrophoretic bands for three genera, *Microcebus, Mirza* and *Cheirogaleus,* some common bands only for *Mirza* and *Cheirogaleus,* and numerous specific bands for each genus. Thus, it was possible to propose a cladogram showing the emergence of *Cheirogaleus, Mirza* and *Microcebus* successively (Rumpler and Crovella, 1992). The systematic position of *Allocebus* was more difficult to determine precisely. Nevertheless, using different probes (one consisting of *Microcebus* fragments and another of *Cheirogaleus* fragments) it appeared that *Allocebus* is more closely related to *Microcebus* than to the other Cheirogaleidae.

b) Branching of the Lemuridae. The evolution of Lemuridae is characterized by a short trunk bearing two common rearrangements. This interpretation differs from our previous one because we erroneously proposed that *Varecia* emerged first. Reinterpretation of its karyotype shows that it acquired the same rearrangement of chromosome PRI *3* as the other Lemuridae and that this chromosome was further translocated and inverted. It now appears that all Lemuridae diverged from a common point by a trichotomy:

-one branch leads to *Varecia* with at least ten rearrangements.

-another branch is common to *Lemur catta* and *Hapalemur* species. On this branch, a dichotomy could be reconstructed, which allowed a clear separation of all species and subspecies of *H. griseus.*

-a final branch constitutes a short common trunk to all *Eulemur.*

The karyotype of *E. fulvus* represents an ancestral form from which all *Eulemur* other karyotypes diverged. A common trunk was observed only for *E. f. collaris* and *E. f. albocollaris*. It is impossible to propose a branching scheme for the various *Eulemur* species. This is a limitation of the cytogenetic studies and points out the need for other approaches. Additional data were provided by molecular studies of HR-DNA. These demonstrate the existence of dichotomic events for these species (Figure 5) (Jung *et al.*, 1992). Comparison of the data obtained from these two approaches provides apparently contradictory interpretations. *E. fulvus,* which has kept the most ancestral karyotype, appears to have undergone the most complex evolution as interpreted from restriction enzymes of patterns of HR-DNA. This provides another example of the difficulty in establishing a strict parallel between chromosomal and molecular changes and proposing relationships from the study of a single biological parameter.

The necessity of considering multiple parameters is also amplified by the comparison of *Varecia* with other lemurs. Cytogenetic studies do not allow us to branch Varecia with either the *Hapalemur/L. catta* group or *Eulemur.* Study of HR-DNA lead to the same interpretation (Montagnon *et al.*, 1993). However, data on βhemoglobin chains led Duffy and Coppenhaver (1986) to propose a common trunk between *Varecia* and *L. catta.* This interpretation does not contradict ours, but suggests that *L. catta* and *Varecia* underwent a common evolution during which no chromosomal or HR-DNA modifications occurred (Figure 5), although point mutations obviously did.

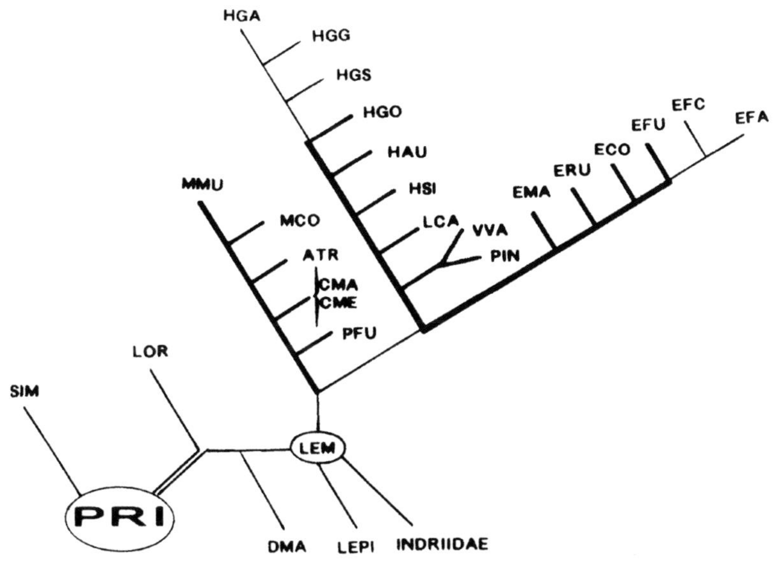

Figure 5. Phylogenetic tree of Cheirogaleidae and Lemuridae based on molecular biology data (thick line). The subspecies names are indicated by the same 3-letter codes as in Figure 2.

Finally the same molecular biology technique has also been fruitfully used for extinct lemurs. HR-DNA studies of *Pachylemur insignis* revealed that *Pachylemur* is closer to *Varecia* than to *Eulemur*. After successive hybridizations of a *Pachylemur* blot with *E. fulvus* and *Varecia* HR-DNA probes, the *Pachylemur* blot only gave a signal with the *Varecia* probe (Crovella *et al.*, in press).

CONCLUSIONS

The existence of a common chromosomal trunk for all Prosimii remains uncertain. The present data argue for an independent origin of Lorisiformes and Lemuriformes. The relationships between *Daubentonia* and other Malagasy lemurs are more firmly established. All share a common but fairly short evolution from their last common ancestor, which suggests a monophyletic origin of the settlement of Malagascar by lemurs. However, the branch leading to *Daubentonia* is distinct from the long trunk common to all other lemurs. Thus, *Daubentonia* was separated early, which may justify its classification among either Lemuriformes or in the separate sub-order Daubentoniformes. By comparison, the trunk common to all Lorisiformes is much longer and the separation between Galagidae and Lorisidae is clearly defined.

Most cytogenetic interpretations are compatible with or were confirmed by other approaches, principally molecular methods. However, in a limited number of cases, the cytogenetic approach has reached its limits. Fortunately, other methodologies will obviously fill some of these gaps.

ACKNOWLEDGEMENTS

The authors thank L. Alterman for his critical and fruitful comments on some aspects on the phylogeny interpretation and for stylistic changes of the manuscript, K. Izard. We are grateful to G. Cadiou for his photographic work, and to M. Lavaux for her secretarial assistance.

REFERENCES

Couturier, J., & Dutrillaux, B. (1986). Evolution chromosomique chez les Carnivores, *Mammalia, 50,* 124-162.

Crovella S., Montagnon D., Rakotosamimanana B., Rumpler Y. Molecular biology and systematics of an extinct lemur: Pachylemur insignis. Primates (in press).

Duffy, L.K. & Coppenhaver, D.H. (1986). Molecular evolution of prosimian hemoglobins, *Annals of the New York Academy of Sciences,* 254-257.

Dutrillaux, B., Couturier J., Muleris, M., Rumpler Y., & Viegas-Péquignot, E. (1986). Relations *chromosomiques entre sous-ordres et infra-ordres, et schéma évolutif général des Primates, Mammalia, 50,* 108-121.

Ishak, B., Warter, S., Dutrillaux, B., & Rumpler, Y. (1988). Phylogenetic relation between Lepilemuridae and other Lemuriform families, *American Journal of Primatology, 15,* 275-280.

Jung, K.Y., Crovella, S., & Rumpler, Y. (1992). Phylogenetic relationships among Lemuriform species determined from restriction genomic DNA banding patterns, *Folia Primatologica, 58,* 224-229.

Montagnon, D., Crovella, S., & Rumpler, Y. (1993). Comparison of sequences of highly repeated DNA of some Lemuridae and taxonomic implications. *Cytogenetics and Cell Genetics, 63,* 131-134.

Rumpler, Y., Warter, S., & Dutrillaux, B. (1988). Chromosomal evolution of Malagasy lemurs. XI. Phylogenetic position of *Daubentonia madagascariensis, Folia Primatologica, 50,* 124-129.

Rumpler, Y. & Dutrillaux, B. (1990). *Chromosomal Evolution and Speciation in Primates.* RBC, University of the Basque Country (Ed.), (vol. 23), New York/Berlin: Springer Verlag.

Rumpler, Y. & Crovella., S. (1992). Importance of comparative cytogenetics for primate systematics and phylogeny, Limits and alternatives, *Karger Gazette, 54,* 8-9.

Viegas-Péquignot, E., Petit, D., Benazzou, T., Prod'homme, M., Lombard, M., Hoffschir, F., Descailleaux, J., & Dutrillaux, B. (1986). Evolution chromosomique chez les Rongeurs. Phylogénie chromosomique chez les Sciuridae, Gerbillidae et Muridae, et étude d'espèces appartenant à d'autres familles de Rongeurs, *Mammalia, 50,*164-202.

PHYLOGENETIC STUDIES OF GLUTATHIONE-METABOLIZING ENZYMES IN PRIMATE LENSES

William B. Rathbun

Department of Ophthalmology
Lions Research Building
University of Minnesota
Minneapolis, Minnesota 55455

INTRODUCTION

Morphological characteristics of the nose are aids which have been used to differentiate between the strepsirhines and haplorhines and also between the catarrhines and platyrrhines. However, application of characteristics of a second sense organ, the eye, to primate classification does not seem to have occurred. In part, this may be due to the minute fund of comparative data that presently exists. The lens is the simplest ocular organ. It varies greatly between species regarding gross characteristics such as color, shape and size. The variation in size is in both actual values and relative values as a proportion of the body mass. Of the many tissues and organs which make up the primate eye, the clear ocular lens has perhaps been the most charismatic, its fibrillar structure even serving as a subject for Leeuwenhoek's microscopic studies.

All cells formed during the life of a lens are retained; the older cells are overgrown by younger ones, which form overlying layers reminiscent of the layered makeup of an onion. The cells of the inner cortex and nucleus differ from the outer germinative epithelium in that they have lost their nuclei and mitochondria and contain certain characteristic structural proteins, such as α-crystallin. Some metabolic pathways vary with cell age. For example, the citric acid cycle, contained in the mitochondria, is only operational in the outermost cell layer and therefore forms only a minority of the required ATP. Proteins and enzymes, may be retained for long periods and be subjected to post-translational modification.

Many biochemical characteristics set the lens apart from other tissues. Among these are its unique high level of the sulfur-containing antioxidant tripeptide, glutathione. The high concentration of glutathione is apparently required for lens clarity, as low glutathione con-

Creatures of the Dark: The Nocturnal Prosimians
Edited by L. Alterman *et al.*, Plenum Press, New York, 1995

centrations are invariably followed by cataract development. Accordingly, the lens expends at least 11% of its ATP derived from glycolysis (or 8% of total ATP) on glutathione biosynthesis (Rathbun and Wicker, 1973).

The biosynthesis of glutathione is performed by two enzymes reacting sequentially. These are γ-glutamylcysteine synthetase (E.C. 6.3.2.2), which catalyzes reaction I, and glutathione synthetase (E.C. 6.3.2.3), which catalyzes reaction II.

$$\text{glutamate} + \text{cysteine} + \text{ATP} \xrightarrow{Mg^{++}} \gamma\text{-glutamylcysteine} + \text{ADP} + P_i \quad \text{(I)}$$

$$\gamma\text{-glutamylcysteine} + \text{glycine} + \text{ATP} \xrightarrow{Mg^{++}, K^{+}} \underset{\text{(glutathione)}}{\gamma\text{-glutamylcysteinylglycine}} + \text{ADP} + P_i \quad \text{(II)}$$

Although glutathione is a reactant in a variety of biochemical systems, one function is of special interest in the lens: its participation in the detoxification of hydrogen peroxide. Interaction of ultraviolet light with ascorbic acid, which occurs at very high concentrations in the aqueous humor, forms hydrogen peroxide (Pirie, 1965a, 1965b). This oxidant rapidly invades the lens and, if uncontrolled, can reach toxic levels. The concentration is controlled by an intralenticular enzymic reaction with glutathione. Ultraviolet light also interacts directly with various molecules within the lens to form toxic free radicals, some of which are neutralized by direct action with glutathione. Glutathione also has a vital function in maintenance of membrane components. If any portion of the glutathione defense system fails, membrane damage occurs, followed by failure of water extrusion, which changes local refractive indices to create light scatter and thus cataracts.

Detoxification of hydrogen peroxide is carried out via the glutathione redox cycle. The cycle consists of two enzymes reacting sequentially: glutathione peroxidase (E.C. 1.11.1.9), which catalyzes reaction III, and glutathione reductase (E.C. 1.6.4.2), which catalyzes reaction IV to regenerate the reduced glutathione.

$$2 \text{ reduced glutathione} + H_2O_2 \longrightarrow \text{glutathione disulfide} + 2\ H_2O \quad \text{(III)}$$

$$\text{glutathione disulfide} + \text{NADPH} + H^{+} \longrightarrow 2 \text{ reduced glutathione} + \text{NADP}^{+} \quad \text{(IV)}$$

It is generally thought that the primitive primate habit was that of nocturnality. Diurnal habit may be a derived condition. Increased sharpness of vision is commensurate with adaptive survival for primates under diurnal conditions. This was achieved with development of a fovea and the ability to deform the lens for accommodation, some degree of color vision (achieved with cone cells in the macula) and possibly increased binocularity. Additionally, it was essential that the diurnal primate have protection from the ultraviolet portion of the solar spectrum to avoid blindness due to cataract development before puberty.

Alteration of the several glutathione-related enzyme systems may differ between species in adaptation to a sunlit environment. The enzyme molecules may vary among species in such elements as primary structure, quantity and stability. This laboratory has investigated the activity of four glutathione-related enzymes (γ-glutamylcysteine synthetase, glutathione synthetase, glutathione peroxidase and glutathione reductase) in the ocular lenses of a variety of mammalian species, including the primates. This paper will review the published and some previously unpublished findings of our laboratory regarding primates in a phylogenetic context. It is the purpose of this communication to show that variation of several parameters regarding these enzymes is instructive of and consistent with accepted phylogenetic grouping of the primates.

MATERIALS AND METHODS

Because the only comparative data available is from the author's laboratory, a brief description of methods common to all the work follows. All lenses for interspecies comparison were young to middle-aged adult primates. The lenses for aging studies included every available age group. Eyes from non-human primates were harvested shortly after sacrifice, frozen and stored at -74° C. They were then shipped on dry ice via overnight air express to this laboratory and stored at -80°C until used. During an eight-year period, eyes were obtained from the following Regional Primate Research Centers: Delta, New England, University of Oregon, University of Washington, University of Wisconsin and Yerkes. Additional sources of lenses were: the Southwest Foundation for Biomedical Research, Duke University Primate Center, the Brookfield Park Zoo of Chicago, the National Zoo, the Como Park Zoo of St. Paul and the University of Minnesota Lions Eye Bank. Non-primate sources can be found in the individual references.

Wet lens weights and activities of the four enzymes were determined on the same lenses. γ-Glutamylcysteine synthetase and glutathione synthetase activities were determined by measurement of inorganic phosphate formation from ATP, using standardized preparation, assay conditions and media for each enzyme (Sethna *et al.*, 1982/1983). Units of activity were as earlier defined (Rathbun *et al.*, 1983). Glutathione peroxidase and glutathione reductase activities were determined by rate measurements of NADPH oxidation, using standard assay conditions and unit definitions (Bergad *et al.*, 1982; Rathbun *et al.*, 1983).

Units of enzyme activities must be expressed in terms of some parameter. When using lenses, comparison of enzyme activities between species is best expressed by using units/g wet weight and sometimes specific activity (units/mg soluble protein) when specimens of roughly the same stage in life are compared. Data so generated usually yield the tightest grouping for plots. However, aging studies may be better expressed as units/lens to compensate for low enzyme activities sometimes found in the inner cortical and nuclear regions. Comparisons between clear and cataractous lenses of the same age group are also best expressed as units/lens to correct for soluble protein loss and water invasion of the lens. This review will use both units/g lens and units/lens, as appropriate. The trends observed in aging are usually similar despite different methods of data expression, but some exceptions exist. To aid in brevity, manuscript reviewers often insist upon use of a single expression, usually units/g lens, as in the aging study of Holleschau and Rathbun (1994). Some data of that study will be expressed herein as units/lens. The only differences in trends shown by the two methods of data presentation are noted in Table 1.

Abbreviations for enzymes used in text and figures are as follows: γ-GCS, γ-glutamylcysteine synthetase; GSHS, glutathione synthetase; GSHPx, glutathione peroxidase; GSSGR, glutathione reductase.

RESULTS: REVIEW OF BOTH PUBLISHED AND UNPUBLISHED WORK

A. Glutathione Biosynthesis: γ-Glutamylcysteine Synthetase and Glutathione Synthetase

1. Effect of species variation on γ-GCS and GSHS activities. The activities were reported for γ-GCS and GSHS in the lenses of 35 species, representing 8 mam-

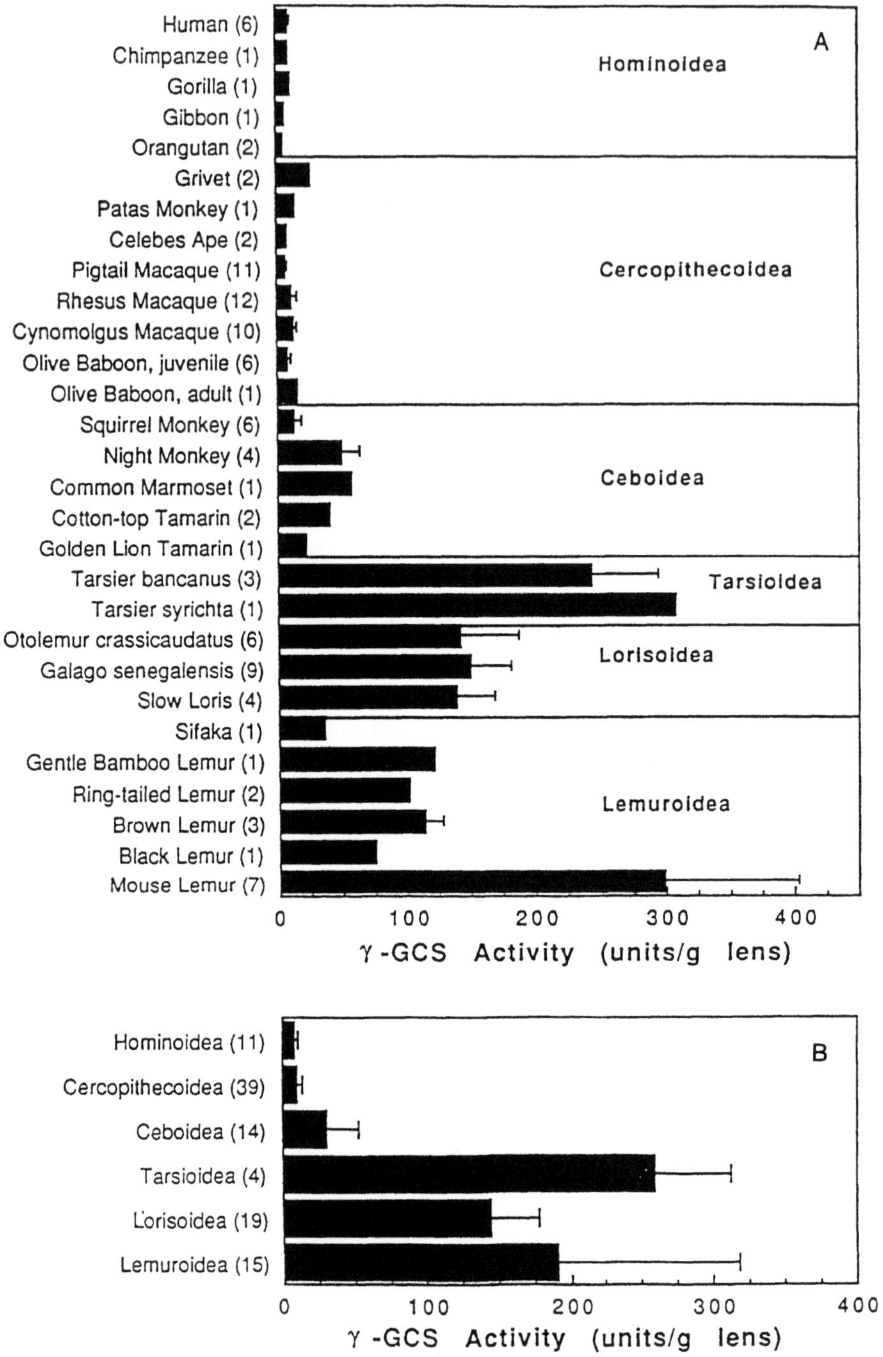

Figure 1. Lenticular γ-GCS activity (units/g lens) in primates. A: average activity in lenses of 28 species. The number in parentheses after the species name indicates the number of specimens analyzed while the error bars represent one standard deviation. B: average enzyme activity in six primate superfamilies. Redrawn from Rathbun, Holleschau and Alterman (1994).

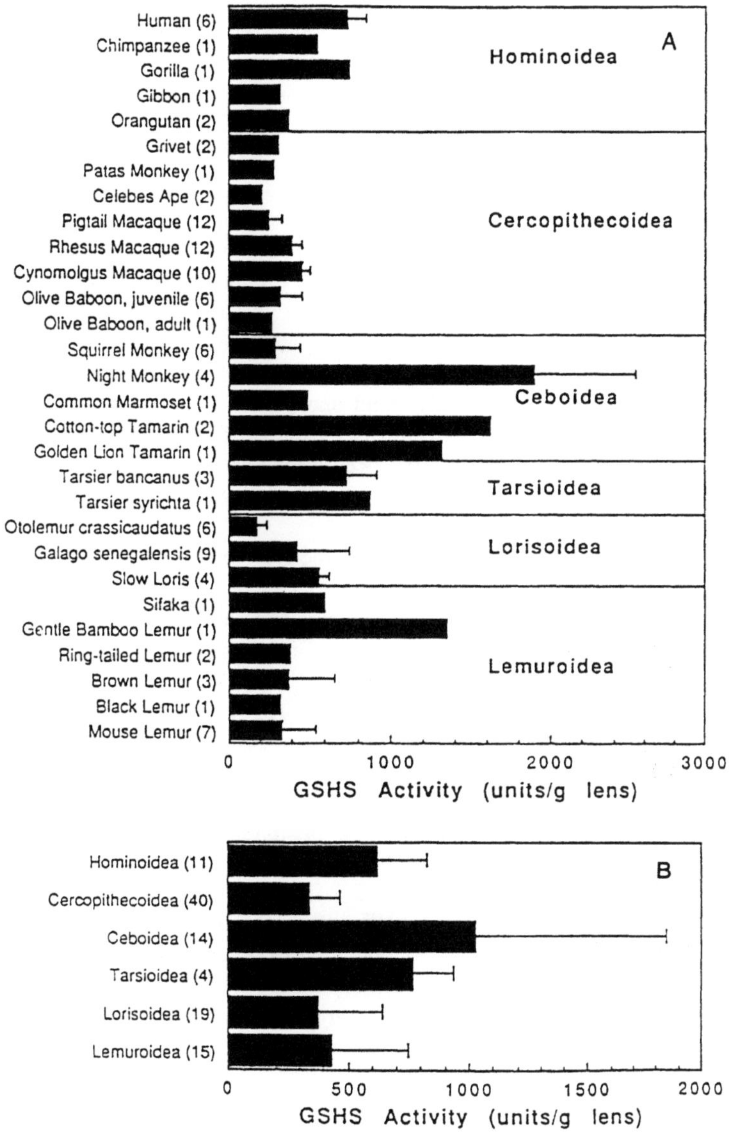

Figure 2. Lenticular GSHS activity (units/g lens) in primates. A: average activity in lenses of 28 species. B: average enzyme activity in six primate superfamilies. Redrawn from Rathbun, Holleschau and Alterman (1994).

malian orders (Rathbun *et al.*, 1986a). Although γ-GCS activity varied widely from less than 10 to 220 units/g lens, the values for most orders varied between 50 to 150 units/g lens. The pattern for primates was striking in that the activity value for *Otolemur crassicaudatus*, 180 units/g lens, was approximately 20-fold that of the two macaques and two hominoids (*Homo sapiens* and *Pan troglodytes*) that were examined. In contrast, a random pattern characterized the enzyme activity values of GSHS assayed from the same lenses.

An investigation of both enzymes in primate lenses was recently reported (Rathbun *et al.*, 1994). In that work data from adults of 28 species, representing 23 genera, were divided into the six basic primate groups suggested by Martin (1990): Hominoidea, Cercopithecoidea, Ceboidea, Tarsioidea, Lorisoidea and Lemuroidea. The pattern of γ-GCS activity was high in prosimian lenses, similar to the patterns observed for many non-primates (Fig. 1A). Among the prosimians the highest values were found in the nocturnal species and the lowest in the diurnal sifaka. In great contrast, the data for the platyrrhini were considerably lower. The lowest values were found in the catarrhini. Average values for the six groups are shown in Figure 1B, in which the differences in activities of the enzyme between the prosimians and simians are very marked.

Low γ-GCS activity was consistently found among adult anthropoids and represented a critical difference when compared to prosimians in expression of the enzyme. Its low activity caused γ-GCS to be termed an Achilles' heel of the Old World simian lens (Rathbun, 1983). Additional loss of activity was shown to be proportional to the degree of subcapsular cataract in humans (Rathbun *et al.*,1993).

GSHS activities were highest in lenses of some New World monkeys (ie, *Aotus* and the tamarins) and lowest in the Old World monkeys (Fig. 2A). Some apparently random variation of activity was found in the New World monkeys and lemurs. The average values for activity of this enzyme in the six primate superfamilies show lemurs, lorisids and Old World monkeys to be about the same (Fig. 2B).

The relative activities of the two enzymes required for glutathione synthesis show marked differences among various phylogenetic groups. The ratios of GSHS:γ-GCS activities per lens were commonly between 1:2 to 1:5 in non-primates and prosimians (Rathbun *et al.*,1986a; Rathbun *et al.* 1994; Fig 3). In contrast, the ratio averaged 1:45 in the New and Old World monkey lenses and 1:75 in hominoid lenses. The high ratios show that γ-GCS activity is rate-limiting for glutathione synthesis in the Anthropoids.

2. Effect of age on γ-GCS and GSHS activities. γ-GCS activity (units/lens) increased with age three-fold in rabbit and mouse lenses (Rathbun *et al.*, 1983; Rathbun *et al.*, 1986c), and a smaller increase was shown by the mouse lemur lens (Fig. 4A). A remarkable increase with age occurred in *Otolemur crassicaudatus* (Fig. 4B). In contrast, there was no change in activity with age in dog lenses (Rathbun *et al.*, 1983, Rathbun and Holleschau, 1992).

The aging patterns of γ-GCS activity in lenses of the cercopithecoids and hominoids are quite different from those of *Otolemur* and the mouse lemur. A logarithmic loss of enzyme activity was shown for olive baboons, pigtail (Fig. 4C) and rhesus monkeys, humans and orangutans (Rathbun and Holleschau, 1992; Rathbun *et al.*, 1986b; Sethna *et al.*, 1983). The temporal loss was greatest in human lenses (sixteen-fold, units/g lens or five-fold, units/lens) in an age span from 3.5 days to 84 years (Sethna *et al.*, 1983).

Reduction of GSHS activity associated with aging was observed in mouse lemur, *Otolemur crassicaudatus*, olive baboon, pigtail macaque, rhesus monkeys, humans and orangutans (Rathbun and Holleschau, 1992; Rathbun *et al.*, 1986b; Sethna *et al.*, 1983).

Loss of this enzyme activity with age is apparently a characteristic shared by all the primates.

3. Effect of species variation on γ-GCS and GSHS thermal stability. Enzymes that exhibit differing thermal inactivation curves have been thought to differ in their primary structures (Northrop and Kunitz, 1933; Ohrloff, 1978). This comparison is made by maintaining an enzyme solution for a standard time at variable temperatures, followed by rapid cooling and measurement of residual enzyme activity. Such comparisons require a large homogeneous source of tissue, which has rarely been available in primate investigations of lens enzymes.

A few comparative studies were made among the primates and representatives of six other mammalian orders. γ-GCS activities of rhesus monkey and sheep lenses were the most unstable to elevated temperatures. In comparison, the enzyme activities from human lenses and *Otolemur crassicaudatus* were more stable to elevated temperatures, and the thermal patterns were similar, but not identical, to each other. Lens preparations from other non-primates (noted below) yielded variable, non-identical curves compared to those of the above primates (Rathbun and Schmidt, unpublished).

GSHS activity was much more stable to elevated temperatures than was γ-GCS activity. The thermal curves of *Otolemur crassicaudatus*, rhesus monkey, humans and rabbits were essentially superimposable and slightly different from sheep but distinct from the thermal profiles of rat, horse, kangaroo and spotted leopard (Rathbun and Schmidt, unpublished).

B. The Glutathione Redox Cycle: Glutathione Peroxidase and Glutathione Reductase

1. Effect of species variation on GSHPx and GSSGR activities. The continual battle against hydrogen peroxide derived from the aqueous humor (Pirie, 1965a, 1965b) is first fought by the reaction catalyzed by lenticular GSHPx. In a survey of species from eight mammalian orders, the activity of this enzyme was very low in two marsupials, cat, guinea pig and *Otolemur crassicaudatus*. In contrast, exceedingly high values were found in three Old World monkeys (Rathbun *et al.*, 1986a).

An enlarged study of the redox cycle enzyme activities was carried out in primate lenses (Rathbun *et al.*, 1994). The activity levels of GSHPx were remarkably low in the tarsier and lorisid groups and variable, but low, in the lemurs and hominoids. Very high values were found in the cercopithecoids (Fig. 5A). Averages of these values clearly distinguish the Old World monkeys from all other living primate superfamilies, as shown by Figure 5B.

GSSGR activity in the previously mentioned eight mammalian orders was medium to low. In the primates, the very low GSSGR activity of *Otolemur crassicaudatus* contrasted with high activities in six catarrhine species (Rathbun *et al.*, 1986a).

The enlarged study of primate lenticular enzymes showed that GSSGR activity was very low in lenses of all prosimians, except for the gentle bamboo lemur (Fig. 6A). The enzyme activity was many-fold higher in all anthropoids except for *Aotus*. Averages of these values (Fig. 6B) show that the hominoids had the highest glutathione reductase activity, clearly setting hominoidea apart from the other five primate superfamilies.

2. Effect of age on GSHPx and GSSGR activities. Aging had little effect upon GSHPx activity (units/lens) in dog, rabbit or mouse lenses (Rathbun *et al.*, 1983; Rathbun *et al.*, 1986c). The enzyme activity doubled in the mouse lemur lens with aging (Fig. 7A), but decreased in the *Otolemur crassicaudatus* lens (Fig. 7B).

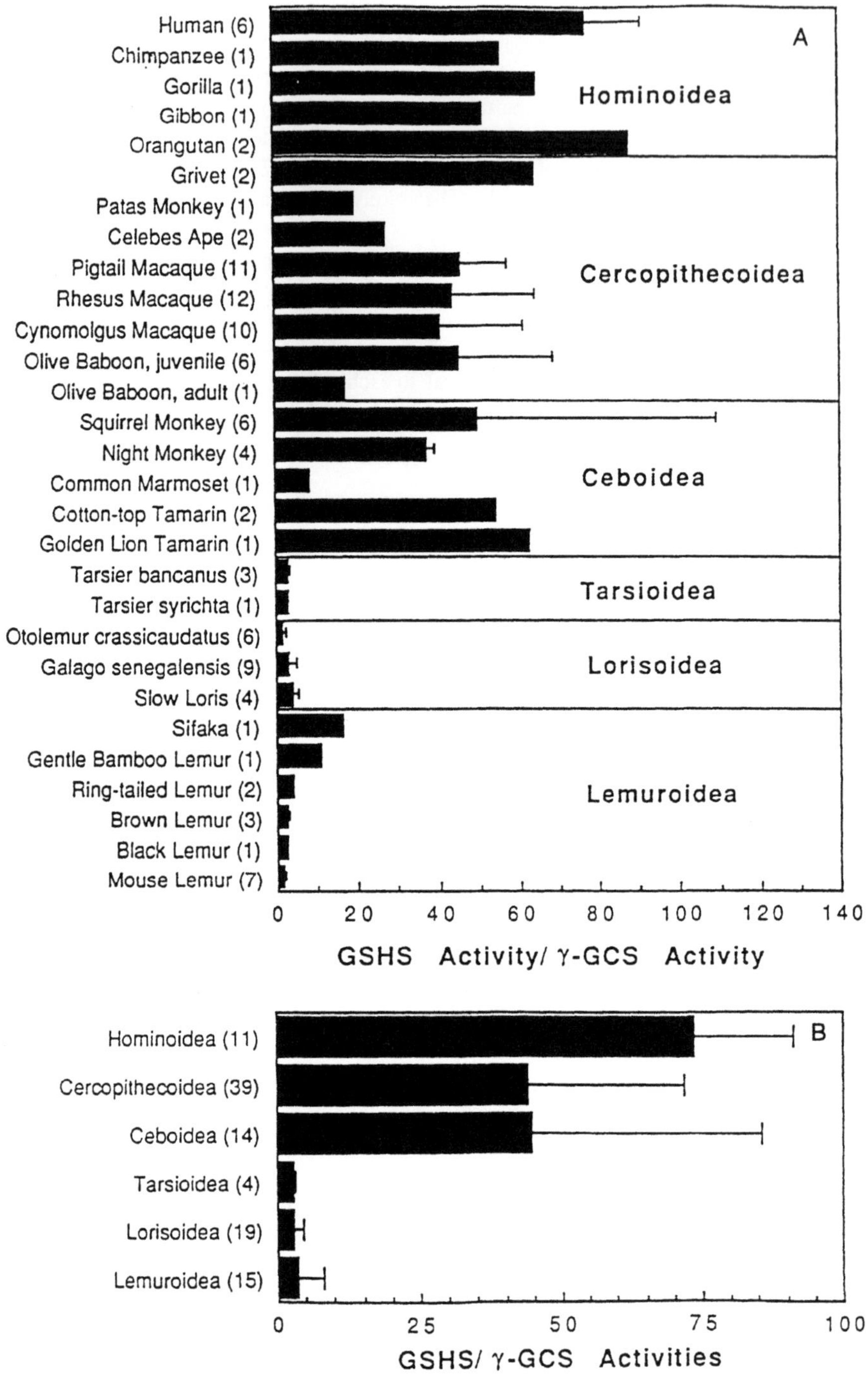

Figure 3. Ratio of the average GSHS:γ-GCS activity of primate lenses. A: activity ratio in lenses of 28 species. B: activity ratio in six primate superfamilies. Redrawn from Rathbun, Holleschau and Alterman (1994).

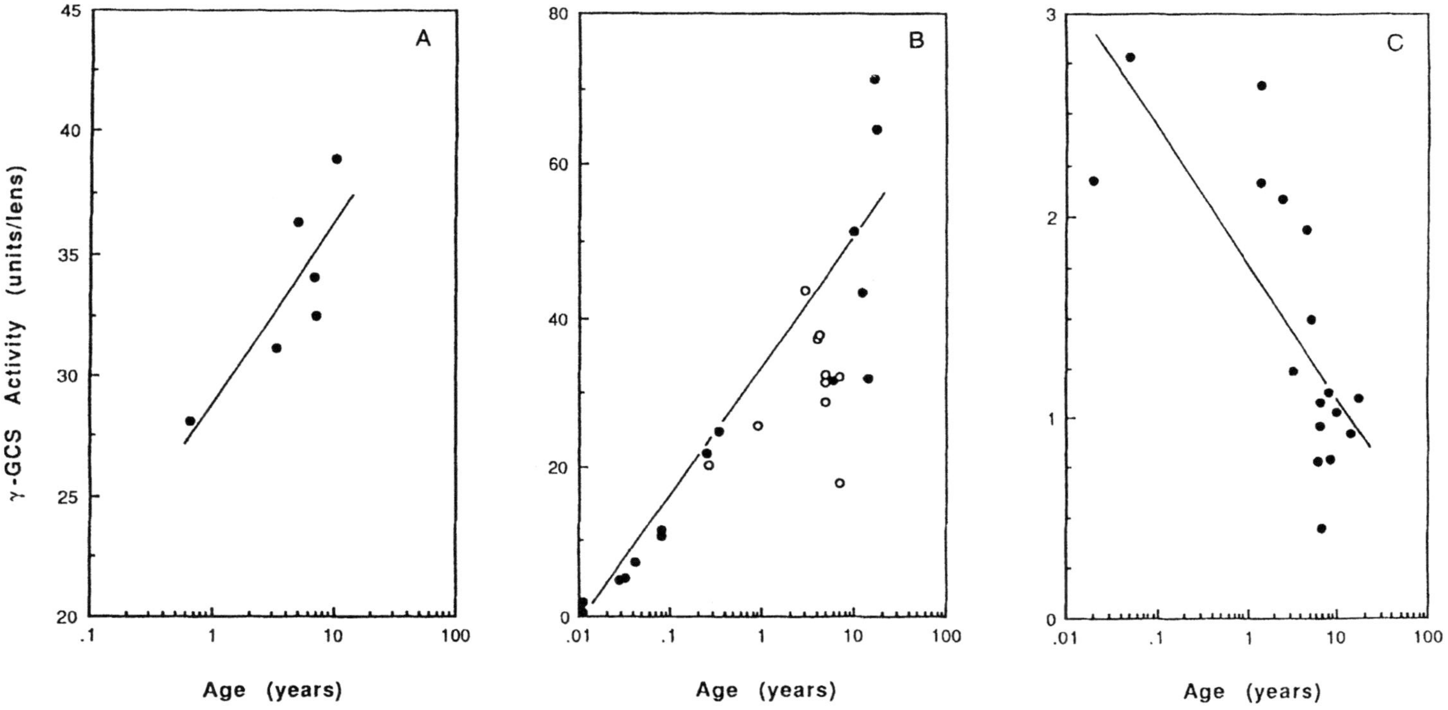

Figure 4. Effect of age on γ-GCS activity in primate lenses. Enzyme activity is expressed as units/lens; each plot includes a logarithmic regression line. A: mouse lemur, *Microcebus murinus*. B: galagos; symbols: ●, *Otolemur crassicaudatus;* ○, *Galago senegalensis*. The logarithmic regression line refers only to data of *Otolemur crassicaudatus*. C: pigtailed monkey, *Macaca nemestrina*. A loss of enzyme activity similar to that shown in C was demonstrated for the following Anthropoids: *Papio anubis, Macaca mulatta, Pongo pygmaeus* and *Homo sapiens*. Redrawn from Rathbun and Holleschau (1992).

In contrast to the *Otolemur* pattern, GSHPx activity of all catarrhine lenses increased with aging. The aging pattern of the pigtail monkey is an example (Fig. 7C). Other species in which the GSHPx activity increased with aging were the olive baboon, rhesus monkey, orangutan and human. In both rhesus monkey and human lenses GSHPx activity rose to a peak by late maturity, then declined to form a bell-shaped curve (Holleschau and Rathbun, 1994; Rathbun *et al.*, 1986b; Rathbun and Bovis, 1986).

GSSGR activity (units/lens) was unchanged with aging in the lenses of dog or rabbit. The enzyme activity rose to a peak, then declined in the mouse lens (Rathbun *et al.*, 1983; Rathbun *et al.*, 1986c). GSSGR activity tripled with age in mouse lemur (Fig. 8A) and increased nine-fold in *Otolemur* (Fig. 8B) lenses. It is significant in data expression (see Methods) that the enzyme activity of *Otolemur crassicaudatus* showed a slight decrease with age when expressed as units/g lens. The values for the mouse lemur GSSGR activity, expressed as units/g lens, were not published by Holleschau and Rathbun (1994) due to data point scatter.

The Old World monkey pattern contrasted sharply with those of the two prosimians shown in Figure 8. GSSGR activity declined logarithmically with age in lenses of rhesus monkeys (Rathbun *et al.*, 1986b), the olive baboon and pigtail monkey (Holleschau and Rathbun, 1994). Figure 8C illustrates the decline in the pigtail monkey lens, when expressed as units/lens. These aging profiles distinguish the cercopithecoids from the prosimians in the effect of aging on GSSGR activity. The response of GSSGR activity to aging by the hominoids' lenses was variable. The enzyme activity declined in the orangutan, but remained unchanged with aging in the human (Holleschau and Rathbun, 1994; Rathbun and Bovis, 1986).

3. Effect of species variation on GSHPx and GSSGR thermal stability. Thermal inactivation studies were carried out on both GSHPx and GSSGR from rhesus monkeys, humans and *Otolemur crassicaudatus*. Thermal inactivation profiles were comparable in rat and *Otolemur* lenticular GSHPx. Enzyme activities from both rhesus monkey and human lenses were very unstable, and the two patterns were essentially superimposable and easily distinguishable from *Otolemur* and rat (Holleschau and Rathbun, 1991).

GSSGR activity from all species was extremely stable to heat. The most stable form was from human lenses, although those from both rhesus monkey and rat were almost as stable. The pattern for *Otolemur crassicaudatus* GSSGR activity was easily distinguished from those of the foregoing, as that enzyme activity was the least stable to elevated temperature (Holleschau and Rathbun, 1991).

C. Summary of Distinguishing Differences

Table 1 summarizes and qualitatively compares several enzyme parameters of γ-GCS, GSHPx and GSSGR. Various combinations of these parameters substantiate the division of the primates into six groups, following Martin's scheme (1990). A fourth enzyme, GSHS, did not appear to yield useful systematic data for primates and was not included in Table 1.

DISCUSSION

Rate-limiting Enzyme Reactions as Selection Agents: A Hypothesis

On theoretical grounds, in order for an enzyme to be an effective target for selection, it must be one that catalyzes a rate-limiting reaction in a metabolic sequence. That is, it must form a "bottle neck" of a metabolic pathway. Therefore, effectiveness of the metabolic pathway will be proportional to the activity of this rate-limiting enzyme. Of the four enzymes that form the subject of this presentation , γ-GCS was shown to be rate-limiting in glutathione synthesis (Gander *et al.*, 1983) and GSSGR was most likely rate-limiting in the glutathione redox cycle (Rathbun *et al.*, 1983). Indeed, major differences in characteristics of these two enzymes were shown among the living primates.

Table 1. Summary of enzyme parameters

Group	Enzyme activity (units/g lens)	Effects of Aging (units/lens)	Thermal Stability
	γ-Glutamylcysteine Synthetase Activity		
Lemuroidea	very high	increase	----
Lorisoidea	very high	large increase	similar to human
Tarsioidea	very high	----	----
Ceboidea	low	----	----
Cercopithecoidea	very low	loss	rhesus least stable
Hominoidea	very low	large loss	similar to galago
	Glutathione Peroxidase Activity		
Lemuroidea	medium	increase	----
Lorisoidea	very low	loss	similar to rat
Tarsioidea	medium	----	----
Ceboidea	medium	----	----
Cercopithecoidea	very high	increase, then loss	unstable like
human			
Hominoidea	medium	increase, then loss	unstable like
rhesus			
	Glutathione Reductase Activity		
Lemuroidea	low	increase	----
Lorisoidea	low	large increase[1]	least stable
Tarsioidea	low	----	----
Ceboidea	high	----	----
Cercopithecoidea	high	loss	very stable
Hominoidea	very high	variable	most stable

[1] A slight loss with aging when plotted as units/g lens was reported by Holleschau and Rathbun (1994).

Enzyme Characteristics Differentiating Suborders Prosimii and Anthropoidea

The differences between γ-GCS activity of the prosimians and simians is quite striking (Fig. 1). Low activity of this enzyme in simians, especially of the Old World as compared to any other primate or non-primate group, appears to be a defining characteristic. The large ratio of GSHS/γ-GCS shows that not only is the γ-GCS reaction rate-limiting in glutathione biosynthesis based on substrate concentration as well as product inhibition mechanisms (Gander *et al.*, 1983), but the paucity of γ-GCS activity in lenses was also rate-limiting among the Anthropoids (Fig. 3).

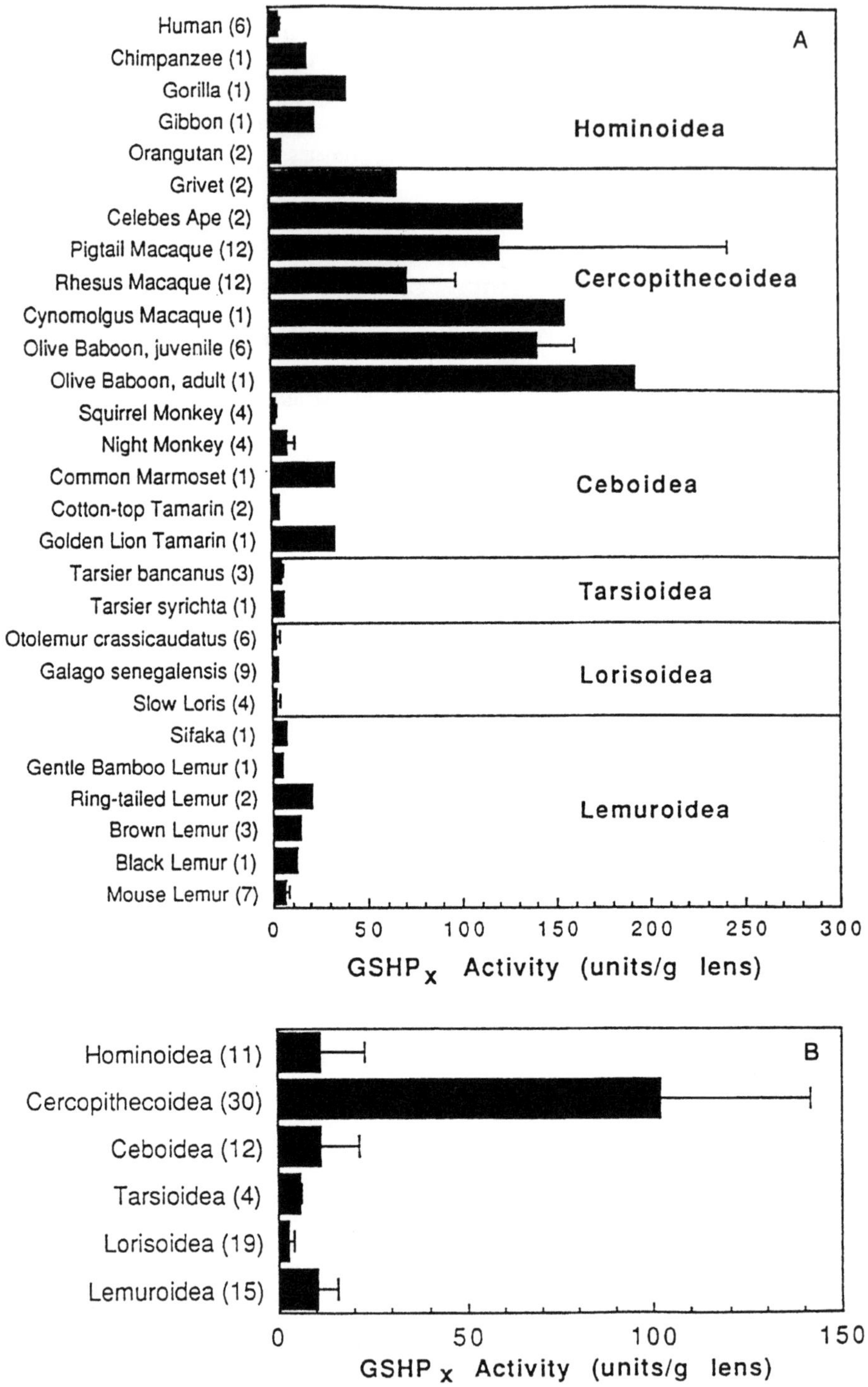

Figure 5. Lenticular GSHPx activity (units/g lens) in primates. A: average activity in lenses of 27 species. B: average enzyme activity in six primate superfamilies. Redrawn from Rathbun, Holleschau and Alterman (1994).

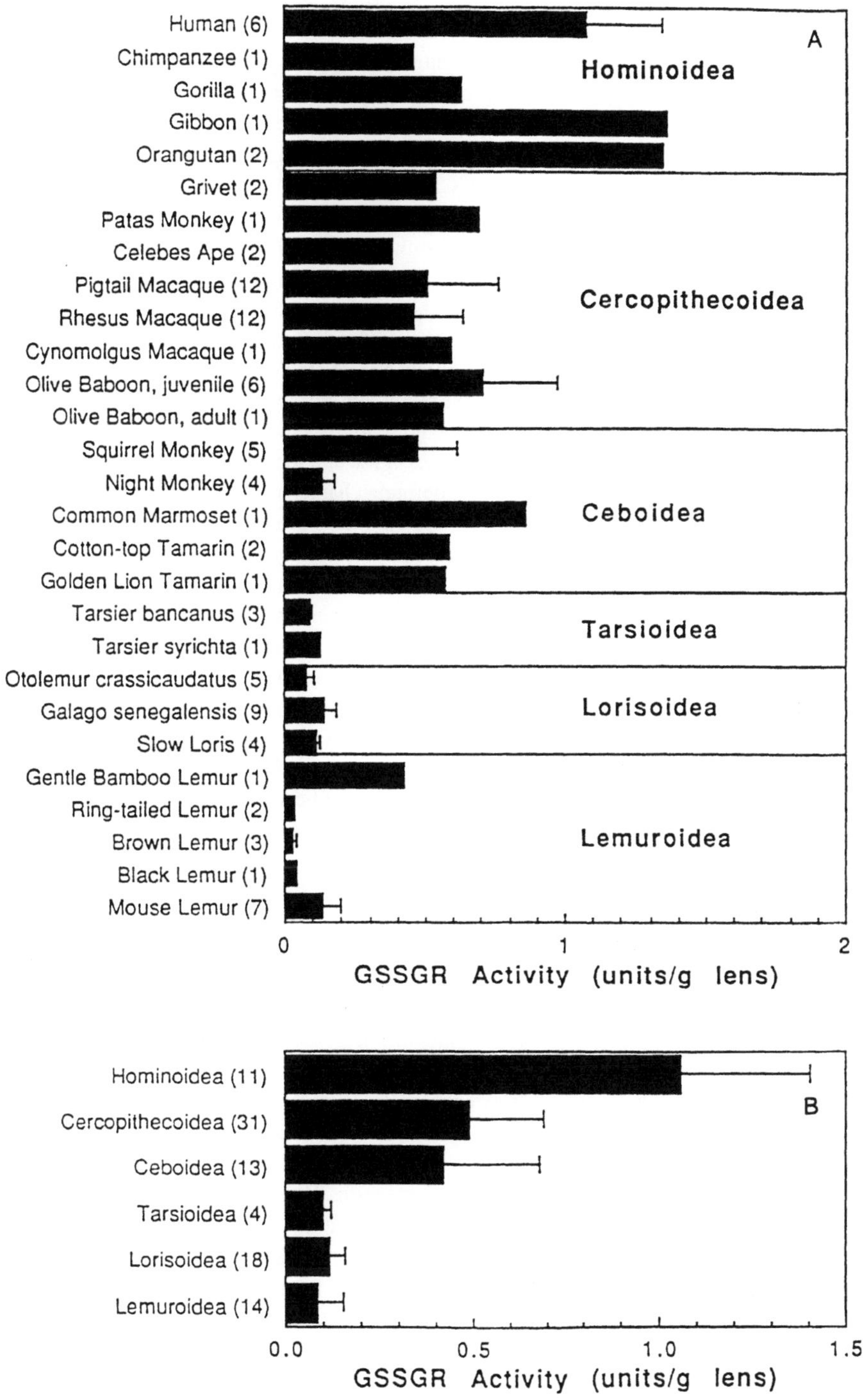

Figure 6. Lenticular GSSGR activity (units/g lens) in primates. A: average activity in lenses of 27 species. B: average enzyme activity in six primate superfamilies. Redrawn from Rathbun, Holleschau and Alterman (1994).

A consistent and significant loss of γ-GCS activity with aging characterizes the Old World simians. This loss (example: Fig. 4C) contrasts with the gain of activity with age in mouse lemur lenses and especially in *Otolemur* lenses (Fig. 4A, 4B). The activity decline is probably not related to innate enzyme instability to heat as *Otolemur* and *Homo* thermal inactivation profiles were quite similar and different from the rhesus monkey. Therefore, another factor must be responsible for activity loss with aging. These data are consistent with the possibility that low activity of γ-GCS may have first appeared in an ancestral Anthropoid, following the branching from the prosimians' precursor.

Differences between the prosimians and simians in GSSGR activity are almost as striking (Fig. 6). The activity in the lenses of Anthropoids was high, but decreased with aging in all simians tested except *Homo*, where there was no change. In contrast, GSSGR activity was very low in the prosimians, but increased with age in both *Otolemur* and mouse lemur. Moreover, the enzyme activity was most stable to elevated temperatures in lenses from rhesus monkey and humans but relatively unstable in *Otolemur*. The highest enzyme activity was found in three of the five hominoid species, suggesting that a vigorous glutathione reductase activity may be important in the ape lenses. High lenticular GSSGR activity, decreasing with age, appears to be a defining characteristic of Anthropoids.

In the event that only a small amount of glutathione is available to a cell, it is important that the maximum amount possible be maintained in the biologically active form, ie, reduced glutathione. The loss of γ-GCS activity (five-fold, units/lens; 16-fold, units/g lens) in the human lens, which largely occurs before puberty, is associated with a parallel loss of glutathione content. During the reproductive ages 95% to 98% of the glutathione is maintained in the reduced state by the GSSGR-catalyzed reaction (Rathbun and Murray, 1991). It is interesting that among the simians, in which the glutathione synthesis enzymes were greatly reduced, GSSGR activity occurred at very high levels. The only simian genus that was the exception to exhibiting high GSSGR activity was the nocturnal *Aotus*. It is this author's hypothesis that high glutathione reductase activity might be a compensatory mechanism to offset low γ-GCS activity and so aid survival in a sunlit environment.

Tarsius Membership in Haplorhini or Prosimii; Comparison with *Aotus*

Two characteristics of gross eye morphology, considered by themselves, lend support to inclusion of *Tarsius* with the Anthropoidea to form the Haplorhini. The shared characteristics include existence of a small fovea and lack of a tapetum.

There are many characters of the eye that *Tarsius* does not share with the simians. The visual receptor cells of *Tarsius* are entirely rod cells; cone cells as found in the retina of the Anthropoidea have not been shown (Castenholz, 1984). Such rod cells are much more sensitive to low light levels than are cone cells. Therefore, the *Tarsius* retina is especially suitable for a nocturnal environment.

The lens of *Tarsius* is almost spherical (Castenholz, 1984) and therefore unlikely to be readily flattened to allow for variable focus, as observed in diurnal simians. The spherical shape, typical of nocturnal species such as in Rodentia, is especially useful for spreading light rays across the entire retina rather than to a discrete focus such as a fovea. The lens of *T. bancanus* (589 mg) was the largest listed in the primate survey of Rathbun *et al.* (1994). Its size also makes this lens efficient for light-gathering through a wide visual angle.

Within the *Tarsius* lens the activities of γ-GCS, GSHPx and GSSGR, as well as the activity ratio of GSHS:γ-GCS, are all similar to the prosimians and totally unlike those of the simians. These enzyme data, together with the shape and size of the lens and the pure rod cell character of the retina, do not lend support to the concept that the *Tarsius* ancestral stock was originally diurnal and therefore a member of the Haplorhini.

Comparison with the nocturnal *Aotus* is of interest. This species has only a rudimentary tapetum and ill-defined fovea (Martin, 1990). The lens size, 303 mg, was the largest among the simians reported in the survey by Rathbun *et al.* (1994). None of the other Ceboidea exceeded 76 mg, while lenses in the larger-bodied chimpanzee and white-handed gibbon were 230 mg, second only to the night monkey among the simians. Therefore, with respect to size *Aotus* exhibited a large light gathering-dispersing optical element perhaps comparable to that of *Tarsius*.

Activity of γ-GCS from the night monkey was comparable to other New World monkey species, and six-fold less than *Tarsius*. Unfortunately, age-related data are not available for either *Tarsius* or *Aotus*. The ratio of GSHS:γ-GCS activities of *Aotus* was 37, compared to 3 for *Tarsius*. GSHPx activity was not definitive, because the low activities in *Tarsius* and *Aotus* were comparable to activities found in various species of all major living primate groups with the exception of the Old World monkeys (Fig. 5A).

GSSGR is the only glutathione-metabolizing enzyme in which the *Aotus* lens differed from all other Anthropoid species. GSSGR activity of *Aotus* and *Tarsius* lenses were almost identical. This was the sole lens enzyme characteristic that was shared and which is consistent with the concept that a high activity level of GSSGR in the lens is not required for a diurnal habit. However, this character by itself does not offer enlightenment as to whether either or both *Aotus* and *Tarsius* might be secondarily nocturnal.

Other Relationships

All six species of the Old World monkeys demonstrated higher GSHPx activity than any other primate species surveyed by Rathbun *et al.* (1994). When compared to the lorisids, the enzyme activity difference was 50-fold. This consistent difference characterized the Old World monkeys from any other primate and non-primate group.

These data show no significant differences between the slow loris and the galagos in any of the enzyme activities. However, the γ-GCS activity of the mouse lemur lens was more than double that of other lemurs and indeed, of all other strepsirhines. Additionally, GSSGR activity of this species was two to three-fold greater than other lemurs except for the gentle bamboo lemur. Additional data is required to determine whether differences in this enzyme might be significant between the cheirogalids and other lemurs.

Possible Application

It is apparent that compared to the prosimians, all simians have a defective gene regarding the γ-GCS molecule. This enzyme is critical to maintenance of a high level of glutathione which is required for a diurnal environment. Perhaps the emerging technique of gene therapy could be used to alleviate this defect. *Otolemur* might provide an appropriate source of a better quality gene for the simians because its enzyme activity increases dramatically during life as opposed to the marked decrease characteristic of the simians. Additionally, the *Otolemur* enzyme is relatively thermally stable.

ACKNOWLEDGMENTS

The author thanks Ann Holleschau for use of her data concerning GSHPx and GSSGR activities expressed as units/lens which were previously unpublished in this form. This work was supported in part by National Institutes of Health research grant EY-01197-19.

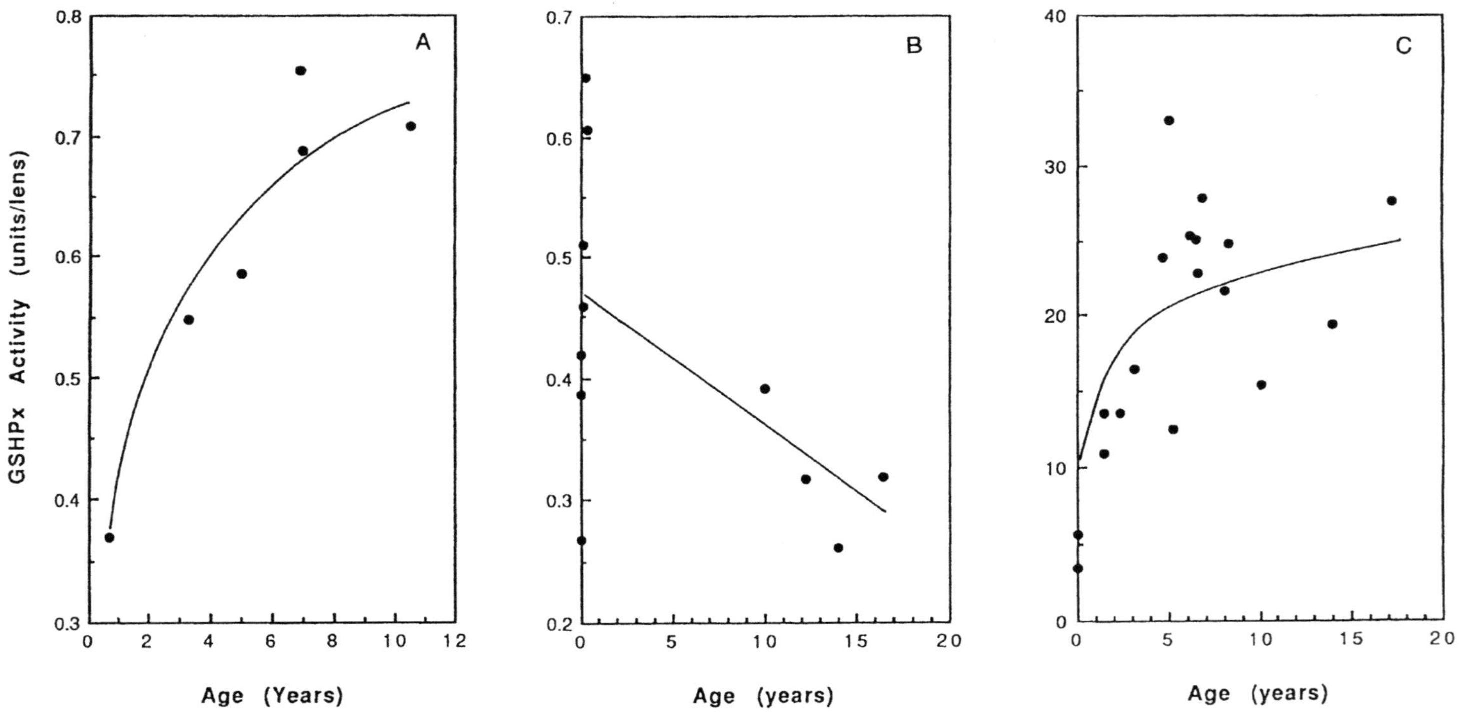

Figure 7. Effect of age on GSHPx activity in primate lenses. Enzyme activity is expressed as units/lens; each plot includes a logarithmic regression line. A: mouse lemur, *Microcebus murinus*. B: *Otolemur crassicaudatus*. C: pigtailed monkey, *Macaca nemestrina*. A gain of enzyme activity was also demonstrated for the following Anthropoids: *Papio anubis, Macaca mulatta, Pongo pygmaeus* and *Homo sapiens*. Previously unpublished data.

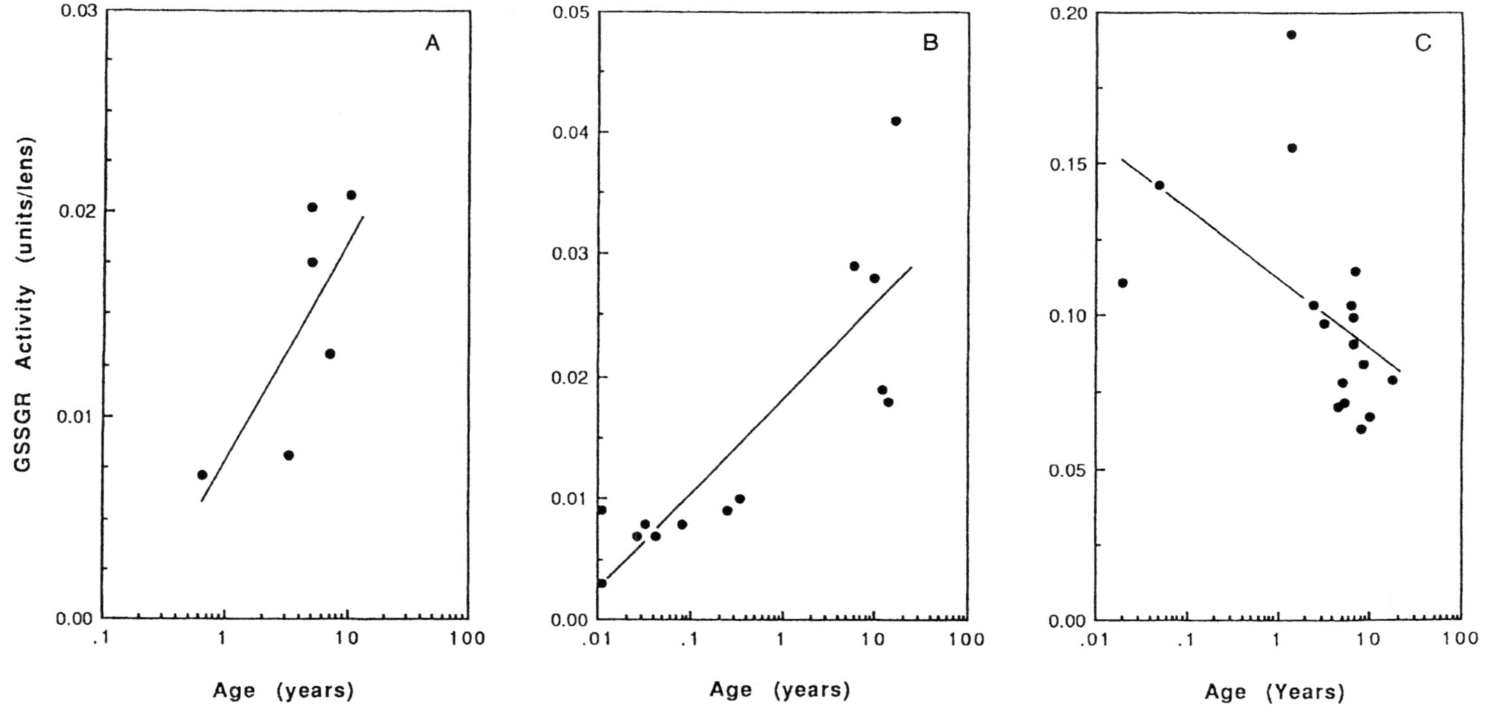

Figure 8. Effect of age on GSSGR activity in primate lenses. Enzyme activity is expressed as units/lens; each plot includes a logarithmic regression line. A: mouse lemur, *Microcebus murinus*. B: *Otolemur crassicaudatus*. C: pigtailed monkey, *Macaca nemestrina*. A loss of enzyme activity similar to C was demonstrated for the following Anthropoids: *Papio anubis*, *Macaca mulatta* and *Pongo pygmaeus*. Previously unpublished data.

and unrestricted funds from Research to Prevent Blindness, Inc. W.B.R. is a Research to Prevent Blindness Senior Scientific Investigator.

REFERENCES

Bergad, P.L., Rathbun, W.B. & Linder, W. (1982). Glutathione peroxidase from bovine lens: a selenoenzyme. *Experimental Eye Research*, 34, 131-144.

Castenholtz, A. (1984). The eye of tarsius. In "Biology of Tarsiers", (Niemitz, C., ed.), Gustav Fisher Verlag, Stuttgart, New York.

Gander, J.E., Sethna, S.S. & Rathbun, W.B. (1983). Bovine lens γ-glutamylcysteine synthetase: inhibition by glutathione and adenine nucleotides. *European Journal of Biochemistry*, 133, 635-640.

Holleschau, A.M. & Rathbun, W.B. (1991). Thermal inactivation study of glutathione peroxidase and glutathione reductase activities in lenses of primates and non-primates. *Current Eye Research*, 10, 221-229.

Holleschau, A.M. & Rathbun, W.B. (1994). The effects of age on glutathione peroxidase and glutathione reductase activities in lenses of Old World simians and prosimians. *Current Eye Research*, in press.

Martin, R.D. (1990). "Primate Origins and Evolution: a Phylogenetic Reconstruction", Princeton University Press, Princeton, N.J.

Northrop, J.H. & Kunitz, M. (1933). Isolation and properties of crystalline pepsin. *Ergebn. Enzymforsch.*, 2, 104-117.

Ohrloff, C. (1978). Age changes of enzyme properties in crystalline lens. *Interdisciplinary Topics in Gerontology*, 12, 158-179.

Pirie, A. (1965a). A light-catalyzed reaction in the aqueous humour of the eye. *Nature*, 205, 500-501.

Pirie, A. (1965b). Glutathione peroxidase in lens and a source of hydrogen peroxide in aqueous humour. *Biochemical Journal*, 96, 244-253.

Rathbun, W.B. (1986a). Glutathione synthesis in evolution: an Achilles' heel of human and other Old World simian lenses. *Ophthalmic Research*, 18, 236-242.

Rathbun, W.B. (1986b). Activity of glutathione synthesis enzymes in the rhesus monkey lens related to age: a model for the human lens. *Current Eye Research*, 5, 161-166.

Rathbun, W.B. & Bovis, M.G. (1986). Activity of glutathione peroxidase and glutathione reductase in the human lens related to age. *Current Eye Research*, 5, 381-385.

Rathbun, W.B., Bovis, M.G. & Holleschau, A.M. (1986a). Species survey of glutathione peroxidase and glutathione reductase: search for an animal model of the human lens. *Ophthalmic Research*, 18, 282-287.

Rathbun, W.B., Bovis, M.G. & Holleschau, A.M. (1986b). Glutathione peroxidase, glutathione reductase and glutathione-S-transferase activities in the rhesus monkey lens as a function of age. *Current Eye Research*, 5, 195-199.

Rathbun, W.B. & Holleschau, A.M. (1992). The effects of age on glutathione synthesis enzymes in lenses of Old World simians and prosimians. *Current Eye Research*, 11, 601-607.

Rathbun, W.B., Holleschau, A.M. & Alterman, L. (1994). Glutathione metabolism in primate lenses: a phylogenetic study of glutathione synthesis and glutathione redox cycle enzyme activities. *American Journal of Primatology*, 33, 101-120.

Rathbun, W.B., Kuck, J.F.R. & Kuck, K.D. (1986c). Glutathione metabolism in lenses of Emory and cataract-resistant mice: activity of five enzymes. *Current Eye Research*, 5, 189-194.

Rathbun, W.B., Sethna, S.S, Skelnik, D.L. & Bistner, S.I. (1983). Glutathione metabolism in lenses of dogs and rabbits: activities of five enzymes. *Experimental Eye Research*, 36, 845-858.

Rathbun, W.B., Sethna, S.S. & Van Buskirk, G.E. (1977). Purification and properties of glutathione synthetase from bovine lens. *Experimental Eye Research*, 24, 145-158.

Rathbun, W.B., Schmidt, A.J. & Holleschau, A.M. (1993). Activity loss of glutathione synthesis enzymes associated with human subcapsular cataract. *Investigative Ophthalmology and Visual Science*, 34, 2049-2054.

Rathbun, W.B. & Wicker, K. (1973). Bovine lens γ-glutamyl transpeptidase. *Experimental Eye Research*, 15, 161-171.

Sethna, S.S., Holleschau, A.M. & Rathbun, W.B. (1982/1983). Activity of glutathione synthesis enzymes in human lens related to age. *Current Eye Research*, 2, 735-742.

SPECIES DIVERSITY IN LORISIDS: A PRELIMINARY ANALYSIS OF *ARCTOCEBUS, PERODICTICUS,* AND *NYCTICEBUS*

Jeffrey H. Schwartz and Jeremy C. Beutel

Department of Anthropology
University of Pittsburgh
Pittsburgh, PA 15260

INTRODUCTION

In recent decades, the strepsirhine family, Lorisidae[1], has been thought of as consisting taxonomically of only four extant genera, of which two are considered to be monotypic: the Asian *Loris* (which is now apparently confined to the island of Sri Lanka), and the African *Perodicticus potto.* The African genus, *Arctocebus,* apparently comprises two species, *A. calabarensis* and *A. aureus.* And the Asian genus *Nycticebus* also is supposed to embrace only two species, *N. pygmaeus* and the somewhat larger *N. coucang. N. pygmaeus* is distributed roughly throughout the mainland of southeast Asia. *N. coucang* is sympatric with *N. pygmaeus* on the mainland, but also occurs on the islands of Borneo, Sumatra, Java, and, purportedly, the Philippines.

The question of species recognition, however, is not a simple one. For example, consider *Nycticebus.* In the most recent taxonomic evaluation of the genus, Groves (1971: 45) distinguished *N. pygmaeus* from *N. coucang* on the basis of: 1) smaller size, 2) larger ears, 3) very flattened occiput, 4) anteriorly extended premaxilla, 5) diastemata on either side of the upper anterior premolar, 6) upper molars subequal in size, 7) palate extending to the distal side of M^3, 8) fur fine, wavy, reddish or orange in color, and 9) dorsal stripe not dark and terminating in lumbar region. *N. coucang* is characterized not only by its larger size, but by having 1) a diastema only between the upper anterior premolar and the upper canine, 2) upper molars that decrease in size from front to back, 3) a palate that does not extend as far as the distal side of M^3, 4) darker, coarse, crisp fur, and 5) dark dorsal stripe that often extends to the sacrocaudal region. Although Groves recognized that "[t]here is considerable geographic variability in *N. coucang,*" he represented as subspecies or races specimens that earlier taxonomists had proposed as separate species in a nominal genus for which *Nycticebus* is the senior objective synonym.

While not suggesting that the suppression of all of these previously proposed species of *Nycticebus* is unwarranted, it nevertheless should be pointed out that there are differences between specimens included within the species *N. coucang* that might be reflective of the potential existence of morphs, which, in turn, could be considered as representing separate species. Thus, there are some *N. coucang* that develop and erupt two pairs of upper deciduous incisors that are replaced by two pairs of permanent successors (2 + 2); others that

[1] Recently, Jenkins (1987) has presented the case that, although the family names Lorisidae and Galagidae have been employed consistently since early in this century, these are not the first names applied to these groups of prosimians. Rather, Loridae Gray, 1821 and Galagoninae Gray 1825, respectively, are. Thus, in spite of familiarity, we should more properly use Loridae and Galagonidae as respective family names. Nevertheless, in light of the facts that one of us (JHS) is preparing an appeal to suppress these latter names and that Lorisidae and Galagidae are widely understood among taxonomists, we shall here use the more familiar terms of taxonomic reference.

develop and erupt two pairs of upper deciduous incisors that are replaced by one pair of permanent successors (2 + 1); and yet other *N. coucang* that develop and erupt only one pair of upper deciduous incisors which are replaced by one pair of permanent successors (1 + 1) (Schwartz, 1974). This would appear to represent a morphocline of derivedness, the first state of which is primitive for primates, in general. In addition, not all *N. coucang* can be characterized by the same sequence of dental growth and eruption (Schwartz, 1974). However, subsets of *N. coucang* can be identified by unified sequences of dental growth and eruption, and these subsets appear to correspond to upper incisor-based morphs. Finally, while compiling descriptions of strepsirhine teeth for another study (Schwartz and Tattersall, 1985), one of us (JHS) noticed that those *N. coucang* with the primitive upper incisor character states (2 + 2) had cheek teeth with puffier cusps and less crisp cresting than those with the more derived upper incisor states. The latter had cheek teeth with taller, more pointed cusps and more crisply defined cresting. Given that very few strepsirhines possess puffy to low, flat-cusped cheek teeth (i.e. *Perodicticus, Cheirogaleus,* and *Daubentonia),* one could suggest that similarly characterized *N. coucang* would be displaying the more derived condition.

Although Groves (1971) cited differences in upper incisor number in the descriptions of his subspecies of *N. coucang,* he did not consider these differences sufficient to warrant recognition of these morphs at the level of the species. However, given the additional information that could affect the systematics of *Nycticebus,* one would think that a reevaluation of the criteria upon which species could be recognized would be in order. Furthermore, insofar as the other lorisid genera have long been considered either monotypic or species-restricted, a reevaluation of the systematics of all extant lorisid genera might prove fruitful. This contribution, in which we present our initial studies on *Arctocebus, Perodicticus,* and *Nycticebus,* represents the first stage of such a reassessment.

METHODS AND MATERIALS

The specimens of *Arctocebus* (n = 32), *Perodicticus* (n = 44), and *Nycticebus* (n = 45 for nondental measurements; n = 18 additional specimens for dental measurements) used in this study are housed in the collections of: the Natural History Museum, London (BM[NH]); the Anthropological Institute and Museum, Zürich (AMZ); the Adolph Schultz collection of the Anthropological Institute and Museum, Zürich (AMZ-AS); the National Museum of Natural History, Washington, DC (USNM); and the American Museum of Natural History, New York City (AMNH). Details of craniodental morphology were recorded for all specimens. Skins of *Nycticebus* were available and also included in the analysis. The following cranial measurements were taken on samples of each genus (when not precluded by juvenality of, or damage to, specimens): L (maximum cranial length); NL (maximum length of nasal bones); BO (also Biorb; maximum horizontal distance between outer margins of orbital rims); OD (minimum interorbital distance); BP (width from porion to porion); BC (maximum breadth measured on the maxilla from upper canine to upper canine); TL (minimum distance between temporal lines). Upper canines and upper and lower P2s-M3s of 19 specimens of *N. coucang* (but not those on which cranial measurements were taken) were measured. Maximum length of the right humerus as well as mesiodistal (MD) length (L) and buccolingual (BL) width (W) of upper and lower molars were also recorded for specimens of *Arctocebus* and *Perodicticus.* All specimens were described and measured in mm by JHS; the statistical analyses were carried out by JCB. Because detailed descriptions of dental morphology are available elsewhere (Schwartz and Tattersall, 1985), they will not be repeated here.

For the sake of brevity, we accept the hypotheses that *Arctocebus, Perodicticus,* and *Nycticebus* are cladistically 1) strepsirhine (or, more broadly, prosimian), 2) members of a galagid + lorisid clade, and 3) members of a lorisid clade (e.g. see reviews by Groves, 1974; Schwartz, 1986; Schwartz and Tattersall, 1985).

RESULTS

Prior to investigating the systematics of each genus, one must evaluate the hypothesis that each taxon regarded as a genus (whether or not it ultimately should be regarded as a genus, a species, or some other taxonomically-interpreted morphological unit) does appear to be distinguished by at least one autapomorphy. In the context of this study, the taxon that

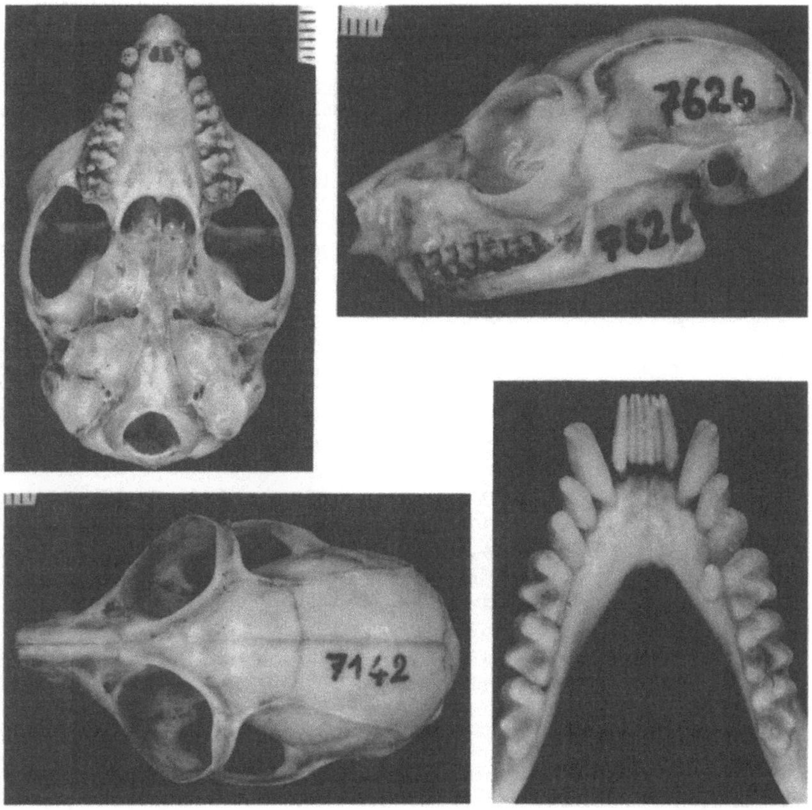

Figure 1. *Arctocebus calabarensis.* Clockwise from upper left: AMZ 7142, AMZ 7626, AMZ 7665 (not to scale; note supernumerary tooth lingual to left P_4 and M_1), and AMZ 7141. When present, scale is in mm. (© J. Schwartz)

most readily presents itself as a distinct morphological unit is *Arctocebus calabarensis.*

I. Arctocebus (Fig. 1)

A dental feature that distinguishes *Arctocebus* from other lorisids is its deeper and more expansive hypoflexid notch, especially on M_1. In fact, the straight cristid obliqua contacts the metaconid on M_1, may do so on M_2, and is quite medially emplaced on M_3. Although many Paleogene primates had an *arcuate* cristid obliqua on M_1 that contacted the metaconid, few (e.g. *Smilodectes, Necrolemur)* were similar in detail to *Arctocebus* on all lower molars (see Schwartz, 1986). Craniodentally, the *Arctocebus* specimens used in this study (most are housed in the AMZ) presented themselves as a morphologically uniform sample. Attempts at morphologically distinguishing specimens by sex using subsets of the sample that were already identified as male or female were unsuccessful. Specimens were even uniformly similar in displaying little or no dental attrition. Morphologically, therefore, this *Arctocebus* sample presented itself as a single morph, which, given the nature of binomial nomenclature, could reasonably be represented taxonomically as the genus and species, *A. calabarensis.*

The uniformity of the sample also was tested statistically. First, all craniodental measurements and right humeral length were subjected to an unpaired, two-tailed t-test in order to evaluate the degree to which sexual dimorphism was expressed in the sample. Table 1 ranks the measurements in descending order of degree of overlap of the male and females subsets. Only three of the 24 measurements fell below 5% ($p \leq .05$ that individuals were from the

Table 1. *Arctocebus calabarensis.* Ranked, unpaired, two-tailed t-test of hypothesis that males and females are from the same population.* Measurements in mm.

Rank	Measurement	Probability	Male $\bar{x}$	n	Female $\bar{x}$	n
1.	L	.945	54.62	11	54.57	6
2.	rM_1L	.8923	3.29	17	3.3	5
3.	lM_2W	.827	2.55	13	2.54	5
4.	TL	.8211	12.14	11	11.97	6
5.	lM_1L	.8169	3.22	12	3.24	5
6.	OD	.8066	5.03	11	4.98	6
7.	BC	.8022	9.28	9	9.33	6
8.	BP	.6732	24.22	11	24.38	6
9.	NL	.6497	14.24	10	13.98	6
10.	rM_2W	.5749	2.49	14	2.52	5
11.	rM_1W	.3108	2.49	14	2.54	5
12.	Biorb	.2723	30.16	11	30.87	6
13.	Hum(r)	.258	57.97	10	59.68	6
14.	lM_1W	.2033	2.52	12	2.58	5
15.	lM^2L	.1507	3.26	14	3.35	6
16.	lM^1W	.1427	4.12	14	4.22	6
17.	rM^1W	.1246	4.11	13	4.22	6
18.	rM^2W	.1107	4.3	13	4.48	6
19.	rM^2L	.0934	3.26	13	3.37	6
20.	rM_2L	.0714	3.35	14	3.42	5
21.	lM_2L	.0596	3.31	13	3.42	5
22.	lM^1L	.0376	3.25	14	3.35	6
23.	rM^1L	.0186	3.26	13	3.45	6
24.	lM^2W	.0139	4.17	14	4.38	6

*21 measurements: > 5% overlap; 3 measurements: < 5% overlap.

same population); remarkably, 14 measurements fell above 20%, reflecting 80% and higher overlap between the two sexes ($p \geq .80$ that individuals were from the same population). Clearly, as has also been demonstrated for the extant indriids on the basis of dental measurements (Gingerich and Ryan, 1979), sexual dimorphism in *Arctocebus* is negligibly reflected in the dental (as well as in the cranial and humeral) metric data collected for this study.

Table 2. *Arctocebus calabarensis.* Length and width (in mm) and summary statistics of upper and lower first and second molars. Combined sample of specimens (male, female, and sex unknown).

		Maxillary Dentition					*Mandibular Dentition*				
		n	$\bar{x}$	Range	s	CV	n	$\bar{x}$	Range	s	CV
Right											
M1	L	28	3.31	3.1-3.7	.18	5.412	28	3.29	3.1-3.6	.12	3.583
	W	28	4.13	3.7-4.7	.19	4.655	28	2.54	2.3-2.8	.13	4.951
M2	L	27	3.26	3.1-3.6	.13	4.09	28	3.35	3.1-3.7	.12	3.489
	W	27	4.33	4.0-4.7	.21	4.871	28	2.52	2.3-2.8	.11	4.497
Left											
M1	L	29	3.28	3.1-3.6	.12	3.617	26	3.25	3.0-3.6	.14	4.377
	W	29	4.16	3.9-4.7	.17	3.98	26	2.57	2.3-3.0	.12	4.769
M2	L	28	3.25	3.1-3.6	.14	4.229	27	3.34	3.0-3.7	.14	4.084
	W	28	4.24	4.0-4.5	.17	4.027	27	2.56	2.5-3.0	.10	3.963

With the exception of the coefficient of variation (CV) for right M^1 length (CV = 5.412), CVs for right and left (mesiodistal) lengths and (buccolingual) widths for upper and lower M1-2 fell below 5 (Table 2), as did the those for skull length and biorbital, biporion, and bicanine breadths (Table 3). The most variable feature, not only for *Arctocebus* but for the entire study sample, was minimum separation of temporal lines (CV = 10.506 for *Arctocebus).*

Table 3. *Arctocebus calabarensis.* Cranial and right humeral measurements (in mm) and summary statistics. Combined sample (male, female, sex unknown).

	n	$\bar{x}$	Range	s	CV
Right humeral length	21	57.87	48.5-63.9	3.52	6.081
Skull length	23	54.52	50.6-57.4	1.65	3.029
Nasal length	22	14.22	12.6-15.9	1.04	7.305
Biorbital breadth	22	30.33	28.2-32.0	1.27	4.189
Interorbital breadth	23	5.0	4.4-5.6	.32	6.401
Biporion breadth	23	24.30	22.7-25.9	.82	3.384
Bicanine breadth	21	9.37	8.5-10.2	.44	4.718
Min. temporal line sep.	23	12.27	9.8-14.6	1.29	10.506

II. Perodicticus

The sample of *Perodicticus* is housed in the AMZ. It was sufficiently variable to allow for the delineation of morphs, which were identified as Morphs A-E. Collectively these morphs are characterized by an extraordinary amount of cheek tooth attrition and enamel crazing, especially of the upper canines, lower P2s, and lateral teeth of the toothcomb (33 of 36 specimens analyzed), and an overall chunky or robust appearance, e.g. relatively short, deep face and mandible, deep zygoma, broad snout with bulging canine root, expanded goneal region (Fig. 3-7). The potential distinctiveness of this sample (regardless of the potential reality of the morphs contained therein) is borne out by comparison with specimens assigned to *P. potto* that have been illustrated elsewhere as representative of the taxon (e.g. see illustrations in Elliot, 1912; Gregory, 1922; Schwartz and Tattersall, 1985; also compare specimens in Fig. 2, 4-7). We shall refer to the latter as the "textbook" potto (e.g. see Fig. 2) and Morphs A-E as the "Zürich" potto.

The textbook potto (Fig. 2, 4-7) is distinguished from the Zürich potto (Fig. 3-7) in that, overall, the former is craniofacially more gracile and its teeth are less bulky. Specifically, the face and mandible of the textbook potto are longer and slenderer; when viewed from above,

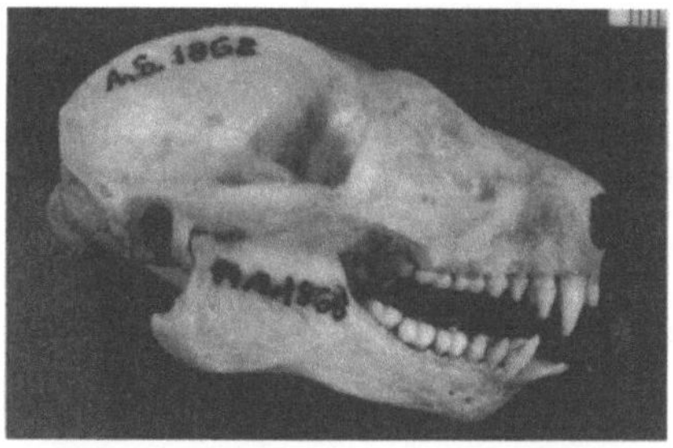

Figure 2. "Textbook" Potto (AMZ-AS 1868). Note descending gradient of crown height from $\underline{C}$-P^3 and P_2-P_4. See text for further discussion. Scale is in mm. (© J. Schwartz)

the mandible tapers anteriorly more noticeably, forming a more elongate and narrower "V" shape (Fig. 5). Dentally, and in contrast to the Zürich potto, the textbook potto lacks a diastema between the upper canine and anterior premolar (Fig. 2 and 3); the upper canine is also taller and its root does not inflate the snout as noticeably. In the textbook potto, there is a gradient of decrease in crown height from the upper canine to P^2 to P^3 (Fig. 2). In the Zürich potto, the P^2, although slightly taller than the premolar behind it, is noticeably smaller than the upper canine (Fig. 3). In the textbook potto, the upper incisors are separated by slight gaps; the gaps are further accentuated because these teeth are splayed out, rather than being orthally implanted and close together (as in the Zürich potto). In the lower dentition of the textbook potto (Fig. 2), the lower anterior premolar is not as tall as in the Zürich potto (in which this tooth is distinguished by its height) (Fig. 3); also in the textbook potto, the premolars form a gradient of decreasing crown height from the first to the last (thereby mirror-

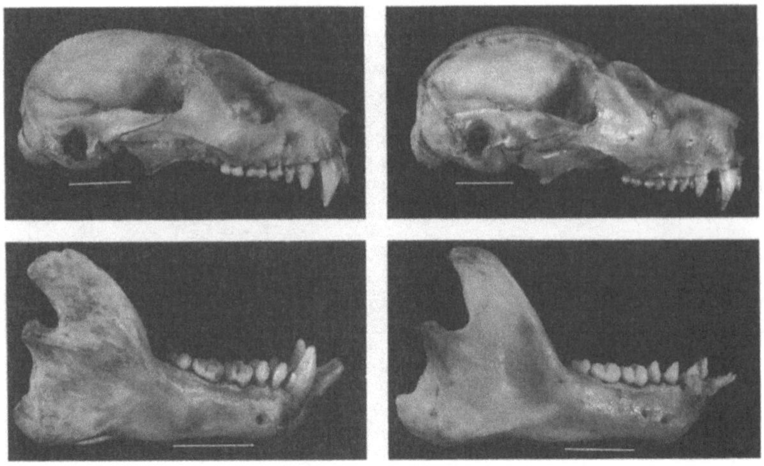

Figure 3. Zürich Potto. Left, above and below: skull and mandible of Zürich (Morph A) potto (AMZ 7191). Right, above and below: skull and mandible of Zürich (Morph B) potto (AMZ 6620). Note in both steep decrease in crown height from $\underline{C}$ to P^2. An equally steep decrease in height from P_2 to P_3 (as seen in AMZ 7191) also characterizes Zürich potto; the specimen of Morph B illustrated here has (probably antemortem) damage to its lower P2s, but was chosen because other aspects of its dentition are clearly preserved. See text for further discussion. Bar = 1 cm. (© J. Schwartz)

ing the gradient from the upper canine to the middle upper premolar). In general aspects of the skull (e.g. bulkiness, depth of zygoma) and teeth (e.g. verticality of upper incisors, diastema between the upper canine and P^2, shortness of P^2), the Zürich potto compares more favorably with *Nycticebus* than does the textbook potto (cf. Fig. 9, 11, 12).

It is fortunate that the potential existence of two major groups of such morphologically distinct pottos did become apparent during study of the Zürich collection because our consideration here of the issue of a single species of *Perodicticus* would have been even more tentative. Although we intend to study the textbook potto more extensively, the obvious morphological differences between it and the Zürich potto signal the possibility not only that *Perodicticus* may not be monotypic, but that, once the systematics of potto-like primates is better understood, the recognition of additional genera may be warranted. As it stands, however, we are confident that we can discuss the Zürich potto sample as a potential taxonomic entity, in light of which the possibility that this entity might itself embrace more than one definable morph becomes even more intriguing.

As mentioned above, the sample of Zürich *Perodicticus* appeared to be morphologically subdivisible into what we identified as Morphs A-E. Morph A (n = 10) can be distinguished

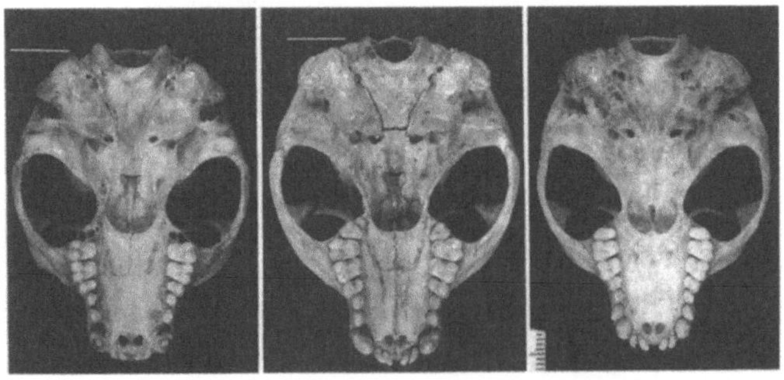

Figure 4. Basicranial views of "textbook" potto (AMZ-AS 1868; left), Zürich (Morph A) potto (AMZ 7191; middle), and Zürich (Morph B) potto (AMZ 6620, right). In the textbook potto, note splay and separation of upper incisors and contact between $\underline{C}$ and P^2. In the Zürich potto, note orientation of upper canine. See text for further discussion. Scale is in mm; bar = 1 cm. (© J. Schwartz)

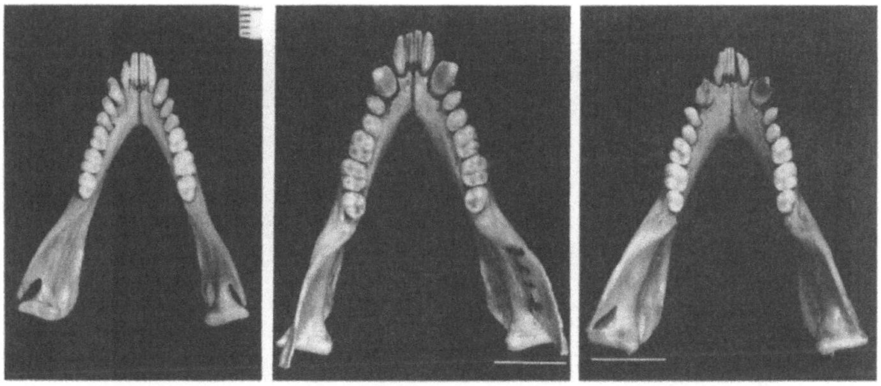

Figure 5. Mandibles of "textbook" potto (AMZ-AS 1868; left), Zürich (Morph A) potto (AMZ 7191; middle), Zürich (Morph B) potto (AMZ 6620, right). In the textbook potto, note the slenderer, more tapering V-shaped mandible and relatively smaller P_2. See text for further discussion. Scale is in mm; bar = 1 cm. (© J. Schwartz)

from the other morphs by a large P^2, protoconids on M_{1-3} that bulge buccally, and, overall, by appearing to have larger teeth (Fig. 3-7). In addition, the upper dentition of Morph A is characterized by: large upper canines that are rotated somewhat anteriorly (so that their buccal sides face obliquely forward); a P^2 that is taller than P^3 and which bears sharp mesial and distal margins; a P^3 that is conical to subconical in outline and which bears a marked lingual cingulum; a very buccolingually transverse P^4 with a large paracone and marked pre- and postcingula (which are reminiscent of protocristae); M^{1-2} that are broad across the protocone region and which bear a marked, perhaps beaded or wrinkled, lingual cingulum that wraps around the base of the protocone and then broadens in the region of the hypocone. In its lower dentition, all premolars (of which P_2 is the tallest and P_4 is expanded somewhat lingually) sport a band of lingual cingulid and the M_2 protoconid is markedly swollen buccally.

The teeth of Morph B *Perodicticus* (n = 18; Fig. 3-7)—especially the upper and lower middle and posterior premolars—appear to be smaller than those of Morph A. M^{1-2} bear a less complete lingual cingulum, and, consequently, have a more discretely delineated (although smaller) hypocone. The lower P4, which may bear a teardrop-shaped posterior fovea, is distinguished by a vertical ridge that courses down its lingual surface. The P_3 is obliquely set in the mandible (i.e. its mesial edge faces somewhat buccally).

The three other potential morphs are represented by fewer specimens and thus are discussed here more tentatively (and not illustrated). Similar to Morph A *Perodicticus,* Morph C (n = 4) has large, bulbous upper molars, but, as in Morph B, the cingulum does not surround the protocone lingually. In further similarity with Morph B, the P_4 is not enlarged. In contrast to both Morph A and B, however, the lower dental arcade of potential Morph C is not angularly "V"-shaped but swells out along the sides. The teeth of Morph D (n = 2) are more gracile overall compared to the other morphs and the inferior surface of the mandible is flattened in the region of the symphysis. In addition, the premolars are generally smaller,

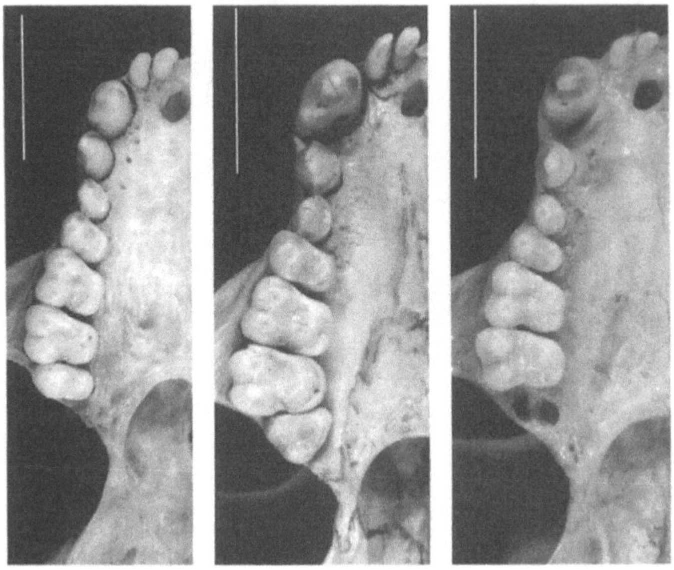

Figure 6. Close-up of occlusal morphology of upper dentition of "textbook" potto (AMZ-AS 1868; left), Zürich (Morph A) potto (AMZ 7191; middle), and Zürich (Morph B) potto (AMZ 6620, right). Note differences in orientiation of upper canine, as well as premolar and molar morphology (especially lingual cingulum of Morph A and the small hypocone of Morph B pottos). See text for further discussion. Bar = 1 cm. (© J. Schwartz)

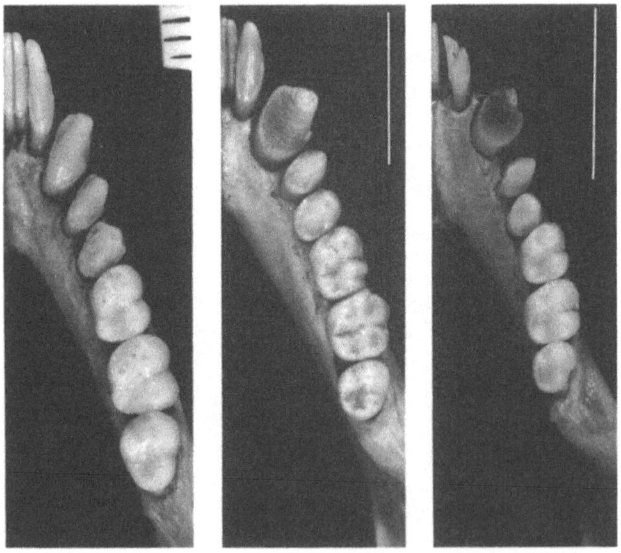

Figure 7. Close-up of occlusal morphology of lower dentition of "textbook" potto (AMZ-AS 1868; left), Zürich (Morph A) potto (AMZ 7191; middle), and Zürich (Morph B) potto (AMZ 6620, right). Note difference in P_2 height in textbook potto. See text for further discussion. Scale is in mm; bar = 1 cm. (© J. Schwartz)

P^{2-3} are subequal in size, and the upper incisors and canine are noticeably more slender. Morph E (n = 1) has a straighter occipit (when viewed from above) and M^{1-2} are narrower mesiodistally, with M^1 being "waisted" and its hypocone being slender and distolingually distended.

Table 4. *Perodicticus potto:* Ranked, unpaired, two-tailed t-test of hypothesis that males and females are from the same population.* Measurements in mm.

Rank	Measurement	Probability	Male $\bar{x}$	n	Female $\bar{x}$	n
1.	OD	.9958	9.24	14	9.24	8
2.	lM_1W	.9807	3.46	15	3.46	7
3.	rM^2W	.98	5.39	15	5.4	7
4.	rM_2L	.9753	4.26	15	4.26	7
5.	BC	.9707	17.39	14	17.37	8
6.	lM^1W	.9364	5.04	15	5.03	7
7.	lM_2W	.8939	3.98	15	4.00	6
8.	Biorb	.8826	39.98	14	39.89	8
9.	rM^1L	.806	4.95	15	4.9	7
10.	TL	.7405	14.39	14	14.71	8
11.	rM^1L	.7249	4.15	15	4.19	7
12.	lM^2W	.643	5.45	15	5.37	7
13.	lM^2L	.5549	4.01	15	3.94	7
14.	rM^2L	.5357	4.04	15	3.97	7
15.	lM^1L	.4532	4.15	15	4.09	7
16.	NL	.3896	16.4	14	17.02	8
17.	rM_2W	.3181	3.86	15	3.97	7
18.	BP	.2811	37.13	14	37.91	8
19.	lM_2L	.2585	4.2	15	4.3	6
20.	Hum(r)	.2399	70.92	12	72.71	8
21.	rM_1W	.2043	3.39	15	3.27	7
22.	lM_1L	.1719	4.04	15	3.93	7
23.	rM_1L	.1645	4.07	15	3.9	7
24.	L	.0448	63.07	14	64.63	8

*23 measurements > 5% overlap; 1 measurement < 5% overlap.

As with the *Arctocebus* sample, morphological differences between male and female *P. potto* in either cranial or dental features were not visually detectable. Nevertheless, the entire sample of Zürich potto was subjected to an unpaired, two-tailed t-test to investigate whether

Table 5. *Perodicticus potto.* Length and width (in mm) and summary statistics of upper and lower first and second molars. Combined sample of specimens (male, female, and sex unknown).

		Maxillary Dentition					*Mandibular Dentition*				
		n	x̄	Range	s	CV	n	x̄	Range	s	CV
Right											
M1	L	33	4.17	3.6-4.8	.26	6.234	33	4.01	3.4-4.7	.27	6.639
	W	33	4.95	4.0-5.9	.41	8.267	33	3.32	2.3-3.9	.28	8.476
M2	L	33	4.00	3.0-4.5	.31	7.787	33	4.22	3.4-4.8	.27	6.344
	W	33	5.37	3.9-6.2	.55	10.173	33	3.88	2.6-4.5	.35	8.985
Left											
M1	L	34	4.13	3.6-4.8	.04	6.325	34	3.99	3.6-4.7	.22	5.407
	W	34	4.98	4.0-5.7	.06	7.469	34	3.38	2.5-4.2	.33	9.669
M2	L	34	3.96	3.0-4.5	.32	8.069	32	4.18	3.6-4.8	.26	6.297
	W	34	5.35	4.0-6.2	.47	8.849	32	3.92	2.6-4.7	.44	11.096

Table 6. *Perodicticus potto.* Cranial and right humeral measurements (in mm) and summary statistics. Combined sample (male, female, sex unknown).

	n	x̄	Range	s	CV
Right humeral length	31	71.23	57.6-79.7	4.11	5.767
Cranial length	35	63.89	59.3-77.6	2.93	4.586
Nasal length	34	16.58	12.8-19.0	1.52	9.178
Biorbital breadth	35	39.93	36.7-42.1	1.36	3.413
Interorbital breadth	35	9.30	7.3-11.0	.87	9.396
Biporion breadth	35	37.57	33.8-43.2	2.03	5.41
Bicanine breadth	34	17.29	15.1-20.9	1.06	6.139
Min. temporal line sep.	35	15.10	10.4-19.3	2.04	13.48

any metric variability could be attributed to sexual dimorphism rather than potential taxonomic dimorphism. As can be seen in Table 4, skull length was the only measurement that could reliably distinguish females from males. with skulls of females being slightly longer than those of males.

With regard to lengths and widths of upper and lower M1-2s, CVs for the pooled Zürich potto sample consistently ranged above 5 (Table 5). This is in contrast to the *Arctocebus* sample in which CVs for dental measurements all fell below 5 (Table 2). Similar to *Arctocebus*, the Zürich potto sample had low CVs for skull length and biorbital breadth (CV < 5) but otherwise it was characterized by having CVs that were consistently higher in almost every other case (Table 6). In most cases, CVs for dental measurements calculated separately for Morph A and Morph B (Table 7) were lower than for the pooled sample (Table 5). CVs for dental measurements were usually lower for Morph B than Morph A, although, with the exception of length of lower left M1, CVs for *Arctocebus* were always lower (Table 2). With regard to cranial and humeral measurements, Morph B CVs were consistently lower than those for the pooled Zürich potto sample, whereas those for Morph A often were considerably higher (Table 8). These data suggest that Morph B, at least, might be a distinct entity. When, however, Morph A is compared to other Zürich potto in an unpaired, two-sample t-test, its distinctiveness (i.e. belonging to a different population) is more strongly indicated (for 16 of 24 measurements, $p < .05$) (Table 9). In contrast, in a similar test, only three measurements hinted at the possibility that Morph B represented a different population ($p < .10$) (Table 10). With the exception of nasal length, Morph A was on average larger than other Zürich potto, as had been suspected from visual observation.

Table 7. *Perodicticus potto:* Length and width (in mm) and summary statistics of upper and lower first and second molars of Morphs A and B. Combined samples (male, female, sex unknown).

		Maxillary Dentition					Mandibular Dentition				
		n	x̄	Range	s	CV	n	x̄	Range	s	CV
Morph A											
Right											
M1	L	8	4.38	3.9-4.8	.31	7.098	8	4.2	3.9-4.7	.28	6.734
	W	8	5.27	4.7-5.9	.44	8.341	8	3.54	3.1-3.9	.24	6.747
M2	L	8	4.26	4.0-4.5	.21	5.006	8	4.39	4.2-4.8	.20	4.629
	W	8	5.76	5.3-6.2	.38	6.62	8	4.14	3.7-4.5	.29	6.95
Left											
M1	L	8	4.31	3.9-4.8	.36	8.263	8	4.18	3.9-4.7	.26	6.239
	W	8	5.19	4.7-5.7	.39	7.532	8	3.59	3.3-3.9	.20	5.661
M2	L	8	4.14	3.9-4.5	.22	5.317	7	4.4	4.0-4.8	.26	5.868
	W	8	5.67	5.1-6.2	.42	7.461	7	4.37	4.0-4.7	.30	6.953
Morph B											
Right:											
M1	L	18	4.13	3.9-4.3	.17	4.139	18	3.96	3.4-4.3	.24	5.945
	W	18	4.85	4.0-5.4	.35	7.151	18	3.3	2.8-3.6	.18	5.401
M2	L	18	3.94	3.6-4.3	.19	4.937	18	4.22	3.9-4.7	.19	4.399
	W	18	5.21	3.9-6.1	.50	9.626	18	3.85	3.4-4.2	.24	6.315
Left:											
M1	L	18	4.1	3.7-4.3	.17	4.098	18	3.96	3.6-4.3	.16	4.165
	W	18	4.98	4.5-5.4	.25	5.089	18	3.41	3.1-4.2	.25	7.313
M2	L	18	3.95	3.4-4.3	.23	5.776	17	4.15	3.9-4.5	.16	3.912
	W	18	5.32	4.8-6.2	.35	6.65	17	3.9	3.6-4.2	.16	4.154

Table 8. *Perodicticus potto:* Cranial and right humeral measurements (in mm) and summary statistics for Morphs A and B. Combined samples (male, female, and sex unknown).

	n	x̄	Range	s	CV
Morph A					
Right humeral length	9	72.23	68.4-79.7	4.03	5.574
Cranial length	10	64.69	62.0-77.6	4.71	7.284
Nasal length	10	16.34	13.7-19.0	1.68	10.278
Biorbital breadth	10	40.37	38.0-41.7	1.14	2.812
Interorbital breadth	10	9.42	7.5-11.0	1.19	12.608
Biporion breadth	10	38.66	35.5-43.2	2.58	6.663
Bicanine breadth	10	18.0	16.5-20.9	1.19	6.61
Min. temporal line sep.	10	15.12	10.4-19.2	2.46	16.26
Morph B					
Right humeral length	16	70.91	67.0-75.7	2.59	3.657
Cranial length	17	63.46	60.2-65.9	1.37	2.161
Nasal length	16	16.78	14.2-18.6	1.42	8.463
Biorbital breadth	17	39.6	37.5-42.0	1.29	3.261
Interorbital breadth	17	9.13	7.9-9.8	.60	6.547
Biporion breadth	17	36.98	35.0-38.8	1.20	3.254
Bicanine breadth	17	16.97	15.5-18.3	.82	4.822
Min. temporal line sep.	17	14.84	11.0-19.3	2.18	14.724

Table 9. *Perodicticus* Morph A versus Other *P. potto:* Ranked, unpaired, two-sample t-test of hypothesis that Morph A and other *P. potto* are from the same population. Measurements in mm.

Rank	Measurement	Probability	A $\bar{x}$	n	Other $\bar{x}$	n
1.	lM_2W	.0011	.43	7	.38	25
2.	lM_1L	.0049	.418	8	.394	26
2.	rM^2L	.0049	.426	8	.392	25
4.	rM^1W	.008	.527	8	.485	25
5.	rM^1L	.0092	.438	8	.411	25
6.	BC	.0097	1.8	10	1.7	24
7.	lM_2L	.0104	.44	7	.412	25
8.	rM_1W	.0115	.354	8	.326	25
9.	rM_2W	.0121	.414	8	.379	25
10.	rM^2W	.0169	.576	8	.524	25
11.	rM_1L	.0172	.42	8	.395	25
12.	lM^1L	.0233	.431	8	.408	26
13.	lM^2W	.0254	.567	8	.525	26
14.	lM_1W	.0405	.359	8	.332	26
15.	rM_2L	.0416	.439	8	.417	25
16.	BP	.0426	3.866	10	3.713	25
17.	lM^1W	.074	.519	8	.492	26
18.	lM^2L	.0804	.414	8	.391	26
19.	Biorb	.234	4.037	10	3.976	25
20.	L	.3167	6.469	10	6.358	25
21.	Hum(r)	.393	7.223	9	7.082	22
22.	NL	.5547	1.634	10	1.669	24
23.	OD	.6147	.942	10	.925	25
24.	TL	.9754	1.512	10	1.51	25

Table 10. *Perodicticus* Morph B versus other *P. potto.* Ranked, unpaired, two-sample t-test of hypothesis that Morph B and other *P. potto* are from the same population. Ranks with more than a 36% chance of being from the same population are not listed. Measurements in mm.

Rank	Measurement	Probability	B $\bar{x}$	n	Other $\bar{x}$	n
1.	rM^2W	.0668	.521	18	.556	15
2.	BC	.0778	1.697	17	1.761	17
3.	BP	.0977	3.698	17	3.812	18
4.	rM^1W	.1202	.485	18	.507	15
5.	Biorb	.1653	3.96	17	4.024	18
6.	rM^2L	.2001	.394	18	.408	15
7.	rM_1L	.2632	.396	18	.407	15
8.	OD	.2679	.913	17	.946	18
9.	rM^1L	.2839	.413	18	.423	15
10.	lM_1L	.3524	.396	18	.403	16

III. Nycticebus

Insofar as the impetus for this systematic reinvestigation of extant lorisids arose from observations of gross differences among specimens of *N. coucang,* we have reinitiated the study of thie genus. We are still in the process of gathering metric and detailed morphological dental data, but we can here discuss some metric dental data (n = 19) as well as cranial data for *N. coucang* (n = 45 for morphological study; n = 37 for metric analysis) and *N. pygmaeus* (n = 9 for both morphological and metric cranial analyses)

Although the genus *Nycticebus* is often cited as being highly variable, it is difficult to tease from the literature any definitive diagnostic information by which one could identify any particular specimen as belonging to the taxon. It usually is the case that *Nycticebus* is compared to *Loris*, the only other Asian primate with a six-toothed toothcombed, and distinguished from the latter taxon primarily because of its different geographic distribution and

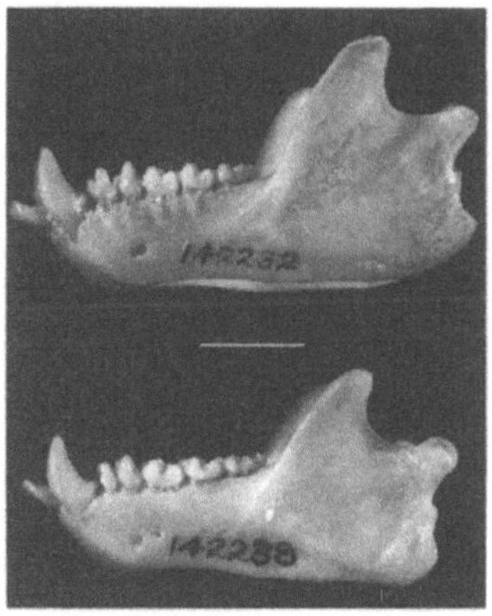

Figure 8. Mandibles of *Nycticebus (N. coucang borneanus).* Above: NMNH 142232; below: NMNH 142238. Note variably developed postcondylar tubercle. Bar = 1 cm. (© J. Schwartz)

the fact that it is larger, skeletally more robust, and dentally less crisply cusped and crested. In pelage, *Loris* and *Nycticebus* appear to be primitively similar to one another in that their fur is generally brown to ruddy and speckled with gray and their eyes are circum-scribed by a band of dark hair (Schwartz, pers. observ.). Differences between individuals attributed to *Nycti-cebus* in pelage, number of incisors, or any other feature are thus relevant only within the context of having accepted the taxonomic reality of the larger taxon. If, however, a diagnosis is supposed to convey something distinctive about a potential taxon—i.e. something autapomorphic—we must strive first to identify at least one feature that distinguishes a morph (which we might call *Nycticebus)* from other lorisids.

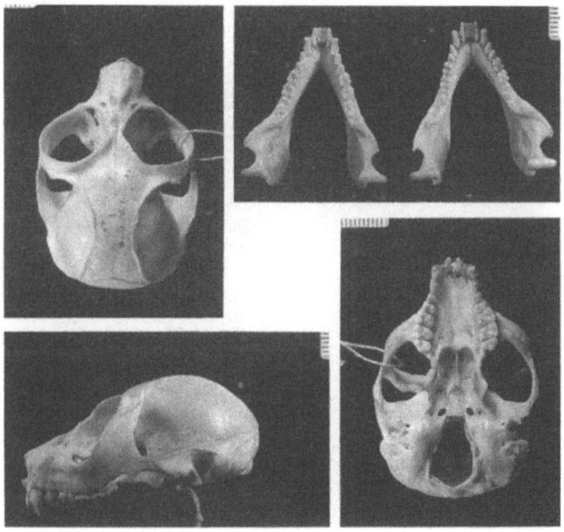

Figure 9. *Nycticebus pygmaeus.* The mandible closer to the scale is BM(NH) 26.10.4.9. All other illustrations are of BM(NH) 10.3.10.1. See text for discussion. Scale is in mm. (© J. Schwartz)

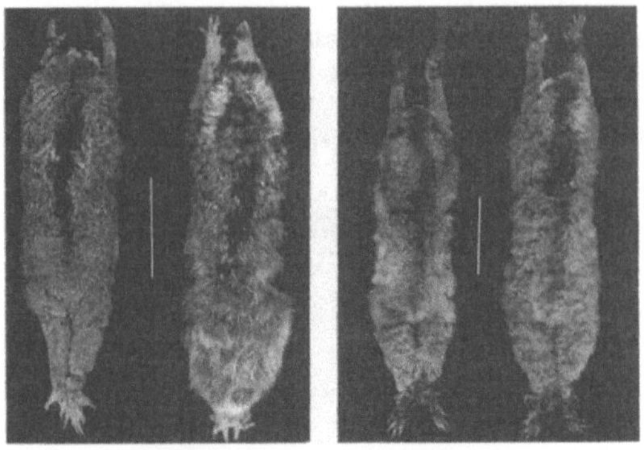

Figure 10. Skins of *Nycticebus.* Left pair (bar = 10 cm): *N. c. javanicus* (left; NMNH 521836) and *N. pygmaeus* (right; NMNH 256913). Right pair (bar = 10 cm): *N. c. borneanus* (left, female, NMNH 142238; right, male, NMNH 142232). (© J. Schwartz)

In comparison to *Loris, Perodicticus,* and *Arctocebus, Nycticebus* is distinctive in having the deepest mandible and the most markedly enlarged and expanded goneal region. Among the four genera, *Perodicticus* is distinguished by its somewhat "puffy" or inflated snout, which is produced by stout upper canine roots, while *Nycticebus* is delineated further by having a markedly "puffy" snout and stouter canine roots. Uniquely among lorisids (and, apparently among all primates with the exception of the orangutan and *Sivapithecus), Nycticebus* possesses a rugose goneal region that is surmounted by a postcondylar rugosity, which, in some individuals, is enlarged into a tubercle (Fig. 8).

This, of course, should be taken only as a preliminary list of autapomorphies that might delineate the taxon *Nycticebus.* But if they are accepted as defining the genus, we can then proceed to the next question: Are *N. coucang* and *N. pygmaeus* "real" species?

In terms of the craniodental features cited by Groves (1971) as being distinctive of *N. pygmaeus* (Fig. 9), it would appear from comparison with other lorisids, with galagids, and more broadly with other strepsirhines, that a long snout (= anteriorly protruding premaxilla), a long palate (extending to the distal side of M^3), and large M^3's (subequal upper molars) are primitive features for the clade (e.g. compare illustrations in Schwartz and Tattersall, 1985). *N. pygmaeus* also appears to retain a primitively colored and patterned pelage (Fig. 10).

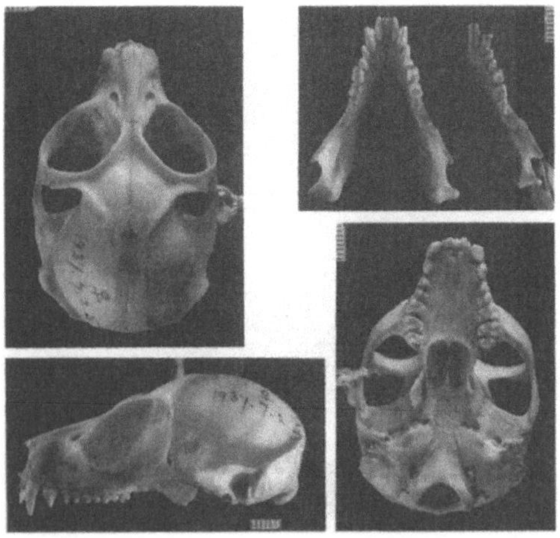

Figure 11. *Nycticebus coucang* (Java). The half-mandible closer to the scale is BM(NH) 67.4..12.75. All other illustrations are of BM(NH) 1937.2.5. Only I^1 is present in this *N. coucang.* See text for further discussion. Scale is in mm. (© J. Schwartz)

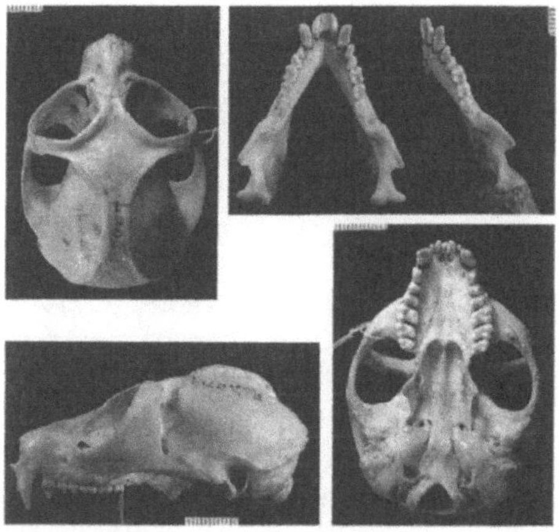

Figure 12. *Nycticebus coucang* (Laos, Thailand, and Assam). Clockwise from upper left: skull (Laos, BM[NH] 26.10.4.8), mandible (Assam, BM[NH] 43.63), half-mandible (Laos, BM[NH] 26.20.4.8), basicranium (Laos, BM[NH] 26.20.4.8), and skull (lateral view, Thailand, BM[NH] 14.6.18.7; note unusual sagittal crest). Note primitive number of upper incisors but deep nasal fossa. See text for further discussion. Scale is in mm. (© J. Schwartz)

The development of diastemata on either side of the upper canine is difficult to sort out in terms of polarity. It could be argued that these features are primitive as much as they are derived In accordance with Groves's metric observations, our sample of specimens attributed to *N. pygmaeus* is on average smaller for all cranial measurements than the sample of specimens attributed to *N. coucang* (Table 11). We are not able to comment on larger ear size or occipital flattening in *N. pygmaeus*.

Distinguishing a "*coucang*" subclade within *Nycticebus* on the basis of anything but larger size is difficult. Groves (1971) suggested that *N. coucang* could be characterized as having a diastema between the upper canine and P^2 but not between the upper canine and I^2. However, the latter configuration would appear to be primitive in that *N. pygmaeus* and a plethora of strepsirhines also lack a diastema between the upper canine and I^2. Although some *N. coucang* lack a diastema between the upper canine and I^2, those that have thin I^2's

Table 11. *Nycticebus coucang* and *N. pygmaeus:* Cranial measurements (in mm) and summary statistics. Combined samples (male, female, and sex unknown).

	n	x̄	Range	s	CV
N. coucang					
Skull length	37	60.01	54.4-69.8	4.75	7.907
Nasal length	35	15.51	12.6-20.6	2.07	13.351
Biorbital breadth	34	37.94	34.7-43.5	2.31	6.085
Interorbital breadth	37	4.38	1.9-6.0	.81	18.411
Biporion breadth	37	35.79	32.0-40.3	.37	6.307
Bicanine breadth	37	13.37	11.4-20.0	1.57	11.713
Min. temporal line sep.	35	6.27	.3-12.0	2.94	46.943
N. pygmaeus					
Skull length	9	51.87	48.3-54.9	2.54	4.893
Nasal length	8	12.89	10.6-15.5	1.85	14.388
Biorbital breadth	9	34.23	32.2-36.6	1.54	4.49
Interorbital breadth	9	2.92	2.4-3.6	.47	16.212
Biporion breadth	9	32.19	30.1-34.5	1.67	5.199
Bicanine breadth	9	10.54	9.8-11.6	.48	4.526
Min. temporal line sep.	9	11.26	8.2-13.7	2.02	17.941

Table 12. *Nycticebus coucang.* Ranked, unpaired, two-tailed t-test of hypothesis that males and females are from the same population.* Measurements in mm.

Rank	Measurement	Probability	Male $\bar{x}$	n	Female $\bar{x}$	n
1.	Biorb	.6114	38.18	13	37.7	11
2.	BC	.5716	13.07	13	13.3	13
3.	OD	.5483	4.55	13	4.36	13
4.	BP	.5365	35.34	13	35.86	13
5.	L	.4993	59.47	13	60.72	13
6.	NL	.446	15.04	12	15.65	12
7.	TL	.2361	6.59	13	5.11	12

*All measurements > 23% overlap.and those lacking this tooth altogether could be described as having a diastema in front of the

upper canine. The posterior extent of the palate also varies among specimens, as does the length of the dorsal stripe. We agree with Groves (1971) that having reduced M^3's distinguishes *N. coucang* from *N. pygmaeus.* We would also include as distinctive of *N. coucang* reduced M_3's, which are primitively long in *N. pygmaeus* (as in the majority of extant strepsirhines/prosimians) but shorter in *N. coucang* because the hypoconulid is either smaller or absent.

Table 13. *Nycticebus coucang.* Dental measurements (in mm) and summary statistics for all specimens (i.e. all subspecies of *N. coucang).* Combined sample (male, female, and specimens of unknown sex).

		Maxillary Dentition					*Mandibular Dentition*				
		n	$\bar{x}$	Range	s	CV	n	$\bar{x}$	Range	s	CV
Right											
C	L	18	2.84	2.3-3.8	.39	13.662					
	W	18	1.95	1.6-2.6	.28	14.122					
P2	L	17	2.18	1.3-2.6	.29	13.306	18	2.63	2.1-2.9	.21	7.911
	W	17	1.70	1.5-1.9	.12	7.139	18	1.75	1.5-2.1	.16	8.935
P3	L	17	1.57	1.3-2.0	.19	11.954	16	1.91	1.4-2.7	.34	17.818
	W	17	1.86	1.2-2.5	.30	15.835	16	1.03	0.8-1.3	.16	15.535
P4	L	18	1.81	1.5-2.2	.24	13.022	18	1.98	1.5-2.4	.25	12.658
	W	18	2.36	2.0-3.1	.25	10.768	18	1.31	1.1-1.6	.14	10.751
M1	L	17	2.63	2.1-3.1	.26	9.923	18	2.28	1.8-2.7	.21	9.351
	W	17	3.12	2.4-3.9	.38	12.086	18	2.05	1.8-2.5	.20	9.729
M2	L	17	2.36	1.9-2.9	.23	9.746	18	2.34	1.9-2.7	.21	8.809
	W	17	3.16	2.6-3.8	.33	10.384	18	2.02	1.7-2.5	.20	9.729
M3	L	16	1.85	1.6-2.1	.14	7.822	13	2.55	2.1-3.0	.22	8.762
	W	16	2.59	2.3-3.2	.24	9.322	13	1.72	1.5-2.1	.16	9.180
Left											
C	L	18	2.79	2.3-3.9	.38	14.013					
	W	18	2.01	1.6-2.7	.28	14.051					
P2	L	18	2.31	2.0-2.6	.19	8.187	17	2.64	2.5-2.9	.11	4.083
	W	18	1.66	1.5-1.9	.12	7.085	17	1.89	1.6-2.9	.30	15.935
P3	L	17	1.56	1.2-2.0	.19	12.331	18	1.83	1.4-2.5	.25	13.610
	W	17	1.83	1.2-2.3	.28	15.262	18	1.14	0.8-1.9	.29	25.556
P4	L	18	1.78	1.5-2.1	.18	10.133	17	1.89	1.7-2.2	.17	9.113
	W	18	2.40	1.8-3.0	.26	11.029	17	1.36	1.0-1.6	.15	11.113
M1	L	18	2.64	2.2-3.1	.25	9.346	18	2.27	1.9-2.7	.20	8.952
	W	18	3.13	2.4-3.7	.29	9.299	18	1.96	1.2-2.4	.28	14.077
M2	L	17	2.41	2.0-2.8	.20	8.440	18	2.30	1.9-2.7	.22	9.415
	W	17	3.09	2.5-3.7	.28	9.048	18	2.00	1.6-2.4	.21	10.363
M3	L	16	1.83	1.6-2.2	.17	9.133	13	2.44	1.9-2.8	.25	10.340
	W	16	2.58	2.2-3.1	.21	8.216	13	1.74	1.5-2.3	.24	13.763

Table 14. *Nycticebus coucang.* Ranked, unpaired, two-tailed t-test of hypothesis that males and females are from the same population; (all subspecies combined). Ranks above 32% not included. Measurements in mm.

Rank	Measurement	Probability	Male $\bar{x}$	n	Female $\bar{x}$	n
1.	lP_4W	.0140	1.24	6	1.42	9
2.	lM_1W	.0368	1.77	7	2.06	9
3.	lM_1L	.0598	2.15	7	2.35	9
4.	lP^3L	.0678	1.45	6	1.65	9
5.	rP_4L	.0719	1.91	7	2.11	9
6.	lP_2W	.0808	2.05	7	1.77	8
7.	rM_1L	.1412	2.20	7	2.36	9
8.	rM^2W	.1465	3.03	7	3.28	9
9.	rP_3W	.1850	0.95	5	1.08	9
10.	lP_4L	.2029	1.84	6	1.96	9
11.	lP^2W	.2128	1.70	7	1.62	9
12.	lP^4W	.2241	2.32	7	2.49	9
13.	lP_2L	.2455	2.69	7	2.62	8
14.	lM^1W	.2553	3.04	7	3.22	9
15.	lM^2W	.2588	3.00	7	3.17	9
16.	rP_3W	.2608	1.25	7	1.34	9
17.	lM^1L	.2683	2.56	7	2.71	9
18.	rM^1L	.2735	2.54	6	2.69	9
19.	rM^1W	.3006	2.99	6	3.22	9
20.	rM^3W	.3168	2.53	6	2.66	9

With regard to overall pelage coloration, Javanese and Bornean *N. coucang* have the lightest brown hair and Sumatran *N. coucang* the darkest. Perhaps each intensity represents a derived state relative to the generally present brown to ruddy color. *N. coucang* from mainland southeast Asia (Thailand, Laos, Vietnam, Burma, and Assam) are characterized by having silver/gray hair at least along the flanks, arms, and neck—which might represent a derived condition. *N. coucang* from Malaysia are similar to *N. pygmaeus* in retaining primitively ruddy brown hair sprinkled with gray flecks (Fig. 10).

The possibility that subgroups which are hinted at by pelage patterning constitute real subclades within *N. coucang* is indicated further by various craniodental features. For example, as pointed out earlier, some *N. coucang* erupt only one pair of permanent upper incisors; the tooth missing is I^2. This seemingly derived state for incisor number is seen in specimens from Java and Borneo. Sumatran *N. coucang* do erupt a lateral upper incisor, but, compared to the same tooth in other, especially mainland, *Nycticebus*, this I^2 is extremely slender. As such, we might hypothesize a morphocline within *N. coucang* of I^2 size reduction-to-absence, which, in turn, leads to the suggestion that Sumatran *N. coucang* could be the sister taxon of a Javanese-Bornean sister group. This potential three-morph clade within *N. coucang* is indicated further by the observation that specimens from these three islands are distinguished from other *N. coucang* (and strepsirhines, in general) in that their infralateral orbital margins are conspicuously more laterally flared (Fig. 11).

Another feature of possible significance in delineating morphs within *N. coucang* is the development on the "face" of a deep depression or fossa anterior to the lacrimal canal; we shall refer to this as the nasal fossa. Because a nasal fossa of shallow to moderate depth is common among strepsirhines, similarly configured *N. coucang* would be primitive in this regard. In contrast, a distinctly deep nasal fossa would emerge as potentially derived and, when shared by at least two groups, suggestive of their relatedness. This latter condition is present in *N. coucang* from Laos, Burma, and Assam. It is thus of interest to note that specimens from Laos and Assam, at least, are distinguished among *N. coucang* in that the large hypocone on M^2 is narrow and essentially extends directly lingually, rather than being broad and characteristically distended distolingually (Fig. 12).

To the extent of our preliminary assessment of *Nycticebus,* it appears that reevaluation

Table 15. *Nycticebus coucang coucang:.* Ranked, unpaired, two-tailed t-test of hypothesis that males and females are from the same population. Ranks above 28% not included. Measurements in mm.

Rank	Measurement	Probability	Male $\bar{x}$	n	Female $\bar{x}$	n
1.	lP^2W	.0413	1.77	2	1.59	4
2.	rP^4W	.0444	2.18	2	2.47	4
3.	lP^4W	.0528	2.29	2	2.55	4
4.	rP_3L	.0733	2.46	1	2.00	4
5.	lP_4L	.0771	1.83	2	1.97	4
6.	rM^1W	.0773	2.92	1	3.19	4
7.	rP_4W	.0890	1.15	2	1.30	4
8.	rP_4L	.1105	1.88	2	2.17	4
9.	lP_2L	.1130	2.82	2	2.61	4
10.	lM^1W	.1237	3.12	2	3.23	4
11.	rM^2W	.1336	3.05	2	3.21	4
12.	rP^2L	.1374	2.44	2	2.25	4
13.	rM_3L	.1435	2.81	2	2.54	4
14.	lP_2W	.1607	1.88	2	1.72	4
15.	rM^1L	.1819	2.53	1	2.70	4
16.	rP^4L	.2110	1.66	2	1.89	4
17.	lP_4W	.2200	1.26	2	1.38	4
18.	rM^3W	.2483	2.47	2	2.66	4
19.	rM_3W	.2678	1.67	2	1.77	4
20.	rP_2W	.2717	1.84	2	1.68	4

of the taxonomic unity of at least *N. coucang* is in order. On morphological grounds, a few, seemingly derived features (pelage coloration, depth of nasal fossa, M^2 hypocone distension), may be seen as reflecting the existence of a clade within *N. coucang* that consists of specimens from Laos, Assam, and Burma. Other features (I^2 size reduction, more marked infralateral orbital flare) distinguish Sumatran, Bornean, and Javanese *N. coucang.* Within this latter potential clade, a sister group relationship between Bornean and Javanese *N. coucang* is indicated by the absence (= developmental suppression) of I^2 and the development of the lightest brown hair.

Although Groves (1971) more than adequately demonstrated the high degree of variability in *N. coucang,* we calculated CVs for the cranial measurements on our sample (Table 11). All CVs were above 5, and CVs for three measurements (not including minimum separation of temporal lines) were above 10. An unpaired, two-tailed t-test of the cranial measurements using specimens of known sex indicates that it is highly unlikely that sexual dimorphism is contributing to this high degree of variability within the sample (Table 12).

Coefficients of variation on dental measurements (buccolingual widths and mesiodistal lengths of upper C-M3 and lower P2-M3) also reflect a high degree of variability within *N. coucang.* (Table 13). For example, of these 52 dental measurements, the CV of only one (mesiodistal length of left P_2) fell below 5; and CVs for 47 measurements (90%) were higher than 8. The effects of sexual dimorphism on the variability in the sample was tested by an unpaired, two-tailed t-test using specimens of known sex (Table 14): only two of the 52 measurements (buccolingual widths of left P_4 and left M_1) appear to distinguish males from females ($p < .05$). The negligible expression of sexual dimorphism (at least in dental metrics) in *N. coucang* was tested further by an unpaired, two-tailed t-test on an *N. c. coucang* subset of the sample (Table 15). Although the number of males and females in this subset is quite small, the results were comparable to those based on the entire *N. coucang* sample: only two of the 52 tooth measurements (in this case, buccolingual widths of left P^2 and right P^4) appear to distinguish males from females.

Conclusion

Although the preceding represents an initial investigation into the question of species diversity among extant lorisids, the results appear to be suggestive. There seems to be considerable morphological as well as metric variability within some species currently recognized. This variability does not appear to be correlated with sexual dimorphism, which is negligibly expressed in the samples analyzed here.

Groves (1971) demonstrated a high degree of variability within *N. coucang,* but he argued that this was not an artifact of lumping several species together. Although he did focus in a similar fashion on the variability of *N. pygmaeus,* Groves did subsume in this species *N. intermedius,* which Dao (1960) had suggested was a viable species from northern Vietnam. Recently, Zhang *et al* (1993) studied mitochondrial (mt) DNA restriction site variability in *Nycticebus.* They found that even when three species—*N. coucang, N. pygmaeus,* and *N. intermedius*—were recognized, there was a large amount of variability (polymorphism) in each. Nonetheless, because they thought that their reconstructed date of separation between *N. intermedius* and *N. pygmaeus* was too recent (0.5 Ma) for speciation to have occurred, Zhang *et al* (1993) concluded that *N. intermedius* should not be recognized as a separate species. Thus *N. coucang* and *N. pygmaeus* are species that are remarkable for their variability.

Inasmuch as a major factor in our study was the discovery that *Arctocebus calabarensis* was minimally variable both morphologically and metrically, and sexual dimorphism was not significantly expressed in this taxon, we investigated the issue of variability from the perspective of chance. We asked the question, "Compared to *A. calabarensis,* is the variability that we see in *P. potto* and *N. coucang* likely to have been random (i.e. due to chance)?" We

Table 16. Summary Table of Coefficients of Variation for Humeral, Cranial, and Dental Data.

	P. potto	Potto(A)	Potto(B)	Arcto.	Nyct. all	N. couc.	N. couc.	N. pyg.
n=	31-35	7-10	16-18	21-29	43-46	34-37	13-18	8-9
Hum(r)	5.767	5.574	3.657	6.081				
L	4.586	7.284	2.161	3.029	9.349	7.907		4.893
NL	9.178	10.278	8.463	7.305	15.047	13.351		14.388
Biorb	3.413	2.812	3.261	4.189	7.099	6.085		4.490
OD	9.396	12.608	6.547	6.401	23.211	18.411		16.212
BP	5.410	6.663	3.254	3.384	15.435	16.475		5.199
BC	6.139	6.610	4.822	4.718	14.141	11.713		4.526
TL	13.480	16.260	14.724	10.506	47.022	46.943		17.941
rM^1L	6.234	7.098	4.139	5.412			9.923	
rM^1W	8.267	8.341	7.151	4.655			12.086	
rM^2L	7.787	5.006	4.937	4.090			9.746	
rM^2W	10.173	6.620	9.626	4.871			10.384	
lM^1L	6.325	8.263	4.098	3.617			9.346	
lM^1W	7.469	7.532	5.089	3.980			9.299	
lM^2L	8.069	5.317	5.776	4.229			8.440	
lM^2W	8.849	7.461	6.650	4.027			9.048	
rM_1L	6.639	6.734	5.945	3.583			9.351	
rM_1W	8.476	6.747	5.401	4.951			9.729	
rM_2L	6.344	4.629	4.399	3.489			8.809	
rM_2W	8.985	6.95	6.315	4.497			9.729	
lM_1L	5.407	6.239	4.165	4.377			8.952	
lM_1W	9.669	5.661	7.313	4.769			14.077	
lM_2L	6.297	5.868	3.912	4.084			9.415	
lM_2W	11.096	6.953	4.154	3.963			10.363	

employed an approach developed by Gaulin and Robbins (1991: 67) to test "[t]he chance of exactly s successes in t trials given only two possible outcomes, success and failure, and a p chance of success on any given trial." For our purposes here, we substituted "higher CV" for "success," "lower CV" for "failure," and "measurement" for "trial." We used the following formula (Gaulin and Robbins, 1991):

$$(t!/s![t-s]!)p^s(1-p)^{(t-s)}$$

where s = number of higher CVs, t = number of measurements, and p = probability (chance) of a higher CV for any given measurement. According to Gaulin and Robbins (1991: 67), "To compute the chance of at least s successes [higher CVs], this formula must be iterated for s, s+1, s+2 [to] t successes, and the results of these iterations must be added."

Table 17. Relative Ranking of Coefficients of Variation for Dental Measurements.

	P. potto (all)	Potto (A)	Potto (B)	*Arctocebus*
Right:				
M^1W	2	1	3	4
M^1L	2	1	4	3
M^2W	1	3	2	4
M^2L	1	2	3	4
M_1W	1	2	3	4
M_1L	2	1	3	4
M_2W	1	2	3	4
M_2L	1	2	3	4
Left:				
M^1W	2	1	3	4
M^1L	2	1	3	4
M^2W	1	2	3	4
M^2L	1	3	2	4
M_1W	1	3	2	4
M_1L	2	1	4	3
M_2W	1	2	3	4
M_2L	1	2	4	3

To begin, we chose to compare the CVs for dental measurements of the combined *P. potto* sample, the Morph A potto sample, and the Morph B potto sample to the *A. calabarensis* sample. Table 16 summarizes the CVs for cranial, humeral, and dental measurements. Thus, for any dental measurement, four CVs are compared and ranked relative to one another from "highest" to "lowest." These rankings are presented in Table 17. We chose "1" to represent the highest CV (= most variation) and "4" the lowest CV (= least variation). As is obvious from the comparison, *A. calabarensis* is always less variable than the total *P. potto* sample. Using the formula above, we calculated that p = .000015258 that all 16 CVs for *P. potto* were higher than those for *A. calabarensis* due to chance (i.e. randomness) alone. Curiously, when the Morph A potto sample is considered, it, too, is more variable than the *A. calabarensis* sample. This is unlikely to have been by chance (i.e. p = .00015258 that all 16 CVs would be higher in the former than the latter due entirely to chance). The Morph B potto sample is less variable than the Morph A potto sample and the total *P. potto* sample, but somewhat more variable than the *A. calabarensis* sample. This, too, is not likely to have been by chance (p = .161391794 that 13 of 16 CVs in Morph B potto would be higher than in *A. calabarensis).* Insofar as there are discernible morphological differences between Morphs A and B, we interpret these calculations as indicating that it is reasonable to recognize each morph as a separate species within the genus *Perodicticus.* Continued study will hopefully clarify the difference between Morph A and Morph B in within-morph variability and test the suggestion also advanced here that yet another morph—the textbook potto—can be distinguished within the genus. This morph might also represent a separate species of *Perodicticus.*

Table 18. Relative Ranking of Coefficients of Variation for Cranial Measurements. See text for abbreviations.

	P. potto	*A. calabarensis*	*N. coucang*	*N. pygmaeus*
L	3	4	1	2
NL	3	4	2	1
Biorb.	4	3	1	2
OD	3	4	1	2
BP	2	4	1	3
BC	2	3	1	4
TL	3	4	2	3

Because we do not yet have comparable dental measurements for *N. pygmaeus*, our relative ranking of CVs for *Nycticebus* was based on cranial data (Table 18). As in the rankings above, *A. calabarensis* was the least variable taxon overall. Interestingly, the *P. potto* sample (which we did not subdivide into morphs for this test), although comparatively more variable than the *A. calabarensis* sample, was *less* variable than the collective *N. coucang* and *N. pygmaeus* samples. Again, compared to *A. calabarensis*, it appears unlikely that the higher degrees of variability in the other taxa is due to chance alone: i.e., $p = .015625$ that six of seven CVs would be higher in *P. potto* and *N. pygmaeus* than *A. calabarensis*, and $p = .0078125$ that all CVs would be higher in *N. coucang* than in *A. calabarensis*. In conjunction with our initial morphological study, we interpret these results as indicating that variability in *N. coucang* may be greater than would be expected of a single species. The CV rankings of *N. pygmaeus* also seem to indicate that a more rigorous investigation of the monotypy of this taxon be undertaken. In the case of mtDNA restriction site analysis, *N. coucang* and *N. pygmaeus* are, respectively, considerably more variable than Zhang *et al* (1993) would have expected of a species. Rather than let the application of hypothesized rates of mtDNA change guide the analysis, we suggest that these additional data are potentially informative with regard to species diversity within the genus *Nycticebus*.

Although the focus of this investigation has been a small group of strepsirhine primates—a group we intend to continue studying—we think that the questions raised in this contribution are generally viable systematic issues. Clearly, there is still much work to be done on the systematics of extant taxa.

Acknowledgments

We thank L. Alterman and M. K. Izard for inviting us to contribute to their symposium. For assistance and access to specimens, we are indebted to G. Musser and W. Fuchs (AMNH), R. Thorington (USNM), P. Jenkins and P. Andrews (BM[NH]), and R. D. Martin and Fr. Grüber (AMZ). S. J. C. Gaulin generously offered suggestions for analyzing the metric data as did the editors and anonymous reviewers for revising the original manuscript. Some research was supported by a Provost's faculty grant to JHS from the University of Pittsburgh. The study of the collection in Zürich was made possible by a fellowship from the Adolph Schultz fund of the Anthropological Institute and Museum, Zürich-Irchel, to JHS; thanks to all at the Institute for their assistance and especially to R. D. Martin for his continual encouragement and support of this project. A list of specimens used in this study will be made available upon request.

REFERENCES

Dao, V. T. (1960). Sur un nouvelle especes *Nycticebus au Vietnam*. *Zoologisches Anzeiger, 164*, 240-243.

Elliot, D. G. (1912). *A Review of the Primates, volume 1*. New York: The American Museum of Natural History.

Gaulin, S. J. C., & Robbins, C. J. (1991. Trivers-Willard effect in contemporary North American society. *American Journal of Physical Anthropology, 85*, 61-69.

Gingerich, P. D., & Ryan, A. S. (1979). Dental and cranial variation in living Indriidae. *Primates, 20*, 141-159.

Gray. J. E. (1821). On the natural arrangement of vertebrose animals. *London Medical Repository, 15*, 296-310.

Gray, J. E. (1825). Outline of an attempt at the disposition of the mammalia into tribes and families with a list of the genera apparently appertaining to each tribe. *Annals of Philosophy, 26, new series 10,* 337-344.

Gregory, W. K. (1922). *The Origin and Evolution of the Human Dentition.* Baltimore: Williams and Wilkins.

Groves, C. P. (1971). Systematics of the genus *Nycticebus. Proceedings of the 3rd International Congress of Primatology, Zürich, 1970, volume 1,* 44-53.

Groves, C. P. (1974). Taxonomy and phylogeny of prosimians. In R. D. Martin, G. A. Doyle & A. C. Walker (Eds.), *Prosimian Biology* (pp. 449-473). London: Duckworth Press.

Jenkins, P. D. (1987). *Catalogue of Primates in the British Museum (Natural History). Part IV: Suborder Strepsirrhini, including the subfossil Madagascan lemurs and Family Tarsiidae.* London: British Museum (natural History).

Schwartz, J. H. (1974). Dental development and eruption in the prosimians and its bearing on their evolution. Ph.D. thesis, Columbia University, New York City.

Schwartz, J. H. (1986). Primate systematics and a classification of the order. In D. R. Swindler and J. Erwin (Eds.), *Comparative Primate Biology, volume 1: Systematics, Evolution, and Anatomy* (pp. 1-41). New York: Alan R. Liss.

Schwartz, J. H., & Tattersall, I. (1985). Evolutionary relationships of living lemurs and lorises (Mammalia, Primates) and their potential affinities with European Eocene Adapidae. *Anthropological Papers of the American Museum of Natural History, 60,* 1-100.

Zhang, Y., Chen, Z. & Shi, L. 1993. Phylogeny of the slow lorises (Genus *Nycticebus):* an approach using mitochondrial DNA restriction enzyme analysis. *International Journal of Primatology, 14,* 167-175.

THE SUBSPECIES OF *LORIS TARDIGRADUS* AND THEIR CONSERVATION STATUS: A REVIEW

Helga Schulze and Bernhard Meier

Ruhr-Universität Bochum
Institut für Anatomie
Abteilung. Funktionelle Morphologie
44780 Bochum, Germany

INTRODUCTION

This paper constitutes a review of the subspecies classification of *Loris tardigradus*, their distribution, and conservation status in the field and in captivity. Present descriptions of the subspecies of the slender loris are largely based on small samples and are distinguished by minor variations in size and pelage color (Osman Hill, 1933, 1953; Jenkins, 1987). Obviously, in some cases the inaccessibility of type specimen localities has played a major role. Furthermore, according to Osman Hill and Phillips (1932), and after 13 years experience of breeding this species in captivity, there are large individual and age-related variations in coat color and body size, which therefore are not sufficient for distinguishing races or species. Some *Loris* subspecies might turn out to be identical (Jenkins, 1987) or to be representatives of a clinal variation. *L. t. tardigradus* (Sri Lanka) and *L. t. malabaricus* (India), for instance, are both small, brown and restricted to western rainforests, whereas two larger grey subspecies, *L. t. nordicus* (Sri Lanka) and *L. t. lydekkerianus* (India), are found in dry-zone shrub jungle. During the last glacial period, when India and Sri Lanka were connected, they probably belonged to only one brown and one grey form. Forms intermediate between some subspecies have been found (Osman Hill, 1933). Type and specimen locations are shown in Figs. 1 and 2. Additional data are presented in the Specimen List below.

Research with advanced methods would be necessary to verify the present taxonomy. It might show that undescribed subspecies or even different species exist within the genus *Loris*. Thus far, no behavioral differences of features such as vocalizations have been described within the genus.

Creatures of the Dark: The Nocturnal Prosimians
Edited by L. Alterman *et al.*, Plenum Press, New York, 1995

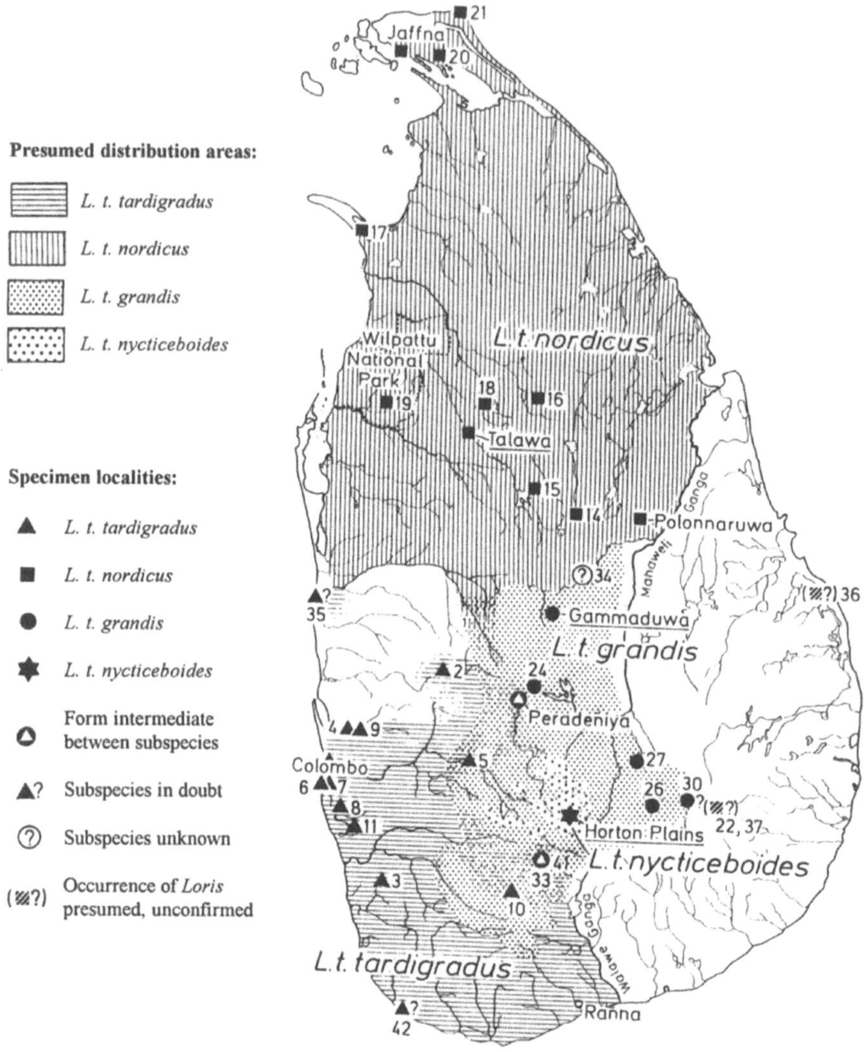

Figure 1: The distribution of slender loris subspecies in Sri Lanka is redrawn from Osman Hill (1953), and corrected with regard to specimen localities, altitude and vegetation. The numbers added to the specimen locality symbols refer to corresponding numbers in the Specimen List below. Names of type localities are underlined. Some information about *Loris* distribution in Sri Lanka by Osman Hill seems to be guesswork, based on unconfirmed verbal information, altitude, climate, vegetation and occurrence of sympatric simian primate species. During surveys in 1979 no lorises were found in the south-eastern parts of the island, east of the Mahaweli Ganga river. This result is in accordance with Osman Hill's distribution map; it was also confirmed by local people. The chief-gamewarden of Yala National Park confirmed the lack of lorises east of the Walawe Ganga river (Meier, Nieschalk, unpublished).

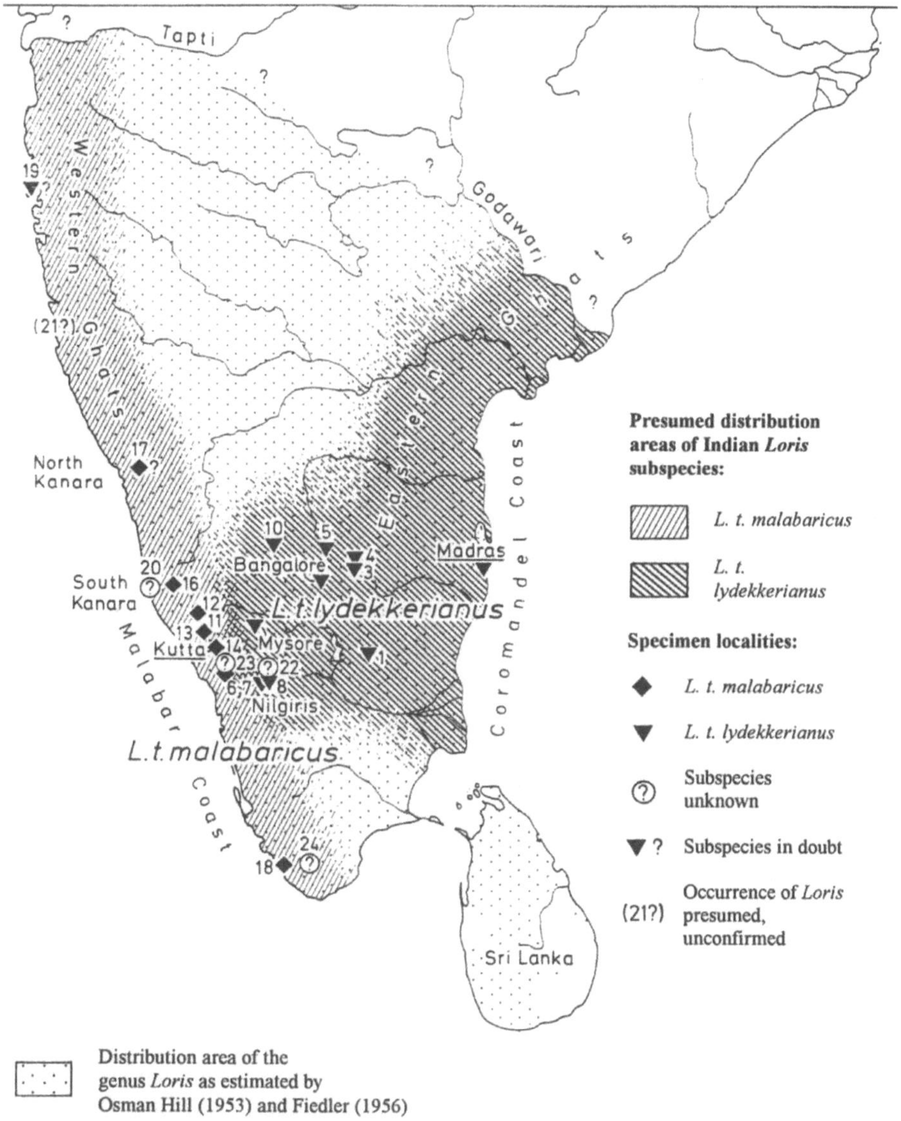

Figure 2: The distribution of slender loris subspecies in India is based on distribution areas described by Osman Hill (1933, 1953), as well as on the origin of museum specimens collected between 1889 and 1929 (Jenkins, 1987) and on specimen localities mentioned by other authors. Some of Osman Hill's information about *Loris* distribution is apparently guesswork, based on habitat quality and distribution of sympatric simian primate species. Numbers added to the specimen locality symbols refer to corresponding numbers in the Specimen List below with additional information (source, date of capture or other available data). Names of type localities are underlined.

SUBSPECIES OF *LORIS*

For body measurements and weight data see Table 1.

L. t. tardigradus **Linnaeus, 1758, Slender loris** (Figs. 4a-f). Smallest subspecies. Pelage: red-brown with or without indistinct dorsal stripe, ventral color similar to dorsal color or greyer. Circumocular patches dark red-brown; white interocular stripe sometimes bifurcated above, but not extended around the circumocular patches as in *L. t. grandis* (Osman Hill, Phillips, 1932). Preauricular hairs pale brown, never white. Short, scanty fur on forearms, hands and feet. Ears yellowish, dusky at margin (Osman Hill, 1933, 1953). Intergrades with *L. t. grandis* in the low hills towards Kandy (Phillips, 1935).

(***L. gracilis zeylanicus*** **Lydekker, 1905**: Status in doubt (Jenkins, 1987). Smaller, browner than *L. t. nordicus*; probably identical with *L. t. tardigradus* (Osman Hill, 1933).

L. t. malabaricus **Wroughton, 1917, Malabar slender loris.** Pelage: light red-brown or grey-brown with or without white frosting on dorsum and head. Dorsal stripe usually absent, indistinct if present. Ventrally buff or dirty white; red-brown circumocular patches. Cutaneous pigmentation unrecorded (Osman Hill, 1953). Type specimen with a tail of 7 mm length (Shortridge, quoted by Wroughton, 1917).

L. t. grandis **Hill and Phillips, 1932, Highland slender loris** (Figs. 4g, h). Pelage: dark grey or grey-brown with an indistinct dark dorsal stripe, usually frosted. Ventral hairs distally white or light buff, basal half black (Osman Hill, 1933). Circumocular patches black, brown or red-brown. The white interocular stripe bifurcates above, forming a pale zone around the dark circumocular patches separating two darker zones. Ears dusky throughout (Osman Hill, 1933, 1953). Coat thicker, extremities more heavily covered than in *L. t. tardigradus*, especially forearms and dorsum of hands are well covered. Hairless skin only on the palms of hands, soles of feet and tip of the rhinarium. Wooly coat mixed with more ordinary hair than in the velvety coat of other subspecies. In some respect intermediate between the lowland subspecies of Sri Lanka and *L. t. nycticeboides*. Intergrades with "one or other of the lowland races" at lower altitude (Phillips, 1935).

L. t. nycticeboides **Hill, 1942.** Medium size, but looking much larger because of the thick fur (hair length up to 25 mm, ventrally 30 mm). Pelage: brown without dorsal stripe or frosting; ventrally buff. Circumocular patches very dark brown, almost black. Ears completely hidden, covered with greyish-brown fur, slightly pigmented towards the margin (Osman Hill, 1933, 1942). Limbs, especially hind limbs, rather short, stout and thickly furred; superficially resembling a slow loris. First described as an extreme highland form of *L. t. grandis* because of some similarities (Nicholls, 1939). More distinct from the lowland races than from *L. t. grandis*.

L. t. nordicus **Hill, 1933, Northern Ceylonese slender loris** (Figs. 5a-e). Largest subspecies of Sri Lanka. Pelage: grey, grey-brown or buff-brown, with or without white frosting on dorsum and head, usually with dark dorsal stripe. Ventrally white or light buff, ventral hairs, except from the throat hairs, usually (not always) with a grey base (main difference to *L. t. lydekkerianus*). Circumocular patches dark grey (almost black), grey or grey-brown, preauricular hairs white or light grey. Pigmentation variable (Osman Hill, 1953), ears black (Osman Hill, 1933) or yellow.

In the *L. t. nordicus* from Polonnaruwa bred at the Ruhr-University, juvenile animals show a reddish or yellowish pelage with light reddish circumocular patches and pigmented ear rim whereas in adults the pelage is usually grey, the circumocular patches are grey to almost black, and the ears are always unpigmented. A few adult animals remain yellowish with indistinct, light brownish-grey circumocular patches. In old animals, the circumocular patches become lighter grey and less distinct. Body size, too, varies considerably in the captive-bred animals at Ruhr-University.

***L. t. lydekkerianus* Cabrera, 1908, Mysore slender loris** (Figs. 5f-h). Largest subspecies. Pelage: buff or grey-buff, with or without white frosting on dorsum and head, with or without dark dorsal stripe. Ventrally white or buff, ventral hairs usually without grey or occasionally with a tiny grey base (main difference to *L. t. nordicus*). Circumocular patches grey, dark brown or almost black, preauricular hairs white. Ears black (Osman Hill, 1953). According to Seth (1963), 12 specimens from the type locality Madras were dorsally of blackish-grey or occasionally a deep fawn color without dorsal stripe; they had white cheeks and were ventrally white, sometimes with a light brownish tinge. Their extremities were fawn, sometimes mixed with grey, on the lateral surface.

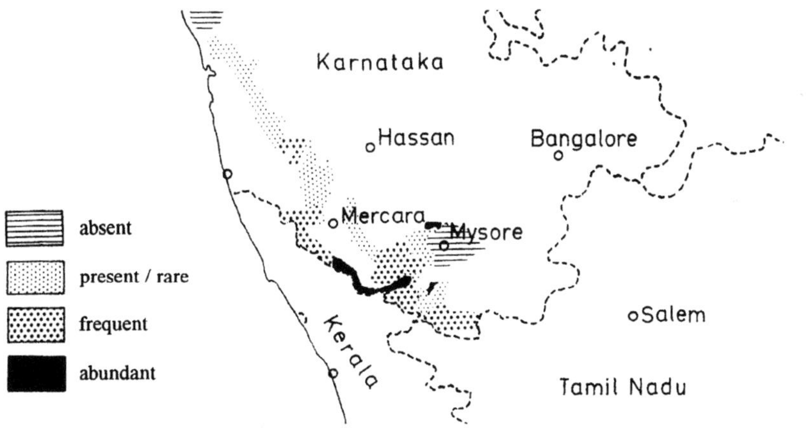

Fig. 3: *Loris* distribution and abundance in certain forests of Karnataka State, India, redrawn and modified from Devaraj Sarkar *et al.* (1981), extensive population study of *Loris* (subspecies not distinguished) in the major forests of the coastal region of Karnataka and the Mysore Plateau.

CONSERVATION AND CAPTIVE STATUS

In India, the moist evergreen forest, habitat of *L. t. malabaricus*, has been reduced to isolated patches (Eudey, 1987). According to Devaraj Sarkar *et al.* (1981) an extensive population study in Karnataka State showed that slender lorises were still frequent in some

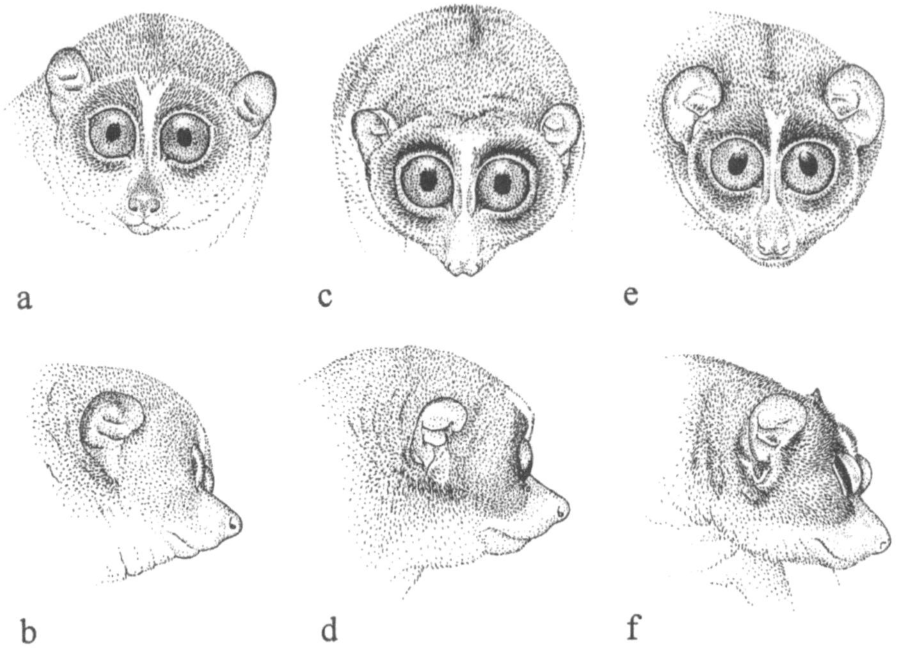

Figure 4: slender lorises of the small reddish or brown rainforest subspecies and highland subspecies
a-f: *L. t. tardigradus?* Small brown or reddish specimens, stock imported from Sri Lanka. Exact origin unknown. **a, b:** wildcaught female, imported from Colombo Zoo, presently kept at Ruhr-University.
c, d: captive-bred male "1964 Ganesh", Duke University Primate Center. Ganesh and most of the other slender lorises at DUPC are small reddish or brownish lorises with the body-weight of Indian *L. t. malabaricus* (Izard and Rasmussen, 1985), but the wildcaught founders were all imported from Sri Lanka.
e, f: wildcaught male, imported from Colombo Zoo, presently kept at Ruhr-University. This specimen resembles other small brown lorises in size and color, but differs from them in some facial characteristics, including ear size. Compare with g, h: possibly an intermediate *tardigradus / grandis* specimen?

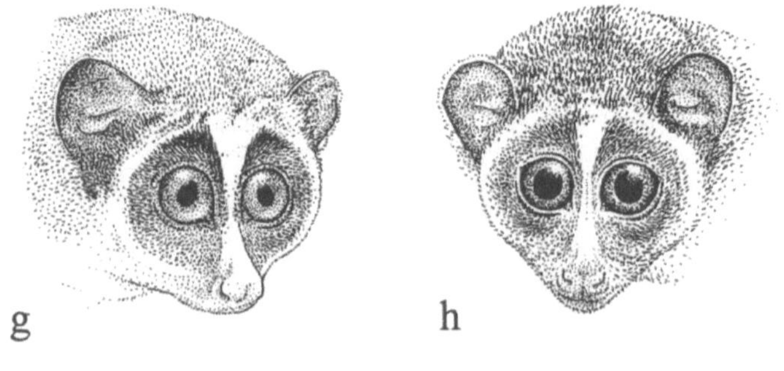

g, h: *L. t. grandis* from the type locality, Gammaduwa. **g:** female, **h:** male. Both redrawn after photos in Osman Hill (1933)

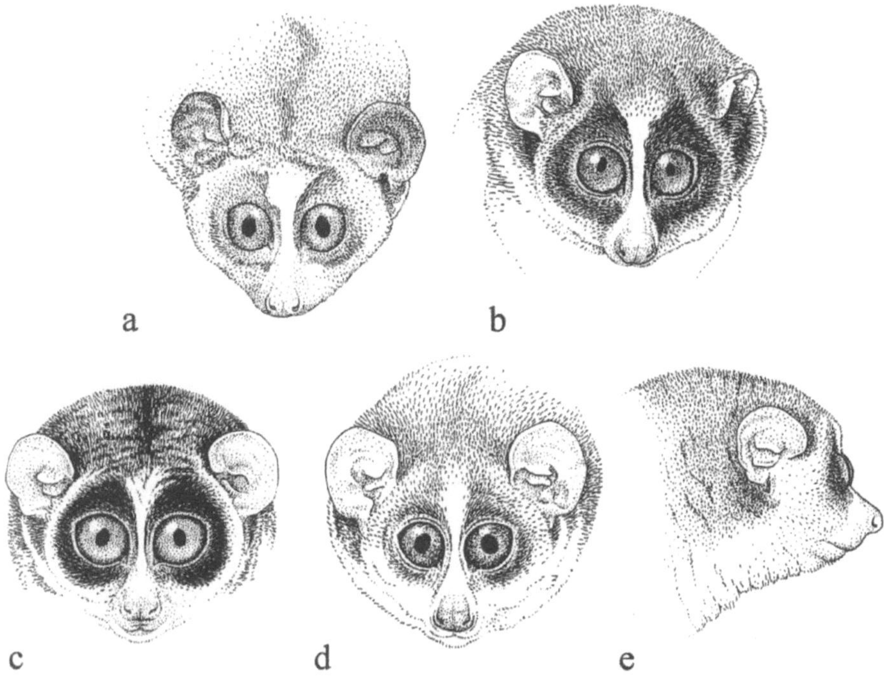

Figure 5: slender lorises of the large grey dry zone subspecies
a-d: *L. t. nordicus*, Sri Lanka: variation within one subspecies. **a:** male with dark ears from the type locality, Talawa. Redrawn from a photo in Osman Hill (1933). **b:** female, probably from Wilpattu. Redrawn from a photo from Skansen-Akvariet, Stockholm. **c-e:** specimens from Polonnaruwa kept at Ruhr-University. **c:** adult female, dark grey with black circumocular patches and heavy frosting. **d:** adult female with light-coloured yellowish pelage (usually a colour typical for subadult animals) and light, indistinct circumocular patches. **e:** *L. t. nordicus* male from Polonnaruwa.

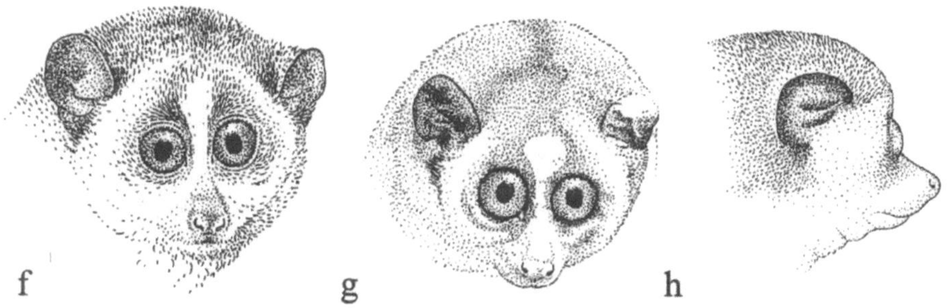

f-h: *L. t. lydekkerianus* from India. **f:** specimen from the type locality, Madras. Drawn after a photo of a dead specimen in Seth (1963). **g:** female from the Bidigiriranjans, Nilgiri Hills. Redrawn from a photo in Osman Hill (1933). **h:** female from Bangalore. Redrawn from photos in Swayamprabha and Kadam (1980).

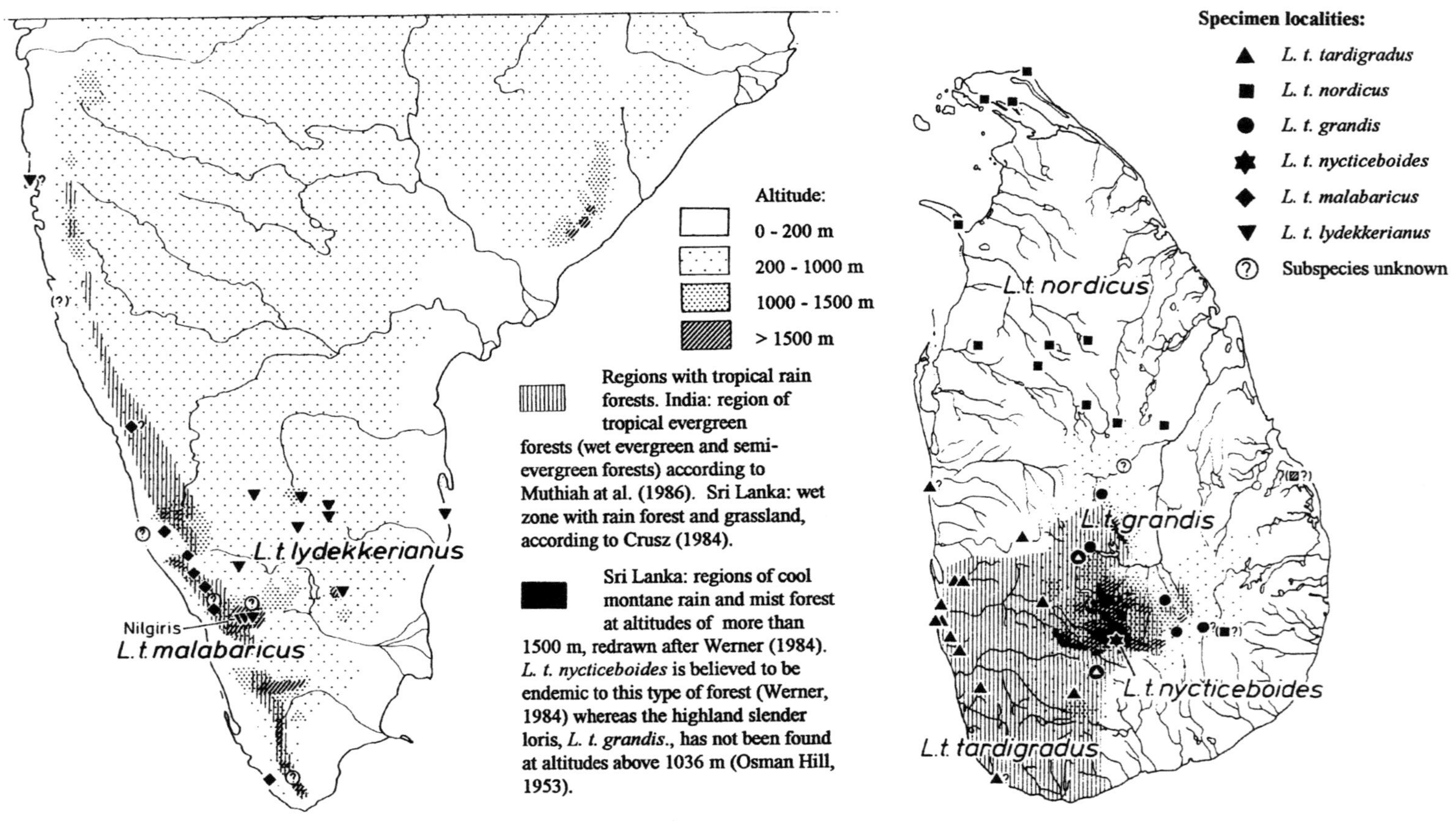

Figure 6: Distribution of Loris subspecies in relation to altitude and vegetation.

Table 1: Body measurements (adults). Lengths rounded to full mm.
Data based on Osman Hill (1933) with additional data from other authors.

	L.t.tardigradus	*L.t.nordicus*	*L.t.grandis*	*L.t.nycticeboides*	*L.t.lydekkerianus*	*L.t.malabaricus*
Length,				Osman Hill, 1942:		
head and body	18.2-20.6 cm	20.5-23.8 cm	-	20.4-21.3 cm	23.6-26 cm	18-22 cm
n =	10	6	-	2	4	4
Average (male)	18.9 cm	22.7 c.	21 cm	20.4 cm	26 cm	22 cm
n =	7	3	4	1	1	1
Average (female)	20.2 cm	21.7 cm	20.2 cm	21.3 cm	24.2 cm	21.8 cm
n =	3	3	4	1	3	2
Max. cranial length	4.8-5 cm	5-5.4 cm	5.1-5.3 cm	5.3, 5.4 cm	5.2-5.5 cm (Jenkins, 1987)	4.8 (Jenkins, 1987) -5.2 cm
Average	4.9 cm	5.3 cm	5.2 cm		5.4 cm	5 cm
n =	7	4	3	2	9	5
Foot length	3.5-4.4 cm	4.2-5.4 cm	-	4.7	4.5-5.3 cm	4.4-4.5 cm
n =	10	6	-	2	4	4
Average (male)	4.1 cm	4.8 cm	4.5 cm	4.7 cm	5.3 cm	4.4 cm
n =	7	3	4	1	1	1
Average (female)	4.2 cm	5.8 cm	4.5 cm	4.7 cm	4.8 cm	4.5 cm
n =	3	3	4	1	3	2
Ear length	1.9-2.3 cm	2.6-3 cm	-	2.5-2.6 cm	2.6-3.1 cm	2.5 cm
n =	10	6	-	2	4	2
Average (male)	2.2 cm	2.8 cm	2.4 cm	2.6 cm	3.1 cm	-
n =	7	3	4	1	1	-
Average (female)	2.1 cm	2.8 cm	2.6 cm	2.5 cm	2.9 cm	2.5 cm
n =	3	3	4	1	3	2
Weight	85-127 g	198.1-339.6 g	205-240 g	140, 190 g	226-347 g	about 170 g
Average	102.4 g	235.8 g	222.5 g	165 g	283 g	(Ryley, 1913)
n =	4	4*	2 (preserved)	2	4	

**L. t. nordicus*: Wilachchiya specimens: 566-679 g (Mayor, quoted by Osman Hill, 1933, as "almost certainly an error and ... probably computed purely by guesswork"), Wildcaught *L. t. nordicus* specimens from Polonnaruwa: 228-285 g (4 males), 238-287 g (5 females); some unusually large captive-bred specimens weigh up to 450 g without being fat (unpublished data).

regions on the Mysore Plateau and in parts of the coastal forest (Fig. 3) but, in general, "lorises belong to India's most endangered species"; they are decreasing in number due to hunting as they are believed to be a potent love charm and of medicinal importance in the treatment of eye diseases (Devaraj Sarkar *et al.*, 1981).

Osman Hill (1933) noted that *Loris* was rare in Sri Lanka. Today, the habitats are increasingly being destroyed, especially in the wet zone. Distribution seems to be patchy. An unsystematic 12-week search in Sri Lanka in 1979 (Meier and Nieschalk, unpublished) suggested that in most parts of the country lorises appeared to be absent even when the habitat appeared to be suitable. This situation is aggravated by habitat fragmentation leading to loss of genetic diversity if not extinction. Killing of lorises in Sri Lanka due to local superstition is also reported.

The description of *L. t. nycticeboides* is based on only two adult specimens, found in 1938 in the forests below the Horton Plains, and their captive offspring, none of which survived. Since that time no new data have been published on the occurrence of this thickly furred highland subspecies but, in 1982, local rangers claimed that slender lorises still occurred in the Horton Plains where temperature may fall below 0°C (Dmoch, pers. comm.).

Field surveys are needed, and for *L. t. malabaricus* a plan for *in situ* management is necessary (Stevenson *et al.*, 1991). A recommended captive program, "Nucleus I", suggests maintaining of a "captive nucleus (50-100 individuals) to always represent 98% of the wild gene pool", into which new animals should be periodically introduced to maintain the desired level of genetic diversity (Stevenson *et al.*, 1991).

Table 2: Conservation status of *Loris tardigradus* subspecies
Data from the Global Captive Action Plan for Primates, Discussion edition, 15 September 1991 by Stevenson, Foose and Baker:

Origin	Subspecies	Category of threat: Red data*[1]	Mace/Lande*[2]	Recommended captive programme
Sri Lanka:	*L. t. tardigradus*:	vulnerable	vulnerable	Nucleus I
	L. t. nordicus:	vulnerable	vulnerable	Nucleus I
	L. t. grandis:	vulnerable	vulnerable	Nucleus I
	L. t. nycticeboides:	vulnerable	vulnerable	Nucleus I
India:	*L. t. malabaricus:*	endangered	vulnerable	Nucleus I
	L. t. lydekkerianus:	-	-	-

*[1] IUCN threatened species categories *[2] categories according to Mace and Lande (1991)

CAPTIVE BREEDING

The susceptibility of slender lorises to psychic stress was already mentioned by Osman Hill (1953) and Manley (1966); our experience showed that especially quarantain after transfer of animals to other institutions may result in a high level of mortality. Breeding success is still low or absent in some captive colonies. The American Regional Studbook for Asian Prosimians (Fitch-Snyder, 1993) includes a considerable number of slender lorises, but reliable information about subspecies is lacking in most and crossbreeding might occur. A captive population of purebred *L. t. nordicus*, descended from an original breeding population of nine specimens from Polonnaruwa (Meier, 1989), now consists of 69 animals in several breeding facilities. A studbook for this population, including husbandry information with special emphasis on stress problems, may be obtained from the authors.

DISTRIBUTION OF SLENDER LORIS SUBSPECIES: SPECIMEN LIST

The following is a review of the literature and various sources of information about the distribution of *Loris* populations. The following data are included:

Origin: place of observation/capture or origin of museum specimens as noted on labels. The places are numbered. On the distribution map, either the number or the name of the place is noted next to the symbol indicating the place of origin.

Column 1: place numbers. Locations in close proximity are listed under one number.

Column 2: names of localities and some additional information.

Abundance, if known, and other additional information:

Column 2

Altitude:

Column 3: The altitude of habitat is given in meters.

Date:

Column 4: date of observation, capture or importation of wild-caught specimens. If authors provide no date but indicate the present-day occurrence of lorises, the date of publication is mentioned - in parentheses. If the time of occurrence is not mentioned, "prior to (date of publication)" is noted. If it appears that the information was published much later, no date is mentioned. More recent dates (from 1960 onwards) are printed bold and underlined.

References:

Column 5

The general distribution of *Loris* subspecies in South India and Sri Lanka is contained in Figs. 1 and 2. Anderson (1881, quoted by Osman Hill, 1933) mentions a specimen in the Calcutta museum which was obtained from a Burmese and was said to have come from the Shan States in Burma, but the occurrence of lorises in Burma has never been confirmed.

Sri Lanka

L. t. tardigradus: Distribution: "Ranging sparsely over the wet lowlands of S.W. Ceylon from Colombo in the north to about Ranna on the south coast; inland into the foothills where it intergrades, e. g., at Peradeniya, with the next race" (Osman Hill, 1953). Probably well distributed throughout the low-country wet zone, but nowhere common (Phillips, 1931). Above 300 m grades into *L. t. grandis* (Osman Hill, 1933, Phillips, 1935). According to Phillips (1935), in the north and south-east of its range *L. t. tardigradus* probably meets and intergrades into *L. t. nordicus*.

Type locality: unknown

Other locations:

1	2	3	4	5 [column number]
1	Colombo, 06^{0} 56^{1} N 79^{0} 51^{1} E,	0-6 m	1903, 1937 + ?	(Osman Hill, 1933, Jenkins, 1987)
2	Polgahawela, 07^{0} 20^{1} N 80^{0} 19^{1} E,		(1924)	(Jenkins, 1987)
	(Phillips, 1924: "seems to be found more often in parts of the Kelami valley and around Polgahawela than anywhere else")			
3	Matugama	30 m	prior to 1933	(Osman Hill, 1933, Phillips, 1935)
4	Henaratgoda (several specimens)		prior to 1933	(Osman Hill, 1933)
5	Kitulgala	210 m	prior to 1933	(Osman Hill, 1933, Phillips, 1935)
6	Wellawatta	270 m	prior to 1933	(Osman Hill, 1933)
7	Nugegoda		prior to 1933	(Osman Hill, 1933)
8	Kesbewa	15 m	prior to 1933	(Osman Hill, 1933, Phillips, 1935)
9	Gampaha		prior to 1933	(Osman Hill, 1933)
10	Madampe	120 m	prior to 1935	(Phillips, 1935)

11 Bandaragama 15 m prior to 1935 (Phillips, 1935)

Location not clear or not found on the map:

Rayigama Korale (W. P.) prior to 1935 (Phillips, 1935)

Bulangoda? (Jenkins, 1987); = Bulatgangoda?, 07^0 07^I N 80^0 02^I E (Jenkins, 1987)

(= Balangoda? See below: Intermediate types *L. t. tardigradus*/*L. t. grandis*)

Cotta? (= Kotte = Sri Jayawardanapura near Colombo?) (Osman Hill, 1933)

L. t. nordicus: Distribution: Lowland dry zone of north and east Ceylon, including the Jaffna peninsula (Osman Hill, 1953). Distribution likely to prove "almost identical with that of the Ceylonese race of ..... *Pithecus entellus thersites*" (Osman Hill, 1933).

Type locality: Talawa, North Central Province, 08^0 13^I N 80^0 21^I E

13 m prior to 1931 (Phillips, 1931, Osman Hill, 1933)

Other locations:

12	Polonnaruwa		1970	(Petter + Hladik, 1970; Hladik + Petter, 1970;
			1980	Meier, 1989)
	dry lowland zone including the Jaffna peninsula		prior to 1933	(Osman Hill, 1933)
13	Jaffna	4 m	prior to 1931	(Phillips, 1931, Osman Hill, 1933)
14	Sigiriya (C.-P.)	200 m	prior to 1931	(Phillips, 1931, Osman Hill, 1933)
15	Kekirawa (N.-C.-P.)	120 m	prior to 1931	(Phillips, 1931, Osman Hill, 1933)
16	Tammanewa	30 m	prior to 1931	(Phillips, 1931; Mayor, quoted by Osman Hill, 1933)
17	Mannar (N.-W.-P.)	4 m	prior to 1931	(Phillips, 1931, Osman Hill, 1933)
18	Anuradhapura (N.-C.-P.), 08^0 21^I N 80^0 23^I E	87 m	1913	(Phillips, 1931, Osman Hill, 1933, Jenkins, 1987)
19	Wilpattu National Park: "distributed throughout Wilpattu" (Eisenberg, Lockhart, 1972), but local rangers said to one of us (B. M.) that they were very rare.	< 152 m	1972	(Eisenberg, Lockhart, 1972)
19	Wilpattu		1984	(Verner-Carlsson, pers. comm.)
20	Chavakachheri, 09^0 39^I N 80^0 09^I E		1933	(Jenkins, 1987)
21	Point Pedro, 09^0 50^I N 80^0 14^I E			(Jenkins, 1987)
22	Uva (Monaragala), 07^0 09^I N 81^0 07^I E	<150 m	1939	(Jenkins, 1987)

? (Lowland, not within estimated distribution area of *L. t. nordicus*; oral reports indicate that *Loris* is common around Monaragala are mentioned by Osman Hill (1933), but without information of the subspecies)

Location not clear or not found on the map:

23	Wilachchiya (North-Central-Province)	prior to 1931	(Phillips, 1931, Osman Hill, 1933)
	(= ?) Willichia (North Central Province) ?	1914	(Jenkins, 1987)

L. t. grandis: Distribution: "hills of the Central Province at 900 m average altitude. Uncommon and sparsely distributed throughout its range (Osman Hill, Phillips, 1932). Intergrades with *tardigradus* at lower altitudes but not known to meet range of *nycticeboides*" (Osman Hill, 1953) because no lorises have been found at altitudes between 1036 m and 1829 m.

Type locality: Mousakande, Gammaduwa, 07^0 34^I N 80^0 43^I E

671-1036 m prior to 1931 (Phillips, 1931, Jenkins, 1987)

(According to Phillips (1931), typical specimens of *L. t. grandis* had only been obtained from the type locality whereas in lower parts of the hills specimens intergrade with other subspecies)

Other locations:

24	D'Udawattakele (forest reserve near Kandy)		1969-1970	(Petter + Hladik, 1970)
25	Gammaduwa District		prior to 1933	(Osman Hill, 1933)
26	Namunucula (village or mountain with the same name?), Tonacombe Estate, 06^0 53^I N 81^0 07^I E	329 m		(Jenkins, 1987)
27	Badulla District (?)		prior to 1933	(Osman Hill, 1933)

(Phillips, quoted by Osman Hill, 1933, in the description of *L. t. grandis*: large form known in the Badulla district)

Location not clear or not found on the map:

28 Opalgalla (?) prior to 1933 (Osman Hill, 1933)

(Osman Hill, Phillips, 1932: specimen resembling a *L. t. tardigradus* female in general color, but definitely diagnosed as *L. t. grandis* by size, character of fur, well-clothed hands and facial appearance)

29 near Bandarawela (?) prior to 1933 (Osman Hill, 1933)

(Phillips, quoted by Osman Hill, 1933, in the description of *L. t. grandis*: large form known near Bandarawela)

(*L. t. grandis*?)

30 Badalkumbura, near Nakkala, Uva Province, 06^0 54^I N 81^0 14^I E, 152 m 1955 (Jenkins, 1987)

31 Moneragala District, East Ceylon, 07^0 06^I N 80^0 38^I E (?) (Jenkins, 1987)

L. t. nycticeboides: Distribution: known only from the type locality (Osman Hill, 1953). Restricted to the montane rain and mist forests growing at an altitude of more than 1500 m in Sri Lanka (Werner, 1984).

Type locality: Horton plains 1829 m (0,1)

regarded as rare (Osman Hill, 1942)2134 m (1,0) 1938, (Osman Hill, 1942, Jenkins, 1987)

In 1982, rangers said that slender lorises still occur on the Horton Plains where temperature may fall below 0^{0}C **1982 ?** (Dmoch, pers. comm.)

Intermediate types *L. t. tardigradus*/*L. t. grandis*

32 Peradeniya, 07^0 15^I N 80^0 36^I E 469 m prior to 1933 (Osman Hill, 1933)

(Jenkins, 1987: specimen labelled as *L. t. tardigradus*)

33 Balangoda 533 m prior to 1931 (Phillips, 1931, Osman Hill, 1933)

(Osman Hill, 1933: intermediate size; facial aspects of *L. t. grandis* with white throat and cheeks, upper parts dull fawn with little frosting,)

(Intermediate type *L. t. nordicus*/*L. t. grandis*?)

34 Elahera-Tal, north of Gammaduwa prior to 1931 (Osman Hill, 1933)

(Osman Hill, 1933: the specimen should be intermediate between *L. t. nordicus* and L. t. grandis, but it is almost identical with the *tardigradus/grandis*-specimen from Balangoda)

(*L. t. tardigradus*/*L. t. nordicus*?)

35 Chilaw: red specimen described by Tennent prior to 1861 (Osman Hill + Phillips, 1932)

"belonging to the usual red variety" (*L. t. tardigradus* or possibly an intermediate form between *L. t. nordicus* and *L. t. tardigradus*?). According to Osman Hill and Phillips (1932), the specimen might also have been a juvenile *L. t. nordicus* with the reddish color of immature animals.

Sri Lanka; no information about subspecies given:

1 near Colombo, 06^0 56^I N 79^0 51^I E 1879 (Jenkins, 1987)

1 Colombo, 06^0 56^I N 79^0 51^I E 1906 (Jenkins, 1987)

36 Batticaloa:"said to be common" (Osman Hill, 1933. In Osman Hill, 1953, the whole area is regarded as free of lorises. Meier, Nieschalk, unpublished: no lorises found during an unsystematic search in 1979)

37 Monaragala ("common"), possibly *L. t. nordicus* prior to 1933 (Osman Hill, 1933)

(Meier, Nieschalk, unpublished: no lorises found during an unsystematic search in 1979)

38 Wellassa district prior to 1933 (Osman Hill, 1933)

39 Tamankaduwa district prior to 1933 (Osman Hill, 1933)

40 Bintenne district prior to 1933 (Osman Hill, 1933)

41 Balangoda (see also no. 33) 533 m prior to 1924 (Phillips, 1924)

42 Galle prior to 1933 (Osman Hill, 1933)

India

Osman Hill, 1933, 1953: *Loris tardigradus* probably occurring at the west coast north up to theTapti river; east coast: "distribution ...... possibly nearly ...... the same as that of ..... *Macaca radiata*, i.e. covering the whole of South India as far north as the Godavari river" (Osman Hill, 1933).

L. t. lydekkerianus: General distribution: south and east India. Precise range to north unknown (Osman Hill, 1953). "It is not known whether it reaches the east coast or not" (Osman Hill, 1933). Regarded as a dry zone form (Osman Hill, 1933), but, according to Devaraj Sarkar (1981), also occurring in moist evergreen forest. Specimens have been found occurring at altitudes of 2000 und 3000 ft. (Shortridge; Elliott, 1913; quoted by Osman Hill, 1933) and 4700 ft (Kinloch 1898, quoted by Osman Hill, 1933).

Type locality: Madras (Osman Hill, 1933) **1962** (Osman Hill, 1933, Seth, 1963)

Osman Hill (1933) mentioned that the type specimen might have been not from Madras, but brought there for sale because no lorises have been recorded from there "in recent times". But in 1962, several *L. t. lydekkerianus* were caught in the vicinity of Madras (Seth, 1963).

Other locations:

	Coromandel region of south India		prior to 1935	(Osman Hill, 1933, 1953)
	Eastern Ghats (precise range to north unknown):		prior to 1933	(Osman Hill, 1933)
1	Chettiri range (Chittiri hills), Salem district, Eastern Ghats, $11^{0}\,50^{I}$ N $78^{0}\,29^{I}$ E,	610 m	1912, 1929	(Jenkins, 1987)
	Western Ghats		**(1981)**	(Devaraj Sarkar, 1981)
2	Mysore, Mysore plateau and attendant ranges:			(Osman Hill, 1953)
2	"Mysore": said to be common"		prior to 1933	(Osman Hill, 1933)
3	Kolar town, East Mysore, $13^{0}\,08^{I}$ N $78^{0}\,08^{I}$ E,	849-1227 m		(Jenkins, 1987).
3	Kolar district: "evident abundance"		(1913)	(Ryley, 1913 a)
	(Shortridge, quoted by Osman Hill, 1933: Loris "said to be very common" in the Kolar district)			
3	Kolar district		**(1981)**	(Devaraj Sarkar, 1981).
4	Malur, Kolar District, Mysore, $13^{0}\,20^{I}$ N $78^{0}\,10^{I}$ E, 1912			(Osman Hill, 1933; Jenkins, 1987)
5	Nandedroog, (Nandidrug, Nundidroog), N.-W. Kolar District, East Mysore$13^{0}\,22^{I}$ N $77^{0}\,42^{I}$ E,	910 m		(Osman Hill, 1933, Jenkins, 1987)
6	Nilgiris: "said to be common"		prior to 1933	(Osman Hill, 1933)
7	Bidigiriranjans (in the Nilgiris)		prior to 1933	(Osman Hill, 1933, 1953)
8	Kotagiri (in the Nilgiris)	1430 m	prior to 1898	(Kinloch, 1898, quoted by Osman Hill, 1933)).
9	Bangalore, $12^{0}\,59^{I}$ N $77^{0}\,35^{I}$ E	950 m	1919 1959-**(1965)** **(1980)** **(1984)**	(Osman Hill, 1933; Ramaswami, Anand Kumar, 1962; Swayamprabha + Kadam, 1980; Sarma + Kadam, 1984, Manjula, 1984)
	(apparently abundant; numerous animals caught and killed for research in the past 30 years)			
	Devaraj Sarkar (1981): *L. t. lydekkerianus occurring* in the Bangalore district..			
	Jenkins (1987): *L. t. lydekkerianus?* Subspecies of specimen in question.			
10	Tumkur District		**(1981)**	(Devaraj Sarkar, 1981)

Location not clear or not found on the map:

(Malur, Kolar District, East Mysore, $13^{0}\,44^{I}$ N $75^{0}\,20^{I}$ E ?) 1912 (Jenkins, 1987)

Coordinates wrong? Not located within Kolar District.

L. t. malabaricus: General distribution: Wet coastal tract (Malabar tract) of west India, including western Ghats and Wynaad plateau (Osman Hill, 1953). Eastern limits of distribution unknown (Osman Hill, 1933). *L. t. malabaricus* is regarded as a wet zone form (Osman Hill, 1953), but, according to Devaraj Sarkar (1981), also occurs in other types of forest. According to Osman Hill (1933), in the distribution area of L. t. *malabaricus* specimens have been found at considerable altitudes: 2000, 3000 ft in South Coorg (museum specimens labelled *L. malabaricus* or *L. t. gracilis*) whereas the similar-looking *L. t. tardigradus* of Sri Lanka seems to be confined to the lowland.

Type locality: Kutta, South Coorg 1913 (Shortridge, quoted by Osman Hill, 1933)

Other locations:

11 Coorg (especially south Coorg): (Wroughton, 1917)
12 Mercara (1981) (Devaraj Sarkar, 1981)
13 Virajpet (Virarajendrapet), 12° 12' N 75° 48' E, 610, 910 m 1913 (Shortridge, quoted by Osman Hill, 1933)
14 Huvinakadu Estate, Kutta, 12° 01' N 75° 58' E 867 m 1913
15 Wynaad, Wynaad-Plateau (South Coorg) (= Wayanad district, Kerala) 1903 (Shortridge, quoted by Osman Hill, 1933)
16 South Canara (1981) (Devaraj Sarkar, 1981)
17 North Kanara ? (Ryley, 1913)
"The Loris is said to be rare as far as North Kanara" (Osman Hill, 1933, mentioned in the description of *L. t. malabaricus*; from Shortridge's notes in Ryley, 1913)
18 Trevandrum, Travancore State, 08° 29' N 76° 55' E 1889 (Jenkins, 1987)
18 Travancore: "Specimens from here are referrable to this race" prior to 1918 (Wroughton, 1918, quoted by Osman Hill, 1933).
(Travancore: former British-Indian state with the capital Trevandrum, now southern Kerala)

? India; subspecies not clear:

19 Bombay, Sassoon 14 dock, 18° 58' N 72° 50' E 1914 (Jenkins, 1987)
(Museum specimen labelled "*L. t. lydekkerianus*"; correct?)
(Wroughton's "Bombay collection" of three *L. t. malabaricus* (skins and skulls only), used for the description of the subspecies by Osman Hill (1933), includes specimens from other localities; Osman Hill does not mention Bombay as a locality where *L. t. malabaricus* has been found).
20 Mangalore, 12° 52' N 74° 53' E prior to 1849
"The record of *L. t. lydekkerianus* from Mangalore is apparently based on a poorly preserved specimen .... the subspecific identification of which can no longer be established" (Jenkins, 1987)
21 West coast near Ratnagiri given as a "possible locality" for *Loris* (Blanford 1888)
(Osman Hill, 1933: Location mentioned in connection with *L. t. lydekkerianus*)
22 Bandipur National Park: "thickly distributed" (1981) (Devaraj Sarkar, 1981)
23 Nagarhole National Park: "thickly distributed" (1981) (Devaraj Sarkar, 1981)
24 Mundanthurai wildlife sanctuary, Tirunelveli district, Tamil Nadu (single individuals, "at certain points") (1984) (Johnson, 1984)

Location not clear or not found on the map:

25 Thittimati (= forest reserve, Mysore Plateau): "thickly distributed" (1981) (Devaraj Sarkar, 1981)
26 Anekadu (= forest reserve, Mysore Plateau): "thickly distributed" (1981) (Devaraj Sarkar, 1981)

ACKNOWLEDGEMENTS:

We are grateful to H. Preuschoft who provided the facilities for keeping and breeding our stock and other support, to H. Fitch-Snyder (Zoological Society of San Diego) for sending us a copy of the American Regional Studbook for Asian Prosimians, to K. Weisenseel and J. Drake (DUPC) for help in taking photos, and to the staff of the British Museum of Natural History. In particular, we thank the editors for careful revision of the manuscript.

REFERENCES

Anderson, J. (1881). Catalogue of Mammalia in the Indian Museum (pp.97-98). Calcutta.
Blanford, W.T. (1888). Fauna of British India, Mammals (p.47). London.

Crusz, H. (1984). Parasites of endemic and relict vertebrates: a biogeographical review. In C.H. Fernando, (Ed.), *Ecology and Biogeography in Sri Lanka* (pp. 321-351). Dr. W. Junk Publishers, The Hague.

Devaraj Sarkar, H.B., Murali, S., Prasad, D.T., Shekarappa, B.M., & Vijayalakshmi, V. (1981). The population and distribution of the slender loris (*Loris tardigradus*) in Karnataka State. *Tigerpaper, 8 (1)*, 7-10. Bangkok.

Eisenberg, J.F., & Lockhart, M. (1972). An ecological reconnaissance of Wilpattu National Park, Ceylon. *Smithsonian Contributions to Zoology, 101*, 1-119. Smithsonian Press, Washington.

Eudey, A.A. (1987). Action Plan for Asian Primate Conservation 1987-91. IUCN Primate Specialist Group.

Fiedler, W. (1956). Übersicht über das System der Primaten. In H. Hofer, A.H. Schultz, & D. Starck (Eds.), *Primatologia 1* (pp. 321-351). Basel.

Fitch-Snyder, H. (1993). American Regional Studbook for Asian Prosimians. San Diego Zoological Society.

Hladik, C.M., & Petter, J.J. (1970). Le loris tardigrade. Observations de terrains effectuées à Ceylan. *Science et Nature, 101*.

Izard, M.K.; Rasmussen, D.T. (1985). Reproduction in the slender loris (*Loris tardigradus malabaricus*). *American Journal of Primatology, 8*, 153-165.

Jenkins, P.D. (1987). "Catalogue of Primates in the British Museum (Natural history)." Part IV: Suborder Strepsirrhini. British Museum (Natural history), London.

Johnson, J.M. (1984). Diurnal activities of the slender loris, *Loris tardigradus*, in the Mundanthurai Sanctuary, Tamil Nadu (India). In M.L. Roonwal, S.M. Mohnot, & N.S. Rathore (Eds.), *Current Primate Researches* (pp. 389-395). Jodhpur.

Kinloch, A.M. (1898). Distribution of Slender Loris. *Journal of the Bombay Natural History Society, 11*, 736.

Mace, G.M.; Lande, R. (1991). Assessing extinction threats: toward a reevaluation of IUCN threatened species categories. *Conservation Biology, 5 (2)*, 148-157.

Manjula, A. (1984). Changes in levels of proteins in the male reproductive system of the slender loris, *Loris tardigradus lydekkerianus* (Cabr.). In M.L. Roonwal, S.M. Mohnot, & N.S. Rathore (Eds.), *Current Primate Researches* (pp. 441-447). Jodhpur, India.

Manley, G.H. (1966). Prosimians as laboratory animals. *Symposia of the Zoological Society of London 17*, 11-39.

Meier, B. (1989). Genetic management of captive prosimian populations, a report of experience with *Loris tardigradus*. *Human Evolution 4 (2-3)*, 223-229.

Mutiah, S., Vasanthra Kumaran, T., Ramaswami, G., & Subbiah, S. (Eds.) (1987). *A Social and Economic Atlas of India*. Oxford University Press, Delhi.

Nicholls, L. (1939). Period of gestation of *Loris*. *Nature, 143*, 246. London

Osman Hill, W.C. (1933). A monograph on the genus *Loris*. *Ceylon Journal of Science (B) 18 (1)*, 89-129.

Osman Hill, W.C. (1942). The slender loris of the Horton Plains, Ceylon, *Loris tardigradus nycticeboides* subsp. nov.. *Journal of the Bombay Natural History Society, 43*, 73-78.

Osman Hill, W.C. (1953). Primates: Comparative Anatomy and Taxonomy. Vol. I, Strepsirhini. Edinburgh University Press, Edinburgh.

Osman Hill, W.C., & Phillips, W. W. A. (1932). A new race of slender loris from the Highlands of Ceylon. *Ceylon Journal of Science/Spolia Zeylanica, 17 (2)*, 109-123.

Petter, J.J., & Hladik, C.M. (1970). Observations sur le domaine vital et la densité de population de *Loris tardigradus* dans les forêts de Ceylan. *Mammalia, 34 (3)*, 394-409

Phillips, W.W.A. (1924). A guide to the mammals of Ceylon. *Ceylon Journal of Science/Spolia Zeylanica, 13 (1)*, 261-283.

Phillips, W.W.A. (1931). The food of the Ceylon slender loris in captivity. *Spolia Zeylanica, 16*, 205-208.

Phillips, W.W.A. (1935). Manual of the Mammals of Ceylon. Colombo Museum, Ceylon.

Ramaswami, L.S., & Anand Kumar T.C. (1962). Reproductive cycle of the slender loris. *Naturwissenschaften, 49*, 115-116.

Ryley, K.V. (1913). Mammal survey of India - *Loris lydekkerianus*, Cabr., The Mysore Slender Loris. *Journal of the Bombay Natural History Society, 22*, 283-287.

Sarma, R., & Kadam, M. (1984). Study of Germ-Cell Morphology and the Spermatogenic Cycle in the Slender loris, *Loris tardigradus lydekkerianus* (Cabr.). In M.L. Roonwal, S.M. Mohnot, & N.S. Rathore (Eds.), *Current Primate Researches* (pp.167-175). Jodhpur, India.

Seth, P.K. (1963). Preliminary observations on slender lorises. *Anthropologist, 7*, 45-51. Delhi.

Stevenson, M., Foose, T.J., & Baker, A. (1991). Global Captive Action Plan for Primates, Discussion edition, 15 September 1991. Captive Breeding Specialist Group (CBGS), IUCN-Species Survival Commission (SSC), Primate Specialist Group + Conservation International.

Swayamprabha, M.S., & Kadam, K.M. (1980). Mother-infant-relationship in the slender loris (*Loris tardigradus lydekkerianus*). *Primates, 21 (4)*, 561-566.

Werner, W.L. (1984). Die Höhen- und Nebelwälder auf der Insel Ceylon (Sri Lanka). Tropische und subtropische Pflanzenwelt 46. Steiner, Wiesbaden.

Wroughton, R.C. (1917). The slender loris of Malabar, *Journal of the Bombay Natural History Society, 25*, 45.

Wroughton, R.C. (1918). Summary of results from the Indian mammal survey, *Journal of the Bombay Natural History Society, 25*, 563.

FLUORESCENCE *IN SITU* HYBRIDIZATION REVEALS HOMOLOGIES AMONG TARSIER, GALAGO, AND HUMAN KARYOTYPES

Kim Coleman Healy

Department of Pharmacology
Yale University School of Medicine
47 College Street, Suite 227
New Haven, CT 06510

INTRODUCTION

The current majority position among morphologists regards *Tarsius* and Anthropoidea as sister taxa. This grouping, suborder Haplorhini, is supported by shared derived characters of the orbit, retina, placenta (Luckett, 1982) and basicranium (MacPhee and Cartmill, 1986) in addition to the derived nasal morphology from which the clade takes its name. *Tarsius* ' s highly derived body plan complicates the interpretation of tarsier-anthropoid similarities, however, leading other morphologists to reject them as nonhomologous consequences of tarsier specializations. Schwartz and Tattersall (1987) propose a tarsier-lorisiform clade in lieu of a tarsier-anthropoid one; Simons and Rasmussen (1989) regard the most salient tarsier-anthropoid correspondences in cranial anatomy as convergent.

Since *Tarsius*'s morphological traits permit divergent interpretations, many systematists have looked to molecular characters to help resolve its position. Both amino-acid-sequence data (Baba *et al.*, 1982; de Jong and Goodman, 1989; Czelusniak *et al.*, 1990) and DNA-sequence data (Koop et al, 1989a and b) appear to support haplorhine monophyly; however, maximum-parsimony trees joining *Tarsius* and anthropoids frequently are only a few mutations shorter than contending trees in which Haplorhini is not a clade.

Poorman *et al.* (1985) undertook the first G-banding analysis of a tarsier karyotype. Their comparison of *Tarsius bancanus* and *Galago crassicaudatus monteiri* karyotypes revealed apparent galago counterparts for some two-thirds of the *T. bancanus* chromosomes. These correspondences presented an evolutionary dilemma. If homologous and derived, they would contradict anatomical and molecular evidences for the phyletic sibship of *Tarsius* and Anthropoidea. Indeed, the Poorman *et al.* (1985) results were invoked by Schwartz and Tattersall (1987) in support of a tarsier-lorisiform clade, and by Simons and Rasmussen (1990) as an example of dissent over the postulated synapomorphies of haplorhines in non-fossilizable characters. On the other hand, if the tarsier-galago chromosome correspondences proved homologous and primitive, they would represent conservation of chromosome morphology over some 60 million years of primate divergence.

This situation prompted a reexamination of tarsier and galago karyotypes with increased-resolution techniques (Healy, 1992). While this work was in progress, Dutrillaux and Rumpler (1988) published an R-banded karyotype of *T. syrichta*, asserting that this karyotype shares no recognizable similarities with any other known primate karyotype. The sharp contrast between their results and those of Poorman *et al.* (1985) had several possible explanations: major karyotypic differences between *T. syrichta* and *T. bancanus*, nonequivalence of G- and R- bands as indicators of intertaxon homology, or differences in pattern

Creatures of the Dark: The Nocturnal Prosimians
Edited by L. Alterman *et al.*, Plenum Press, New York, 1995

interpretation between the two laboratories. Further comparison of *T. syrichta, T. bancanus,* and *G. c. monteiri* karyotypes (Healy, 1992) was therefore undertaken with both G and R-banding at increased resolution (technique of Ikeuchi *et al.*, 1984).

Chromosome banding is an *indirect* indicator of genetic homology. Banding homology is an *hypothesis* inferred from the *similarity* of chromosomal patterns. As such, banding-based homology determination is less reliable between more divergent karyotypes; isolated similarities, outside a context of corresponding sets of basic segments, may easily represent homoplasy. For comparisons beyond the useful limits of banding analysis, a more direct technique of homology recognition is desirable.

Fluorescence *in situ* hybridization methods are useful in the investigation of highly diverged karyotypes or in cases where banding correspondences are ambiguous, because *in situ* hybridization can indicate genetic homology directly without relying on pattern recognition to mediate homology recognition. A particularly valuable technique for investigating chromosome evolution is chromosomal *in situ* suppression hybridization, or chromosome painting (Lichter *et al.*, 1988). This technique uses DNA "libraries" derived from the entire content of a given human chromosome to label counterparts to that chromosome in the karyotype of another species (see Weinberg *et al.* [1990] and Jauch *et al.* [1992] for examples). The source of isolated chromosomes for probe construction may be human-rodent somatic cell hybrids containing only one human chromosome pair, or isolated chromosomes prepared *in vitro* by fluorescence-activated flow cytometry. DNA fragments from the target chromosome are cloned in phage vectors and subcloned into plasmids (Collins et al, 1991).

Probes are labeled by nick-translation with biotinylated or digoxigenin-labeled nucleotides. Since cloned chromosome libraries contain both chromosome-specific sequences and interspersed repetitive sequences which are distributed throughout the genome, the interspersed repeats are suppressed by prehybridizing the probe with unlabeled repetitive fractions of human DNA. Labeled probes are then hybridized overnight to heat-denatured chromosomes and detected by immunofluorescent staining with conjugated avidin or anti-digoxigenin antibodies. The ionic strength and concentration of denaturing formamide in the hybridization mixture and subsequent washing solutions are adjusted to stabilize binding of labeled probe sequences to homologous target sequences while minimizing their binding in other regions.

This work utilizes chromosome painting with human chromosome libraries to complement G- and R-banding in the reassessment of tarsier-galago chromosome correspondences.

MATERIALS AND METHODS

Metaphase spreads were prepared by the method of Ikeuchi *et al.* (1984) from *T. syrichta* (TSY) and *G. c. monteiri* (GCM) skin fibroblast cultures. The one *G. c. monteiri* male cultured was born at the Duke University Primate Center (DUPC); *T. syrichta* subjects were either adults wild-caught on Leyte, Philippines, and resident at the DUPC, or infants born at the DUPC to such adults. Table 1 summarizes the available life-history data. G- and R-banding were done on the first five individuals listed; *in situ* hybridization was done on "Cotabatu's son" and on the last female listed.

G-banding was performed by a standard Giemsa-trypsin technique, using trypsin concentrations adapted to the extreme sensitivity of TSY and GCM chromosomes. R-banding followed the procedure of Verma and Babu (1988).

Human chromosome probes used for *in situ* hybridization were the plasmid libraries pBS 1, pBS 4, pBS 6, pBS 7, pBS 9, and pBS 11 (Collins *et al.*, 1991). Slides received RNAse and pepsin pretreatment, formalin postfixation, and ethanol dehydration as described in Wiegant *et al.* (1991). Just before hybridization, they were denatured in 70% formamide, 2 X standard saline-citrate (SSC) at 80°C for 1 min; denaturation was halted by passage through a second ethanol series at –20°C.

Probe labeling and *in situ* hybridization followed the basic method of Wienberg *et al.* (1992), modified in stringency to increase hybridization between divergent human and prosimian DNAs. Libraries were labeled by nick translation (Sambrook *et al.*, 1987) with biotin-11-dUTP or digoxigenin-11-dUTP. 500–600 ng of labeled DNA was ethanol-precipitated with 15 μg of the Cot-1 human repetitive DNA fraction (Bethesda Research Labs) as competitor and 7 μg salmon sperm DNA as carrier. DNA was resuspended in 10 μl of 25% formamide, 1 X SSC and 10% dextran sulfate, denatured for 10 min at 75°C and allowed to reanneal at 37°C for 30 min before being applied to slides and sealed under an 18 x 18 mm coverslip with rubber cement. (The 25% formamide hybridization mixture was used for galago and some tarsier preparations. Other tarsier preparations received a 50% formamide, 2XSSC, 10% dextran sulfate hybridization mixture.) Hybridization proceeded overnight at 37°C in a humid chamber.

Table 1. *T. syrichta* individuals karyotyped in this study.

Animal	Provenance	Date of death
Adult female "Kiomba"	Wild-caught on Leyte, Philippines; lived 2 yrs., 5 mos. in captivity at DUPC	10-30-87
1-day-old female "Cotabatu's daughter"	Born at DUPC to mother wild-caught on Leyte	10-31-87
Infant female "Gobo's daughter"	Born at DUPC to mother wild-caught on Leyte	5-22-88
Infant male "Cotabatu's son"	Born at DUPC	5-22-88
Adult male "Bohol"	Wild-caught on Leyte; lived at DUPC 5 yrs.	6-18-90
Adult female	Wild-caught on Leyte; lived at DUPC 8 yrs.	12-3-92

After overnight incubation, coverslips were removed and slides washed 3 X 5 min at 42°C in a solution containing the same formamide and SSC concentrations as the respective hybridization mixture. Preparations with the 25% formamide (low-stringency) hybridization mixture then were washed 3 X 5 min in 0.1 X SSC at 42°C, while 50% formamide (moderate stringency) preparations were washed in 0.5 X SSC at 50°C. Nonspecific protein binding sites were blocked by incubation in 4 X SSC, 3% bovine serum albumin, 0.1% Tween-20 for 30-60 min at 37°C. Biotin-labeled probes were detected with fluoresceinated avidin; digoxigenin-labeled probes were detected with rhodamine-conjugated anti-digoxigenin. After immunostaining for 30 min at 37°C, slides were washed 3 X 5 min at 42° C in 4 X SSC containing 10% Tween-20, counterstained in 0.2 µg/ml 1,4- diamidinophenylindole (DAPI), rinsed in 4 X SSC + 10% Tween-20, and mounted in diazabicyclooctane-glycerol antifade solution (Johnson *et al.*, 1982).

Metaphases were imaged by epifluorescence on a Nikon Axiophot microscope equipped with a Photometrics Nu200 cooled CCD camera. The following Zeiss filters were used: for DAPI, 4880XX-01 (excitation 362–368 nm, dichroic 395 nm, emission >397 nm); for fluorescein isothiocyanate, 4880XX-10 (excitation 450–490 nm, dichroic 510 nm, emission 515–565 nm); for rhodamine, 4880-XX-15 (excitation 540–552 nm, dichroic 580 nm, emission >590 nm). Captured images were enhanced with NIH Image or Adobe Photoshop, and hybridization signals were merged with DAPI images using the Gene Join program (developed by T. C. Rand in the laboratory of D. C. Ward). Full details of the imaging and optics are discussed in Ballard and Ward (1993).

RESULTS AND DISCUSSION

The G-banded karyotype of *T. syrichta* is shown in Figure 1, which presents four examples of each chromosome at different extensions in order to illustrate the reproducibility of banding patterns and exemplify their range of preparative variations. The order of the karyotype is based on measured relative lengths of 13 G-banded karyotypes. Equivalences to the nomenclature of Dutrillaux and Rumpler (1988) are provided in Table 2.

Table 2. TSY chromosome nomenclature.

Poorman (1982)	Rumpler *et al.* (1983)
1	1
2	3
3	4
4	5
5	2
6	25
7	26
8	6
9	8
10	7
11	9
12	16
13	12
14	11
15	15
16	13
17	17
18	10
19	14
20	18
21	19
22	27
23	21
24	20
25	28
26	22
27	23
28	29
29	30
30	24

The discrepancy between the Poorman *et al.* (1985) G-band results and the Dutrillaux and Rumpler (1988) R-band results reflects neither nonequivalence of G- and R-banding patterns nor major karyotypic changes between *T. syrichta* and *T. bancanus* (Healy and Poorman, unpubl. observ.). Increased-resolution G- and R-band comparison of the *T. syrichta* and *G. c. monteiri* karyotypes shows that the vast majority of G-band correspondences proposed by Poorman *et al.* (1985) appeared to have been fortuitous similarities occasioned by chromosome condensation (Healy and Poorman, unpubl. observ.), confirming Dutrillaux and Rumpler's (1988) conclusion that *T. syrichta* has a highly derived karyotype with few correspondences to *G. c. monteiri* or other primate karyotypes.

There remain two possible correspondences between *T. syrichta* and *G. c. monteiri* chromosomes (Figure 2, top). TSY 19 (18 in Dutrillaux and Rumpler [1988]) resembles the distal two-thirds of GCM 8 (nomenclature of Poorman, 1982); TSY 26 (19 in Dutrillaux and Rumpler, 1988) may be related to GCM 13 by means of a whole-arm paracentric inversion.

It is difficult to assess the significance of such isolated similarities of relatively simple banding patterns. However, the correspondence between TSY 19 and GCM 8 invite further

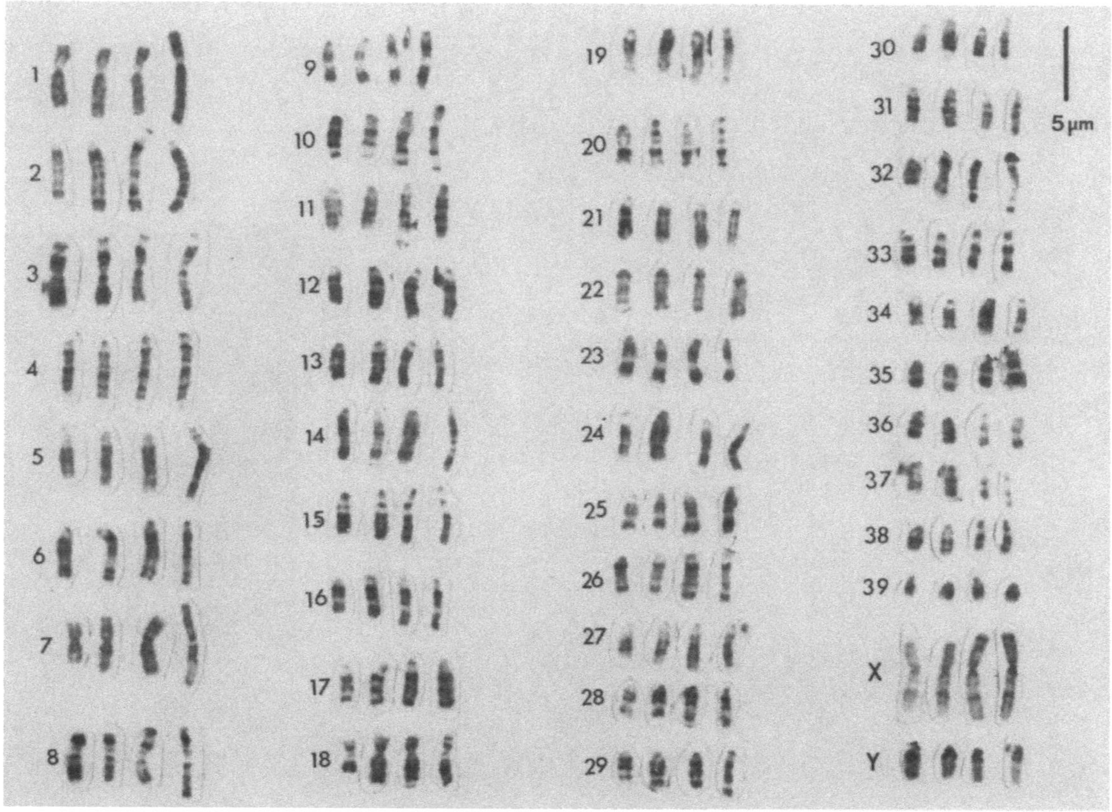

Figure 1. G-banded chromosomes of *T. syrichta*. The reproducibility of the banding patterns and the effect of chromosome extension are shown by the collocation of four examples of each chromosome from different cells.

attention because Dutrillaux et al. (1986) regard GCM 8 as a rearranged homologue of the element that became HSA 1p; by a different rearrangement, the same primitive element gave rise to *Cheirogaleus medius* (CME) chromosome 2 (Dresser and Hamilton, 1979; Dutrillaux *et al.*, 1986; Rumpler and Dutrillaux, 1986). (See Figure 2, bottom.)

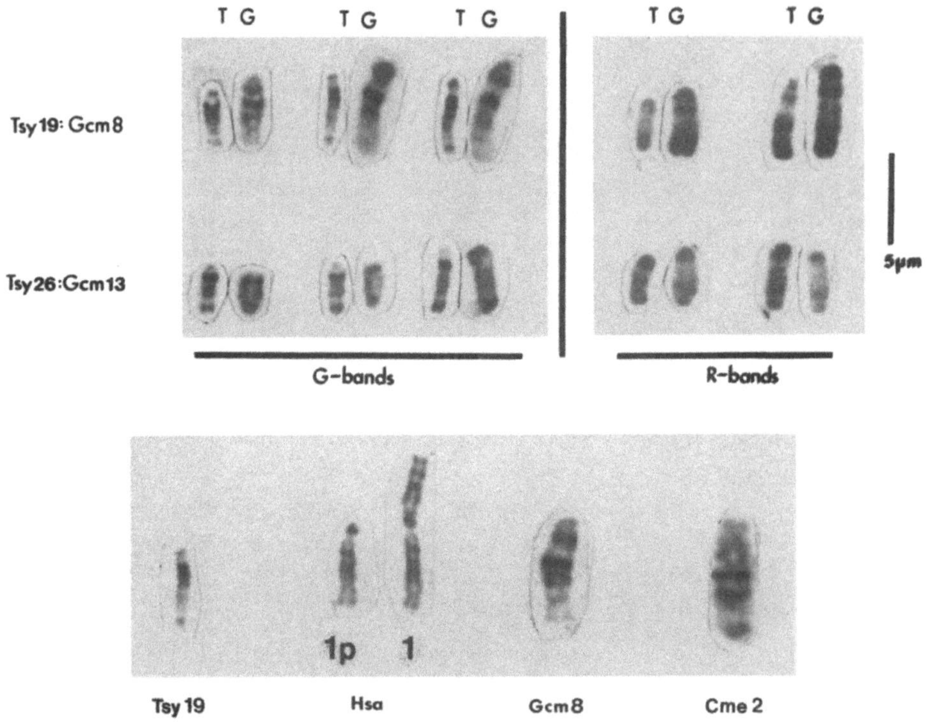

Figure 2. Top panel: G- and R- band similarities between TSY 19 and the distal two-thirds of GCM 8 (nomenclature of Poorman, 1982). Bottom panel: Relationship of TSY 19 and GCM 8 to HSA 1p and CME 2 (Dresser and Hamilton, 1979).

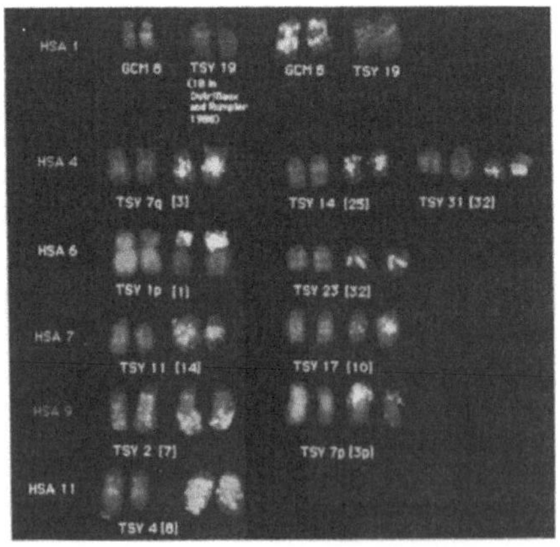

Figure 3. Fluorescence *in situ* hybridization of human chromosome libraries to TSY and GCM chromosomes.

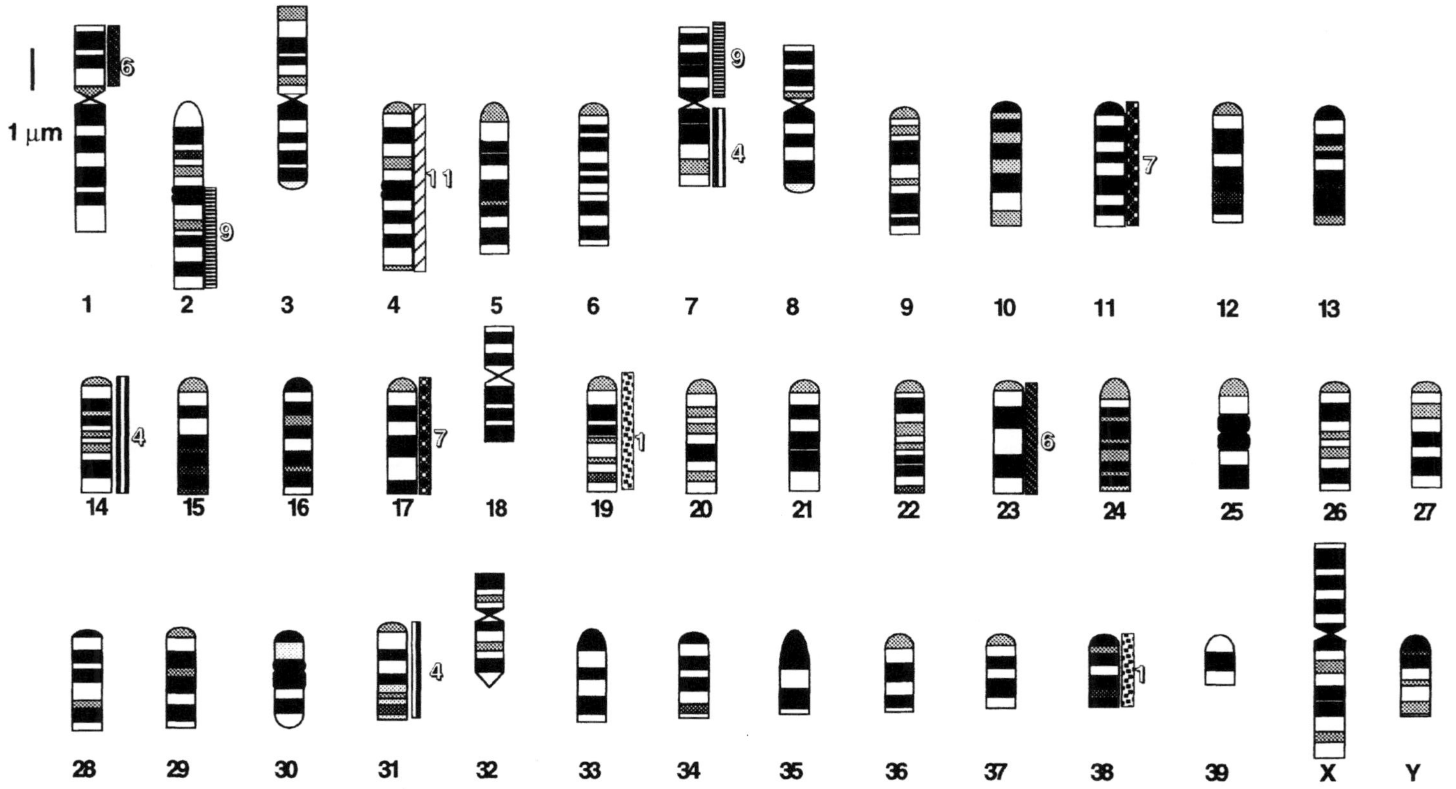

Figure 4. Human homologies of TSY chromosomes determined by *in situ* hybridization, indicated on the complete TSY G-band idiogram. Black numbers below chromosomes are TSY numbers; white numbers at the right of chromosomes are human chromosome numbers.

Fluorescence *in situ* hybridization results are shown photographically in Figure 3 and idiogrammatically in Figure 4. Hybridization of a human chromosome library, pBS1, to TSY and GCM chromosomes labels TSY 19 and 38, and GCM 8 and 21. The conjunction of banding similarity and common hybridization between TSY 19 and GCM 8 makes a stronger case for their homology than banding alone would. The hypothesis of homology could be further tested by hybridization of single-copy probes specific to the human 1p region. If in fact this element is homologous among these taxa, it represents a chromosomal segment conserved from the common ancestor of Primates.

Further tarsier-human homologies are identified by hybridization to other human libraries. HSA 4 hybridizes to TSY 7q, 14, and 31; HSA 6, to TSY 1p and 23; HSA 7, to TSY 11 and 17; HSA 9, to the distal portion of TSY 2 and to TSY 7p; and HSA 11, to TSY 4. In all of these cases, the tarsier banding patterns are sufficiently rearranged that the homologies revealed by hybridization could not have been predicted. This result and the similar study of Jauch *et al.* (1992) on gibbon karyotypes, which are drastically rearranged with respect to the primitive anthropoidean karyotype, illustrates the potential of chromosome painting for the identification of genetic homology in chromosomes with divergent banding patterns. The distribution of HSA 4, HSA 6, HSA 7 and HSA 9 sequences to multiple TSY chromosomes is not surprising in view of the large diploid-number differential between the two species. Indeed, *T. syrichta* exceeds the chromosome number (2n = 56) of the primitive primate karyotype reconstructed by Dutrillaux *et al.*, 1986; it is therefore probable that the evolution of the TSY karyotype from the primitive karyotype involved chromosomal fission.

CONCLUSIONS

G- and R-banding comparison of extended *T. syrichta* and *G. c. monteiri* chromosomes shows neither the widespread correspondences cited by Poorman *et al.* (1985) nor the total absence of similarity cited by Dutrillaux *et al.* (1988). The majority of the Poorman *et al.* (1985) similarities appear to have been artifacts of chromosome condensation (Healy and Poorman, unpubl. observ.); therefore, the karyotype does not constitute a source of evidence against tarsier-anthropoid sibship. Tarsier-galago G-band comparisons gave radically different results at the different levels of chromosome extension and different sample sizes represented by Poorman *et al.* (1985) and the present study, illustrating the importance of thorough characterization of the normal range of banding patterns for all chromosomes for the validity of intertaxon comparisons. Comparison of isolated examples of chromosomes may lead to spurious similarities; homologous chromosomes must have equally similar banding patterns throughout their ranges of extension.

A single primitive retention is visible in the *T. syrichta* karyotype. TSY 19 shows G- and R-band similarity to portions of GCM 8 and HSA 1p; *in situ* hybridization results are consistent with the hypothesis that these elements are homologous, and have therefore been retained from the ancestral primate karyotype.

These results illustrate the limitations of banding as an indicator of homology across wide phyletic distances and the utility of chromosome painting as an alternative indicator of homology. Further work with prosimian karyotypes can test hypotheses of homology based on banding comparison and establish fluorescence *in situ* hybridization as a new modality for chromosomal systematics.

ACKNOWLEDGMENTS

I am grateful to the staff of the Duke University Primate Center for tissue sampling, to J. H. Tepperberg and G. Briscoe for shipping assistance, to D. C. Ward in whose laboratory I performed the *in situ* hybridization work, and to M. D. Healy and G. Warren for commenting on the manuscript. This work was supported by NIH grant R01 HG00246 to David C. Ward and is publication # 612 of the Duke University Primate Center.

REFERENCES

Baba, M.L., Weiss, M.L., Goodman, M., & Czelusniak, J. (1982).The case of tarsier hemoglobin.*Systematic Zoology*, *31*, 156–165.

Ballard, S.G. & Ward, D.C. (1993) Fluorescence *in situ* hybridization using digital imaging microscopy. *Journal of Histochemistry and Cytochemistry, 41*, 1755–1759.
Collins, C., Kuo, W.-L., Segraves, R., Fuscoe, J., Pinkel, D., & Gray, J.W. (1991). Construction and characterization of human chromosome-specific plasmid libraries by subcloning inserts from Charon 21a libraries made by the National Laboratory Gene Library Project.*Genomics, 11*, 997–1006.
Czelusniak, J., Koop, B., Tagle, D., Shoshani, H., Goodman, M., Braunitzer, G., Kleinschmidt, T., de Jong, W.W., & Matsuda, G. (1988).Perspectives from amino acid and nucleotide sequences on cladistic relationships among higher taxa of Eutheria. In H. H. Genoways (Ed.) *Current Mammalogy* (pp. 545–572). New York: Plenum Press.
de Jong, W.W. & Goodman, M. (1988). Anthropoid affinities of *Tarsius* supported by lens α-A crystallin sequences. *Journal of Human Evolution , 17*, 575–582.
Dresser, M. E. & Hamilton, A. E. (1979). Chromosomes of Lemuriformes. V. A comparison of the karyotypes of *Cheirogaleus medius* and *Lemur fulvus fulvus. Chromosoma, 24*, 160–167.
Dutrillaux, B. & Rumpler, Y. (1988). Absence of chromosomal similarities between tarsiers (*Tarsius syrichta*) and other primates. *Folia primatologica, 50*, 130–133.
Dutrillaux, B., Couturier, J., Muleris, J., Rumpler, Y., & Viegas-Pequignot, E. (1986). Relations chromosomiques entre sous-ordres et infra-ordres, et schema evolutif general des Primates. *Mammalia, 50*, 108–123.
Healy, K. C. (1992). Chromosome banding as a source of phylogenetic information: a critique based on the case of *Tarsius*. Unpublished Ph. D. dissertation, Duke University.
Ikeuchi, T. (1984). Inhibitory effect of ethidium bromide on mitotic chromosome condensation and its application to high-resolution chromosome banding. *Cytogenetics and Cell Genetics , 38*, 56–61.
Jauch, A., Wienberg, J., Stanyon, R., Arnold, N., Tofanelli, S., Ishida, T., & Cremer, T. (1992). Reconstruction of genomic rearrangements in great apes and gibbons by chromosome painting. *Proceedings of the National Academy of Sciences USA, 89*, 8611–8615.
Johnson, G.D., Davidson, R.S., McNamee, K.C., Russell, G., Goodwin, D., & Holbrow, E.J. (1982) Fading of immunofluorescence during microscopy: A study of the phenomenon and its remedy. *Journal of Immunological Methods, 55*, 231–242.
Koop, B.F., Siemieniak, D., Slighton, J.L., Goodman, M., Dunbar, J., Wright, P.C.,. Simons, E.L. (1989). *Tarsius* δ- and β-globin genes: conversions, evolution, and systematic implications. *Journal of Biological Chemistry, 264*, 68–79.
MacPhee, R.D.E. & Cartmill, M. (1986). Basicranial structures and primate systematics. In D.R. Swindler & J. Erwin (Eds.) *Comparative Primate Biology, Volume 1: Systematics, Evolution, and Anatomy* (pp. 219–275). New York: Alan R. Liss, Inc.
Poorman, P.A. (1982). Banded chromosomes of *Galago crassicaudatus monteiri, G. c. garnettii*, and a subspecific hybrid. *Cytogenetics and Cell Genetics, 34*, 296–304.
Poorman, P.A., Cartmill, M., MacPhee, R.D.E., & Moses, M.J. (1985). The G-banded karyotype of *Tarsius bancanus* and its implications for primate phylogeny. *American Journal of Physical Anthropology, 66*, 215. (Abstract.)
Rumpler, Y. & Dutrillaux, B. (1986). Evolution chromosomique des Prosimiens. *Mammalia, 50*, 83–107.
Sambrook, J., Frisch, E.F., & Maniatis, T. (1987). *Molecular Cloning: A Laboratory Manual.* Volume 2. Cold Spring Harbor Laboratory Press.
Schwartz, J.H. & Tattersall, I. (1987). Tarsiers, adapids and the integrity of Strepsirhini. *Journal of Human Evolution, 16*, 23–40.
Simons E. L. & Rasmussen, T. (1989). Cranial morphology of *Aegyptopithecus* and *Tarsius* and the question of the tarsier-anthropoidean clade. *American Journal of Physical Anthropology, 79*, 1–23.
Verma, R.S. & Babu, A. (1989). *Human Chromosomes: A Manual of Basic Techniques.* New York: Pergamon Press.
Wienberg, J., Jauch, A., Stanyon, R., & Cremer, T. (1990). Molecular cytotaxonomy of primates by fluorescence *in situ* hybridization. *Genomics, 8*, 347–350.
Wiegant, J., Galjart, N.Y., Roop, A.K., d'Asso, A. (1991). The gene encoding human protective protein (PPG B) is on chromosome 20. *Genomics, 10*, 345–349.

BEHAVIOR OF CAPTIVE *LORIS TARDIGRADUS NORDICUS*: A QUALITATIVE DESCRIPTION, INCLUDING SOME INFORMATION ABOUT MORPHOLOGICAL BASES OF BEHAVIOR

Helga Schulze and Bernhard Meier

Ruhr-Universität Bochum
Institut für Anatomie
Abtlg. Funktionelle Morphologie, MA 0/46
44780 Bochum, Germany

INTRODUCTION

In 1980, a captive breeding program for *Loris tardigradus nordicus* was started at the Ruhr-University (Meier, 1989). Experience showed that successful husbandry and breeding of slender lorises is not difficult if the behavioral needs of the animals are taken into account. On the other hand, the species is very susceptible to improper captive conditions which easily lead to reproductive failure or deaths. This paper was written to provide knowledge about slender loris behavior which might also contribute to better husbandry conditions.

MATERIAL

Nine wild-caught slender lorises (*Loris tardigradus nordicus* from Polonnaruwa, Sri Lanka) and about 70 animals up to the fourth captive generation were observed. The animals are kept alone or in family groups of up to eight animals. The cages are connected by passages, measure 2.5-40 m^3 and are densely equipped with natural branches of various diameters and orientations. Throughout the year day length is kept constant (13 hours, neon light). Frequent cleaning of the branches is avoided. Slender lorises are very cautious and excitable animals, but we found that too thorough protection from disturbance increases their susceptibility to psychic stress. Therefore, we habituated our animals to a certain amount of unfamiliar stimuli such as visitors, a radio automatically switched on at intervals and some "environmental enrichment".

Creatures of the Dark: The Nocturnal Prosimians
Edited by L. Alterman *et al.*, Plenum Press, New York, 1995

METHODS

More than 150 hours of *Loris* behavior were recorded with a remote-controlled infrared video camera, and slow motion video analysis was employed to recognize typical behavioral patterns. A standard ethogram for computerized behavioral comparison was developed and used to recognize stereotyped features of behavior. Typical examples of all observed behavioral patterns were documented by photos. The results were re-examined by systematic and routine observations after habituating the animals to human observers. The illustrations are redrawn from still video pictures and slides.

THE BEHAVIORAL REPERTOIRE OF SLENDER LORISES

Adaptations to Arboreal Life, Use of the Hands and Feet

Slender lorises are well adapted to grasping locomotion on a variety of substrates (Preuschoft *et al.*, this volume). Fig. 1 illustrates the use of hands and feet of *Loris*. The strong feet (Figs. 1a-f) are forceps-shaped. The strong first digit and the opposed digits 3-5 are used for locomotion and clinging to the substrate. The second digit can be extended independently (Fig. 1f). It has a toilet claw which is used for scratching and for cleaning of the auditory canal. During locomotion, the toilet claw usually does not touch the branch.

Figs. 1g-m show the use of hands in *Loris*. The fingers close synchronously. Precision grips with the use of single fingers, as seen in simian species dealing with small objects (Preuschoft *et al.*, 1993; Christel, 1993), do not occur. Whole-hand control can be explained by the finger flexor tendons descending from common muscle bellies; Osman Hill (1953) gives a detailed description of digital tendons. The thumb and the opposed digits 3-5 can be clasped around the substrate like a forceps (Fig. 1i).

Prey and other objects are seized in a stereotyped manner, which indicates the existence of a fixed action pattern. The object is first fixated with the eyes. Activation of the behavioral pattern obviously depends on the presence of an optical stimulus (Jolly 1964a). Before or in the beginning of the grasping movement, the ears are often laid back. If prey is removed after the start of the action, the grasping movement is finished, the empty fist is drawn back, and often the mouth opens before the behavior is stopped. Very similar, highly stereotyped grip patterns in *Galago senegalensis* were analyzed by Bishop (1962, 1964) who also observed that, when mealworms were offered on a rotating turntable, the animals were unable to compensate for the movement of the bait, but struck the place where it had been in the beginning of the movement. In *Loris*, the grip may be performed in a very precise way or fiercely and with little precision, both with the same movement pattern. Difficult manipulatory tasks may be solved by combined use of both hands and sometimes the mouth.

The prey-catching grip is used for a variety of other purposes such as eating of nutritious liquids (baby pap), exploratory and play behavior. During locomotion, food is carried in the mouth (Fig. 1l). Occasionally the hands are used for carrying of infants which do not properly cling to the mother's fur (Fig. 1m).

The mobility of limbs and trunk contributes to safe locomotion in the trees. The legs of the Lorisidae are very well vascularized by retia mirabilia (Miyake *et al.*, 1991) and are, in spite of their slender appearance, of considerable strength and stamina. Slender lorises have an unusually low metabolic rate, specialized for energy-saving (Müller *et al.*, 1984). For longer-lasting behaviors and for locomotion they prefer to stay on top of more or less horizontal substrates (Fig. 2a) or, to a lesser degree, suspended postures (Figs. 2b, c). Clinging to vertical branches consumes more energy and is mainly used for locomotion upwards and downwards. For resting, sitting postures are most common; suspended

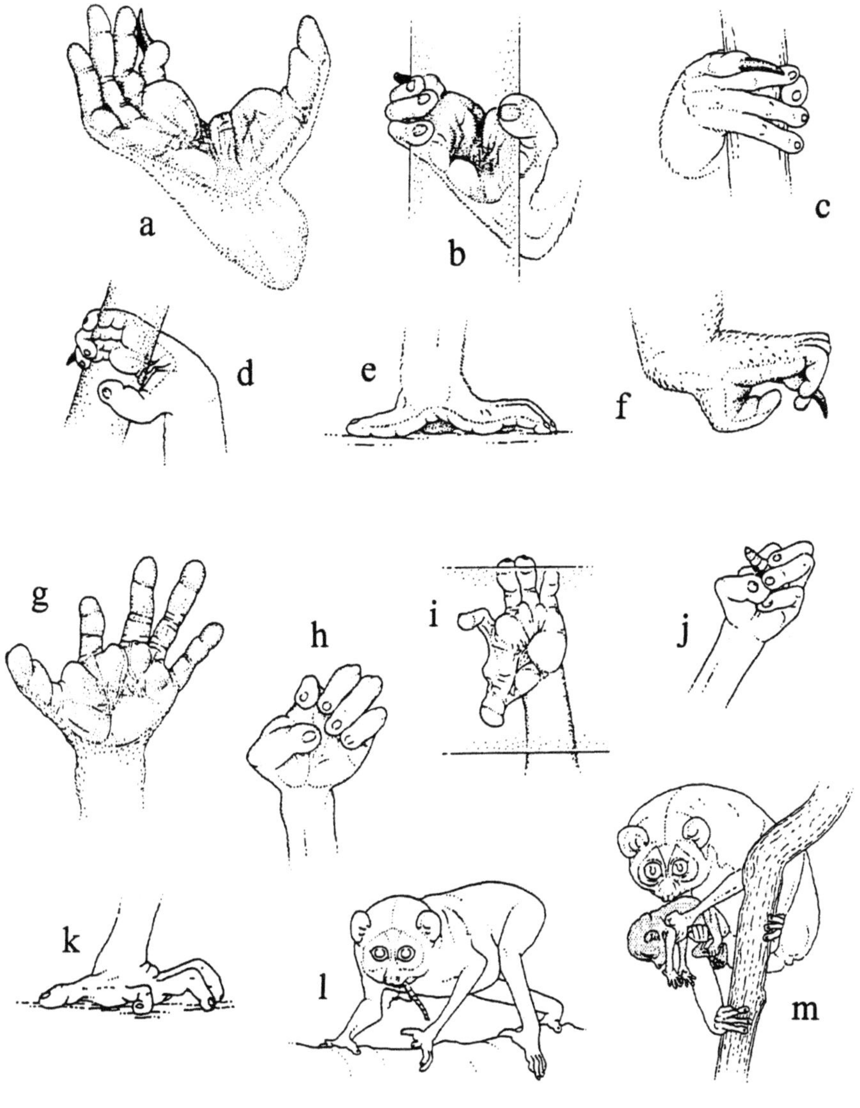

Figure 1: use of the hands and feet. a: loris foot showing the forceps-shape. **b, c:** forceps grip of the foot during locomotion (b: plantar side seen through a glass tube). The toilet claw of the second digit is not used for locomotion, it is often laid over the third digit. **d:** hook-shape of the foot employed for suspended postures. **e:** foot during locomotion on flat ground. **f:** for scratching, the second digit with the toilet claw can be extended while digits 3 to 5 are flexed. The first digit is usually flexed, but may also be extended. **g:** loris hand, redrawn after a photo in Petter and Hladik (1970). **h:** the fingers close synchronously with the thumb opposed to digits 3-5. **i:** forceps grip of the hand during locomotion (palmar side seen through a glass tube). Reduction of length of the second digit by involution and flexion diminishes bending stress when the hand is rolled off (Nieschalk and Demes, 1993). **j:** a mealworm has been seized with the typical whole-hand grip. **k:** hand during locomotion on flat ground. **l, m:** carrying of objects. During locomotion, food is carried in the mouth (a). If a loris infant is too weak to cling to its mother's fur, it may be abandoned, but the mother may also carry it with one or both hands for hours (b), using the feet and only one hand for locomotion. Use of the mouth for carrying infants has never been observed.

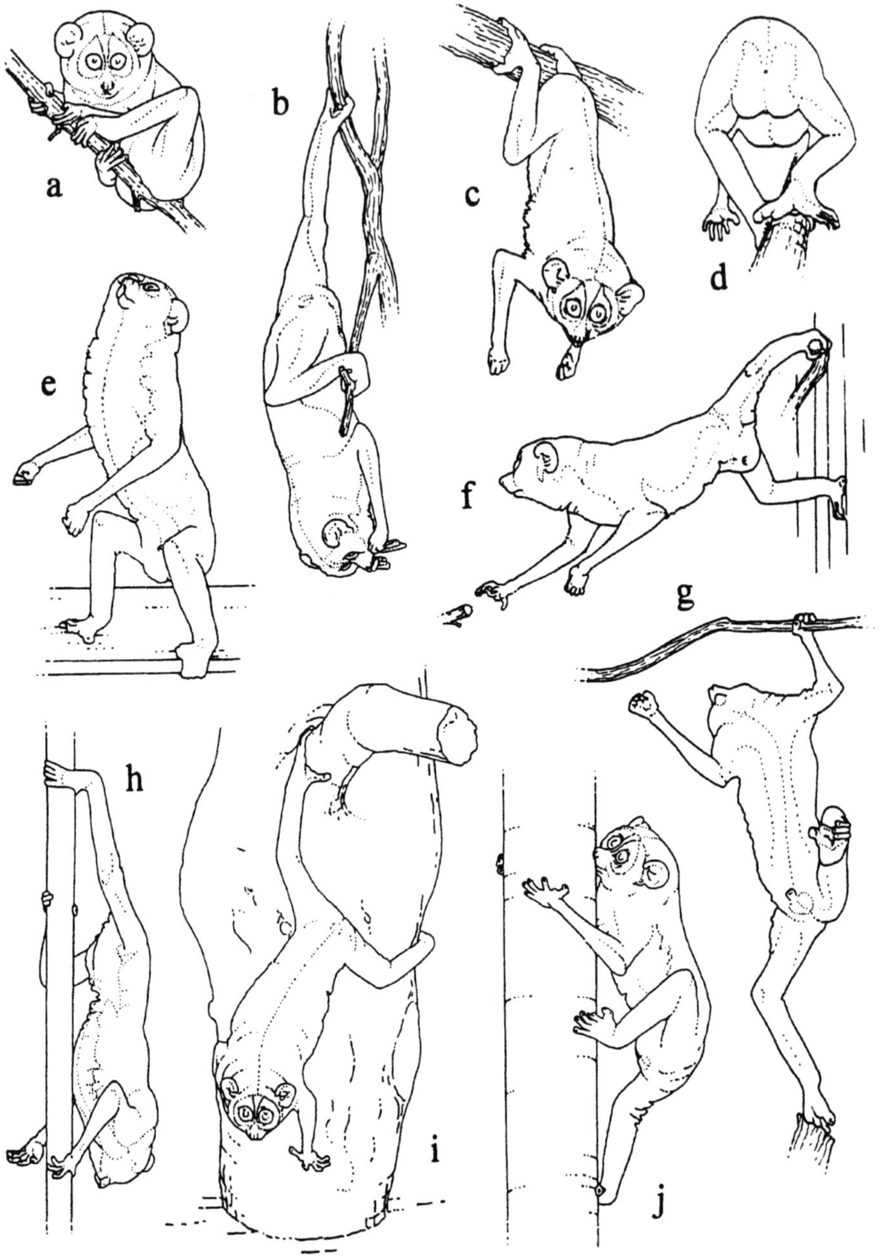

Figure 2: postures and locomotion. a: sitting on top of the substrate is the usual posture for resting. **b, c:** suspended postures, with both feet attached to the substrate, are frequent. **d:** walking on top of more or less horizontal substrate is the preferred way of moving. **e:** bipedal stand allows use of the hands for handling of objects. The animal pictured here clearly disliked standing upright on a horizontal board without bending the toes of at least one foot around the edge for safe hold (redrawn after a photo by Tardieu). **f:** Bipedal postures also allow reaching for distant branches or prey. **f, g:** "bridging" over substrate gaps. **h-j:** vertical substrates are mainly used for locomotion upwards and downwards (both head first). The animals prefer thin branches which allow a safe grip (h). When climbing up and down large trunks (i), the animals need a rough surface to which they can cling.

postures are often used for comfort behavior and play. Clinging to the substrate merely with the hands occurs only exceptionally. Bipedal postures, both suspended and standing, are frequent. They allow the use of the hands for object-manipulation, enable the animals to cross gaps in the substrates (Figs. 2f, g) and to reach distant prey. For horizontal bridging, starting from a vertical branch, a loris can use a horizontal posture (Fig. 2f), suspended with the upper hindlimb and using the lower one as a brace (see Preuschoft *et al.*, this volume). This posture allows swinging of the body in various directions and can be maintained for a considerable time. For horizontal bridging, starting from a horizontal branch, the animal has to employ a fast movement resembling a jump, with the feet remaining attached; if it fails to reach the branch it aimed for, it falls back into a bipedally suspended posture.

Slender lorises rely on camouflage rather than on rapid flight. They are specialized in deliberate grasping locomotion with silent, inconspicuous movements. Jumping or locomotor phases without contact to the substrate do not occur. For locomotion, our lorises prefer twigs of a diameter which allows a safe grip, especially when the substrate is more or less vertical (Fig. 2h). But they also show a considerable ability to climb up and down vertical trunks with a diameter up to 10 cm (Fig 2j), even if the surface is smooth. Locomotion is then slow and cautious. Subramoniam (1957) observed that captive slender lorises, which possibly belonged to the smaller subspecies *L. t. malabaricus*, avoided vertical substrates with diameters of more than 2.5 cm and were unable to climb smooth trunks of 10 cm diameter. Walking on top of branches is the preferred way of moving. The long, slender limbs allow an average stride length of 43 cm (Nieschalk, 1991; Nieschalk and Demes, 1993), and beside the usual flowing, noiseless locomotion, a fast, almost trot-like "running" without airborn phases occurs, especially on continuous horizontal substrates. Quadrupedal locomotion suspended under branches is employed for short distances when an obstacle does not allow walking on top of the substrate. Male lorises especially often show a considerable amount of locomotor activity; some very active specimens developed stereotyped locomotor patterns even in cages of more than 10 m^3, furnished densely with branches. Correspondingly, Rasmussen (1986) found that in captive *Nycticebus* "... most of the traveling observed ... did not seem to be directed to a specific goal; usually lorises circuited the room just for the sake of locomotion". Females are usually less active.

The Sensory World of *Loris*: Relevant Stimuli and Communication

Our lorises are strictly nocturnal although some activity may occur at the end of the daylight period when the animals are hungry. Diurnal activity in *L. t. lydekkerianus* in the wild was reported by Johnson (1984). Lorisoids show typical adaptations for nocturnal life. Their large eyes are provided with a tapetum (Osman Hill, 1953); the retina possesses narrow attenuated rods and lacks cones (Detwiler 1940; Osman Hill, 1953). The overlapping visual fields of the forward-oriented eyes indicate a good stereoscopic vision (Osman Hill, 1953). Detwiler (1940) found that in slow lorises the ciliary muscles are unusually large in relation to the lens, and he concluded "that the accomodation mechanism is far superior to that of man".

The number of groups of tactile vibrissae is largely reduced in lorisoids, as compared to the Malagasy lemurs (Osman Hill, 1953). In our slender lorises few short mystacial and supraorbital vibrissae are left, and the eyes are protected by true eyelashes. Single straight hairs can be found, distributed all over the limbs and body. These project about 6-10 mm over the silky, undulated usual fur hairs (Fig. 3a) and obviously have a tactile function. A very old *Loris* male at Ruhr-University, blind for several years, shows a distinct reaction when these hairs are touched. In spite of his blindness, this animal still moves with considerable speed without touching obstacles, an ability which can be explained by the combined use of memory, tactile stimuli and olfactory orientation. In cases of loss of

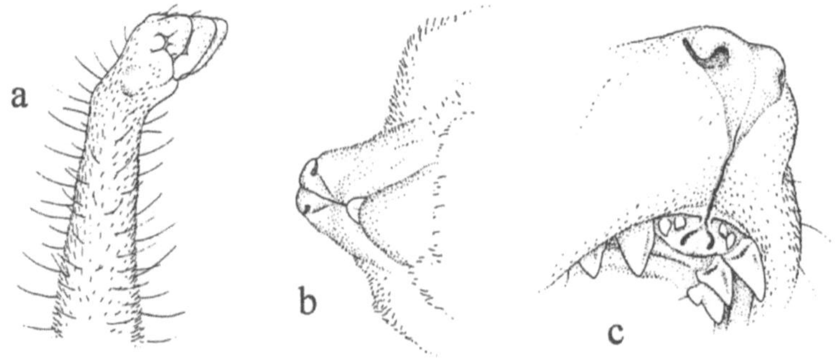

Figure 3: Sensory abilities of the slender loris. a: fine straight hairs which protrude over the usual fur might have a tactile function, for instance for locomotion in the dark. **b:** during close approach to objects for exploration and during naso-genital approach, often a hardly visible, lizard-like forward-stretching of the tongue in quick succession can be observed. Some animals also stretch out their tongue continuously for several seconds. **c:** the moist rhinarium and the entrances of the vomeronasal organ are connected by a furrow which descends from the philtrum and clefts the upper lip. Lip and gum are connected by two folds of labial skin which pass the diastema between the median incisors on either side of the furrow (Hofer, 1977).

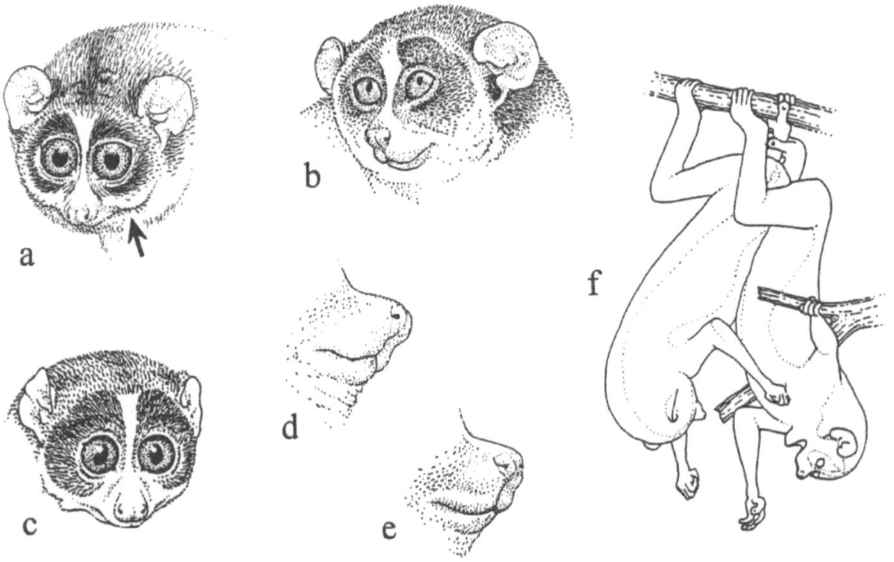

Figure 4: facial expressions and ear movements. a: ears in normal position. When an animal perceives something unfamiliar; fur gaps may become visible in the corners of the mouth (indicated by arrow), sometimes extending to the ears. They possibly indicate stress. **b:** facial expression indicating severe social, or sometimes environmental, stress; the ears are drawn down to the sides of the head. **c:** Laying back of the ears is regularly seen in the initial phase of the prey-catching movement and sometimes during severe stress; it seems to be a protective behavior rather than a signal (here: animal caught for tattooing, with wide-open, protruding eyes indicating fear). **d:** in calm animals, the lip cleft is s-shaped. **e:** in stress situations, the lips may look tense, the lip cleft almost straight. **f:** open-mouth-face during playwrestling is a playful biting intention movement of short duration and fairly inconspicuous, its function as an optical signals is doubtful.

orientation, for instance after sleeping, he employs sniffing and feeling around with the hands. According to Pariente (1979), behavior of wild *Microcebus coquereli* and other leaping prosimians "implies that they possess considerable specialization of the higher nervous centers for the storage of a large body of information about the environment"; and "even the non-leaping prosimians, such as the lorisines, seem to memorize features of their environment". This is certainly true in our lorises who react with signs of stress as soon as they notice the slightest changes of their environment. A study of pathway selection by slender lorises like those examining pottos and angwantibos (Charles-Dominique, 1977), has not yet been done in slender lorises. However, their long-range vision would certainly be sufficient for optical recognition of possible detours from a distance.

For olfactory exploration, the object of interest is approached and sometimes touched with the moist rhinarium; audible sniffing and biting of the object occur. An inconspicuous forward-stretching of the tongue (Fig. 3b) in quick succession, as in reptiles, may be observed during exploration at close proximity. This behavior may be connected to the accessory olfactory system (see Schilling and Evans, this volume). Some animals often keep the tip of their tongue stretched forward for several seconds when interested in a (sometimes distant) stimulus. The purpose of this behavior might be an enlargement of the moist body surface (rhinarium) which collects soluble traces of substances from the air. Chemoperception in *Loris* includes a variety of behaviors: substances are inhaled together with the air by sniffing or are collected in the liquid on the rhinarium or tongue; from here the liquid may be transported to the papilla palatina by movement of the tongue or through the moist median furrow of the rhinarum (Fig. 3d), tested by taste buds on the tongue or the papilla palatina and then transported into the vomeronasal organ. In unsystematic tests, strongly smelling substances (perfume, fresh carnivore feces) sometimes caused signs of disgust, but curiosity rather than caution. Scent plays an important role in territorial and sexual behavior. Olfactory searching with the nose close to the substrate during locomotion is observed in connection with urine-marking and exploration of unfamiliar cages. Olfactory searching for food occurs in some specimens whereas others rely more on optical stimuli (see section on food consumption). Moving prey behind glass provokes immediate prey-catching behavior; audible, but hidden objects are usually ignored. Only small objects which fall down and emit audible signals often cause considerable interest and sometimes pursuit over a longer distance; in this case, however, optical stimuli cannot be excluded. Noise caused by moving of twigs by the keeper, for instance during cage cleaning, usually causes symptoms of stress or flight reactions.

Optical stimuli in communication consist of variation of movement speed, different postures and changes of orientation towards conspecifics. Abrupt, jerky movements indicate an aggressive mood. Normal movements are often fast, but flowing and noiseless. Very slow, hesitating movements and frequent freezing to immobility are signs of environmental stress. Suspended locomotion under twigs may be a signal in sexual behavior (the male often reacts with immediate attempts to mount and copulate). When waiting for tidbits from the keeper, the animals often show suspended postures, sometimes in connection with body-wiggling. Forward-stretching of the arm during allogrooming, a behavior by which the brachial scent gland region is presented, might be an optical as well as an olfactory signal. Conspecifics usually react with eager allogrooming. Some facial changes and ear movements can be observed (Fig. 4), but their function as signals is not clear.

Acoustic Communication

Descriptions of vocalizations do not always allow a clear identification. Therefore, sonograms from *L. t. nordicus* from our colony have been included with kind permission of H. Helmich (1987, unpublished). Table 1 gives a review of synonyms of *Loris* vocalizations used in literature. Some vocalizations of *Nycticebus* have been included for comparison because acoustic communication in this species has been studied in detail (Zimmermann 1985, 1989). Vocalizations in slow lorises, however, include more and different calls than *Loris* vocalizations, and similar calls may be uttered in different contexts.

Table 1: synonyms for slender loris vocalizations (German synonyms in parentheses). A question mark behind a synonym indicates that context and description are similar, but vocalizations might be not identical. Similar *Nycticebus* calls, included for comparison, partly occur in a different context.

***Loris* vocalizations** Authors	Rasmussen, 1986	Helmich, unpubl.	Subramoniam, 1957	***Nycticebus* voc.**
whistle (Pfiff)	whistle	(Pfeifen)	whistle	whistle; II.1-II.3[1]
chitter (Meckern)	chitter	(Schnattern)	distress sound;	long kecker; II.5[1]
screeching (Zirpen)	chitter	(Schnattern)	screeching ?	
monosyllabic chitter (Abwehr-Laut)	-	-	- -	short kecker[2] ?
zic call (Zick-Ruf)	infant click	-	-	zek-call[1,2]
-	-	-	-	ultrasound ?[2]
krik call[3] ("Kcht")	-	(Schniefen)	-	krik call[3], ek-call: II.4 ? [1]
growl (Fauchen)	pant-growl	(Fauchen)	growl	snarl: II.7[1]; grunt: II.8[1] ?
-	-	-	warning sound	
scream (Angstgeschrei)	scream	(Schreien)	S.O.S.	scream[1], II.6[1]

[1] Zimmermann 1985; [2] Zimmermann 1989

[3] We adopted the designation "krik call" from Rasmussen (1986), who used it for a comparable *Nycticebus* call emitted in a similar context (see below).

The Whistle is a loud, usually monosyllabic, occasionally di- and trisyllabic vocalization, each syllable ending with a decrease of frequency. The duration ranges from short low squeaks to sequences of loud, clear whistles lasting more than 5 seconds. Frequencies up to 14 kHz, with a maximum intensity between 6 and 9 kHz, have been recorded (Helmich, unpublished). Whistles are uttered in aggressive excitement, for instance if an adult unrelated conspecific of the same sex approaches the cage of the animal. They may be emitted by both sexes, but are much more common in males. If quarreling occurs within a family group, one dominant animal may whistle, often in connection with noisy locomotor display (see below), while inferior group members show signs of social stress and avoidance behavior. Whistling may cause immediate aggression in neighboring groups; animals in other cages may also begin to whistle. Petter and Hladik (1970) observed that whistles of wild lorises were emitted two or three times per night and could be heard over a distance of 100 m by humans. In *Nycticebus*, whistling in connection with sexual interest or as a contact seeking call has been described (Rasmussen, 1986; Zimmermann, 1985). In our adult slender lorises, no whistling occurs in such a context. Low whistle-like squeaks by abandoned infants (in connection with zic calls, see below) might be a similar vocalization, but occur only rarely.

Chitter consists of a series of short rhythmic clicks which may last up to 4 minutes. With increasing intensity, chitter turns to voiced, cricket-like screeching. Durations of 50 ms of the single clicks, with breaks of 12,5-30 ms, and frequencies up to 20 kHz have been recorded (Helmich, unpublished). Chitter is an intraspecific defensive vocalization, often connected with defensive staring (turning of the face towards the conspecific) or pushing

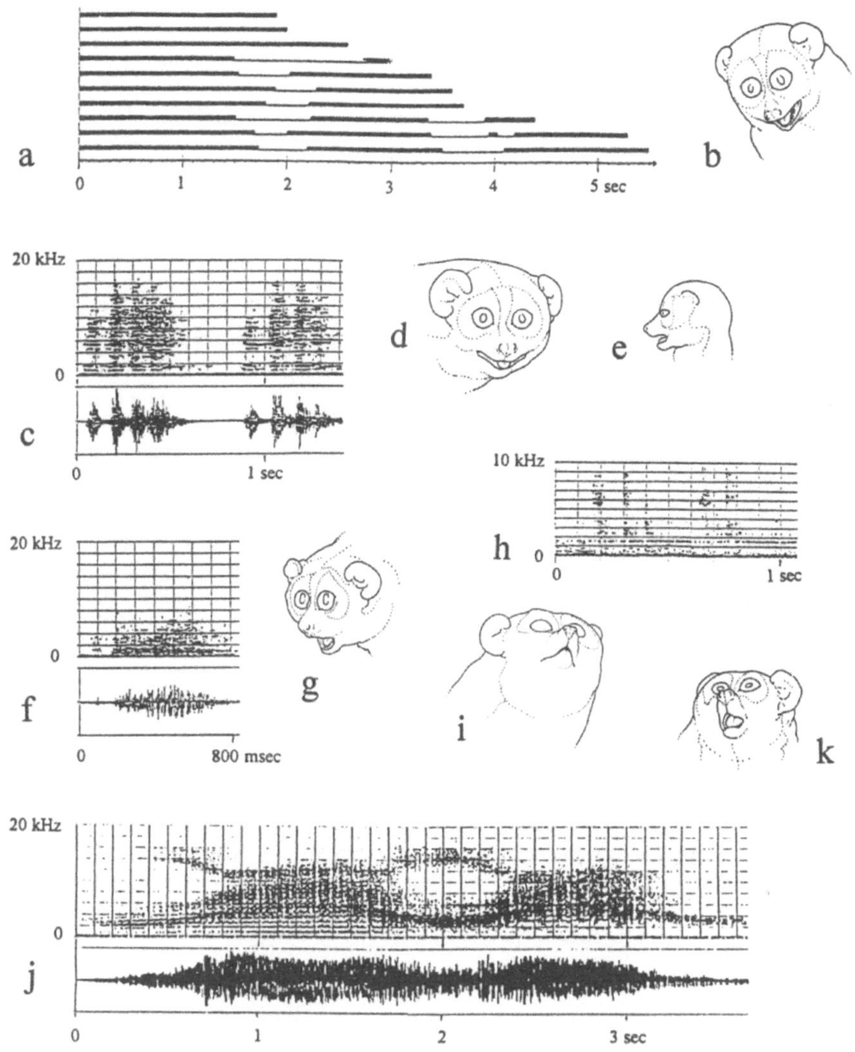

Figure 5: vocalization. a: duration of mono-, di- and trisyllabic whistles. **b:** facial expression during whistling **c-k:** sonograms and facial expression during vocalization: **c, d:** chitter. **e:** infant zic call. **f, g:** growl. **h, i:** krik call. **j, k:** scream.

away of an opponent with one or both hands. Chitter as a low intensity threat is frequent in females; males may screech during severe quarreling, but rarely chitter. Screeching occurs in connection with abrupt movements, sometimes urination and defecation indicate considerable excitement. According to Subramoniam (1957), chitter and a "warning sound" (repeated "krichit-krichit" in quick succession) were uttered when lorises were seized by a human or threatened by a dog. In our animals chitter directed toward humans occurs rarely. No other rhythmic vocalization in the described context was ever heard.

Monosyllabic Chitter is a single sharp smacking or crunching noise uttered by females when approached by a male. It is probably identical with the higher, but otherwise very similar monosyllabic zic call of *Loris* infants (see below). The male usually reacts with retreat, turning away of the head or with an appeasing krik call (see below).

The Zic Call is a high, sharp, usually very short monosyllabic call uttered by infants. With increasing intensity, it turns to voiced disyllabic and trill-like calls of longer duration. In rare cases a single squeak, resembling a low short whistle, may be emitted between two zic calls. Zimmermann (1989), who described "zek-calls" in *Nycticebus* infants, also recorded pure ultrasound calls, "presumably being high frequency variants zek calls". In *Loris*, no study of the emission of ultrasounds has been made. The zic call obviously signals unwellness, for instance when an infant dislikes grooming or has been separated from the mother; high intensity zic calls occur when an infant is abandoned for a longer time or hurt. The mother usually reacts by approaching and picking up the infant, or she may wait until it has mounted her on its own. She may, in addition, try to appease the infant by krik calls. Other group members, too, may react by approaching and handling the infant. Continuous high intensity zic calls cause considerable excitement, not only in the mother who may start to growl (in one case, even sham-attacks against the infant were observed), but also in animals in neighboring cages.

The Krik Call is a low short vocalization, single or repeated at irregular intervals, which sounds like a combination of a crackling noise and a hiss. With increasing intensity it turns to a voiced, whining or squeaking sound which resembles a human baby's voice. Durations of 50-75 ms and frequencies up to 4-9 kHz have been recorded (Helmich, unpublished). We adopted the designation for this call from Rasmussen (1986) who used it for a "string of short soft clicks" by *Nycticebus* observed in a similar context, but did not hear it in *Loris*. Sonograms of similar *Nycticebus* calls ("calls type II.4", "ek-calls") published by Zimmermann (1985) resemble the *Loris* sonogram in Fig. 5h.

The krik call is an appeasing vocalization regularly heard from males when they approach a female for sniffing. During estrus, males frequently emit krik calls while pursuing females; in a few cases male and female were heard duetting. A blind old male, attacked by a female he approached during estrus, uttered loud voiced krik calls and high calls closely resembling infant zic calls; the attack was immediately stopped. When group members quarrel, infants may become so excited that they threaten or attack their mother. In such cases, females sometimes use the krik call, combined with allogrooming, for appeasing the infant.

Growl is an unvoiced expiratory vocalization, resembling the spit of a cat. It may include voiced sounds. Growl may be repeated with every exhalation, alternating with a voiced sound during inspiration, as long as a perceived threat continues. Loudness varies with the intensity of the stimulus. Growl is mainly a defensive threat against predators; in intraspecific encounters it is used only occasionally when the animal is very excited. Growling is often connected with bipedal defensive postures, attacks or mock attacks or abrupt lateral swaying of the body. Fierce biting or emission of an odor from brachial gland exudate may occur. Still (1905), who kept both slender lorises and cobras in captivity, observed that a loris threatened by a cat gave a "perfect imitation" of cobra threat behavior, including the "regular breathing sound", now and then a "cobra's hiss", and an erect bipedal posture in connection with lateral swaying.

Scream: with increasing intensity of fear, growl may suddenly turn to long, loud screaming like the screams of a cat. Screams are an expiratory vocalization and, like growling, may be repeated as long as the perceived threat continues, only interrupted by short periods of inhaling. Duration of screams was always more than 0.5 seconds; frequencies up to 18 kHz were recorded (Helmich, unpublished). Screaming occurred in rare cases when an animal was caught and handled, indicating extreme fear. It is usually accompanied by the emission of odor from brachial gland exudate. Screaming during intraspecific attack fights, as reported by Rasmussen (1986), was never heard in our animals.

Besides vocal communication, rattling of twigs is apparently used as an acoustic signal. If quarreling occurs in a group, dominant aggressive animals show a locomotor display in the course of which they produce a continuous rattling of twigs. Most of this noise is caused by rapid locomotion with conspicuous, jerky movements which looks clearly different from the usual silent and flowing movements. Several other techniques are employed to produce noise, for instance brushing against twigs with the body or pulling along twigs with the feet which, when finally released, spring back. During running, the animals may repeatedly "fall" down from the branch by releasing the grip of the forelimbs while both feet remain firmly attached, both body and branch abruptly swinging down. This behavior looks unintended, but it has repeatedly been observed in this context and does not occur during rapid locomotion in a different context. An aggressive animal has been observed standing, firmly holding a twig with both hands and shaking it by rapid flexion and extension of both arms; in this case intention was evident. A noisy locomotor display may also be shown in playful mood, and some animals regularly cause considerable noise with twigs or dry leaves when impatiently waiting for food or tidbits. Rattling of twigs, caused by the keeper, causes flight reactions or signs of stress; when animals are disturbed, sudden flight often causes an initial rattling of twigs which might possibly be a substitute for vocal alarm calls.

Urine Deposition, Territorial Marking and other Scent Production

In captivity, feces appear to have no olfactory function in *Loris*, they fall to the ground and are ignored by conspecifics. The olfactory function of urine, however, is evident. In captive *L. t. lydekkerianus*, Seitz (1968) measured an average daily quantity of urine of 6.3 ml in a female (n=9 days) and 8 ml in a male (n=11 days). In our animals, males usually save their urine for more frequent marking with small quantities whereas emission of larger quantities at a time is observed more often in females, especially after sleeping period. Urine-marking is performed in the same way by both sexes. For dot marking (Fig. 6a), locomotion is slowed down or stopped for a moment, and a small quantity of urine is deposited, often with one leg raised. Dots of urine are deposited in irregular intervals, often on parts standing out of the substrate surface. Rhythmic micturition as described by Ilse (1955) does not occur in our animals. On smooth horizontal surfaces, the penis or clitoris are often dragged over the substrate during locomotion, leaving a continuous trace of urine (trail marking, Figs. 6b, c). Dot and trail marking may be combined. Some shy and cautious individuals avoid cages urinemarked by strange conspecifics for hours, but most of our animals enter without hesitation and immediately start sniffing and dot marking branches. In cages which contain new branches without previous urine-marks, dot marking is also increased. Seitz (1968) indicated that in *Nycticebus*, urine-marks serve for both olfactory orientation in the dark and territorial marking. He also found sexual differences in the spatial distribution of urine in this species. Allomarking with urine occurs rarely. Occasionally a male dot-marked his mate or infant. Urine-marking on infants sometimes caused threat reactions (staring, pushing with the hand, chitter) by the infants or their mothers. In one exceptional case, an adult female who met her mother after months of separation showed "passing-over" as described for angwantibos (Manley, 1976): she repeatedly passed back and forth over the mother, each time trail-marking until the mother's back was completely wet. This action was finished by intense mutual grooming and playwrestling. Seitz (1968) describes a case of passing-over (a female marked her mate) in his captive *L. t. lydekkerianus*, and Rasmussen (1986) reports passing-over as a rare behavior in slender and slow lorises. Genital-scratching grooming as described for the potto (Manley, 1976) was not observed in *Loris*.

Urine-washing (Fig. 6d-f) is performed in a stereotyped way by animals of either sex: the animal raises hand and foot on one side; penis or clitoris are touched by the rhythmically moving hand. The urine emitted is then distributed over palms and soles by "washing" movements of both hand and foot. (In contrast to the description of Andrew and Klopman, 1976, the sole of the foot is pointing downwards during "washing", which might indicate some differences among prosimian species). This procedure is repeated with hand and foot on the contralateral side. Up to four changes of side occur. Very young animals may occasionally start urinewashing as in Fig. 6d, but then continue with scratching of the fur instead of the "washing" movement (Figs. 6e, f). Urine-washing in our animals is usually restricted to more or less horizontal substrates on which they can stand; this supports the observations of Ilse (1955) in *Loris* and of Doyle (1975) in *Microcebus* and several galago species whereas the lorises at the Duke University Primate Center, according to Rasmussen (1986), urine-washed throughout their rooms. Harcourt (1981) mentions several possible purposes of urine-washing. Moistening of the skin for better climbing is one (Welker, 1973), thermoregulation and olfactory marking are also discussed. The latter is unlikely in our lorises as the frequency of urine-washing is not remarkably increased in unfamiliar cages. Urine-washing is regularly seen when activity starts after sleeping periods. Besides, males also urinewash with nervous-looking movements while in pursuit of females in estrus, and during some unsystematic tests with musk-perfume, several animals urinewashed immediately, appearing agitated and sometimes with signs of disgust (flinging of food, abrupt shaking of hands, avoidance), as soon as they perceived the smell. Our impression that, in excited slender lorises, urine-washing sometimes serves as a displacement behavior is shared by Schilling (1979) who observed urine-washing in situations eliciting attention, curiosity, fear etc. in several prosimian species. Osman Hill (1938) and Andrew and Klopman (1976) mention urine-washing in prosimians in connection with increased micturition due to fright or excitement.

Olfactory communication includes direct sniffing of conspecifics. Osman Hill (1933, 1953) mentions large glands found in the female clitoris which, during estrus, is frequently inspected nasaly by males. Genital scent gland regions in both sexes with a female "pseudoscrotum", as described in pottos (Sanderson, 1940; Charles-Dominique, 1966; Manley, 1976) and anwantibos (Manley, 1976) do not occur in *Loris* although Montagna and Ellis (1960) mention some unspecified glands in the scrotal skin. Inspection of the male genital region by females occurs only exceptionally. Unfamiliar conspecifics may be cautiously sniffed on a part of the body within reach; no ritualized form of sniffing was observed in such situations. The brachial organ, a specialized glandular region on the median side (flexor surface) of the forearms, has been discovered by Osman Hill (1956) and described in detail by Montagna and Ellis (1960). It is presented to the conspecific, by forward-stretching of the arm, during social grooming (see section on comfort behavior) and is eagerly allogroomed. If our animals are frightened by catching or other human activities, secretion of a badly smelling fluid can be observed in the region of the brachial organ (see also Alterman, this volume), but no similar smell was noticed in cases of social stress.

A nasal scent secretion like that of *Nycticebus* (Zimmermann *et al.*, 1979) is not known in *Loris*; facial rubbing on the substrate as a regular reaction to smell of unfamiliar conspecifics (observed in *Nycticebus* by Rasmussen, 1986) was neither seen in our slender lorises nor in the *Loris* specimens Rasmussen observed. Rubbing of the brachial organ on branches is sometimes observed, but conspecifics afterwards show no noticeable interest in sniffing the brachial region or the branch used for rubbing. Sledging (rubbing of the anal

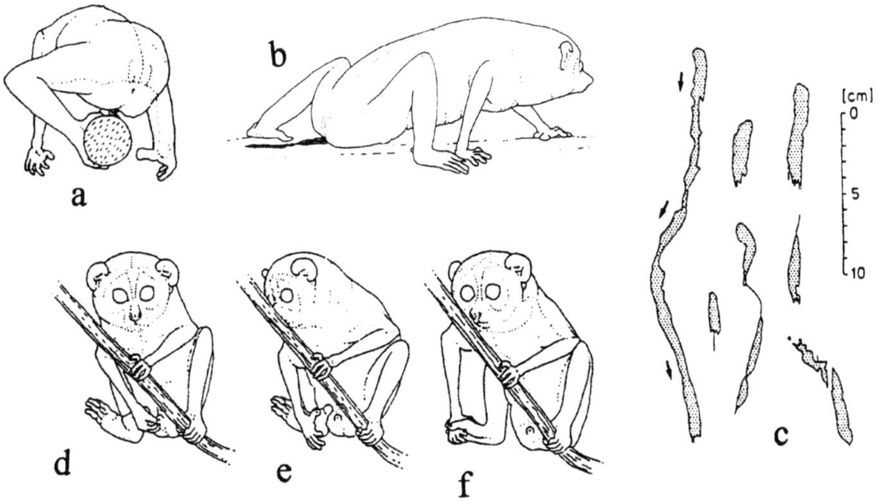

Figure 6: behavior connected with deposition of urine. **a:** for dot marking, locomotion is stopped for a moment. **b:** trail marking during locomotion. **c:** urine left on a horizontal glass surface during trail-marking. **d-f:** urine-washing: the animal urinates into the palm of the hand (d) and then wipes the sole of the foot with alternating movements of both hand and foot (e). During "washing", both limbs are moved further laterally (f).

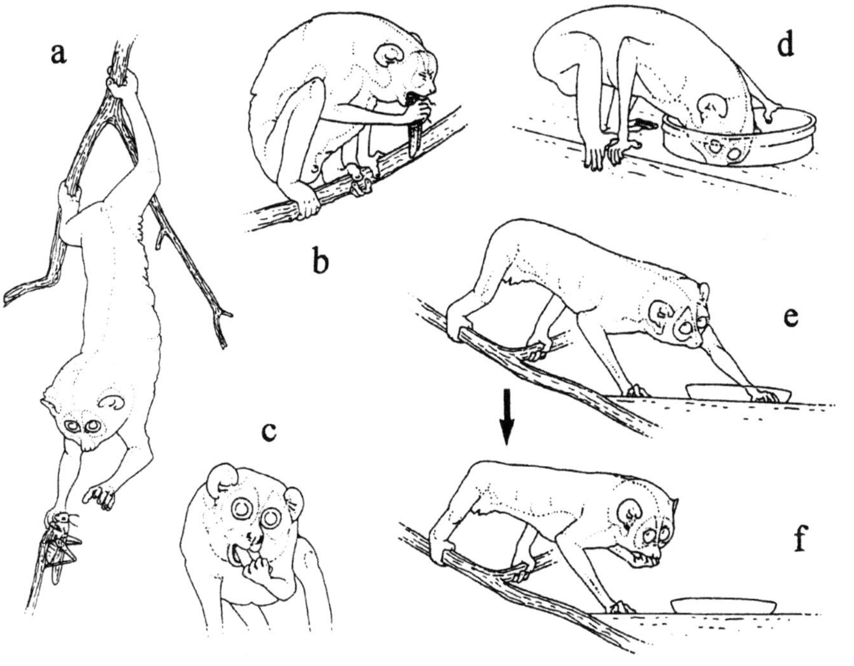

Figure 7: food consumption. **a:** food is seized with one or both hands. **b:** larger, struggling prey are killed by crushing their heads with the teeth. **c:** for chewing, food items are pushed between the molars. **d:** water is licked up in a dog-like fashion. **e, f:** nutritious liquids such as baby milk pap are often seized with one hand with typical prey-catching movement (e) and licked off the palm (f).

region on branches) rarely occurs. We interpret it as rubbing without any olfactory meaning.

Food Consumption

Lorisids are more or less omnivorous and in their natural habitat feed on insects, other small prey and various vegetable food. They may be specialized for catching insect species other sympatric mammals will refuse, such as hairy caterpillars, very small, badly smelling or tasting and poisonous prey. Such fairly abundant food is not preferred, but tolerated (Hladik and Hladik, 1972; Charles-Dominique, 1976; Rasmussen and Izard, 1988). Analysis of feces from our newly captured animals consisted almost exclusively of chitinous remains of small insects. Analyses of feces and stomach contents, previously published by Hladik and Petter (1970), Petter and Hladik (1970), and Hladik and Hladik (1972), also showed that slender lorises in the wild are extremely insectivorous, eating mainly very small prey. Small geckos (probably *Hemidactylus frenatus*, a species abundant in Ceylonese houses) were the first food items accepted by our freshly caught animals. Local people reported that they have observed lorises eating eggs and nestlings of small birds. Killing of larger animals has been reported at least from captive lorises. Kinnear (1919) for instance observed that captive *L. t. lydekkerianus* killed and ate two gerbils. Feeding on fruits, leaves, young sprouts and flowers was occasionally observed in the wild (Johnson, 1984). In publications about food preferences of captive lorises, leaves have usually been neglected, but our lorises take some sorts of green salad even more readily than sweet fruits, and parts of the leaves and smaller twigs of rubber trees (*Ficus benjamina*) were chewed off.

Olfactory examination is used to identify food-items. Food-searching is more variable. In the wild, systematic searching of twigs and liana for prey, with the nose close to the substrate, was observed (Hladik and Petter, 1970), a behavior described by Rasmussen (1986) as "nose-down search". Our *L. t. nordicus* seem to react primarily to optical perception of movements. Goonan (1983) also observed that motionless prey was not recognized by his captive lorises (*L. t. tardigradus* and *L. t. grandis*). Our animals prefer moving small prey. Both animal prey and immobile food particles are seized with the stereotyped prey-catching movement with one or both hands; sometimes in addition the teeth are used. Small animals like mealworms are often eaten alive; larger prey (locusts, crickets) are immediately killed by crushing their heads with the teeth (Fig. 7b).

Our animals drink water only occasionally; it is lapped up with the tongue in dog-like fashion (Fig. 7d). Milk or baby pap may be licked up like water, or the animal makes the typical movement of seizing solid food and then licks the liquid off the palm (Figs. 7e-f). Subramoniam (1957) observed that captive lorises licked up dew drops from the leaves.

Resting and Sleeping

For resting and sleeping, lorises usually assume sitting postures on top of more or less horizontal substrates (Fig. 8a). Sleeping sites which allow comfortable leaning against a lateral support are preferred. Under conditions of environmental stress, animals may occasionally sleep in a suspended posture, in a place as high as possible (Fig.8b).

Lorisids are generally believed to lack nest-building behavior. Cowgill (1969), however, mentions a "nest" built in sawdust on the cage floor by a female potto. In *Loris*, some arranging of substrates occurs. In our colony, in two cases, animals pulled a piece of cloth, attached to the wiremesh, into the cage and draped it so that they could sleep hidden behind it. Subramoniam (1957) observed a slender loris sitting down surrounded by foliage, "... pulling down one of the adjoining leaves and taking a seat with the leaf immediately below the haunches". Sleeping on the ground, wrapped up in paper, is regularly observed in some captive slow lorises and pottos (Schweigert, pers. comm.), and wild slow lorises are, according to native people (Orang Asli), sometimes found sleeping on the ground, hidden under dead foliage (Wiens and Zitzmann, pers. comm.). In our lorises, sleeping on the floor has never been observed; our cage floors, however, do not provide any substrate for hiding.

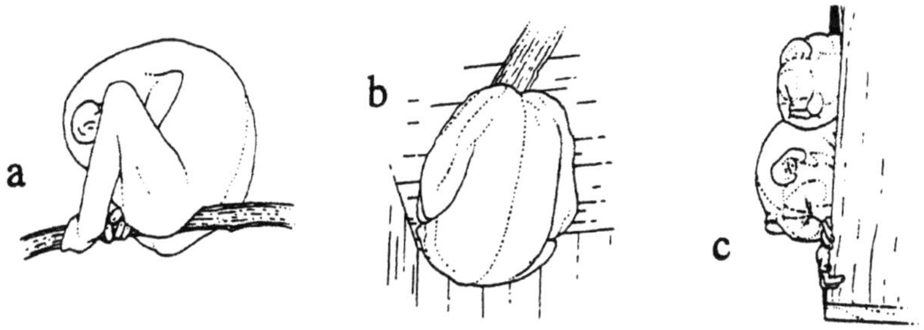

Figure 8: Sleeping. a: usually lorises sleep rolled up, sitting on more or less horizontal substrates. b: suspended sleeping posture under conditions of psychic stress. Usually our animals sleep in the open; wooden boxes are sometimes used by shy animals or after disturbance (c).

Stretching, Yawning and Comfort Behavior

After sleeping, our lorises begin their activity period with yawning, wiping of face and ears with the hands or forearms and licking of hands and arms. Grooming and licking of the fur follow, first in parts of the body which can easily be reached in the sleeping posture. Later the limbs and body are increasingly stretched and moved before the animal gets up for its nocturnal activity.

During resting periods, mutual and self-grooming is frequent. Licking is usually self-directed whereas conspecifics are rather groomed with the tooth-comb; newborn infants are licked by the mother. Fig. 9a shows the typical prosimian tooth-comb of *Loris*, which is used for grooming the fur. Loose hairs are probably removed from the tooth-comb by the sublingua (Fig. 9b-d), a structure attached to the muscular tongue. It consists of cartilage and connective tissue, but does not contain any muscle fibers (Bolk *et al.*, 1938), which means that it can only be moved passively by the tongue. The whole sublingua surface is said to be corneous (Bolk *et al.*, 1938). The hairs are swallowed and then excreted with the feces, which often consist mainly of hair. One old animal in our colony died of a large gastric trichobezoar. Deaths due to trichobezoars have also been reported from angwantibos and slow lorises (Manley, 1966; Osman Hill, 1953). The characteristic pushing movements of the head during grooming with the toothcomb (Figs. 9f, h, i) are easily distinguishable from the similar, but more flowing movements during licking (Fig. 9e). Allogrooming not only helps the animals to keep their fur clean in parts of the body they cannot groom or lick themselves (Fig. 11b), but also is meaningful as an appeasement behavior and for the establishment of social relations when, for instance, adults of different sex meet for the first time. Scent might play an important role in this context; the brachial organ is presented to the conspecific for grooming and is in fact groomed more eagerly than other parts of the body.

For scratching, the toilet claw of the foot is used (Fig. 10c). Cleaning of the ear with the toilet digit (Fig. 10e) is sometimes started with a scratching movement before the toilet claw is cautiously introduced into the auditory canal. After scratching or cleaning of the ear, the toilet claw is always cleaned by licking (Fig. 10f). Lying on a branch, upside down (Fig. 10g), is a behavior which mainly occurs when needs have been satisfied and which apparently indicates wellbeing. It is often associated with rubbing of the back or arm-pit or

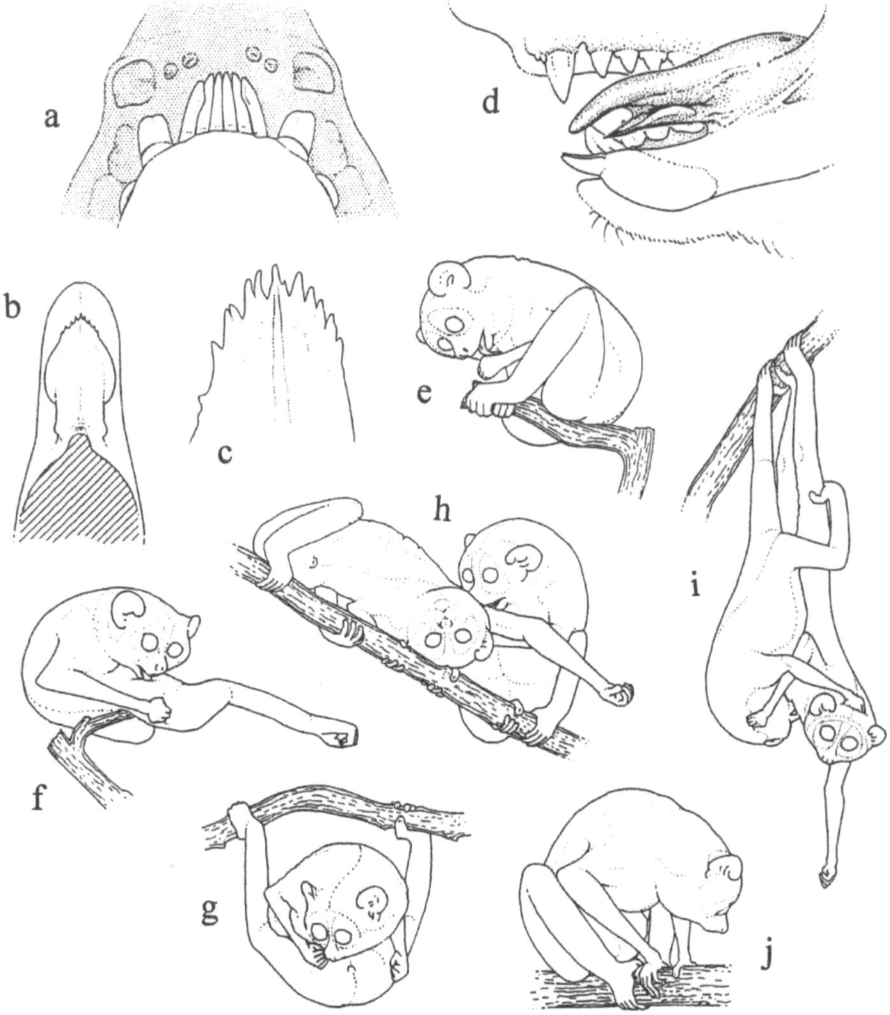

Figure 9: comfort behavior with use of the tongue, tooth comb and sublingua. a: Loris toothcomb, used for grooming of the fur (view from the ventral side, upper jaw stippled grey). **b:** tongue and sublingua, seen from below. The sublingua, a structure attached to the muscular tongue, is regarded as an instrument for cleaning of the tooth-comb. **c:** tip of a sublingua, original size: 5 mm. **d:** lateral view on the tongue and sublingua in the mouth. **e-i:** use of toothcomb or tongue for comfort behavior: **e:** licking. **f:** autogrooming with the toothcomb. **g:** for licking or grooming of the ano-genital region, often a bipedal hanging posture is employed. **h, i:** allogrooming. In animals which intend to be groomed, the region of the brachial organ (arm-pit) is presented to the conspecific by forward-stretching of the arm; usually eager allogrooming is the reaction. **j:** if a conspecific approaches from cranial direction, lowering of the head may be used as a grooming-solicitation (a similar posture by *Nycticebus* is shown in a figure by Ehrlich and Musicant, 1977).

with solitary play. For sledging, one or both feet are raised, the hands pull the body forward, the anal region is dragged along the branch surface. This behavior rarely occurs in our lorises.

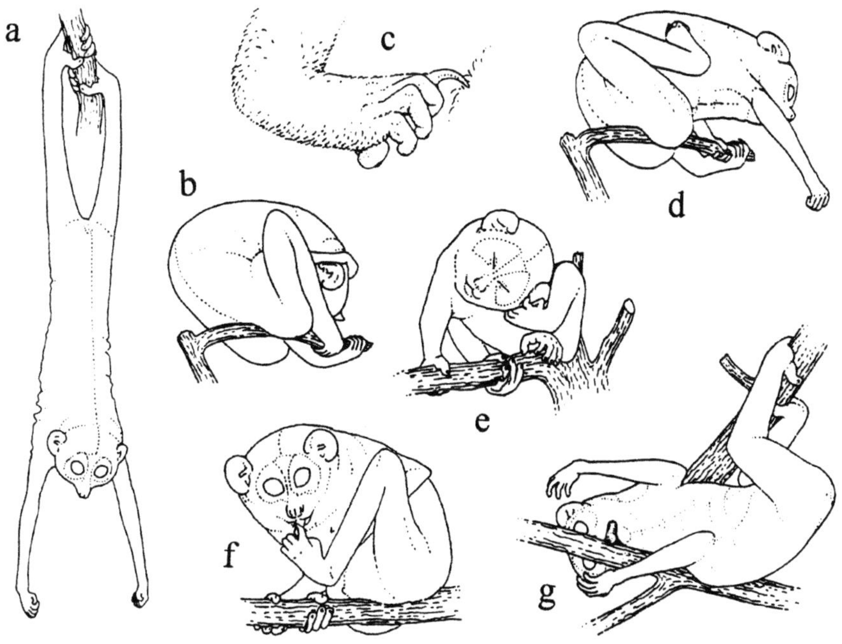

Figure 10: other comfort behavior. **a:** stretching. **b:** hand and forearm are used for wiping head or ears. Wiping is immediately followed by licking the hand. **c, d:** scratching with the toilet claw. **e:** use of the toilet claw for cleaning of the ear. **f:** after use, the toilet claw is always cleaned by licking. **g:** lying on a branch, upside down, often occurs after satisfaction of needs (feeding, sexual behavior). We therefore interpret it as a sign of wellbeing. It is often connected with stretching, rubbing of the back or arm-pit on the substrate or, as in this figure, with play behavior.

Social Behavior

In their natural habitat in Sri Lanka, slender lorises were found in colonies in areas otherwise free of lorises (one of us, B. M., together with Nieschalk); Petter and Hladik (1970) found these colonies to consist of very small adjacent territories. Sizes of home ranges may vary dependent on the food resources and may, for instance, comprise 0.9 ha (Johnson, 1984) or little more than one tree (Petter, Hladik 1970). Free-living slender lorises covered distances of about 20-30 m per night (Petter, Hladik 1970). Slender lorises in the wild usually move alone or by twos, rarely in "groups" (Petter and Hladik, 1970; Hladik and Petter, 1970; Devaraj Sarkar *et al.*, 1981; Johnson, 1984). Of 15 animals observed for some time in the wild in Sri Lanka, six were seen together with a conspecific. Sleeping huddled with conspecifics has been observed (Bayart, pers. comm.). Social behavior in the wild has not yet been studied. It may resemble the social system of pottos in which small territories of several females are overlapped by a larger male territory (Charles-Dominique, 1971, 1990). This hypothesis is supported by the fact that our captive males show significantly more locomotor activity than the females.

During the first 20 months of captivity, some of our wild-caught females, kept together with a mate, concentrated their activity in the lower parts of the cage while the males spent more time on the upper branches. We first misinterpreted the different preference of altitude as a differential use made of the habitat by the sexes. Later, the

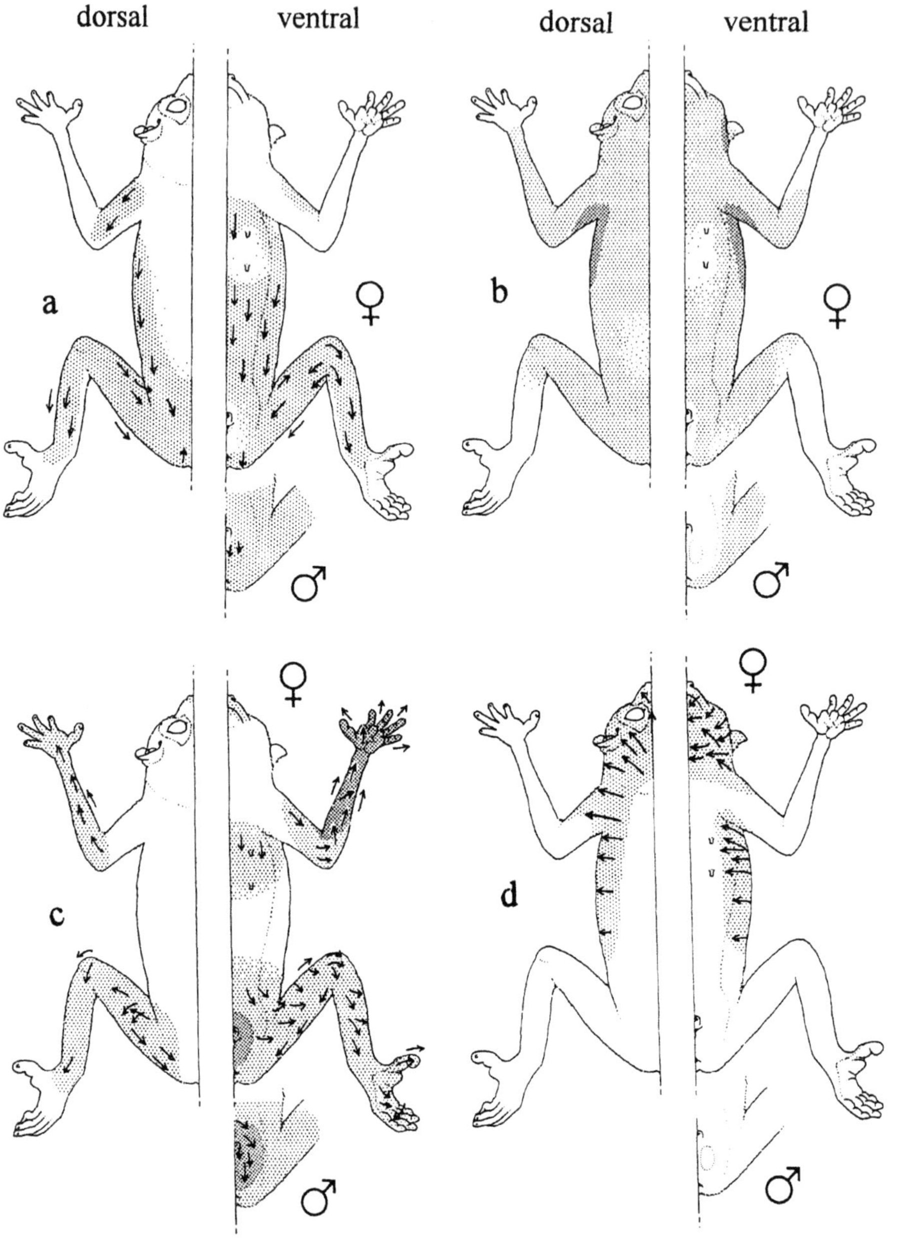

Figure 11: use of different grooming behavior for the body regions. Groomed areas are stippled grey; areas groomed with above-the-average intensity are stippled dark grey. The arrows indicate the observed direction of grooming movements. **a:** areas autogroomed with the toothcomb. **b:** areas allogroomed. Besides social function, allogrooming provides care for regions of the body which cannot be groomed by the animal itself. The observed intense grooming of the arm-pit might indicate a social meaning of the scent gland exudates of the brachial organ. **c:** areas licked. In most parts of the body, both licking and grooming may occur. Hands, genital region and toilet claw are licked. **d:** areas scratched with the toilet claw.

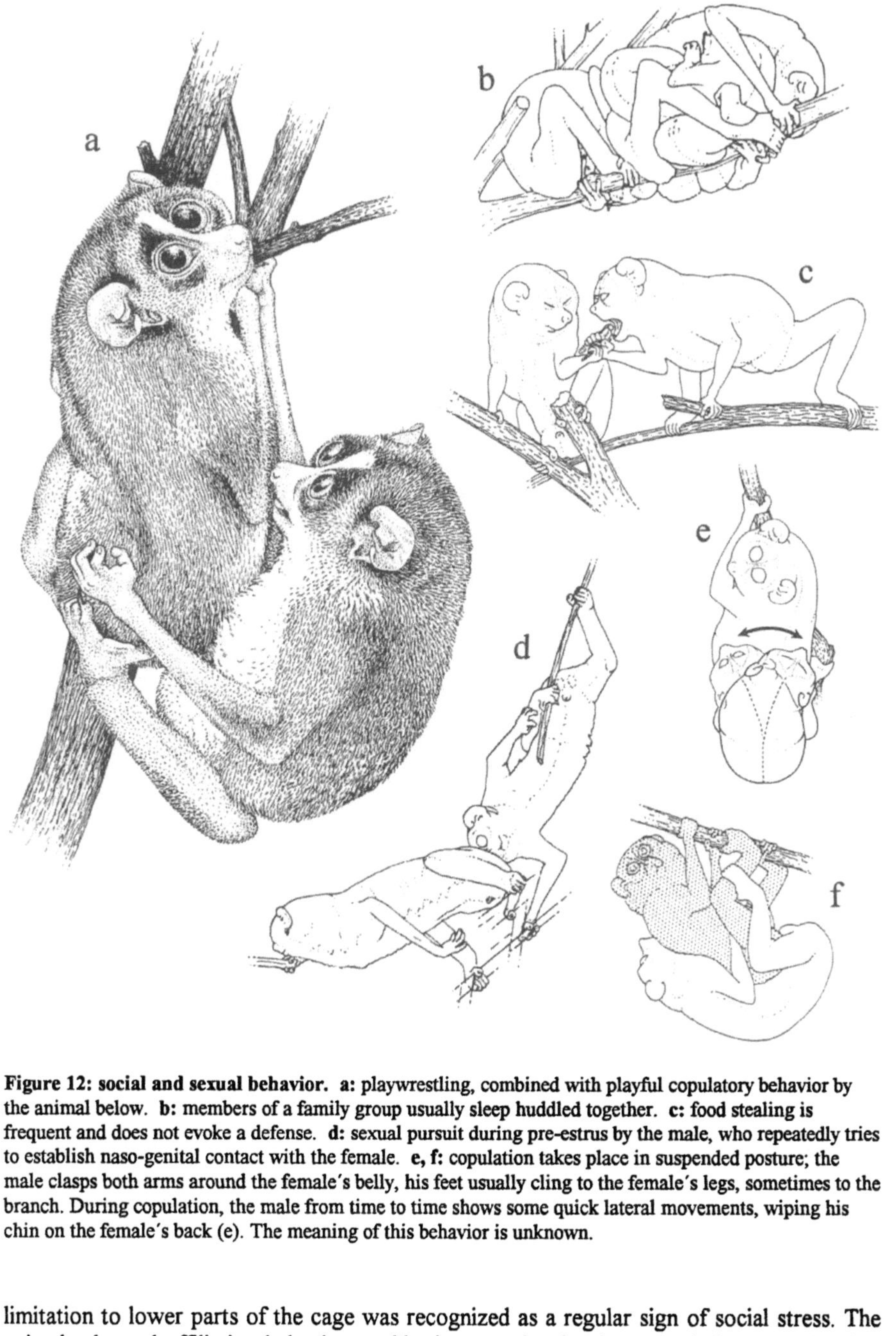

Figure 12: social and sexual behavior. a: playwrestling, combined with playful copulatory behavior by the animal below. **b:** members of a family group usually sleep huddled together. **c:** food stealing is frequent and does not evoke a defense. **d:** sexual pursuit during pre-estrus by the male, who repeatedly tries to establish naso-genital contact with the female. **e, f:** copulation takes place in suspended posture; the male clasps both arms around the female's belly, his feet usually cling to the female's legs, sometimes to the branch. During copulation, the male from time to time shows some quick lateral movements, wiping his chin on the female's back (e). The meaning of this behavior is unknown.

limitation to lower parts of the cage was recognized as a regular sign of social stress. The animals showed affiliative behavior at this time, staying in close proximity to each other during 20 % of the activity period and sleeping closely huddled together.

Our animals usually react with whistling and attacks upon the approach of unfamiliar individuals of the same sex. The frequency of urine-marking is increased in such situations.

Territorial defense seems to depend more on the perception of an opponent than on the place where the animals meet. Offensive fighting of animals in the cage of the opponent has been observed. Territorial behavior in our colony is continuous and not dependent on seasonality.

In contrast to our experience, Rao (1932) kept "a large number of" *L. t. lydekkerianus* in only two cages for "long periods". He observed that the animals slept huddled together in one corner of their cage. No fighting between them is mentioned, although Rao described slender lorises as animals whose life is "marked by mutual intolerance" with the exception of the breeding seasons. Goonan (1993) describes a "coalition" of two unrelated adult *Loris* females which included affiliative behavior and moving around together. According to Subramoniam (1957), lorises, released in a tree, always followed a leading conspecific and then crowded near each other. Subramoniam also observed peaceful coexistence and even affiliative behavior between adults put together in one cage, in spite of some vocalizations (whistle, chitter) which in our animals occur in an agonistic context (Subramoniam assumed these vocalizations to be affiliative). In other cages all animals fought against each other (Subramoniam 1957). In our colony, the degree of aggressiveness has clearly declined in the past years. The reasons are not clear, but availability of food throughout the activity period may play a role. According to Petter and Hladik (1970), the size of territories in the wild may depend on availability of food, an hypothesis which is confirmed by the increased occurrence of aggressive behavior of our animals when they are hungry.

Our animals regularly sleep huddled with members of their family group (Fig. 12b). They usually move around independently, peacefully share their food and show no behavior indicating the existence of a rank order. If a female rejects a sexually interested male with a defensive threat, the latter usually reacts with appeasing vocalization, retires or turns away his head whereas, during quarreling, males are often dominant over females. Social play can regularly be seen in our breeding groups. It consists of hand-grabs aimed at hands or feet of conspecifics, seizing each other's fur, playwrestling or playful biting. It is often connected with playful assumption of a copulation posture (Fig. 12a) or allogrooming. Infants and subadult animals spend more time playing than adults, but even very old animals playwrestle. Stealing of food from conspecifics (Fig. 12c) is common and never leads to defensive behavior. Peaceful food-sharing was also reported for *Loris* by Rasmussen (1986) and Goonan (1993) and for several prosimian species by Doyle (1974). Food is not intentionally offered to group members, but most animals do not mind if even a favorite food item is taken from hand or mouth. Occasionally an animal may try to run away with a piece of food if group members approach with the apparent intention to take it. Infants often steal food from the parents. In doing so, young animals apparently learn about edible substances. Food is also stolen from offspring by their parents.

When an experienced male is introduced to a female, usually affiliative or sexual behavior is observed. When meeting for the first time, both partners may initially act as though they were not interested in the conspecific, moving around and sniffing twigs before approach. The first peaceful contact is usually established by sniffing some part of the conspecific (the initiative may be with either sex); then allogrooming may follow. A ritualized greeting behavior was never observed. Problems may occur if animals are young and unexperienced or have been kept solitary for a longer period. In several cases, unexperienced males approached a new mate without appeasement behavior and tried to mount almost immediately for copulation. This behavior was successful with some very tolerant females, but in other cases the males were attacked and injured by the female. After being kept solitary for several months, three males showed signs of excitement when introduced to a female and repeatedly attacked her. When a tranquilizer was added to the food prior to the introduction, the same individuals remained peaceful, showing appeasing and affiliative behavior.

Mutual defense against predators has never been observed in adults, but when an infant, found lying on the ground, is picked up by a keeper and held up towards the mother, even very tame females may approach with growling, perform mock attacks or bite the human hand. As the animals are not aggressive towards humans in other situations, this behavior can be interpreted as a defense of the infant.

Sexual behavior. Estrus is not seasonal in our captive *L. t. nordicus* and in the captive slender lorises at Duke University Primate Center which belong to a small reddish subspecies (Izard and Rasmussen, 1985). A study by Ramaswami and Anand Kumar (1962) indicates that in wild *L. t. lydekkerianus* an annualperiodic breeding seasonality may exist. Olfactory communication during estrus has already been described above. Copulation may take place after playwrestling with mounting, but more often a longer-lasting pursuit of the female by the male precedes copulatory behavior (Fig. 12d), also described by Rasmussen (1986) under the synonym "sexual monitor sequence". The duration of pursuit depends on the behavior of the female. It may last more than five hours or may be stopped and repeated several times if the female is aggressive. During pursuit, the male emits appeasing krik calls in irregular intervals. The female may first react with defensive threats. Later she increasingly shows suspensory locomotion, and the male often reacts with approach and attempts to mount; a suspensory posture appears to signal that the female is ready for copulation. The copulation itself also takes place in suspended posture (Figs. 12e, f). Poor captive reproduction of *Loris* in zoos may be due to the lack of horizontal substrate which allows quadrupedal hanging. This is confirmed for pottos by Cowgill (1969) and and for slender lorises by Drake (pers. comm.). Copulation may last 2-16 minutes. The male, clinging to the quadrupedally suspended female, at intervals shows a lateral wiping of the chin on the female's back (Fig. 12e). Copulation is ended by low-intensity threat and wiggling by the female; both animals carefully lick their genital region afterwards, a behavior which does not occur after play-mounting. Sometimes a vaginal plug, consisting of a hardened liquid which appears to be produced by the male (Manley, 1967), closes the vagina after copulation. It may prevent the sperm from dripping out and certainly is an obstacle to further copulations by other males. The plug is later removed and at least partly eaten by the female.

Maternal Behavior and Relation of other Group Members to the Offspring. The few births observed in our colony took place during the daylight period and in the open, even in cages with nest boxes. Births in sitting and hanging postures occurred. In the first weeks of life, infants cling to the mother's belly throughout the activity period (Figs. 13a, b). During the day, they sleep supported by the mother (Fig. 13c). Infant parking as described for the other lorisid species (Charles-Dominique, 1966; Cowgill, 1969; Blackwell and Menzies 1968; Osman Hill, 1937; Tenaza and Fitch, 1984) is an uncommon behavior in our animals. The mother shows no behavior which indicates her intention to deposit the infant. When the infant starts to climb on twigs, usually at the end of the sleeping period, the mother may occasionally get up and leave it alone until it begins to emit "zic" calls. If an infant falls down, it also reacts with "zic" calls. Hearing zic vocalization, the mother usually approaches the infant and waits until it has mounted her again. If the infant does not cling to her properly, the mother may abandon it or pick it up with her hands (Fig. 13e) and try to carry it to higher parts of the cage: she seizes a thin part, for instance the belly or a leg, and carries the infant often with the head hanging down (Fig. 13f), sometimes banging it against twigs during locomotion. The infant in this situation desparately clings to the twigs, and the mother repeatedly has to stop and to loosen its grip.

First locomotion of the infant on branches is usually observed at the age of 5-7 weeks. The duration of intervals in which an infant is carried is determined by the offspring, not by the parents, and differs individually. It rapidly decreases when the juveniles are 2-3 months old (Meier, Preuschoft, 1985). Four-months-old and sometimes older juveniles still return

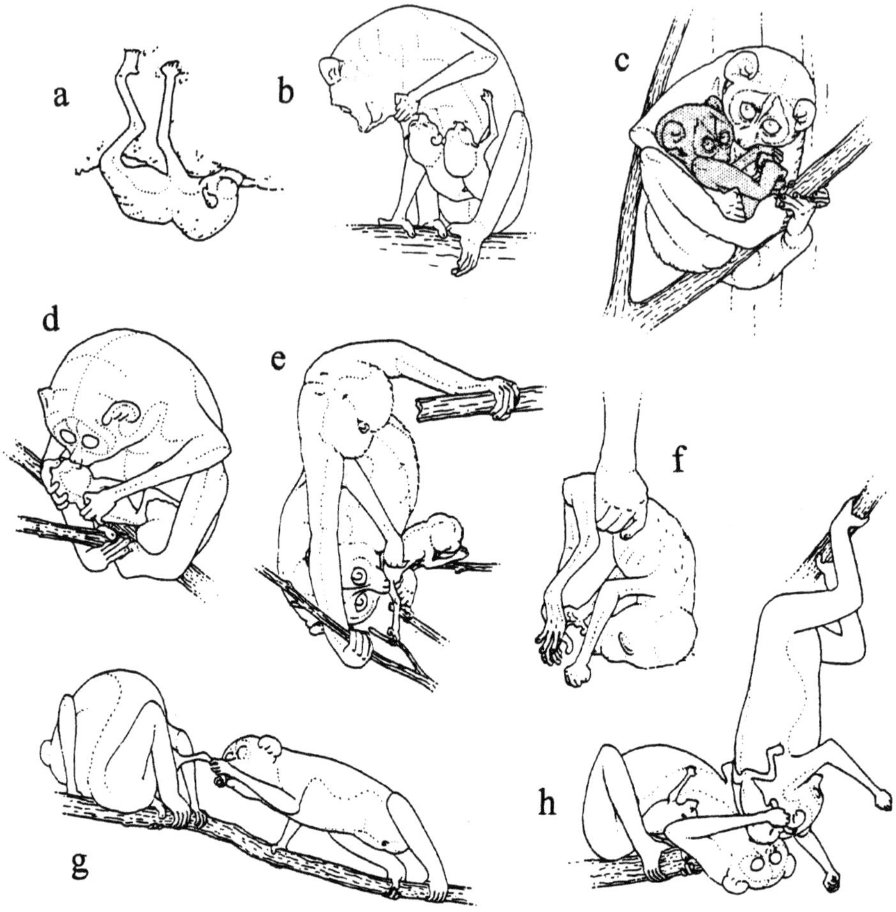

Figure 13: maternal care and relation between infants and group members. **a:** during the activity period, infants cling to the mother's belly. **b:** twins cling to the mother side by side. As the mother has two pairs of mammary glands, the infants may suckle in staggered positions, allowing better clinging to the mother's fur. **c:** during daytime, the infant sleeps cradled by the mother's arms and legs, often with hands and feet loose. **d:** the infant is carefully licked by the mother. **e:** the mother tries to pick up an infant which had fallen down and is now clinging to a twig. **f:** for carrying an infant which does not cling to her properly, usually the mother seizes a thin part of it, often a leg or the belly. **g:** after birth, group members show considerable interest in the infants and often try to take one for examination. When the mother does not defend her offspring, an infant may be torn away from her belly. **g:** the infant usually gets hold of the "thief" and clings tightly to his fur, a situation which may cause nervous efforts of the latter to get rid of the unfamiliar burden. The mother tries to get her infant back while the other twin still clings to her belly.

to the mother toward the end of the nocturnal activity period, are carried for some time and sleep cradled on her legs. The mother does not threaten juveniles in order to prevent them from clinging to her, although females with fully-grown twins sometimes had considerable problems with locomotion. Prior to parturition, in several cases females have tried to get rid of juveniles still clinging to them by running around and repeatedly passing between narrow twigs where the offspring was sometimes ripped off. Females may also closely roll up in a sleeping posture, leaving no space for the offspring to squeeze into. Similar postures were

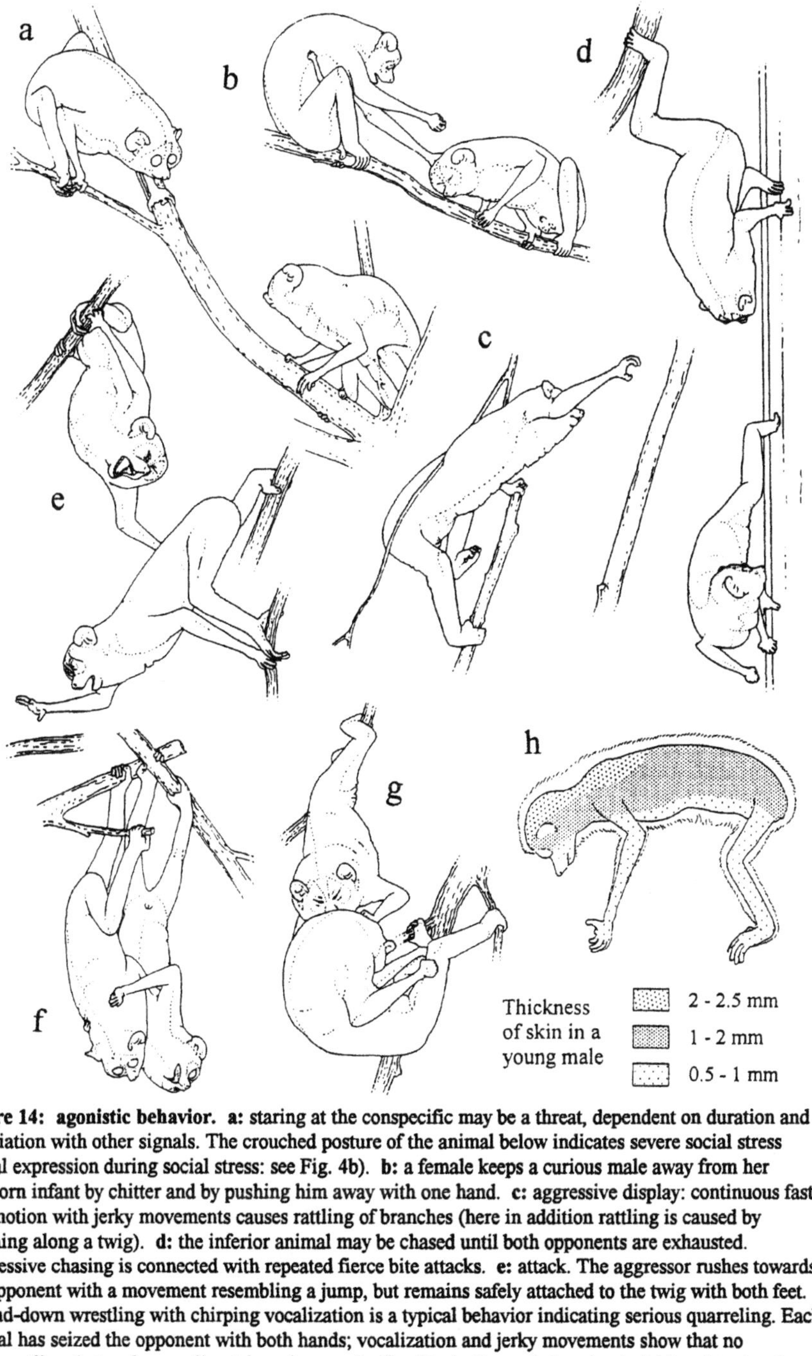

Figure 14: agonistic behavior. a: staring at the conspecific may be a threat, dependent on duration and association with other signals. The crouched posture of the animal below indicates severe social stress (facial expression during social stress: see Fig. 4b). **b:** a female keeps a curious male away from her newborn infant by chitter and by pushing him away with one hand. **c:** aggressive display: continuous fast locomotion with jerky movements causes rattling of branches (here in addition rattling is caused by brushing along a twig). **d:** the inferior animal may be chased until both opponents are exhausted. Aggressive chasing is connected with repeated fierce bite attacks. **e:** attack. The aggressor rushes towards the opponent with a movement resembling a jump, but remains safely attached to the twig with both feet. **f:** head-down wrestling with chirping vocalization is a typical behavior indicating serious quarreling. Each animal has seized the opponent with both hands; vocalization and jerky movements show that no playwrestling, but real quarreling takes place. **g, h:** the skin in the nuchal region is thicker than in other parts of the body and provides some protection against bite wounds. When attacked in a situation in which an escape is impossible, the animals may turn this region towards the aggressor.

assumed as a reaction to attempts of older offspring to suckle. Sometimes a mother used her hands to keep a subadult away from her teats. No threat behavior occurred in this context. After birth, females usually defend their infants against conspecifics, including older offspring. In this situation some low intensiy threat against the offspring may occur.

Other group members show a strong interest in newborn infants and may take one for examination or playing if the mother does not react with a threat (Fig. 13g). Some males regularly carry one infant of a pair of twins for some time while the mother carries the other sibling. The mother may actively try to get her infant back (Fig. 13h) or just stay close to it, often grooming the male, until it crawls over to her on its own. Occasionally, an infant may be dropped by a group member and get hurt. In few cases, especially in periods of environmental stress, infants were killed by conspecifics. Cannibalism is rare. Males were seen chewing small parts off a dead infant which apparently was not freshly killed. Attacks against older infants by adults were not observed. Even when an infant, excited by fighting in the group, attacked and bit an adult, the latter reacted with appeasing behavior.

Agonistic Behavior. Quarreling within family groups usually starts with threat behavior (Figs. 14a-c). In captivity, the subordinant opponent, unable to leave, may be chased to complete exhaustion (Fig. 14d) unless the animals are separated by the keeper. Appeasing behavior has little protective effect when the excitement of the aggressor escalates. One of our males, fiercely chasing his mate, attacked and bit a broom held between them. Severe injuries due to fighting have occurred in several breeding facilities; in some cases animals were wounded so badly that they had to be euthanized. Continued social stress may also lead to the death of animals. Aggression may be directed against a certain individual or against all adult conspecifics. If quarreling occurs in groups with male offspring close to maturity, aggression between father and son may escalate rapidly and may lead to severe fighting within a few hours. In periods of severe social stress, the subordinant animal either runs around, trying to escape, or stays close to the ground of the cage where it may try to hide. It usually shows a characteristic posture with crouched back (Fig. 14a, animal below) and facial expression (see above, Fig. 4b), quietly staring upwards or towards the opponent.

Anti-Predator Behavior and Stress

Even slight changes within the familiar territory of slender lorises cause signs of excitement. During environmental stress, the animals show silent, cryptic movements and repeated bouts of immobility. Inconspicuous retreat or rapid flight upwards may occur. The animal may try to hide behind some cover or just remain motionless, often hiding the face with the conspicuous dark and white optical markers (Fig. 15a, b). Falling from the substrate as a means of flight, especially as a reaction to the presence of snakes, was reported for slow lorises (Tenaza, Fitch, 1984) and pottos (Charles-Dominique, 1977). In our lorises it was not observed as a regular behavior although tests with live snakes were made. In the wild, a preference of the region of thin twigs was observed (Petter, Hladik 1970; Hladik, Petter, 1970); here the animals are usually hidden by foliage, and heavier predators cannot follow them.

If an escape is impossible, defensive behavior (Figs. 15c, d) may be shown. Optical enlargement by stretching of the legs and arching of the back (Fig. 15c) and production of a bad smelling brachial gland exudate (see Alterman, this volume) may occur as a defensive threat against humans. A subadult female, disturbed by the keeper's hand during feeding, lowered the head between the forelimbs and fiercely knocked forward with the nuchal region several times as described for pottos by Charles-Dominique (1971, 1977). Rasmussen (1986) reports similar postures and behaviors (defensive shielding, body-butting) for *Loris* and *Nycticebus*. The nuchal region in *Loris* is not equipped with

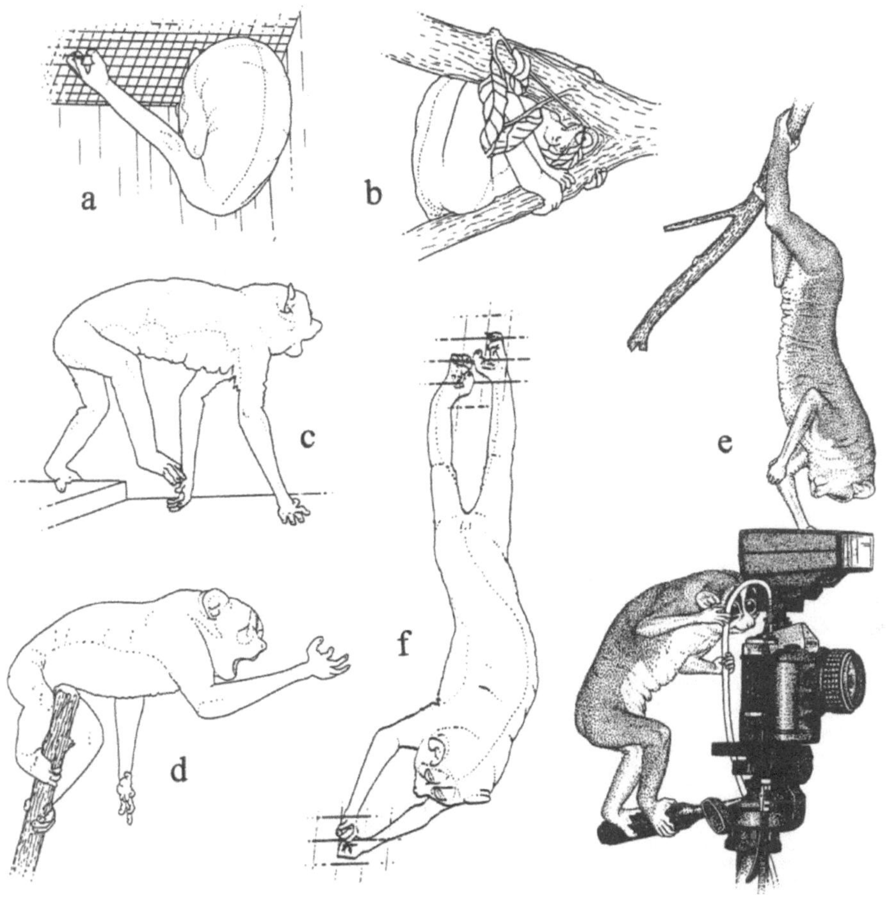

Figure 15: behavior connected with unfamiliar stimuli: stress, defense and curiosity. a: frightened by the approach of a human, a loris has reacted with flight up to the ceiling of the cage. Turning away the conspicuously colored face (**a**) is also a common reaction when a loris is frightened by a human. A frightened animal usually tries to hide behind some cover (**b**) or just freezes to motionlessness in the open, sometimes for hours. If hiding or escape seem impossible, the animal may show defensive threat behavior as for instance an extension of the legs and cat-like arching of the back in connection with loud growling (**c**). A bipedal defensive posture allows quick attacks (**d**). Less frightening stimuli usually have a different effect: the animals approach the object, look at it, sniff it and repeatedly touch it with the prey-catching movement pattern (**e**). Examination of smaller objects, like play behavior, is often connected with body-wiggling (**f**) and other behavior indicating playful mood.

prolongued spinal processes as in *Perodicticus* (Charles-Dominique, 1971, 1977), but its skin is fairly thick (Fig. 14h).

Environmental stress may lead to epileptiform seizures with rhythmic convulsions and saliva dripping from the mouth, and in rare cases animals may feign death. If a severe disturbance continues for longer periods, the lasting stress may lead to organ lesions, such as bleeding gastric ulcera, to rejecting of food, starvation and finally death.

Exploratory Behavior, Learning and Adaptability

Slender lorises are very curious. Large objects are usually first observed from a safe distance, sometimes in connection with lateral swaying of the body or headcocking. After cautious approach, small objects are examined by repeated grasping, sniffing and biting. Among young animals, playing with small non-edible objects for 20 minutes and more was observed in connection with body-wiggling and other signs of a playful mood. Epileptic seizures, however, may occur during clearly playful exploration, indicating that it is connected with considerable excitement. Curiosity, resulting in a thorough knowledge of the familiar territory, seems to be one of the main protective strategies in slender lorises.

In learning and intelligence tests with lorises, psychic stress due to environmental changes leads to poor performance. Gorter (1937) writes that "one should have much patience in investigating their (*Nycticebus*) behaviour, otherwise one runs the risk to undervalue their capacities". We agree. In addition, learning success in *Loris* may remain unrecognized by the researcher. Our animals have repeatedly maintained a certain behavior for months, showing no sign of learning, and then suddenly developed completely new habits, which we can only understand as a consequence of previous learning (increasing tameness, for instance, was often shown with delay). Ehrlich and Musicant (1976) made similar observations in some slow lorises, who, in discrimination tests with two-dimensional black-and-white patterns, showed poor learning success before performance suddenly "went to near perfect", whereas others were "fast learners" and showed learning success immediately. Lorises apparently learn better if a task allows for their inherited behavior, for instance eliciting the prey-catching movement. According to Ehrlich (1968), *Nycticebus*, in bar pressing for a food reward, worked readily and "maintained a high level of response" when the number of presses necessary to obtain the reward was increased, but if the number of 25 necessary presses was exceeded, performance rapidly declined. According to Johnson (1984), fortune tellers in Tamil Nadu use tame lorises to pick out cards from packs spread out on the ground. Learing and problem-solving tests in slender or slow lorises by Gorter (1937), Jolly (1964a, b), Ehrlich (1968), Ehrlich and Musicant (1976), Ohta (1983) and Rozman and Pöppelmann (unpublished) describe tests which required performance of the prey-catching grip. More variable experimental manipulatory tasks successfully solved by simians apparently exceed the motivation or abilities of lorises. Slender lorises are said to be unable to open doors (Subramoniam, 1957), and in fact, most of our animals show no interest in solving such problems. However, some of our individuals succeeded in opening various types of doors and shutters even if considerable strength was needed.

Some long-term behavioral changes have occurred in our captive population, possibly in connection with improved husbandry conditions, environmental enrichment, and learning. Aggressive and territorial behavior became rare in the past years, and an extreme susceptibility to psychic stress is increasingly replaced by tameness and curiosity. In captivity, slender lorises can adapt to their environment within certain species-specific limits. If the latter are exceeded for a longer period, or if the necessary time for habituation is missing, severe suffering and possibly death of animals will be the consequence.

ACKNOWLEDGEMENTS

We are grateful to Dr. Moritzen and Dr. Rößler from the library of Ruhr-University for search for literature, to H. Helmich for unpublished data and sonograms, to T. Eickhoff, G. Fellenberg, J. Rozman, D. Pöppelmann, B. Grütering, E. Strothdrees and others who at times did the work of keepers without being paid, to G. Rao and other members of our institute for their support, to Prof. Preuschoft for discussion and revision of the text and to the editors (in particular to K. Izard) for patience and careful revision.

REFERENCES

Andrew, R.J., & Klopman, R.B. (1976). Urine-washing: comparative notes. In R.D. Martin, G.A. Doyle, & A.C. Walker (Eds.), *Prosimian Behavior* (pp.303-312). Duckworth, London.

Bishop, A. (1962). Control of the hand in lower primates. *Annals of the New York Academy of Sciences, 102*, 316-337.

Bishop, A. (1964). Use of the hands in lower primates. In J. Buettner-Janusch (Ed.), *Evolution and Genetic Biology of Primates* (pp.133-223). Academic Press, London.

Blackwell, K., & Menzies, J.I. (1968). Observations on the biology of the potto (*Perodicticus potto*, Miller). *Mammalia, 32*, 447-451

Bolk, L., Göppert, E., Kallius, E., & Lubosch, W. (Eds.) (1938). *Handbuch der vergleichenden Anatomie der Wirbeltiere*. Urban und Schwarzenberg, Berlin.

Charles-Dominique, P. (1966). Glandes préclitoridiennes de *Perodicticus potto*. *Biologia Gabonica, 2*, 355-359

Charles-Dominique, P. (1971). Éco-éthologie des prosimiens du Gabon. *Biologia Gabonica, 7*, 121-228.

Charles-Dominique, P. (1976). Ecology and feeding of five sympatric lorisids in Gabon. In R. D.Martin, G. A. Doyle,, & A. C. Walker, (Eds.), *Prosimian Behaviour* (pp.131-150). Duckworth, London

Charles-Dominique, P. (1977). Ecology and Behavior of Nocturnal Primates. Prosimians of Equatorial West Africa. Duckworth, London.

Charles-Dominique, P. (1990). Structures sociales chez les prosimiens. In J. J. Roeder, & J. R. Anderson (Eds.), *Primates, Recherches Actuelles* (pp.63-77). Masson, Paris.

Christel, M. (1993). Grasping techniques and hand preferences in hominoidea. In H. Preuschoft, & D. Chivers (Eds.), *Hands of Primates* (pp.91-108). Springer, Wien, New York.

Cowgill, U. (1969). Some observations on the prosimian *Perodicticus potto*. *Folia Primatologica, 11*, 144-150.

Detwiler, S.R. (1940). The eye of *Nycticebus tardigrada*. *Anatomical Record., 76 (3)*, 295-301.

Devaraj Sarkar, H.B., Murali, S., Prasad, D.T., Shekarappa, B.M., & Vijayalakshmi, V. (1981). The population and distribution of the slender loris (*Loris tardigradus*) in Karnataka State. *Tigerpaper 8 (1)*, 7-10. Bangkok.

Doyle, G.A. (1974). Behavior of prosimians. In A. M. Schrier, & F. Stollnitz (Eds.), *Behavior of Nonhuman Primates: Modern Research Trends*, Vol. 5 (pp.155-353). Academic Press, New York.

Doyle, G.A. (1975). Some aspects of urine-washing in four species of prosimian primate under semi-natural laboratory conditions. In S. Kondo, H. Kawai, & A. Ehara (Eds.), *Contemporary Primatology, 5th Int. Congr. Primat., Nagoya 1974* (pp.232-237). Karger, Basel.

Ehrlich, A. (1968). Food-motivated behavior in prosimians. *Folia Primatologica, 8*, 66-71.

Ehrlich, A., & Musicant, A. (1976). Visual discrimination learning and memory in nocturnal prosimians. *Animal learning and behavior, 4 (4)*, 431-435.

Ehrlich, A., & Musicant, A. (1977). Social and individual behaviors in captive slow lorises. *Behaviour, 60*, 195-220.

Goonan, P.M. (1993). Behaviour and reproduction of the slender loris (*Loris tardigradus*) in captivity. *Folia Primatologica, 60*, 146-157.

Gorter, F.J. (1937). Experiments on the learning and problem solving of *Nycticebus tardigradus* L. *Archives Neerlandaises de Zoologie., 2*, 95-111.

Harcourt, C. (1981). An examination of the function of urine-washing in *Galago senegalensis*. *Zeitschrift für Tierpsychologie, 55*, 119-128.

Hladik, C.M., & Hladik, A. (1972). The food supplies and vital territories of primates in Ceylon. *Terre vie, 26*, 149-215

Hladik, C.M., & Petter, J.J. (1970). Le loris tardigrade. Observations de terrains effectuées à Ceylan. *Science et Nature, 101*.

Hofer, H.O. (1977). The anatomical relations of the ductus vomeronasalis and the occurrence of taste buds in the papilla palatina of *Nycticebus coucang* (Primates, Prosimiae) with remarks on strepsirrhinism. *Gegenbaurs morphologisches Jahrbuch, 123*, 836-856.

Ilse, D.R. (1955). Olfactory marking of territory in two young male loris, *Loris tardigradus lydekkerianus*, kept in captivity in Poona. *British Journal of Animal Behaviour, 3*, 118-120.

Izard, M.K., & Rasmussen, D.T. (1985). Reproduction in the slender loris (*Loris tardigradus malabaricus*). *American Journal of Primatology, 8*, 153-165.

Jolly, A. (1964 a). Prosimians' manipulation of simple object problems. *Animal behaviour, 12*, 560-570.

Jolly, A. (1964 b). Choice of cue in prosimian learning. *Animal behaviour, 12*, 571-577.

Johnson, J.M. (1984). Diurnal activities of the slender loris, *Loris tardigradus*, in the Mundanthurai Sanctuary, Tamil Nadu (India). In M.L. Roonwal, & N.S. Rathore (Eds.), *Current Primate Researches* (pp.389-395). Jodhpur.

Kinnear, N.B. (1919). Notes on the Malabar slender loris, *Loris lydekkerianus. Journal of the Bombay Natural History Society, 26*, 836-837.
Manley, G.H. (1966). Prosimians as laboratory animals. *Symposia of the Zoological Society of London* 17, 11-39.
Manley, G.H. (1967). Gestation periods in the Lorisidae. *International Zoo Yearbook, 7*, 80-81.
Manley, G.H. (1976). Functions of the external genital glands of *Perodicticus* and *Arctocebus*. In R. D.Martin, G. A. Doyle,, & A. C. Walker, (Eds.), *Prosimian Behaviour* (pp.311-329). Duckworth, London.
Montagna, W., & Ellis, R.A. (1960). The skin of the slender loris. *American Journal of Physical Anthropology, 18*, 19-43
Meier, B., Preuschoft, H. (1985). Behavioral ontogeny and maternal care in slender loris (*Loris tardigradus nordicus* HILL, Prosimiae). *Verhandlungen der Deutschen Zoologischen Gesellschaft, 78*, 341.
Meier, B. (1989). Genetic management of captive prosimian populations, a report of experience with *Loris tardigradus. Human Evolution, 4 (2-3)*, 223-229.
Miyake, S., Ikeda, A., Ohta, S., & Matsumoto, S. (1991). Three-dimensional analysis of the peculiar arterial patterns of the extremities in Lorisidae: the rete mirabile. In A. Ehara (Ed.), *Primatology Today* (pp.569-570). Elsevier.
Müller, E.F., Nieschalk, U., & Meier, B. (1984). Thermoregulation in the slender loris (*Loris tardigradus*). *International Journal of the Primatological Society*, July 1984, Nairobi, Kenya.
Nieschalk, U. (1991). Fortbewegung und Funktionsmorphologie von *Loris tardigradus* und anderen kleinen quadrupeden Halbaffen in Anpassung an unterschiedliche Habitate. Dissertation, Bochum.
Nieschalk, U., & Demes, B. (1993). Biomechanical determinants of reduction of the second ray in Lorisinae. In H. Preuschoft, & D.J. Chivers. (Eds.), *Hands of Primates* (pp.225-234). Springer, Wien, New York.
Ohta, H. (1983). Learning set formation in slow lorises (*Nycticebus coucang*), *Folia Primatologica, 40*, 256-267.
Osman Hill, W.C. (1933). A monograph on the genus *Loris. Ceylon Journal of Science, (B), 18, Pt.1*, 89-129.
Osman Hill, W.C. (1937). On the breeding and rearing of certain primates in captivity. *Ceylon Journal of Science, 20*, 385-389
Osman Hill, W.C. (1938). A curious habit common to lorisoids and platyrrhine monkeys. *Ceylon Journal of Science (B) 11, Pt.1*, 65.
Osman Hill, W.C. (1953). Primates: Comparative Anatomy and Taxonomy. Vol. I, Strepsirhini. Edinburgh University press, Edinburgh.
Osman Hill, W.C. (1956). Body odour in lorises. *Proceedings of the Zoological Society of London, 127*, 580
Pariente, G. (1979). The role of vision in prosimian behaviour. In Martin, R.D., & Doyle, G.A. (Eds.), *The Study of Prosimian Behavior* (pp.411-459). Academic Press, New York.
Perry, J.M., Izard, M.K., & Fail, P.A. (1992). Observations on reproduction, hormones, copulatory behavior, and neonatal mortality in captive *Lemur mongoz. Zoo Biology 11*, 81-97.
Petter, J.J., & Hladik, C.M. (1970). Observations sur le domaine vital et la densité de population de *Loris tardigradus* dans les forêts de Ceylan. *Mammalia, 34 (3)*, 394-409
Preuschoft, H., Godinot, M., Beard, C., Nieschalk, U., & Jouffroy, F.K. (1993). Biomechanical considerations to explain important morphological characters of primate hands. In H. Preuschoft, & D.J. Chivers (Eds.), *Hands of Primates* (pp.245-256). Springer, Wien, New York.
Ramaswami, L.S., & Anand Kumar T. C. (1962). Reproductive cycle of the slender loris. *Naturwissenschaften, 49*, 115-116.
Rao, C.R.N. (1932). Observations on the habits of the slow loris, *Loris lydekkerianus. Journal of the Bombay Natural History Society, 32*, 206-208
Rasmussen, D.T. (1986). Life history and behavior of slow lorises and slender lorises: implications for the lorisine-galagine divergence. Dissertation, Duke University.
Rasmussen, D.T., & Izard, M.K. (1988). Scaling of growth and life history traits relative to body size, brain size, and metabolic rate in lorises and galagos (Lorisidae, Primates). *American Journal of Physical Anthropology, 75*, 357-367.
Sanderson, I.T. (1940). The mammals of the north Cameroon forest area. *Transactions of the Zoological Society of London, 14*, 623-725
Schilling, A. (1979). Olfactory communication in prosimians. In G.A. Doyle, & R.D. Martin (Eds.), *The Study of Prosimian Behavior* (pp.461-542). Academic Press, New York.
Seitz, E (1968). Die Bedeutung geruchlicher Orientierung beim Plumplori *Nycticebus coucang* Boddaert 1785 (Prosimii, Lorisidae). *Zeitschrift für Tierpsychologie, 26 (6)*, 73-103.
Still, J. (1905). Notes: On the loris in captivity. *Spolia Zeylanica, 3*, 155-157.

Subramoniam, S. (1957). Some observations on the habits of the slender loris, *Loris tardigradus* (Linné). *Journal of the Bombay Natural History Society, 54*, 387-398.

Tenaza, R., Fitch, H. (1984). The slow loris. *Zoo nooz, 57*, 10-12

Welker, Ch. (1973). Ethologische Bedeutung des Harnwaschens von *Galago crassicaudatus* E. Geoffroy, 1812 (Lorisiformes: Galagidae). *Folia Primatologica, 20*, 429-452.

Zimmermann, E., Zimmermann, P., & Zimmermann, A. (1979). Soziale Kommunikation bei Plumploris (*Nycticebus coucang*). *Zeitschrift des Kölner Zoo, 22*, 25-36.

Zimmermann, E. (1985). Vocalizations and associated behaviours in adult slow loris (*Nycticebus coucang*). *Folia Primatologica, 44*, 52-64.

Zimmermann, E. (1989). Reproduction, physical growth and behavioral development in slow loris (*Nycticebus coucang*, Lorisidae). *Human Evolution, 4 (2-3)*, 171-179.

PERSPECTIVES ON PERCUSSIVE FORAGING IN THE AYE-AYE (*DAUBENTONIA MADAGASCARIENSIS*)

Carl J. Erickson

Psychology Department - Experimental
and Duke University Primate Center
Duke University
Durham, NC 27706 U.S.A.

INTRODUCTION

In 1858 Richard Owen learned that H.M. Sandwith would be traveling to Mauritius as the new Colonial Secretary. He urged Sandwith to visit neighboring Madagascar and bring back an aye-aye, dead or alive. Eventually the British Museum was to receive its specimen "immersed in a keg of colourless spirit" but not before Sandwith had examined the behavior of this peculiar animal in considerable detail. In March of 1859 he wrote to Owen of his observations:

"......bending forward his ears, and applying his nose close to the bark, he rapidly tapped the surface with the curious second (sic) digit, as a Woodpecker taps a tree, though with much less noise, from time to time inserting the slender finger into the wormholes as a surgeon would a probe. ....I watched these proceedings with intense interest, and was much struck with the marvellous adaptation of the creature to its habits, shown by his acute hearing, which enables him aptly to distinguish the different tones emitted from the wood by his gentle tapping; his evidently acute sense of smell, aiding him in his search; ....the curious slender finger, unlike that of any other animal. . . he used alternately as a pleximeter, a probe, and a scoop." (Owen, 1866; p. 38).

Although it is more than 135 years since Sandwith described this tapping, probing and extraction behavior, it is only recently that the sensory-motor and tactical aspects of this unusual foraging activity have become the subject of systematic scrutiny. The questions are many: What food items are the object of such activity? Is echolocation and/or a cutaneous sense involved in their localization? How sensitive is the detection mechanism? Are there strategic aspects to prey detection? A series of studies conducted both at the Duke University Primate Center (DUPC) and in Madagascar has begun, not only to provide some specific answers to these and other questions, but to reveal the importance of the aye-aye as a model for addressing broad issues of perceptual and inferential processes. In general, I am interested in understanding the natural history of percussive foraging, but I also hope to use the behavior as a window upon the aye-aye's broader behavioral capabilities. The aye-aye has the largest brain of all the prosimians, and one suspects that its brain size and unusual foraging behavior are somehow linked. The studies described here were designed with the intention of examining the aye-aye's detection and tactical capabilities using its tap-scanning behavior.

TAP-SCANNING AND THE STIMULUS TO EXCAVATION

The aye-aye feeds on a wide variety of foods, including sap, nectar, fungus, fruit, ants, insect larvae and nuts (c.f. Pollock *et al.*, 1985; Iwano and Iwakawa, 1988; Winn, 1989; Sterling, 1993). Percussive foraging seems to be largely focused on woodboring insect larvae and the inquiline residents of their mines, although it may also be used to locate the locules of canarium nuts (Sterling, 1993) and the milk level in coconuts (Winn, 1989). Aye-ayes excavate and ingest large lepidopteran and coleopteran woodborers (Sterling, 1993), and I have found some of the latter in foraged trees to reach 8 g and 8.5 cm in length. Aye-ayes may also feed on the larvae of buprestid, tenebrionid, and elaterid members of the coleoptera (Pollock *et al.*, 1985; Erickson, in press). Owen (1866) proposed that prey may generate audible sounds revealing their location, and Feller (pers. comm.) notes that the subsurface activities of large cerambicids are sometimes audible to humans. Because the aye-aye has large, bat-like pinnae, it seems likely that the movements of prey are indeed detectable; moreover, the tapping behavior might stir quiescent prey into audible activity. Detection can be further aided by any small pores and channels providing olfactory access to the larval galleries below.

In my initial studies I compared the responses of aye-ayes when presented with subsurface features of various kinds (Erickson, 1991). The standard procedure involved the presentation of a spruce woodblock 35 cm long, 6 cm wide, and 3.5 cm thick. (See Fig. 1.) Six different stimulus conditions were placed along the midline of the woodblock: (1) *Condition LO*. A subsurface cavity of 3 cm diameter contained six small, living mealworms (i.e. *Tenebrio* larvae). Four small apertures connected cavity and surface, allowing olfactory access. (2) *Condition L*. A cavity was prepared and provisioned as in condition LO, but in this case the cavity was not connected to the surface by apertures. (3) *Condition D*. This condition was similar to that in condition L except that the mealworms had been diced into four or five pieces to prevent audible movement. (4) *Condition E*. Cavities in this condition were empty; they had neither mealworms nor surface apertures. (5) *Condition F*. Four 3 mm holes were drilled into a section of the upper surface of the woodblock, but they did not connect to a cavity. (6) *Condition B*. A blank section of woodblock served as a control. It contained no cavity, mealworms or surface holes. The positions of these conditions were varied randomly from one woodblock to another. Surface thickness over the four cavities in each woodblock was the same, but surface thicknesses varied from woodblock to woodblock, ranging from 7 to 17 mm. Animals were presented with 10 woodblocks at each of six different surface thicknesses.

As expected, across conditions excavation frequency decreased as surface thickness increased (See Erickson, 1991 for details). Quite unexpected was the finding that empty cavities were excavated much more often than were blank sections of block or false cavities (conditions B and F). Thus the cavities themselves, independent of any prey they may contain, have stimulus qualities capable of evoking excavation behavior (see Fig. 2) Although cavities containing live mealworms were opened slightly more often than those that were empty or those that contained diced mealworms, the differences were not large. However, it should be noted that mealworms are much smaller than many of the large larvae on which the aye-aye normally feeds, and it is possible that responses to conditions L and LO would have been higher

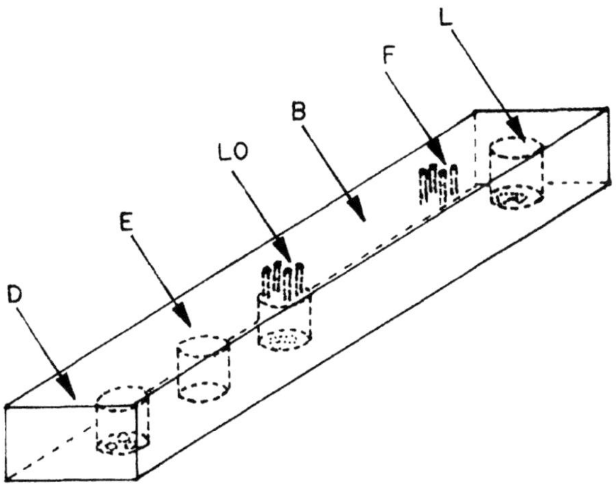

Figure 1. Typical arrangement of cavities, olfactory access holes, and blank sections in standard spruce woodblocks. The sequence of conditions was varied in each woodblock.

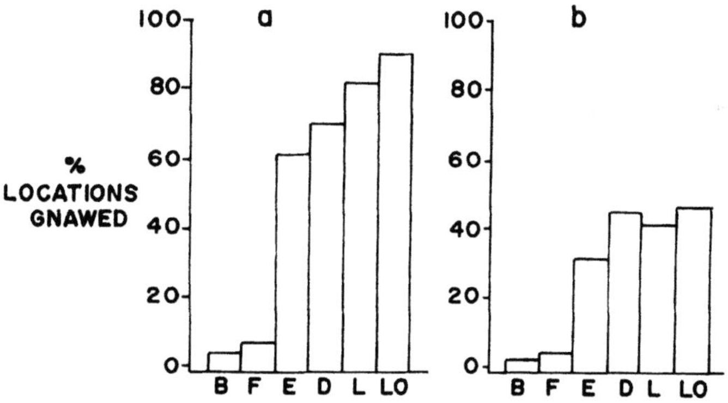

Figure 2. Distribution of investigatory behavior under six stimulus conditions. B:blank; F:false; E:empty; D:diced mealworms; L:live mealworms, no access holes; LO:live mealworms, access holes present. Measurements are for pairs of animals: (a) Samantha and Nosferatu, (b) Poe and Annabelle.

had large woodboring larvae been provided. In short, the study illustrates the important role of the cavities themselves in eliciting excavation; the full significance of prey sounds remains unknown.

If empty cavities are a potent stimulus to excavation, how are they detected? Do subsurface hollow spaces have specific, signature characteristics that elicit excavation? For example, does the aye-aye attend to the resonance unique to air-filled cavities? In a preliminary experiment Stephen Nowicki and I presented woodblocks containing four cavities; a blank section again served as a control. The surface over all cavities was 5 mm and no perforations connected any of the cavities with the surface. One, two or three of the cavities was provisioned with live mealworms. These cavities -empty or provisioned- were included simply to maintain the interest of the animals. A fourth cavity in each block was filled with acoustical foam, a material used commercially to absorb sound energy; in this case we attempted to dampen echoes within the cavity and interfere with cavity detection and excavation. Our results were not what we had expected. In each of ten woodblocks the aye-ayes detected and excavated the cavities containing the acoustical foam. Their responses suggested that the stimulus to excavation is not unique to air-filled cavities.

Consequently, the hypothesis was revised to suggest that any subsurface area containing material of lower density than the surrounding wood is of interest to the aye-aye. In order to test this new hypothesis we designed a more elaborate study. Cavities were backfilled with materials of lower, higher or equal density than the surrounding spruce. Balsa and air (low density), cast iron and maple (higher density) and spruce from a scrap end of the woodblock itself (equal density) were used. Five different surface thicknesses, ranging from 3mm to 15 mm, were included. The surface thickness and position of the various cavity conditions varied from one woodblock presentation to the next. Again our results were contrary to expectation (Fig. 3). The density of subsurface materials had no apparent effect on excavation frequency; in fact, the frequency with which metal and balsa plugs (i.e. high and low density materials) were uncovered was identical for each series tested. Moreover, even when cavities were backfilled with spruce plugs from the woodblock itself, they were readily detected and excavated.

Although this study suggested that the stimulus to excavation is not unique to air-filled spaces (or even low density areas) within the wood, there remained a slight chance that minor air spaces trapped around the plugs had been responsible for the excavation behavior. As a control for this possible artifact, another study was performed in which woodblocks contained cavities packed either with fine sand or sawdust. These woodblocks also contained a section in which as many as 16 small holes, each of which was 2.5 mm in diameter, were drilled in a circular pattern 2.5 cm in diameter. These holes were drilled to a depth equivalent to that of the backfilled cavities. The purpose of this condition was to simulate minor air spaces that could have been trapped around the plugs in the study described above. The results revealed that cavities packed with sand or sawdust were excavated much more often than areas where minor airspaces could be found. Together, these studies support the conclusion that it is not the acoustical properties unique to air-filled "hollow" or "empty" subsurface cavities that stimulate the excavation behavior of the aye-aye; cavities with and without prey as well as those tightly backfilled with materials of widely varying density elicit the behavior.

These studies tell us much more about what the stimulus is not than what it is. However, the common element in the studies performed thus far seems to be the presence of an interface. Perhaps the aye-aye simply responds to any departure from the acoustical properties of intact subsurface wood. It is possible that such detection is enhanced by properties peculiar to the tropical trees on which the aye-aye feeds. Many of the excavated trees that we examined in Madagascar have a thin, rindlike bark and a homogeneous wood that is free of the growth rings so characteristic of temperate species. These sources might provide the aye-aye with a higher signal-to-noise ratio than it would encounter in knotty seasonal growers. It must be

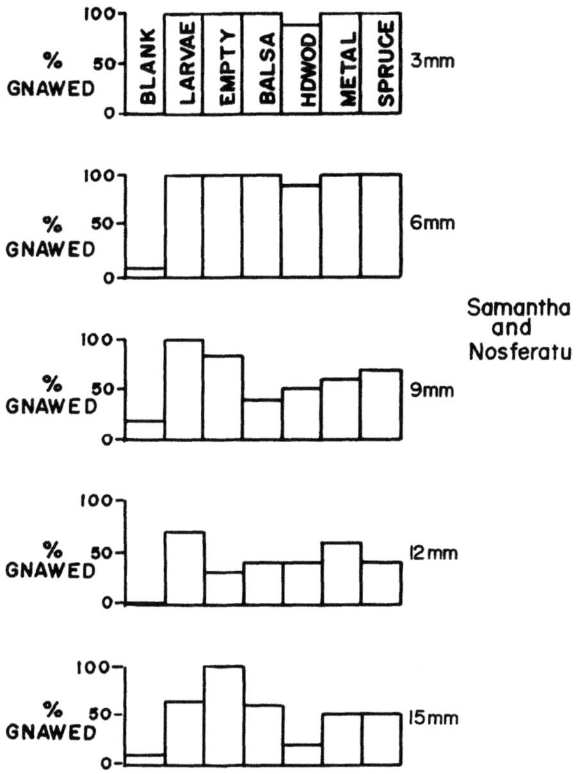

Figure 3. Woodblocks contained cavities plus a blank section. The cavities either were provisioned with mealworms, remained empty, or were backfilled with balsa, hardwood, metal, or a plug made from the spruce woodblock material itself. Surface thicknesses over the cavities varied from 3mm to 15 mm. Data for one pair of animals are shown, but other animals performed similarly.

emphasized that these studies provide no insight into the aye -aye's discrimination capabilities. Although a wide variety of subsurface features are excavated with approximately equal frequency, it remains quite possible that the aye-aye is readily detecting the differences among them; the failure to exhibit differential excavation may be due to motivational indifference rather than any deficiency in discrimination capability. Insight into this issue must await studies in which motivation is experimentally manipulated.

THE EXPLOITATION OF LARVAL MINES

Petter (1977) and Ganzhorn and Rabesoa (1986) have remarked on the apparent inefficiency of aye-aye extractive foraging. Our demonstration that aye-ayes readily open cavities devoid of prey seems to underscore such an impression. Although some of the cerambicid larvae are very large (Erickson, in press) and nutritious (Sterling, 1993), and are probably capable of compensating for much time and effort spent in the search for them, they often occupy mines of considerable length (Fig. 4) I estimated that mines in some foraged trees coursed through the bole for a meter or more, but I suspect that some are much longer. Ilka Feller has measured the length of mines generated by some New World cerambicids at more than four m. How then does an animal efficiently find prey within such extended cavities? Until we know more about the natural history of the woodboring larvae on which the aye-aye feeds, some ambiguity is sure to remain. However, some of the following hypotheses regarding foraging tactics can be suggested.

It is likely that a woodboring larva is most often found at the terminus of its mine. Once a mine is found, the aye-aye can tap-scan its way along the surface, tracing its course until an end is located. Since the mine has two ends, the aye-aye has a 50-50 chance of payoff, but of course the chance of pay-off will be markedly enhanced if larvae are known to move selectively in one direction. For example, larvae might be more easily located if they are guided by a geotaxis, tending to move selectively up or down the tree. Moreover, the cross-sectional dimensions of the mine probably increase as the larva grows; and this increase, if detectable to the aye-aye, could also serve as a cue to the end in which prey are to be found. My analysis of excavation locations in my field study did, in fact, confirm that aye-ayes often locate their entry points over the terminal portions of mines. In two instances the remains of captured larvae were clearly visible; however, I found no evidence of geotaxic behavior in the larvae.

The influence of frass on aye-aye foraging is not clear. Unlike some of the xylophagous hymenoptera and isoptera that eject sawdust from their galleries, woodboring cerambicids backfill their mines with the masticated detritus of shredded wood as they move along. This becomes so tightly packed in the mine as to make the mine almost invisible when the bole is crosscut. In early studies I packed artificial cylindrical cavities with sawdust and found that aye-ayes can readily identify them in spite of such fill. In more recent, preliminary work I have attempted to simulate the frass-filled mines more closely, and I find a similar result. However, more research is needed before the effects of frass on mine detectability can be adequately estimated. If frass makes mines undetectable, the aye-aye would be unable to follow their course by tap-scanning. On the other hand, if frass makes all but the area immediately surrounding the larva undetectable, we can infer that once the aye-aye has found a frassless sector, it has also found its prey.

Interpretation of these issues is complicated by the fact that the distribution of entry points observed in foraged trees exhibits much variability. Not only are excavated pits found at the terminus, they can be found at other locations as well; and it is not uncommon for a single mine to be entered at several locations. In some trees more than half the excavations opened onto no mines at all! Without direct observation of aye-aye tap-scanning patterns followed by pit and mine analysis, our interpretation of foraging tactics must remain somewhat speculative. Experiments with captive animals are helpful, however.

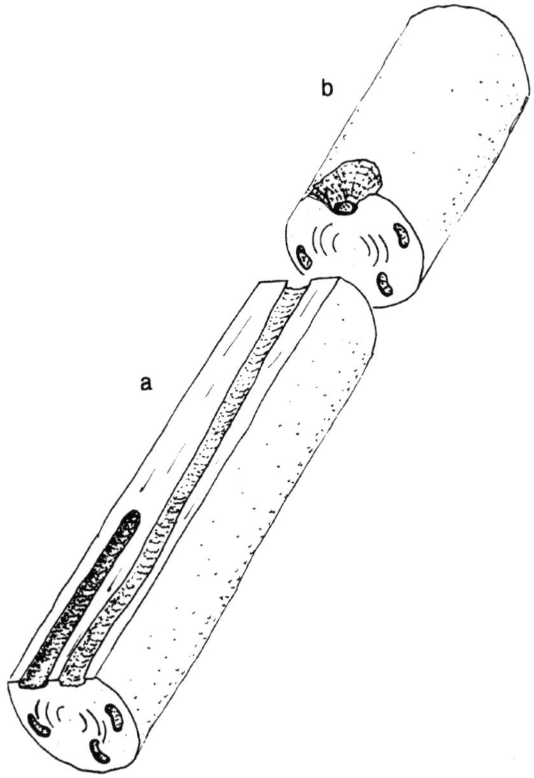

Figure 4. Larval mines often course long distances beneath the wood surface. The surface of Section *a* has been planed away to reveal the terminal end of one mine and thc configuration of another. Section *b* has been cross-cut through an aye-aye excavation revealing a mine.

Recently I have presented animals with crude simulations of larval mines by using a router to incise channels into woodblocks. These "mines" were 20, 30 or 40 cm long. The aye-ayes readily found the mines. Moreover, they placed excavation pits at regular intervals along their course, approximately 7.5 cm apart. It appears that once a cavity is detected, excavations are distributed according to some strategic pattern. Although these initial trials involved straight mines containing no frass, more recent investigations using meandering mines containing frass have suggested similar results.

Why should aye-ayes probe mines at regular intervals if woodboring larvae are most likely to be found at the terminus? Such probing may represent a search for inquilines. For example, beetle larvae of the family Elateridae are known to prey upon woodboring cerambicids. I found several large elaterids in the mines of cerambicids; and when we fed them to the aye-ayes at the zoological park in Ivoloina, Madagascar, they were readily consumed. The cerambicid mines also contained an assortment of scarabs, tenebrionids and buprestids as well as some millipedes, scorpions and a crab; but logistical difficulties prevented us from testing their acceptability to the aye-ayes. Pollock *et al.* (1985) also suggest that aye-ayes eat elaterids. They examined galls on an *Afzelia bijuga* (Leguminosaceae) tree where aye-ayes were seen feeding, and found not only elaterids, but a variety of dermapteran, hymenopteran, and lepidopteran species as well. In one instance they even found a tree frog, *Anodonthyla rouxae* (Microhylidae: Cophylinae) behind a gall, and they suggest that it too might be included in the aye-aye's diet. In general, we find that the trees on which the aye-aye forages contain complex communities of organisms. Additional research is required to census the full range of woodborers and inquiline residents on which the aye-aye feeds and to record the full repertoire of foraging techniques used in their capture.

SOME CONCLUSIONS AND ADDITIONAL HYPOTHESES

Many of the observations reviewed here seem to reinforce the common impression that aye-aye extractive foraging is inefficient at best. Aye-ayes gouge pits more than 3 cm deep, removing as much as 146 cc of wood in the process; yet this effort may be expended when no prey occupy the subsurface cavities. Even when mines do contain prey and are readily located, the prey may occupy only a small sector of a very large cavity and require extensive search before capture. The fact that aye-ayes also open mines that have been tightly backfilled with materials varying widely in density--even cavities that have been plugged with stock from the woodblocks themselves-- suggests a remarkable lack of specificity in the features that elicit excavation. How then can this behavior be profitable?

A precise cost-benefit analysis of percussive foraging is not possible, but our studies suggest a number of features that may make it more effective than it at first appears.

1. Many woodboring larvae are large and nutritious, and they probably provide a substantial payoff when found. Furthermore, in some instances their activities may be directly audible.

2. Once mines have been excavated and woodboring larvae have been removed from them, they may become home to inquiline residents such as frogs. These may provide substantial payoff with little additional search cost and excavation effort.

3. Aye-ayes may have search tactics that increase the efficiency of finding prey. Mines usually run parallel to the main axis of the tree. By tapping around the bole, aye-ayes are more likely to encounter them. Woodboring larvae are more often to be found at the terminal sections of their mines, and excavation at these locations is more likely to be successful. Moreover, If mine size increases as larvae progress or if larvae are geotaxic, their location within extended mines should be even more easily identified.

ACKNOWLEDGMENTS

I owe special thanks to K. Izard who urged me to observe the aye-ayes and without whose help the research would have never begun. I am grateful to L. Alterman who also encouraged these first tentative steps into primatology. Bill Hess deserves special mention for his care and concern for the aye-ayes at the DUPC. I am grateful to Donna Brown who provided important comments on the manuscript. Andrea Katz, Charles Welch, Patricia Wright, Eleanor Sterling, Ilka Feller, Warren Steiner, Benjamin Andriamihaja and my guides, Albert Telo and Tsirindahy all contributed in various but important ways. I thank Emily Houde for helping me with the illustrations. Supported by NSF Grant NSF-DIR-89-13211 to the DUPC and a Duke University Research Council Grant to C.E. DUPC Publication 575.

REFERENCES

Erickson, C.J. (1991). Percussive foraging in the aye-aye, (*Daubentonia madagascariensis*). *Animal Behaviour*. *41*,793-801.

Erickson, C.J. (in press). Feeding sites for extractive foraging by the aye-aye, (*Daubentonia madagascariensis*). *American Journal of Primatology*.

Ganzhorn, J.U. & Rabesoa,J. (1986). The aye-aye (*Daubentonia madagascariensis*) found in the eastern rainforest of Madagascar. *Folia Primatologica, 46*, 125-126.

Iwano, T. & Iwakawa, C. (1988). Feeding behaviour of the aye-aye, (*Daubentonia madagascariensis*) on nuts of ramy (*Canarium madagascariensis*). *Folia Primatologica, 50*, 136-142.

Owen, R. (1866). On the aye-aye (*Chiromys Cuvier*). *Transactions of the Zoological Society of London, 5*, 33-101.

Petter, P.P. (1977). The aye-aye. In H.S.H. Prince Rainier III, & Bourne, G.H. (Eds.), *Primate Conservation* (pp. 37-57). New York: Academic Press.

Pollock, J.I., Constable,I.D., Mittermeier,R.A., Ratsirarson,J. & Simons,H. (1985). A note on the diet and feeding behavior of the aye-aye (*Daubentonia madagascariensis*), *International Journal of Primatology, 6,* 435-447.

Sterling, E.J. (1993). Behavioral ecology of the aye-aye, (*Daubentonia madagascariensis*) on Nosy Mangabe, Madagascar. Unpublished Ph.D. Dissertation, Yale University.

Winn, R.M. (1989). The aye-ayes, (*Daubentonia madagascariensis*) at the Paris Zoological Garden: Maintenance and preliminary behavioural observations, *Folia Primatologica, 52*, 109-123.

RIGHT HAND PREFERENCE AND PERFORMANCE BIASES IN THE FORAGING BEHAVIOR OF THE AYE-AYE

Garrett W. Milliken

Department of Neurobiology and Anatomy
University of Texas Medical School at Houston
P.O. Box 20708
Houston, TX 77225

INTRODUCTION

Research on handedness in primates has revealed a leftward bias for prosimians (Forsythe, Milliken, Stafford, & Ward, 1988; Forsythe & Ward, 1988; Milliken, Forsythe, & Ward, 1989; Milliken, Stafford, Dodson, Pinger, & Ward, 1991; Sanford, Guin, & Ward, 1984; Stafford, Milliken, & Ward, 1993; Subramoniam, 1957; Ward, Milliken, Dodson, Stafford, & Wallace, 1990) and a rightward bias for anthropoid species (Bresard & Bresson, 1983; Beck & Barton, 1972; Deuel & Dunlop, 1980). The leftward preference seen in prosimians is related to the basic prehensile type of manual control whereby all fingers move in concert, and the right preference of anthropoids reflects a refinement in manual function that involves independent control of each finger.

MacNeilage, Studdert-Kennedy, and Lindblom (1987) have proposed that the origins of behavioral lateralization were propagated by the need for a division of labor between the hands associated with unimanual insect capture while maintaining stability in the arboreal substrate. They view lateralization as an evolving dynamic where the primary form (left hand preference for visually guided insect capture, right hand for postural support) formed the foundation for later developing intricacies associated with fine motor control. These authors assert that the left hand specialization for whole hand (prehensile) grasping relegated fine motor tasks to the right hand when species changed their feeding niche to include fruits that require manipulation to remove the edible portions from the protective covering. The change in feeding niche modified the basal manual division of labor wherein the left hand retained the function of prehensile grasping and the right hand developed refined motor skill to effect feeding.

In interpreting data on lateral biases it is important to distinguish between preference and performance measures of lateralization. Preference

Creatures of the Dark: The Nocturnal Prosimians
Edited by L. Alterman *et al.*, Plenum Press, New York, 1995

measures of lateralization are seen as quantitative indices of the overall tendency to respond with the left or right hand. The preference measure is the most widely used characterization in the non-human primate laterality literature and is generally concerned with simple food reaching. Performance measures of lateralization are seen as more qualitative and focus on efficiency differences between the hands. It has been argued that performance measures are better indicators of underlying cerebral asymmetries than simple preference measures (Bryden, 1982). Performance measures have been used to assess laterality in black-capped capuchin monkeys (Fragaszy & Mitchell, 1990), rhesus macaques (Preilowski, 1979), and the lesser bushbaby (Dodson & Ward, in press). Fragaszy and Mitchell (1990) examined latency of prehension in grasping a variety of shaped knobs. They found no relationship between hand preference on the experimental task and preferences observed in spontaneous acts, and that use of the preferred hand was associated with faster latencies. Preilowski (1979) used a force transducer to test the ability of rhesus macaques to produce isometric pressure of a specific duration between the fingertips of left and right hands. He found that his subjects were able to perform better with the right hand than with the left. In an ongoing study of the lesser bushbaby, Dodson et al. are finding a stable left hand advantage for ballistic prey capture latency in this prosimian.

The aye-aye is a prosimian that has developed the ability to move the third finger independently of the other fingers (Milliken, Ward, & Erickson, 1991). Aye-ayes use this ability for two purposes in foraging for wood-boring insect larvae. The first involves rapid tapping to locate the galleries where the larvae live (Erickson, 1991); the second involves extraction of the larva from the cavity. This paper presents data that pertain to the extraction of larvae from cavities. The limits of extractive ability were tested using apparatuses that simulate the galleries of wood boring insects. These apparatuses permitted examination of the fingers during retrieval of grubs from various depths (vertical cavities) and from cavities cut in differently angled orientations (horizontal cavities). The apparatuses also allow comparisons to made between preference and performance measures of lateralization. The results that follow will show that aye-ayes can tactilely detect the presence of a grub in a cavity, and that when removing grubs from cavities the right finger is preferred and is more efficient than the left finger. These results are taken as evidence that right hand preference and performance biases are linked to independent finger control and are not necessarily confined to anthropoid primates.

METHODS

Subjects

Data for this project were collected at the Duke University Primate Center in the summers of 1989 and 1990. Four wild-caught aye-ayes *(Daubentonia madagascariensis)* served as subjects for these experiments. The 2 male subjects were received in 1987 and had ages estimated as mature (Nosferatu) and less than 1 year old (Poe). A mother-daughter pair arrived in 1988 and had ages estimated as mature (Samantha) and less than 6 months (Annabellee). All subjects were maintained on reversed 12/12 light cycles and pair-housed in enclosures approximately 40 m^3. The rooms contained wood nest boxes, as well as trees and lianas for climbing. The aye-ayes were provisioned midway through the dark cycle with an assortment of fruit, vegetables, and insect larvae presented in a dish. The larval species regularly offered (*Tenebrio molitor, Sarcophaga bullata, Galleria mellonella)* were also used in testing.

Materials

To examine methods used by the aye-aye to extract grubs from wood cavities, two types of apparatus were constructed. The apparatus for determining effective reach had vertical cavities, whereas that for evaluating digit flexibility had horizontal cavities. Both apparatuses were attached to an internally illuminated plywood camera box (30 x 13 x 13 in.), and consisted of cavities that were cut into hardwood. The cavities had a Plexiglas window facing the camera that permitted the actions of the finger to be videotaped when inside the cavity. It was necessary to protect the top surface of the apparatuses with a steel plate to prevent the aye-ayes from chewing the wood. The steel plates had 1 cm. holes cut to allow access to the cavities.

The vertical cavity apparatus (VCA) tested the abilities of the aye-ayes to extract grubs from vertical cavities that were of different depths. The VCA (Figure 1) required the aye-aye to probe downward to retrieve a grub through a metal hole 1 cm diameter into a vertical cavity that was 1.5 cm wide and either 6, 7.5, 9, or 10 cm. deep.

The second apparatus (Figure 2) was designed to test extractive ability in horizontally oriented cavities. The horizontal cavity apparatus (HCA) had cavity channels (1 cm wide, 1.25 cm deep) cut in straight, right, obtuse, and acute angle orientations. The angled cavities were located at the end of a straight runway that extended from the entrance hole to the angled cavity branching point.

Procedure

Data were collected midway through the animals' active period at feeding time. It was necessary to videotape from both inside and outside the apparatuses in order to view the finger inside the cavity, and externally, to identify the subject and hand used. The tapes were synchronized during analysis to permit evaluation of the movements of the finger inside the cavity with respect to the hand used. The procedures for analysis of each apparatus will be presented separately.

For the vertical cavity apparatus (VCA), videotapes were analyzed using the dependent variables of *probes*, *lifts*, and whether or not the subject retrieved a grub *(success)*. A *probe* was defined operationally by the appearance of the finger inside the cavity opening. During a probe, the aye-aye moved the third or fourth finger inside the cavity. This movement involved flexion and adduction of the finger, and resembled a digging motion. When a grub was balanced on the fingertip, the finger was lifted. A *lift* was defined by the distal phalanx moving upward beyond the halfway mark indicated on the cavity. The number of lifts and the result of the probe (success or failure in extracting a grub) was recorded for each probe sequence. Also indicated was the hand (left or right) and finger(s) (D3, D4, D3 & D4) used to enact the probe.

For the horizontal cavity apparatus (HCA), the dependent variables for comparing performance in the various cavity angles were the number of *flexions* made by the distal phalanx when inside a particular cavity, the amount of time *(duration)* that the finger was inside the cavity, and whether or not the subject successfully removed a grub from the cavity *(success)*. The *flexion* movement is illustrated in Figure 3 and was defined by the terminal phalanx flexing and extending. While the finger was inside the cavity, the number of flexions were counted. A *duration* sequence commenced when the fingertip entered the proximal end of the cavity, and concluded when the finger was removed from the cavity. *Success* was defined by the larva being removed completely from the cavity. It should be noted that data was recorded only when the finger was inside an *angled* cavity and not when the finger was in the runway leading to or from the cavity. The *success* measure in the horizontal cavity apparatus differs from that used in the vertical cavity apparatus in that the grub had to be removed from the cavity itself and not from the runway and aperture. In the vertical cavity experiment, subjects removed a grub from a particular cavity depth with one fluid motion. In the horizontal cavity apparatus, a retrieval usually involved at least three motions: (1) removal of the grub from the cavity, (2) removal of the grub from the runway, and (3) removal of the grub from the apparatus through the aperture.

The HCA also provided a measure of tactile sensitivity. Measurement of tactile sensitivity was obtained by comparing duration and activity measures of finger function when a grub was inside the cavity to that when the cavity was vacant.

The targeted sample size for each subject on the HCA was 160 cavity entries (40 in each cavity angle; 20 with grub present, 20 with vacant cavity). This sample was obtained for both of the male subjects (Nosferatu and Poe), but was not obtained for the females: Annabellee made 135 cavity entries, and Samantha made 105 entries. The low response in the female subjects was caused by two factors. In the case of the juvenile female, Annabellee, her low response was caused by her roommate, Poe, who would seldom allow her the opportunity to probe the apparatus. Samantha sustained an injury after the second day of data collection, whereby the fingernail of her right hand was torn off as she was being captured for medical treatment. This injury effectively removed her from laterality analysis, but she was included in all other analyses.

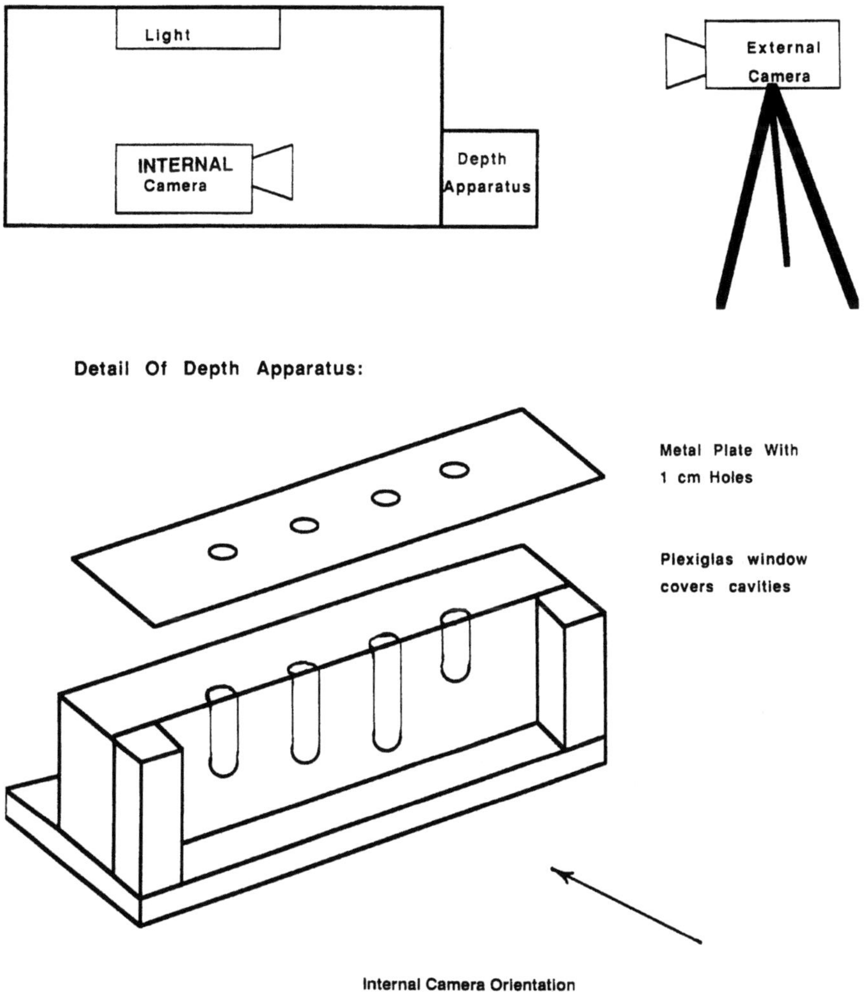

Figure 1. Vertical cavity apparatus (VCA). Cavity channels were 6, 7.5, 9, and 10 cm deep.

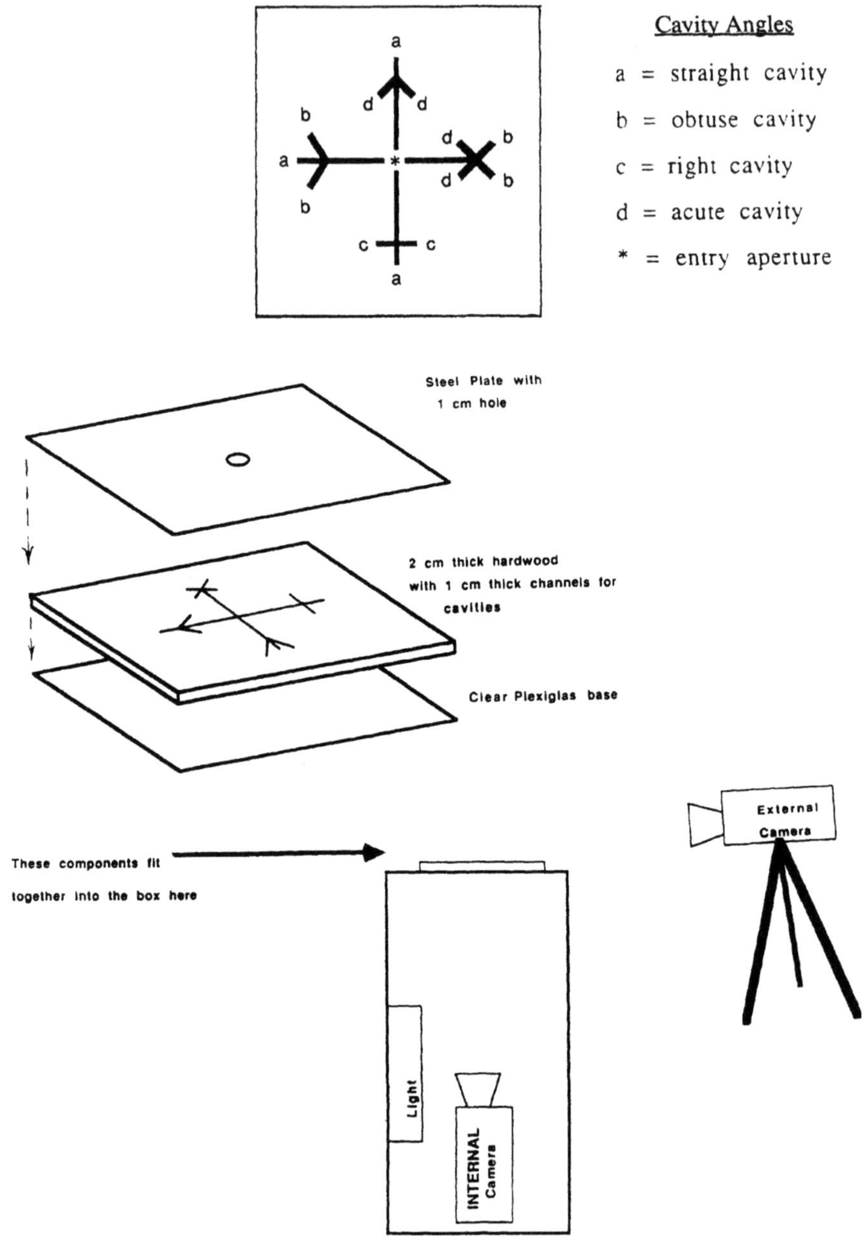

Figure 2. Horizontal Cavity Apparatus (HCA) for determining extractive ability from differently angled cavity channels.

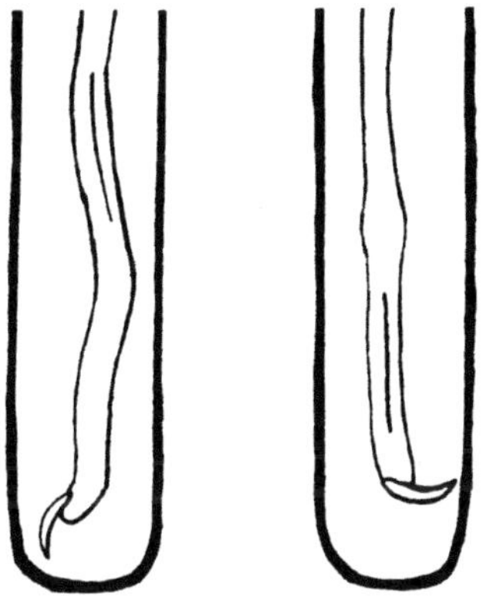

Figure 3. Flexion movement of the terminal phalanx.

Statistics

Binomial z-scores were used to evaluate lateral preferences with respect to frequency of left and right finger use for individual subjects (Edwards, 1963). T-tests were used in group comparisons of performance and lateral preference.

To compare performance in the different cavity depths and different cavity angles, within-subjects one-way analyses of variance were employed using the different cavities of each apparatus as repeated measures. Use of this statistic is deemed appropriate because each aye-aye was exposed to each type of cavity in both experimental conditions. Significant ANOVAS were followed up with Tukey's honestly significant difference tests to specify performance differences by comparing the means from each cavity.

RESULTS

Vertical Cavity Apparatus

Finger Use. Three strategies were observed for probing in the vertical cavity apparatus. These strategies are specified as probing with the third finger, fourth finger, and third and fourth finger in unison. Figure 4 shows that all subjects favored use of the third finger for probing the cavities, with the female aye-ayes (Samantha and Annabellee) using the fourth finger and third and fourth finger combination strategies more than the males. The higher frequency of response of the females with the longer, larger, but less coordinated fourth finger may reflect the difference in finger size between the males and females. The longer fingered males had a total of only seven probes using the fourth finger. The limited use of the fourth finger, and third and fourth finger combination probe strategy by all subjects, is reason to consider only third finger function in the subsequent analyses.

General Responses Characteristics: Cavity Depths. When the third finger probes for all subjects were combined (Figure 5) there was an overall tendency for probe frequency to diminish with cavity depth. There was also a tendency for each individual to respond more to the shallow cavities that were more proportionate to the length of the third finger than to the deeper cavities. A repeated measures one-way analysis of variance was used to test this trend but differences in frequency of responses to the different cavity depths were non-significant (F [3, 12] = 2.61, ns). This result is probably due to the low response by Samantha to the 6 cm cavity and could have been influenced by her high success rate for this cavity. A high success rate translates into fewer probes because the cavity is rapidly depleted, conversely the probe frequency is amplified if there are more unsuccessful attempts. The highest proportion of success in any cavity by any subject was by Samantha in the 6 cm cavity.

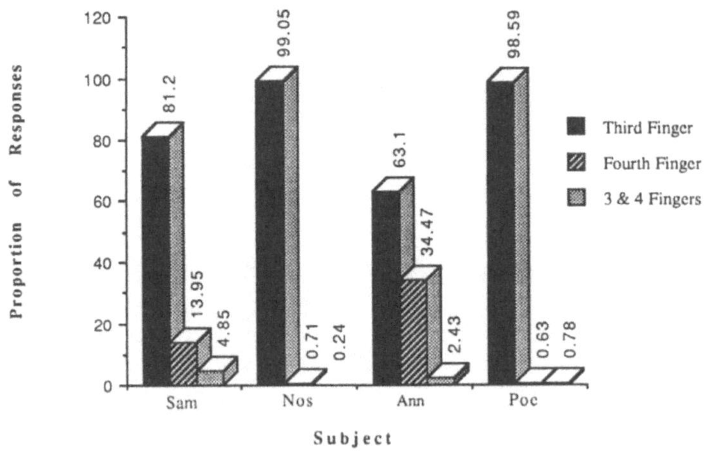

Figure 4. Proportion of third finger, fourth finger, and combination third and fourth finger probe strategies.

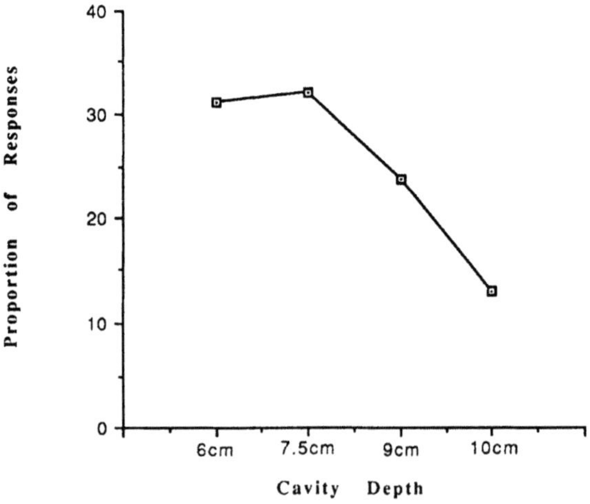

Figure 5. Combined proportion of responses as a function of cavity depth.

Activity. The activity of the third finger inside the cavity was analyzed in terms of the number of lifts that occurred during a probe sequence. Table 1 shows the number of probes and number of lifts for each subject in each cavity depth. The table also shows a derived measure of activity, a lift/probe average, derived by dividing the total number of lifts by the number of probes. The higher the lift/probe average, the more lifts per probe. The data from Table 1 show that there were individual differences in lift/probe averages between cavities. When lift/probe averages were combined for all subjects, there were no differences that were related to cavity depth. The lift/probe averages also did not differ between successful and non-successful retrieval attempts t (15) = .405, p = .691, ns, meaning that activity was not a predictor of success in obtaining a grub.

Success. Successful retrieval of grubs was most associated with depth of cavity. Table 2 shows the number of successful and unsuccessful probes for each subject in each cavity. These results are illustrated in Figure 6 as combined proportions by cavity. The figure shows that the rate of success diminished between the shallow (6 cm, 7.5 cm, & 9 cm) and deepest (10 cm) cavities, a result that was confirmed by a repeated measures one-way ANOVA, F (3, 12) = 4.81, p = .029. A Tukey's post hoc test revealed significant differences between the deepest cavity (10 cm) and the 6 cm and 9 cm cavities. The mean comparison between the 10 cm and 7.5 cm cavity performance approached significance.

Table 1. Frequency of probes and lifts for each subject in each cavity.

Subject Cavity	Probes	Lifts	Lift/Probe Average
Samantha			
6	75	111	1.48
7.5	130	156	1.20
9	126	169	1.34
10	88	107	1.22
Nosferatu			
6	123	144	1.17
7.5	133	156	1.17
9	83	109	1.31
10	80	126	1.58
Annabellee			
6	146	202	1.38
7.5	82	142	1.73
9	24	28	1.17
10	8	8	1.00
Poe			
6	193	241	1.25
7.5	208	320	1.54
9	178	244	1.37
10	50	52	1.04

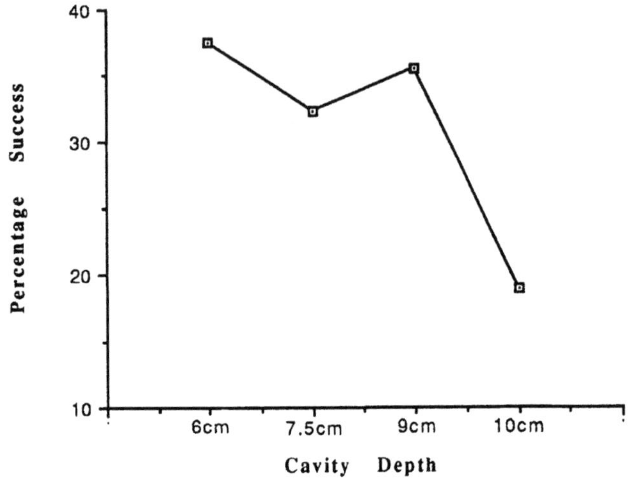

Figure 6. Combined percentage of successful retrievals as a function of cavity depth.

Table 2. Percentage of successful retrievals for each subject in each cavity.

Subject Cavity	Failed Attempts	Successful Attempts	Percentage Success
Samantha			
6	27	48	64
7.5	52	78	60
9	55	71	56
10	53	35	40
Nosferatu			
6	69	54	44
7.5	102	31	23
9	46	37	45
10	54	26	32
Annabellee			
6	126	20	14
7.5	68	14	17
9	22	2	8
10	8	0	0
Poe			
6	138	55	28
7.5	147	61	29
9	119	69	33
10	48	2	4

Lateral Preference. All subjects were right preferent in probing, with the strongest lateral biases appearing in the shallow (6 and 7.5 cm) cavities (Figure 7). The frequency of left and right probes using the third finger, the resultant z-score for each cavity, and the overall lateral preference for probing are presented in Table 3. A within subjects repeated measures ANOVA using the z-scores from each cavity approached significance, F (3, 12), 3.42, p = .066, lending some support to the assertion that strength of hand preference diminished with cavity depth.

The finding of a right finger preference in probe frequency in the shallow cavities may be explained by the analysis of probe persistence, defined as the tendency for subjects to remain at, and repeatedly probe in the same cavity, thus forming a probe sequence. After initiating a probe into a given cavity, the aye-aye could withdraw the finger and probe the same cavity again, or change .cavities. The initial probe represents the first probe in a sequence. To evaluate probe persistence, the frequency of responses to a given cavity were divided by the number of initial entries into that cavity. The resultant number is the average number of probes that made up the sequence. Table 4 gives the probe frequency, number of initiated sequences, and the average number of probes that were included in a sequence. The data are presented by subject and cavity and show that the number of probes in a sequence diminished with cavity depth, a result confirmed by a repeated measures one-way ANOVA, F (3, 12) = 7.91, p = .0067. Longer probe sequences were associated with shallow cavities (6 cm and 7.5 cm), and shorter probe sequences were found in the deeper cavities (9 cm and 10 cm). A Tukey's post hoc test found significant differences between the 6 cm cavity and the deeper cavities (9 cm and 10 cm).

This analysis indicates that the aye-ayes were more likely to

Table 3. Hand preference analysis: Overall frequency of left and right third finger probing as a function of cavity depth.

Cavity	Left	Right	Z	Cavity	Left	Right	Z
Samantha				Nosferatu			
6	16	59	4.97*	6	61	62	0.09
7.5	54	76	1.93	7.5	36	97	5.29*
9	71	55	-1.43	9	44	39	-0.55
10	45	43	-0.21	10	37	43	0.67
Total	186	233	2.30*	Total	178	241	3.08*
Annabellee				Poe			
6	45	101	4.63*	6	2	191	13.60*
7.5	46	36	-1.10	7.5	26	182	10.82*
9	12	12	0.00	9	106	72	-2.55*
10	5	3	-0.71	10	12	38	3.67*
Total	108	152	2.73*	Total	146	483	13.44*

Note. Positive z-score denote a rightward bias, negative z-scores a leftward bias.
* p < .05

persist with subsequent probes when they initiated a probe in a shallow cavity. The analysis of persistence shows that the right probe preferences associated with shallow cavities could have resulted from more persistent probing with the right finger, and the diminishment of right preference in the deeper cavities shows a shift toward greater use of the left hand. The number of probes in a sequence was greater for the right hand than the left, t (15) = 3.08, p = .008. Table 4 gives the average left and right sequence length and again it is apparent that the largest disparity between the sequences was in the shallow cavities.

Table 4. Analysis of probe persistence

	Probe Frequency			Initiated Sequences						Changes	
Cav	Left	Right	Probe Total	Init Total	Avg Seq	Left	Right	Avg Left	Avg Right	L=>R	R=>L
Samantha											
6	16	59	75	10	7.50	3	7	5.33	8.43	2	5
7.5	54	76	130	20	6.50	10	10	5.40	7.60	5	3
9	71	55	126	30	4.20	20	10	3.55	5.50	10	9
10	45	43	88	38	2.32	21	17	2.14	2.53	8	7
Nosferatu											
6	61	62	123	5	24.60	3	2	20.33	31.00	3	1
7.5	36	97	133	14	9.50	4	10	9.00	9.70	4	3
9	44	39	83	11	7.55	7	4	6.29	9.75	6	4
10	37	43	80	20	4.00	8	12	4.63	3.58	5	8
Annabellee											
6	45	101	146	13	7.77	4	9	11.25	11.22	9	9
7.5	46	36	82	17	4.82	11	6	4.18	6.00	6	5
9	12	12	24	12	2.00	8	4	1.50	3.00	4	0
10	5	3	8	7	1.14	4	3	1.25	1.00	0	0
Poe											
6	2	191	193	16	12.06	1	15	2.00	12.73	1	0
7.5	26	182	208	25	8.32	8	17	3.25	10.71	7	2
9	106	72	178	35	5.09	25	10	4.24	7.20	11	8
10	12	38	50	45	1.11	9	36	1.33	1.06	0	3

Note. Init. Total column denotes the frequency of probe sequences initiated. Av. Seq represents the average number of probes that made up the sequence. Avg. Left and Avg. Right represent the average sequence length for each finger. Changes reflect the number of right-to-left, and left-to-right finger changes during a probe sequence.

Table 4 also includes a breakdown of the number of *changes* that occurred within probe sequences. A change refers to the subject changing from the right finger to the left, or left finger to the right. When the aye-ayes changed fingers during a probe sequence, the change was more often from left to right, but this trend was not statistically significant, t (15) = 1.50, ns. In the deep cavities most subjects showed left to right changes indicating that more sequences were initiated with the left hand.

The analysis of probe sequences and finger changes help to amplify the idea that the right finger preference most prevalent in the shallow cavities was likely to break down in the deeper cavities. Recalling the finding that success decreased with cavity depth, presented in the section on response characteristics, it is likely that the decrease in right finger usage derives from the aye-aye seeking alternative approaches to the difficult cavities.

Success. Measurement of success in terms of laterality can be conceptualized in two ways. The first is termed *productivity* and is defined by comparing the number of grubs obtained by the left and right fingers. Productivity is an adaptively significant measure of lateralization because it compares the nutritional contribution of each finger. Productivity is analogous to preference measures of lateralization whereby the frequency of success of the left and right fingers is compared with z-scores. The second conceptualization is related to performance measures of lateralization and is termed *efficiency*, and is derived by dividing the number of successful retrievals by the number of responses. For lateral assessment, the efficiency measure compares percentage of left and right success. Any lateral differences detected in success measures (productivity and efficiency) can be taken as indices of motor control advantage because there was no difference in activity when the lift/probe averages were compared between left and right fingers, $t(15) = 0.28$, ns.

The productivity measure of success revealed no left finger advantage in obtaining grubs. Figure 8 depicts the frequency of left and right finger responding and the associated frequency of success for each subject and each cavity. The figure shows two trends related to the depth of the cavity. In shallow cavities, there appear to be more successful retrievals associated with the right finger than the left, but in the deeper cavities there is a slight increase in left finger success. Table 5 gives the results of both measures of success (productivity and efficiency). For the frequency based productivity measure it can be seen that all of the significant z-scores occurred in the shallow cavities, and that the right finger had more successful retrievals than the left finger. The analysis of combined left and right success shows that two of the subjects had a right finger advantage, and that the other two subjects had no difference between the fingers for productivity.

Results from the efficiency measure of success differ from those for the productivity measure in that no subject displayed a finger efficiency bias. Table 5 also shows the results of the efficiency measure of success. Z-scores were calculated for each cavity depth by comparing the percentage of success with the right hand to the percentage of success with the left hand. It must be noted that z-scores were calculated only if a subject made twenty responses

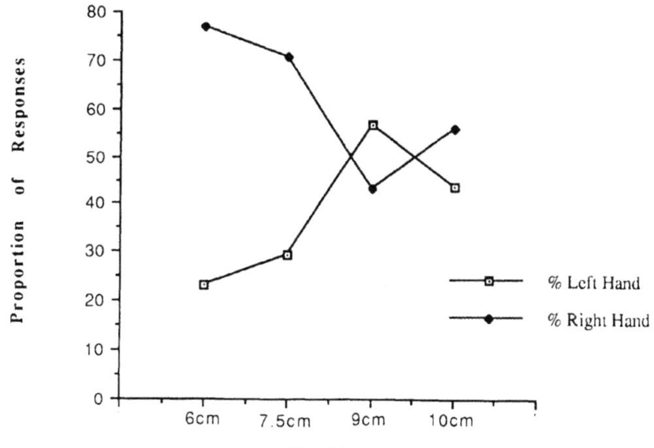

Figure 7. Combined proportions of lateral responses as a function of cavity depth.

Table 5. Laterality analyses of success.

Cavity	Productivity: Unsuccessful Left	Unsuccessful Right	Successful Left	Successful Right	Z	Efficiency % Success: Left	Right	Z
Samantha								
6	5	22	11	37	3.76*	--.--	62.71	--.--
7.5	27	25	27	51	2.72*	50.00	67.11	1.58
9	33	22	38	33	-0.59	53.52	60.00	0.61
10	31	22	14	21	1.18	31.11	48.84	1.98*
Total					3.41*			1.20
Nosferatu								
6	33	36	28	26	-0.27	45.90	41.94	-0.42
7.5	27	75	9	22	2.34*	25.00	22.68	-0.34
9	22	24	22	15	-1.15	50.00	38.46	-1.23
10	20	34	17	9	-1.57	45.95	20.93	-3.06*
Total					-0.33			-1.51
Annabellee								
6	39	87	6	14	1.79	13.33	13.86	0.10
7.5	36	32	10	4	-1.60	21.74	11.11	-1.85
9	10	12	2	0	-1.41	--.--	--.--	--.--
10	5	3	0	0	0.00	--.--	--.--	--.--
Total					0.00			-0.91
Poe								
6	1	137	1	54	7.14*	--.--	28.27	--.--
7.5	20	127	6	55	6.27*	23.08	30.22	0.98
9	70	49	36	23	-1.69	33.96	31.94	-0.25
10	12	36	0	2	1.41	--.--	5.26	--.--
Total					6.84*			-0.23

Note. Positive z-scores reflect a right bias, negative z-scores a left bias.
* $p < .05$.

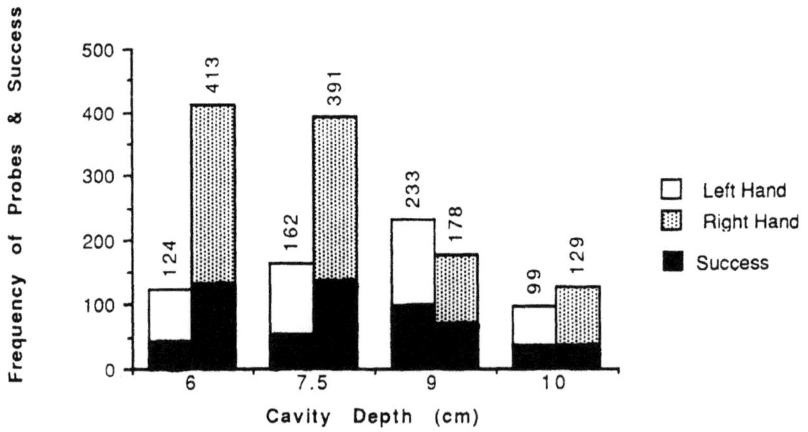

Figure 8 Productivity analysis. Successful retrievals are blackened inside the frequency of left and right finger probe responses.

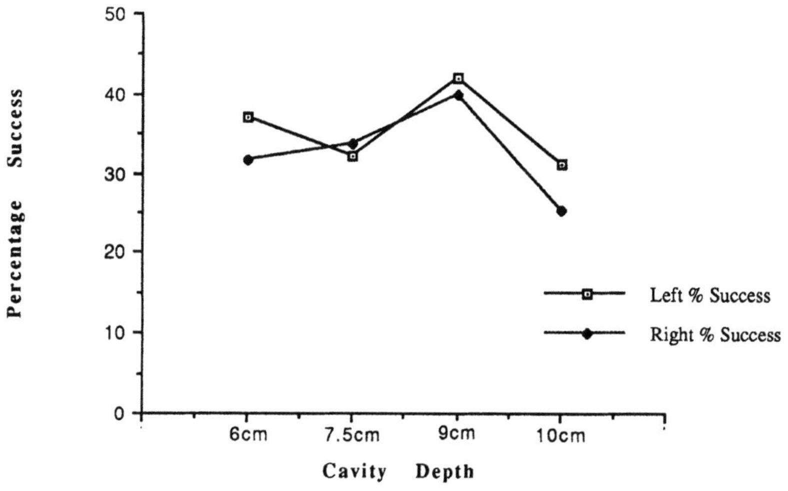

Figure 9. Efficiency analysis of success. The combined percentage success of left and right fingers are plotted as a function of cavity.

by each finger in a cavity, as this was deemed sufficient for analysis. Significant z-scores derived from the two older subjects (Samantha and Nosferatu) and occurred in the deepest (10 cm) cavity. Samantha displayed a right bias in efficiency, and Nosferatu a left bias. When the efficiency percentages were combined for all subjects, it can be seen (Figure 9) that there is slightly more success associated with the left hand than the right in all but the 7.5 cm cavity. These lateral differences were not statistically significant when evaluated with z-scores. This result indicates that both the left and right fingers were equally efficient in extracting grubs from vertical cavities.

To summarize, the right finger was used more for probing and was more productive than the left in extracting larvae. However, when the left finger was used to extract grubs, it was as efficient as the right.

Horizontal Cavity Apparatus

The horizontal cavity apparatus (HCA) utilized performance measures to determine the influence of cavity angle on the ability to extract larvae. The performance measures included: overall rate of success, number of flexions, and duration of time inside the cavity. These measures were also utilized in assessment of lateral preference. The number of flexions and duration of time in a cavity were used to ascertain tactile sensitivity by comparing performance when there was a grub present in the cavity to when the cavity was vacant. The analysis of responses to differently angled cavities differs from the previous experiment in that only the first 20 cavity entries were scored. This number was chosen because it provided a close approximation to the responses that were obtained for all four subjects. A somewhat matched sample enabled comparison of performance in the different angled cavities.

Finger Use. The third finger was used by all subjects exclusively. Use of the third finger for this apparatus was necessitated by task requirements where digit flexibility was paramount to entering the cavities. The third finger is thinner and has more joint flexibility than the other fingers. The rigidity and thickness of the fourth finger probably prevented it from entering the apparatus, and no aye-aye attempted to fit it into the aperture.

General Response Characteristics: Cavity Angle

Duration. Duration scores for each angled cavity were derived from the amount of time the finger was inside the cavity. Figure 10a shows the overall average duration for all subjects in each cavity angle. It can be seen that there was a decrease in duration from acute to right to obtuse, but an increase in the straight alley. The increase in straight alley duration could have resulted from the location of the straight cavity. The straight cavities were located at the end of the runway, and by position were usually entered first. If an aye-aye was attempting to probe an angled cavity that was located off to the side at the termination point of the runway, it is possible that overextension into the straight cavity would prevent entry to the targeted angled cavity. A repeated measures oneway ANOVA found overall average durations did not vary as a function of cavity, F (3, 12) = 1.35, ns.

Activity. The activity measure for the horizontal apparatus was digit flexion. A flexion was defined as a dorsal-ventral movement of the terminal phalanx of the third finger. This type of measure was not possible to obtain in the vertical cavity apparatus because the cluster of grubs in the bottom of the cavity obstructed the view of finger flexions. The flexion measure is potentially unique in that the aye-aye is the only species known to exhibit phalangeal control, that is, movement of the distal phalanx that is independent of movement in the proximal segments (Milliken et al., 1991).

The activity measure of flexion is presented as overall averages for the population in Figure 10b where it can be seen that flexions diminished as the cavity angle straightened. A repeated measures one-way ANOVA affirmed the difference in average flexions as a function of cavity angle, F (3, 12) = 14.69, p < .001. A Tukey's post hoc test found significant differences between the difficult (acute & right) and the less extreme cavities (obtuse and straight). There were no differences between the acute and right angle means, or the obtuse and straight means.

Success. A successful retrieval was defined by the grub being removed completely from the cavity. It should be noted that success in the horizontal cavity apparatus differs from the previous experiment in that success in the vertical cavity apparatus was based on the aye-aye obtaining and consuming a grub, but in the horizontal cavity experiment success was measured in terms of whether the grub was removed from the cavity. The percentage of successful retrievals in each cavity for the all subjects combined are presented in Figure 11. The figure shows that the acute cavity had the fewest successful retrievals, and most successes occurred in the straight and right angle cavities. The largest disparity in success rate was seen in the most difficult cavity (acute), wherein juveniles (Poe & Annabellee) had few successful retrievals when compared with adults. A repeated measures oneway ANOVA between success and cavity angle was significant F (3, 12) = 3.828, p = .05, but a Tukey's post hoc test revealed that only the means of the acute and straight cavities differed.

To summarize the results of the above measures: there were no duration differences that were angle dependent, but there was a greater amount of activity in the acute and right angle cavities when compared with the flexions that occurred in the obtuse and straight cavities. The only differences detected in the analysis of success were between the acute and straight cavity, with fewer successful retrievals deriving from the acute cavity.

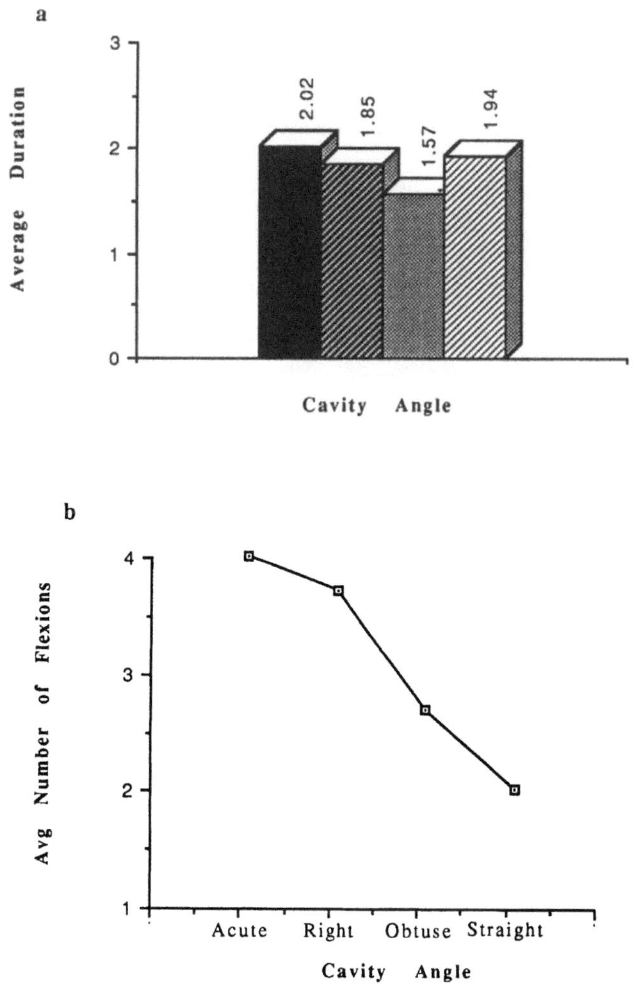

Figure 10. (a) Overall average duration of time spent in each angled cavity. (b) Overall average number of flexions for the population as a function of cavity angle.

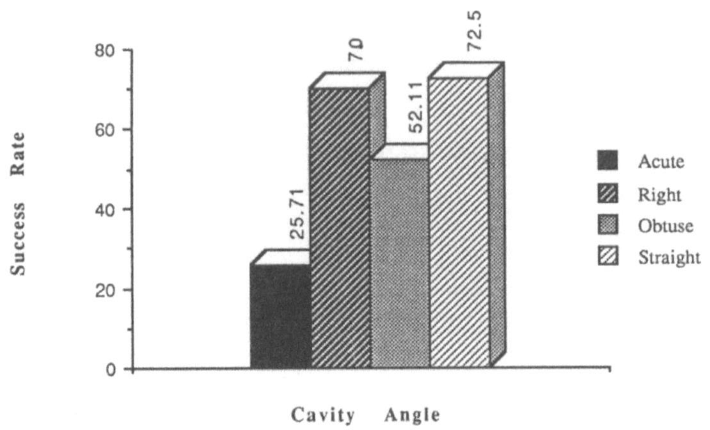

Figure 11. Percentage of successful retrievals in each angled cavity for all subjects combined.

Tactile Sensitivity

To evaluate tactile sensitivity, duration and activity measures were compared between probes into cavities with a grub present and probes into vacant cavities.

Duration. Figure 12a compares the overall average duration for each subject when there was a grub present or when the cavity was vacant. It can be seen that all of the aye-ayes spent more time in the baited cavities than in cavities that were vacant. A within subjects t-test between the average durations from baited (M = 2.36s) and vacant (M = 1.41s) cavities was significant, t (15) = 5.08, $p < .001$, demonstrating that the aye-ayes were spending more time in the baited cavities.

Activity. The second measure used to discern differences between empty and baited cavities was the average number of digit flexions when inside the cavity. Figure 12b depicts the overall average number of flexions for each subject when there was a grub present or when the cavity was vacant. The figure shows that all subjects showed a greater number of flexions when there was a grub in the cavity. This activity bias was confirmed with a within subjects t-test that found more flexions to occur in cavities with grubs (M = 3.52) than in vacant cavities (M = 2.71), t (15) = 2.59, p = .02.

These results, when combined with the results from the duration analysis indicate that the aye-ayes were able to tactilely discriminate between the baited and vacant cavities, and spent more time and energy in cavities that were baited.

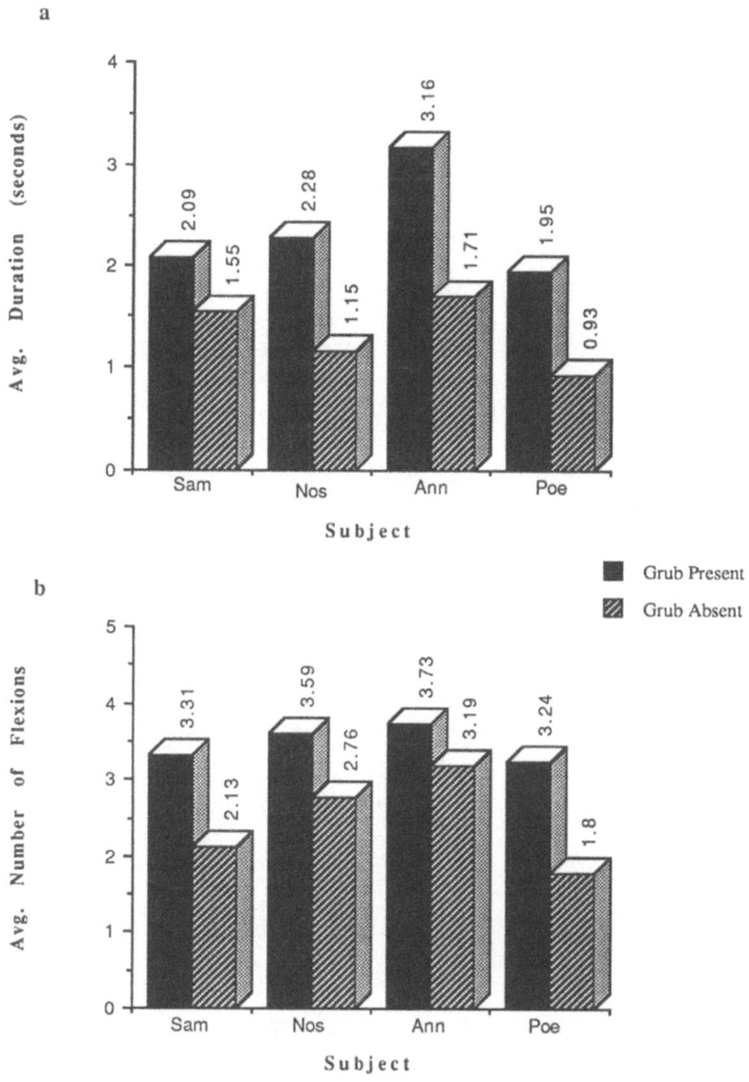

Figure 12. (a) Comparison of average duration of time spent in baited versus vacant cavities for each subject. (b) Comparison of overall average number of flexions for each subject in vacant and baited cavities.

Laterality

Measures of lateral preference in the horizontal cavity apparatus differ from those for the vertical cavity apparatus in that the horizontal apparatus study includes a duration measure that compares the left and right fingers on time spent inside the cavities, and a measure of activity that is termed digit flexion.

Because Samantha lost the fingernail of the right middle finger after the second day of data collection, she was intact for only 52 of 94 responses. Her results were included in the above analyses, but her scores were not used in assessment of lateral preferences because of the biasing nature of her injury.

Frequency of Probing. As in the vertical cavity apparatus, the frequencies of left and right finger probing were evaluated with z-scores. The central location of the aperture to the apparatus, and the hidden nature of the angled cavities below, permits a simplified presentation of probe preference as a combined measure for all cavities. Table 6 lists the frequency of left and right finger responding for the three subjects, along with the resultant z-score. Only one subject (Poe) had a significant preference and that was with the right hand. The reason probe preferences did not occur was possibly due to the complex configuration of the cavities routed off a single aperture. If a subject chose to use one finger preferentially to probe, he or she would have to reposition the body to reach certain cavities. While on the apparatus the aye-ayes frequently switched fingers, rather than repositioning the entire body, to access cavities that were not accessible in the initial probe. It was fortuitous that the aye-ayes did not display strong probe preferences because this assured a somewhat matched comparison sample of left and right responses for efficiency measures of lateralization.

Table 6. Probe preference analysis in horizontal cavity apparatus

Subject	Left Probes	Right Probes	Z Score
Nosferatu	87	73	-1.11
Annabellee	69	66	-0.26
Poe	66	94	2.21*

*p < .05

Duration. The amount of time the left and right finger spent inside each type of cavity was compared in the duration analysis. Figure 13a combines duration averages from all cavity angles, and reports the left and right latencies for each subject. It can be seen that there was a leftward bias in duration for all subjects. The average time spent in each cavity angle by the left and right fingers for all of the subjects combined are depicted in Figure 13b. The figure shows that there was a pronounced duration difference between the left and right finger in the acute cavity, with a less dramatic leftward bias also occurring in the obtuse angle cavity. In the right and straight cavities there was a slight rightward bias.

Activity. Results derived from the activity measure of digit flexion differ from the laterality findings seen above in the duration analysis. Figure 14a shows the combined average number of flexions for the left and right fingers for each subject. It can be seen that for two subjects there was a leftward bias in activity, with Poe having a substantial left bias and Nosferatu showing a somewhat less substantial bias. Annabellee had a very slight bias in flexions with the right finger. The average number of flexions that occurred in each cavity angle by the left and right fingers for all of the subjects combined is depicted in Figure 14b. The figure shows that in the more difficult cavities (acute and right angle) there were marked differences in flexions between the left and right fingers with the left finger exhibiting greater average flexions. In the easier cavities (obtuse and right angle) there was a shift to a right bias for activity. A within subjects t-test between the average number of flexions by the left (M = 3.62) and right (M = 3.36) fingers in each cavity was not significant, t (11) = .75, ns.

Success. A measure of lateral efficiency was obtained by comparing the rate of success between the left and right fingers in extraction of larvae from cavities. Figure 15a shows the overall success rates for each subject as a function of left and right hand. The male subjects had the strongest right biases, while young Annabellee had only a slight rightward bias. Figure 15b presents the combined averages of all subjects in terms of left and right success for each cavity angle. The graph shows that there was a right finger advantage in extracting grubs from each angled cavity. A t-test comparing the percentage of successful retrievals between left and right fingers for each cavity angle revealed that the right finger (M = 65.58) was more successful than the left (M = 48.42) in extracting larvae from cavities, t (11) = 2.44, p = .033. Evaluation of efficiency of retrieval with the left and right fingers for each cavity angle and each subject are presented in Table 7. The z-scores compared the percentage success of the right finger with the left finger. The table shows that the only left advantage was exhibited by Poe in the obtuse cavity, while all other significant z-scores favored a right bias. The more difficult cavities elicited the strongest right biases. Poe was the only subject to show a significant z-score in the analysis of combined left and right percentage, and this was a right finger advantage. For the other two subjects, total efficiency of left versus right finger comparisons were in the right finger direction, but were non-significant.

To summarize the laterality findings from the horizontal cavity apparatus: (1) There were no real differences in duration of time spent in the differently angled cavities, though all subjects spent more time with the left hand in most of the cavities. (2) The activity measure found more flexions associated with the left hand for all subjects, with most of the leftward tendency coming from the more difficult angled cavities (right and acute), but more right hand flexions in the easier cavities (obtuse and straight). (3) Success was more pronounced with the right finger than the left for all subjects in all cavity angles. The largest margins of right success occurred in the difficult cavities.

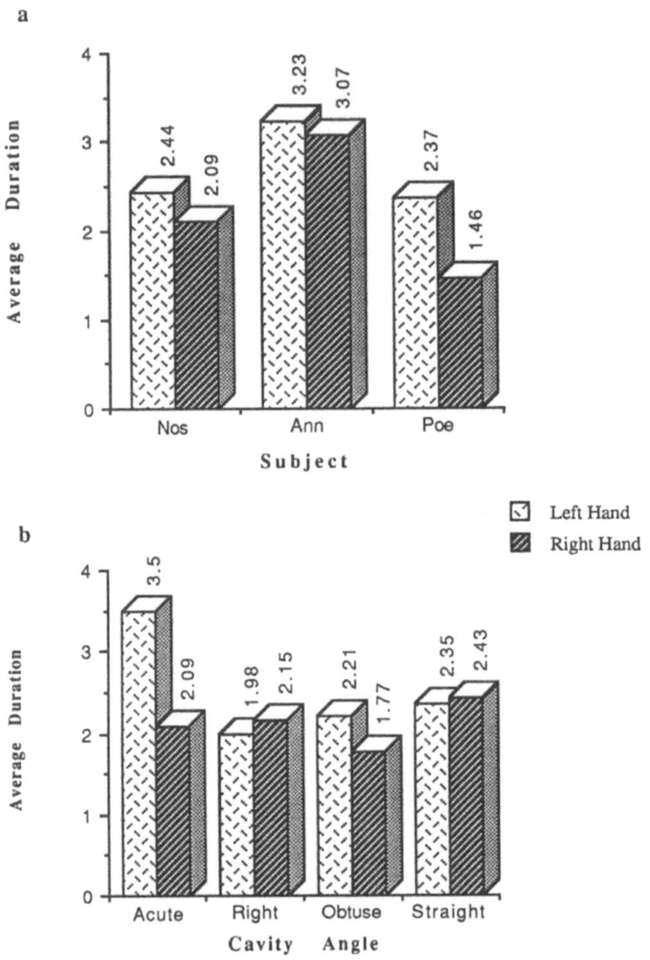

Figure 13. (a) Average duration of time spent by left and right finger inside cavities for each subject. (b) Overall duration of left and right finger inside each angled cavity.

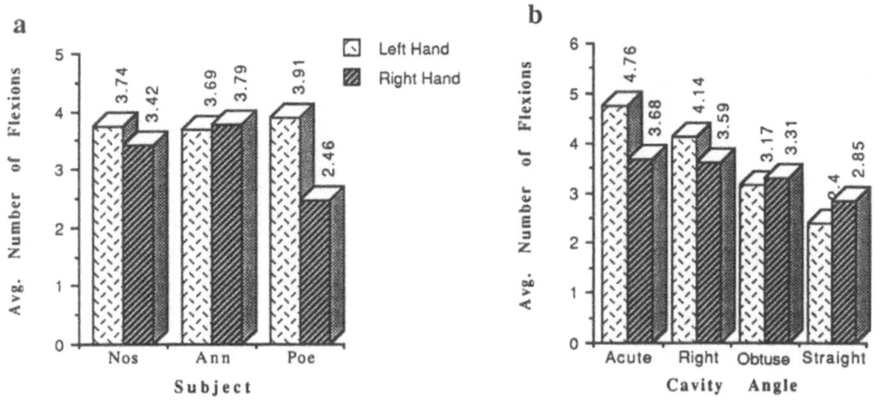

Figure 14. (a)Overall average number of flexions by left and right fingers for each subject. (b) Overall average number of flexions of left and right fingers as a function of cavity angle.

Table 7. Efficiency measure of success as z-scores for each subject in each cavity

Subject	Cavity Angle				Total
	Acute	Right	Obtuse	Straight	
Nosferatu	0.62	0.62	0.82	1.99*	1.43
Annabellee	-1.83	5.00*	1.49	1.87	0.41
Poe	8.19*	3.17*	-3.67*	0.23	2.34*

*p < .05.

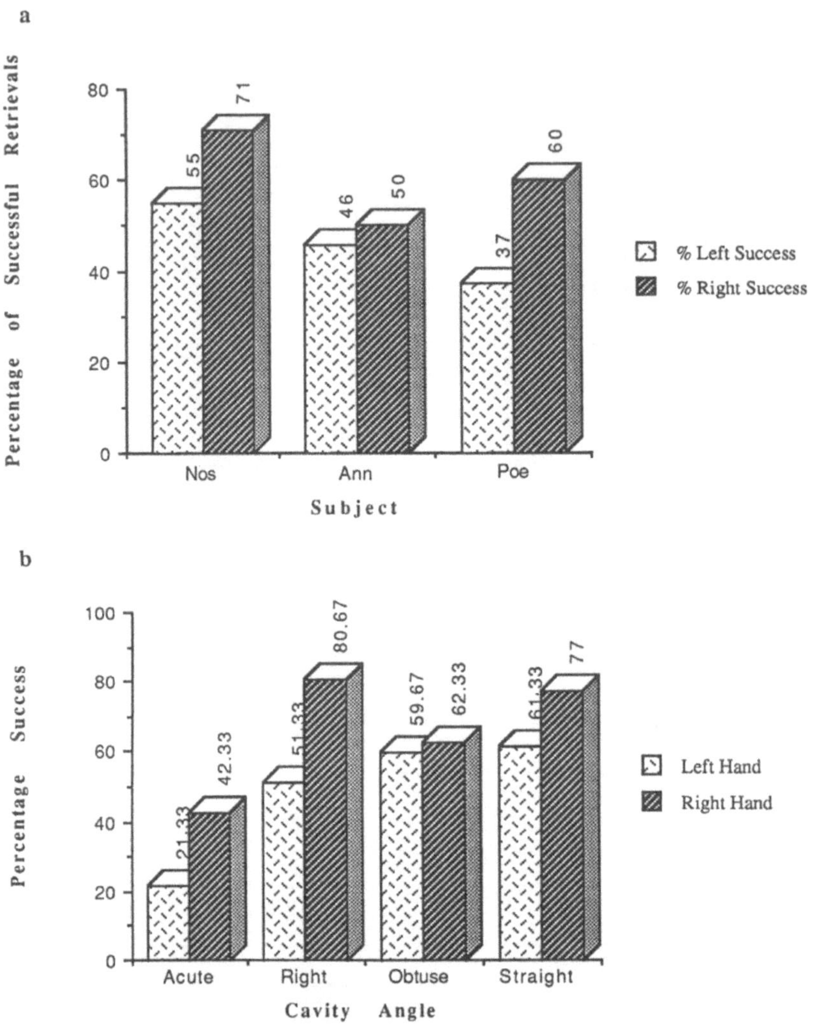

Figure 15. (a) Percentage of left and right finger successes for each aye-aye. (b) Overall percentage success of left and right fingers as a function of cavity.

Comparison of Horizontal and Vertical Cavity Apparatuses

The vertical cavity apparatus (VCA) required the aye-aye to make a probe into a cavity that was straight and wide, to hook and balance a grub, and to lift it through the aperture and into the mouth. The finger was held straight and the positional finger movements resembled a stirring motion. The hooking motion of the finger involved the distal phalanx adducting, then flexing. To retrieve a grub from the horizontal cavity apparatus (HCA), the subject had to probe through the aperture into a shallow channel that required the finger to flex into a right angle (at the metacarpophalangeal articulation) as it was inserted into the runway to the angled cavities. A further bend in the finger was required to enter any of the angled cavities. The HCA elicited the same type of hooking motion as the VCA, but balancing the grub on the claw was not required. Instead the grub was removed by hooking, encircling, and pulling it first from the angled cavity, then from the runway, and finally from the apparatus through the aperture. Given the description of the task requirements for each apparatus, it is apparent that the HCA required more finger movements and contortions than the VCA, and may be construed as a more difficult task.

Comparison of efficiency in grub extraction between the two apparatuses provides evidence that the HCA was the more difficult task. Efficiency is a measure of success that is derived by dividing the number of successful

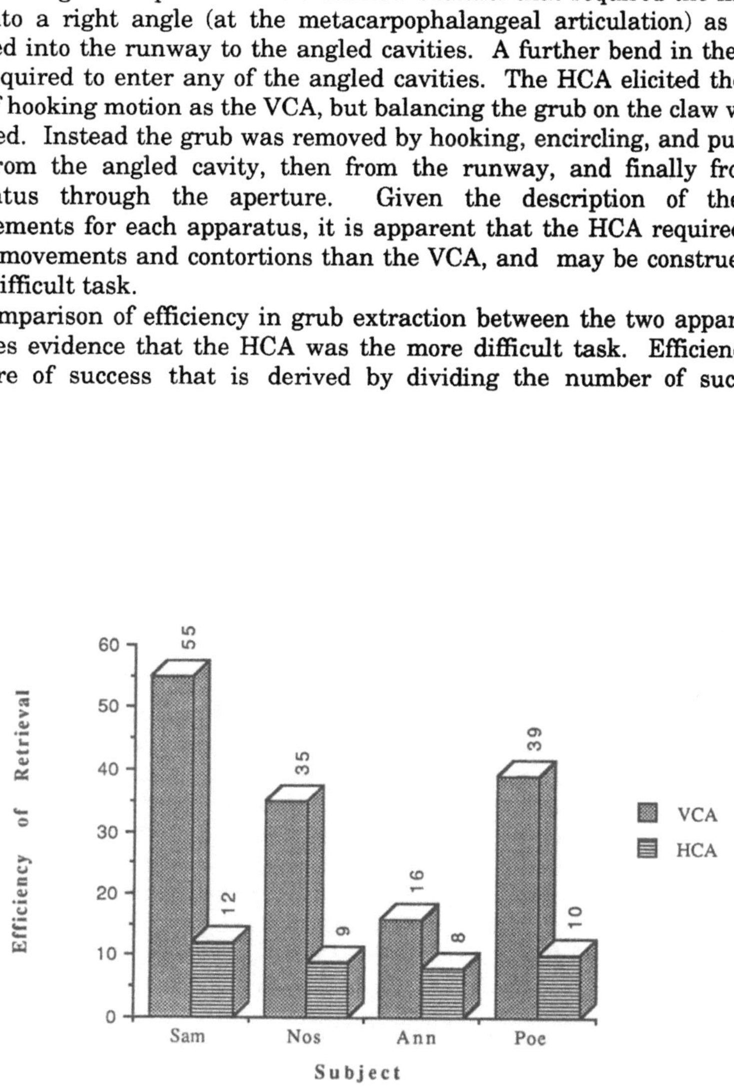

Figure 16. Comparison of retrieval efficiency between the vertical cavity apparatus (VCA) and horizontal cavity apparatus (HCA).

retrievals by the frequency of attempts. For the HCA, the measure of success differs from the analysis presented previously in the section on cavity comparisons. In the cavity comparison section success was defined by the grub being removed from a *cavity*, but in the comparison analysis of success between apparatuses success is defined as removing a grub from the *apparatus*. Figure 16 compares the efficiency of each subject for both apparatuses. The figure shows that the aye-ayes were more efficient at obtaining grubs from the VCA (M = 36.25) than the HCA (M = 9.75), t (3) = 3.68, p = .035. These results illustrate the fact that the HCA presented more of a challenge to the subjects than the VCA. Finding that the two tasks differed in difficulty helps to explain differences found between apparatuses in preference and performance measures of laterality.

DISCUSSION

The horizontal and vertical cavity apparatuses permitted evaluation of physical parameters of cavity depth and angle on the ability of the fingers to extract grubs. The aye-ayes used in the present study were able to extract grubs from all cavity depths and all of the angled orientations. The aye-ayes did not need inducements to learn the probing tasks. Their curious nature led them to explore new additions to their room, and upon encountering an aperture that afforded probing, they were quick to investigate. These experiments were run during the regularly scheduled feeding time, when the aye-ayes were provisioned with fruits, vegetables, egg, and a dish of easily accessible insect larvae. When the apparatus was placed in the room the aye-ayes immediately left the feeding stand and started to probe the apparatus. This preference for species-typical behavior was also seen when the food arrived late and data collection was interrupted. Rather than leave the apparatus to eat the grubs provided with the food, the aye-ayes preferred the challenge of removing grubs from the apparatus; a context that more closely approximates the natural foraging condition.

Because all subjects used only the third finger in the horizontal cavity apparatus, the vertical cavity apparatus (6, 7.5, 9, 10 cm depths) provided the only opportunity to compare extractive ability between the third, fourth, and third and fourth finger combination probe strategies. The preference for the group was to probe with the dexterous third finger, which given the unique abilities of independent digit control and phalangeal control is best suited for the task. The fourth finger exhibited a small degree of phalangeal control, but the flexions of the terminal phalanx were mirrored by the third finger that was outside of the cavity. Similar shadowing was not seen in the fourth finger when the phalanx of the third finger was flexing and extending in the cavity. The fourth finger was seldom used by the longer fingered males, and was used more so by the females and especially in the deeper cavities. The use of the fourth finger by the females could have been a means of compensating for their smaller finger size. This trend was evident within females as the juvenile Annabellee had a smaller finger than Samantha and shifted to use of her fourth finger in the 7.5 cm deep cavity, while Samantha's threshold was the 9 cm cavity. The females also displayed the greatest amount of combination probing where the third and fourth fingers were used together in probing. It would have been interesting to include an 11 cm cavity to see if males would demonstrate the shift to fourth or combination finger use. Examination of success percentages for the different probe strategies showed that the third finger strategy was the most effective. Annabellee had minimal success with all probe strategies, and was the only subject to display greater success with fourth (16%) and combination (20%) probing than third finger probing (14%). Her lack of success could reflect her lack of neuromuscular maturation, or her unfamiliarity with grub extraction from wood cavities. Annabellee was an infant when she was captured with her mother, and probably never had an opportunity to forage in the wild. All adult subjects had more success when probing with the third finger than the fourth or combination strategies. This is likely due to the superior motor control that was displayed by this finger in comparison to the fourth finger.

There was a tendency for the frequency of responding to diminish with cavity depth. All subjects showed more responding to the 6, 7.5, and 9 cm deep cavities that were more approximate to the length of their fingers, than to the 10 cm cavity that was slightly deeper. This implies that the aye-ayes were probing cavities that were more likely to elicit a food reward, and/or the aye-ayes were able to remember the location of the "difficult" cavity where they were less frequently (if ever) rewarded. These hypotheses receive some support when success rates are considered. Most subjects displayed the worst retrieval performance from the 10 cm cavity, and had higher percentages for success in the shallow cavities. This differential success was, however, not related to less effort or activity in the deeper cavity as there were no differences in activity (lift/probe average) as a function of cavity depth.

In the horizontal cavity apparatus, a dimension of measurement was added in the form of duration of time spent in each cavity. When durations were averaged for all subjects, there was a trend for longer durations as the cavity angles became more difficult (more bending by the finger required for entry). An exception to this trend was that durations were greatest in the straight cavity. An inflation of straight cavity duration could have resulted from extension of the finger past the termination point of the runway. This seems likely when it is recalled that entry points for the more extreme angled cavities were located off to the side of the runway. A consideration of the results of the flexion measure of activity supports the idea that the duration scores from the straight cavity were inflated by finger overextension. The number of flexions that were executed by the terminal phalanx were counted during the time that the finger was inside a cavity angle, and the number of flexions did diminish as the cavity angle became less extreme. Because flexion of the distal phalanx is taken as an intention movement to retrieve a grub, and unlike durations, flexions did not increase like duration in the straight cavity, it could be argued that the longer durations scored in the straight cavity were a result of overextension. It seems likely that the aye-ayes were trying to reposition the finger to a side cavity with lateral movements of the finger originating from the metacarpal extension rather than the distal phalanx. Thus, although longer durations were recorded for the straight cavity, evidence of grub retrieval attempts in the form of finger flexions was not inflated for this cavity.

The finger flexion measure of activity is a more veridical characterization of the phalangeal activity of the finger than the activity measure from the VCA where the measure of lift was based on gross movements that involved the hand and arm as well as the finger. In the HCA, differences in the number of flexions were most pronounced between more difficult (acute and right angle) and less difficult (obtuse and straight) cavities. This activity measure seems to bear some relationship to success in grub retrieval wherein the aye-ayes were most successful in retrieving grubs from the straight and right angled cavities and were least successful in the acute angle cavity. The only significant difference found between mean success rates was between the acute and straight cavities. The fact that the aye-ayes displayed more activity in the acute angled cavity and the least number of flexions in the straight cavity suggests that activity may be the inverse to successful retrieval and not necessarily a predictor of whether or not the aye-aye obtained a grub. More likely, however, activity is tied to the difficulty in cavity angle where increased activity in the more difficult cavities was a means of exploring and confirming the presence or absence of grubs.

The horizontal cavity apparatus provided the opportunity to test tactile sensitivity in both duration and activity measures by comparison between probes into vacant cavities and cavities that contained grubs. Both duration and activity were greater when the cavity was baited indicating that the aye-ayes were using tactile cues to determine if the cavity contained a grub. It is likely that the aye-ayes could not use their normal mode of echolocation by tapping because the cavities were covered with a steel plate. Tapping diminished after the initial exposure to the apparatus suggesting that auditory information was not useful for determining cavity location. Because auditory cues were not useful, the aye-ayes had to rely on tactile cues to sense the presence of larvae.

The ability to detect a grub in contact with the fingertip is evidence of tactile sensitivity. This tactile sensitivity functions as a motor activation system. When the fingertip came into contact with a grub, the distal phalanx would flex and the small claw would encircle the larva between the middle phalanx and cavity wall. The hand and finger would then be withdrawn and the grub delivered to the mouth. This behavioral sequence displays the interplay between sensory and motor systems, and may be thought of as a rudimentary form of manipulation where the dynamic interaction between touch and control are used by one finger to effect feeding. The differences in duration and activity between the vacant and baited cavities are presented as evidence that the aye-ayes were relying on tactile cues to ascertain the presence of a grub.

The third finger is so specialized for probing cavities that it is never used in locomotion or postural support. The finger is either held adducted or crossed over the fourth finger during locomotion and climbing. Disuse of the medius in locomotor behavior could be explained by the double-jointed metacarpo-phalangeal articulation whereby high flexibility does not contribute to postural support. Another explanation for disuse could be that the aye-aye is protecting the distal phalanx from undue wear that could foster injury or calluses and decrease tactile sensitivity. Regardless of the motives, the aye-aye takes great caution to protect the finger and uses it for effective rather than positional purposes.

Laterality is a confusing phenomenon because of the definitions that are used to describe functional differences between the two hands. The dichotomy between hand preference and performance was presented in the introduction, wherein hand preference was defined as consistent lateralized usage on relatively simple tasks such as food reaching; and performance concerns differences between the two hands on a qualitative variable such as reaction time or performance efficiency. Handedness (preference) is further defined as consistent lateralized usage on familiar, highly practiced, and relatively simple tasks, whereas manual specialization (performance) is defined as consistent lateralized usage on novel, unpracticed, relatively complex tasks (Young, Segalowitz, Corter, & Trehub, 1983). Fagot and Vauclair (1991) affirm that both hands could possess the basic abilities to perform a specific task when it requires a low level of manipulative skill, but beyond that skill level, only one hand would be able to perform it, since the capacity of the other would be exceeded. Skills that require increased manipulative ability would fall under the category of manual specialization and would be evaluated with performance measures, while simple forms of hand bias would be categorized as handedness and would enlist preference measures.

Qualitative performance differences between the hands were assessed with efficiency measures based on percentage of successful retrievals. The problem with efficiency measures is relating lateralized differences to adaptive significance. If one hand becomes highly specialized for a particular task, how would the animal manage if it sustained injury to the hand? As mentioned previously, Samantha lost the fingernail of her right hand after testing in the vertical cavity apparatus. The results from the vertical cavity apparatus show that the right finger was preferred by Samantha for probing. Her success ratio for the horizontal cavity apparatus (using her non-preferred hand) was directly comparable to all of the intact subjects. Given this example, manual specialization is not seen as crucial to the survival of the animal.

There is also the problem of categorizing behavior to fit the definitions of laterality. For example in the grub extraction experiments, what the animal might view as simple food reaching, the experimenter might perceive as skillful manual dexterity in moving the finger and lifting a grub from a cavity. It is therefore necessary to test our assumptions about definitional differences in lateralization phenomena by careful analysis and interpretation of the data.

The horizontal cavity apparatus (HCA) was more challenging than the vertical cavity apparatus (VCA) as evidenced by retrieval efficiency differences. The fact that the retrieval tasks were graded in complexity infers that each task was measuring a different aspect of lateralization. Because the measures of preference in the VCA revealed stable significant right preferences for all subjects, and measures of performance detected no lateral biases, it could be implied that the VCA was a simpler task and therefore measured lateral preference. The analysis of probe persistence indicated that in the shallow cavities where most preference was detected and where most success occurred, the aye-ayes displayed longer probe sequences than the deeper and more difficult cavities. These behavioral characteristics of probing in shallow cavities fit the definition of preference in that the task was simple and repetitive. The lack of preference in the deeper cavities illustrates the point that lateral preference diminishes as the task becomes more difficult. As preference drops out as a factor, finger efficiency becomes more important. For the more difficult task, the HCA, no stable preferences were detected, but a right performance efficiency advantage was found when the success rates of all aye-ayes were compared. The greater use of the non-preferred finger in both the VCA and HCA as difficulty increased may be indicative of the aye-ayes seeking alternative strategies when the reinforcement ratio is too low.

In the VCA, the right preference can be conceptualized in terms of adaptive significance if success is interpreted as productivity. The productivity measure of lateralization is a comparison of the frequency of success by the left and right fingers. Productivity has a component of adaptive significance because it accounts for the animal's goal in the foraging task, that is to obtain a grub and consume it. In the analysis of success the productivity measure detected a right finger advantage in obtaining grubs and this occurred only in the shallow cavities. Recalling the result of a right preference for probing in the shallow cavities, a relationship is established between frequency of responding and productivity.

Success in the VCA was also analyzed in terms of efficiency where the percentage of success was compared between the left and right hands. When percentage success was collapsed across cavities, there were no lateral differences in performance between the left and right hands. When the analyses of productivity and efficiency are compared, the emergent result becomes a right finger advantage for productivity in the shallow cavities, and no difference between the hands for efficiency. To extrapolate these findings to definitions presented above: the right finger preference for probing translated directly to increased productivity with the right finger. Performance between the left and right fingers did not differ, indicating that both hands were equally efficient at extracting grubs from vertical cavities.

The nature of the probing task in the HCA required the switching of fingers in order to reach certain cavities relative to the position of the animal and single aperture. In the HCA, only Poe showed a preference for probing and this was a right bias. It was fortunate that there were no strong probe preferences because this experiment focused on performance measures of lateralization, by comparing duration, flexion, and success between the two hands.

Results from the duration and activity measures of lateralization had in common a leftward bias in difficult cavities, and a right tendency for simpler cavities. For activity there were more flexions in the acute and right angled cavities, and a greater amount of right finger activity in the easier cavities (obtuse and straight). These results from performance measures are somewhat symmetrical with results obtained from the VCA in that there was a right finger advantage for the simpler tasks, and a left bias for difficult tasks. Measurement of efficiency based on rate of success revealed a more consistent trend that was a right finger advantage for each subject. The finding that the right advantage in success was consistent for each angled cavity and most pronounced in the difficult cavities was not what would be expected given the left advantage seen in performance measures of duration and activity. The emergent picture from the HCA indicates that although the left finger displayed longer duration and more activity, this energy was not converted into success in extracting the grub. Instead the right finger displayed an advantage in success, and could be thought of as more efficient than the left finger.

The findings of stable preference patterns and unstable performance patterns of lateralization leaves one to decide which measures are most useful for categorizing underlying asymmetries. It has been argued that performance measures are the best means of assessing underlying cerebral asymmetries. If this is so, then one would predict that aye-ayes have symmetrical brains and the preference measures of lateralization are based only on external setting conditions (type of apparatus, lighting, etc.). Because the right advantage was manifested in two different apparatuses, using different measures, (probing and productivity in the VCA and efficiency in the HCA) it seems unlikely that setting conditions are responsible for the bias. Preference measures are equally useful to performance measures in categorizing behavioral lateralization if not superior, and neurological asymmetries could occur at several levels that may be addressed using different measures. It is for this reason that research into laterality phenomena should utilize a range of tasks to discern different levels and degrees of preference and performance.

The study of finger lateralization in the aye-aye has provided results that help to affirm the theory of the evolution of behavioral lateralization expressed by MacNeilage et al. (1987). The aye-aye represents a transitional prosimian form that has evolved rudimentary manipulative skills in one finger. That the aye-aye exhibits a right preference in productivity (VCA), and efficiency (HCA), is evidence that the specialization of the left hemisphere for tasks involving dexterous finger work are related to the manual abilities of the species.

Acknowledgments

The author would like to thank the staff of the Duke University Primate Center for making this project possible, particularly, Kay Izard, Bill Hess, and Fred Stewart. Further, the author is indebted to Carl Erickson and the members of his dissertation committee for their constructive criticisms of this project. He would especially like to thank Dr. Jeannette P. Ward for her friendship and support throughout his graduate training period. This project was supported by a Center of Excellence grant from the State of Tennessee to the Psychology Department at Memphis State University and NS09366.

REFERENCES

Beck, C. H. M., & Barton, R. L. (1972). Deviation and laterality of hand preference in monkeys. *Cortex,* 8: 339-363.

Bresard, B. & Bresson, F. (1983). Handedness in *Pongo pygmaeus* and *Pan troglodytes. J. Hum. Evol.,* 12: 659-666.

Bryden, M. P. (1982). "Laterality and Functional Asymmetry in the Intact Brain," Academic Press: New York.

Deuel, R. K., & Dunlop, N. L. (1980). Hand preference in the rhesus monkey: Implications for the study of cerebral dominance. *Arch. Neurol.,* 37: 217-221.

Edwards, A. L. (1963). "Statistical Methods for the Behavioral Sciences," Holt, Reinhart, & Winston: New York.

Erickson, C. J. (1991). Percussive foraging in the aye-aye, *Daubentonia madagascariensis. Animal Behaviour,* 41: 793-802.

Fagot, J., & Vauclair, J. (1991). Manual laterality in nonhuman primates: A distinction between handedness and manual specialization. *Psych. Bull.,* 109: 76-89.

Forsythe, C., Milliken, G. W., Stafford, D., K., & Ward, J. P. (1988). Posturally related variations in the hand preferences of the ruffed lemur *(Varecia variegata variegata)*. *J. Comp. Psych.,* 102: 248-250.
Forsythe, C., & Ward, J. P. (1988). Black lemur *(Lemur macaco)* hand preference for food reaching. *Primates,* 29: 369-374.
Fragaszy, D. M., & Mitchell, S. R. (1990). Hand preference and performance on unimanual and bimanual tasks in capuchin monkeys *(Cebus apella)*. *J. Comp. Psych.,* 104: 275-282.
Geschwind, N. F., & Galaburda, A. M.(1985). "Cerebral Lateralization: Biological Mechanisms, Associations, and Pathology," MIT Press: Cambridge, MA.
MacNeilage, P. F., Studdert-Kennedy, M. G., & Lindblom, B. (1987). Primate handedness reconsidered. *Behav. Brain Sci.,* 10: 247-303.
Milliken, G. W., Forsythe, C., & Ward, J. P. (1989). Multiple measures of hand-use lateralization in the ring-tailed lemur *(Lemur catta)*. *J. Comp. Psych.,* 103: 262-268.
Milliken, G. W. (1990). "Analysis of the structure, function, and evolution of primate hands." Unpublished major area paper: Memphis State University, Memphis, TN.
Milliken, G. W., Ward, J. P., & Erickson, C. J. (1991). Independent digit control in foraging by the aye-aye *(Daubentonia madagascariensis)*. *Folia Primatol,* 56: 219-224.
Milliken, G. W., Stafford, D. K., Dodson, D. L., Pinger, C. D., & Ward, J. P. (1991). Analyses of feeding lateralization in the small-eared bushbaby *(Otolemur garnettii)*: A comparison with the ring-tailed lemur *(Lemur catta)*. *J. Comp. Psych.,* 105: 274-285.
Milliken, G. W. (1991). An examination of hand use in foraging by the aye-aye *(Daubentonia madagascariensis)* (Doctoral dissertation, Memphis State University, 1991). *Diss. Abst Int.* 52-12: 6695B. (University Microfilms # 9214868).
Preilowski, B. (1979). Performance differences between hands and lack of transfer of finger posture and sensory-motor skill in intact rhesus monkeys: Possible model for the origin of cerebral asymmetry. *Neurosci. Letters,* Supplement 3, S89.
Sanford, C., Guin, K., & Ward, J. P. (1984). Posture and laterality in the bushbaby *(Galago senegalensis)*. *Brain, Behav.,Evol,* 25: 217-224.
Stafford, D. K., Milliken, G. W., & Ward, J. P. (1990). Lateral bias in feeding and brachiation in *Hylobates. Primates,* 31: 407-414.
Stafford, D. K., Milliken, G. W., & Ward, J. P. (1993). Patterns of hand and mouth lateral biases in bamboo leaf shoot feeding and simple food reaching in the gentle lemur *(Hapalemur griseus)*. *Amer. J. of Primatol,* 29: 195-207.
Stephan, H. (1972). Evolution of primate brains: A comparative anatomical investigation. In R. H. Tuttle (Ed.), "The Functional and Evolutionary Biology of the Primates," Aldine Press: Chicago.
Subramoniam, S. (1957). Some observations on the habits of the slender loris *(Loris tardigradus)*. *J. of the Bombay Nat. History Soc.* 54:386-398.
Ward, J. P., Milliken, G. W., Dodson, D. L., Stafford, D. K., & Wallace, M. (1990). Handedness as a function of sex and age in a large populaton of *Lemur. J. Comp. Psych.* 104: 167-173.
Young, G., Segalowitz, S. J., Corter, C. M., & Trehub, S. E. (1983). Manual specialization and the developing brain: An overview. In G. Young, S. J. Segalowitz, C. M. Corter, & S. E. Trehub (Eds.), "Manual Specialization and the Developing Brain," Academic Press: New York.

LATERALITY IN AFRICAN AND MALAGASY PROSIMIANS

Jeannette P. Ward

Department of Psychology
The University of Memphis
Memphis, TN 38152

INTRODUCTION

The comparison of species, whether with respect to physical structure or behavioral capacity, is a powerful strategy in the attempt to gain insight into the systematics of the development of lifeforms. This strategy has, however, been more extensively and effectively employed in the study of the structure and physiology of organisms than of behavior. Most studies of animal behavior reported in the literatures of biology, ethology, neuroscience and psychology evaluate the behavior of a single species in terms of the adaptive significance for that species or in terms of its suitability as an animal model for human behavior. Even the two journals dedicated to species comparison (i.e.,*The Journal of Comparative Psychology* and *The International Journal of Comparative Psychology*) have characteristically published more single species reports than those comparing the behavior of two or more species. In the study of primate behavior, there is currently an increasing incidence of reports of direct comparison of species along specific behavioral dimensions.

The comparative approach involves seeking patterns of similarity or difference in the behaviors of organisms as a function of species membership. The "natural experiments" which have resulted in the formation of different species may be employed to study the commonalities of behavior that exist within a taxon as a consequence of a shared phylogenetic history as well as differences in behavior that may have arisen in adaptation to divergent ecological niches. Taxa which are particularly suitable for comparative study are those which contain a large number and variety of species and thereby offer many dimensions along which comparisons may be made. In this respect, prosimians constitute a most appropriate group for comparative study.

The prosimian group is one of two major groupings within the Order Primates and contains three of the six major primate classifications (Martin, 1990). These three prosimian classifications -- lemurs, loris group, and tarsiers -- include 24 genera and 50 species which inhabit a variety of ecological niches in tropical and subtropical regions of Madagascar, Africa, South Asia, and South-East Asia. There is great species variation along a number of dimensions. Body weights range from a minimum of 60 g. to a maximum of more than 6 kg. Dietary specializations include frugivory, folivory, and insectivory. Although many species are nocturnal in habit, some are diurnal and others have recently been categorized as cathemeral (Tattersall, 1987), that is, without a clear circadian pattern of activity. Prosimian social habits range from usually solitary to living in relatively large troops, and preferred habitat varies from upper or lower canopy of an arboreal environment to more frequent terrestriality. The postural orientation of the body and characteristic manner of locomotion also differ among prosimian species and include quadrupedal climbing, running and leaping as well as bipedal clinging and leaping. Even from this very brief overview of species variation it is apparent that the prosimian group offers many opportunities for behavioral comparison.

Prosimians are also of special interest to primate phylogeny because of their primitive morphology relative to the simians and because of evidence of very early origins of prosimian species in primate phylogeny. Understanding the course of phylogenetic development of the primates, or of any primate characteristic, should begin with prosimians. For this reason alone, it is difficult to understand why the total research effort devoted to prosimian species so greatly lags behind that expended on the simians. This is most true with regard to controlled laboratory studies of behavior. Many of the behavioral studies that employ primates in experimental research evaluate hypotheses that are directly related to the concerns of humans. The preferred employment of monkeys and apes in these studies likely reflects the fact that these are perceived as more relevant to humans because of their greater similarity and more recent common ancestry with humans. This constitutes a serious oversight which neglects the "founding fathers" of the primate order. The study of laterality in prosimians has been an attempt to remedy this oversight by investigating a topic that originated in questions about an important characteristic of human behavior.

WHY STUDY LATERALITY?

The presence of lateral bias in behavior does not follow our expectations of congruence between structure and function. For most organisms, the dorsal-ventral and anterior-posterior planes of organization are asymmetrical both in structure and in function. However, across the midline, in the lateral plane, there is an apparent essential symmetry of structure. This bilateral structural symmetry makes any lateral asymmetry in behavioral function a fascinating biological puzzle.

Human handedness, defined as a population-level bias toward use of the right hand, was undoubtedly the first phenomenon to be recognized as a manifestation of laterality. There are several Biblical references to handedness. One of the first nature/nurture controversies was between the followers of Aristotle and Plato regarding the origin of handedness. The early beginnings and historical course of development of ideas about human handedness have been thoroughly explored by Harris (1980; 1990). Documentation of the possibility of motoric lateralization in hand use for nonhuman primates only began in the 1700's (Harris, 1993).

In the nineteenth century, a second manifestation of the human laterality phenomenon began to emerge. The concept of functional specialization of the cerebral hemispheres can be dated to Broca (1861). The issue of human brain asymmetry gradually became a matter of debate until in this century Sperry's studies with "split brain" patients captured the imagination of the scientific community (see for example Sperry, 1968; 1974). Research into functional specialization of the human cerebral hemispheres has been vigorously pursued and evidence amassed that many human cognitive functions are lateralized to one brain hemisphere or the other; in particular, language comprehension and production are, in most humans, lateralized to the left hemisphere. The fact that the left hemisphere also controls the motor functions of the species-typical dominant right hand has constituted a sort of "Holy Grail" for those predisposed to find those structures and functions that could be held to be entirely unique to human primates (see Corballis, 1991). Laterality as functional specialization of the cerebral hemispheres and human handedness came to have much the same status as symbolic language as evidence of the clear separation of human from nonhuman primates. However, this view was based on a body of nonhuman primate evidence that was miniscule as compared to the mass of human research.

In the past 10 years, much evidence of motoric laterality in both prosimian and simian primates has been published (for reviews see Bradshaw and Rogers, 1993; MacNeilage *et al.*, 1987; 1988; 1991; Ward and Hopkins, 1993), and there is now also considerable evidence of cerebral functional specialization in several simian species (Hamilton and Vermeire, 1991; Heffner and Heffner, 1986; Hopkins *et al.*, 1993; Hopkins *et al.*, 1992; Morris and Hopkins, 1993; Vauclair *et al.*, 1993). This chapter reviews the evidence for motoric laterality in prosimians. Consistently findings of lateral biases in several response systems supports the view that laterality was an early development in primate phylogeny and is characteristic of most, perhaps all, primates.

THE STUDY OF PROSIMIAN LATERALITY: METHODS

The first problem to be surmounted in the study of prosimians was to obtain access to sufficient populations of subject species. When my students and I began this study about 10 years ago, we had only a modest laboratory colony of galagos (*Galago senegalensis*). In 1985 we obtained a breeding nucleus of four male and six female bushbabies (*Otolemur garnettii*) which has since grown into a colony of over 70 animals. All except two females of this foundation stock were wildborn. In addition, galagos (*G. moholi*), mouse lemurs (*Microcebus murinus*), and dwarf lemurs (*Cheirogaleus medius*) were obtained from the Duke University Primate Center (DUPC), Durham, NC. Thus, these five nocturnal species were studied in my laboratory at The University of Memphis (UM). Data were obtained from other prosimians resident at the DUPC and at Memphis Zoological Gardens (MZG), Memphis, TN, and Louisiana Purchase Gardens and Zoo (LPGZ), Monroe, LA.

Data were collected both in laboratory and free-ranging studies by videotaping ongoing behaviors and scoring various responses for lateral bias from the videotaped records. This added to the accuracy of scoring, and permitted convenient assessment of accuracy from interrater reliabilities, and the scoring of types of responses and response sequences that would otherwise be very difficult or impossible. For example, a stopwatch function on the videotapes permitted the scoring of duration of holding with the right or left hand.

The question of adequate sample size was an early matter of concern. Because lateral bias for an individual animal is a function of the proportion of responses made from the left side as compared to those from the right side, the assessment is based on probability, and the issue of sample size is thus particularly important. In most studies, 100 discrete response samples have been required and in some, lateral biases have been estimated for individual subjects based on several hundred responses. Many studies in the literature simply set a percentage criterion to establish hand preference, a procedure which does not permit a probability estimate of error in the determination of whether an individual subject does in fact have a hand preference. For the prosimian studies, the use of the z-score was adopted. This statistic permits the evaluation of lateral bias against the expected binomial distribution of total responses in terms of a selected confidence interval. The calculation is $z = (X - M)/$ (square root of $N \times p \times q$); wherein $X =$ the number of either right or left hand responses, $N =$ the total number of responses, $M = N/2$, and both p and $q = 0.5$. The confidence interval for any set of z-scores was selected in accordance with the number of subjects to be evaluated. For example, if 100 prosimians were evaluated and categorized with respect to hand preference by z-score for a particular study, the use of a criterion score of $z > 1.96$ (a two-tailed confidence interval of 0.05) would risk the miscategorization of as many as five prosimians as hand preferent when they were in fact ambipreferent, whereas a $z > 3.3$ (a two-tailed confidence interval of 0.001) for this population would reduce that risk to one in one thousand or considerably less than one animal in the evaluation set of 100. In the evaluation of prosimian lateral biases a stringent confidence interval was selected in order to keep the probability of error in evaluation to less than one subject per study.

One response consideration that has recently concerned some researchers is the question of bouts (King and Landau, 1993; Lehman, 1991; 1993; McGrew and Marchant, 1992). It is apparent that if a small sample of reaching responses is scored on one occasion from an animal that is sitting still in one position and reaching repeatedly with the same hand (a "bout"), this sample is not adequate to determine a lateral bias for that subject. None of the prosimian research has determined lateral bias under these limited conditions. In every study, large response samples were taken over many daily sessions. For some studies of free-ranging subjects, food was thrown some distance from each subject so that locomotion to each piece of food was required. In other studies when food plates were used, plates were placed in a different location on each test day. Also, the fact that the feeding was in social groups promoted movement about the food plate. Most of the laboratory studies have involved discrete trials with the subject returning to a startbox between reaches which precluded the possibility of repetitive bouts of reaching.

Lateral bias in hand use has been a principal focus of our studies. However, hand preference has been assessed in several ways: 1) as the relative frequency of simple food reaching by the right and left hands, 2) as relative duration of food holding by each of the two hands, and 3) as the relative frequency of manipulation of food between the mouth and the right or left hand. When these multiple measures have been used, they have also been analyzed within subjects to determine individual consistency of lateral hand bias across measures. We have also used videotaped records to study patterns of hand laterality in the species-typical foraging sequence of the gentle lemur (*Hapalemur griseus griseus*). The specific details of test procedures employed in each study are available in the reports listed in Table 1.

Lateral bias in whole-body turning has also been studied in several species with remarkable results. In a study by Larson *et al.*, (1989) which required that galagos run discrete trials from a startbox to a foodcup about 30.5 cm away to retrieve a mealworm (*Tenebrio* larva), it was observed that many of the animals had a consistent lateral turning bias when making the 180^0 turn to return to the startbox. When this behavior was quantified, a strong leftward population-level turning bias was found for this species. Subsequently, whole-body turning bias was investigated for all of the prosimian species resident in the UM colonies. *G. moholi* (Dodson *et al.*, 1992) and *O. garnettii* (Stafford, 1993) were evaluated for lateral bias in whole-body turning in a discrete trials procedure similar to the initial study. *M. murinus* was tested for lateral turning bias in a circular test apparatus that measured free

Table 1. Bibliography of prosimian laterality studies by species.

NOCTURNAL PROSIMIAN SPECIES	
African Prosimians	
Galago (*G. senegalensis*)	Sanford *et al.*, 1984
Galago (*G. senegalensis*)	Larson *et al.*, 1989
Galago (*G. moholi*)	Dodson *et al.*, 1992
Bushbaby (*O. garnettii*)	Milliken *et al.*, 1991
Bushbaby (*O. garnettii*)	Stafford, 1993
Magalasy Prosimians	
Mouse lemur (*M. murinus*)	Dodson *et al.*, 1992
OTHER MAGALASY PROSIMIAN SPECIES	
Ruffed lemur (*V. v. variegata*)	Forsythe *et al.*, 1988
Black lemur (*E. macaco*)	Forsythe and Ward, 1988
Ring-tailed lemur (*L. catta*)	Milliken *et al.*, 1989
Lemur species	Ward *et al.*, 1990
Gentle lemur (*H. griseus*)	Stafford *et al.*, 1993

-responding rotational bias during nightly 6 hr test periods (Dodson *et al.*, 1992). *C. medius* was evaluated for turn bias in a modified discrete trials procedure that required running from one end of a 30.5 x 172.7 cm rectangular plexiglas box to the other end to obtain a food reward, then turning 180^0 to travel back to the original position once more to obtain another food reward, and so on, back and forth for 10 turning trials per session over six sessions (unpub.).

Analysis of Strength and Direction of Lateral Bias

As our work progressed in these studies, it became apparent that there were two dimensions along which these motoric responses might be scored. Our initial interest was in the directional bias of these behaviors, whether to the right or left. In the analysis of these biases the strength or consistency of the biases became an issue in the question of whether the strength of bias warranted the categorization of the animal as left or right preferent. It soon became apparent to us that both left and right preferent individuals varied greatly in their strength of bias. This led us to wonder whether strength of lateral bias irrespective of direction might not be an important and separate phenomenon. Thus, our results have been analyzed in terms of both directionality and strength of lateral bias in subject populations.

THE STUDY OF PROSIMIAN LATERALITY: RESULTS

In the study of laterality, the unit of analysis must always be the individual organism, with statements about population-level bias of a group derived from the collective data of the subject sample. Although the logic of this statement is obvious, the concept of directional population-level lateralization is often employed as though this were the superordinate construct, as in the frequently encountered proposition that the laterality of human primates is different than that of nonhuman primates because humans have a population-level right handedness. I doubt that this is intended as a politically incorrect handedism, but it does call into question the status of the 10-15% of humans who are sinistral or ambidextrous. At present I know of no percentage value for populations that represents a generally agreed upon criterion for a judgment of species-typical laterality.

Direction of Lateral Hand Bias

At the present time, 345 individuals representing 13 prosimian species have been evaluated for lateral hand preference, with the result that 82.6% of this total sample were characterized by statistically significant *z*-scores as hand preferent whereas 17.4% were ambilateral. The summary of these results by species and direction of hand preference is given in Table 2. Although it is certain from these data that lateral bias in hand use is characteristic of a substantial proportion of prosimians, it is not possible to ascertain whether any species should be regarded as having a left or right directional bias. The small number of individuals representing most of the species evaluated and the possibility of founder effects, even for those species represented by a larger sample such as the ring-tailed lemurs (*L. catta*) and the bushbabies (*O. garnettii*), make speculations about species-typical direction of lateral bias premature. Several of the larger species groups shown in Table 2 have a substantial proportion of left preferent individuals (*G. senegalensis*, *L. catta*, *E. f. collaris*, and *E. macaco*); on the other hand, with 49 individuals now categorized, the *O. garnettii* in our laboratory seem to constitute a right biased group.

When the subtotals for nocturnal prosimians shown in Table 2 are considered with respect to all three categories of hand use, the distribution does depart from chance ($\chi^2 = 13.2$, $df = 2$, $p < 0.01$) which supports the contention of lateral bias in hand use. However, a comparison of the number of left and right preferent nocturnal individuals does not support a directional bias for this group ($z = 0.92$, n.s.). When the subtotals for Other Magalasy Prosimians are analyzed similarly, there is evidence of a biased distribution both across the three hand use categories ($\chi^2 = 34.94$, $df = 2$, $p<0.01$) and between left and right preferent individuals ($z = 2.47$, $p<0.05$) such that there is a greater proportion of left preferent animals than would be expected by chance. Analysis of Totals for the entire group of 345 animals shows differential distribution into the three hand categories ($\chi^2 = 42.16$, $df = 2$, $p<0.01$) but no statistically significant bias to employ the right or left hand preferentially ($z = 1.48$, n.s.). A further analysis shows that group hand bias evaluations may have been affected by the gender composition of the various groups.

A pattern of directional bias related to gender emerged in several prosimian studies (Milliken *et al.*, 1989; Milliken *et al.*, 1991; Ward *et al.*, 1990). Although this gender bias in directionality has not been found in every study, the recurrence of the pattern of greater incidence of left hand preference in males as compared to right

Table 2. Summary of number of individuals characterized as left, right, or ambilateral in hand preference for simple food reaching as a function of species.

	LEFT	AMBI	RIGHT	TOTALS
NOCTURNALS				
African Prosimians				
G. moholi	5	0	11	16
G. senegalensis	21	6	8	35
O. garnettii	12	12	25	49
Magalasy Prosimians				
M. murinus	3	1	4	8
C. medius	2	2	4	8
Subtotals (%)	43 **(37)**	21 **(18)**	52 **(45)**	116
OTHER MAGALASY PROSIMIANS				
H. griseus	8	0	5	13
E. coronatus	6	1	4	11
E. f. albifrons	6	4	5	15
E. f. collaris	15	4	5	24
E. f. fulvus	3	3	5	11
E. f. rufus	8	3	3	14
E. f. sanfordi	3	8	6	17
E. macaco	20	1	12	33
E. mongoz	9	5	12	26
E. rubriventer	1	0	4	5
L. catta	28	10	17	55
V. v. variegata	5	0	0	5
Subtotals (%)	112 **(49)**	39 **(17)**	78 **(34)**	229
TOTALS (%)	155 **(45)**	60 **(17)**	130 **(38)**	345

hand preference in females provoked my interest (Ward *et al.*,1993), particularly in view of the fact that similar associations in motoric laterality have been reported for other primate species such as hylobatids (Stafford *et al.*, 1990), chimpanzees (Bard *et al.*, 1990; Hopkins and Bard, 1993), and humans (Annett, 1972, 1978; Porac and Coren, 1981; Seltzer *et al.*, 1990). A summary of species directional biases by gender is given in Table 3. In this table the number of males and females assessed for each species is shown as well as the proportion of each sex biased to the left or right. Inspection of the totals shows the percentage of ambilateral animals to be quite similar in males and females. By contrast, the male group has a greater proportion of left hand preferent animals than right preferent animals ($z = 2.82, p<0.01$); these left and right proportions are not different in the female group ($z = 0.69$, n.s.). Thus, in consideration of all the prosimians evaluated to the present time, it is concluded that there may be a population-level bias towards the left hand in male prosimians. Perhaps when a larger number of females have been sampled, the apparent tendency toward right bias in females may become clearer.

Table 3. Percentage of lateral hand preference for each species by gender.

		%LEFT	%AMBI	%RIGHT
NOCTURNAL SPECIES				
African Prosimians				
G. moholi	Male (n=8)	50.0	0	50.0
	Female (n=8)	*12.5*	*0*	*88.5*
G. senegalensis	Male (n=7)	85.7	0	14.3
	Female (n=3)	*33.3*	*0*	*66.7*
O. garnettii	Male (n=29)	24.1	27.6	48.3
	Female (n=20)	*25.0*	*20.0*	*55.0*
Malagasy Prosimians				
M. murinus	Male (n=5)	20.0	20.0	60.0
	Female (n=3)	*67.0*	*0*	*33.0*
C. medius	Male (n=4)	50.0	50.0	0
	Female (n=4)	*75.0*	*25.0*	*0*
OTHER MALAGASY PROSIMIANS				
H. griseus	Male (n=8)	50.0	0	50.0
	Female (n=5)	*80.0*	*0*	*20.0*
E. coronatus	Male (n=6)	66.7	0	33.3
	Female (n=5)	*40.0*	*20.0*	*40.0*
E. fulvus	Male (n=10)	40.0	30.0	30.0
albifrons	Female (n=5)	*40.0*	*20.0*	*40.0*
E. fulvus	Male'(n=14)	71.4	14.3	14.3
collaris	Female (n=10)	*50.0*	*20.0*	*30.0*
E. fulvus	Male (n=6)	33.3	33.3	33.3
fulvus	Female (n=5)	*20.0*	*20.0*	*60.0*
E. fulvus	Male (n=8)	50.0	12.5	37.5
rufus	Female (n=6)	*66.7*	*33.3*	*0*
E. fulvus	Male (n=10)	20.0	40.0	40.0
sanfordi	Female (n=7)	*14.3*	*57.1*	*28.6*
E. macaco	Male (n=16)	81.3	6.3	12.5
	Female (n=13)	*46.2*	*15.4*	*38.5*
E. mongoz	Male (n=15)	46.7	20.0	33.3
	Female (n=11)	*18.2*	*18.2*	*63.6*
E. rubriventer	Male (n=3)	0	0	100.0
	Female (n=2)	*50.0*	*0*	*50.0*
L. catta	Male (n=35)	57.2	17.1	25.7
	Female (n=20)	*40.0*	*20.0*	*40.0*
V.v.variegata	Male (n=5)	100.0	0	0
	Female (n=0)			
TOTALS	Male (n=189)	50.3	17.5	32.2
	Female (n=127)	*37.8*	*18.9*	*43.3*

Direction of Lateral Whole-Body Turn Bias

The discovery of a lateral bias in prosimian whole-body turning was unexpected but more so the fact that many animals had a directional bias to turn to the left.

Only the nocturnal prosimians resident in the UM laboratories have been evaluated for this response bias. The results of this testing are shown in Table 4. It may been seen from this table that the left turning effect was most strongly manifested in the galagos [10 *G. senegalensis* from Larson *et al.*(1989) and 16 *G. moholi* from Dodson *et al.*(1992)]. However, for each species tested, the number of individuals categorized as left turning exceeded those preferring to turn right. The distribution across the three turning categories of all prosimians tested was different from chance (χ^2 = 27.7, $df = 2, p<0.001$). The number of animals that turned left consistently was also greater than expected by chance ($z = 4.52, p<0.001$). I consider this left whole-body turning bias to be a manifestation of a more pervasive lateral bias in posture and postural mechanisms.

A left turning bias has also been reported for children (Gospe *et al.*,1990) and adult humans (Seltzer *et al.*, 1990). Bracha *et al.* (1987) found that the direction of the adult bias was contingent on gender and lateralization in other systems. For example, males who were consistently right-sided in hand, foot, and eye preference had a right-turning bias, whereas females consistently right-sided and males with mixed dominance turned left preferentially. Research with laboratory rats by Glick and his colleagues supports the view that these turning biases are related to neurochemical imbalances (Glick, 1993; Glick *et al.*, 1977). The finding of left whole-body turning biases in both prosimians and human primates suggests that this directional postural bias may be characteristic of most primates.

This postural turning bias may also bear some relationship to a locomotor directional lateral bias that has been reported for primates: leading out in locomotion from the right side. Right leading limb has been reported for the black and white ruffed lemur, *V. v. variegata*, (Forsythe and Ward, 1986), four species of apes (Heestand, 1986), pigmy chimpanzees (Hopkins *et al.*, 1993) and humans (Seltzer *et al.*, 1990). I propose that there is an essential pattern of lateral bias in the posture and locomotion of primates which is most pronounced in those primates whose posture is most characteristically bipedal or upright in orientation of the long axis of the body. This includes a tendency to carry weight to the left side of the body (Seltzer *et al.*, 1990), to lead out in action from the right side of the body, and to rotate the body to the left in turning.

Table 4. Direction of whole-body turn bias as number of individuals and percentage of species population (%) for nocturnal prosimians.

	LEFT (%)	AMBI (%)	RIGHT (%)	TOTALS
Mouse lemur (*M. murinus*)	6 (75)	2 (25)	0 (0)	8
Dwarf lemur (*C. medius*)	3 (30)	5 (50)	2 (20)	10
Galago (*G. spp.*)	23 (88)	1 (4)	2 (8)	26
Bushbaby (*O. garnettii*)	7 (44)	5 (31)	4 (25)	16
TOTALS	39 (65)	13 (22)	8 (13)	60

Strength of Lateral Bias and the Within-Subject Effect of Postural Change

The first study of lateral bias reported for any prosimian (Sanford *et al.*, 1984) recorded the effect of posture on lateral bias. Galagos, who as a group were somewhat indefinite in hand preference when reaching downward from a quadrupedal posture for a mealworm, were greatly strengthened in the consistency of hand preference when required to stand upright in a bipedal posture before reaching forward for the same prey item. In a follow-up study by Larson *et al.* (1989) which employed a different set of tasks with a new population of galagos, a similar effect was found. Eight food cups were constructed that varied visibility of prey, required angle of reaching, and elevation of the target prey and permitted testing the effects of these variables in a completely counterbalanced design. Only the degree of elevation of the target prey was found to affect consistency of lateral hand bias in food reaching. In tests using the four variants of the task that required fully upright stance prior to reaching, all galagos were more consistent in use of the preferred hand than for comparable reaching from a quadrupedal posture.

Five black and white ruffed lemurs tested at the MZG showed that postural effects on reach strength did not necessarily require a bipedal posture (Forsythe *et al.*, 1988). When the five brothers reached during casual foraging about a moated island, only one manifested a lateral hand preference, a 71% left hand bias. When food (banana) was thrown to the surface of the island and away from the location of the lemurs such that they were required to locomote to the food item and position themselves prior to reaching, all five lemurs revealed a left hand bias with an average bias of 96.2%. In a final test condition, banana slices were thrown into the water of the moat next to the concrete ledge that surrounded the island. In this test, the average left hand bias for the five lemurs was 99.8%; only one of 515 total reaches into the moat was made with the right hand, and this by the youngest of the five siblings. To reach into the moat, it was necessary for the lemurs to rest their hindquarters on top of the concrete ledge, then lean over the side of the ledge into the moat, bracing their upper body with one hand and reaching for the banana slice with the other. Both the postural preadjustment promoted by the discrete presentation of food items on the surface of the island and the extreme postural preadjustment required for reaching into the moat strengthened lateral hand bias.

Similar effects of postural alteration on strength of hand bias have been reported recently for squirrel monkeys, *Saimiri sciureus*, and cotton-top tamarins, *Saguinus oedipus*, (Roney and King, 1993); for gorillas, *Gorilla gorilla gorilla*, and gibbons, *Hylobates lar*, (Olson *et al.*, 1990); for chimpanzees, *Pan troglodytes*, and orangutans, *Pongo pygmaeus*, (Hopkins, 1993); and for pygmy chimpanzees, *Pan paniscus*, (Hopkins *et al.*, 1993). In all of these studies, the lateral bias of individual primates has been strengthened by an extreme, usually upright and bipedal posture. This within-subject change in the consistency of manifested lateral bias with the assumption of a vertical orientation for the long axis of the body may be thought of as a state effect. When the postural state of a particular individual is altered, the consistency of that individual's lateral hand use bias is also altered.

The Postural Orientation of Species and Strength of Laterality

Could there also be a between-subjects species-typical effect of posture? In the section that follows, evidence is presented to support the proposal that the typical postural orientation of the long axis of the body of various species constitutes an enduring trait of that species which contributes to the greater or lesser strength of lateral bias characteristic of that species.

The first study that brought species differences in strength of lateral bias to my attention was a study of 194 individuals of six species and five subspecies (*E. fulvus sp.*) of lemurs resident at the DUPC (Ward *et al.*, 1990). It was, in fact, an observation about the propensity of younger lemurs to pick up food with the mouth rather than the hand that drew me into consideration of species differences in strength of lateral bias. Lemurs younger than one year were not included in this study, but it soon became apparent that even one- and two-year-old juveniles favored the mouth in picking up food. When the average "mouthedness" of all juvenile lemurs (*Mean* = 29.4%) was compared to that of all adults, those 3-30 years of age, (*Mean* = 11.8%), the age effect on mouth preference was clear ($t = 8.74, p<0.001$). It had also been observed that when the juveniles did use their hands, they were less consistent in lateral hand preference than the adults. However, statistical comparison of the strength of bias for all adults as compared to that for all juveniles did not exceed chance because of the extreme variability of scores in both groups. Because this variability appeared to be linked to species membership, a further analysis of both mouthedness and strength of lateral bias was conducted on adult and juvenile scores matched by species. For each species, without exception, the juveniles of the species were more mouth biased and less strongly lateralized in hand preference than the adults of the same species. A comparison of juvenile mouth bias by species averages (*Mean* = 32.7%) with those for adults (*Mean* = 13.2%) confirmed the age difference in mouth bias ($t = 7.61, p<0.001$). A paired comparison by species also showed adults to be more strongly lateralized in hand use (*Mean* = 84.1%) than juveniles (*Mean* = 72.6%, $t = 4.22, p<0.002$). It was also observed that in some cases the juveniles of strongly lateralized species had higher hand bias scores than the adults of less lateralized species.

In subsequent studies whenever possible, both mouth and hand retrievals of food were recorded and analyzed for strength of bias. A summary of results for the nocturnal prosimians tested in our laboratory and all ring-tailed lemurs tested is given in Fig. 1 and 2. The species averages were derived from adult animals. The species shown have been arranged from left to right in these figures in accordance with my appraisal of the extent to which the species characteristic axial orientation is more horizontal or more vertical. The quadrupedal mouse and dwarf lemurs (*M. murinus and C. medius*) with long torsos and short legs are typically observed with a horizontal long body axis. The galagos (*G. senegalensis* and *G. moholi*) with a short back and relatively long hind legs both locomote by vertical clinging and leaping and also frequently rest or feed with the hind legs deeply flexed and the torso vertical. The bushbabies (*O. garnettii*) and ring-tailed lemurs (*L. catta*) are intermediate along this dimension with the lemurs ranked as somewhat more upright in seated posture than the bushbabies. This ranking by usual long axis orientation is now based only on observations and should be quantified in future studies to permit direct correlation of postural orientation with other behavioral variables.

Fig. 1 illustrates that the average mouth bias of adults of the species evaluated decreases with upright posture. By contrast, Fig. 2 shows the strength of lateral hand bias to increase with more upright species posture. Observe that Fig. 2 does not illustrate a function that is reciprocal to that of Fig. 1. What is shown in Fig. 2 is not the propensity to use the hand in lieu of the mouth, but rather the *strength of the tendency to use a particular hand repeatedly*. A final measure analyzed by species posture is the strength of lateral whole-body turn bias. Fig. 3 illustrates that the strength of lateral bias in whole-body turning is enhanced with the more vertical body orientation.

The tendency of an animal to use the mouth in lieu of the hand, when the mouth is typically closer to the substrate where food is found, is readily understood in terms

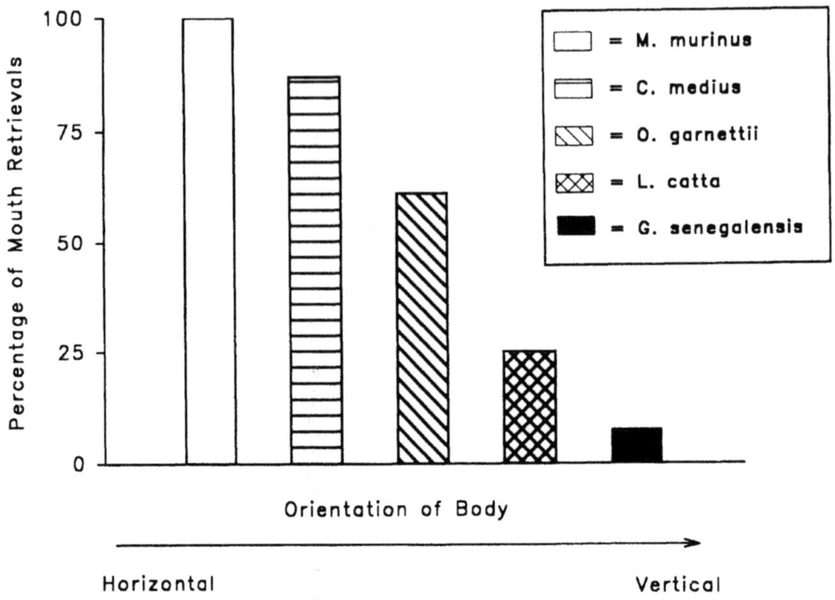

Figure 1. Strength of mouth bias as average percentage of mouth retrievals of food from total food retrievals (with mouth and hands) for five species ranked from left to right in terms of species characteristic long axis orientation from horizontal to vertical. Number of individuals contributing to species averages: 8 *M. murinus*, 10 *C. medius*, 23 *O. garnettii*, 55 *L. catta*, and 16 *G. moholi*.

of speed and ease of feeding. However, the reason for a more consistent lateral bias in the use of a preferred hand or in the execution of whole-body turns by an animal with an upright posture is not so intuitively obvious.

SOME THOUGHTS ABOUT POSTURE AND LATERAL BIAS IN BEHAVIOR

Several studies have now demonstrated that reaches executed by an individual from an unstable posture are more consistent with respect to lateral hand bias than those made from a stable posture by that same individual. Evidence has also been presented to illustrate the greater consistency of lateral bias in both reaching and turning associated with the degree of upright posture characteristic of species. Might these within- and between-subject postural effects be related phenomena, perhaps mediated by a common neural substrate?

The actions of every organism must be conducted within the earth's gravitational field, a silent but powerful factor which shapes and limits all patterns of movement. Recumbent or seated animals or those in a quadrupedal or tripedal stance are secure

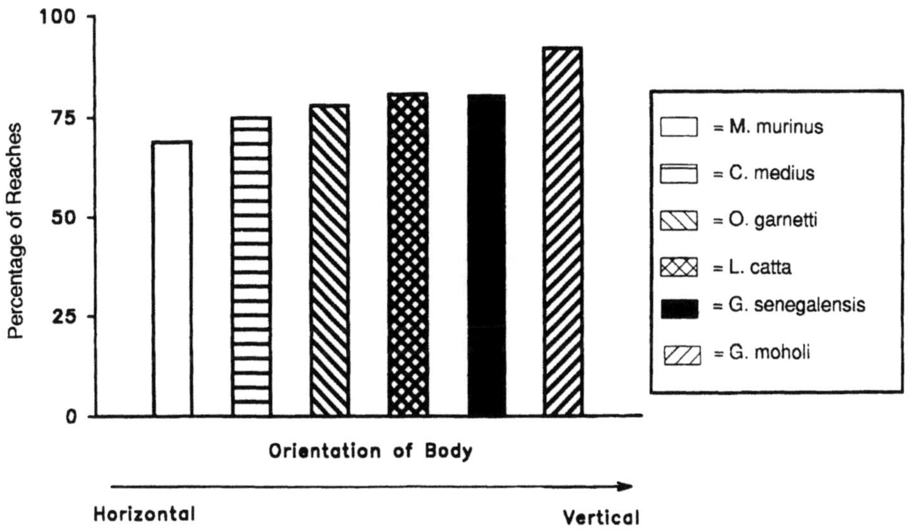

Figure 2. Strength of lateral reach bias as average percentages of food reaches with the preferred hand (from hand reaches only) for six species ranked left to right in terms of species characteristic long body axis orientation from horizontal to vertical. Number of individuals averaged for each species: 8 *M. murinus*, 10 *C. medius*, 49 *O. garnettii*, 55 *L. catta*, 35 *G. senegalensis*, 16 *G. moholi*.

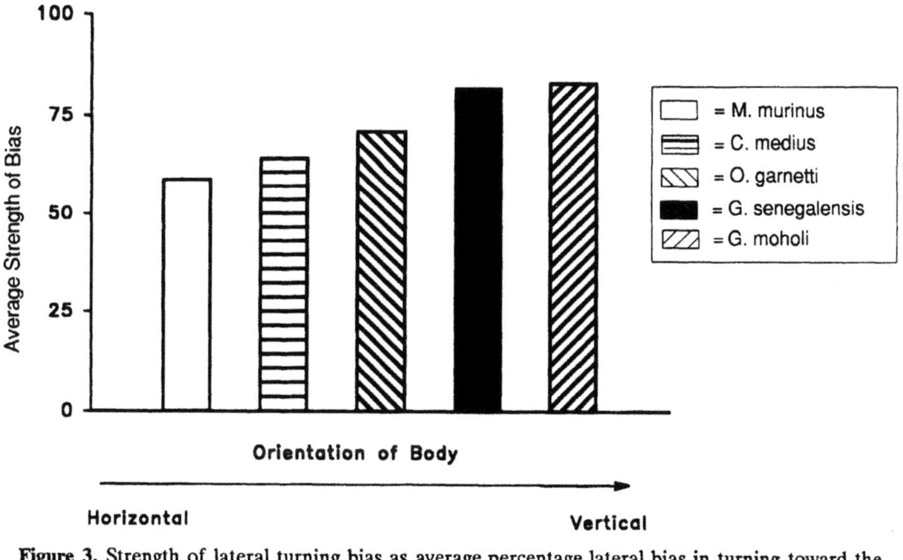

Figure 3. Strength of lateral turning bias as average percentage lateral bias in turning toward the preferred side for each of five species ranked left to right in terms of species characteristic long body axis orientation from horizontal to vertical. Number of individuals contributing to average for each species: 8 *M. murinus*, 10 *C. medius*, 16 *O. garnettii*, 10 *G. senegalensis*, 16 *G. moholi*.

in opposition to the downward pull of gravity. (We recognize this principle of stability when chairs and tables are constructed with three or four supporting legs.) When, in the service of specific needs, an animal rises in the vertical plane, elevating the center of gravity, and balancing weight over two points of support, a challenge is posed to the neuromuscular system that mediates this action. For example, a primate, whether human or nonhuman, standing quietly on two hindlegs has its' center of gravity elevated and balanced over and between the two hindlimbs. A simple act such as reaching out to the side or forward with one arm demands a global readjustment in the musculature of the trunk and lower limbs to effectively maintain that upright posture. The maintenance of balance during the execution of patterns of action that continually shift the body's weight laterally requires complex and efficient reflexes. Although located for the most part in the spinal cord, these reflexes are modulated by intersegmental spinal influences and by descending neural influences from several regions of the brain. During actions that invoke lateral imbalance, a postural modulating system adjusts the unstable bipedal posture to a transient but balanced monopedal posture such that weight is supported by the trunk and leg of one side of the body while the other side of the body is engaged in leading out in action. This bilateral reciprocation serves to maintain postural balance against the force of gravity during various activities of the organism and also may constitute a factor that influences the strength of lateral motoric biases.

The assumption of upright posture may set into motion a general systemic arousal that serves to intensify the integration of motor subsystems and thereby enhance whatever inherent lateral biases are present. This proposed mechanism could apply equally well to the transient effect in the within-subject manipulation of postural state or as a chronic condition of species-typical postural orientation. Although this is a hypothetical construct, neural mechanisms have been demonstrated that illustrate the principles involved. Ito and Sanada (1965) found righting reflexes of monkeys to be activated by simple alteration of the orientation of the body axis with respect to gravity. They further demonstrated that this effect was mediated by increased neural activity originating from pressure receptors (pacinian corpuscles) located in the visceral organs. These visceral afferents are known to project to the cerebellum and reticular activating system (Ito and Sanada, 1965) and also to the primary and secondary somatosensory and orbital areas of neocortex (Amassian, 1951; Downman, 1951; Korn, 1969). In consideration of these results, Talbott (1979) suggested that visceral afferents may increase the overall integration of postural mechanisms. Perhaps changes in posture with respect to gravity serve to alter the integration of lateral biases in neural systems mediating movement and thereby influence the manifestation of lateral bias in response.

PROSIMIANS AND THE STUDY OF LATERALITY: A REPRISE

The comparative study of prosimians has made a substantial contribution to the understanding of laterality as a behavioral phenomenon. A large proportion of prosimians representing a variety of species native to both Africa and Madagascar have been shown to have consistent lateral biases in behavior. These findings support the view that laterality is a general characteristic of primates and likely has been since the origins of this group. The variations in postural habit among prosimians has made possible the discovery and exploration of the intriguing effect of posture on the expression of lateral bias, a phenomenon that was not previously recognized. It is hoped that the potential of prosimians as subjects for comparative study will continue to be explored in the context of a variety of questions concerning primate behavior.

ACKNOWLEDGMENTS

The research described here was supported in part by National Science Foundation Grant BNS-8707754 to the author. Support for the research and for this report was also received from a Centers of Excellence grant awarded to the Department of Psychology, UM, by the State of Tennessee. The author would also like to express appreciation to the DPC, Durham, NC, to the MZG, Memphis, TN, and to the LPGZ, Monroe, LA, for access to their subject populations, and to the many student collaborators who by their diligent efforts made this report possible.

REFERENCES

Amassian, V.E. (1951). Cortical representations of visceral afferents. *Journal of Neurophysiology, 14*, 433-444.

Annett, M. (1972). The distribution of manual asymmetry. *British Journal of Psychology, 63*, 343-358.

Annett, M. (1978). Genetic and nongenetic influences on handedness. *Behavior Genetics, 8*, 227-249.

Bard, K.A., Hopkins, W.E., & Fort, C. (1990). Lateral bias in infant chimpanzees (*Pan troglodytes*). *Journal of Comparative Psychology, 104*, 309-321.

Bracha, H.S., Seitz, D.J., Otemaa, J., & Glick, S.D. (1987). Rotational movement (circling) in normal humans: Sex difference and relationship to hand, foot and eye preference. *Brain Research, 411*, 231-235.

Bradshaw, J.L., & Rogers, L.J. (1993). *The Evolution of Lateral Asymmetries, Language, Tool Use, and Intellect*. San Diego, CA: Academic Press.

Broca, P. (1861). Remarks on the seat of the faculty of articulate language, followed by an observation of aphemia. In G. von Bonin (Ed.), *The Cerebral Cortex* (pp. 49-72). Springfield, IL: C.C.Thomas, 1960.

Corballis, M.C. (1991). *The Lopsided Ape*. New York: Oxford University Press.

Dodson, D.L., Stafford, D.K., Forsythe, C., Seltzer, C.P., & Ward, J.P. (1992). Laterality in quadrupedal and bipedal prosimians: Reach and whole-body turn in the mouse lemur (*Microcebus murinus*) and galago (*Galago moholi*). *American Journal of Primatology, 26*, 191-202.

Downman, C.B.B. (1951). Cerebral destination of splanchnic afferent impulses. *Journal of Physiology, 113*, 434-441.

Forsythe, C., Milliken, G.W., Stafford, D.K., & Ward, J.P. (1988). Posturally related variations in the hand preferences of the ruffed lemur (*Varecia variegata variegata*). *Journal of Comparative Psychology, 102*, 248-250.

Forsythe, C., & Ward, J.P. (1986). The lateralized behavior of the ruffed lemur (*Varecia variegata*) [Abstract]. *American Journal of Primatology, 12*, 342.

Forsythe, C., & Ward, J.P. (1988). Black lemur (*Lemur macaco*) hand preference in food reaching. *Primates, 29*, 369-374.

Glick, S.D. (1993). Rotational behavior in children and adults. In J.P.Ward & W.D.Hopkins (Eds.) *Primate Laterality: Current Behavioral Evidence of Primate Asymmetries* (pp. 307-317). New York: Springer-Verlag.

Glick, S.D., Jerussi, T.P., & Zimmerburg, B. (1977). Behavioral and neuropharmacological correlates of nigrostriatal asymmetry in rats. In S.Harnad, J.Jaynes, L. Goldstein, R.W.Doty, & G. Krauthamer (Eds.), *Lateralization in the Nervous System* (pp. 213-249). New York: Academic Press.

Gospe, S.M., Mora, B.J., & Glick, S.D. (1990). Measurement of spontaneous rotational movement (circling) in normal children. *Journal of Child Neurology, 5*, 31-34.

Hamilton, C.R., & Vermeire, B.A. (1991). Functional lateralization in monkeys. In F.L.Kitterle (Ed.), *Cerebral Laterality: Theory and Research* (pp. 19-34). Hillsdale, NJ: Lawrence Erlbaum.

Harris, L.J. (1980). Left-handedness: Early theories, facts and fancies. In J. Herron (Ed.), *Neuropsychology of Left-handedness* (pp. 303-329). New York: Academic Press.

Harris, L.J. (1990). Cultural influences on handedness: Historical and contemporary evidence. In S. Coren (Ed.), *Left-handedness: Behavioral Implications and Anomalies* (pp.195-258). Amsterdam: Elsevier.

Harris, L.J. (1993). Handedness in apes and monkeys: Some views from the past. In J.P.Ward & W.D.Hopkins (Eds.), *Primate Laterality: Current Behavioral Evidence of Primate Asymmetries* (pp.1-41). New York: Springer-Verlag.

Heestand, J.E. (1986). *Behavioral lateralization in four species of apes*. Unpublished Ph.D. dissertation, University of Washington, Seattle.

Heffner, H.E., & Heffner, R.S. (1984). Effect of unilateral and bilateral auditory cortex lesions on the discrimination of vocalizations by Japanese macaques. *Journal of Neurophysiology, 56*, 683-701.

Hopkins, W.E. (1993). Posture and reaching in chimpanzees (*Pan troglodytes*) and orangutans (*Pongo pygmaeus*). *Journal of Comparative Psychology, 107*, 162-168.

Hopkins, W.D., & Bard, K.A. (1993). Hemispheric specialization in infant chimpanzees (Pan troglodytes): Evidence for a relation with gender and arousal. *Developmental Psychobiology, 26*, 219-235.

Hopkins, W. D., Bennett, A. J., Bales, S. L., Lee, J., & Ward, J. P. (1993). Behavioral laterality in captive bonobos (*Pan paniscus*). *Journal of Comparative Psychology, 107*, 403-410.

Hopkins, W. D., Fagot, J., & Vauclair, J. (1993). Solution of mirror-image and mental rotation problems by baboons (*Papio papio*): Unilateral input enhances performance. *Journal of Experimental Psychology: General, 122*, 61-72.

Hopkins, W.D., Morris, R.D., Savage-Rumbaugh, E.S., & Rumbaugh, D.M. (1992). Hemispheric priming by meaningful and nonmeaningful symbols in language-trained chimpanzees: Further evidence of a left hemisphere advantage. *Behavioral Neuroscience, 106*, 575-582.

Ito, T., & Sanada, Y. (1965). Location of receptors for righting reflexes acting upon the body in primates. *Japanese Journal of Physiology, 15*, 235-242.

King, J.E., & Landau, V.I. (1993). Manual preference in varieties of reaching in squirrel monkeys. In J.P. Ward & W.D. Hopkins (Eds.), *Primate Laterality: Current Behavioral Evidence of Primate Asymmetries* (pp. 107-124). New York: Springer-Verlag.

Korn, H. (1969). Splanchnic projection of the orbital cortex of the cat. *Brain Research, 16*, 23-38.

Larson, C.F., Dodson, D.L., & Ward, J.P. (1989). Hand preferences and whole-body turning biases of lesser bushbabies (*Galago senegalensis*). *Brain, Behavior and Evolution, 33*, 261-267.

Lehman, R.A.W. (1991). Hand choice by monkeys is governed not only by preference but also by perseveration. In A. Ehara, T. Kimura, O. Takenaka, & M. Iwamoto, (Eds.), *Primatology Today* (pp. 265-268). Amsterdam: Elsevier.

Lehman, R.A.W. (1993). Manual preference in prosimians, monkeys and apes. In J.P. Ward & W.D. Hopkins (Eds.). *Primate Laterality: Current Behavioral Evidence of Primate Asymmetries* (pp. 149-182). New York: Springer-Verlag.

MacNeilage, P.F., Studdert-Kennedy, M.G., & Lindblom, B. (1987). Primate handedness reconsidered. *Behavioral and Brain Sciences, 10*, 247-303.

MacNeilage, P.F., Studdert-Kennedy, M.G., & Lindblom, B. (1988). Primate handedness: A foot in the door. *Behavioral and Brain Sciences, 11*, 737-746.

MacNeilage, P.F., Studdert-Kennedy, M.G., & Lindblom, B. (1991). Primate handedness: The other theory, the other hand, and the other attitude. *Behavioral and Brain Sciences, 14*, 344-349.

Martin, R.D. (1990). *Primate Origins and Evolution: A Phylogenetic Reconstruction*. Princeton, NJ: Princeton University Press.

McGrew, W.D., & Marchant, L.F. (1992). Chimpanzees, tools, and termites: Hand preference or handedness? *Current Anthropology, 33*, 114-119.

Milliken, G.W., Forsythe, C., & Ward, J.P. (1989). Multiple measures of hand-use lateralization in the ring-tailed lemur (*Lemur catta*). *Journal of Comparative Psychology, 103*, 262-268.

Milliken, G.W., Stafford, D.K., Dodson, D.L., Pinger, C.D., & Ward, J.P. (1991). Analyses of feeding lateralization in the small-eared bushbaby (*Otolemur garnettii*): A comparison with the ring-tailed lemur (*Lemur catta*). *Journal of Comparative Psychology, 105*, 274-285.

Morris, R. D., & Hopkins, W.D. (1993). Perception of human chimeric faces by chimpanzees (*Pan troglodytes*): Evidence for a right hemisphere asymmetry. *Brain and Cognition, 21*, 111-122.

Olson, D.A., Ellis, J.E., & Nadler, R.D. (1990). Hand preferences in captive gorillas, orang-utans and gibbons. *American Journal of Primatology, 20*, 83-94.

Porac, C., & Coren, S. (1981). *Lateral Preferences and Human Behavior*. New York: Springer-Verlag.

Roney, L.S., & King, J.E. (1993). Postural effects on manual reaching laterality in squirrel monkeys (*Saimiri sciureus*) and cotton-top tamarins (*Saguinus oedipus*). *Journal of Comparative Psychology, 107*, 380-385.

Sanford, C., Guin, K., & Ward, J.P. (1984). Posture and laterality in the bushbaby (*Galago senegalensis*). *Brain, Behavior and Evolution, 25*, 217-224.

Seltzer, C., Forsythe, C., & Ward, J.P. (1990). Multiple measures of motor lateralization in human primates (*Homo sapiens*). *Journal of Comparative Psychology, 104*, 159-166.

Sperry, R.W. (1968). Hemispheric deconnection and unity in conscious awareness. *American Psychologist, 23*, 723-733.

Sperry, R.W. (1974). Lateral specialization in the surgically separated hemispheres. In F. O. Schmitt & F. G. Worden (Eds.), *The Neurosciences: Third Study Program* (pp.5-19). Cambridge, MA: MIT Press.

Stafford, D.K. (1993). *Effects of Sex and Handedness on Visuospatial Memory in the Small-eared Bushbaby (Otolemur garnettii).* Unpublished Ph.D dissertation, Memphis State University, Memphis, TN.

Stafford, D.K., Milliken, G.W., & Ward, J.P. (1990). Lateral bias in feeding and brachiation in Hylobates. *Primates, 31*, 407-414.

Stafford, D.K., Milliken, G.W., & Ward, J.P. (1993). Patterns of hand and mouth lateral biases in bamboo leaf shoot feeding and simple food reaching in the gentle lemur (*Hapalemur griseus*). *American Journal of Primatology, 29*, 195-207.

Talbott, R.E. (1979). Ferrier, the synergy concept, and the study of posture and movement. In R.E. Talbott & D.R. Humphrey (Eds.), *Posture and Movement* (pp. 1-12). New York: Raven Press.

Tattersall, I. (1987). Cathermeral activity in primates: A definition. *Folia Primatologica, 49* (3-4), 200-202.

Vauclair, J., Fagot, J., & Hopkins, W.D. (1993). Rotation of mental images in baboons when visual input is directed to the left cerebral hemisphere. *Psychological Science, 4*, 99-103.

Ward, J.P. (1991). Prosimians as animal models in the study of neural lateralization. In F.L.Kitterle (Ed.), *Cerebral Laterality: Theory and Research* (pp.1-17). Hillsdale, NJ: Erlbaum.

Ward, J.P., & Hopkins, W.D. (Eds.), (1993). *Primate Laterality: Current Behavioral Evidence of Primate Asymmetries*. New York: Springer-Verlag.

Ward, J.P., Milliken, G.W., Dodson, D.L., Stafford, D.K., & Wallace, M. (1990). Handedness as a function of sex and age in a large population of *Lemur*. *Journal of Comparative Psychology, 104*, 167-173.

Ward, J.P., Milliken, G.W., & Stafford, D.K. (1993). Patterns of lateralized behavior in prosimians. In J.P.Ward & W.D.Hopkins (Eds.), *Primate Laterality: Current Behavioral Evidence of Primate Asymmetries* (pp.43-74). New York: Springer-Verlag.

ACOUSTIC COMMUNICATION IN NOCTURNAL PROSIMIANS

E. Zimmermann

German Primate Center
Laboratory of Biocommunication
Kellnerweg 4
37077 Göttingen, FRG

INTRODUCTION

Research on nonhuman primate vocal communication began about three decades ago when the first sound analyzing machines became available. Vocalizations recorded on tape could be displayed visually as sonagrams. They could be characterized acoustically in the spectral and temporal domain and the size and shape compared objectively. In 1963, the first sonagrams of nocturnal prosimians were published (Andrew, 1963). In the following decade, however, interest in the study of prosimian vocal behavior remained rather limited (Klopfer, 1977). Thus, it was not until 1979 that the first comparative survey of vocalizations in all of the main taxonomic groups of prosimians was presented (Petter and Charles-Dominique, 1979). Since then, sound analyzing and synthesizing techniques have become much more sophisticated. A significant amount of descriptive information on the structure and function of vocalizations of nocturnal prosimians has become available. Compared to simian primates, however, analytical and experimental studies designed to reveal information conveyed by the acoustic signals and their use in prosimian primates (messages and meanings as defined by Smith, 1977), as well as information about constraints and patterns in the evolution of their calls, are far more rare.

In this paper, the current state of the art in research on acoustic communication in nocturnal prosimians will be reviewed by focusing on three major topics. Firstly, the descriptive information which is now available on vocalizations in the main taxonomic groups will be briefly summarized stressing problems which have to be taken into account when comparing vocal repertoires of different species studied by different investigators using different methods. Secondly, the role played by vocalizations in social life will be addressed. Finally, constraints and patterns in the evolution of social calls will be considered.

DEFINITIONS

To avoid confusion, two terms must be clarified. Firstly, the term "nocturnal prosimians", as used in this paper, is restricted to nocturnal strepsirhines. According to Nash *et al.* (1988), Martin (1990) and Rumpler (1992), this group consists of the African and Asian Lorisidae with 16 species, the Malagasy Cheirogaleidae with six species, the Malagasy Daubentonidae with one species, the Malagasy Lepilemuridae with six species and the Malagasy Indridae with two species. At least four species of the Malagasy Lemuridae (*Hapalemur aureus*, *H. griseus*, *H. simius* and *Eulemur mongoz*) show a cathemeral activity pattern. They are excluded from this review. To date, there is no consensus about the genera name in the bushbabies (Nash *et al.*, 1988). The genera name "Galago" is therefore used throughout this paper. Secondly, the term "communication" has many meanings. Here, the definition most commonly used in the field of eco-ethology is applied. "Communication is a process in which actors use specially designed signals or displays to modify the behavior of reactors" (Krebs and Davies, 1993).

A SURVEY OF NOCTURNAL PROSIMIAN VOCALIZATIONS

A necessary prerequisite for any sophisticated experimental and analytical study is the description of the vocal repertoire of the respective species. Studies that have determined at least part of the vocal repertoires of adult nocturnal prosimians are listed in Table 1. Of 31 known extant species, information on 23 species is now available. Most detailed studies have been conducted on African and Asian lorisids and Malagasy cheirogaleids.

Three major problems arise, however, when comparing the size and shape of vocal repertoires of different species studied by different investigators with different methods. They make it to date extremely difficult to reveal general trends or peculiarities in their vocal repertoires. These are:

1. Differences in the quality of recording and analyzing techniques. Particularly in the early days of bioacoustics, little attention was paid to the sound recording and analyzing equipment and the effect the environment may have on the quality of

sonagrams and oscillograms. Frequency bands may appear suppressed or enhanced in an artificial way if the recording equipment does not have a flat frequency response for the entire recorded range. Artificial noise or artificial frequency bands or sidebands may result when the intensity of a sound is too high during recording or as an input to the analyzing system. In addition, if field recordings are used or if recordings are made in small rooms without sound-dampening, excessive contamination in both frequency and temporal domain may occur as a result of environmental acoustics. For example, refraction and reflection of frequencies may lead to smeared frequencies in rapidly modulated calls and to longer call duration (for review, see Michelsen and Larsen, 1983 and Brown and Gomez, 1992).

2. Differences in sampling accuracy. As Snowdon (1986) emphasized, there may be a bias in field recording towards those vocalizations that are loud enough to be heard over some distance, such as alarm or advertisement calls. Calls used for short distance social interactions are difficult to record with proper sound quality. In captivity it is often much easier to record all calls. The captive environment, however, may limit the full range of calls due to the absence of social or other environmental stimuli. Furthermore, captive born monkeys may show an inappropriate use of particular calls, such as alarm or food calls if they have not had the opportunity to learn the appropriate sound-object relations from experienced adults. My own studies of bushbabies and mouse lemurs reveal that, in contrast to wildborn animals, captive born animals reared by inexperienced adults do not display any form of predator avoidance behavior toward snakes; rather they try to approach and sniff them.

3. Differences in the size of the vocal repertoire in the same species. There have been various attempts of categorizing the different sound classes within the repertoire of a given species, often with slightly different results for the same species. Thus, for example, Daschbach *et al.* (1981) described the vocal repertoire of the adult slow loris (*Nycticebus coucang*) as containing four call classes, whereas Zimmermann (1985b) classified eight call classes based on finer resolution of call structures and calls not observed by the other authors. Acoustically distinct call classes may be discriminated by means of cluster or discriminant analysis. This approach was recently applied by Stanger (1993) for classifying the vocal repertoire of Coquerel's dwarf lemur (*Mirza coquereli*). The most widely used approach to differentiating call classes in the vocal repertoires of nocturnal prosimians is to correlate subtle differences in call structures with differences in their usage (Petter and Charles-Dominique, 1979; Zimmermann *et al.*, 1985a, b; Zimmermann *et al.*, 1988; Macedonia and Stanger, submitted). The third and most reliable, but time-consuming, approach is to ask the animals themselves how they perceive and differentiate their own vocalizations. By using an operant conditioning technique, Petersen *et al.* (1984) discerned that Japanese macaques (*Macaca fuscata*) discriminate one social call group, the "coo"-group, as two separate call classes on the basis of the position of the peak frequency. Humans would treat these calls as one.

Table 1. List of species for which there is sonagraphic information

Taxa		Environment	Data	References
Lorisidae				
Galaginae				
Galago	alleni	Lab, Field	Li	2,13,21,25
	crassicaudatus	Lab, Field	De	1,11,21,25
	demidoff	Lab	De	1,2,13,14,21,25
	elegantulus	Field	Li	2,13
	gallarum	?	?	
	garnettii	Lab, Field	De	11,21,25
	matschiei	Field	Li	21
	moholi	Lab, Field	De	10,23,25
	senegalensis	Lab	De	1,16,18,19,21,23
	thomasi	?	?	
	zanzibaricus	Field	Li	4,19, 21, 25
Lorisinae				
Arctocebus	calabariensis	?	?	
Loris	tardigradus	Lab	De	7,25
Nycticebus	coucang	Lab	De	1,5,8,16,18,20,21
Nycticebus	pygmaeus	Lab	Li	21, 25
Perodicticus	potto	Lab	Li	2,13,21,25
Cheirogaleidae				
Allocebus	trichotis	?	?	
Cheirogaleus	medius	Lab,Field	De	3,13, 14,25
Cheirogaleus	major	Field	Li	13, 25.
Mirza	coquereli	Lab, Field	De	12,13,14,25
Microcebus	murinus	Lab, Field	De	1,3,6,9,13,14,22,24
	rufus	Lab	Li	6
Phaner	furcifer	Field	Li	13, 25
Lepilemuridae				
Lepilemur	dorsalis	?	?	
	edwardsi	Field	Li	13
	leucopus	?	?	
	mustelinus	?	?	
	ruficaudatus	Field	Li	13, 25
	septentrionalis	Field	Li	13
Indridae				
Avahi	laniger	Field	Li	13
	occidentalis	?	?	
Daubentonidae				
Daubentonia	madagascariensis	Field	Li	13

(1) Andrew, 1963;(2) Charles-Dominique, 1977;(3) Cherry et al., 1987; (4) Courtenay and Bearder, 1988; (5) Daschbach et al., 1982; (6) Glatston, 1979; (7) Helmich, 1987; (8) Horr, 1969; (9) Kuhn, 1989; (10) Mascagni and Doyle, 1993; (11) Masters, 1991; (12) Pages, 1980; (13) Petter and Charles-Dominique,1979; (14) Stanger, 1988; (15) Stanger, 1993; (16) Zimmermann, 1981; (17) Zimmermann, 1985a; (18) Zimmermann, 1985b; (19) Zimmermann, 1989a; (20) Zimmermann, 1989b; (21) Zimmermann, 1990; (22) Zimmermann, 1991; (23) Zimmermann et al., 1988; (24) Zimmermann and Lerch, 1993; (25) Zimmermann, pers. observ. Abbreviations: Li = limited, De = detailed.

Seyfarth *et al.* (1980) conducted play-back experiments in the field and were able to show that vervet monkeys (*Cercopithecus aethiops*) could perceive different variants of an onomatopoetically similar call class as functionally different, solely on the basis of slight variations in vocal structure. Similar results were obtained for various monkeys (e.g. marmosets: Snowdon, 1986; macaques: Gouzoules *et al.*, 1984; Hopp *et al.*, 1992). To date, however, there is no information on how nocturnal prosimians categorize species-specific calls.

THE ROLE OF VOCALIZATIONS IN THE SOCIAL LIFE OF NOCTURNAL PROSIMIANS: A CASE STUDY ON THE ACOUSTIC BEHAVIOR OF THE LESSER MOUSE LEMUR (*Microcebus murinus*)

Vocalizations in simian primates are known to encode information about sex, age, individual identity and information about internal states and behavioral intentions (e.g. contact, aggression, defense or alarm), as well as information about objects (such as food and predators) and information about social relations, including dominance-subordinance relations (for reviews see Snowdon *et al.*, 1982; Todt *et al.*, 1988; Cheney and Seyfarth, 1990; Zimmermann, 1992). Nocturnal prosimian vocalizations are believed to convey information mainly about internal states and intentions (Petter and Charles-Dominique, 1979).

Nocturnal prosimians have limited vision (Pariente, 1979) relative to diurnal species, live in a dense forest environment and have a pregregarious social structure (see Richard, 1987, for review). In these species, selection should favor the evolution of mate advertisement and recognition patterns in the acoustic communication channel.

Indeed, some nocturnal prosimian species, like the African bushbabies, have developed loud or advertisement calls with species-specific acoustic patterns (Zimmermann *et al.*, 1989; Zimmermann, 1990). It is suggested that they play a major role in the species-specific mate recognition system. To date, however, no experimental study in any nocturnal primate species has been conducted which confirms a major role for vocalization in the mating context. We investigated the vocal and social behavior of a nocturnal Malagasy prosimian species, the lesser mouse lemur (*Microcebus murinus*), with a view to answering the following questions: Is there a vocalization within the vocal repertoire which is produced at a high rate mainly during the mating season? To what extent is this vocalization, if it exists, altered by sexual competition? Does a female discriminate among males in a mating situation? If so what may be the discriminative behavioral clues?

The lesser mouse lemur is a suitable model for such a study since it has already been described as highly vocal (Martin, 1972) and it is relatively easy to maintain in captivity where it breeds freely. Furthermore, this species has a seasonally restricted breeding period which may be triggered artificially by photoperiodic modification (Petter-Rousseaux, 1975; Perret, 1992). It is the smallest known primate species and it has retained the most archaic, primate pattern in terms of its morphology and karyology (Martin, 1972; Rumpler and Dutrillaux, 1990).

Mouse lemurs live in the dry forest areas of western Madagascar where they occupy the fine branch niche (Harcourt and Thornback, 1990). Acoustic signals should be of great importance for communication because of their secretive nocturnal habits within the dense and bushy regions of these forests.

The type of vocalizations adult mouse lemurs exhibit and the context in which they are emitted are summarized by representative sonagrams in Figure 1 (for methods of recording and analyzing see Zimmermann and Lerch, 1993) and Table 2. Acoustic characteristics are displayed in Table 3.

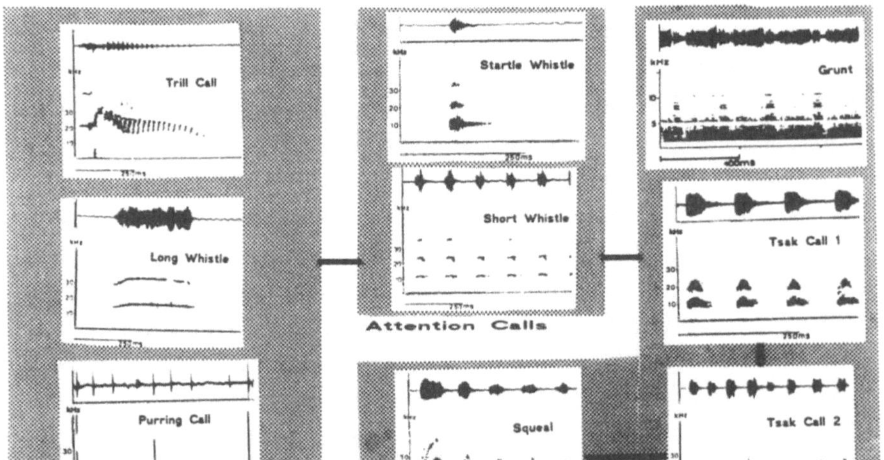

Figure 1. Vocal repertoire of the adult lesser mouse lemur. Lines indicate possible transitions between vocalizations

Table 2. Synopsis of mouse lemur calls

Vocalization	Stimulus (sender)	Behavior of sender	Acoustic coding	Behavior of recipient	Vocal response of recipient
Trill I	unknown c, estrous f, spontaneously (m)	marking, approach, sniffing , mounting	combination of whistle + sweep + fm pulse series, harmonic	observing s, attack, withdrawl or tolerance	sometimes trill calling
Trill II	male (estrous f)	marking, observing e	?	searching for m	sometimes trill calling
Trill III	unknown or sub c (f,m)	marking, sniffing, attack	?	withdrawl or attack	sometimes trill calling
Trill IV	infant (f) nursing	approach, grooming,	combination of sweep + separated fm pulse series, harmonic	approach, grooming, suckling	twitter,peep purr
Long whistle	unspecific, unknown objects, onset or offset of activity period (f,m)	observing, marking, approach o	slow fm pulses, narrowband, harmonic	unspecific, scanning e	sometimes long whistles
Short whistle	unspecific, disturbances (f,m), delivery (f), in tame animals when held by keeper (f,m)	scanning e	rapid fm pulses, narrowband, harmonic	unspecific, scanning e	sometimes short whistles
Startle whistle	sudden movement of external o (f,m)	fleeing to nest	single, fm pulse harmonic	fleeing to nest	
Tsak	disturbances from c (f,m)	aggressive threat	rapid fm pulses, broadband, harmonic	withdrawl or attack	sometimes trill or tsak
Squeal	unspecific, sudden unexpected noises (f,m), after repeated contact refusals by estrous f (m)	observing e	medium fm pulses, unspecific broadband, harmonic	unspecific	sometimes squeal in neighbours
Purr	familiar c, keeper in tame animals (f,m)	approach, grooming	rhythmically repeated short pulses, noisy	grooming	purr
Grunt	cornered by c or keeper	defensive threat	irregularly repeated pulses of variable duration, noisy	withdrawl	

f=female, m=male, c=conspecific, e=environment, o=object, s=sender, sub=subordinate, fm=frequency modulate

Table 3. Acoustic characteristics of vocalizations in adult lesser mouse lemurs

Vocalization	n	FAN kHz	FEN kHz	FP kHz	UD ms	IUI ms	M
Trill I	8	13-25	5-10	23-36	440-1100		single call
Long whistle	16	8-19	9-19	9-26	50-800	50-6000	series
Short whistle	16	5-16	9-15	8-18	8-50	30-300	series
Startle whistle	3	8-16	12-18	12-18	25-50		single call
Tsak	16	5-25	5-17	8-28	14-50	20-100	series
Squeal	5	4-11	3-12	9-19	32-100	180-900	series
Purr	16	1-3	2-7	5-30	5-30		series
Grunt	16	0.5-2	8-18	50-70	5-400		series

n=number of animals; FAN=onset frequency of fundamental; FEN=end frequency of fundamental; FP=peak frequency of fundamental; UD=unit duration; IUI=interunit interval; M=miscellaneous; Minimum and maximum frequencies and durations are presented, respectively

Eight major, structurally different vocal patterns can be discriminated on the basis of computer spectrography. They are produced by adults irrespective of sex and rank. The sound communication range extends from about 0.5 kHz to about 36 kHz. Vocal patterns are categorized according to the general behavioral context; contact calls, attention calls, and aggressive and defensive calls. A high degree of variability in vocal morphology and vocal usage can be detected within most of the eight vocal patterns. The trill advertisement call of adult males demonstrates the role this variation plays in social life .

The "trill" advertisement call is an example of a contact call with a complex acoustic design. It is used very frequently by sexually active males during the mating season (Zimmermann and Lerch, 1993). Two types of experiments were designed in which social interactions and acoustic behavior were analyzed in order to clarify its message and meaning (Kugel, 1993). In the first experiment a group of two unfamiliar males was established. There could, therefore, be no competition for a female. In the second experiment a group consisting of the same males with a female in estrus was established. The female was unfamiliar with either both or with only one of the males. If intrasexual competition has an effect on trill calling behavior, then trill calling activity should be low in the first experiment and greatly enhanced in the second. If females discriminate mates on the basis of relatedness/familiarity and try to avoid incest, they should exhibit a higher level of aggression towards closely related than towards unrelated males in a mating situation. If trill calls encode acoustic information about individual identity and calling activity about quality , the acoustic pattern of this call should have a higher variability among different males than within the same male and, its occurrence should be rank dependent.

Table 4 illustrates the results for dyads of the two males before and after the inclusion of an estrous female. Male groups were tested on three consecutive days, each for 90 min at the beginning of their activity period. Males developed a linear rank order as shown by the percentage of agonistic behaviors (chasing, retreat and defensive behaviors) in both situations. Dominant males exhibited the highest number of aggressive and the least number of defensive behaviors. Sexual competition increased the frequency of aggression among males without altering the established rank. Trill calling activity was rank dependent and increased considerably during sexual competition.

Table 4. Agonistic encounters and trill calling activity of male mouse lemur during the breeding period with and without access to an estrous female collected and analyzed for 90 min over three consecutive days (n= 2 groups/experiment)

	Males without female			Males with estrous female		
Behavioral Category	Dominant male	Subordinate male	N	Dominant male	Subordinate male	N
Chasing	100%	0%	9	63%	27%	122
Retreat	8%	92%	194	22%	78%	238
Defensive behaviors	14%	86%	7	42%	58%	55
Trill calling activity	73%	27%	15	65%	35%	390

Table 5 shows the behaviors of a female toward dominant and subordinate and related and unrelated males of the same rank. The estrous female was tested over three consecutive days for 90 min at the beginning of the activity period after being introduced to two males. Females were tested with pairs of males, in one pair both being unfamiliar to her and, in another pair, one being her son with whom she had been housed until the first breeding season.

Females displayed a higher level of aggression toward dominant males rather than subordinate males. Aggression in females correlated with the rate of contact approaches of males toward females. Subordinate males exhibited an equal rate of contact approaches to females irrespective of their relatedness to the female. Females, however, reacted much more aggressively toward their sons than toward subordinate, unfamiliar males.

Table 5. Aggressive behavior of an estrous female towards males of different ranks and of different relatedness/familiarity over three consecutive days during the first 90 min of the activity period (n=2 groups/experiment)

Behavior of Female toward	Dominant male	Subdominant male	N	Related male	Unrelated male	N
Chasing	66%	34%	189	74%	26%	65
Aggressive Threat	55%	45%	426	68%	32%	193

These results imply that females do indeed distinguish between males in a mating situation and exercise active mate choice. Do trill advertisement calls convey acoustic information about individual identity which provide females with clues for discriminating between potential mates? To investigate this the complex acoustic pattern of trill calls produced by these males was analyzed by means of computer spectrography and multivariate statistics. Cluster analysis revealed that trill calls of adult male mouse lemurs provide reliable acoustic information about individuality. A stepwise discriminant function analysis showed that a set of only three different variables (onset frequency of whistle, bandwidth of third frequency modulation and end frequency of seventh frequency modulation) provides maximum separation between the trill calls of different males. Based on this analysis, almost 90% of the trill calls could be correctly assigned to the individual who produced them. A principal component analysis showed that this set of variables was unaffected by differences of age, weight and context (Zimmermann and Lerch, 1993).

More recent results indicate that the stereotypic, individually specific vocal patterns remain constant over at least two consecutive breeding seasons. Trill advertisement calls, therefore, provide a reliable basis both for assessing the quality of a respective male through its calling rate and for individual recognition by its vocal structure. Calls showing these characteristics are probably the most important vocalizations in the repertoire of nocturnal prosimians with regard to mate advertisement and recognition, sexual selection and speciation. They can help to identify cryptic species within monomorphic species groups, to clarify the status in polymorphic species or to develop inferences about phylogeny.

CONSTRAINTS AND PATTERNS IN THE EVOLUTION OF ADVERTISEMENT CALLS IN NOCTURNAL PROSIMIANS

In all nocturnal prosimians studied thus far, calls have been described as being emitted by either sex primarily during the mating season. These calls were often associated with sitting on exposed sites, marking behavior or displacing or chasing a competing individual of the same sex. They evoke either mate attraction and acceptance or antiphonal responses from visually isolated conspecifics. Since they seem to signal the presence of a breeding animal over long distances for the purpose of mating, they are termed here "advertisement" calls. Although no experimental studies, other than those referred to above, have affirmed the adaptive significance of these calls, it can be postulated that they play an integral role in mate advertisement and recognition. According to Paterson (1980) and Littlejohn (1981), the mate advertisement and recognition system has evolved to achieve efficient syngamy. Species distinctiveness in the acoustic pattern of their calls may, therefore, have evolved particularly in those species which rely largely on the acoustic communication channel, as in forest dwelling, nocturnal prosimians.

Various factors have been discussed which are likely to influence the evolution of acoustic signals in such species. Environmental effects, such as attentuation and degradation, may mask the shape of a call over long distances in a forest environment and may distort effective transmission of acoustic information. Distortion is much greater in calls with high frequency fundamentals and rapidly repeated pulses than for those with low frequency fundamentals and slowly repeated pulses. According to the acoustic adaptation hypothesis (Wiley and Richards, 1978; Krebs and Davies, 1993), an animal living in a tropical forest environment can overcome these problems if it produces advertisement calls in the low frequency range which contain either pure notes, as opposed to trills, or trills with widely spaced notes.

The advertisement calls of all the galagines (Zimmermann, 1990) and one Malagasy cheirogaleid, *Phaner furcifer* (Figure 5), closely fit these predictions. Their advertisement calls consist of a series of rhythmically repeated, stereotypic notes for which the main energy lies in the lower frequency range between 0.3 and 1.5 kHz. All Asian and African lorisines and all the other Malagasy cheirogaleines, however, emit advertisement calls in the higher frequency range with a clear harmonic structure and narrow frequency bands (see Figure 3 and Table 1 for detailed references). According to the acoustic adaptation hypothesis, such calls do not seem to be optimally adapted for acoustic advertising over longer distances.

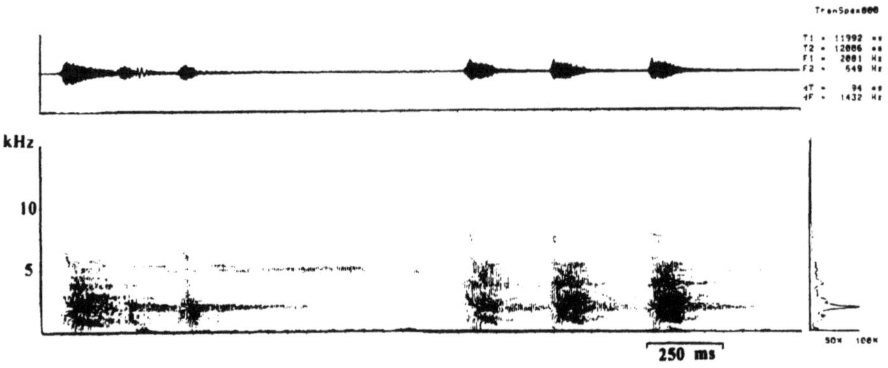

Figure 2. Advertisement call in Phaner furcifer displayed as oscillogram (above) and sonagram (below)

What could be the explanation for this apparent contradiction? There might be a theoretical optimum to which an acoustic signal could evolve to minimize attentuation and degradation. Vocal production, however, may be further constrained by ambient noise and predation risk. Thus, it is known from studies on vocal advertising in frogs and birds, that their predators, such as barn owls and bats, may track them by listening for their calls (Alcock, 1989; Konishi, 1973). A comparison of the acoustic structure of the advertisement calls of closely related taxa, with different strategies for avoiding predation, the African and Asian lorisines on the one hand, and the African galagines on the other, demonstrates how predation pressure appears to affect sound production.

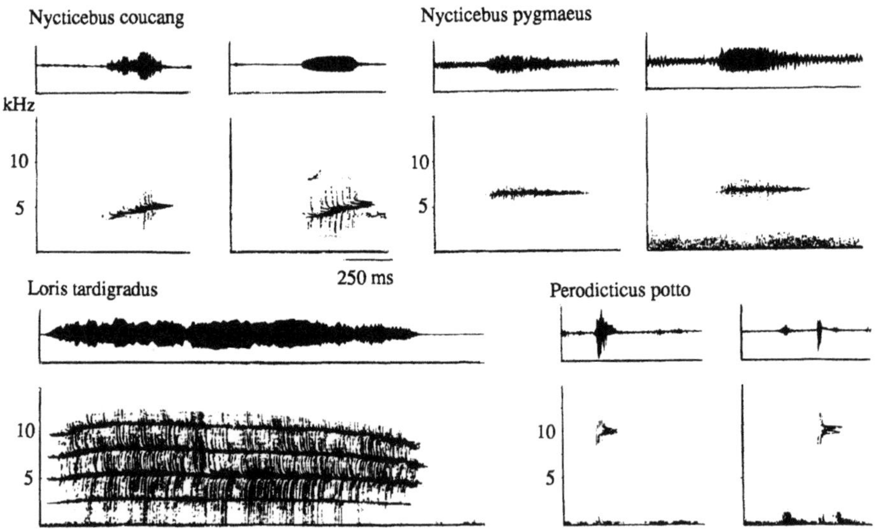

Figure 3. Comparison of the advertisement call in the Lorisinae displayed as oscillograms and sonagrams: Whistle call

The African and Asian lorisines are slow-moving prosimians which are confronted with the problem of having "to appear conspicuous to their own species and inconspicious to predators" (Marler 1955). Confronted by potential predators they are not able to move rapidly, compared to the fast leaping and hopping African galagines or the Malagasy cheirogaleid *Phaner furcifer*. Indeed, the acoustic pattern of their advertisement calls seems to reflect a compromise. Calls have high frequency fundamentals and are emitted singly or at an irregular repetition rate (see Figure 3 and Zimmermann, 1985b). Such acoustic characteristics minimize the probability of being located by predators. At least two of the lorisines, the slow loris (*Nycticebus coucang*) and the potto (*Perodicticus potto*), are highly sensitive to that range (Heffner *et al.*, 1969), suggesting that both their vocal and auditory system evolved together.

The question arises, however, as to why small, fast moving Malagasy cheirogaleids also display high frequency fundamentals in their sexual advertisement calls. A possible hypothesis emerges when one compares two ecologically similar species groups. These are the Malagasy lesser mouse lemur (*Microcebus murinus*) and the African dwarf bushbaby (*Galago demidoff*), being house-mouse sized ecological counterparts, and the Malagasy Coquerel's dwarf lemur (*Mirza coquereli*) and the African Senegal bushbaby (*Galago senegalensis*), being grey squirrel-sized ecological counterparts. All four species are known to inhabit similar forest environments and have similar social structures (for reviews see Bearder, 1987; Richard, 1987; Martin , 1990).

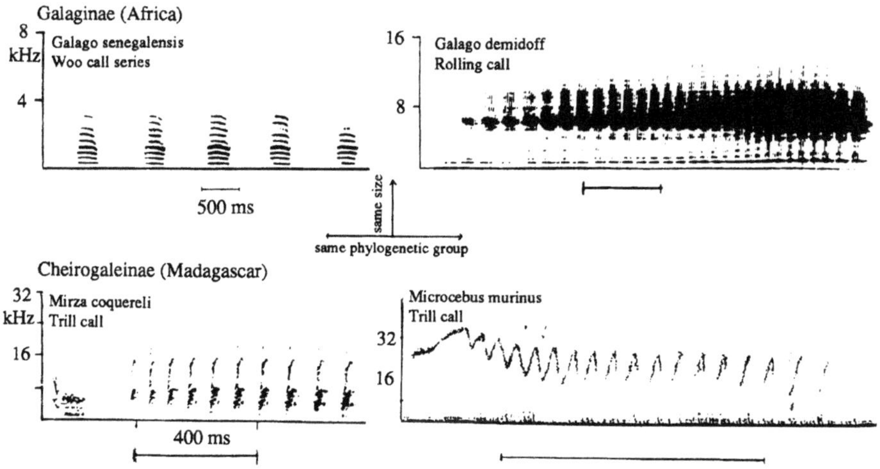

Figure 4. Comparison of the advertisement call in galagines and cheirogaleines displayed as sonagrams

Size-correlated differences should emerge by comparing the two groups in terms of body-weight differences and phylogenetically correlated differences by comparing the African with the Malagasy forms.

Differences with regard to fundamental frequency as well as unit and call duration within the African galagos and the Malagasy lemurs are indeed an effect of size differences. The smaller species in each group exhibit a higher fundamental frequency, a broader bandwidth, a shorter call duration and a higher unit repetition rate than the larger species (Figure 4; Zimmermann, 1990; Stanger 1993; Lerch and Zimmermann, 1993). These effects are due to differences in body mass and larynx morphology. In general the fundamental frequency of a call is determined by the mass and tension of the vocal cords as well as air pressure through the larynx. The more massive the vocal cords, the lower the frequency of the call. In addition, the fundamental frequency of the call can be actively modified by changing the tension and length of the vocal cords by activating different laryngeal muscles. Vocal cord mass as well as air capacity of the lungs increases with body size (see, for example, Lieberman, 1984; Sutton and Jürgens, 1986). A divergence in body size, in response to ecological selection, may, therefore, lead to a corresponding divergence of fundamental frequency and duration of calls and probably to the subsequent premating isolating mechanism.

Comparing the overall acoustic shape of these calls of the bushbabies with the lemurs indicates, however, that it is much more similar within the most closely related group phylogenetically than between the two ecologically similar groups. Thus, bushbabies have developed a stereotyped "bark" series with low frequency fundamentals while the lemurs have developed "trill" calls with high frequency fundamentals. This implies bushbaby- and lemur-specific sound-producing mechanism. Divergent predation pressures appear to have favored different strategies in the evolution of their vocal signalling systems which, in turn, impose phylogenetic constraints on further evolution.

Predation on small nocturnal lemurs by owls constitutes a major threat to their survival in present-day Madagascar (Goodman *et al.*, 1993). Predation pressure from nocturnal predatory birds, like owls, appears to be less imposing in Africa than in Madagascar. Thus, in South-African bushbabies predation rate was estimated to be about 15% of the population per year (Cheney and Wrangham, 1987) whereas in the Malagasy mouse lemur predation by barn owls was found to be greater than 25% of the population per year (Goodman *et al.*, 1993). Bushbabies have, therefore, been able to evolve optimally habitat-adapted loud, low frequency advertisement calls which can carry reliably over long distances (several hundred meters under field conditions). Because of the scarcity of small rodents in Madagascar, risk of predation by owls is much higher for small nocturnal Malagasy primate species than for their ecological counterparts in Africa. It is likely that this highly increased predator pressure forced small nocturnal Malagasy lemurs to evolve advertisement calls in the high frequency and ultrasonic range, a range which is undetectable for nocturnal predatory birds and quite unusual in other primate species.

Last, but not least, the question arises as to how advertisement calls evolve within a closely related taxonomic group. By means of evolutionary systematics, analyses were made of the calls of the bushbabies.The advertisement calls of each of the eight species studied were characterized by a set of acoustic characters (Zimmermann, 1990). Variations in the advertisement call in all species were classified as particular character states. Character states were compared to the infant isolation calls to assess their degree of conservatism. The infant isolation call is a precursor of the adult advertisement call. Irrespective of body size this call displays a similar acoustic pattern in all bushbaby species (Zimmermann, 1990). Interestingly, a contextually and structurally similar call was found in infants of the African and Asian lorisids, implying a common phylogenetic origin. A hypothetical tree of advertisement call evolution is shown in Figure 5.

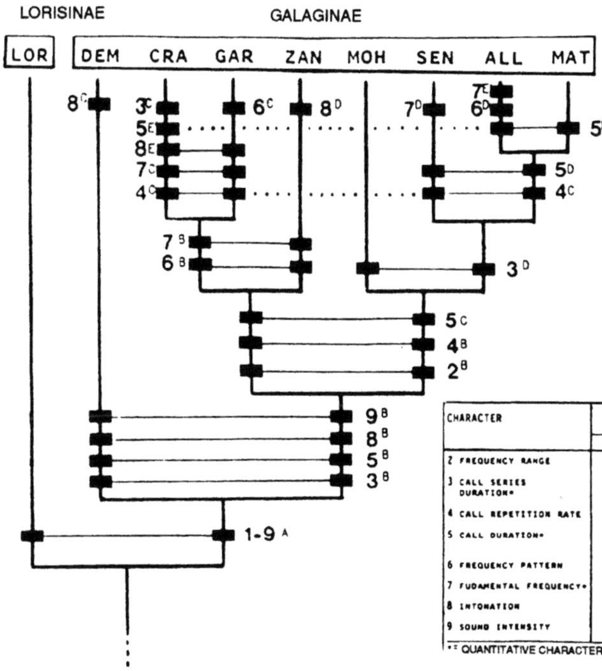

CHARACTER	CHARACTER STATE A	B	C	D	E
2 FREQUENCY RANGE	BROAD	NARROW			
3 CALL SERIES DURATION*	VERY SHORT	SHORT	MEDIUM	LONG	
4 CALL REPETITION RATE	HIGH	MEDIUM	SLOW		
5 CALL DURATION*	VERY SHORT	SHORT	MEDIUM	LONG	VERY LONG
6 FREQUENCY PATTERN	CF	FM	MFM	MCF	
7 FUDAMENTAL FREQUENCY*	MEDIUM	MEDIUM-HIGH	LOW-MEDIUM	LOW	MEDIUM+HIGH
8 INTONATION	IR	333	123	321	121
9 SOUND INTENSITY	LOW	HIGH			

*= QUANTITATIVE CHARACTERS

Figure 5. Evolution of the advertisement call in the galaginae based on shared acoustic characters.

Some interesting facts emerge. *Galago demidoff* shows the least change of its advertisement call structure. Its advertisement call differs from the infant isolation call and the outgroup in only four of the nine characters and one autapomorphy and, therefore, branches first. Its call consists of a definite number of pulses with relatively low fundamental frequencies and a broad bandwidth due to the large number of harmonics. Sound intensity is low initially and increases gradually to its maximum towards the end.

The remaining seven bushbaby species share three derived characters as synapomorphies. Two differentiations emerge. Frequency modulated pulses produced in a species-specific dynamic pattern characterize the *garnetti-zanzibaricus-crassicaudatus* group and distinguishes it from the medium-sized bushbabies. In *Galago zanzibaricus* autapomorphies were the dynamic temporal pattern, along with the intonation of pulses within call series which decrease towards the end, in *Galago garnettii* the unique multiple frequency modulated pulses and in *Galago crassicaudatus* the highly derived pulse and pulse series duration.

All medium-sized bushbabies display stereotypically repeated pulses which may form a series lasting up to an hour with almost similar intonation of pulses in the entire series. The most striking differences among them were, in *Galago moholi*, the occurrence of single, double and triple pulses in series, in *Galago senegalensis*, the extremely low fundamental frequency in single pulse series, in *Galago alleni*, the noisy overlapped harmonic spectra of the pulses and, in *Galago matschiei*, the highly derived character states. To explain the reasons for differences in the acoustic patterns, more information about the eco-ethology of the respective species and the underlying mechanisms of sound production and perception is needed.

CONCLUSION

Recent research on vocal behavior in nocturnal prosimians has revealed an unexpectedly rich array of differently structured sounds used for communication. The most elaborate studies on vocal repertoires have been carried out on cheirogaleids and lorisids. There is a particular lack of information on all of the other nocturnal Malagasy lemur species.

Experimental and analytical studies assessing the role of vocalizations in social life are just beginning. Thus far, they have concentrated on advertisement calls in the lesser mouse lemur (*Microcebus murinus*). They have implied an integral role these calls play in mate recognition, sexual selection and speciation. Future studies should be conducted on a larger variety of populations of the same and of different species. They will help to identify cryptic species within monomorphic species groups, to clarify the species status in polymorphic species and to develop inferences about phylogeny. These studies should also focus on the differentiation of these calls during ontogeny and on their perception by conspecifics.

Scant attention has been paid thus far to the evolution of social calls. Environmental, morphological and physiological factors which appear to constrain the evolution of acoustic patterns used for long distance communication in tropical forests are discussed. An approach was introduced by which patterns in the evolution of advertisement calls in closely related species could be reconstructed. To illuminate mechanisms which have caused changes in acoustic character states, a much better knowledge of the psycho-physiology of call production and perception and the eco-ethology of the respective species has to be obtained. Such an interdisciplinary approach will provide a new understanding of the evolution of primate vocal communication.

ACKNOWLEDGEMENTS

I thank the following researchers from whom I obtained recordings: H. Fitch-Snyder (pygmy slow loris) and H. Helmich and H. Preuschoft (slender loris). I am grateful to the German Primate Center and to H. and M. Zimmermann for providing space for conducting research on galagos, lorises and mouse lemurs and to H.-J. Kuhn, U. Juergens, E. Sterling and G. Doyle for comments on earlier versions of this manuscript. Financial support was provided by several grants of the German Research Council (DFG AZ Zi 345/ 1; 3; 5; 8).

REFERENCES

Alcock, J. (1988). Animal Behavior. Sinauer Ass., Sunderland.

Andrew, R.J. (1963). The origin and evolution of the calls and facial expressions of the primates. *Behaviour, 20,* 1-109.

Bearder, S.K. (1987). Lorises, busbbabies and tarsiers: Diverse societies in solitary foragers. In B.B. Smuts, D.L. Cheney, R.M. Seyfarth, R.W.Wrangham and T.T. Struhsaker (Eds.), *Primate Societies,* (pp. 11-24). Chicago: Univ. Chicago Press.

Brown, C.H. and Gomez, R. (1992). Functional design features in primate vocal signals: The acoustic habitat and sound distortion. In T. Nishida, W. C. McGrew, P. Marler, M. Pickforf, F. de Waal. (Eds.) , *Topics in Primatology,* Vol. 1 (pp. 177-198). Tokyo, Univ. Tokyo Press.

Charles-Dominique, P. (1977). Ecology and Behaviour of Nocturnal Primates. New York Columbia Univ. Press, New York.

Cherry, J.A., Izard, M.K. and Simons, E.L. (1987). Description of ultrasonic vocalizations of the mouse lemur (*Microcebus murinus*) and the fat-tailed lemur (*Cheirogaleus medius*). *American Journal of Primatology, 13,* 181-185.

Cheney, D. L. and Seyfarth, R.M. (1990). How Monkeys See the World. Univ. Chicago Press, Chicago.

Cheney, D.L. and Wrangham, R.W. (1987). Predation. In B.B. Smuts, D.L. Cheney, R.M. Seyfarth, R.W.Wrangham and T.T. Struhsaker (Eds.), *Primate Societies,* (pp. 11-24). Chicago:Univ. Chicago Press.

Courtenay, D.O. and Bearder, S.K. (1988). The taxonomic status and distribution of bushbabies in Malawi with emphasis on the significance of vocalizations. *International Journal of Primatology 10, 17-34.*

Daschbach, N., Schein, M.W. and D.E. Haines (1981). Vocalizations of the slow loris (*Nycticebus coucang*). *International Journal of Primatology , 2,* 71-80.

Glatston, A. (1979). Reproduction and Behaviour of the Lesser Mouse Lemur in Captivity. PhD thesis, University of London, unpubl.

Goodman, S.M., O'Connor, S. and Langrand, O. (1993). A review of predation on lemurs: Implications for the evolution of social behavior in small nocturnal primates. In P.M. Kappeler , J.U. Ganzhorn (Eds.), *Lemur Social Systems and their Ecological Basis* (pp. 51-66). New York: Plenum Press.

Gouzoules, S., Gouzoules, H. and Marler, P. (1984). Rhesus monkey (*Macaca mulatta*) screams: Representational signalling in the recruitment of agonistic aid. *Animal Behaviour, 32,* 182-192.

Harcourt, C. and Thornback, J. (1990). Lemurs of Madagascar and the Comores. IUCN, Gland.

Heffner, H.E. and Masterton, B. (1969). Hearing in primitive mammals. *Journal of Comparative Physiology and Psychology, 71,* 175-182.

Helmich, H. (1987). Das Lautrepertoire adulter Schlankloris (*Loris tardigradus*) im Verhaltenskontext in Gefangenschaft. Schriftliche Hausarbeit, University of Bochum, unpubl.

Hopp, S.L., Sinnot, J.M., Owren, M. J. and Petersen, M. (1992). Differential sensitivity of japanese macaques (*Macaca fuscata*) and humans (*Homo sapiens*) to peak position along a synthetic coo call continuum. *Journal of Comparative Psychology* , 106, 128-136.

Horr, D.A. (1969). Communication and Behavior of a Cage-Colony of Slow Loris (*Nycticebus coucang*). PhD thesis, Harvard University, unpubl.

Klopfer, P. H. (1977). Communication in prosimians. In T.A. Seboek (Ed.), *How Animals Communicate* (pp. 841-850) Bloomington: Indiana University Press.

Krebs, J.R. and Davies, N.B. (1993). An Introduction to Behavioural Ecology. Blackwell Scientific Publ., Oxford.

Konishi, M. (1973). Locatable and nonlocatable acoustic signals for barn owls. *American Naturalist, 107,* 775-785.

Kuhn, M. (1989). Verhaltensbiologische Untersuchungen zur Ontogenese von Mausmakis (*Microcebus murinus*) mit besonderer Berücksichtigung der akustischen Kommunikation. Diplomarbeit, University of Stuttgart-Hohenheim, unpubl.

Kugel, S. (1993). Verhaltensbiologische und bioakustische Untersuchungen zur strukturellen Variabilität des Soziallauts "Triller" beim Mausmaki (*Microcebus murinus*). Diplomarbeit, University of Stuttgart-Hohenheim, unpubl.

Lieberman, P. (1984). The Biology and Evolution of Language. Harvard Univ. Press, Cambridge.

Littlejohn, M.J. (1981). Reproductive isolation: A critical review. In W.R. Atchley and D.S. Woodruff (Eds.), *Evolution and Speciation* (pp. 298-334). Cambridge: Cambridge Univ. Press.

Macedonia, J.M. and Stanger, K.F. (submitted). Phylogeny of the Lemuridae revisited: Evidence from communication signal structure, *Folia Primatologia.*

Mascagni, O. and Doyle, G.A. (1993). Infant distress vocalizations in the South African Lesser bushbaby (*Galago moholi*), *International Journal of Primatology, 14,* 41-60.

Martin, R.D. (1972). A preliminary field-study of the Lesser Mouse Lemur (*Microcebus murinus*), *Advances in Ethology* , *9,* 43-90.

Martin, R.D. (1990). Primate Origins and Evolution. Chapman and Hall, London.

Marler, P. (1955). Characteristics of some animal calls. *Nature* , 176, 6-8.

Masters, J.C. (1991). Loud calls of *Galago crassicaudatus* and *G. garnettii* and their relation to habitat structure. *Primates* , *32,* 153-167.

Michelsen, A. and Larsen, O.N. (1983). Strategies for acoustic communication in complex environments. In F. Huber and Markl, H. (Eds.), *Neuroethology and Behavioral Physiology (pp. 321-331)*. Heidelberg : Springer.

Nash, L.T, Bearder, S.K. and Olson, T.R. (1989). Synopsis of Galago species characteristics. *International Journal of Primatology, 10*, 57-80.

Pages, E. (1980). Ethoecology of *Microcebus coquereli* during the dry season. In P. Charles-Dominique, H.M. Cooper, A. Hladik, C.M. Hladik, E. Pages, G. Pariente, A. Petter-Rousseaux and A. Schilling (Eds.), *Nocturnal Malagasy Primates* (pp. 97-116). New York: Academic Press.

Pariente, G.F. (1979). The role of vision in prosimian behavior. In G.A. Doyle and R.D. Martin (Eds.), The Study of Prosimian Behavior (pp. 411-459). New York: Academic Press.

Patersen, H.E.H. (1980). The recognition concept of species. In E.S. Vrba (ed.), *Species and Speciation* (pp. 21-29). Pretoria: Transvaal Museum Monograph No. 4.

Perret, M. (1992). Environmental and social determinants of sexual function in the male Lesser Mouse Lemur (*Microcebus murinus*). *Folia primatologica, 59*, 1-25.

Petersen, M.R., Beecher, M.D.; Zoloth, S.R., Green, S., Marler, P., Moody, D.B. and Stebbins, W.C. (1984). Neural lateralization of vocalizations by Japanese macaques: Communicative significance is more important than acoustic structure, *Behavioral Neuroscience, 98*, 779-790.

Petter, J.-J. and Charles-Dominique, P., 1979, Vocal communication in prosimians. In G.A. Doyle and R.D. Martin (Eds.), The Study of Prosimian Behavior (pp. 247-305). New York: Academic Press.

Petter-Rousseaux, A. (1975). Activite sexuelle de *Microcebus murinus* soumis a des regimes photoperiodiques experimentaux, *Annales de Biolgie Animale Biochimie Biophysique, 153*, 503-508.

Richard, A.F.(1987). Malagasy prosimians: Female dominance. In B.B. Smuts, D.L. Cheney, R.M. Seyfarth, R.W. Wrangham and T.T. Struhsaker (Eds.), *Primate Societies*, (pp. 25-33). Chicago:Univ. Chicago Press.

Rumpler, Y. and Dutrillaux, B. (1990). Chromosomal evolution and speciation in primates. *Cell Biological Reviews, 23*, 1-112.

Rumpler, Y. (1992). Die Klassifikation der Lemuren. In V. Ceska, H.-U. Hoffmann and K.-H. Winkelsträter (Eds.) *Lemuren im Zoo (pp. 71-77). Berlin:* Verlag Paul Parey.

Seyfarth, R.M., Cheney, D.L. and Marler, P. (1980). Vervet monkey alarm calls: Semantic communication in a free-ranging primate. *Animal Behaviour, 28*, 1070-1094.

Smith, W.J. (1977). The Behavior of Communicating. Harvard Univ. Press, Cambridge.

Snowdon, C.T. (1986). Vocal Communication. In G. Mitchell and J. Erwin (Eds.) *Comparative Primate Biology* (pp. 495-530). New York: Alan R. Liss.

Snowdon, C.T., Brown, C.H. and Petersen, M.R. (1982). Primate Communication. Cambridge Univ. Press, Cambridge.

Stanger, K.F. (1988). Lautrepertoire und begleitendes Verhalten von nachtaktiven Halbaffen. Diplomarbeit, University of Tübingen, unpubl.

Stanger, K.F. (1993). Structure and function of vocalizations of nocturnal prosimians. PhD thesis, University of Tübingen, unpubl.

Sutton, D. and Jürgens, U. (1986). Neural control of vocalization. In H.D. Steklis and J.Erwin (Eds.), *Comparative Primate Biology Vol. 4* (pp. 625-648). New York: Alan R. Liss Inc.

Todt, D., Goedeking, P. and Symmes, D. (1988). Primate Vocal Communication. Springer Verlag, Heidelberg.

Wiley, R.H. and Richards, D.G. (1978). Physical constraints on acoustic communication in the athmosphere: Implications for the evolution of animal vocalizations. *Behavioral Ecology and Sociobiology, 3*, 69-94.

Zimmermann, E. (1981). First record of ultrasound in two prosimian species. *Naturwissenschaften, 68*, 531.

Zimmermann, E. (1985a). Vocalizations and associated behaviours in adult Slow Loris (*Nycticebus coucang*). *Folia primatologia, 44*, 52-64.

Zimmermann, E.(1985b). The vocal repertoire of the adult Senegal bushbaby (*Galago senegalensis senegalensis*). *Behaviour , 94*, 211-233.

Zimmermann, E. (1989a). Reproduction and behavioral development in slow loris (*Nycticebus coucang*), *Human Evolution , 4*, 171-179.

Zimmermann, E. (1989b). Aspects of reproduction, behavioral and vocal development in Senegal bushbabies (*Galago senegalensis*). *International Journal of Primatology, 10*, 1-16.

Zimmermann, E. (1990). Differentiation of vocalizations in bushbabies (Galaginae) and the significance for assessing phylogenetic relationships. *Z.eitschrift fuer zoologische Systematik und Evolutionsforschung, 28* , 217-239.

Zimmermann, E. (1991). Ontogeny of acoustic communication in prosimian primates. In A. Ehara, T. Kimura, O., Takenaka and M. Iwamoto (Eds.) *Primatology today* (pp. 337-340). Amsterdam, Elsevier Science Publ.

Zimmermann, E. (1992). Vocal communication by nonhuman primates. In R.D. Martin, D. Pilbeam and J.S. Jones (Eds.), *The Cambridge Encyclopedia of Human Evolution* (pp. 124-128). Cambridge: Cambridge Univ. Press.

Zimmermann, E., Bearder, S.K., Doyle, G.A. and Andersson, A. (1988).Variations in vocal patterns of Senegal and South African lesser bushbabies and their implications for taxonomic relationships. *Folia primatologia, 51*, 87-105.

Zimmermann, E. and Lerch, C. (1993). The complex acoustic design of an advertisement call in male mouse lemurs (*Microcebus murinus*) and sources of its variation. *Ethology , 93*, 211-224.

SPECIES DIVERSITY AMONG GALAGOS WITH SPECIAL REFERENCE TO MATE RECOGNITION

Simon K. Bearder, Paul E. Honess and Lesley Ambrose

Anthropology Unit
Department of Social Sciences
Oxford Brookes University
Oxford OX3 0BP, U.K.

INTRODUCTION

This paper argues that the number of species of nocturnal primates has been seriously underestimated. It has been traditional to separate species largely on the basis of the physical characteristics of museum specimens. This is satisfactory when the animals' themselves recognize one another by sight, as in the majority of birds and day-living primates, but where scent and sound play the predominant role in male-female recognition, species boundaries can easily be overlooked. As Tinbergen has pointed out, closely related species do not ordinarily interbreed because: 'the various signals serving attraction, persuasion, appeasement, and synchronisation, are so very different from one species to another.' (Tinbergen, 1953, p.36). Here we use one conspicuous element in the communication system of galagos, which are all nocturnal, to demonstrate that similar-looking populations frequently belong to different species which have yet to be investigated.

Galagos (also known as bushbabies) are distributed throughout most of Africa south of the Sahara in forests, woodlands and savannahs. Between 1977 and 1986 most authorities revised their estimates of the number of species from six to eleven, largely on the basis of the work by Olson on museum specimens (Olson, 1979; Nash *et al.*, 1986). Since then, field and laboratory studies of a number of populations indicate that this figure remains an underestimate. Our data confirm that there is some way to go before the number of species can be stated with confidence.

We aim to demonstrate the use of 'vocal fingerprinting' as a practical means of distinguishing among species in the wild. Data on calling patterns collected over the past 26 years are summarized to illustrate the most conspicuous differences among species and to provide a baseline for the future identification of newly-found populations. We discuss ways in which our results can be verified, and a guide to productive areas for future fieldwork.

Since no two classifications agree on the specific and generic naming of galagos we have taken two of the most recent interpretations as a starting point against which to test our conclusions (see Table 1 and Nash *et al.*, 1989). We begin by dividing the galagos into six groups based on calls given in the context of reassembly at dawn, which are unambiguously related to mutual attraction. We then demonstrate a set of the loudest calls in the repertoire of well-established species remains more-or-less constant between widely separated populations. These calls can therefore be used to define a vocal fingerprint typical of each species. A comparison of vocal fingerprints for all populations that have been studied acoustically reveals qualitatively different repertoires in 16 taxa. We suggest that the technique provides a robust and convenient means of censusing many nocturnal primates for taxonomic, phylogenetic and conservation purposes.

Table 1. Recent classifications of the galagos. Common names given are those used in the text. Species identified by this study are also listed.

Olson (1979)	**Groves (1989)**
1. *Otolemur crassicaudatus* (Large-eared Greater)	1. *Otolemur crassicaudatus*
2. *O. garnettii* (Garnett's)	2. *O. garnettii*
3. *Galago elegantulus* (Elegant/Needle-clawed)	3. *Euoticus elegantulus* (Southern elegant)
4. *G. matschiei* (Eastern Needle-clawed)	4. *E. pallidus* (Northern elegant)
5. *G. senegalensis* (Senegal)	5. *Galago matschiei*
6. *G. gallarum* (Somali)	6. *G. senegalensis*
7. *G. moholi* (Mohol/Southern Lesser)	7. *G. gallarum*
8. *Galagoides alleni* (Allen's)	8. *G. moholi*
9. *G. zanzibaricus* (Zanzibar)	9. *G. alleni*
10. *G. thomasi* (Thomas's)	10. *Galagoides zanzibaricus*
11. *G. demidoff* (Demidoff's)	11. *G. demidovii*
	Additional species (this study)
	1. Makande Allen's
	2. Newala small/Kalwe small
	3. Matundu dwarf
	4. Rondo dwarf
	5. Amani dwarf

METHODS

This account is based on tape recordings of galagos in the wild. Table 2 summarizes the locations that have been sampled on a country by country basis from South to North and lists the taxa that have been identified. Additional information is available from the literature and from preliminary sound recordings which are not sufficiently loud for sonagraphic display. Even poor quality tape recordings have been used to help identify the caller by comparison with a reference library of calls from known species, but only good quality recordings are shown here. Uher (4200 Report Monitor) and Marantz (PMD 222) tape recorders were used with Sennheiser (MKH 816T) and Audio-technica (AT 9350) directional condenser microphones.

When using recordings from the field, only the loudest calls in the repertoire are suitable for display because of generally high levels of background noise. These calls have been given sound-descriptive labels and divided into three broad categories for each study population, based on the context in which they were heard: A) 'Advertising' calls facilitating contact, gathering and display; B) 'Mild alarm' when disturbed, for example by a conspecific, an observer or a change in the environment; C) 'Intense alarm' when extremely agitated in the presence of danger. These categories are not intended to imply a hard and fast function to the calls in question (see Discussion). Qualitative differences among calls are illustrated on oscillograms produced with a Digital Sound Analyzer (MC 6800, A. Zimmermann, 1985; E. Zimmermann, 1990). For direct comparison, all calls are shown on the same scale using a 1 kHz display. No attempt is made to analyze the call structure quantitatively at this stage, although this work is in progress. The purpose of this initial treatment is to highlight gross structural differences in the call repertoires.

We employ the following working definitions of terms:

Species: Members of a species share a repertoire of common loud calls that differs qualitatively from the repertoires of related species.

Call: A sound or series of sounds of a distinct type.

Unit: The basic element of a call that is represented as a continuous tracing along the temporal axis of an oscillogram.

Phrase: A group of units that is separated from other similar groups by a time interval greater than the time interval separating the units within a phrase.

Where we are unsure about the scientific name associated with a particular population, or where no name exists, we use the name of the first study site where tape recordings were made, together with a size category (dwarf <90 g; small 90-300 g; medium 300-600 g; large >600 g).

Table 2. Sites at which populations of galagos have been sampled acoustically (* = tape recordings not suitable for sonagraphic display).

Key to species: 1. Elegant; 2. Makande Allen's; 3. Korup Allen's; 4. Eastern needle-clawed; 5. Senegal; 6. Southern lesser; 7. Large-eared greater; 8. Garnett's; 9. Matundu dwarf; 10. Rondo dwarf; 11. Amani dwarf; 12. Newala small; 13. Zanzibar; 14. Kalwe small; 15. Thomas's; 16. Demidoff's.

Country	Area	Map Reference	Species Code (see above)
South Africa	Eshowe, Natal	23°53'S; 31°28'E	7
	Ofcolaco, Transvaal	23°80'S; 30°15'E	7, 6
	Naboomspruit, Transvaal	24°35'S; 26°27'E	6
Zimbabwe	Mutare	18°90'S; 32°05'E	7*
Malawi	Lilongwe	13°59'S; 33°47'E	6*
	Kalwe, Nkata Bay	11°36'S; 34°18'E	7, 14
Tanzania	Mtopwa, Mtwara	10°40'S; 39°23'E	7, 8, 12, ?
	Ziwani, Mtwara	10°20'S; 40°15'E	8, 10
	Mtwara	10°15'S; 40°20'E	7, 8, 12
	Narunyu, Lindi	10°07'S; 39°35'E	7
	Rondo, Lindi	10°05'S; 39°25'E	8, 10, 12
	Litipo, Lindi	10°15'S; 40°20'E	7, 8, 12
	Lindi	10°00'S; 39°40'E	7, 8
	Matundu, Kilombero	08°15'S; 36°15'E	9
	Nambiga, Ulanga	08°35'S; 36°30'E	12
	Bagilo, Morogoro	06°55'S; 37°40'E	? *
	Kiamboza, Morogoro	07°00'S; 37°45'E	8, 9
	Kiono/Zaraninge	06°10'S; 38°35'E	?
	Msumbugwe	05°32'S; 38°45'E	9
	Amani West, Muheza	05°07'S; 38°35'E	8, 11*
	Moshi	03°20'S; 37°25'E	8
Kenya	Diani, Mombasa	04°15'S; 39°36'E	8, 13
	Kibwezi, Tsavo	02°25'S; 37°58'E	5*
Uganda	Bwindi	01°00'S; 29°58'E	4, 15, ?
	Kibale, Fort Portal	00°30'N; 30°11'E	4, 15
	Semliki, Bundabugio	00°49'N; 30°09'E	15, 16
Gabon	Franceville	01°40'S; 13°35'E	15, 16
	Makande	00°45'S; 11°55'E	1, 2, 15, 16
	Makokou	00°30'N; 12°50'E	1, 2, 15, 16
Bioko	Grand Caldera	03°35'N; 08°45'E	3, 15
Cameroon	Korup, Mundemba	04°55'N; 08°53'E	3, 15, 16
	Korup, Nguti	05°20'N; 09°25'E	1, 3, 15, 16
	Mt. Kupe	05°05'N; 09°50'E	1, 3, 15, 16

RESULTS

Sound recordings were obtained from 68 galago populations. The animals within each population give a diversity of call types, of which three or four are usually sufficiently loud to allow graphic display. We have divided these populations into six groups based on structural distinctions in the calls associated with contact and reassembly (Table 3). Examples of calls within each group are illustrated in Figure 1.

Table 3. A key to identifying galagos based on the structural characteristics of their 'advertising' calls, in order of increasing complexity (see Figure 1).

Group 1: Click calls - Elegant galagos

Very brief, single units given at irregular intervals with no obvious pattern (Fig.1.1).

Group 2: Repetitive Calls - Lesser galagos

Single units or short phrases arranged at regular intervals into a series of highly variable length, from tens to hundreds of repetitions. Units within the series may be given at higher or lower intensity (volume) depending on the excitation of the caller (Figures 1.2-1.6).

Group 3: Trailing Calls - Greater galagos

A patterned sequence of units arranged into a phrase which has a loud beginning and trails away over the last few units. Calls are given singly (Figures 1.7-1.8).

Group 4: Rolling Calls - Southern dwarf galagos

Phrases in which single, double or triple units gradually increase in intensity and trail away. Calls are generally given in shorter bouts than in the case of Repetitive Calls. Phrases may be repeated a variable number of times and the first phrase is generally the longest (Figures 1.9-1.10).

Group 5: Incremental calls - Zanzibar galagos

A patterned phrase in which the units are arranged into distinct groups or 'unit sequences' which tend to get longer as the call progresses. Calls are given singly (Figures 1.11-1.13).

Group 6: Crescendo Calls: - Dwarf galagos

Phrases in which the units increase in intensity and speed of repetition leading to a marked crescendo. Each crescendo is given singly or repeated in a patterned sequence (Figures 1.14-1.15).

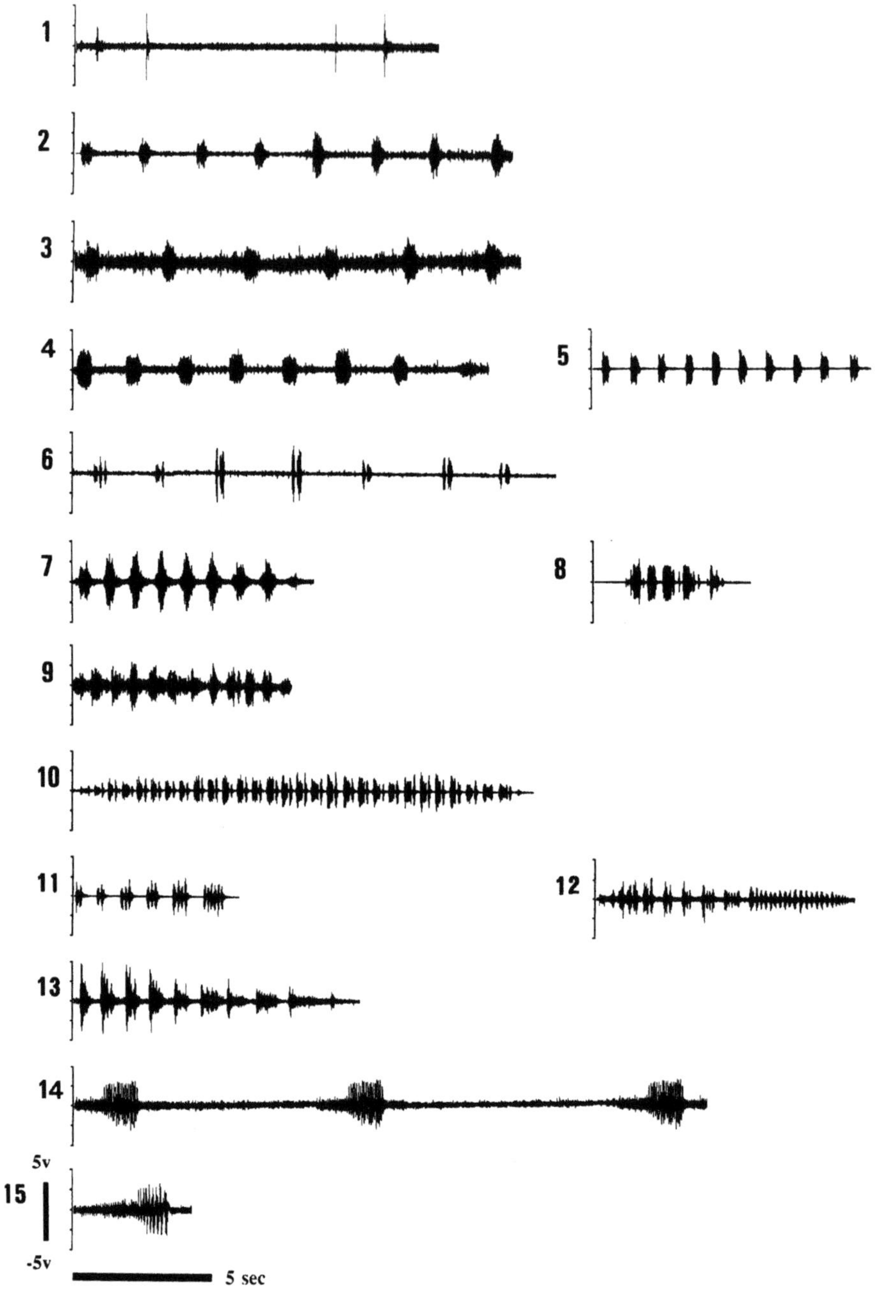

Figure 1. Oscillograms of the vocal advertisements of different galago species (Table 3). Key to species: 1. Elegant; 2. Makande Allen's; 3. Korup Allen's; 4. Eastern needle-clawed; 5. Senegal; 6. Southern lesser; 7. Large-eared greater; 8. Garnett's; 9. Matundu dwarf; 10. Rondo dwarf; 11. Newala small; 12. Zanzibar; 13. Kalwe small; 14. Thomas's; 15. Demidoff's.

The Loud-call Repertoires of Different Taxa

Figures 2-15 illustrate the loud call repertoires of a number of different populations for which recordings are available from the wild. In three cases we have included repertoires from different parts of a species geographical range to illustrate within-species variation. Species and proposed species having qualitatively different vocal repertoires are given below. Within each group, taxa are listed in order of body size (from large to small).

Group 1: Galagos with Click Calls (1 species). Elegant or Needle-clawed galagos can be distinguished from all other species by their reassembly call, a high pitched click or 'tsic' of very brief duration. The call repertoire from the Korup National Park in Cameroon is illustrated in Figure 2. Preliminary results indicate that the calls of populations in Gabon are very similar to those in Cameroon.

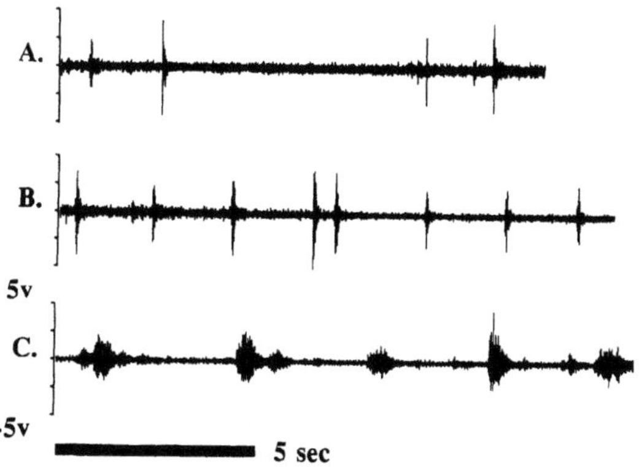

Figure 2. Oscillograms of the loud calls of Elegant galago, Korup, Cameroon: A) Click or 'tsic'; B) Shrill Yap; C) Screech yap.

Group 2: Galagos with Repetitive Calls (5 species). This group includes two species of Allen's galagos (Makande, Gabon and Korup, Cameroon), the Eastern Needle-clawed, and the Senegal and Mohol/Southern lesser galagos (Figures 3-6). A sixth species allied to this group (the Somali galago) has not been sampled. The calls of the two Allen's galagos are very different from each other with the exception of the 'advertising' calls which are broadly similar. This is contrary to the general rule whereby such calls tend to be highly characteristic.

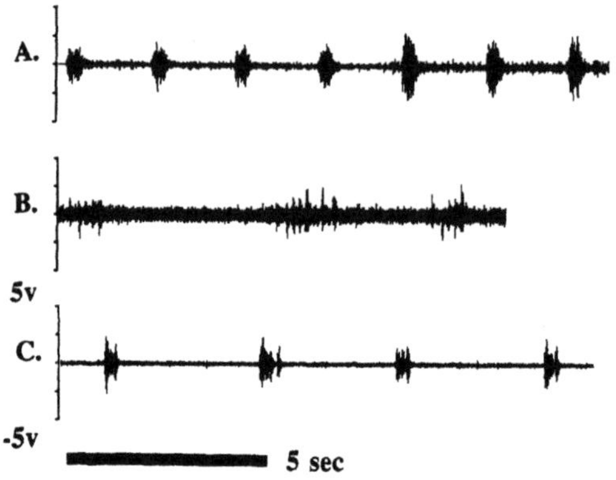

Figure 3. Oscillograms of the loud calls of 'Allen's' galago, Makande, Gabon: A) Croaks; B) Piping rattle (recording unconfirmed, see Charles-Dominique, 1977); C) Unit whistle.

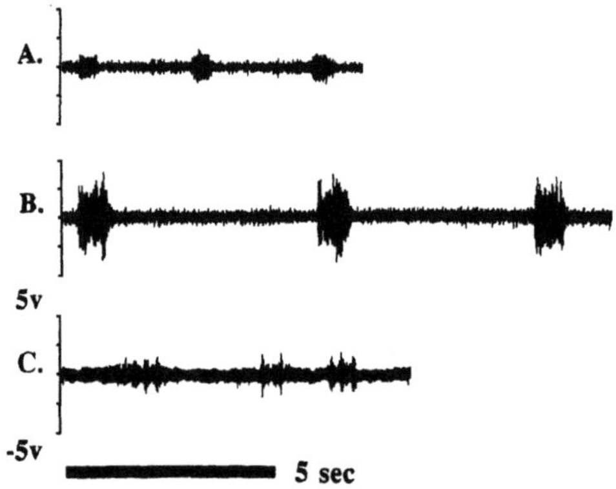

Figure 4. Oscillograms of the loud calls of Allen's galago, Korup, Cameroon: A) Croak; B) Mew whistle; C) 'Quok' yap.

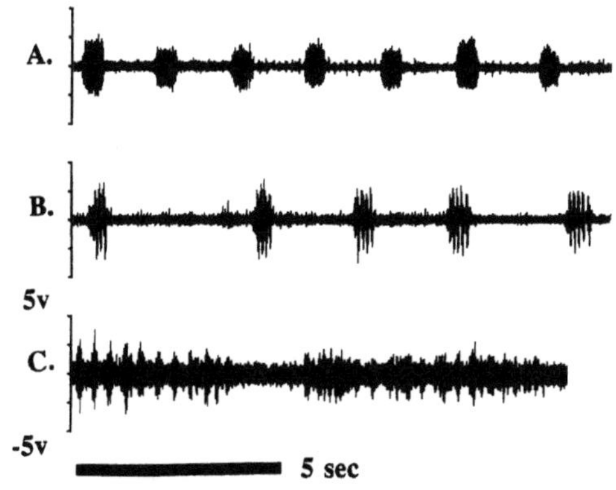

Figure 5. Oscillograms of the loud calls of Eastern Needle-clawed galago, Kibale, Uganda: A) Bark; B) Grunt-yap; C) Screech yap.

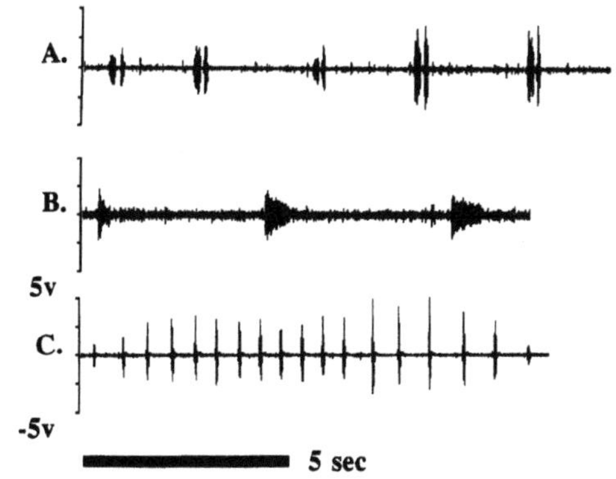

Figure 6. Oscillograms of the loud calls of Southern lesser galago, Naboomspruit, S. Africa: A) Bark; B) Whistle; C) Yap.

Group 3: Galagos with Trailing Calls (2 species). Both the Large-eared Greater and Garnett's galago have similar vocal repertoires. Their geographical ranges overlap over a large part of East Africa and they can be most easily distinguished by their contrasting 'advertising' calls (Figures 7 and 8). Our surveys have revealed the presence of a pygmy form of large-eared greater galagos living in southern Tanzania in the same region as individuals of normal size, but we cannot detect a qualitative difference in the vocalisations of the two forms (Crouse, 1988).

Group 4: Galagos with Rolling Calls (3 species). Recent fieldwork in Tanzania has revealed the presence of three kinds of Southern dwarf galagos that merit separation into three species on the basis of their divergent vocal repertoires (Figures 9 and 10). These are the Matundu, Rondo and Amani dwarf galagos. They differ from previously known dwarf species in lacking a crescendo call (see Group 6). Their different call repertoires correlate with differences in behavior and habitat preference but little is yet known about their anatomy. The calls of the Amani population are audibly distinct from the other two but very high levels of background noise preclude graphic display.

Group 5: Galagos with Incremental Calls (2/3 species). This group includes three populations of small galagos surveyed: Newala in South East Tanzania; Diani near Mombasa, Kenya; and the Kalwe Forest near Nkata Bay, Malawi (Figures 11-13). It is probable that the Diani population are Zanzibar galagos, although we do not have call samples from Zanzibar island to verify this. The other two populations also exhibit incremental calls but of a somewhat different kind. Their calls of anxiety and alarm are similar and their separation may prove to be unwarrented.

Group 6: Galagos with Crescendo Calls (2 species). Two species are presently known for this group: Thomas's and Demidoff's dwarf galagos. They frequently live together in the same habitat and are so similar in size and appearance that they have usually been lumped together as one species (Charles-Dominique, 1977; Corder, 1989). In fact, Demidoff's dwarf galago is somewhat smaller and prefers secondary growth within forest or forest edge, seldom moving above five metres high, whereas Thomas's generally remains above five metres in primary vegetation. We have been able to demonstrate that Demidoff's galagos maintain a relatively constant vocal fingerprint within an extensive geographical range (Figure 15) but we are less certain about variability among populations presently attributed to Thomas's galagos. Calls from Kibale Forest in Uganda are illustrated here (Figure 14). The most conspicuous differences in comparison with Demidoff's calls are: 1) an additional reassembly call (trill); 2) a multiple crescendo call in which the first crescendo is generally the loudest; and 3) absence of the distinctive buzz calls of Demidoff's galagos.

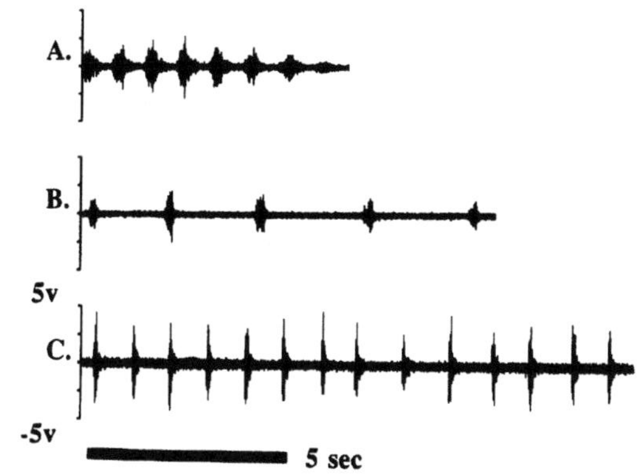

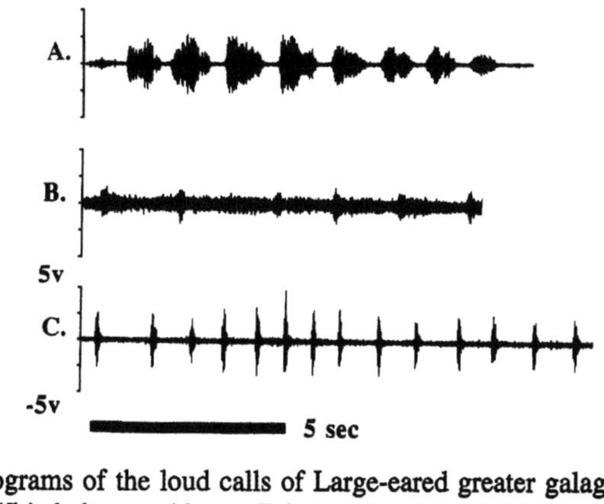

Figure 7. Oscillograms of the loud calls of Large-eared greater galago: A) Cry; B) Squawk; C) Whistled-yap. Above: Eshowe, S. Africa. Below: Ofcolaco, S. Africa.

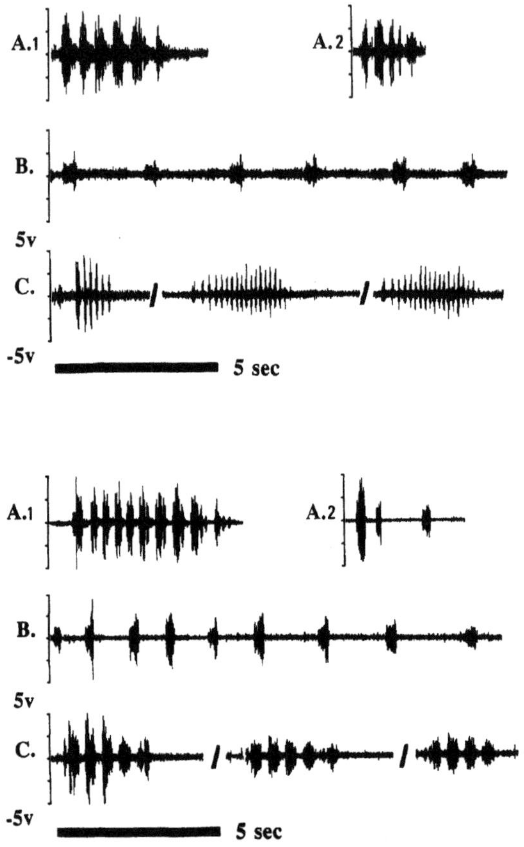

Figure 8. Oscillograms of the loud calls of Garnett's greater bushbabies: A1) Trailing call; A2) Short-squawk; B) Squawk; C) Cackle. Above: Rondo, Tanzania. Below: Diani, Kenya.

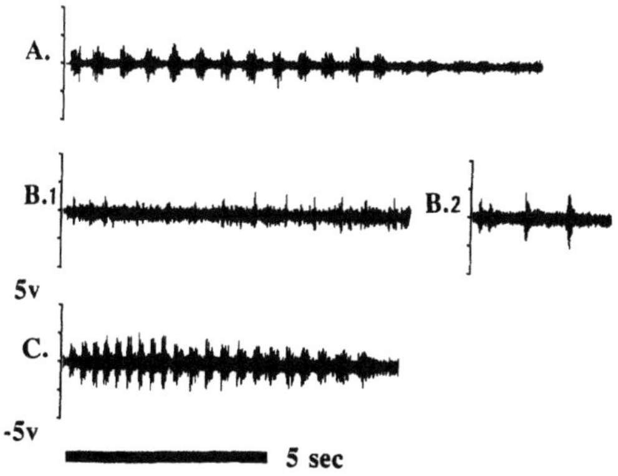

Figure 9. Oscillograms of the loud calls of Matundu southern dwarf galagos, Tanzania: A) Single unit rolling call; B1) Yap; B2) Shrill buzz; C) Descending shriek.

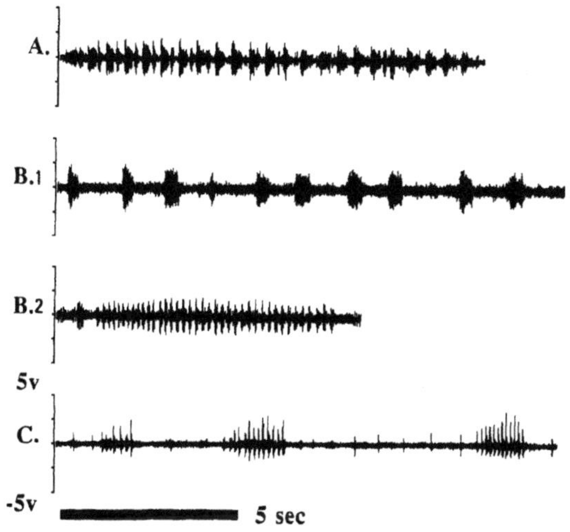

Figure 10. Oscillograms of the loud calls of Rondo southern dwarf galagos, Tanzania: A) Double unit rolling call; B1) Screech-grunt; B2) Buzz and Rapid whistle; C) Yaps and Shivering twitter.

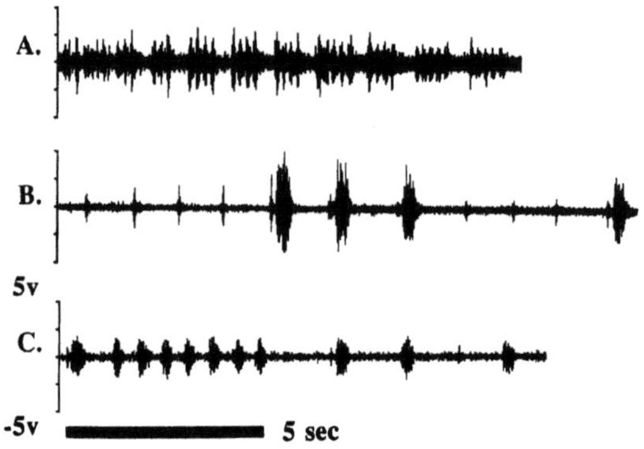

Figure 11. Oscillograms of the loud calls of Newala small galagos, Tanzania: A) Incremental call; B) Yap and Screech; C) Descending Screech.

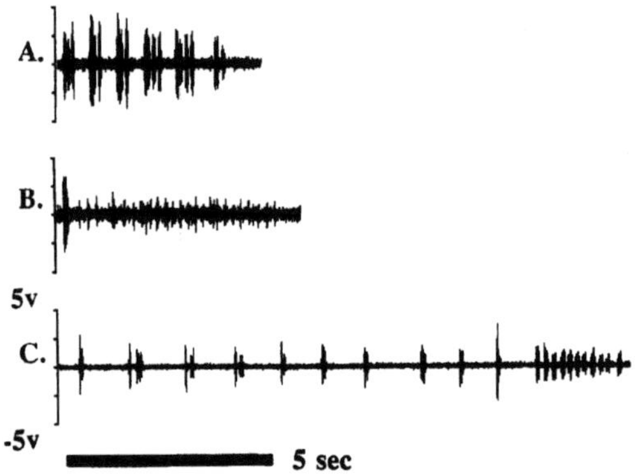

Figure 12. Oscillograms of the loud calls of Diani small (Zanzibar) galagos, Kenya: A) Incremental call; B) Buzz and Rapid chatter; C) Yap and Chirrup.

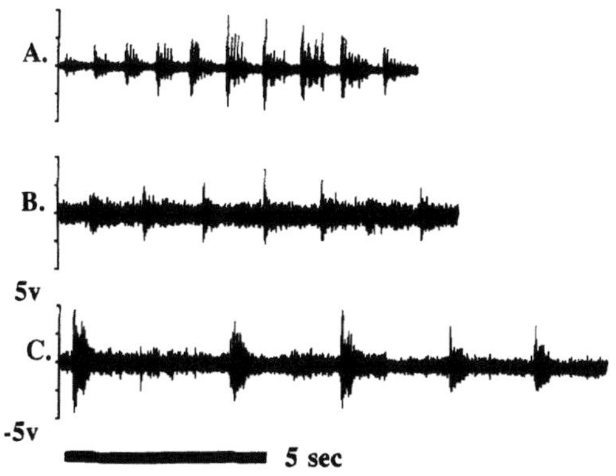

Figure 13. Oscillograms of the loud calls of Kalwe small galagos, Malawi: A) Incremental call; B) Yap; C) Screech.

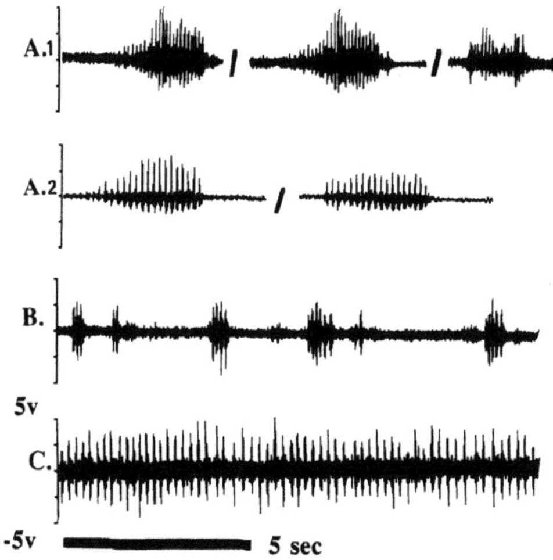

Figure 14. Oscillograms of the loud calls of Thomas's dwarf galago, Kibale, Uganda: A1) Multiple crescendo; A2) Trill; B) Screech-yap; C) Rapid yap.

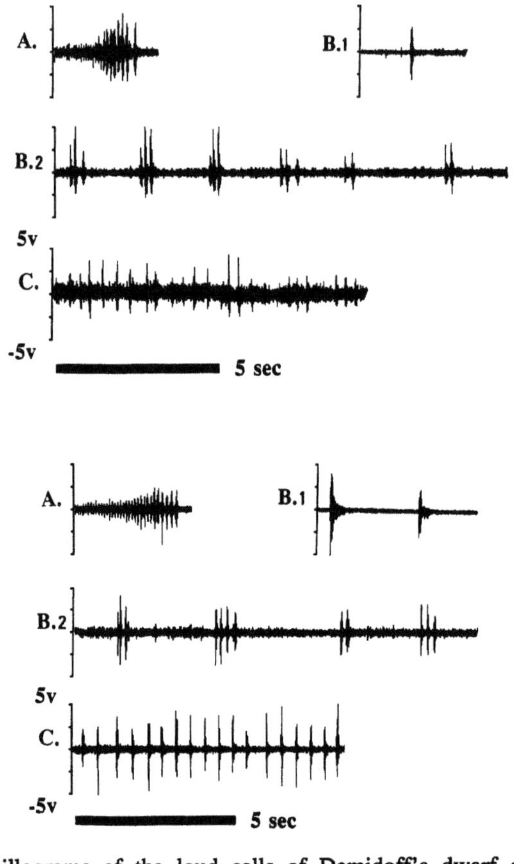

Figure 15. Oscillograms of the loud calls of Demidoff's dwarf galagos: A) Single crescendo; B1) Buzz; B2) Unit yap; C) Rapid yap. Above: Semliki, Uganda. Below: Korup, Cameroon.

DISCUSSION

There is little controversy about six of the species listed in Table 1 (Olson, 1979: 1, 2, 4, 5, 7 and 11). Taking these species as a starting point, it is clear that their loud-call repertoires vary in a species-typical way, with relatively little divergence among different parts of the species geographical range. It follows that qualitative differences in calling patterns can be used to establish the validity of those taxa about which there is still debate and to indicate the existence of previously unrecognized species. This discussion focuses on four aspects of the use of vocalizations to identify species: the problems encountered with fieldwork; the theoretical basis for relying on vocal fingerprints; how species should be named and grouped; and a review of future avenues of research which can be used to extend and verify our initial conclusions.

The Advantages and Shortcomings of Fieldwork on Vocalizations

There are practical reasons why it can be difficult to compare vocal repertoires recorded under different conditions, even when using identical equipment. In the field it is rarely possible to ensure that there are no obstructions between the animal and the microphone and background noise is almost inevitable, especially in forests. The distance between the caller and the microphone usually means that the higher frequencies of the call are not recorded, making the frequency range (and the number of harmonics) quite different from calls taped under ideal laboratory conditions. Nevertheless, the variety and number of calls heard in the constantly changing conditions of the wild provides good opportunities for obtaining recordings and for interpreting the context of the calls in comparison with captivity. In addition, the number of populations *of known origin* that can be sampled is considerably greater. It is for these reasons that we have chosen to compare only broad features of the loudest calls in the repertoire of each species and to use only calls recorded from documented sites in the wild.

The most obvious practical advantage of using vocalizations is that they can be collected for later analysis by anyone with a tape recorder without disturbing or harming the animals. In our experience, while some species of galago are easily trapped, others are almost impossible to capture. In these cases it is difficult to obtain body measurements and samples (skin, hair, blood) for comparative study along other lines, and tape recordings provide a useful alternative.

A problem encountered when trying to compare the calls of different species is deciding how to divide the repertoire. Galago calls are complex, with discrete calls and graded series that can be mixed together in highly variable sequences in terms of volume and speed of repetition, which are associated with changing levels of arousal. This is related to poorly understood aspects of the anatomy and physiology of sound production. For example, galagos are capable of making a call when breathing in as well as when breathing out, which can give the effect of two animals calling instead of one. The division of calls by function is also difficult on occasions. For example, calls made in the context of disturbance by a predator also appear to carry information about the identity and position of the caller. They can induce counter-calling in other galagos and evidently have a dual function. We have therefore chosen to avoid terms that imply hard and fast functions to each call. Nevertheless, calls can be divided into recognizable structural categories and grouped according to their predominant context for the purposes of comparison.

The Validity of Vocal Fingerprinting

Paterson's concept of a species highlights the importance of a mutually tuned system of communication shared by the males and females of each species (Paterson, 1985, see Vbra, 1985; Masters, 1989; 1993). Among galagos, which are nocturnal and highly vocal, a significant part of their Specific-Mate Recognition System (SMRS) evidently consists of vocalization. Zimmermann and Stanger, among others, have shown that vocal structure provides valuable clues to the phylogeny of galagos and cheirogaleids (Zimmermann, 1990; Stanger, 1993; chapters in this volume) but there is relatively little information on species-specific repertoires and virtually nothing on local dialects (Tattersall, 1989). This paper is a contribution in this direction.

The galagos represent a group of 'cryptic species' that are relatively difficult to separate on the basis of their morphology. They recognise potential mates predominantly by sound and/or scent with sight playing a secondary role. Consequently, they may show considerable within-species variation in size and coloration (as in the case of Large-eared greater galagos which weigh between 600 g and 2 kg, and may be brown, grey, silver or black). Conversely, species may look so alike that it is extremely difficult to tell them apart, even on close examination of living animals (as in the case of Demidoff's and Thomas's dwarf galagos which live together in the same habitat). When calls play a major role in the communication system of such species they usually provide an excellent means of identification, and there are many well-established examples among insects (Alexander, 1968; Toms, 1993), amphibians (Littlejohn and Oldham, 1968; Ralin, 1977), birds (Lanyon, 1969; Marler, 1975) and mammals, including both nocturnal and diurnal primates (Zimmermann *et al.*, 1988; Stanger, 1993; Gautier, 1988; Oates and Trocco, 1983). One problem has been a reluctance among some traditional taxonomists to accept the validity of specific distinctions based on behavioral data. Yet it is behavior in the form of relatively stereotyped displays and signalling systems that ensures that members of a species are brought together to reproduce. Furthermore, components of the sound signals are determined by structures that appear to be species-specific (e.g. vocal cords, vocal tract and neuronal pathways) (Evans, 1982; Liebermann, 1984). Differences in these structures are reflected in the characteristics of the calls which can be illustrated and measured on a computer sonagraph.

Among galagos, all populations studied to date have one or more 'advertising' calls which play an important role in social coordination (Petter and Charles-Dominique, 1979). These calls may be given by both sexes shortly after leaving the sleeping place, when gathering to sleep in groups at dawn and when contacting companions during the night. The same calls are often given when interacting with rivals of the same sex, sometimes involving long bouts of counter-calling. They are particularly common at times of mating and birth, when social interactions are paramount. Not surprisingly, the structural characteristics of 'advertising' calls appear to provide a good guide to species identity, and we have used these calls in the first instance. In some cases, however, it can be difficult to decide at what point the variation between the advertising calls reflects a specific boundary as opposed to a local dialect. It is for this reason that we have extended the comparison to include qualitative differences between all loud calls within the repertoires of each population, providing a more rigorous guide to the extent of separation.

The Naming of Species and their Phylogeny

We have not yet assigned scientific names to the newly uncovered species, not because we lack confidence in their validity, but to avoid repeating the mistake of applying labels that may later prove misleading. For example, the names *orinus* and *granti* have been associated with dwarf and small galagos in East Africa (Lawrence and Washburn, 1936; Thomas and Wroughton, 1907). It is possible that the populations that we have referred to as Newala small and Kalwe small galagos will turn out to be Grant's galago (*Galagoides granti*), but this remains to be confirmed. At present there are three contenders for the name *orinus* (the Matundu, Rondo and Amani dwarf galagos) or it may turn out to be none of these (our visit to Bagilo, where the type specimen was collected, proved inconclusive). A total of 58 names have been applied to galago populations in the past and our data show that a major new classification will be necessary in the future. The time for that is still some way off.

Finally, we have avoided drawing inferences about the evolutionary relationships of galagos based on their calls. Our six groups of 'advertising' calls conform, to some extent, with suggested genera (*Euoticus, Galago, Otolemur, Galagoides*) but in view of the imminent use of DNA analysis it seems preferable to wait for more precise information before speculating further about galago phylogeny (Brown *et al.*, 1982; Hillis and Moritz, 1990). One initial result that is of interest is a broad correlation between species diversity and niche width. Galagos which are able to survive in a range of habitats appear less likely to undergo speciation (and therefore have fewer close relatives) than those which are prone to isolation (see Crompton, 1989; Tattersall, 1989). This is true, for example, for Garnett's galago in comparison with species of the Zanzibar group living in the same areas.

Vocal Fingerprinting in the Broader Context

Now that there is serious doubt about the existing descriptions of different species it is important to examine other lines of evidence that can confirm or reject the divisions based on the patterns of calling. The newly discovered species should co-vary in their chromosome arrangement, details of anatomy and reproductive biology, and in their biochemical structure, revealed by studies of the DNA. Senegal and Southern lesser galagos, for example, have been shown to vary in their calls, chromosome number and reproductive biology (Nash *et al.*, 1989; Izard and Nash, 1989) despite the fact that they are almost indistinguishable in the field, even to a trained observer. It seems likely that many nocturnal primates and other mammals will have been lumped together although they are unable to interbreed. Vocalizations provide a convenient clue to boundaries between species, but many mammals (including lorises and pottos) are largely silent, and it will probably take much longer to determine the true extent of speciation in these groups. Our research is expanding to include mitochondrial DNA analysis on hair and tissue samples from the field and museums in collaboration with the Institute of Zoology, London. In this way it will soon be possible to measure genetic distance among species and populations and to better assess their phylogenetic affinities (Paabo, 1989; Thomas *et al.*, 1990). The widespread distribution of a large number of closely related galago species will provide useful information on how new species may have arisen and help to clarify our knowledge of the expansion and contraction of African forests. But the work is of more than academic interest; many of the remaining forest 'islands' are shrinking at an alarming rate due to human activities, and species may be forced to extinction before they are known to exist. There is an urgent need to highlight this problem and to identify areas where the risks are most severe.

SUMMARY AND CONCLUSIONS

Scientists, like galagos, appear to remain in the dark. It is perhaps not surprising that species have been overlooked, given our previous reliance on visual cues. For human beings, seeing is believing and it can be difficult to come to terms with the fact that animals may look very similar yet not recognize one another as members of the same species. For species with dispersed social systems that are active at night the senses of sound and smell play an important role in communication (Bearder, 1987). Here we have demonstrated that the loudest calls made by galagos can be used to define a species-specific vocal fingerprint. The call repertoires remain more-or-less constant between dispersed populations of the same species, but separate species have qualitatively different repertoires. We have identified six groups of species on the basis of their 'advertising' calls (the Elegant, Lesser, Greater, Southern dwarf, Zanzibar and Dwarf groups). To date our fieldwork, and the generous help of people from many parts of Africa, has produced recordings suitable for analysis from 10 of the galago species classified by Olson and Groves (see Table 1). *Euoticus pallidus* in Cameroon could not be distinguished as a separate species from *E. elegantulus* in Gabon at this stage of the analysis. No field recordings have yet been obtained for *Galago senegalensis* or *G. gallarum.* The first of these species has been shown to have a specifically distinct call-repertoire in captivity (Zimmermann *et al.*, 1988; Zimmermann, 1989), but we do not know of any recordings of the Somali galagos, and they cannot, therefore, be confirmed as a separate species in this analysis. In addition, we have provided evidence that there are a further five or six species that have not been recognized previously, and there are good reasons to believe that several more species remain to be uncovered. Our current fieldwork is designed to characterize species in the Dwarf and Zanzibar groups in Tanzania, and to compare the four (or more) galagos of Gabon with those of Cameroon. Other potentially important projects on nocturnal primates include:

1) Pygmy greater galagos in S.E. Tanzania (Lindi and Newala Districts).

2) An unidentified small species in the region of Mount Marsabit in northern Kenya.

3) Detailed examination of populations allied to Elegant, Allen's, Thomas's and Demidoff's galagos in the central African forest block.

4) Follow-up studies of populations in central and northern Malawi (Courtenay and Bearder, 1989).

5) Lesser galagos related to *Galago senegalensis* from Senegal in the west to northern Kenya and Somalia in the east (including *G. gallarum;* Nash *et al.*, 1989).

6) Collection of hair, skin, blood and tissue samples, scent gland secretions and measurements/photographs, especially from lorises which lack loud vocalizations. (Note that penile morphology appears to correlate well with species identity (Dixson, 1989).

ACKNOWLEDGEMENTS

This work has been supported by grants from Oxford Brookes University, the Boise Fund, the Wenner-Gren Foundation, the Leverhulme Trust (Grants to Institutions), the Fauna and Flora Preservation Society and the Wildlife Conservation Society, New York. Our recent field work has been made possible through the Commission for Science and Technology of the Republic of Tanzania. We thank the organizers and participants of Creatures of the Dark for their help. P. Honess is affiliated with the Wildlife Conservation Research Unit, Oxford University.

Our warmest thanks go to the many people who have helped with our research programme over the years. We would particularly like to thank those who have provided direct help in designing and implementing the main body of this study. These include: N. and L. Baker, P. Boyd, M. Bruford, N. Burgess, T. Butynski, R. Corder, K. Crouse, A. Dixson, F. Dowsett-Lemaire, J. Faldborg, T. and R. Fison, T. Fraser, A. Perkin, K. Halberg, C. Harcourt, L. Hill, P. Hills, K. Howell, P. Jenkins, G. Jones, J. Kingdon, J. Lagunda, V. Lyamuya, R. Martin, T. Olson, U. Petry, D. Pomeroy, P. Rodewald, D. Shaaf, K. Stanger, T. Struhsaker, F. Vincent, A. Walker, A. and E. Zimmermann.

REFERENCES

Alexander, R.D. (1968). Arthropods. In T.A. Seboek (Ed.), *Animal Communication* (pp. 167-216). London: Indiana University Press.

Bearder, S.K. (1987), Lorises, bushbabies and tarsiers: Diverse societies in solitary foragers. In D. Cheney, R. Seyfarth, B. Smuts, T. Struhsaker, R. Wrangham (Eds.), *Primate Societies* (pp. 11-24). Chicago: University of Chicago Press.

Brown, W.M., Prager, E.M., Wang, A. & Wilson, A.C. (1982). Mitochondrial DNA sequences of primates: tempo and mode of evolution, *Journal of Molecular Evolution, 18*, 225-239.

Charles-Dominique, P.(1977). *Ecology and Behaviour of the Nocturnal Primates.* London: Duckworth.

Corder, R.F. (1989). *The Taxonomic Status of Dwarf Galagos (Galagoides): A Preliminary Analysis on the basis of Vocalisation and Morphological Data.* Unpublished masters dissertation, Durham University (Great Britain).

Courtenay, D.O. & Bearder, S.K. (1989). The taxonomic status of bushbabies in Malawi with emphasis on the significance of vocalisations. *International Journal of Primatology, 10*, 17-34.

Crompton, R.H., 1989, Mechanisms for speciation in Galago and Tarsius. *Human Evolution, 4*, 105-116.

Crouse, K.E. (1988). *The Distribution and Taxonomic Status of a Small Southern Tanzanian Galago (Primates: Galagidae).* Unpublished masters dissertation, Oxford University (Great Britain).

Dixson, A.F. (1989). Effects of sexual selection upon the genitalia and copulatory behavior in male galagos, *International Journal of Primatology, 10*, 47-55.

Evans, E.F. (1982). Basic principles and psychophysics of sound, In H.B. Barlow & J.D. Mollon (Eds.), *The Senses* (pp. 239-250). Cambridge: Cambridge University Press.

Gautier, J-P. (1988). Interspecific affinities among guenons as deduced from vocalisations. In A. Gautier -Hion (Eds.), *A Primate Radiation - Evolutionary Biology of the African Guenons* (pp. 194-226). Cambridge: Cambridge University Press.

Groves, C.P.(1974). Taxonomy and phylogeny of prosimians, In R.D. Martin, G.A. Doyle & A.C. Walker (Eds.), *Prosimian Biology.* London, Academic Press.

Hillis, D.M. & Moritz. C. (eds.) (1990), *Molecular Systematics.* Massachusetts: Sinauer Associates.

Izard, K. & Nash, L.T. (1989). Contrasting reproductive parameters in *Galago senegalensis braccatus* and *G. s. moholi. International Journal of Primatology, 9*, 519-527.

Lanyon, W.E., 1969, Vocal characters and avian systematics. (pp.291-310) In R.A. Hinde (Ed.), *Bird Vocalizations.* Cambridge: Cambridge University Press.

Lawrence, B. & Washburn, S.L. (1936). On a new race of *Galago demidovii. Occasional Papers Boston Society of Natural History, 8*, 255-266.

Liebermann, P. (1984). *The Biology and Evolution of Language*. Cambridge: Harvard University Press.
Littlejohn, M.S. & Oldham, R.S. (1968). *Rana pipens* complex: Mating call structure and taxonomy. *Science, 62*, 1003-1004.
Marler, P. (1975). Specific distinctiveness in the communication signals of birds. *Behaviour, 11*, 13-39.
Masters, J.C. & Spencer, H.G. (1989). Why we need a new genetic species concept. *Systematic Zoology, 38*, 270-279.
Masters, J.C. (1993). Primates and paradigms: Problems with the identification of genetic species. (pp. 43-64) In W.H. Kimbel & L.B. Martin (Eds.), *Species, Species Concepts, and Primate Evolution*. New York: Plenum.
Nash, L.T., Bearder, S.K. & Olson, T. (1989). A synoptic overview of galago species differences, *International Journal of Primatology, 8*, 45-53.
Oates, J.F. & Trocco, T.F. (1983). Taxonomy and phylogeny of black and white colobus monkeys. Inferences from an analysis of loud call variation. *Folia Primatologica, 40*, 83-113.
Olson, T.R. (1979). *Studies on Aspects of the Morphology of the Genus Otolemur Coquerel, 1859.* Doctoral Dissertation, University of London, (Great Britain). Ann Arbor: University Microfilms International.
Paabo, S. (1989). Ancient DNA: extraction, characterization, molecular cloning and enzymatic amplification. *Proceedings of the National Academy of Sciences, 86*, 1939-1943.
Paterson, H.E.H. (1985). The recognition concept of species. In E.S. Vrba (Ed.), *Species and Speciation* (pp. 21-29). Pretoria: Transvaal Museum.
Petter, J-J. & Charles-Dominique, P.(1979). Vocal communication in prosimians, In G.A. Doyle & R.D. Martin (Eds.), *The Study of Prosimian Behaviour*. (pp. 247-305). London: Academic Press.
Ralin, D.B. (1977). Evolutionary aspects of mating call variation in a diploid-tetraploid species complex of tree frogs (Anura). *Evolution, 31*, 721-736.
Stanger, K.F. (1993). *Structure and Function of the Vocalisations of Nocturnal Prosimians (Cheirogaleidae).* Doctoral dissertation,University of Tubingen. Tubingen: Hans-Joachim Kohler.
Tattersall, I. (1989). The roles of ecological and behavioural observation in species recognition among primates. *Human Evolution, 4*, 117-124.
Thomas, W.K., Paabo, S., Villablanca, F.X. & Wilson, A.C. (1990). Spatial and temporal continuity of kangaroo rat populations shown by sequencing mitochondrial DNA from museum specimens, *Journal of Molecular Evolution, 31*, 565-567.
Thomas, O. & Wroughton, R.C. (1908). The Rudd Exploration of South Africa IX. *Proceedings of the zoological Society of London, 1908*, 164-173.
Tinbergen, N. (1953). *Social Behaviour in Animals*. London: Methuen.
Toms, R.B. (1993). Incidental effects and evolution of sound producing organs in tree crickets (Orthoptera: Oecanthidae). *International Journal of Insect Morphology and Embryology*, 22, 207-216.
Vbra, E.S. (Ed.) (1985). *Species and Speciation*. Pretoria: Transvaal Museum.
Walker, T.J. (1962). Factors responsible for intraspecific variation in the calling songs of crickets. *Evolution, 16*, 407-428.
Zimmermann, A. (1985). Analysesystem fur NF-Signale, *CAL, 1*, 53-55.
Zimmermann, E. (1985). The vocal repertoire of the adult senegal bushbaby (*Galago senegalensis senegalensis*). *Behaviour, 94*, 212-233.
Zimmermann, E. (1990). Differentiation of vocalisations in bushbabies (Galaginae, Prosimiae, Primates) and the significance for assessing phylogenetic relationships, *Journal of zoolical Systematics and Evolution, 28*, 217-239.
Zimmermann, E., Bearder, S.K., Doyle, G.A. & Andersson, A.B. (1988). Variations in vocal patterns of Lesser bushbabies (*Galago senegalensis* and *Galago moholi*) and their implications for taxonomic relationships. *Folia Primatologica, 51*, 87-105.

VOCALIZATIONS OF SOME CHEIROGALEID PROSIMIANS EVALUATED IN A PHYLOGENETIC CONTEXT

Kathrin F. Stanger

Duke University Primate Center
3705 Erwin Road
Durham, NC 27705

INTRODUCTION

Acoustic communication is a widespread mode of information exchange in many social animals, including primates. Social animals may live in highly gregarious groups, or may alternatively spend most of their day in solitude, interacting with only a few individuals (Charles-Dominique, 1978). For small nocturnal primates which tend to forage solitarily and encounter each other infrequently, acoustic signals may play a particularly important role in maintaining and regulating their social relationships.

Since the classic studies of Altmann (1962), Rowell and Hinde (1962), Andrew (1963), and Winter *et al.* (1966), investigations of vocal communication of non-human primates have accumulated rapidly. However, the majority of work investigating vocal communication has addressed only a comparatively small number of primate species (Snowdon, 1986). One major taxonomic gap is the paucity of studies on prosimians. This applies especially to the smaller nocturnal species which are difficult to study in the field and until recently have been rare in captivity. Of 32 extant nocturnal Strepsirhine species (Olson, 1979; Harcourt and Thornback, 1990), the vocal repertoires of only seven species have been documented: *Nycticebus coucang* (Daschbach *et al.*, 1981; Zimmermann, 1985a), *Galago senegalensis senegalensis* (Zimmermann, 1985b), *Galago demidovii*, *Microcebus murinus*, *Cheirogaleus medius* (Stanger, 1988), *Mirza coquereli* (Stanger, 1993), and *Daubentonia madagascariensis* (Stanger and Macedonia, in press). Individual calls of these and a few other species have been reported by Andrew (1963), Charles-Dominique (1977), Petter and Charles-Dominique (1979), Charles-Dominique and Petter (1980), Pages (1980), Cherry *et al.* (1987), Stanger (1989), and Zimmermann and Lerch (1993). In this chapter I discuss the structure of vocalizations in several cheirogaleid species, focussing in particular on how our understanding of the evolution of these vocalizations might be enhanced by employing a phylogenetic perspective.

Vocalizations in Taxonomy and Phylogeny

Since the pioneering work of Struhsaker (1970) on cercopithecines, a number of researchers have used vocalizations as taxonomic or phylogenetic criteria, both in order to produce a better phylogenetic hypothesis and to better understand the evolution of primate vocalizations. The species studied include strepsirhines (Macedonia and Taylor, 1985; Zimmermann *et al.*, 1988; Zimmermann, 1990; Macedonia and Stanger, in press), tarsiers (MacKinnon and MacKinnon, 1980; Niemitz *et al.*, 1991), New World monkeys (Hodun *et al.*, 1981; Snowdon, et al., 1986), Old World monkeys (Struhsaker, 1970, 1981; Oates and Trocco, 1983; Gautier, 1988, 1989), and apes (Marshall and Marshall, 1976).

In the past, vocalizations have been used as evidence for a close phylogenetic relationship between cheirogaleids and galagos (Charles-Dominique and Martin, 1970), but this argument was based on a comparison of sounds by listening and did not explicitly focus on call structure. A preliminary comparison of vocalization structure between cheirogaleids and dwarf galagos (Stanger, 1988, 1989) suggests that these two groups actually produce vocalizations that are very different from each other. This distinction is supported by a recent molecular and morphological study (Yoder, 1992) which groups the cheirogaleids with the Lemuriformes and thus phylogenetically distinct from the Lorisiformes (lorises and galagos). The phylogenetic relationships within the Cheirogaleidae remain unresolved.

Today the cheirogaleids are classified into five genera: *Allocebus*, *Cheirogaleus*, *Microcebus*, *Mirza*, and *Phaner*, which contain seven species. The hairy-eared dwarf lemur, *Allocebus trichotis*, was originally described as *Cheirogaleus trichotis* by Günther in 1875 (Tattersall, 1982), but was assigned to its own genus by Petter and Petter-Rousseaux (1967). Coquerel's dwarf lemur, *Mirza coquereli*, was initially described as a member of *Cheirogaleus* by Grandidier in 1867 (Tattersall, 1982), but in 1868 Schlegel and Pollen (Tattersall, 1982) classified it with *Microcebus* and Gray (1870) assigned it to its own genus: *Mirza*. Based on biological similarities some authors continued to prefer *Microcebus* (Hill, 1953; Napier and Napier, 1967; Petter *et al.*, 1971). Tattersall and Schwartz (1974) reclassified *Mirza* as *Cheirogaleus*, and later, Tattersall (1982) reinstated *Mirza* as a genus name. The fork-marked dwarf lemur, *P. furcifer*, was assigned its own genus by Gray (1870). Petter *et al.* (1977) even consider *Phaner* as the only member of a distinct subfamily Phanerinae within the Cheirogaleidae. For the purposes of the current paper, I distinguish *Allocebus trichotis*, *Cheirogaleus major*, *Cheirogaleus medius*, *Microcebus murinus*, *Microcebus rufus*, *Mirza coquereli*, and *Phaner furcifer* (Tattersall, 1982).

Cheirogaleid Social Systems

The behavioral ecology of many species of Malagasy primates, especially the small nocturnal lemurs of the Cheirogaleid family, remains to be studied. A few field studies in which individual animals were followed or radio-tracked, e.g. *M. murinus* (Martin, 1972), *M. coquereli* (Pages, 1978; 1980), and *P. furcifer* (Charles-Dominique and Petter, 1980), indicate that cheirogaleids live in a social system consisting of more or less overlapping home ranges similar to that described for galagos (e.g. Charles-Dominique, 1977; Bearder, 1987). This social system consists of partially overlapping female home ranges, with mothers sometimes sharing their home range with their subadult female offspring. Females seem to select their home ranges based on their foraging needs, and males, in turn, try to occupy mutually exclusive home ranges which include the home range areas of as many females as possible (Charles-Dominique, 1977). Male home ranges are usually defended against other males, but in some species adult males may share their home range with unrelated younger males (Bearder, 1987). The pattern of overlap between male and female home ranges determines the number of males and females interacting with each other and thus possibly reflects their mating system. The pattern of overlap within and between the sexes varies from species to species (Charles-Dominique and Petter, 1980; Martin, 1972; Pages, 1978, 1980; Bearder, 1987) and is central in establishing the specific composition of their social systems.

This chapter focuses on the vocalizations of four taxa of nocturnal cheirogaleids. All four species: the fat-tailed dwarf lemur (*C. medius*), the mouse lemur (*M. murinus*), Coquerel's dwarf lemur (*M. coquereli*), and the fork-marked dwarf lemur (*P. furcifer*), live sympatrically in the western deciduous forests of Madagasar (Tattersall, 1982; Harcourt and Thornback, 1990). A selection of vocalizations of the four cheirogaleids is presented in the form of sonograms, and the overall acoustic structure of their calls is compared. Due to the difficulty of defining ancestral character states for the vocalization structures found in this group, no phylogenetic hypothesis could be constructed based on vocalization structures alone. Instead, an independent phylogenetic hypothesis was formulated with a cladistic analysis using basicranial and dental data from the literature, as well as behavioral data from my own observations at the Universities of Karlsruhe and Tübingen (Germany), and at the Duke University Primate Center (DUPC). Into this analysis two additional taxa were included: the hairy-eared dwarf lemur (*Allocebus trichotis*), and the greater fat-tailed dwarf lemur (*C. major*), which both inhabit the remnants of the eastern rain forests of

Madagascar. Due to lack of information, the seventh cheirogaleid taxon, the red mouse lemur (*M. rufus*) which occurs both in eastern and western Madagascar (Petter *et al.*, 1977), was not included. Subsequently the evolution of different call structures within the cheirogaleids was investigated by plotting the vocal characters onto this independently acquired phylogeny (McLennan *et al.*, 1988; Brooks and McLennan, 1991).

METHODS

Cladistic Analysis

To investigate the evolutionary history of the call types used by the cheirogaleids, a cladistic analysis was first conducted. Since the ancestral states of vocalization characters cannot be determined with any certainty for this group, I formulated an independent phylogenetic hypothesis using basicranial and dental characters from the literature (e.g. (Cartmill, 1978; Eaglen, 1980; Yoder, 1992). These data were supplemented with behavioral characters (pers. observ.).

The following taxa were included in the cladistic analysis: *Allocebus trichotis*, *Cheirogaleus major*, *Cheirogaleus medius*, *Microcebus murinus*, *Mirza coquereli*, and *Phaner furcifer*.

Morphological and Behavioral Characters used for Cladistic Analysis

1. Os planum. The exposure of the ethmoid in the orbital wall forming an os planum was examined by Cartmill (1978). He examined between 2 and 21 specimens per taxon, and concluded that the os planum was invariably present in *M. coquereli*, *M. murinus*, and *A.trichotis*, but always absent in *C. medius*, *C. major*, and *P. furcifer*. The os planum is usually absent in the Lemuriformes, but it is polymorphic in *Lepilemur*, *Hapalemur*, *Indri* and *Avahi*. The absence of an ethmoid exposure was coded as ancestral.

2. Frontal-palatine contact. In the orbital mosaic of the cheirogaleids, a contact between the frontal and palatine bones has been reported for all *A. trichotis*, *C. major*, *C. medius*, and *P. furcifer*, but not for *M. murinus* or *M. coquereli*. However, in one out of 19 *M. murinus* a frontal-palatine contact occurred, and another specimen showed polymorphism on the two sides of its skull. Because the absence of this character amounted to more than 90% of all cases, *M. murinus* was scored absent for this character. A frontal-palatine contact was absent in most Lorisiformes, but present in most other non-cheirogaleid lemurs as well as tarsiers (Cartmill, 1978). The presence of a frontal-palatine contact was coded as ancestral.

3. Palatine-maxillary suture. The position of the anteriormost point of the palatine-maxillary suture relative to the upper molars was determined by Yoder (1992) who defined the position medial to M^1 as ancestral. This position was documented for *P. furcifer*. In contrast, *A. trichotis*, *C. major*, *M. murinus*, and *M. coquereli* have a suture medial between M^1 and M^2.

4. Posteromedial edge of the palate in relation to M^3. The posteromedial edge of the palate is located medial to M^3 in *P. furcifer*, and posterior to M^3 in all other cheirogaleids studied so far (Yoder, 1992). The medial position was defined as ancestral.

5. Postglenoid process. The postglenoid process may be unattached (*C. major*, *P. furcifer*) or fused with the auditory bulla (*A. trichotis*, *M. murinus*, *M. coquereli*), while the unattached state appears to be ancestral (Yoder, 1992).

6. Medial pterygoid. The medial pterygoid plate protrudes ventrally relative to the lateral pterygoid plate in most Malagassy lemurs including *P. furcifer*. In the other cheirogaleids it does not protrude beyond the lateral plate. This appears to be the derived state (Eaglen, 1980; Yoder, 1992).

7. P^2 higher than P^3. The relative crown height of the upper second and third premolars was used by Eaglen (1980) in his phylogenetic analysis. The widespread distribution of a relatively high crown of P^2 in extant strepsirhines and its occurrence in *Adapis* suggests that this character state is primitive (Eaglen, 1980): *M. murinus* and *M. coquereli* both have higher crowns on P^2 than on P^3, but *C. major, C. medius*, and *P. furcifer* show no difference.

8. P_4 metaconid. This metaconid on the lower forth premolar is widespread among adapids (Gregory, 1920), and in tupaiids (Steele, 1973), suggesting that its presence in *C. major* and *C. medius* is a primitive condition (Eaglen, 1980). It is absent in *M. murinus, M. coquereli*, and *P. furcifer* (Eaglen, 1980).

9. Hypocone $M^{1\text{-}2}$. The upper molar hypocone is present in most strepsirhine outgroups and in most adapids and tupaiines: this evidence suggests that it was present in the last common ancestor of living strepsirhines (Eaglen, 1980), and therefore its presence was coded as ancestral. It is present in *M. murinus*, and *P. furcifer*, but absent in *C. major, C. medius*, and *M. coquereli*.

10. Hypoconulid M_3. A small hypoconulid on the lower third molar is characteristic for early Tertiary primates, and it is widespread among recent strepsirhines (Eaglen, 1980). This suggests that the presence of a small hypoconulid in *M. coquereli* and *P. furcifer* is primitive. *C. major, C. medius*, and *M. murinus* lack this hypoconulid (Eaglen, 1980).

11. Urine washing. Urine washing is a stereotyped behavior pattern consisting of urinating in the hand and subsequently rubbing the hand against the sole of the *ipsi-lateral* hindlimb. This behavior can be repeated on the other side. Urine washing is common among galagos, and *M. murinus* as well as *M. coquereli* show it (pers. observ.), but it is not documented for *C. medius* and for *P. furcifer*, or for any other lemuriform primate. *M. coquereli*, however, shows the movements frequently in a rudimentary form: the limb movements are present, but fore-and hindlimbs do not actually touch. No ancestral state was defined.

12. Fat storage in tail. A seasonal increase in body weight in preparation for the dry season is common among strepsirhines, but the accumulation of fat reserves in the tail is limited to *Cheirogaleus* and *Microcebus*. Meier and Albignac (1991) reported that *A. trichotis* does not store fat for the dry season. No fat-storage has been reported for *P. furcifer*, and the photograph in Tattersall (1982) shows a tail morphology like that of *M. coquereli* where no tissue for fat storage is present. Furthermore, *P. furcifer*, like *M. coquereli* reportedly uses an alternative food source during the dry season (Charles-Dominique and Petter, 1980). No non-cheirogaleid lemuriform primate is known to store fat in its tail, therefore the absence of fat storage in the tail was coded as ancestral.

13. Gular gland. Gular glands are documented for *P. furcifer*, and for the Indridae and *Hapalemur* (Tattersall, 1982) within the strepsirhines, as well as for haplorhines (callitrichids: Hershkovitz, 1977) and for bats (*Molossus:* Heideman *et al.*, 1990). Gular glands are used to deposit scent marks on branches and/or on conspecifics. The distribution of this character indicates the presence of gular glands as ancestral.

Table 1. Character matrix for cladistic analysis: Characters by taxa. ANC: hypothetical ancestor; 0: ancestral character state; 1: derived; ?: unknown.

Character #	1	2	3	4	5	6	7	8	9	10	11	12	13
A. trichotis	1	0	1	1	1	1	?	?	?	?	?	0	1
C. major	0	0	1	1	0	1	1	0	1	1	?	1	1
C. medius	0	0	?	?	?	1	1	0	1	1	0	1	1
M. murinus	1	1	1	1	1	1	0	1	0	1	1	1	1
M. coquereli	1	1	1	1	1	1	0	1	1	0	1	0	1
P. furcifer	0	0	0	0	0	0	1	1	0	0	0	0	0
ANC	0	0	0	0	0	0	0	0	0	0	?	0	0

Thirteen characters were subjected to a cladistic analysis (Table 1). All characters were coded as binary characters with reversible character states, and all characters were weighted equally (Swofford, 1993). Data were analyzed using the "exhaustive search" algorithm in PAUP (Phylogenetic Analysis Using Parsimony, version 3.1, Swofford, 1993). Trees were rooted using Lundberg rooting (Swofford, 1993). This method first computes an unrooted tree for the ingroup taxa only without making any *a priori* assumptions about the polarity of the characters, and a hypothesized ancestor is connected to the tree *a posteriori* to root the tree. The character states of this ancestor were defined based on the character distribution among other mammals (e.g. bats, rodents), other primate taxa, and among early adapids and omomyids for some hard tissue characters (according to the original sources).

The degree to which resulting cladograms are parsimonious is expressed in terms of tree length and consistency index (Rohlf, 1982). Tree length is calculated as the number of steps taken to place all the characters on a tree and reflects the amount of evolutionary change that has occurred in the group being studied. The consistency index (CI) is calculated as the minimum number of possible character changes in the data set divided by the tree length and indicates how closely the tree reflects a single origin for each derived character state (Swofford, 1993). Character state changes of vocalizations were mapped onto the most parsimonious tree.

The cladistic method (Hennig, 1966) is based on the analysis of shared derived characters, or synapomorphies (see also Wiley, 1981). I did not include uniquely derived characters (autapomorphies) because they are not informative for cladistic analysis, and, when included, artificially inflate the consistency index of a proposed phylogenetic hypothesis (Sanderson and Donoghue, 1989). Although I included six taxa in the cladistic analysis, the vocalizations of only three taxa have been studied extensively. Where available, I have included information on individual calls of the other taxa. More detailed studies of the vocal repertoires of these taxa (and the seventh taxon *M. rufus*) will provide tests for the evolutionary scenarios suggested in this chapter.

The Vocalizations of Four Cheirogaleid Species

Vocalizations of captive *C. medius, M. murinus*, and *M. coquereli* were recorded at the Universities of Karlsruhe and Tübingen (Germany) and at the DUPC (Stanger 1988; 1993). Data were collected during more than 500 contact hours with *C. medius* (n= 26 adults), more than 500 hours with *M. murinus* (n= 14 adults), and more than 300 hours with *M. coquereli* (n= 36 adults). Subsets of ten *C. medius*, six *M. murinus*, and ten *M. coquereli* were used for a detailed context analysis of their vocal repertoires. The vocalizations of *P. furcifer* were recorded opportunistically in Madagascar during approximately 40 contact hours (Stanger, unpublished data).

All recordings were made with a Uher SG-631 Logic and a Lennartz tape recorder (*M. murinus* and *C. medius*), as well as a Uher 4400 Report Monitor recorder (*M. coquereli, P. furcifer*). Frequency responses (with reference to a 1 kHz tone) were within ± 3 dB up to 35 kHz (Lennartz 6000/607 with QMC S-100 microphone at 38 cm/sec), 23 kHz (Uher SG-631 Logic with TEAC-TE-50 and Cardoid-CM-310 microphones at 19cm/sec), and 20 kHz (Uher Report Monitor with ECM 10 microphone at 19 cm/sec).

Recordings were analyzed with a Ubiquitous Nicolet-UA-500 and a Kay model 5500 digital real time audiospectrograph. The analyses were conducted with a frequency resolution between 40 and 256 Hz, and a time resolution between 6 and 25 ms. The sonograms presented in this chapter were printed on paper film from a Toennis oscilloscope camera, and with a Kay printer. Sonograms of *P. furcifer* vocalizations were compared by visual inspection to the sonograms published by Niaussat and Molin (1978) and by Charles-Dominique and Petter (1980). Call types were defined based on their acoustic features alone, but for comparative purposes I will also discuss the context in which they were typically (but not exclusively) observed.

RESULTS

The Phylogenetic Relationships within the Cheirogaleids

The cladistic analysis of thirteen characters resulted in one most-parsimonious phylogenetic hypothesis with a tree length of 16 steps and a CI of 0.812. The cladogram, Figure 1, places *P. furcifer* as a sister taxon to all other cheirogaleids. The two *Cheirogaleus* species constitute a sister group to *A. trichotis*, *M. murinus*, and *M. coquereli*. Within the latter group, *M. murinus* and *M. coquereli* are sister taxa, and *A. trichotis* was identified as a sister taxon of these two.

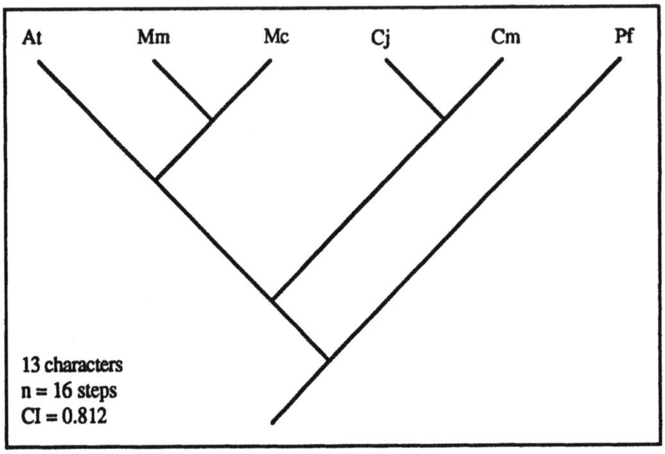

Figure 1. A phylogenetic hypothesis of the cheirogaleids based on morphological and behavioral characters. At = *A. trichotis*, Cj = *C. major*, Cm = *C. medius*, Mm = *M. murinus*, Mc = *M. coquereli*, Pf = *P. furcifer*.

Vocalizations in Captivity: *M. coquereli*, *C. medius*, and *M. murinus*

All three species share many features of their vocalizations. They produce very high-pitched calls with an upper frequency limit for the fundamental of about 32 kHz for *C. medius*, 40 kHz for *M. murinus*, and about 15 kHz for *M. coquereli* (Stanger, 1988; 1993). In *C. medius* and *M. murinus* a number of calls are exclusively emitted in the ultrasonic range with the lowest frequency above 20 kHz.

The three species used between eight and 12 different adult call types in a variety of contexts (Stanger 1988, 1993; Table 2). Some calls of the three species were structurally similar and were emitted in similar contexts and can be considered analogous, if not homologous (but see discussion) as suggested by the monophyletic origin of the cheirogaleids (Yoder, 1992). I discuss a selection of call types here, and describe the associated context.

Trill Calls. Trill calls were absent in *C. medius*, but present in *M. murinus* and *M. coquereli* (Figure 2). In trill calls a whistle (*M. murinus*) or a downward frequency-modulated pulse ("zek", *M. coquereli*) is followed by a rapid succession of up and down frequency modulations, which in *M. coquereli* grades into single upward frequency modulated pulses with a characteristic dynamic pattern of increasing pulse duration and inter-pulse intervals. These calls were emitted in the context of advertisement and in some cases they would elicit reciprocal calling from other groups or separately housed individuals. They were emitted mainly by males, and rarely by females (Stanger, 1993).

Narrow-band Whistles. All three species emitted whistle calls during vocal exchanges between family groups or separately housed individuals (Figure 3). The fundamental was located between 7 and 20 kHz, and calls lasted up to several hundred milliseconds. The bandwidth of an individual call usually did not extend beyond a frequency range of 5 kHz. The whistles of *C. medius* were characterized by a relatively constant frequency and were emitted in a series. Within a whistle series the frequencies of consecutive whistles can be modified, however, and may converge towards the whistle frequency of a reciprocal caller (Stanger, 1993). The whistles of *C. medius* were primarily observed as advertisement calls. The second harmonics of *C. medius* whistles are not visible on this sonogram, Figure 3a. Due to the missing trill element in the advertisement calls of *C. medius*, their whistle calls cannot be unequivocally distinguished from a second type of call emitted between groups as for the other two taxa. In the inter-group context both *M. murinus* and *M. coquereli* used whistles in addition to their trill calls. In *M. murinus* and *M. coquereli* both sexes used whistles during reciprocal calling between groups. Outside the inter-group context *C. medius* and *M. murinus* emitted whistles with a high repetition rate as a response to a disturbance and in the agonistic context. *C. medius* and *M. coquereli* males whistled while sniffing the females' genital areas.

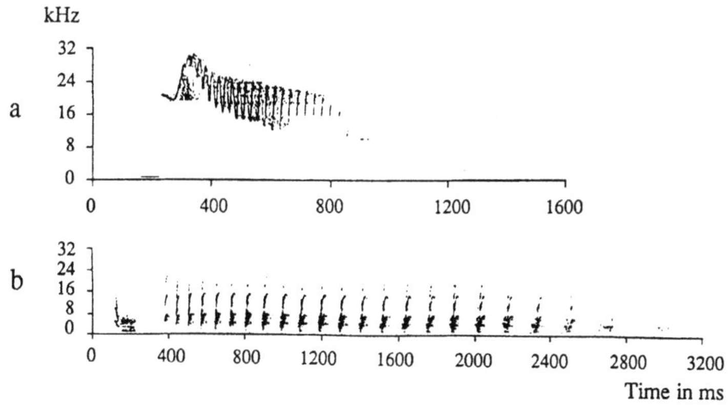

Figure 2. Trill calls of (a) *M. murinus* and (b) *M. coquereli*: trills may be preceded by a whistle.

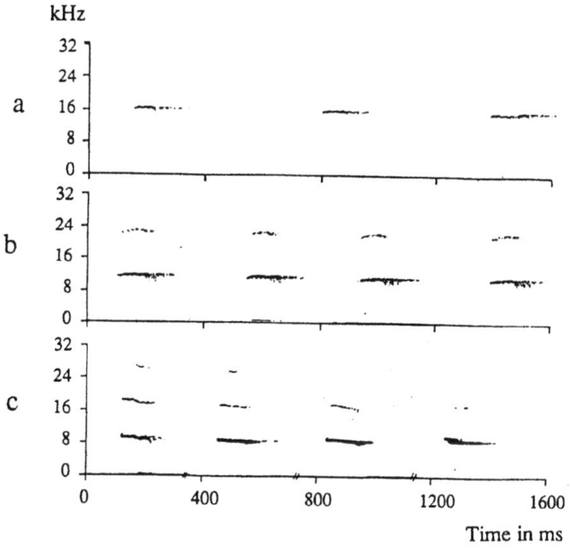

Figure 3. Narrow-band whistle calls of (a) *C. medius*, (b) *M. murinus*, (c) *M. coquereli*.

Frequency-modulated whistles. Among other calls, a downward frequency-modulated whistle was recorded for *C. medius*, *M. murinus*, and *M. coquereli* in the mother-infant context (Figure 4). These whistles were in the ultrasonic range in *C. medius* and *M. murinus*, but audible in *M. coquereli*. These calls were emitted by infants of all three taxa, but in *C. medius* they were also used by adult males approaching (and being approached by) a female.

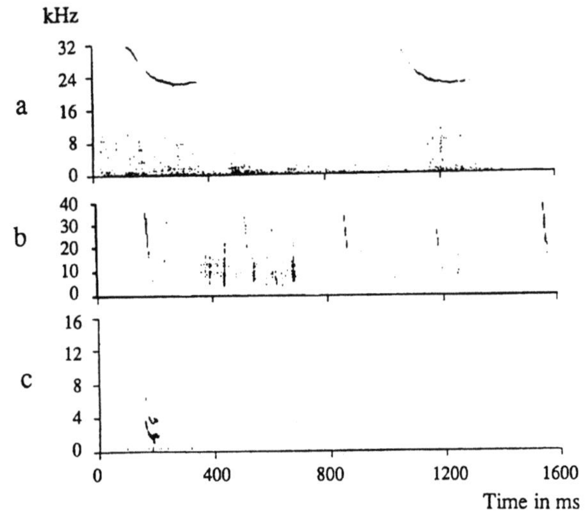

Figure 4. Frequency-modulated whistles of (a) *C. medius*, (b) *M. murinus*, (c) *M. coquereli.*

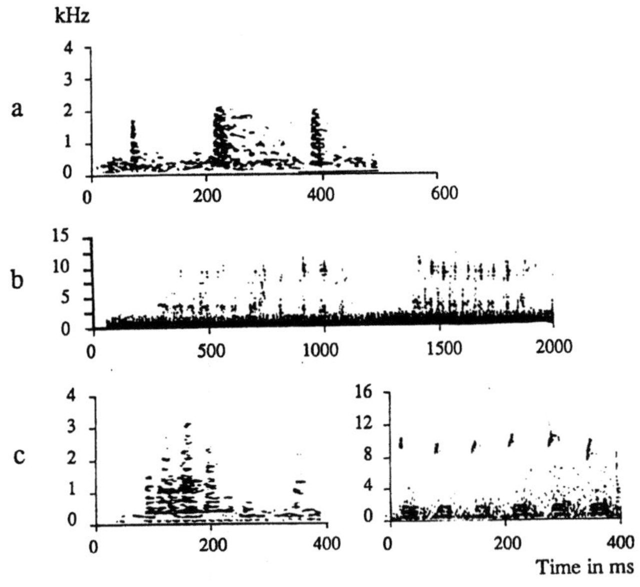

Figure 5. Rolling calls of (a) *C. medius*, (b) *M. murinus*, (c) *M. coquereli.*

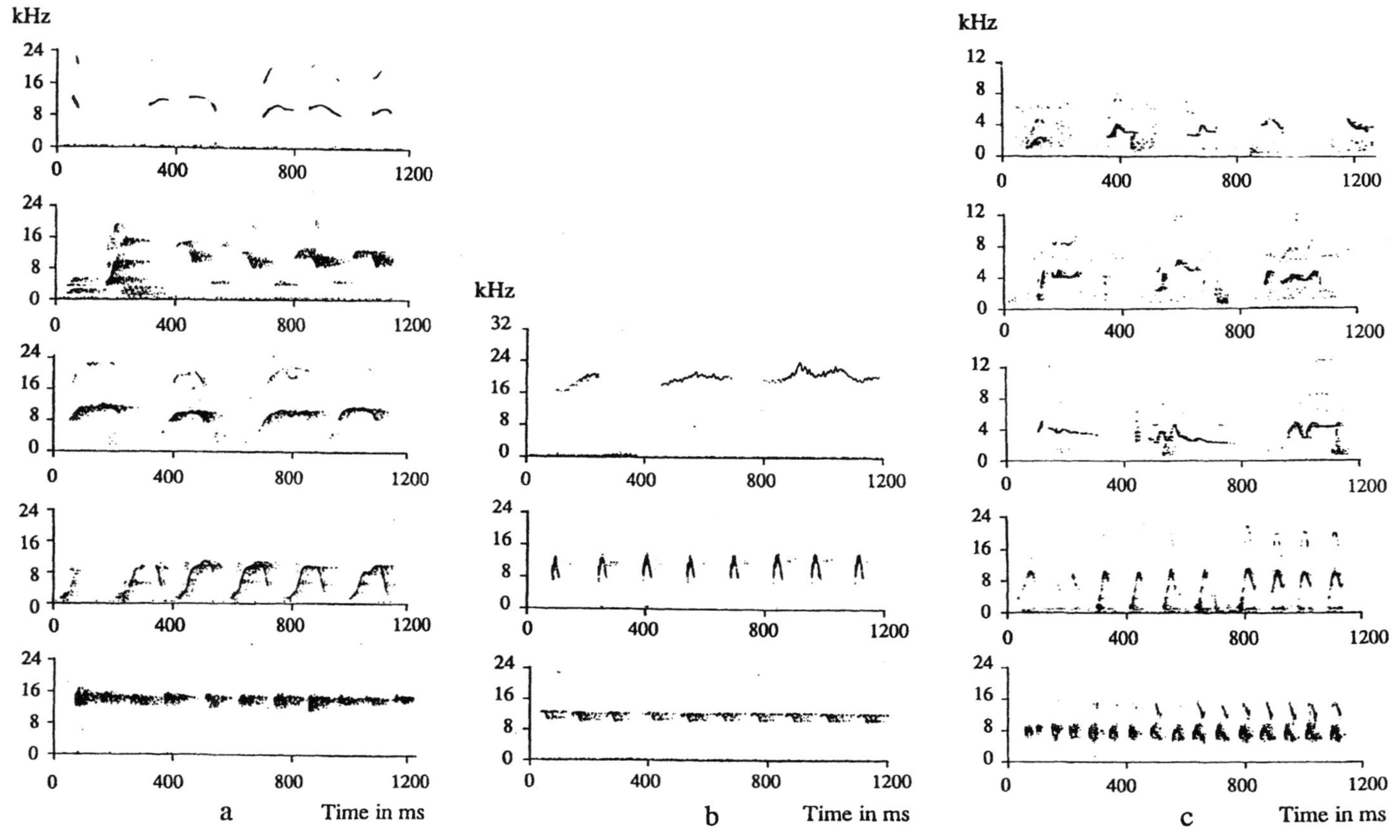

Figure 6. Twitter (top) and Chatter calls (below) of (a) *C. medius*, (b) *M. murinus*, (c) *M. coquereli*.

Rolling Calls. Rolling calls consist of a series of short pulses. Each single pulse lasted less than 90 ms and the main part of the call was located in a frequency range below 5 kHz. In some cases a high frequency component extending up to 15 kHz was added. All three taxa emitted rolling calls (Figure 5) during close contacts between males and females. For *C. medius* only the low frequency part of the call was recorded. *M. coquereli* produced rolling calls with and without high frequency elements, and for *M. murinus* all recordings contain both elements. Rolling calls were emitted during nasogenital contacts and as a submissive signal in *C. medius*, and during mutual grooming in *M. murinus*. In *M. coquereli* they were emitted by contact-seeking males following a female.

Twitter and Chatter Calls. Twitter and chatter calls represent the two ends of what appears to be a gradually changing acoustic structure (Stanger, 1993). Twitter calls consist of variably up- and downward modulated frequencies. They were mainly observed during affiliative interactions. With increasing aggression, the twitter calls were transformed into chatter calls. Chatter calls consist of chevron-shaped, up-down modulated pulses and are generally shorter than twitter calls. The twitter and chatter calls of *C. medius* and *M. coquereli* were more variable in structure than the *M. murinus* calls (Figure 6). Possibly due to the rarity of affiliative interactions between male and female *M. murinus*, only a few twitter call variations were recorded for this species (Figure 6b).

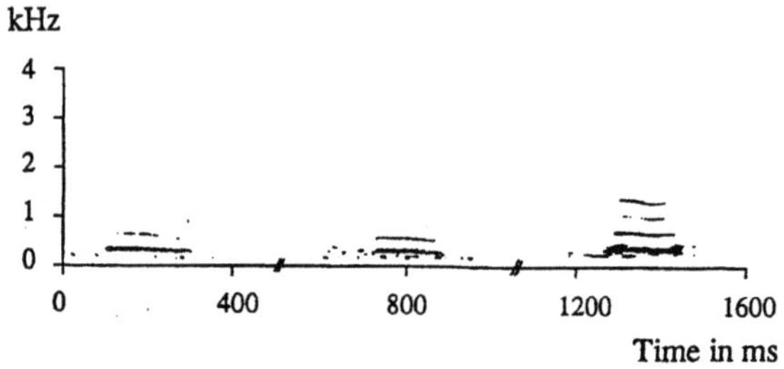

Figure 7. "Hn" calls of *M. coquereli*.

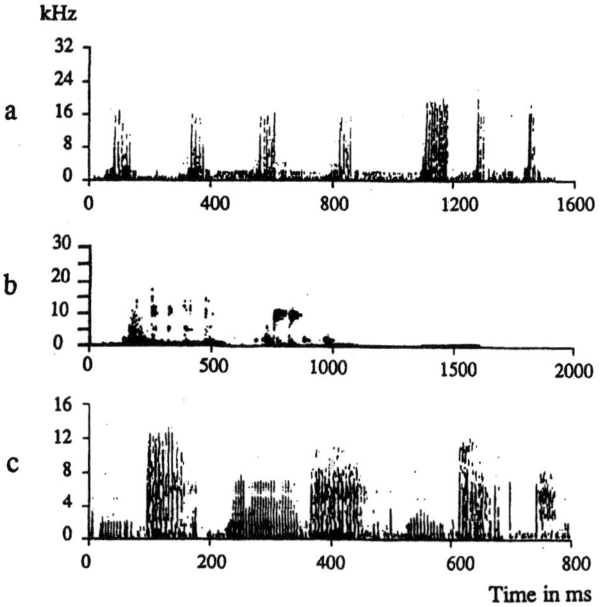

Figure 8. Grunt calls of (a) *C. medius*, (b) *M. murinus*, (c) *M. coquereli*.

"Hn" Calls. "Hn" calls are low frequency sounds with the fundamental located below 500 Hz. They may contain a slight downward frequency modulation. "Hn" calls lasted up to 310 ms. Only *M. coquereli* emitted this call type (Figure 7). These calls were exclusively emitted during locomotion and by mothers approaching their infant. Mothers who failed to emit these calls during approach were invariably attacked when entering the nestbox. The only other cheirogaleid for which "hn" calls are reported is *P. furcifer* and these were also emitted during locomotion (Charles-Dominique and Petter, 1980).

Grunt Calls. Grunt calls are the only call type with a completely noisy call structure. They consist of a series of very short broad-band pulses. All three species emitted grunt calls (Figure 8). Grunts were observed in the agonistic context during fights or as threat calls preceding aggression. This call type can be found in most mammalian taxa, and in reptiles, and represents an archaic call type which, according to Andrew (1963), evolved from an aversive respiration sound.

Squeal Calls. Squeal calls are chevron-shaped in all three species with a variable degree of frequency modulation. In recordings of *C. medius* (Figure 9a), the chevron-shaped calls are obscured by echoes. Squeal calls were generally emitted in distress when animals were handled, but also during fighting and biting in an agonistic context. The calls of *M. murinus* had the highest repetition rate (Figure 9b), and the longest calls were produced by *M. coquereli* (Figure 9c).

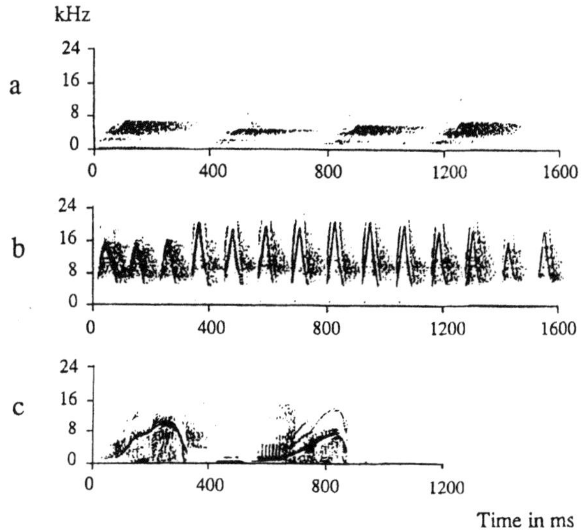

Figure 9. Squeal calls of (a) *C. medius*, (b) *M. murinus*, (c) *M. coquereli*.

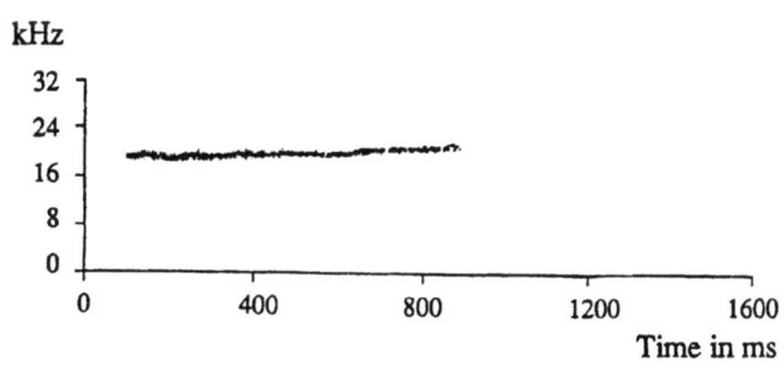

Figure 10. Long high-frequency whistle of *M. murinus*.

Long High-frequency Whistles. *M. murinus* emitted long, high-frequency whistles in the alarm context (Figure 10) while sitting motionless. These whistles were longer and in a higher frequency range than the whistle calls emitted as a response to a disturbance or to the calls of another group. No such call structure was observed in *C. medius* or *M. coquereli.*

"Zek" Calls. *M. coquereli* produced wide-band "zek"calls in the alarm context, as well as in the agonistic context. Depending on the individual, these calls can be upward or downward frequency-modulated, or both (Figure 11). The downward modulated version of the "zek" call resembles the introductory pulse of their trill calls (Figure 2). No such call structure was observed in *C. medius* or *M. murinus*.

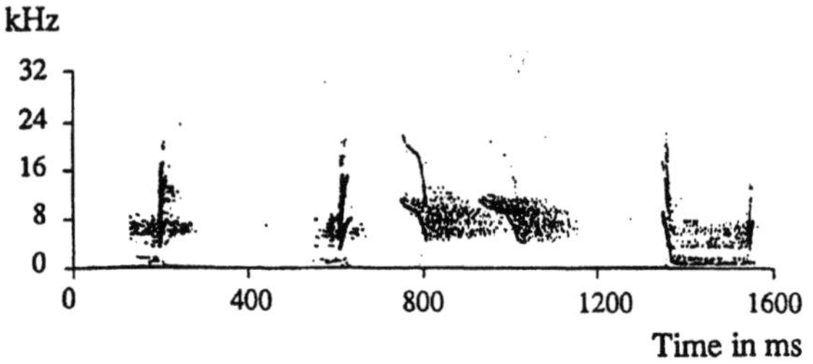

Figure 11. "Zek" calls of *M. coquereli (the sonogram includes the calls of three different individuals).*

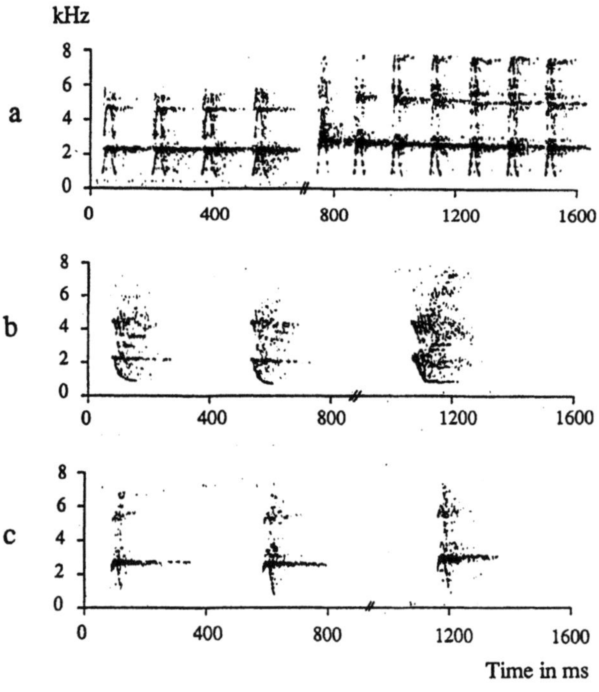

Figure 12. Vocalizations of *P. furcifer*: (a) "kiu" calls, (b) "kea" calls, (c) "ki" calls.

Vocalizations in the Field: *P. furcifer*

The most common vocalizations of *P. furcifer* in the field were series of pulses with variable repetition rates (pers. obs.). Charles-Dominique and Petter (1980) reported two types of advertisement calls: "kiu" calls emitted by both males and females, and "kea" calls emitted exclusively by males. Both call types were wide-band, containing several harmonics (Figure 12 a, b). Most pulse series consisted of up-down frequency-modulated pulses. In some cases the downward modulated part of the pulse was emphasized. When the main energy within such a pulse was located at or close to the peak frequency of the fundamental (Figure 12 c) the pulses sounded like whistles ("ki" calls of Charles-Dominique and Petter, 1980). One call type of *P. furcifer* (Figure 12 b) resembles the downward modulated version of the "zek" calls recorded from *M. coquereli* (Figure 11). No twitter and chatter calls have been heard or reported in the field, but captive studies are needed to determine the absence or presence of these call types in *P. furcifer*. In the field unhabituated *P. furcifer* responded to any approach with a series of wideband pulses ("ki", "kiu", or "kea"calls) and therefore it was not possible to record vocalizations in the undisturbed social context. While travelling, *P. furcifer*, like *M. coquereli*, emit "hn" calls (described as "hon", Charles-Dominique and Petter, 1980), and in distress their squeal calls appear to be almost identical to the squeal calls of *M. coquereli* (Figure 9c).

Table 2. Selected call types of four cheirogaleid taxa. Call types were defined based on their acoustic features alone, but for comparative purposes they are arranged according to the context in which they were typically observed.

Context	*M. coquereli*	*C. medius*	*M. murinus*	*P. furcifer**
Advertisement**	Trill	Whistle	Trill	"Kea"
Between groups***	Whistle	Whistle	Whistle	"Kiu"
Mother-Infant	FM Whistle	FM Whistle	FM Whistle	(?)
Contact seeking	Rolling	Rolling	Rolling	(?)
Affiliative contact	Twitter	Twitter	Twitter	(?)
Agonistic contact	Chatter	Chatter	Chatter	(?)
Locomotion	"Hn"	-	-	"Hn"
Threat	Grunt	Grunt	Grunt	Grunt
Distress	Squeal	Squeal	Squeal	Squeal
Alarm	"Zek"	-	Long Whistle	"Ki"?; "Kea"?

* context information for *P. furcifer* vocalizations from Charles-Dominique and Petter, 1980; ** advertisement calls mainly emitted by males; *** between groups, often accompanied by advertisement calls; -: no call type observed in that context; (?): context rarely observed, call type may or may not be present.

Differences of Vocalization Structures in Comparable Contexts

With the exception of the calls emitted in the alarm context, missing in *C. medius*, and calls associated with locomotion, missing in *C. medius* and *M. murinus*, I found similar call contexts in all three species. However, in some cases the three species emitted structurally different calls in a comparable context (Table 2). In addition to my own data, information on the structure and context of a number of calls of free-ranging *P. furcifer* and *C. major* in the field is available from Petter and Charles-Dominique (1979), and Charles-Dominique and Petter (1980).

Advertisement. With the exception of *C. major* and *C. medius* (Petter and Charles-Dominique, 1979; Stanger, 1988), which emitted tonal and narrow-band whistles, all cheirogaleids studied so far produce wide-band advertisement calls with extensive frequency modulations. The wide-band advertisement calls can be emitted without a characteristic dynamic pattern (*P. furcifer*), or they can be organized in a series of up to 50 pulses with decreasing frequency, increasing pulse lengths and increasing inter-pulse intervals (*M. murinus* and *M. coquereli*).

Number of Calls Between Groups. *C. medius* emitted only whistles during vocal exchanges between groups or distant individuals. *M. murinus* and *M. coquereli* used whistles in the inter-group context in addition to their trills calls. *P. furcifer* males also used two kinds of calls between groups, "kea" calls and "kiu" calls (Petter and Charles-Dominique, 1979; Charles-Dominique and Petter, 1980).

Locomotion. During locomotion "hn" calls were emitted by *M. coquereli* (Stanger, 1993), and by *P. furcifer* ("hon": Petter and Charles-Dominique, 1979, Charles-Dominique and Petter, 1980).

Alarm. Alarm calls are widespread among mammals and common among strepsirhines with three exceptions: Dwarf lemurs *C. major* and *C. medius* (Petter and Charles-Dominique, 1979; Stanger, 1988), from Madagascar, and slow lorises, *Nycticebus coucang* (Zimmermann, 1985b) from Asia lack alarm calls. Besides using no alarm calls at all, different strategies may be applied in the alarm context (Klump and Shalter, 1984). Alarm signals which are hard to localize can be used in combination with cryptic or evasive behavior, or wide-band alarm calls which are localized easily may be used to mob and deter predators. *M. murinus* used ultrasonic whistles with slow and narrow-band frequency modulations. Wide-band and low-pitched alarm calls were documented for *M. coquereli* (Stanger, 1988; 1993), and for *P. furcifer* (Charles-Dominique and Petter, 1980; Stanger, unpubl.)

Character State Changes in Vocalizations

Many of the call types presented above are shared by all cheirogaleid taxa studied so far, and some of them also by other Lemuriformes (Stanger, unpubl.). Therefore it is most parsimonious to presume that these call types were present in the most recent common ancestor of the cheirogaleids (Figure 13). However, several species within the cheirogaleids remain to be studied.

The following vocalization characters varied within the cheirogaleids and are used to retrace the evolution of their character states by mapping them onto the cladogram: (1) the complex acoustic structure of the trill call is produced by *M. murinus* and M. *coquereli;* (2) the acoustic structure of "hn" calls is shared by *M. coquereli* and *P. furcifer*; (3) the long and high-frequency whistle is unique for *M. murinus*; (4) the variable acoustic structure of the "zek" call is unique for *M. coquereli*, but *P. furcifer* emitted one (downward modulated) version of the "zek" call as "kea" calls.

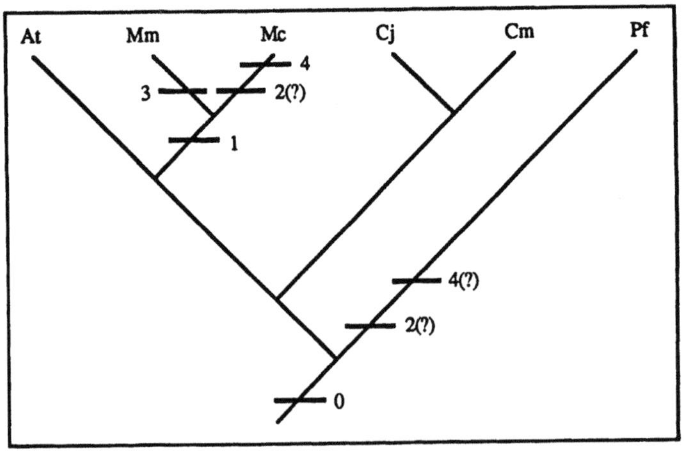

Figure 13. The evolution of vocalization structures within the cheirogaleids. (0) Grunts and squeal calls were presumably present in the most recent common ancestor of the cheirogaleids; (1) the trill call is a shared derived feature of *M. murinus* and *M. coquereli*; (2)"hn" calls are produced by *M. coquereli*, and by *P. furcifer*; (3) a long high-frequency whistle is unique for *M. murinus*, and (4) the different versions of the "zek" call are unique for *M. coquereli*; *P. furcifer produces possibly one version of the "zek" call;* (?) for an alternative hypothesis regarding the evolution of these characters see discussion.

When comparing the acoustic structures of calls that are emitted in comparable contexts the following becomes evident (Figure 14): (1a) in the advertisement context only *C. major* and *C. medius* emit narrow-band advertisement calls; (1b) the advertisement calls of *M. murinus* and *M. coquereli* are not only wide-band but they also consist of a complex temporal sequence with increasing pulse-length and increasing pulse intervals; (2) a second call for inter-group communication is absent in *C. medius*, and probably in *C. major* (Petter and Charles-Dominique, 1979); (3) a special call accompanies locomotion in *P. furcifer* and in *M. coquereli*.; (4) *C. major* and *C. medius* are quiet in the alarm context, whereas (5) *M. murinus* emit narrowband and ultrasonic alarm calls.

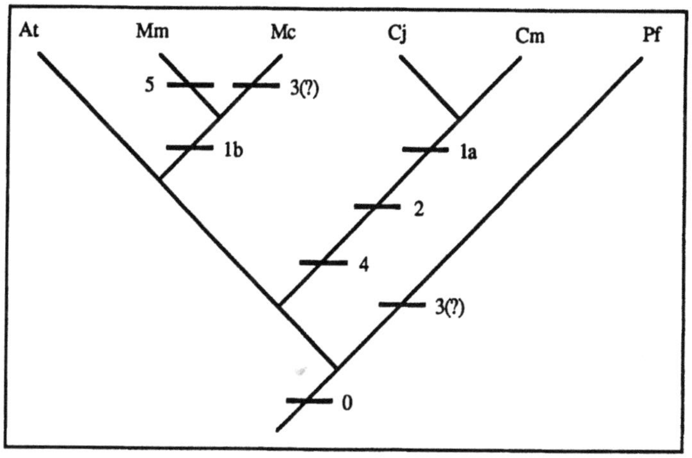

Figure 14. The evolution of vocalization structure by context within the cheirogaleids. (0) In the context of threat and distress, all taxa emitted structurally similar calls, and for long-distance communication between individuals or groups probably two call types were used; (1a) In the advertisement context only *Cheirogaleus* emit narrow-band calls; (1b) the advertisement calls of *M. murinus* and *M. coquereli* contain a characteristic dynamic pattern; (2) second call type for inter-group communication absent; (3) call accompanying locomotion; (4) absence of alarm calls; (5) cryptic alarm call in ultrasonic range; (?) for an alternative hypothesis regarding the evolution of this character see discussion.

DISCUSSION

The Phylogenetic Hypothesis of Cheirogaleid Relationships

The cladistic analysis grouped *P. furcifer* as a sister taxon to all other cheirogaleids. This dichotomy is supported by chromosomal data: *P. furcifer* is the only cheirogaleid with a diploid number of 46 chromosomes including 28 acrocentric autosomes and a fundamental number (number of autosomal arms) of 62. In contrast, all other cheirogaleids studied so far (*C. major*, *C. medius*, *M. murinus*, *M. rufus*, and *M. coquereli*) possess 66 chromosomes, with acrocentric autosomes generating a fundamental number of 66 (Chiarelli, 1973; Rumpler and Dutrillaux, 1979). However, it has been proposed that the primitive karyotype of the strepsirhines consisted exclusively of acrocentric chromosomes and that its fundamental number may have been close to 66 (Martin, 1990). In this case the low number of chromosomes in *P. furcifer* would represent a uniquely derived feature, and thus would not tell us much about evolutionary relationships within the cheirogaleids. Chromosome banding studies, or studies on the level of individual genes (Coleman Healy, this volume) are needed to resolve the relationships among the cheirogaleids more closley.

In the present analysis *M. murinus* and *M. coquereli* are grouped together by one synapomorphy: the lack of a frontal-palatine contact (Cartmill, 1978), and possibly by a second (depending on the ancestral state): the possession of an elaborate behavioral sequence, the urine-washing display (pers. observ.). It remains to be studied whether the urine-washing display is also shared with *A. trichotis*. The presence of an os planum and a postglenoid process fused to the bulla unites *M. murinus* and *M. coquereli* with their sister taxon *A. trichotis* (Cartmill, 1978; Eaglen, 1980; Yoder, 1992). *P. furcifer* and the two *Cheirogaleus* species share premolars with P^2 not higher than P^3 (Eaglen, 1980), but in contrast to *P. furcifer*, in *Cheirogaleus* and in all other cheirogaleids the medial pterygoid does not protrude the ventral plate of the pterygoid (Eaglen, 1980; Yoder, 1992), and they do not possess a gular gland (Tattersall, 1982; Meier and Albignac, 1991). This pattern of phylogenetic relationships resulted as the single most parsimonious solution from a cladistic analysis. The tree length of 16 steps suggested, that on average each character required 1.23 steps to be placed on the suggested phylogenetic tree. A closer analysis revealed that ten characters required only one step to be placed on the tree, whereas three characters (Hypocone M^{1-2}, Hypoconulid M_3, and fat storage in tail) required two steps.

Tattersall (1982) proposed an alternative hypothesis of cheirogaleid relationships, based exclusively on the analysis of dental characters (Tattersall and Schwartz, 1974; 1975). Tattersall (1982) suggested *Allocebus* and *Cheirogaleus* as sister group to *Microcebus*, *Mirza*, and *Phaner*, with *Microcebus* and *Mirza* as sister taxa. Even earlier, based on immunological data, Sarich and Cronin (1976) proposed a sister group relationship of *Cheirogaleus* and *Phaner*, with *Microcebus* as their sister taxon, although they failed to recognize a monophyletic origin of the Malagasy lemurs (Yoder, 1992).

In general, I agree with Cartmill (1978) who stated that we should not use just one single set of characters, but that we should include as many different characters as possible in a phylogenetic analysis. Ideally, all data sets would generate the same phylogenetic hypothesis, but in reality, we find that data are sometimes contradictory and result in conflicting hypotheses. To resolve these conflicts it is necessary that we know more *a priori* about the informative value of each character set for the evolutionary level of interest, but this is rarely the case. As an alternative, we could generate an extensive data set to compensate for contradictions among single characters within the set. Unfortunately, this is difficult at the level of the cheirogaleids. Tattersall (1982) and Tattersall and Schwartz (1974, 1975) did not provide a character matrix for their phylogenetic analyses, and the immunological data presented by Sarich and Cronin (1976) are similarity measures which do not distinguish between homologous and convergent similarities, or between primitive and derived character states. Therefore we are presently left with several hypotheses which are all based on limited character sets instead of working with one comprehensive phylogenetic analysis of cheirogaleid relationships. However, most studies in the past have focused on higher level relationships among the Lemuriformes, and with increasing interest in the cheirogaleids as a family group we undoubtedly will accumulate more characters which will be useful for an extensive cladistic analysis of this group.

For the time being we have to rely on additional evidence supplied by phenetic characters (e.g. chromosomes, immunological data, and vocalizations without information about ancestral character states). As discussed above, chromosomal characters in their present state are not very informative. Unfortunately this is also the case for most molecular data, but for a different reason. Until recently studies have focused on only a few taxa within families, and therefore did not resolve among all the taxa within a family group (e.g. Dene *et al.*, 1976; Sarich and Cronin, 1976; Yoder, 1992). Molecular studies showed, nevertheless, that the cheirogaleids and *Daubentonia* were among the earliest taxa to diverge from a lemuriform ancestor (Dene *et al.*, 1976), and Yoder (1992) documented a monophyletic origin of the Malagasy lemurs. Once there is information available on all taxa within the cheirogaleids, molecular data will be helpful to formulate an independent phylogenetic hypothesis and to test an existing phylogenetic hypothesis by mapping the new data onto the cladogram and evaluating their goodness of fit.

A Note on Homology

One of the most important concepts in cladistics is the detection of phylogenetic relationships by means of shared derived characters or synapomorphies (Hennig, 1966). Shared derived characters can be traced back to a common ancestor and are thus homologous. The homology concept was originally developed by Owen in 1843 (Patterson, 1987; Roth, 1991) to compare similar structures between individuals, and was put into a phylogenetic framework by Darwin (1859). Distinguishing homologous (similar by descent) and analogous (similar by function, and convergent) characters, and showing that a subset of the taxa under investigation uniquely share a derived state of a homologous character is one of the main challenges one faces when establishing a cladistic data set (but see Cartmill, in press).

In morphological studies the likelihood of a morphological convergence can be related to the observed functional convergence of the characters under investigation. In morphological studies we can assess homology and the polarity of characters (primitive versus derived character states) by using fossil evidence, outgroup comparisons, the criterion of complexity (if two complex characters consist of the same combination of simple units they have a high likelihood of being homologous), or ontogenetic information (Andrews, 1987).

In comparison, convergence becomes increasingly more difficult to identify at the molecular level (Andrews, 1987). A large part of molecular data (e.g. chromosomal, immunological, and DNA hybridization data) are basically phenetic. They permit an assessment of overall similarity and differences (calculated as phylogenetic distance) between chromosomes, proteins and DNA, but they cannot by themselves isolate these changes and thus are not able to identify any similarities or differences as homologous, or convergent (Andrews, 1987; Avise, 1994). Protein and DNA sequencing allow a more detailed analysis than the phenetic distance methods, although on most levels of analysis there is rarely a theoretical or empirical way of distinguishing between homologous and convergent changes other than by parsimony (Felsenstein 1981; 1983), or probability alone (Bishop and Friday, 1985; Andrews, 1987).

Homology in Vocalizations

The comparative study of vocalizations poses another but quite similar problem for the assessment of homology. Phylogenetic homology is by definition based on shared structural elements, which can be traced to a common ancestor (Roth, 1984). Vocalizations can be described structurally by means of their frequency and amplitude characteristics and how these change through time. These measurements, however, are indirect measures describing a behavior rather than direct measures of morphological features, for example, changes in vocal tract or larynx. In stating that vocalizations are homologous we are implying that the underlying morphologies and production mechanisms are homologous. Without anatomical and physiological studies we cannot support that notion and homology can only be inferred. In doing so, we assume that similar vocalization structures indicate a common ancestry, that vocalizations are not likely to be highly labile and that they are not subject to selective pressures resulting in similar vocalization structures. Yet this assumption is questionable, and there exists a great body of literature warranting caution (Morton, 1975; Marten and Marler 1977; Marten *et al.*, 1977; Waser and Waser, 1977; Wiley and Richards, 1978, 1982; Gish and Morton, 1981; Michelsen, 1983; Waser and Brown, 1986). Comparative studies are needed to determine the contribution of phylogenetically stable features to primate vocal structure, and to describe features which are more likely influenced by selection pressures through the environment. The features to study, of course, may vary depending on the evolutionary level of interest (e.g. species, genera or families). In most cases it may be easier to determine homologies when studying species just because two species may not have had enough evolutionary time to diverge significantly (beyond recognition) from each other. For the same reason similarities between two taxa are less likely due to convergence when comparing species than they are when comparing families.

Homology by Context

Several (if not most) studies have used context as a criterion to homologize vocalizations (e.g. Gautier, 1989; Zimmermann, 1990; Macedonia and Stanger, in press). This is convenient, because the complete vocal repertoire of a species, and much less those of several species, are rarely known. Comparing calls by their context gives the researcher a basis for comparison. However, in a morphological data set the criterion of context (or function) is the exact criterion used to identify likely analogous data (see above). Furthermore, Nowicki and his coworkers recently argued that there is possibly "an arbitrary realitionship between a signal's function and the motor patterns used to produce it" (Nowicki *et al.*, 1992; see also Smith, in press). If this should also be true for primate vocalizations, the habit of homologizing vocalizations by function would be irreconcilable with the traditional concept of homology by structure and, by inference, production mechanisms.

The likelihood of a correct assessment of homologous vocalizations may depend on the level of analysis, and on how detailed our comparisons are (e.g. overall qualitative features, or quantitative measurements). The more general the descriptions are the more likely it is that analogies may be mistaken for homologies, e.g.whistles in songbirds, rodents, and primates (Marler, 1969; Nowicki and Marler, 1988; Noirot and Pye, 1969; Snowdon *et al.*, 1986). The more detailed the descriptions are, the more likely it is that we will find statistical differences just by chance (Gautier, 1989). As a result we may have to adjust the level of our acoustic analysis to match the level of the phylogenetic resolution we are studying. Within the Lemurifomes the overall vocalization structure and any quantitative measurements are useful only at and below the level of the family group (Stanger, unpubl.). Above that level, the majority of vocalization structures is not directly comparable.

It seems that for a clear assessment of homology, the complete vocal repertoire including context information is needed, and ontogenetic data following changes in structure *and* context from infancy to adulthood are useful if not necessary. Since these data are presently unavailable for the cheirogaleids (and for the Lemuriformes in general) we have to rely on an independent phylogenetic hypothesis (1) to verify the homology of vocalization structures, (2) to study how vocalization structures evolved, and (3) to decide which vocal characters can be used in a phylogenetic analysis. All three points will be addressed below.

Vocalizations: Acoustic Structure and Life Style

The four cheirogaleids share many acoustic features of their calls. However, only *M. coquereli* and *P. furcifer* share "hn" calls, and the production of trill calls is limited to *M. murinus* and *M. coquereli. M. coquereli also* produced highly variable "zek"calls (one version of these "zek" calls is used as the first pulse of their trill calls). A structurally similar call was emitted by *P. furcifer* males as "kea" call. In combination with the "kiu" calls, the "kea" calls of *P. furcifer* may be structural predecessors to the trill calls.

When analyzing the acoustic structures of calls by context several differences between taxa became apparent. *C. major* and *C. medius* produce series of narrow-band whistles in the advertisement context. In contrast, *M. murinus* and *M. coquereli* produce rather long advertisement calls, which consist of a series of wide-band pulses that may function to signal their location in the forest (Wiley and Richards, 1978). *P. furcifer* also emitted wide-band pulses which may be emitted in a long series, but these series lack the temporal organization of the advertisement calls of *M. murinus* and *M. coquereli.* Wide-band structure provides conspecifics with multiple cues for the localization of the caller by inter-aural comparison of the frequency input (Wiley and Richards, 1978). However, these cues may also be utilized by predators, and species especially susceptible to predation may evolve narrow-band advertisement calls in response. In addition to the narrow-band advertisement calls, *Cheirogaleus* ssp. show several unique specializations including the lack of alarm calls that do not give any cues to localize the animals. Although about the same size as *M. coquereli*, *C. major* and *C. medius* are less saltatory and considerably slower, and thus the lack of localizable calls or call features may increase their chance of survival by making them less conspicuous to predators. A lack of alarm calls has also been reported for the slow lorises (*N. coucang*: Zimmermann, 1985a) and is thought to be related to their cryptic life style.

In the context of alarm, *M. murinus* produces narrow-band and high-pitched alarm whistles which are difficult to localize and are outside the range of best hearing of most predators (Konishi, 1973). In contrast, the alarm calls of *M. coquereli* and *P. furcifer* are in a low frequency range and contain rapid frequency modulations over a wider frequency range resulting in wide-band calls, which are more easily localized than the narrow-band calls of *M. murinus*. It could be argued that the high-pitched alarm calls of *M. murinus* are a correlate of their small body size rather than an adaptation to predation pressure. However, the fact that *M. murinus* produce low-pitched calls in other contexts weakens this argument considerably.

M. coquereli and *P. furcifer* emit "hn" calls during locomotion, making themselves more vulnerable to nocturnal predators, which hunt by ear (Konishi, 1973). Calls associated with locomotion have also been reported for *Eulemur* taxa (Macedonia and Stanger, in press) and occur in several haplorhine species, for example, tamarins (Caine and

Stevens, 1990). *P. furcifer* has been documented to travel in pairs (Charles-Dominique and Petter, 1980), and "a loose pair-bond" has also been suggested for *M. coquereli* (Pages, 1980). With "hn" calls *P. furcifer* and *M. coquereli* may be able to monitor their social partners or infants when moving through the forest at night. Furthermore, these calls may enable them to identify any sudden noises in their vicinity as moving conspecifics (when accompanied by "hn" calls), and distinguish these travelling noises from others which may be caused by potential predators. In captivity any locomotor activity of *M. coquereli* is accompanied by "hn" calls (Stanger, 1993), and *P. furcifer* emit these calls whenever they are moving even when alone (Petter and Charles-Dominique, 1979).

The Evolution of Cheirogaleid Call Structures: the Present Evidence

The two *Cheirogaleus* taxa (*C. major* and *C. medius*) share the lack of an alarm call, the lack of a trill call, and as a result a single narrow-band advertisement call. The widespread distribution of wide-band advertisement calls among strepsirhines, and the absence of wide-band advertisement calls in dwarf lemurs, suggests the lack of a wide-band version as derived. Since the cladogram groups the two *Cheirogaleus* species as a monophyletic group, their shared narrow-band advertisment calls (whistles) can be considered homologous. In the context of the phylogenetic hypothesis, the lack of an alarm call and the lack of the wide-band element in their advertisment call constitute a loss of these features since the last common ancestor with *P. furcifer* and the other cheirogaleids.

The presence of a complex advertisement call consisting of a long series of chevron-shaped pulses with increasing pulse lengths and increasing inter-pulse intervals as well as decreasing peak frequencies is a shared derived characteristic of *M. murinus* and *M. coquereli*. With the lack of a frontal-palatine contact (Cartmill 1978), and the possession of an elaborate behavioral sequence, the urine-washing display, this character constitutes the third synapomorphy for this group. The complex structure of these vocalizations and their presence in two sister taxa indicate true homology of these two calls. A second call type for inter-group communication may have been present in the common ancestor of the cheirogaleids and lost in the branch leading to *Cheirogaleus* but the exact location of the loss on the cladogram will be determined by the presence or absence of a second call type in *A. trichotis*.

M. coquereli and *P. furcifer* share "hn" calls and possibly one version of the "zek" call. The sharing of these features may be due to shared production mechanisms due to common descent or to convergence in response to similar selection pressures. It is possible that "hn" and "zek" ("kea") calls evolved independently in *P. furcifer* and *M. coquereli*, possibly as a correlate of the "pair-bond" relations found in both taxa (Pages 1978; Charles-Dominique and Petter 1980). It is equally parsimonious, however, to assume that they were present in the cheirogaleid ancestor and lost in the branches leading to *Cheirogaleus* and to *M. murinus*. The presence or absence of these call types in *A. trichotis* will further support or reject either one of these two hypotheses.

The vocalizations of *A. trichotis* need to be studied, and a more detailed analysis of the vocalizations of *C. major* and *P. furcifer* is needed before a more complete picture of the evolution of cheirogaleid call structures can be inferred. Not only will we learn more about the evolution of calls, we also may be able to test conflicting phylogenetic hypotheses. According to Tattersall's hypothesis (Tattersall, 1982) it should be expected that *P. furcifer* shares more acoustic features with *M. murinus* and *M. coquereli* than either of them with *A. trichotis*. In contrast, the hypothesis presented in the present analysis predicts that *A. trichotis* will be more similar to *M. murinus* and *M. coquereli* than to *C. major* or *C. medius*. In both cases it can be predicted that *A. trichotis* should have whistle calls in its repertoire.

The vocalizations as well as most morphological characters of a seventh cheirogaleid species, *M. rufus*, are still unstudied. As a congener to *M. murinus*, *M. rufus* may be expected to be close to *M. murinus* on the cladogram. If this should be the case it can be predicted that it will have whistles and possibly a trill call in its repertoire.

M. rufus is roughly the same size as *M. murinus* (it is actually a little smaller with a head-body length of about 101 mm compared to 126 mm for *M. murinus*: Petter *et al.*, 1977). For the hypothesis to hold that the alarm call structure in *M. murinus* signifies a cryptic strategy to avoid predators, it is necessary that at least the *M. rufus* from the Western population which are sympatric with *M. murinus*, and thus exposed to the same predators, also emit narrow-band and ultrasonic alarm calls.

In conclusion, with an independently acquired phylogeny we can study the evolution of vocalization structures without any *a priori* knowledge of ancestral character states. By plotting vocal structure onto an existing cladogram we can deduce the evolutionary history of known call structures and also predict the call structures of yet unstudied taxa. By comparing call structures by context, we can trace adaptations of individual taxa to their environments, and we can detect whether comparable call structures are used in a different context today from the one in which they originally may have evolved (Stanger, unpubl.). Future studies of the vocal repertoires of the unknown taxa will provide us with data to test and further modify our present hypotheses. From this process we can gain confidence about the details of the evolution of strepsirhine vocalizations.

ACKNOWLEDGMENTS

I thank R. Foerg-Hoffmann and R. Hoffmann for the opportunity to study *M. murinus* and *C. medius* (1986-1987), and H.-U. Schnitzler for advice and logistical support. J. U. Ganzhorn made my field work in Madagascar (1988) possible. I am grateful to E. L. Simons, scientific director, for his invitation to study *M. coquereli* at the Duke University Primate Center (DUPC), and I thank the DUPC staff for their continuous support. K. Schmidt-Koenig provided the recording equipment, and S. Nowicki has kindly provided the sound analysis facilities. Thanks to S. Nowicki and P. Klopfer for their advise, and to J. Podos for constructive comments on an earlier version of this manuscript. I am grateful to K. Izard, G. A. Doyle, and L. Alterman for the opportunity to participate in this symposium. This study was made possible by DAAD and Landesgraduiertenförderung grants to K.F.S., and by an NSF grant to the DUPC. This is DUPC publication # 580.

REFERENCES

Altmann, S. A. (1962). A field study of the sociobiology of the rhesus monkey, *Macaca mulatta*. *Annals of the New York Academy of Sciences, 102 (2)*, 338-435.

Andrew, R.J. (1963). The origin and evolution of the calls and facial expressions of the primates. *Behavior, 20*, 1-109.

Andrews, P. (1987). Aspects of hominoid phylogeny. In C. Patterson (Ed.), *Molecules and Morphology in Evolution* (pp. 23-53). Cambridge: Cambridge University Press.

Avise, J. C. (1994). *Molecular Markers, Natural History and Evolution*. New York: Chapman and Hall.

Bearder, S. K. (1987). Lorises, bushbabies, and tarsiers: diverse societies in solitary foragers. In B. B. Smuts, D. L. Cheney, R. M. Seyfarth, R. W. Wrangham, T. T. Struhsaker (Eds.), Primate Societies (pp. 11-24). Chicago: The University of Chicago Press.

Bishop, M. J. & Friday, A., E. (1985). Molecular sequences and hominoid phylogeny. In B. Wood, L. Martin, P. Andrews (Eds.), *Major Topics in Primate and Human Evolution* (pp.150-156). Cambridge: Cambridge University Press.

Brooks, D. R. & McLennan, D. A. (1991). *Phylogeny, Ecology, and Behavior. A Research Program in Comparative Biology*. Chicago: The University of Chicago Press.

Caine, N. G. & Stevens, C. (1990). Evidence for a monitoring call in red-bellied tamarins. *American Journal of Primatology, 22*, 251-262.

Cartmill, M. (1978). The orbital mosaic in Prosimians and the use of variable traits in systematics. *Folia Primatologica, 30*, 89-114.

Cartmill, M. (in press). A critique of homology as a morphological concept. American Journal of Physical Anthropology.

Charles-Dominique, P. (1977). *Ecology and Behaviour of Nocturnal Primates*. New York: Columbia University Press.

Charles-Dominique, P. (1978). Solitary and gregarious prosimians: evolution of social structures in primates. In D. J. Chivers & J. Herbert (Eds.), *Recent Advances in Primatology* (pp.139-149). New York: Academic Press.

Charles-Dominique, P. & Martin, R. D. (1970). Evolution of lorises and lemurs. *Nature*, *227*, 257-260.

Charles-Dominique, P. & Petter, J.-J. (1980). Ecology and social life of *Phaner furcifer*. In P. Charles-Dominique, H. M. Cooper, A. Hladik, C. M. Hladik, E. Pages, G. F. Pariente, A. Petter-Rosseaux & A. Schilling (Eds.), *Nocturnal Malagasy Primates* (pp. 75-95). New York: Academic Press.

Cherry, J. A., Izard, M. K. & Simons, E. L. (1987). Description of ultrasonic vocalizations of the mouse lemur (*Microcebus murinus*) and the fat-tailed dwarf lemur (*Cheirogaleus medius*). *American Journal of Primatology*, *13*, 181-186.

Chiarelli, B. (1973). Numerical data on the chromosomes of Tupaioidea, Lorisoidea, Lemuroidea, and Tarsioidea. *Journal of Human Evolution*, *2*, 279-281.

Darwin, C. R. (1859). *On the Origin of Species by Means of Natural Selection*. London, John Murray.

Daschbach, N. J., Schein, M. W. & Haines, D. E. (1981). Vocalizations of the slow loris, *Nycticebus coucang* (Primates, Lorisidae). *International Journal of Primatology*, *2*, 71-80.

Dene, H., Goodman, M., Prychodko, W. & Moore, G.W. (1976). Immunodiffusion systematics of the primates. *Folia primatologica*, *25*, 35-61.

Eaglen, R. H. (1980). The systematics of living Strepsirhini, with special reference to the Lemuridae. Unpublished doctoral dissertation, Duke University (Durham, NC).

Felsenstein, J. (1981). A likelihood approach to character weighting and what it tells us about parsimony and compatibility. *Biological Jornal of the Linnean Society*, *16*, 183-196.

Felsenstein, J. (1983). Statistical inference of phylogenies. *Journal of the Royal Statistical Society*, *146*, 246-272.

Gautier, J.-P. (1988). Interspecific affinities among guenons as deduced from vocalizations. In A. Gautier-Hion, F. Bourlière & J.-P. Gautier (Eds.), *A Primary Radiation: Evolutionary Biology of the African Guenons* (pp. 194-226). Cambridge: Cambridge University Press.

Gautier, J.-P. (1989). A Redrawn Phylogeny of Guenons Based upon their Calls- Biogeographical Implications. *Bioacoustics*, *2*, 11-21.

Gish, S. L. & Morton, E. S. (1981). Structural adaptations to local habitat acoustics in Carolina wren songs. *Zeitschrift für Tierpsychologie*, *56*, 76-84.

Gray, J. E. (1870). *Catalogue of Monkeys, Lemurs, and Fruit-Eating Bats in the Collection of the British Museum*. British Museum, London.

Gregory, W. K. (1920). On the structure and relation of *Notharctus*, an American Eocene primate. Memoirs of the American Museum of Natural History, *3*, 49-241.

Harcourt, C. & Thornback, J. (1990). *Lemurs of Madagascar and the Comores, IUCN Red Data Book*. Switzerland and Cambridge, UK: Gland.

Heideman, P.D., Erickson, K.R. & Bowles, J.B. (1990). Notes on the breeding biology, gular gland, and roost habits of *Molossus sinolae* (Chiroptera, Molossidae). *Zeitschrift für Säugetierkunde*, *55*, 303-307.

Hennig, W. (1966). *Phylogenetic systematics*. Urbana, University of Illinois Press (second edition, 1979).

Hershkovitz, P. (1977). *Living New World Monkeys*. Chicago: University of Chicago Press.

Hill, W. C. O. (1953). *Primates, Comparative Anatomy and Taxonomy I : Strepsirhini*. Edinburgh: At the University Press.

Hodun, A., Snowdon, C. T. & Soini, P. (1981). Subspecific variation in the long calls of the tamarin, *Saguinus fuscicollis*. *Zeitschrift für Tierpsychologie*, *57*, 97-110.

Klump, G. M. & Shalter, M. D. (1984). Acoustic behavior of birds and mammals in the predator context. *Zeitschrift für Tierpsychologie*, *66*, 189-226

Konishi, M. (1973). Locatable and non-locatable acoustic signals for Barn Owls. *American Naturalist*, 107, 775-785.

Macedonia, J. M. & Stanger, K. F. (in press). Phylogeny of the Lemuridae revisited: evidence from communication signals (A preliminary study). *Folia primatologica.*

Macedonia, J. M. & Taylor, L. L. (1985). Subspecific divergence in a loud call of the ruffed lemur (*Varecia variegata*). *American Journal of Primatology*, *9*, 295-304.

MacKinnon, J. R. & MacKinnon, K. S. (1980). The behavior of wild spectral tarsiers. *International Journal of Primatology*, *1*, 361-379,

Maddison, W. P. & Maddison, D. R. (1992). *MacClade*. Sinauer Associates, Inc.

Marler, P. (1969). Tonal quality of bird sounds, in *Bird Vocalizations*. R. A. Hinde (Ed.). Cambridge: Cambridge University Press.

Marshall, J. T. & Marshall, E. R. (1976). Gibbons and their territorial songs. *Science*, *193*, 235-237.

Marten, K. & Marler, P. (1977). Sound transmission and its significance for animal vocalizations I: Temperate Habitats. *Behavioral Ecology and Sociobiology*, 2, 271-290.

Marten, K., Quine, D. & Marler, P. (1977). Sound transmission and its significance for animal vocalization II: Tropical forest habitats. *Behavioral Ecology and Sociobiology*, *2*, 291-302.

Martin, R. D. (1972). Adaptive radiation and behavior of the Malagasy lemurs. *Philosophical Transactions of the Royal Society, London*, *264*, 295-352.

Martin, R. D. (1990). *Primate Origins and Evolution. A Phylogenetic Reconstruction.* London: Chapman and Hall.

McLennan, D.A., Brooks, D. R. & McPhail, J. D. (1988). The benefits of communication between comparative ethology and phylogenetic systematics. A case study using gasterosteid fishes. *Canadian Journal of Zoology*, *66*, 2177-90.

Meier, B. & Albignac, R. (1991). Rediscovery of *Allocebus trichotis* Günther 1875 (PRIMATES) in Northeast Madagascar. *Folia primatologica*, *56 (1)*, 57-63.

Michelsen, A. (1983). Biophysical basis of sound communication. In B. Lewis (Ed.), *Bioacoustics, A Comparative Approach* (pp. 3-38). London: Academic Press.

Morton, E. S. (1975). Ecological sources of selection on avian sounds. *The American Naturalist*, *109*, 17-34.

Napier, J. R. & Napier, P.H. (1967). *A Handbook of Living Primates*. New York, Academic Press.

Niaussat, M.-M. & Molin, D. (1978). Hearing and vocalizations in a Malagasy lemur: *Phaner furcifer*. In D. J. Chivers & J. Herbert (Eds.), *Recent Advances in Primatology* (pp. 821-825). New York: Academic Press.

Niemitz, C., Nietsch, A., Warter, S. & Rumpler, Y. (1991). *Tarsius dianae*: a new primate species from Central Sulawesi (Indonesia). *Folia primatologica*, *56 (2)*, 105-116.

Noirot, E. & Pye, D. (1969). Sound analysis of ultrasonic distress calls of mouse pups as afunction of their age. *Animal Behavior*, *17*, 340-349.

Nowicki, S. & Marler, P. (1988). How do birds sing? *Music Perception*, *5*, 391-426.

Nowicki, S., Westneat, M. & Hoese, W. (1992). Birdsong: motor function and the evolution of communication. *The Neurosciences*, *4*, 385-390.

Oates, J. F. & Trocco, T. F. (1983). Taxonomy and phylogeny of black-and-white *Colobus* monkeys. *Folia Primatologica*, *40*, 83-113.

Olson, T. R. (1979). *Studies on aspects of the morphology and systematics of the genus Otoelmur.* Unpublished doctoral dissertation, University of London (London, UK).

Pages, E. (1978). Home range, behaviour and tactile communication in a nocturnal Malagasy lemur *Mirza coquerli.*. In D. J. Chivers & J. Herbert (Eds.), *Recent Advances in Primatology, Vol 3* (pp. 171-177). New York: Academic Press.

Pages, E. (1980). Ethoecology of *Microcebus coquereli* during the Dry Season. In P. Charles-Dominique, H. M. Cooper, A. Hladik, C. M. Hladik, E. Pages, G. F. Pariente, A. Petter-Rosseaux & A. Schilling (Eds.), *Nocturnal Malagasy Primates* (pp. 97-116). New York: Academic Press.

Patterson, C. (Ed.), (1987). *Molecules and Morphology in Evolution: Conflict or Compromise?* Cambridge: Cambridge University Press.

Petter, J.-J. & Petter-Rousseaux, A. (1967). Contribution a la systematique des Cheirogaleinae (lemuriens malgaches). *Allocebus*, gen, nov. pour *Cheirogaleus trichotis* Gunther 1875. *Mammalia 31*, 574-582.

Petter, J.-J. & Charles-Dominique, P. (1979). Vocal communication in prosimians. In R. D. Martin & G. A. Doyle (Eds.), *The study of Prosimian Behavior* (pp. 247-306). New York: Academic Press.

Petter, J.J., Albignac, R. & Rumpler, Y. (1977). *Faune de Madagascar Mammiferes Lémuriens (Primates Prosimiens).* Paris, ORSTOM CNRS.

Petter, J.J., Schilling, A. & Pariente, G. (1971). Observations eco-ethologiques sur deux lemuriens malgaches nocturnes: *Phaner furcifer* et *Microcebus coquereli. Terre et Vie, 25*, 287-327.

Rohlf, F. J. (1982). Consensus Indices for Comparing Classifications. *Mathematical Biosciences, 59*, 113-144.

Roth, V. L. (1984). On homology. *Biological Journal of the Linnean Society, 22*, 13-29.

Roth, V. L. (1991). Homology and hierarchies: Problems solved and unresolved. *Journal of Evolutionary Biology, 4*, 167-194.

Rowell, T. E. & Hinde, R. A. (1962). Vocal communication in the rhesus monkey (*Macaca mulatta*). *Proceedings of the Zoological Society of London, 138*, 279-294.

Rumpler, Y. & Dutrillaux, B. (1979). Chromosomal evolution in Malagasy lemurs: IV. Chromosomal banding studies in the genuses *Phaner*, *Varecia*, *Lemur*, *Microcebus*, and *Cheirogaleus. Cytogenetic Cell Genetics, 24*, 224-232.

Sanderson, M. J. & Donoghue, M. J. (1989). Patterns of variation in levels of homoplasy. *Evolution, 43 (8)*, 1781-1795.

Sarich, V.M & Cronin, J.E. (1976). Molecular systematics of the primates. In M. Goodman, R. E. Tashian & J. H. Tashian (Eds.), *Molecular Anthropology* (pp. 139-168). New York: Academic Press.

Smith, K. K. (in press). Are neuromotor systems conserved in evolution? *Brain, Behavior and Evolution.*

Snowdon, C. (1986). Vocal communication. In G. Mitchell & J. Erwin (Eds.), *Comparative Primate Biology, Behavior, Conservation, and Ecology* (pp. 495-530). New York: Alan R. Liss, Inc.

Snowdon, C. T., Hodun, A., Rosenberger, A. L. & Coimbra-Filho, A. F. (1986). Long-Call Structure and its relation to taxonomy in lion tamarins. *American Journal of Primatology, 11*, 253-261.

Stanger, K. F. (1988). *Lautrepertoire und begleitendes Verhalten von nachtaktiven Halbaffen. Eine Untersuchung an Galago demidovii (Zwerggalago), Microcebus murinus (Mausmaki) und Cheirogaleus medius (Katzenmaki) in Gefangenschaft.* Unpublished Master thesis, Universität Tübingen (Tübingen, Germany).

Stanger, K. F. (1989). A comparative approach to the organization of vocal signals in nocturnal prosimians. *Human Evolution, 4*, 181-185.

Stanger, K. F. (1993). *Structure and function of the vocalizations of nocturnal prosimians (Cheirogaleidae).* Unpublished doctoral dissertation, Universität Tübingen (Tübingen, Germany).

Stanger, K. F. & Macedonia, J. M. (in press). Vocalizations of the aye-aye (*Daubentonia madagascariensis*). *Folia primatologica.*

Steele, D. G. (1973). Dental variability in the tree shrew (Tupaiidae). *Symposium of the 4th International Congress of Primatology, 3*, 154-179.

Struhsaker, T. (1970). Phylogenetic implications of some vocalizations of *Cercopithecus* monkeys. In J. Napier & P. Napier (Eds.), *Old World Monkeys*(pp.365-444). London: Academic Press.

Struhsaker, T. (1981). Vocalizations, phylogeny and paleogeography of red colobus monkeys (*Colobus badius*). *African Journal of Ecology, 19*, 265-286.

Swofford, D. (1993). *Phylogenetic analysis using parsimony (PAUP)*, version 3.1, Smithsonian Institution.

Tattersall, I. (1982). *The Primates of Madagascar.* New York: Columbia University Press.

Tattersall, I. & Schwartz, J. H. (1974). Craniodental morphology and the systematics of the Malagasy lemurs (Primates, Prosimii). *Anthropological Papers of the American Museum of Natural History, 52 (3)*, 138-192.

Tattersall, I. & Schwartz, J. H. (1975). Relationships among the Malgasy lemurs; the craniodental evidence. In W. P. Luckett, F. S. Szalay (Eds.), *Phylogeny of the Primates* (pp.299-313). New Tork: Plenum Press.

Waser, P. M. & Waser, M. S. (1977). Experimental studies of primate vocalization: specializations for long-distance propagation. *Zeitschrift für Tierpsychologie, 43*, 239-263.

Waser, P. M. & Brown, C. H. (1986). Habitat acoustics and primate communication. *American Journal of Primatology, 10*, 135-154,.

Wiley, E. O. (1981). *Phylogenetics.* New York: John Wiley and Sons.

Wiley, R.H. & Richards, D.G. (1978). Physical constraints on acoustic communication in the athmosphere: implications for the evolution of animal vocalizations. *Behavioral Ecology and Sociobiology, 3*, 69-94.

Wiley, R.H. & Richards, D.G. (1982). Adaptations for acoustic communication in birds: sound transmission and signal detection. In D. E. Kroodsma & E. H. Miller (Eds.), *Acoustic Communication in Birds* (pp. 131-181). New York: Academic Press.

Winter, P., Ploog, D. & Latta, J. (1966). Vocal repertoire of the squirrel monkey. *Experimental Brain Research, 1*, 359-284.

Yoder, A. D. (1992). *The phylogenetic affinities of the Cheirogaleidae: a molecular and morphological analysis.* Unpublished doctoral dissertation, Duke University (Durham, NC).

Zimmermann, E. (1985a). Vocalizations and associated behaviours in adult slow lorises (*Nycticebus coucang*). *Folia Primatologica, 44*, 52-64.

Zimmermann, E. (1985b). The vocal repertoire of the adult Senegal bush baby (*Galago senegalensis senegalensis*). *Behaviour, 94*, 212-233.

Zimmermann, E. (1990). Differentiation of vocalizations in bushbabies (Galaginae, Prosimiae, Primates) and the significance for assessing phylogenetic relationships. *Zeitschrift für zoologische Systematik und Evolutionsforschung, 28*, 217-239.

Zimmermann, E., Bearder, S. K., Doyle, G. A. & Andersson, A. B. (1988). Variations in vocal patterns of Senegal and South African lesser bushbabies and their implications for taxonomic relationships. *Folia Primatologica, 51*, 87-105.

Zimmermann, E. & Lerch, C. (1993). The complex acoustic design of an advertisement call in male mouse lemurs (*Microcebus murinus*, Prosimii, Primates) and source of its variation. *Ethology, 93*, 211-224.

CHEMOCOMMUNICATION IN THE REPRODUCTIVE FUNCTION OF MOUSE LEMURS

Martine Perret

CNRS URA 1183
Laboratoire d'Ecologie Générale
MNHN, 4 avenue du petit Château
91800 Brunoy, France

INTRODUCTION

Prosimians species have retained highly developed chemosensory systems and exhibit various scent-marking behaviors (Schilling, 1979; 1990). They constitute a very convenient group of species with which to test the role of chemocommunication in social behaviour and physiology. However, despite the amount of behavioral data concerned with chemocommunication, particularly studies of marking behaviors, the information that chemical cues convey and the mechanisms underlying their actions are poorly understood.

In nocturnal species, chemical signals are of major importance in social communication because visual signals are obviously far less effective in dense arboreal biotopes. Field studies of marking behaviors in nocturnal prosimians are very few, but it has been hypothesized that scent marks might be used to delineate home-range boundaries or at least determine spacing behaviors (Charles-Dominique, 1977; Schilling, 1980). Moreover, it is clear that chemical signals play a major role in mammalian reproduction (Vandenbergh, 1983).

In the lesser mouse lemur (*Microcebus murinus*), an arboreal and nocturnal Malagasy prosimian, several experiments were conducted on captive animals in an attempt to understand how they communicate and what their chemical cues mean. Studies were facilitated by the small body size and fecundity of this species which remains close to the ancestral primate stock (Dutrillaux, 1979; Rumpler and Dutrillaux,

1991). Because this area of research is very broad, only results on relationships between chemical signals and reproductive function will be presented.

Our breeding colony was established with wild-caught animals from the south-west coast of Madagascar. In this area, environmental resources are limited by the rainy season which lasts from late spring to summer. Behavioral activities and reproductive function of mouse lemurs vary greatly with contrasting seasonal changes in the biotope. During the dry season, both sexes are sexually inactive; they enter a lethargic state and are socially tolerant, nesting in large groups which include both sexes. In contrast, just prior to the rainy season, behavioral activity increases and groups formed during the winter break up. Males establish home ranges overlapping those of philopatric females and the breeding season begins (Martin, 1973; Lebec, 1984; Pages-Feuillade, 1988).

The most interesting time period, in terms of the role played by chemical signals, is that period covering the time when animals emerge from their wintering groups to the formation of spatial organization in home ranges, i.e, from the beginning of sexual activity until the actual mating season. An increase in social communication in both sexes occurs with social competition for home ranges, competition for priority access to receptive females and social competition in terms of maternal investments.

MATERIAL AND METHODS

Photoperiodic Rhythm. Behavioral and physiological circannual rhythms thus far studied in *Microcebus murinus* are dependent on photoperiodic variations (Petter-Rousseaux, 1975; 1980; 1984; Perret, 1980; Perret and Predines, 1984). In captivity, an artificial photoperiodic schedule is imposed consisting of three months of short days (8 hrs of light per day) corresponding to sexual inactivity, followed by five months of long days (14 hrs of light per day) corresponding to breeding season (Perret, 1992).

Social Groups. All subjects were adults, laboratory-born animals from two to five years old. All experimental groups were constituted at the beginning of the long-day period. Males grouped together usually had had no previous contact with one another, whereas females may have had previous contact. Various social group combinations were studied: single-sex groups (SG) which included three males or three females, mixed-sex groups (MG) which included three males and three females living together. In six out of the 12 mixed-sex groups, bulbectomized (BOX) males or females were used to test the role of olfactory communication in social behavior and reproduction. Bulbectomy was performed according to surgical procedures previously described (Perret and Schilling, 1993). Likewise, the role of gonadal activity in chemocommunication has been tested using gonadectomized (GOX) males or females.

Results obtained from these social groups were compared with data from isolated (I) animals. For these animals, olfactory isolation was obtained through continuous individual ventilation (Schilling et al., 1984). To ensure impregnation of single-sex grouped or isolated females, females were placed in a mixed-sex group when they entered pro-estrus. After estrus, all females were isolated.

Behavioral and Physiological Data. In each group, individual frequencies of agonistic, marking or sexual behaviors were recorded using a video tape during the first hour of the nocturnal activity over a period of four weeks, including the occurrence of estrus in females. To assess reproductive function, 100 μl of blood was drawn from males every one or two weeks. Plasma testosterone concentrations were measured using radioimmunoassay procedures previously described (Schilling et al., 1984).

Statistical Analysis. All values are means $\pm$ SEM. Statistical differences in behavioral or physiological data were tested using Mann-Withney tests or analysis of variance for repeated measures. Correlations between specific parameters were evaluated by regression analysis and the G test was used to assert difference in population distribution.

SOCIAL CONTEXT AND CHEMICAL SIGNALS

Mouse lemurs do not exhibit specialized scent glands. They use saliva, feces or urine as chemical signals (Schilling, 1979). Urine is the chemosignal most frequently used and a stereotypical pattern of dispersing urine, urine-washing behavior, constitutes more than 80% of all marking behaviors observed in this species. This behavior is absent or rare under short days and, in both sexes, its frequency increases markedly three to four weeks after photoperiodic stimulation (Schilling and Perret, 1987). This increase in the frequency of marking behaviors is concommittant with changes in reproductive function. In males, gonadal development occurs, with plasma testosterone concentrations reaching up to 50 ng/ml. In females, ovarian activity resumes, with estrus occurring about five to six weeks after the increase of daylength. One might be tempted to think that marking behavior is related entirely to the level of sex hormones, but the frequency of marking in each individual varies greatly according to the social context.

Chemocommunication and Male Reproductive Function

Among males in groups, the number of marking behaviors increased significantly when they competed for priority access to receptive females, as indicated by the increased number of aggressive behaviors in groups with normal animals or males kept with bulbectomized females (MG1 and MG3, Fig. 1). In all other groups without sexual competition, aggressive behaviors were very rare, and the frequency of marking behavior did not differ from that observed for isolated males.

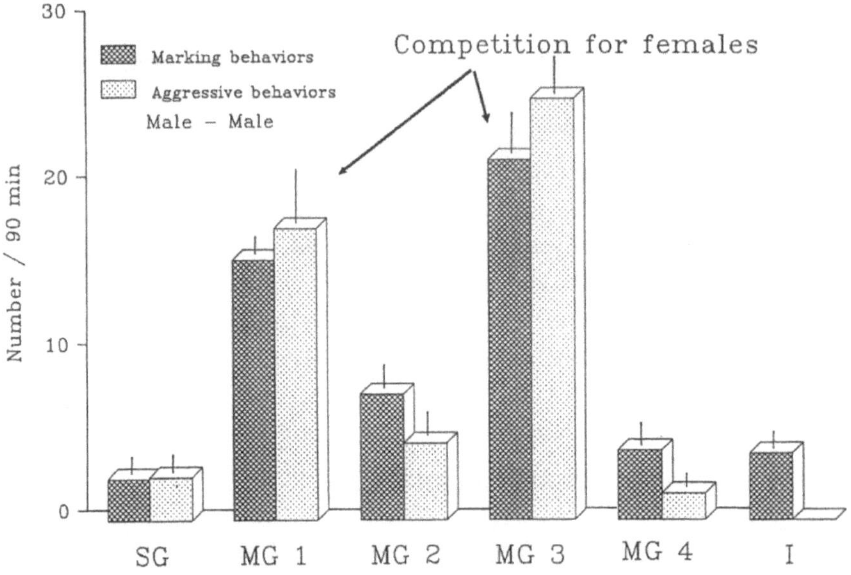

Figure 1. Marking and aggressive behaviors (mean ± sem for 3 groups) in males maintained in different social contexts: single-sex group (SG), mixed-sex groups with normal animals (MG1), with GOX females (MG2), with BOX females (MG3), with BOX males (MG4) or kept isolated (I, N=9).

In the female mouse lemur, the vulva remains closed except for the three-four day-estrous period, a situation which leads to strong competition between males, particularly since estrous periods occur synchronously. In contrast, relatively few marking behaviors were observed for groups with no females, groups with gonadectomized females or for groups including bulbectomized males (SG, MG2, MG4, see Fig. 1). The latter observation strongly suggests that chemosignals from females are a requirement for sexual competition among males.

Males in groups in which sexual competition occurred formed a social hierarchy defined in terms of the direction and number of aggressive encounters between them (Perret, 1992). It appeared that the frequency of marking was rank-related; the highest the rank, the more frequent the marking behavior (Fig. 2). No male social hierarchy developed in groups in which there was no sexual competition, and the frequency of marking behaviors was low and similar for all individuals. The same trends applies to testosterone levels. Compared to the testosterone values recorded in isolated males, only dominant males exhibited high hormonal levels while all other males, either subordinate or unranked, exhibited significantly reduced testosterone levels (* p<0.01; Fig. 2).

Moreover, highest ranking males accounted for more than 60% of all sexual behaviors within a group (sniffing and licking of female genitalia, mounting and

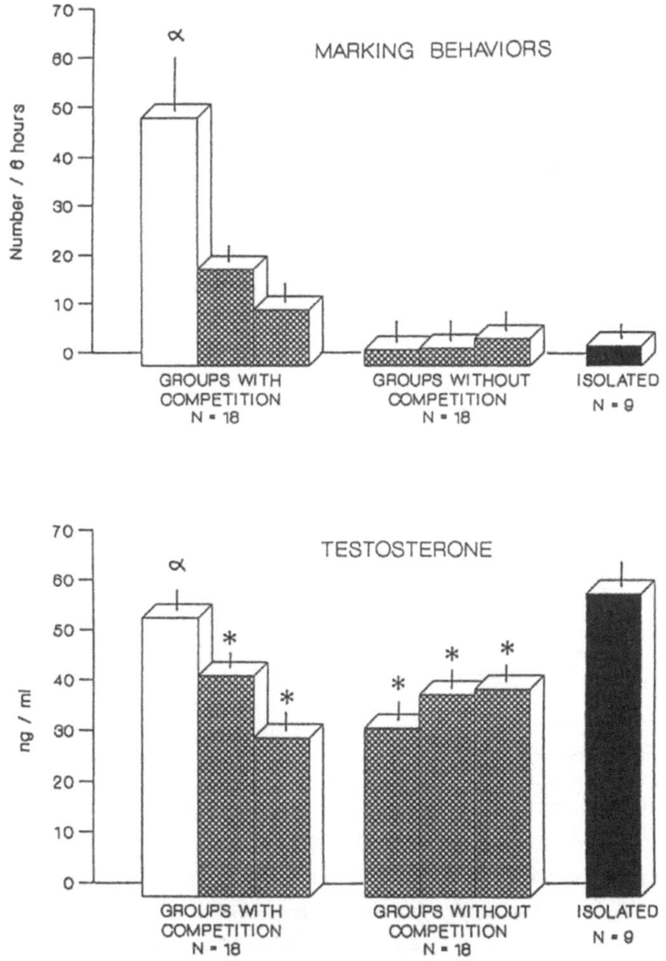

Figure 2. Frequencies of marking behaviors (number/6 hrs) and plasma testosterone levels (ng/ml) according to social context and as a function of rank in groups with sexual competition (α = dominant males). Values compared to that from isolated males (* = p<0.01).

copulation). In mouse lemurs, as has been described for other primates (Fedigan, 1983; Estep et al., 1988; Cowlishaw and Dunbar, 1991), fertility in non-dominant males is limited through behavioral constraints (aggressivity) but also through physiological mechanisms (reduction in sex hormones). Chemocommunication is implicated in these effects since they did not occur when grouped males have been bulbectomized.

Chemocommunication and Female Reproductive Function

In females kept in groups, no social competition was observed. Regardless of social context, aggressive behaviors were very rare and the frequency of marking behaviors remained stable (Fig. 3).

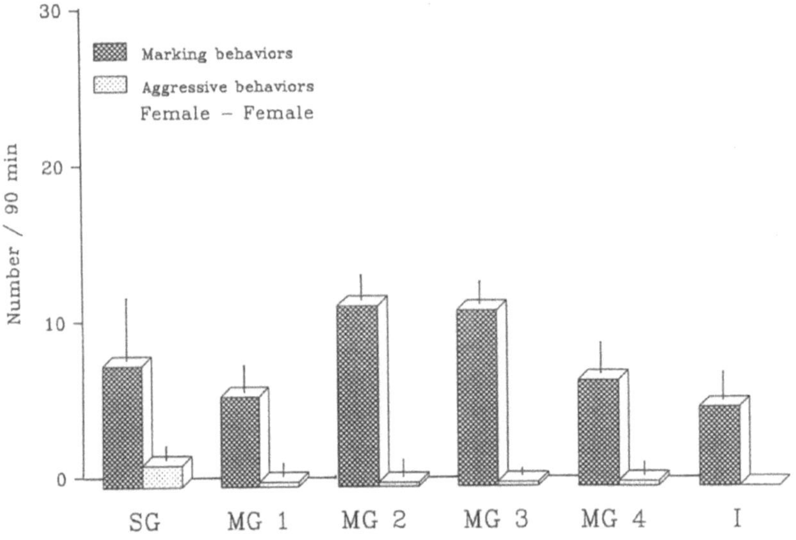

Figure 3. Marking and aggressive behaviors (mean ± sem for 3 groups) in females within the various social groups described. The frequency of marking behaviors did not differ from that observed in isolated females (I, N=5).

However, in all females, a large but temporary increase in marking behaviors, particularly anogenital marking, was observed just prior to vaginal opening. The first seasonal estrus never failed to appear in females and occurred, on average, 40 days after photoperiodic stimulation, regardless of the social context in which the female lived (Fig. 4). Results, however, indicated that the presence of sexually active males enhanced estrous synchrony between females, all estrus within a group occurring over a one week-period. In contrast, estrus occurred over a three week-period in groups in which no sexually active males were present (SG or MG5 groups, see Fig. 4). The importance of chemical stimulation from males for synchrony of estrus within a group is also indicated by the fact that bulbectomized females grouped with males (MG3, Fig.4) behaved in the same way as normal females in single-sex groups. Moreover, the occurrence of estrus was delayed when females were maintained together with no active male stimuli.

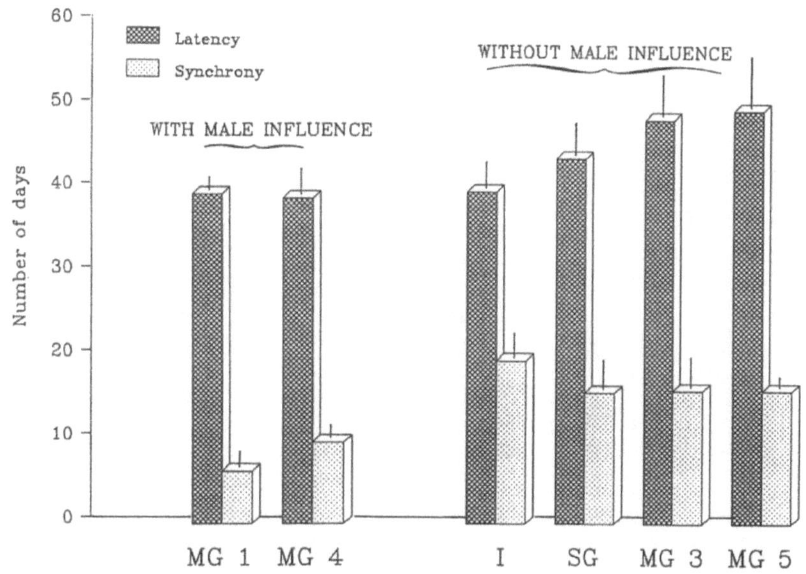

Figure 4. Latency (mean ± sem) after photoperiodic stimulation of estrous occurence in females in relation to group context. Stimuli from sexually active males ensured high estrous synchrony within grouped females.

Fecundity of females can be determined by frequency of pregnancy following mating or by the length of delay in coming into estrus again if not mated (Fig. 5). In contrast to males, no behavioral sexual inhibition occurred within grouped females or females kept in isolation until estrus occurred. Indeed, about 70% females became pregnant irrespective of age, parity or group housing. Likewise, the number of successful pregnancies and the survival of young were not affected by group housing before conception, except for females in single-sex groups in which a high abortion rate was observed (Fig. 5). It was also noted that survival of young was very low in the case of bulbectomized mothers, suggesting that olfactory communication is a necessary component of maternal behaviors. With regard to females which were not impregnated, group living resulted in a significant lengthening of estrous cycle compared with isolated females (Fig. 5; Perret, 1986). Bulbectomized females living in groups reacted in the same way as isolated females, indicating that chemical cues from females themselves might be a factor in the lengthening of the estrous cycle.

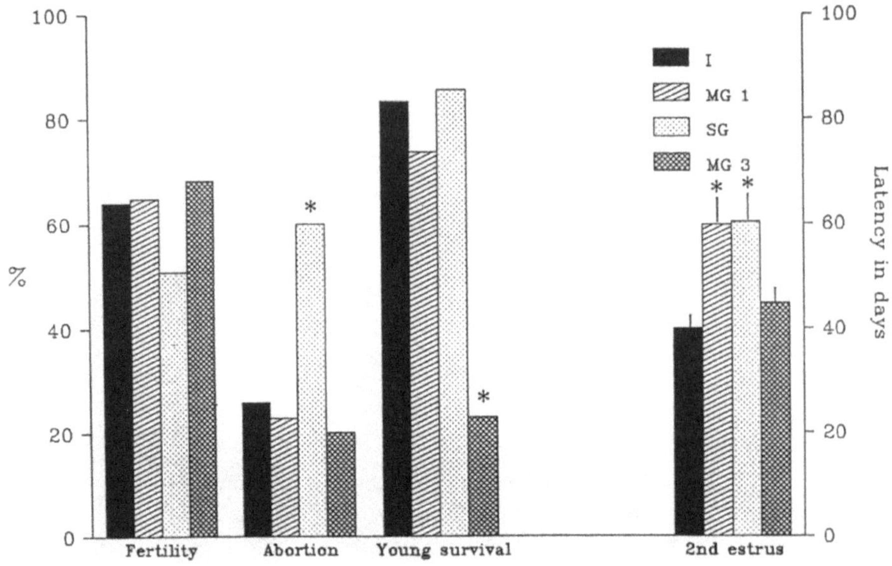

Figure 5. Reproductive capacity of females as a function of social context prior to conception (* = significant differences between group-living and isolated females, $p < 0.01$).

The most significant effect of group housing females concerned the sex ratio of their offspring. The litters of females isolated prior to conception contained 65.2% females, while the litters of females grouped together prior to conception contained 66.7% males (Fig. 6). By contrast, group living prior to conception in bulbectomiized females failed to bias the sex ratio of their litters in favor of males. The shift in sex ratio at birth towards one sex or another was not related to litter size, indicating that there was no selective destruction of embryos. Thus, the bias in sex ratio must have occurred at conception and might be, at least partly attributable to maternal hormones at the time of conception (Perret, 1990).

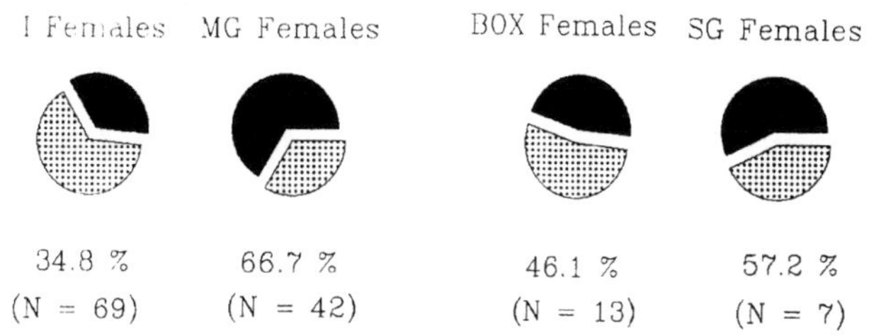

Figure 6. Sex ratio at birth (black = % males in litters and N = number of newborn) in relation to social context prior to conception.

Compared to males, females kept in groups did not establish social hierarchies and their fecundity parameters were not greatly affected, at least in our experimental conditions. Passive competition between females, however, led to a lengthening of the estrous cycle and a bias of sex ratio towards males. As for males, chemosensory systems seemed to be implicated in such effects.

PHEROMONAL EFFECTS OF URINARY SIGNALS

To test for the part played by chemosignals, specifically urinary signals, on reproductive function, animals were kept isolated in air-tight chambers ventilated by fresh air alone or with fresh air combined with urine from various donors.

In males, several pheromonal effects of urine have been demonstrated (Schilling et al., 1984; Schilling and Perret, 1987; Perret, 1992). Urinary stimulation was given simultaneously with photoperiodic stimulation over a period of four consecutive weeks. Significant reductions in testosterone levels occurred during exposure to urine from dominant or isolated males. In contrast, urine from non-dominant males had no effect, while urine from active females enhanced testosterone levels in isolated males (Fig. 7a).

The identity of the critical components in the active urinary signals is not known. However, several attempts have been made to identify the components, in the urine of dominant males, responsible for the testosterone decrease in isolated males (Perret and Schilling, 1987; Perret 1992). Active factors were found to be present in the lipophilic phase of the urine from dominant males, suggesting that they could be steroids. However, urine from a dominant male which had been castrated, was equally effective in depressing testosterone levels in isolated males, thus ruling out sexual steroids. In contrast, urine from dominant males which had their adrenocortical steroids suppressed, failed to produce a decrease in testosterone levels in subject males (Fig. 7b). It appears that, whereas pheromones enhancing reproductive function are known to be under gonadal control, those that depress it may be regulated by the adrenocortical axis.

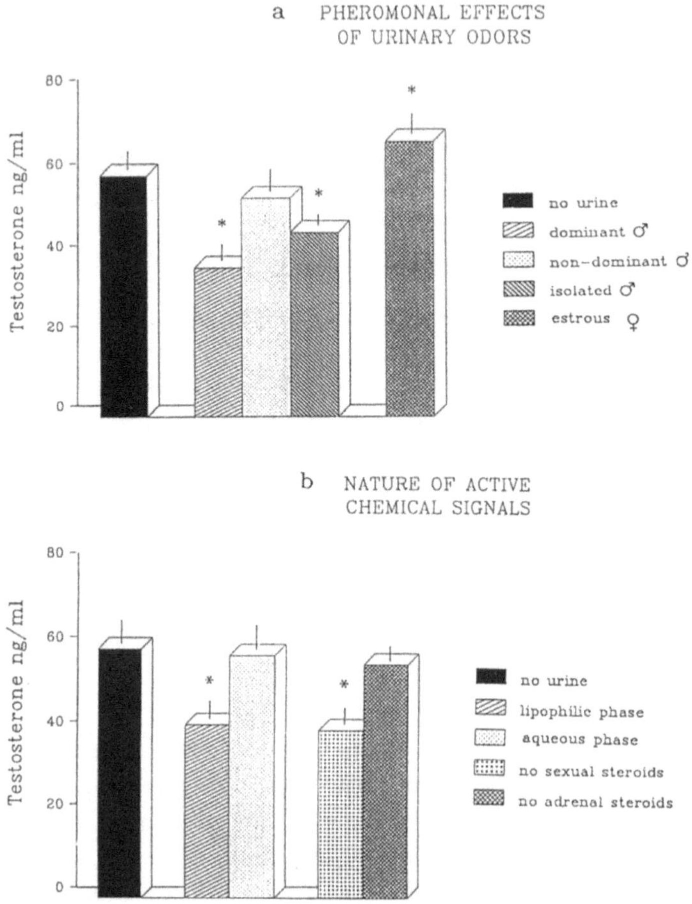

Figure 7. Plasma testosterone concentrations (means ± sem) for 10 isolated males exposed for four weeks to urine from different donors. a: urine from dominant, non-dominant or isolated males and from female in estrus. b: phases of urine from a dominant male. Values are compared to testosterone means for isolated males exposed only to non-odorized air (* = $p<0.01$).

With females, the hypothesis was tested that olfactory signals from females would modify, and, possibly reduce sex hormones in females, as was shown in the case of males. A decrease in sexual hormone concentrations preceding estrus might partially control the preconception bias in sex ratio (James, 1985). Females were isolated and either exposed or not exposed to urine from females for varying duration and at different periods before estrus. A contrasting effect was obtained for the same duration of urinary stimulation; the direction of bias towards females or males depended on the absence or the presence of urine from the beginning of photoperiodic stimulation until the beginning of the follicular phase, i.e. until 20 days before estrus (Fig. 8).

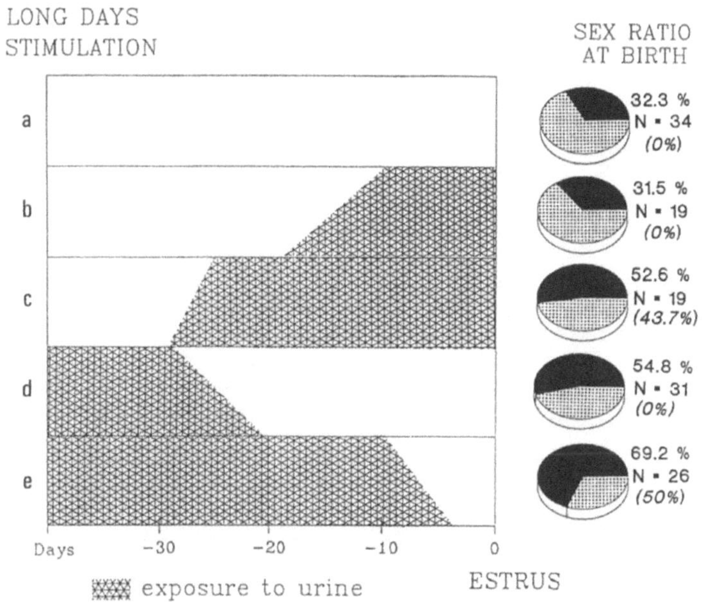

Figure 8. Sex ratio at birth (black = % males in litters, N = number of newborn) and abortion rate (indicated in parentheses) for females isolated before estrus and exposed or not to urine from grouped females for varying durations and at different periods before estrus.

In absence of urinary stimulation during this period mothers produced litters significantly biased toward daugthers (32% males from 53 newborn, lines a-b, Fig. 8). When partially covered by urine, an equilibrated sex ratio was found (54% males from 50 newborn, lines c-d, Fig. 8), while continuous urine stimulation over this period led to a significant bias of sex ratio toward males (69.2% males from 26 newborn, line e, Fig. 8). Results also revealed that abortion rate increased as a function of the duration of urinary exposure ($r=0.56$, $N=74$, $p<0.001$), reaching more than 70% when females were exposed to urine stimulation from the beginning of photoperiodic stimulation until estrus. This corresponds to what has been observed for isosexually grouped females.

Pheromonal effects of urine known in mouse lemurs are included in Table I. Stimulatory effects occur between opposite-sex individuals and include synchrony of estrus and an increase in gonadal hormones as classical effects enhancing reproduction. Inhibitory effects occur between same-sex individuals and include depression of hormonal levels, very apparent in the case of males, and highly probable in the case of females, since the effect of lengthening the estrous cycle as well as the effect of biasing the sex ratio towards males suggests that hormonal levels during the follicular phase have been adversely affected.

Table I: Pheromonal effects of urine in the lesser mouse lemur

	STIMULATORY EFFECTS		INHIBITORY EFFECTS
♂ – ♀	Synchrony of estrus	♂ – ♂	Urine from dominant and isolated males depresses testosterone levels
♀ – ♂	Increase in testosterone levels	♀ – ♀	Lengthening of oestrous cycle (luteal phase defect)
			Bias of sex ratio at birth towards males (preconception mechanism)

POSSIBLE MECHANISMS FOR PHEROMONAL EFFECTS

Mechanisms of pheromonal effects are not understood in mouse lemurs, but several experiments have shown that chemical factors interact with some of the changes initiated by photoperiodic stimulation, more particularly with secretion of prolactin (Perret and Schilling, 1987).

The general pathways that mediate the photoperiodic regulation of the reproductive axis have been extensively studied, mainly in rodents and domestic mammals (Steger et al., 1985; Arendt, 1986; Bronson, 1989). These studies show that, through changes in melatonin secretion, the activity of the suprachiasmatic nuclei is modified, leading to changes in neurotransmitters such as dopamine and the opioids. These regulate the activity of the gonadotrophin pulse generator and prolactin secretion. Under exposure to short or long days, the level of sexual hormones depend on the negative feedback loop. In the mouse lemur, typical of long daylength breeders, prolactin and gonadotrophin secretion is reduced under short photoperiods and the sensitivity of the negative feedback is high, leading to low sexual hormones secretion (Fig. 9). In contrast, exposure to a long photoperiod induces an increase in gonadotrophin secretion and a decrease in negative feedback sensitivity. Prolactin is progressively secreted and stimulates gonadal activity while further reducing the steroid feedback sensitivity. In the male mouse lemur, these endocrinological

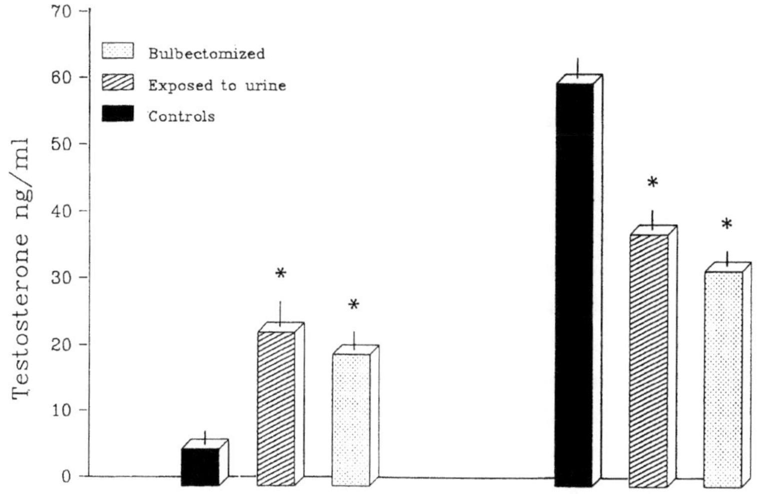

	SHORT DAYS		PHOTOPERIODIC ACTIVATION	
	Control males	exposed to urine or BOX males	Control males	exposed to urine or BOX males
PROLACTIN SECRETION	low	increased	slowly increasing	highly increasing
NEGATIVE FEEDBACK SENSITIVITY	high	decreased	highly decreasing	slowly decreasing
TESTOSTERONE ng/ml	5-10	20-40	50-70	35-45
	sexual rest	resuming sexual activity	rapid sexual activation	reduced sexual activation

Figure 9. Effects of olfactory bulb removal or exposure to urinary factors from a dominant male on testosterone levels in isolated males (N = 10) according to the photoperiodic state. Variations in prolactin secretion and negative feedback sensitivity are indicated (* = significant differences compared to control values).

modifications occur during the first weeks following photoperiodic stimulation and are accompanied by a rapid increase in testosterone secretion (Fig. 9). How the olfactory bulbs intervene in this photoperiodic regulation is unknown.

It has been recently demonstrated that efferent fibers from the olfactory bulbs reach both the suprachiasmatic and hypothalamic nuclei producing gonadotrophin (Mestre et al., 1992; Mick et al., 1993). This suggests that chemical factors may modulate the olfactory bulbs activity which in turn may modulate neurotransmitters turnover.

To test this hypothesis, we compared testosterone responses of urine treated males and bulbectomized males at two contrasting photoperiods, that means at two different neuroendocrine states (Fig. 9). In both photoperiods, testosterone responses of urine-treated and bulbectomized males were similar - their testosterone levels decreased during long days and increased during short days. However, in both photoperiods, these two contrasting responses involved identical mechanisms - an increased prolactin secretion and a corresponding decrease in the sensitivity of the negative feedback. This demonstates that the same chemical signal can act differently depending to the neuroendocrine state of the recipient. Whereas in short days, these changes reactivate the reproductive axis, they reduce the sexual response to long days. It thus appeared that the exposure to the urine from a dominant male depresses the activity of the olfactory bulbs, mimicking bulbectomy. In rodents, several nonsensory effects of bulbectomy on photoperiodic regulation of sexual function have been demonstrated (Clancy et al., 1986; Bittman, 1989; Pieper et al., 1990) and these have been related to the suppression of the tonic influence of the olfactory bulbs on brain nuclei. Such effects have also been found in the mouse lemur (Perret and Schilling, 1993; Schilling and Perret, 1993) and chemosignals might act by modulating the tonic effect exerted by the olfactory bulbs on the hypothalamo-hypophyseal-gonadal axis.

ECOLOGICAL CONCLUSIONS

The question arises as the possible role of chemical factors in wild populations. In the mouse lemur, although the breeding season is controlled primarily by photoperiod, social communication based on urine signals, could be involved in adjusting the reproductive potential of the population to the carrying capacity of the habitat. Stimulatory effects of chemical cues on reproduction are obviously necessary in this solitary species to ensure adequate coordination of sexual behaviors between the sexes. Inhibitory effects of chemical cues, on the other hand, would reduce competition between individuals of the same sex.

Indeed, at the beginning of the breeding season, it is likely that males will attempt to establish their home ranges on sites occupied by females which remain in quiescent state for longer than males. Competition between males would lead to emergence of a core of dominant males through behavioral (aggressiveness, spacing) and physiological mechanisms (pheromonal inhibition) as demonstrated in captivity. Surplus males would be excluded from the core population. Likewise, the onset of sexual activity in females would be influenced by chemical cues from competitive females leading to a bias of sex ratio at birth toward males or a delay in returning to estrus. The bias of offspring toward males would lead, in turn, to a reduction in population, i.e. fewer adult females would be in competition with their mothers (Johnson, 1988) and more males would immigrate. Thus, by means of chemical cues, local social competition which depends largely on food competition, would modulate very precisely behavioral and physiological expressions of seasonal sexual activity for each individual.

REFERENCES

Arendt, J. (1986). Role of the pineal gland and melatonin in seasonal reproductive function in mammals. *Oxford Review of Reproductive Biology*, 8, 266-320.

Bittman, E.L., Crandell, R.G., & Lehman, M.N. (1989). Influence of the paraventricular and suprachiasmatic nuclei and olfactory bulbs on melatonin responses in the golden hamster. *Biology of Reproduction*, 40, 118-126.

Bronson, F.H. (1989). "*Mammalian Reproductive Biology*". University of Chicago Press, Chicago.

Charles-Dominique, P. (1977). "*Ecology and Behaviour of Nocturnal Primates*". Duckworth, London.

Clancy, A.N., Goldman, B.D., Bartke, A., & Macrides, F. (1986). Reproductive effects of olfactory bulbectomy in the Syrian hamster. *Biology of Reproduction*, 35, 1202-1209.

Cowlishaw, G., & Dunbar, R.L. (1991). Dominance rank and mating success in male primates. *Animal Behaviour*, 41, 1045-1056.

Dutrillaux, B. (1979). Chromosomal evolution in primates: tentative phylogeny from *Microcebus murinus* to man. *Human Genetics*, 483, 251-314.

Estep, D.Q., Nieuwenhuijsen, R., Bruce, K.E., Neef, K.J., Walters, P.A., Baker, S.C., & Slob, A.K. (1988). Inhibition of sexual behaviour among subordinate sumptailed macaques, *Macaca arctoides*. *Animal Behaviour*, 36, 854-864.

Fedigan, L.M. (1983). Dominance and reproductive success. *Yearbook of Physical Anthropology*, 26, 91-129.

Johnson, C.N. (1988). Dispersal and the sex ratio at birth in primates. *Nature*, 332, 726-728.

Lebec, A. (1984). Relations entre le comportement agressif du mâle Microcèbe, les autres comportements et la physiologie sexuelle. Doctorat 3ème cycle, Université Paris VI.

Martin, R.D. (1973). A review of the behavior and ecology of the lesser mouse lemur (*Microcebus murinus*). *In Ecology and Behaviour of Primates*, pp. 1-168, R.P. Michael and J.H. Crook, Academic Press, London.

Mestre, N., Petter A., & Bons, N. (1992). Systematisation of the olfactory bulb efferent projection in a lemurian primate (*Microcebus murinus*). *Journal for Hirnforschung*, 2, 173-180.

Mick, G., Cooper, H., & Magnin, M. (1993). Retinal projection to olfactory tubercle and basal telencephalon in primates. *Journal of Comparative Neurology*, 327, 205-212.

Pages-Feuillade, E. (1988). Modalité de l'occupation de l'espace et relations inter-individuelles chez un prosimien malgache. *Folia Primatologica*, 50, 204-220.

Perret, M. (1980). Influence de la captivité et du groupement social sur la physiologie de *Microcebus murinus*. Doctorat Etat, Université Paris XI.

Perret, M. (1986). Social influences on oestrous cycle length and plasma progesterone concentrations in the female lesser mouse lemur (*Microcebus murinus*). *Journal of Reproduction and Fertility*, 77, 303-311.

Perret, M. (1990). Influence of social factors on sex ratio at birth, maternal investment and young survival in a prosimian primate. *Behavioral Ecolology and Sociobiology*, 27, 447-454.

Perret, M. (1992). Environmental and social determinants of sexual function in the male lesser mouse lemur (*Microcebus murinus*). *Folia Primatologica*, 59, 1-25.

Perret, M., & Predines, J. (1984). Effects of long-term grouping on serum cortisol levels in *Microcebus murinus*. *Hormones and Behavior*, 18, 346-358.

Perret, M., & Schilling, A. (1987). Intermale sexual effect elicited by volatile urinary ether extract in *Microcebus murinus* (Prosimian, Primates). *Journal of Chemical Ecology*, 133, 495-507.

Perret, M., & Schilling, A. (1987). Role of prolactin in a pheromone-like sexual inhibition in the male lesser mouse lemur. *Journal of Endocrinology*, 114, 279-287.

Perret, M., & Schilling, A. (1993). Response to short photoperiod and spontaneous sexual recrudescence in the lesser mouse lemur: role of olfactory bulb removal. *Journal of Endocrinology*, 137, 511-520.

Petter-Rousseaux, A. (1975). Activité sexuelle de *Microcebus murinus* soumis à des régimes photopériodiques expérimentaux. *Annales de Biologie Animale, Biochimie et Biophysique*, 15, 503-508.

Petter-Rousseaux, A. (1980). Seasonal rhythms, reproduction and body weight variation in five sympatric nocturnal prosimians in simulated light and climatic conditions. *In "Nocturnal Malagasy Primates"*, pp. 137-152, P. Charles-Dominique *et al.*, eds, Academic Press, New York.

Petter-Rousseaux, A. (1984). Annual variations in the plasma thyroxine level in *Microcebus murinus*. *General and Comparative Endocrinology*, 55, 405-409.

Pieper, D.R., Lobocki, C., Thompson, A., & Subramanian, M.G. (1990). The olfactory bulbs tonically inhibit serum gonadotrophin and prolactin levels in male hamsters on long or short photoperiod. *Journal of Neuroendocrinology*, 2, 707-715.

Rumpler, Y., & Dutrillaux, B. (1991). Chromosomal evolution and speciation in primates. *Cell Biology Review*, 23, 1-115.

Schilling, A. (1979). Olfactory communication in prosimians, *In "The Study of Prosimian Behaviour"*, pp. 461-542, G.A. Doyle and R.D. Martin, eds , Academic Press, New York.

Schilling, A. (1980). The possible role of urine in territoriality in some nocturnal prosimians. *Symposium of the Zoological Society London*, 43, 165-193.

Schilling, A. (1990). Communications par signaux chimiques chez les Prosimiens. *In "Primates: Recherches actuelles"*, pp. 12-143, J.J. Roeder and J.R. Anderson, Masson, Paris.

Schilling, A., Perret, M., & Predines, J. (1984). Sexual inhibition in a prosimian primate: a pheromone-like effect. *Journal of Endocrinology*, 102, 148-151.

Schilling, A., & Perret, M. (1987). Chemical signals and reproductive capacities in a male prosimian primate (*Microcebus murinus*). *Chemical Senses*, 12, 143-158.

Schilling, A., & Perret, M. (1993). Removal of the olfactory bulbs modifies the gonadal responses to photoperiod in the lesser mouse lemur (*Microcebus murinus*). *Biology of Reproduction*, 49, 58-65.

Steger R.W.,Matt, K., & Bartke, A. (1985). Neuroendocrine regulation of seasonal reproductive activity in the male golden hamster. *Neuroscience and Biobehavioral Reviews*, 9, 191-201.

Vandenbergh, J.G. (1983). *"Pheromones and Reproduction in Mammals"*, Academic Press, New York.

THE ACCESSORY (VOMERONASAL) CHEMORECEPTOR SYSTEM IN SOME PROSIMIANS

Charles Evans[1] and Alain Schilling[2]
[1]Glasgow Caledonian University, Glasgow G4 0BA, Scotland U K
[2]Ecologie MNHN, 91800 Brunoy, France

INTRODUCTION

The 35 plus extant species of prosimians have undergone extensive radiation over the estimated 50 million years of their evolutionary history. The range of selective pressures exerted on the group include geographical and climatic isolation; endemic speciation (in Madagascar); niche specialization through dietary and activity pattern changes, as well as social structure adaptations such as male dispersal/female dominance. This radiation has been accompanied by a wide range of anatomical and functional variations, including those of the visual, auditory, and chemical senses.

Of the various nasal chemoreceptor systems, the accessory olfactory system (AOS) has common characteristics found throughout most families of terrestrial mammals. These are principally, a fluid transport system and an enclosed chemoreceptor field contained within the blind-ending sac of the Organ of Jacobson/vomeronasal organ (VNO). The stimulus transfer system is enclosed within the cartilagineous capsule(s) surrounding both the VNO and its accompanying vascular network. The AOS has been most thoroughly studied in rodents, providing strong evidence for an important contribution to central control of reproduction through modulation of PRL and GnRH levels (Wysocki, 1989; Vandenbergh, 1989) although other neurocrines, principally opiates and octapeptides, are also implicated (Labov, *et al* 1988; Bluthe & Danzer, 1993). The primate olfactory system, as in rodents, retains a regenerative capacity eg after tract section (Barber & Raisman, 1978; Graziadei, *et al,* 1980) and hence it can probably maintain continuous neuron replacement throughout life.

A "dual" functioning system has been proposed as a common mammalian mechanism in chemosensory analysis. The separation of the two olfactory pathways - the main and accessory - is largely based on neuro-anatomical evidence of the arrangement of both afferent and efferent tracts. The concept of dual nasal olfactory systems (Raisman, 1972), which are separable but interactive, is now accepted (Keverne, 1979).

The neurosensory apparatus of the AOS comprises bipolar chemoreceptor neurons within the vomeronasal epithelium, the afferent axons comprising the vomeronasal nerve and the vomeronasal primary brain centers (eg accessory olfactory bulb). Other 'modulatory' efferent

Creatures of the Dark: The Nocturnal Prosimians
Edited by L. Alterman *et al.*, Plenum Press, New York, 1995

fibres may also be present as intra-epithelial endings. The free (lumenal) surface of each dendrite bears microvilli which are shorter, less numerous and present a smaller surface area than the ciliated dendrites of the main olfactory receptors. Since receptor cell density is similar in both systems (Schilling, 1987), this reduction in the lumenal surface projections may imply a comparable reduction in odorant trapping efficiency, or possibly a narrower stimulus/response profile of vomeronasal versus main olfactory cells.

The structural criteria on which functional capacity is likely to depend are:

1) a patent duct system, which can be shown to transport fluid from the peripheral intake point into the lumen of the VNO;
2) the presence of mature microvillar neuroreceptors within the epithelium of the VNO, together with precursor cells, immature neurons and their stem (basal) cells;
3) a vasomotor control system (the vascular sinus network); a secretory glandular system, together with their innervation, in part by the trigeminal autonomic motor component of the VNO mucosa;
4) the projection of the axons of the vomeronasal chemosensory cells to their receptor field, demarcated as the glomeruli of the accessory olfactory bulb (AOB). Secondary and tertiary connections of the AOS include specific nuclei of the limbic system in the amygdala and hypothalamus (Scalia & Winans, 1975; Davis,*et al,*1978; Schilling, 1987).

Within the prosimians, in only three species has the vomeronasal organ been fully described: the slow loris and the greater galago (Hedewig, 1980a,b) and the lesser mouse lemur. The latter is the only species in which the functions of AOB have been directly examined (Schilling, 1970; Schilling, *et al,* 1990). The components of the AOS - the VN organ and/or the AOB, have been identified to varying degrees in about 70% of species. The present study considers single species from the main Prosimian families, among the nocturnal forms, both dietary generalists and specialists are compared and then contrasted with a representative diurnal lemur. The range of ecotypes covers gum-feeders, foli/frugivores and partial insectivores. Prosimian olfaction also relates closely to their reproductive cycle(s) (Izard, 1990).

The functional role of the AOS within mammals has been suggested as being related to the particular adaptive requirements of a species, that is, the same structures have differing biological roles (Evans, 1984; Schilling, 1987). Intraspecies variability is slight in most non-primate mammals (Broom, 1897; Cooper & Bhatnagar, 1976), hence single specimens of "diosmic" primates can provide reliable taxon-specific characteristics. Specimens of the following genera were available for the present study: Adults:- (Lemuriformes) - *Phaner, Avahi, Lepilemur ruficaudatus* and *Lemur catta;* Neonates: - *Nycticebus coucang,* and *Microcebus murinus.*

The paraseptal tissues were cut at 7 mm, after formalin fixation, acid decalcification (RDC, Histolabs, U.K) and paraffin embedding. Axon and neurofibril stains were applied to every 20th section. In approximately every 50th section, a selective histochemical reaction for olfactory cells was used to locate and identify the VN neuronal elements (Mendoza & Borisch-Chappuis 1989; Lundh *et al.*, 1989). The specificity of the binding of two lectins (Soybean, SBA & Gorse, UEA-1;Sigma UK) was verified with appropriate control sugars. Other controls were applied to reduce endogenous peroxidase activity and non-specific binding by proteins; a black/brown color reaction with the peroxidase label localized the lectin/sugar complexes (Key and Giorgi, 1986). Control comparisons were made with the respiratory and main olfactory epithelia.

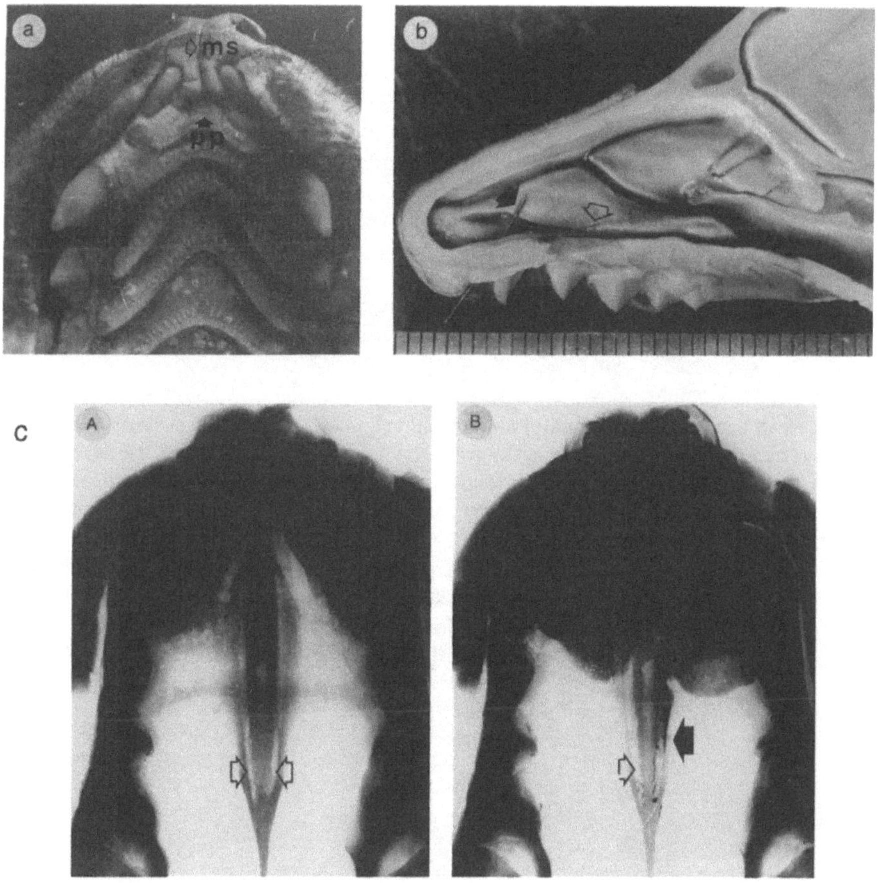

Figure 1 **STIMULUS INTAKE ROUTE**

a *(Phaner)*: Palate and ventral rhinarium; median sulcus (ms), palatine papilla (pp) & sulci X 2.

b *(Lepilemur ruficaudatus)*: hemisection right hand side nasal cavity with partial removal of caudal section of VNO ; bristle(s) inserted into naso-palatine duct (solid arrow) and inside vomeronasal duct (open arrow),.X 1

c *(Lepilemur)*: A. Control radiograph; both VNO lumens empty (open arrows)
B. Post-Injection Radiograph: right VNO lumen (solid arrow) shows ipsilateral uptake of radiopaque tracer - posteriorly as outline to circumference of lumen; left lumen empty (as A - open arrow) ; some spread of tracer into nasal cavity.x1

MORPHOLOGY

In the adults of all species examined, the criteria proposed as basic requirements for functionality were displayed in all species, but with varying degrees of development. Hence only the salient features of each vomeronasal organ will receive specific comment and distinctive structures will be considered in some detail.

A. Cheirogaleidae: *Phaner furcifer* (N= 1 male, adult)

This specimen of the fork marked dwarf lemur has conspicuous dorsal blood vessel(s) at the narrowest point (approx.dia.= 20mm) of the junction of the vomeronasal duct and the nasopalatine duct (NPD, Fig. 3), and, whilst the oral part of the NPD is larger (diameter approx. 60 mm), neither it nor the palatine papilla itself has any prominent associated vasculature. The main (compression) sinuses which run parallel to the VNO's long axis are found initially close to the clearly demarcated non-sensory area on the median side of the VN mucosa. The sinus vessels associated with the VNO and its ducts are always enclosed within cartilagineous capsules. Where the neuroepithelium is maximally developed (i.e in mid-regions of the VNO) and where the non-sensory area is correspondingly reduced (Table 1.), the sinus distribution reverts to the more typical arrangement (Fig. 2b) of a diametrically opposed large lateral sinus and small median sinus(es).

The neuroepithelium shows multiple nuclear stratification of the receptor nuclei with a distinct zonal separation of a band of supporting cell nuclei. In the mid section of the VNO, clusters of receptor cell occur, close to the basal cell layer (Fig. 5b). Lectin localization in this species is confined to receptor cells and while SBA reacts weakly with the cell body (perinuclear) and at the lumenal boundary; strong UEA-1 reactivity identifies a restricted (ventral) receptor zone with long microvilli (Fig. 6).

B. Lepilemuridae: *Lepilemur ruficaudatus* (N = 1 infant, 2 months of age)

The red tailed sportive lemur also presents a wide nasopalatine duct which, as in *Phaner*, is associated with a very large sinus vessel dorsal to the duct (Fig. 3b), again within its own separate cartilage capsule. In the main part of the organ, there is a marked stratification of the supporting and olfactory cell nuclei into distinct layers (Fig. 4). This species had strong localization of lectin staining within the dendritic zone as well as at the perinuclear zone. The SBA reaction reveals a degree of clustering of cell bodies and their dendrites. In the sportive lemur, the microvilli and their associated mucous layer were either non-reactive to SBA or too weakly reactive for resolution at this magnification.

Lemur catta (N=2 male, 2 female adults)

The general appearance of the organ in the ring tailed lemur has been briefly described (Evans, 1984; Fig.1). It is characterized by a complete neurosensory lining (Table 1.) with no demarcation of sensory/non-sensory zones throughout the organ. In cranial (LS) sections the organization of the vascular network comprises an even distribution of like-sized sinus vessels. The NPD/VND junction is again associated with conspicuous sinuses.

D.Indridae: *Avahi laniger*

(N = 1, adult)

The woolly lemur has a VNO which is distinctive in appearance throughout its length. The salient feature is an irregularly shaped lumen, bounded by an epithelium in which the sensory and non-sensory cells are interspersed, showing little or no regional demarcation or nuclear stratification. Possibly correlated with this apparently "scattered" distribution of cell types are the vaso-motor sinuses which are also irregular, both in number and size, but are distributed more or less evenly around the circumference of the organ (ringtail lemur). There is however, no dubiety about the presence of a substantial neuronal component (Fig. 5a), because prominent axon bundles are evident immediately below the basal lamina, while the dendritic layer is consistently narrow. Lectin staining of the neuroepithelium (Fig. 5b) demonstrates clusters of sensory nuclei at an atypical depth and position within the epithelium. There is a less dense staining of the dendritic zone and again minimal uptake at the lumenal surface. The caudal zone of the organ demonstrates a high concentration of sero-mucus glands, with the main entry point of their ducts being dorsal to the organ.

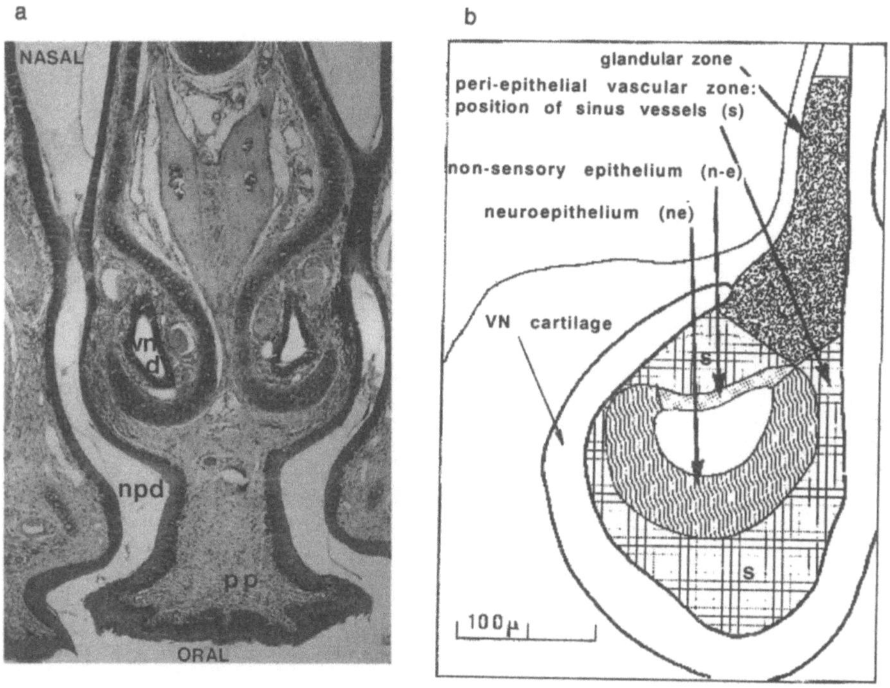

Figure 2 a **ACCESSORY CHEMORECEPTIVE AREA** (*Microcebus murinus*): TS lower septal region showing oral <-> nasal duct = Nasopalatine duct (NPD) & duct of VNO (VND),H & E X 5,

b **VN COMPLEX:** (*Microcebus*) (Schilling '70 Fig. 9) vasomotor sinuses (s); VN cartilage (VNc); neuroepithelium (ne) VN lumen (l), non-sensory zone (NS).

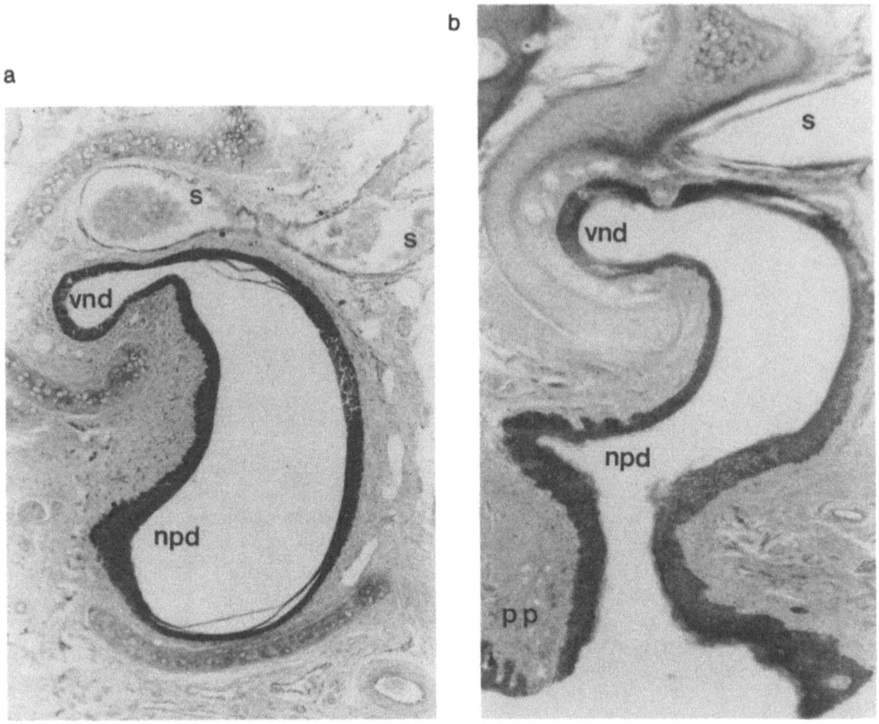

Figure 3 a,b [illegible] PPARATUS a (*Phaner*) & b (*Lepilemur ruficaudatus*): sinus vessels (s) [illegible] D & VND), palatine papilla (pp); H & E, X 40.

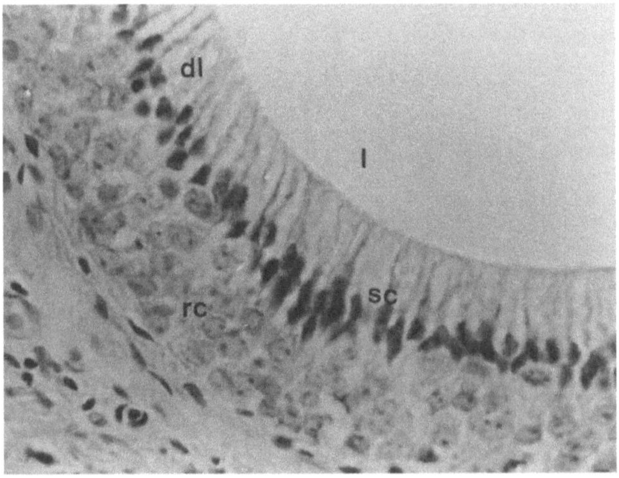

Figure 4 **VN NEUROEPITHELIUM** (*Lepilemur*): stratification of support cell (sc) & receptor cell (rc) nuclei; dendritic layer (dl), lumen (l); H &E X 400.

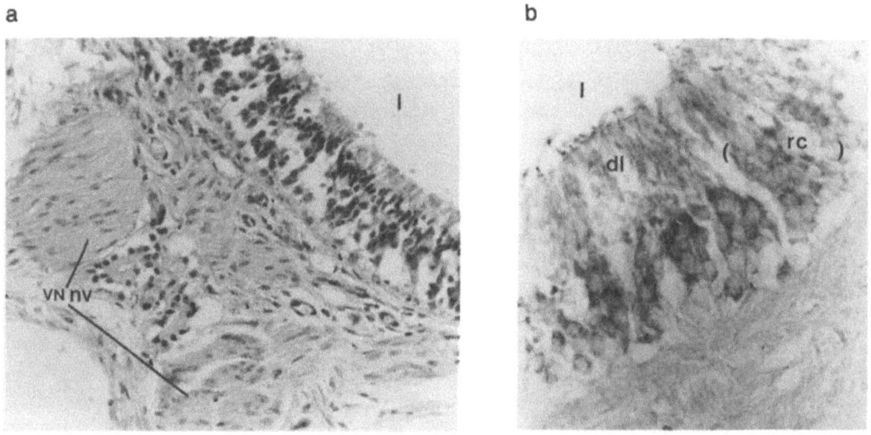

Figure 5 **VNO** *(Avahi)*
a Innervation: vomeronasal nerve (VNnv); H &E X 200.
b Neuronal organization: rc nuclei: - cluster ( ); basal cells (b) SBA Lectin X 400.

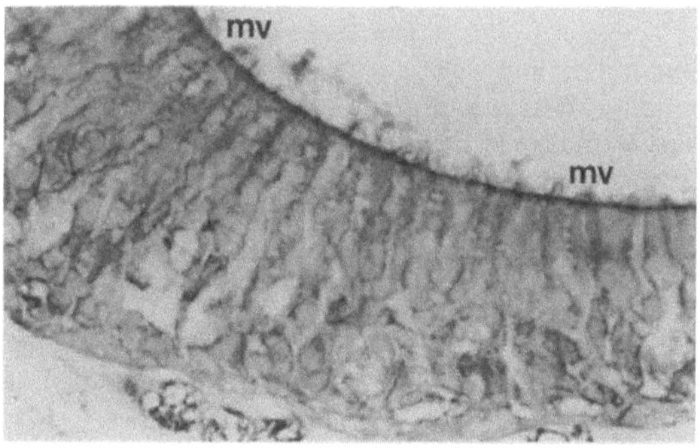

Figure 6 **RECEPTOR SURFACE** *(Phaner)*: microvilli (mv), detected only in ventral zone, showing long processes; UEA Lectin X 200.

ONTOGENY

The ontogeny of the organ was examined in the slow loris and will be compared with the lesser mouse lemur. Relatively few neonates were available in which there was adequate tissue preservation for the examination of histological detail.

A Cheirogalinae: *Microcebus murinus* (N = 1 male, two days old)

The mouse lemur at the neonate stage shows a fully developed VNO: the neuroepithelial height and distribution were all within the adult range (Table 1). In particular, abundant exit axons could be demonstrated in the sub-mucosal layer. The distribution of support and neuronal nuclei did not resemble the very evident stratification in the adult (Schilling, 1970).

Most of the other criteria established as AOS requirements were also fully developed and clearly functional; although not fully quantified in this neonate specimen, the concentration of sero-mucous glands appeared less dense and their distribution more uniform than in the adult. Overall the VN complex was comparable to that of the adult (Fig. 2b and 7a).

B Lorisinae *Nycticebus coucang* (N = 1 male, 11 days old).

In the neonate slow loris, the duct system alone appears to be comparable with that of the adult. There is a well-formed palatine papilla and a clearly patent NP duct leads through to a nasal junction with the VN duct. The papilla, sulci and ducts are lined with pseudostratified epithelium, in which no taste buds could be detected. Vasomotor components were mainly dorsal, and the sparse sero-mucous acini were concentrated only posteriorly. The lumen of the organ was well developed (Table 1) and comparable in relative diameter (176 - 195mm) to that of the adult. The VN epithelium was of uniform thickness (15 - 26mm) while its average depth was approximately one-tenth that of the adult, and much less organized.

General and specific staining of the neonate VN epithelium revealed clusters of cells with irregular shaped and sized dark staining nuclei intermingled with relatively lighter nuclei. In many areas neither clear support cell nor neuronal cell components could be distinguished, and there was, in addition, a relative scarcity of stainable axons.

Given the variety of "intermediate" cell types in the epithelium, it would seem that there is a relatively low level of neurogenesis at this age. Several correlated features suggest this interpretation: a lack of structural organization exemplified by the absence of stratification into distinct nuclear zones; the absence of a dendritic layer and the low axon density as demonstrated by neuron-specific stains; and finally the occurrence of mitotic figures in the basal cell layer. Furthermore, there were small clusters of SBA lectin-stained nuclei but at a low density, plus a very discrete localization of UEA-1 to the apical zone of similar cells; these results suggest that both of the weak staining reactions identified immature neurons only.

There is a full degree of VN organ development in the adult (Hedewig, 1980a), shown by a highly differentiated multi-layered sensory epithelium, plus a discrete non-sensory area. In the neonate, no such distinction between sensory and non-sensory areas was evident. The remaining structures of the VN complex did not show such conspicuous age-related differences, glandular and vascular development being similar in the neonate and the adult.

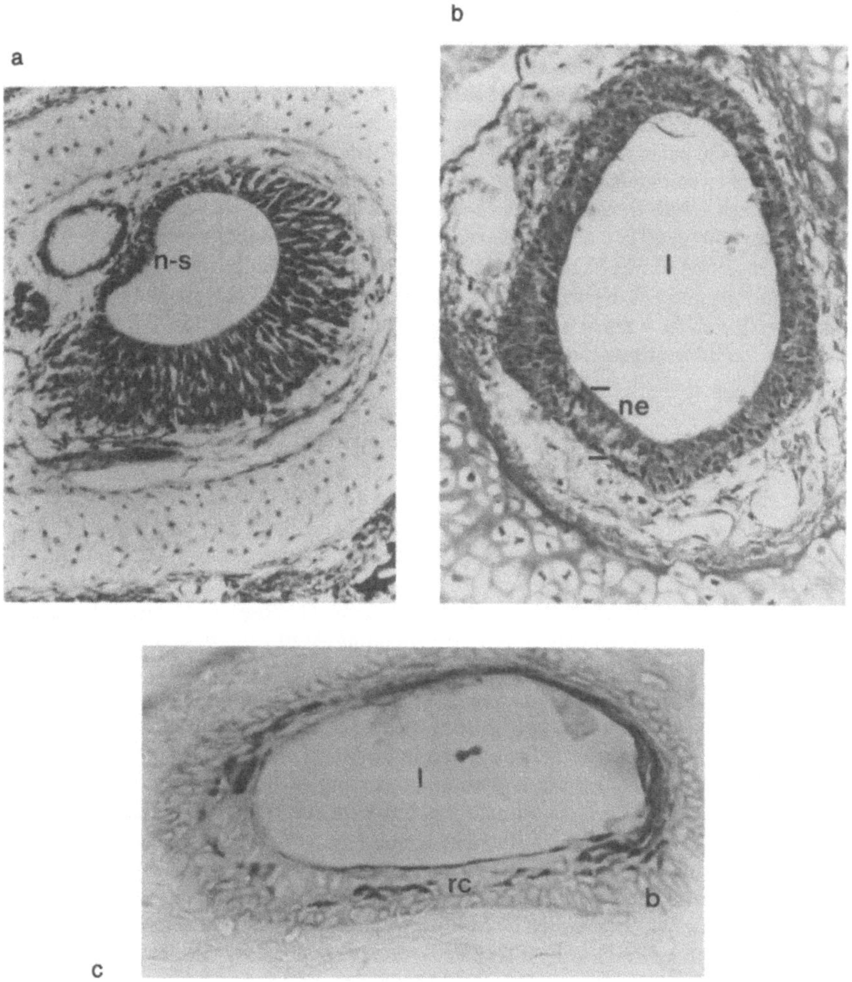

Figure 7 a,b,c **NEONATES** comparison - lesser mouse lemur and slow lLoris
Microcebus : (a) Palmgren Ag X 200; *Nycticebus*: (b)H & E /Aniline Blue X 200& (c) UEA Lectin X 400 neuroepithelium (ne); non-sensory epithelium (n-s)

FUNCTION

If the most likely VN stimulus, that is, a biological fluid (urine, saliva, vaginal exudate), can obtain ready access to the organ, it must also be capable of activating the AOS pathway to the CNS - at least up to the level.of the olfactory bulbs. In several prosimian species, urine is actively or passively dispersed by marking and in any subsequent chemoinvestigation is nosed and lapped by conspecifics (Perret, chapter). The working hypothesis adopted was that urinary chemosignals may be specific stimuli for the vomeronasal system, at least in nocturnal forms which use mainly urine for marking.

It has been shown by radiographic (Schilling, 1980; and unpbl.) techniques that in several prosimian species, uptake of a radiopaque liquid occurs when injected into the oral entry of the VN pathway at the palatine sulci. Fig. 1c shows ipsilateral uptake with radioactivity confined to the lumenal circumference. Furthermore, since the palatine papilla contains elastic fibres, it is possible that tongue movements, such as those described in male *Lemur catta* at the onset of Flehmen (Bailey, 1978), may assist sampling of non/low volatile mixtures. During such chemoinvestigation, fluids could then be sucked into the organ, as occurs after activation of Flehmen during urinalysis in goats (Melese-d'Hospital and Hart, 1985). In anæsthetized mouse lemurs, pressure on the palatine papilla also causes a tongue-lapping reflex; it is therefore likely that during chemoinvestigation partial closure of the oral NPd may then occur (Schilling, 1970; Fig. 35), perhaps to retain sampled fluid during stimulus processing. Alternatively, in species such as *Phaner* and *Lepilemur,* vasomotor control by anterior contraction of the sinuses at the VN duct (Fig. 3a,b) and posterior relaxation of sinuses around the body of the organ, is another possible means of stimulus retention within the VNO lumen.

In the mouse lemur, neurophysiological tracers (e.g WGA-HRP) have been shown to follow or "tag" fluid movement into the VNO lumen, where uptake by vomeronasal microvilli allows intracellular movement of the tracer via the vomeronasal output axons to the glomeruli of the AOB (2nd order neurons). The activated zone of the bulb is shown in Fig. 8. Experiments with metabolic markers of neuronal activity (such as 2-DeoxyGlucose) have shown that oxidative consumption increased markedly in the corresponding AOB (Fig. 9) when liquid urine was deposited in the ipsilateral palatine foramen (Schilling *et al*, 1990). But no activation of the AOB was observed when the stimuli were distilled water or the volatile phase of urine. There was no correlation between the type of urine tested (i.e from anestrus or estrus females and dominant males) and the pattern of activation in the bulb.

During spontaneous chemoinvestigation, processing of stimuli occurs differently. First, "normal" vomeronasal or olfactory stimulation always occurs bilaterally; liquids enter the two palatine foramina and activate the left and right neuroepithelia, but volatiles which enter only one nostril, invariably activate the main olfactory epithelium of each nasal cavity by "re-entry" through the retronasal passage. Second, in these experiments, the AOB was not activated alone. The uptake was always accompanied by MOB activation and hence, it is inferred that biological stimuli do not enter the VNO without prior activation of the olfactory system. Conceivably, any volatile components of a fluid intake could be simultaneously processed by the olfactory or even via nervus terminalis or trigeminal activation.

Experimental studies are not yet available for the majority of prosimians. It is even more difficult in primates than in rodent or ungulate models to define physiological and, just as critically, behavioral responses depending specifically on a vomeronasal mechanism.

In the mouse lemur most chemosignal-linked effects seem to be primarily through the MOS, or at least could be obtained with the volatile phases of male urinary stimuli (Schilling & Perret, 1987). The more central parts of the MOS and AOS have been examined by bulbectomy (that is the removal of both olfactory and VN brain centers), and a comparison made with removal of the vomeronasal organ alone. The latter procedure leads to the disorganization of the layers of the AOB, while leaving the MOB intact. Monitoring of the gonadal responses and of the neuroendocrine cycle of male *Microcebus* was conducted within an established breeding colony. These hormonal changes were correlated with the seasonal variations of the photoperiod and showed that removal of VNO and complete bulbectomy have quite similar effects.

Plasma testosterone levels measured during a complete annual cycle, are significantly higher in operated animals than in controls, except during the period of sexual activity where the situation is reversed. However, such gonadal fluctuations do not inhibit reproduction in the male mouse lemur (Schilling and Perret, 1993). Pilot experiments with males suggest that GnRH-ir can be located in several CNS targets (Schilling *et al.*, 1993), the tertiary projections of the AOS being the same regions of the basal forebrain which regulate reproductive functions.

Comparisons were made of some sexual and social behaviors between intact Mouse Lemurs and those deprived of the VNO; in both cases, sexually inexperienced subjects were used. Vomeronasalectomy appears to have a suppressive effect not only on courtship behavior but also on agonistic interactions. However, these effects are less evident with sexually experienced Mouse lemurs (Schilling & Perret, unpubl.), as with hamsters (Johnson & Pfeiffer,1987) Indeed such animals may well compensate for the lack of VNO input with information from other sensory modalities including other chemosensory systems. Comparable results have been reported after VN duct blockade in cats (Verberne, 1976) and sheep (Blissett, *et al*, 1990).

Although the VNO was fully developed in adults of all species examined, no correlation was apparent with reported estimates of AOB size; there is also little suggestion of any covariation of AOB and MOB volumes (Stephan *et al*, 1982). However, when the interdependence of AOB volume and body weight effects are examined across several groups, the allometric slope is approx. a = 0.57 (Frahm, 1981).

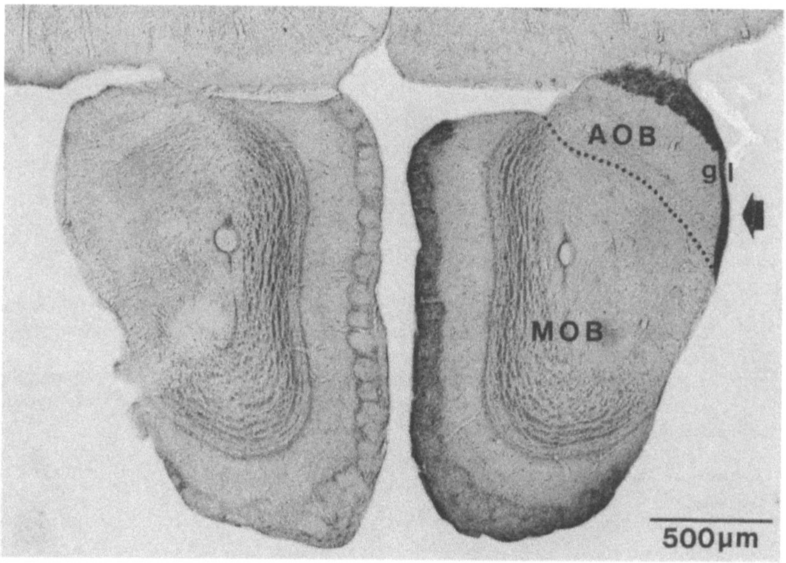

Figure 8 **AOB ~ UNILATERAL ACTIVATION** (*Microcebus*): Differential uptake pattern of anterograde neurotracer (WGA-HRP) - unilateral presentation of conspecific urine via left NPD; reaction product (arrow) restricted to glomerular layer (gl) of left accessory bulb (AOB).

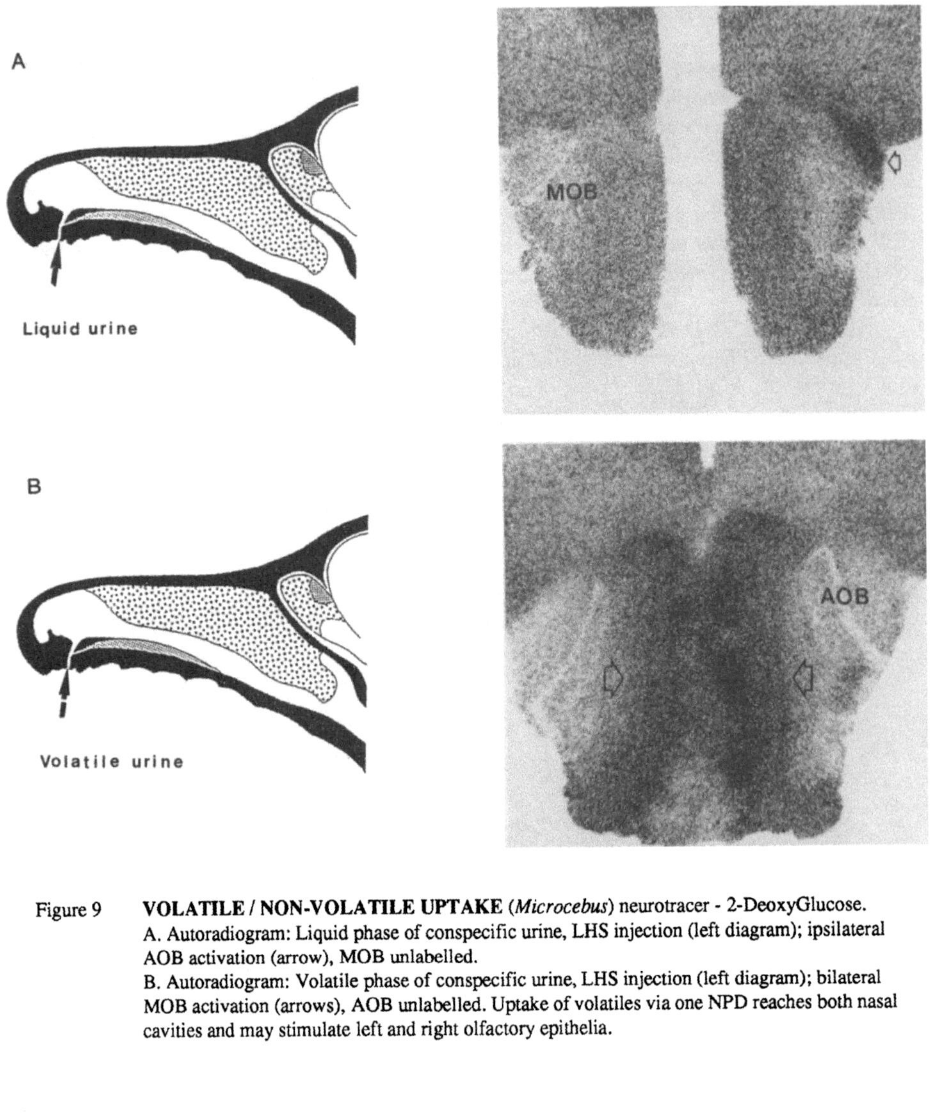

Figure 9 **VOLATILE / NON-VOLATILE UPTAKE** (*Microcebus*) neurotracer - 2-DeoxyGlucose.
A. Autoradiogram: Liquid phase of conspecific urine, LHS injection (left diagram); ipsilateral AOB activation (arrow), MOB unlabelled.
B. Autoradiogram: Volatile phase of conspecific urine, LHS injection (left diagram); bilateral MOB activation (arrows), AOB unlabelled. Uptake of volatiles via one NPD reaches both nasal cavities and may stimulate left and right olfactory epithelia.

Table 1: Comparison of Vomeronasal Organs

	Lepilemur	Phaner	Avahi	Microcebus[1]	L.catta	Nycticebus[2]	Otolemur2
Length of nasal cavity (mm)	29.2	29.1	24.4	17.5	33.8	32	30
Overall length of VNO lumen (mm)	5.5	4.7	4.2	8.5	8.1	10.4	7.0
Length of sensory portion of VNO (mm)	5.3	4.0	4.0	5.4	7.8	5.1	5.3
Lumen dia (min/max) of sensory portion of VNO (μm)	250-400	180-400	200-500	150-225	176-295	120 -220	110 -550
Mean neuro-epithelial height (μm)	75	70	70	70	105	110	100
% of circumference of VNO comprising sensory epithelium	70	75	~90	~70	95	65	55

(From: [1] Schilling, 1970; [2] Hedewig, 1980a)

DISCUSSION

The VNO in one Lorisid species, *Nycticebus coucang*, is not fully developed at birth, although the adult VNO is structurally fully extant and very probably functional. The neuroepithelium shows several features which suggest incomplete neurogenesis in the neonate and hence a post-natal period of growth and differentiation with a consequent slow maturation of accessory olfaction. The period of attainment of functional maturity needs to be determined. This contrasts with the virtual adult status of the organ in neonatal *Microcebus murinus*; while the aye-aye`s early vomeronasal complex shows all the criteria mentioned earlier (Maier, 1986).

The new born slow loris may well not require full VNO function during its suckling period; independence of the VNO is shown by neonate rabbits which perform nipple-search successfully after VNO removal (Hudson & Distel, 1986). In the house mouse, some evidence of post-natal AOS neuronal modelling (Wilson & Raisman, 1980), suggests that altricial/short gestation species do show ontogenic changes in the VN pathway. Hence long gestation/lactation species such as the slow loris, could well show more substantial maturational changes in VN chemoreception. New-born *Lemur* spp., and several other diosmic primates, also seem to be variable in their rate of accessory olfactory development (Evans, unpubl.).

The above findings, together with the absence of other non-volatile (taste) chemoreceptors on the papillary sulci, which occur in the adult (Hofer, 1977), suggest a lower AOS efficiency and/or early dependence on MOS alone. Confirmation in additional specimens is highly desirable for *Nycticebus* and for other Lorisiforms.

Inter-species differences in VN structures (Table 1) include: the degree of development of the sensory neuroepithelium, particularly its height and distribution, and the extent of non-sensory lining present. Few if any structural differences appear associated with activity pattern or other ecological factors. Indeed, two contrasting species show comparable VN receptor distribution, the nocturnal and semi-social woolly lemur compares well with the characteristically diurnal ring tail lemur in that both lack a non-sensory zone. While neuronal density has not yet been established in *L. catta*, the degree of VNO development in the ring tail lemur approaches that of the nocturnal lesser mouse lemur (Table 1). Where there is a specialist intake such as in *Phaner*, and in which there is also a distinctive cellular feature, such as microvillar distribution, the functional linkage is not clear at present. Most structural variants therefore cannot yet be assigned definitive adaptive values.

Vasomotor sinusoids are consistently well developed in all species; hence they all possess a vascular system capable of regulating fluid exchange at key points along the stimulus exchange route (nasopalatine sulci to vomeronasal lumen and back again). Invariant behaviors, such as Flehmen, are probably additional chemoinvestigative patterns, at least in medium to large species. Glandular development in all species further suggests that continuous serous fluid flow - from posterior to anterior - could sustain a passive evacuation of VN lumen contents. Again, reflexive or other control of such features needs to be established.

The stimulus uptake findings suggest that the two main chemosensory systems cooperate in the detection and the processing of conspecifics' chemosignals. Such a "parallel" mode of perception would be expected as a consequence of the operation of dual olfactory systems.

The main nasal olfactory epithelium is presumed continuously to process the volatile phase of all chemosignals together with whichever non-volatile components can reach it. In contrast, vomeronasal activation, probably triggered by the olfactory system, is likely to be involved less often, but more specifically to sample and analyse the components of the liquid phase of the same chemosignals.

It is likely that the VN system selectively detects the high MW component of chemosignal mixtures. A putative carrier protein (Aphrodisin) occurs in hamster vaginal exudate, and similar fractions appear in the scent marks/urine of tamarins (Belcher, *et al*, 1990; Fuchs, *et al*, 1991). Water-soluble and non-volatile residues from labial scent marks elicit licking by male *L.catta* (Evans, 1991; Evans & Pickett, unpubl.). Indeed, 'contact' olfaction at female scent marks seems to provide considerable 'social' information to each sex (Bailey, *et al.*, 1984; Dugmore & Evans, 1990).

Vomeronasal-specific effects have not yet been established for any primate. Although olfactory cues from a dominant female marmoset can suppress cyclicity in subordinates, visual or other cues are also contributory; lesions of either or both olfactory systems do not remove the dominant animal's effect (Barrett, *et al.*,1993). Behavioral and endocrine changes following removal of the VNO in mouse lemurs suggest that the AOS is likely to be involved in the modulation of amygdala/hypothalamic activity, although the degree to which this differs from the functions established in rodents remains unclear. Similar neuroendocrine mechanisms to those of rodents appear to be involved in the modulation of the mouse lemur's central reproductive control mechanisms. The effects of urinary volatiles from dominant males are exerted via MOS/AOS disinhibitory inputs to the hypothalamus, i.e which inhibit prolactin inhibitory factor production, and hence induce prolactin secretion, with consequent hyperprolactinaemia and gonadal suppression in subordinate males (Perret & Schilling, 1987).

Testicular secretion in contrast can be activated by urinary chemosignals from cyclic female mouse lemurs, whereas the suppressive dominant on subordinate effect on testosterone among males may be mediated by the CNS/pituitary/adrenocorticoid system of dominant animals (Perret, Chapter -). The nocturnal primitive prosimians might well be the appropriate candidates to help delineate the role of VN olfaction in primate reproduction and social behaviors.

CONCLUSIONS

The VNO is present in all investigated prosimians and is sufficiently well-developed to be fully functional in the adult. The Cheirogalinae would appear the most primitive in overall development of the AOS, a finding which correlates with cytogenetic and molecular results. Although uptake of urinary or other fluid vehicles seems a likely mode of chemosignal transport, there is as yet no direct evidence that specific molecules are emitted by, transported to and perceived among conspecifics.

The individual chemical territory, fixed in both time and space in nocturnal, solitary prosimians, has probably been transformed into a troop-oriented and more plastic chemical space in the diurnal and gregarious prosimians.

The range of VNO structure encountered across the groups studied shows a considerable uniformity at the cellular level, as expected from other mammalian studies (Cooper & Bhatnagar, 1976). The degree of macroscopic variation did suggest some correlation of olfactory dominance and VNO development, but for any one species, its dietary niche or activity-pattern requirements were more difficult to assign a unique or even a contributory role. Although some of the functions of the AOS may be taxon-specific, these remain to be determined.

The vomeronasal component of the total chemosensory input to the prosimian CNS is much less known than in other groups; chemical signal effects in Mouse Lemurs suggest a substantial influence on the reproductive mechanisms of the adult male and female; these are probably related to the requirements of their social system (Perret, 1992; and chapter -).

The rodent AOS is thought to be concerned with the processing of non-volatile signals involved in feeding, reproductive and other social behaviors (Wysocki, *et al.*, 1985). Within the primate neuroendocrine systems there is likely to be an increase in the complexity of the central integrative mechanisms, with olfaction being involved as one of the factors in the operation of the social/ reproductive axis (Bronson, 1989). Among the prosimians accessory olfaction is probably concerned with similar integrations but with its relative importance being closer to that of non-primate mammals.

ACKNOWLEDGEMENTS

It is a pleasure to record the assistance of Dr E Grigorieva, Mr J Dick. Support by : Leverhulme Trust RF 92/662, Wellcome Trust (CSE); CNRS, URAI/83 (AS).

REFERENCES

Bailey, K. (1978). Flehmen in the Ring Tailed Lemur (*Lemur catta*). *Behaviour,* 65, 309-319.

Barber, P. C., & Raisman, G. (1978). Replacement of receptor neurons after section of the vomeronasal nerves in the adult Mouse. *Brain Research* ,147, 297-314..

Barrett, J., Abbott, D.H.& George L.M. (1993) Sensory cues and the suppression of reproduction in subordinate female marmoset monkeys, *Callithrix jacchus .J ournal Reproduction & Fertility* ,97, 301-310.

Belcher, A..M., Epple, G., Greenfield, K., Richards, L.E, Küderling, I & Smith, A.B. (1990). Proteins: Biologically relevant components of the scent marks of a primate (*Saguinus fuscicollis*).*Chemical Senses ,15*, 431-446.

Blissitt, M., Bland, J.P.& Cottrell, D.F. (1990). Olfactory and vomeronasal chemoreception and discrimination of oestrous and non-oestrous ewe urine odours by rams. *Applied Animal Behaviour*,27, 325-335.

Bluthé, R.M. & Dantzer, R. (1993). Role of the vomeronasal system in vasopressinergic modulation of social recognition in rats. *Brain Research,* 604, 205-210.

Bronson, F.H. (1989). *Mammalian Reproductive Biology*. Chicago UP.

Broom, R. (1897). A contribution to the comparative anatomy of the mammalian Organ of Jacobson. *Transactions Royal Society of Edinburgh,* 39, 231-255.

Cooper, J.G.& Bhatnagar, K.P. (1976). Comparative anatomy of the vomeronasal organ

complex in bats. *J ournal of Anatomy,* 122 , 571-601.
Davis, B. J., Macrides, F., Youngs, W. M., Schneider, S. P., & Rosene, D. L. (1978). Efferents and centrifugal afferents of the main and accessory olfactory bulbs in the hamster. Brain Research Bulletin, 3, 59-72.
Dugmore, S.J., Bailey,K. & Evans, C.S. (1984). Discrimination by male Ring-Tailed Lemurs between the scent marks of male and female conspecifics. *Int ernational J ournal of Primatology,*5, 235-245.
Dugmore, S.J. & Evans, C.S. (1990).Discrimination of conspecific chemosignals by female Ring-Tailed Lemurs *Lemur catta* In:. D. MacDonald (Ed) *Chemical Signals in Vertebrates,V* : (pp 360-366) Oxford: OUP
Evans, C. S. (1984). On the structure and function of accessory chemoreceptive organs. *Acta Zoologica Fennica,* 171, 57-62.
Evans, C.S. (1991) Contact chemocommunication signals in female *Lemur catta* in "Mammalian Semiochemistry" Proc Int Soc Chem Ecol, abs., Dijon.
Frahm, H. D. (1981). Volumetric comparison of the accessory olfactory bulb in Bats. *Acta Anatomia,* 109, 173-183.
Fuchs, E., Rosenbusch, J. & Anzenberger, G. (1991). Urinary protein pattern reflects social rank in male Common Marmosets. *Folia Primatologica,* 57, 177-180.
Graziadei, P.P., Karlan, M., Monti Graziadei, G.A. & Bernstein, J.J .(1980). Neurogenesis of sensory neurons in the primate olfactory system after section of the filia olfactoria. *Brain Research* ,186, 289-300.
Hedewig, R. (1980a). Vergleichende anatomische Untersuchungen an den Jacobsonschen Organen: *Nycticebus coucang ,Morphologisches Jahrbuch,* 126, 543-593.
Hedewig, R. (1980b) Vergleichende anatomische Untersuchungen an den Jacobsonschen Organen *Galago crassicaudatus.Morphologisches Jahrbuch,*126, 676-722.
Hofer, H.O. (1977). The anatomical relations of the ductus vomeronasalis and the occurrence of taste buds in the papilla palatina of *Nycticebus coucang* (Primates, Prosimiae) with remarks on strepsirrhinism. *Morphologisches Jahrbuch,*123, 836-856.
Hudson, R. & Distel, H. (1986). Pheromonal release of suckling in Rabbits does not depend on the vomeronasal organ. *Physiology & Behavior.,* 37:,123-29.
Izard, M.K. (1990) Social influences on the reproductive success and reproductive endocrinology of prosimian primates, In -------??--------(Eds.) *SocioEndocrinology of Primate Reproduction* (pp159-186) New York:Wiley.
Johnson, R. & Pfeiffer, C. (1987). Effect of vomeronasal and main olfactory lesions on communicative behavior & hormonal responses in hamsters. *Chemical Senses,* 12, abs.
Keverne, E.B. (1979). The dual olfactory projections and their significance for behavior. In *Chemical Ecology: Odor communication in animals* F.Ritter (Ed) 75-83. Amsterdam: Elsevier.
Key, B.& Giorgi, P.P. (1986). Soybean agglutinin binding to the olfactory systems of the rat and mouse *Neuroscience Letters* ,69, 131-136.
Labov, J.B., Katz,Y., Wysocki, C.J., Beauchamp, G.K. & Wysocki, L.M. (1988). Levels of immunoreactive ß-endorphin in rostral and caudal sections of olfactory bulbs from male Guinea Pigs exposed to odors of conspecific females *Annals NY Academy of Sciences,* 510, 432-435.

Lundh, B, Brockstedt, U. & Kristensson, K. (1989). Lectin-binding pattern of nasal neuroepithelial cells of mouse *Histochemical J ournal*,21, 33-43.
Maier, W. (1980). Nasal structures in Old and New World Primates. In L. Ciochon & A. Chiarelli (Eds.), *Evolutionary Biology of the New World Monkeys and Continental drift* (pp 219-241).Basel: Karger.
Maier, W. (1986). Nasal capsule and facial skeleton of a fetus of *Daubentonia madagascariensis*. *Primate Report*,.14,127.abs.
Melese-d'Hospital, P.Y & Hart, B.L. (1985). Vomeronasal organ cannulation in male goats: evidence for transport of fluid from oral cavity to vomeronasal organ during Flehmen. *Physiology & Behavior*, 35, 941-944.
Mendoza, A.S. & Borisch-Chappuis, B. (1989). Lectin-binding properties of the vomeronasal organ neuroepithelium, the olfactory epithelium proper and the organ of Masera in mice. *Archives Anatomy Histology & Embryol ogy* ,97, 76-81.
Perret, M. & Schilling, A. (1987). Role of prolactin in the pheromone-like sexual inhibition in the male lesser Mouse Lemur *Journal of Endocrinology* ,114, 279-287.
Perret, M. (1992) Environmental and social determinants of sexual function in the male Mouse Lemur. *Folia Primatologica*, 59, 1-25
Raisman, G. (1972). An experimental study of the projection of the amygdala to the accessory olfactory bulb and its relationship to the concept of a dual olfactory system. *Experimental. Brain Research*,14, 395-404.
Scalia, F. & Winans, S. (1975).The differential projections of the olfactory bulb and accessory olfactory bulb in Mammals *Journal of Comparative Neurology* ,161, 31-56.
Schilling, A. (1970). L'Organe de Jacobson du Lémurien malgache *Microcebus murinus* (Miller, 1777) *Memoires du Muséum Nationale d'Hist oire Naturel* ,61A, 203-280.
Schilling, A. (1980).Olfactory communication in prosimians. In D.M. Stoddart (Ed.), *Olfaction in Mammals* (pp.84-98). London: Acad Pr.
Schilling, A. (1987). L'organe vomeronasale des mammifères. *Journal de Psychologie Normale et Pathologie* ,3-4, 221-278.
Schilling, A. & Perret, M. (1987). Chemical signals and reproductive capacity in a male prosimian primate (*Microcebus murinus*) *Chemical Senses.*, 12, 143-158.
Schilling, A. & Perret, M. (1993). Removal of the olfactory bulbs modifies the gonadal responses to photoperiod in the lesser Mouse Lemur (*Microcebus murinus*). *Biology of Reproduction* ,49, 58-65.
Schilling, A., Serviere, J., Gendrot, G. & Perret, M. (1990).Vomeronasal activation by urine in the primate *Microcebus murinus*: a 2DG study *Experimental Brain Research* ,81, 609-618.
Schilling, A., Peytevin, J. & Martinet, L. (1993). The GnRH neuronal system of the male lesser mouse lemur *Microcebus murinus*: an immunohistochemical study.In'*Hormones, Brain and Behaviour*' Proc. Intl.Conf. 167 abs., Tours.
Stephan, H., Baron, G., & Frahm, H. (1982).Comparison of brain structure volumes in Insectivora and Primates II: Accessory olfactory bulb (AOB) *J ournal der Hirnforschung*, 23, 575-91.
Vandenbergh, J. G. (1989). Coordination of chemical signals and ovarian function during sexual development *J ournal of Animal Science* ,67, 1841-1847.

Verberne, G. (1976).Chemocommunication among domestic cats mediated by the olfactory and vomeronasal senses: 2: The relation between the function of Jacobson's (vomeronasal) organ and Flehmen behaviour. *Zeitschrift der Tierpsychol ogie* ,42, 113-128.
Wilson, K.C. & Raisman, G. (1980). Age related changes in the neurosensory epithelium of the mouse vomeronasal organ. *Brain Research,* 185, 103-113.
Wysocki, C. J. (1989). Vomeronasal Chemoreception: its role in reproductive fitness and physiology. *Neurology and Neurobiology* 50.In J M.Lakoski J.R. Perez-Polo & D.K.Rassin. (Eds):*Neural control of reproductive function* (pp 545-566). New York: Liss.
Wysocki, C.J., Beauchamp, G.K., Reidinger,R.F., & Wellington, J.L. (1985). Access of large and non-volatile molecules to the vomeronasal organ of mammals during social and feeding behaviors. *Journal of Chemical Ecology* ,11, 1147-1159.

TOXINS AND TOOTHCOMBS: POTENTIAL ALLOSPECIFIC CHEMICAL DEFENSES IN *NYCTICEBUS* AND *PERODICTICUS*

Lon Alterman

Department of Biology
Clarke College
1550 Clarke Drive
Dubuque, Iowa 52001-3198

An oral tradition exists in Indonesia that the slow loris or coucang, *Nycticebus coucang*, is venomous (personal observation, also see Wilde, 1972). Few mammals are known to produce toxic compounds, and none is known to routinely use these toxins in allospecific defenses. Yet inhabitants of Kalimantan insist that the coucang protects itself from humans by using its toxic bite. Over the years there have been many accounts of bite-induced shock and prolonged edema of appendages which have been bitten, but the literature is devoid of reports identifying the source of putative toxins. If slow lorises do produce toxins which are used in defense against allospecifics then three sorts of questions might be asked. The most obvious questions are descriptive in nature. What tissue(s) is (are) the source of this venom? What is the chemical nature of putative toxins?

A second line of inquiry is functional. How do lorises use the toxins? Other mammals which have been demonstrated to secrete toxic materials are not believed to use these compounds as defensive weapons. The list of mammals demonstrated to produce toxins includes *Ornithorhyncus*, (platypuses; Tidswell, 1906; Griffiths, 1978); *Blarina*, (shrews; Pearson, 1956); *Neomys* and *Solenodon*, (water shrews; Pucek, 1968). Whereas male platypuses are thought to use toxic materials derived from crural glands and injected by keratinous spurs on the heel to defend lengths of river from other males, eutherians which produce toxins use them to immobilize prey. The eutherian toxins are secreted from submaxillary or submandibular glands and are applied in salivary solution through bites.

An examination of the diet of *N. coucang* suggests that its prey does not include species which require immobilization in order for the loris to ingest them. As a terminal-branch feeder it does not appear that following an impaired insect or chamelion is a practical expenditure of energy. Moreover, the latest surveys of Malaysian populations of *N. coucang* indicate that densities of lorises are extremely low, one or two per kilometer (Barrett, 1981), and that the localization of conspecifics is more of a problem than partition of a limited range as is the case of platypuses. Slow lorises then are unlikely to use putative toxins in a manner similar to other mammals.

An alternative use for poisonous compounds is defense against allospecifics. Postulating such a function seems reasonable for a species whose only defense is thought to be cryptic behavior. Other contributors to this volume (Bearder, Sterling and Richard, Charles-Dominique) have suggested that criteria for species recognition, social structure, space and resource utilization must be adjusted to account for the nighttime behavior of the nocturnal prosimians. Similarly, I suggest that notions of cryptic behavior must be adjusted in discussions of slow lorises.

Creatures of the Dark: The Nocturnal Prosimians
Edited by L. Alterman *et al.*, Plenum Press, New York, 1995

Although slow lorises have been described as quietly moving about branches of either the underbrush (Cartmill, 1972) or terminal branches of the canopy, it may be that these primates are not cryptic. Anthropocentric notions of crypsis include visual inconspicuousness, slowness of movement and the absence of audibly perceptable cues. Forest predators, however, may find the slow loris to be readily detectable. Reptilian and mammalian predators are likely more olfactory and vomerolfactory oriented than humans. Odors from slow lorises can be apparent to humans in the rain forest for at least 50m (pers. obser.). In captivity, slow lorises (and other lorisids) are notorious among keepers for their scent. Potential predators are at least equally likely to detect chemical signals from the coucang.

Whereas nocturnal behavior hides animals from detection by humans it is less significant to nocturnally active animals whose visual apparatus is more oriented towards rhodopsin chemistry than the chromolabes. They see well in conditions of lower light. While I am unable to exhaustively enumerate known predators of slow lorises it appears that the majority of known and suspected predators are night-active and are therefore likely to possess sensitive night vision. In many species visual cues are thought to be dependent on movement of the stimulus. Slowness of the visual target is presumed to confer crypsis, and the very name slow loris is illustrative of locomotor patterns observed by humans. These patterns of locomotion are often noted in artificially high light intensities. Recent night-phase observations of *N. coucang* and *N. pygmaeus* led investigators to ask why these animals were designated "slow" lorises (Alterman, *et al., in prep.*). Slow movement may be an artifact of observations by day-active primates.

While it is apparent that many slow loris vocalizations are largely inaudible to humans because of their frequency range (Zimmerman, 1985), mammalian carnivores are known to detect these frequencies with far greater facility. Potential predators are likely to find their prey far noisier than humans do. Energetic considerations, however, preclude audible signals from a role in the truly long-distance communications between conspecifics suggested by Barrett's (1981) estimates of slow loris density in Malaysia (Michelsen, 1978) .

If slow lorises are far less cryptic than previously thought, possess locomotor behaviors which preclude leaping from branch to branch and manufacture toxins, how do the lorises use these toxins to promote their survival? Is it reasonable to speculate that a loris trapped on a branch will confront, for example, a clouded leopard in an attempt to apply toxins to it? Do the coucangs use their odors to repel attacks by these predators? If so, why would these odorants be toxic? Captive slow lorises have been observed to bite one another during breeding. How does a loris having a toxic bite not endanger the life of its mate(s)?

Each of these questions addresses matters of function and leads us to a final line of inquiry. What can these investigations contribute to our understanding of the roles of chemical defenses in the evolution of the genus *Nycticebus* and the lorisines?

In the remaining sections of this paper I summarize a series of investigations designed to address descriptive and functional questions. In several cases these investigations involve the exposure of glandular exudates to known or putative mammalian predators of either *N. coucang* or *Perodicticus potto.* Potto exudates were investigated to determine if members of *Nycticebus* were unique among lorisines in employing allospecific chemical defenses. Small sample sizes of putative predators of cougang and potto preclude meaningful statistical analyses. As a result, raw data are presented. These results are subsequently used as a basis for this paper's Discussion which therefore must be regarded as speculative. While every effort was made to expand sample size (for example, in the case of *Nandinia binotata* 75% of the world's then captive population was sampled: *n*=3) conclusions which are reached are best regarded as hypotheses for further testing.

EXPERIMENTAL DESIGN AND RATIONALE

Because I was unable to find toxic substances in the saliva of coucangs (despite Wilde's assertions), I concentrated my efforts on the exudates from the brachial gland of slow lorises (Fig. 1). Brachial gland exudates (BGE) are applied to the head and face when the animal appears to be under stress (Fig. 2) and may be mixed with saliva.

A

B

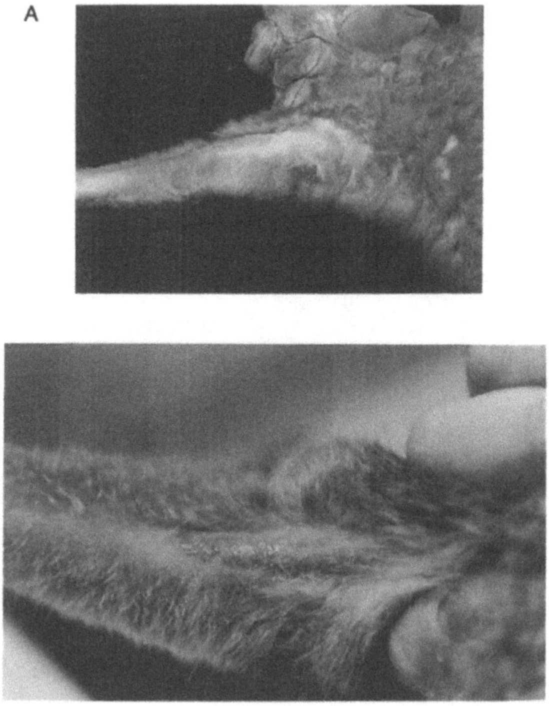

Figure 1. A. Photograph of the right arm of a male coucang with brachial gland proximal to the elbow. The glandular field is seen as a naked patch on the medial surface of the upper arm. B. Detail of the brachial gland with exudate apparent.

Figure 2. Drawing from a photograph of a male coucang applying brachial gland exudate to its head and neck.

Slow lorises from the Duke University Primate Center (DUPC) were restrained by hand while cotton swabs were wiped across one or both brachial glands. In all cases brachial glands were wiped with two cotton swabs simultaneously. In selected cases saliva was withdrawn from the mouth with a ten ml syringe. BGE and saliva samples were placed in sterile glass vials and frozen at -20°C until use.

Frozen BGE swabs were extracted with 2 ml of either 8.8% formic acid (FA) or 1:1 methanol, methylene chloride (MM). These solvents were chosen because they are capable of solvating either hydrophilic compounds in the case of FA or lipids in the case of MM. After 30 min of extraction, solvent vials containing the cotton swab were placed on ice and were assigned numbers whose coding remained unknown to the investigator until the completion of the experiment.

Three hundred µl aliquots were subcutaneously injected into the ventrum of 25 day old male CD mice (Charles River Laboratories, Raleigh, North Carolina). The animals were placed in 29 x 18 x 13 cm translucent mouse cages, and their behavior was monitored for 20 min. Behavior was scored by an observer who was unaware of the identity of the compound being injected.

Investigations described in this section were performed sequentially, i.e. the results of one investigation determined the design of subsequent experiments. It is, therefore, impossible to describe the experiments performed without reference to the results of a previous manipulation. For this reason, portions of this section and those found in Results will be redundant.

In Experiment I, FA and MM extracts (whole extract) from selected lorises (those which secreted copiously) were injected into CD mice. Extraction solvents without BGE were used as controls to assure that none of the expected toxic effects was due to the nature of solvents used.

In Experiment II, duplicate samples of BGE were obtained from animals whose BGE had been observed to kill subject animals in Experiment I. Two hundred µl aliquots of FA or MM extracts of these BGE's were applied to an *n*-octyl high performance liquid chromotography (HPLC) column (SOTA Chromatography, Crompond, New York, model C-8E, 150 x 4.6 mm) and were eluted with linear gradients of water and acetonitriule (ACN) on a Gilson HPLC (Gilson Medical Electronics, Middleton, Wisconsin). Fractions were collected from the HPLC at 3 min intervals. These fractions were applied to mice under conditions identical to those of Experiment I except for the control samples employed. The controls employed in this investigation were homologous 3 min fractions eluted from the HPLC with 200 µl of extraction solvent as the "sample" rather than BGE.

Experiment III employed chromatographic fractions from Experiment II. In this experiment fractions determined to have toxic activity were combined sequentially with other fractions from the same elution which were not toxic.

Experiment IV was designed to elucidate the chemical nature of fractions from Experiment II which had caused mice to die within the parameters of the toxicity tests. In this experiment toxic fractions were digested with either trypsin (Worthington TPCK, I. U. B. 3.4.21.4, Freehold, New Jersey), cyanogen bromide (CNBr) (Pierce Chemical, Rockford, Illinois), or clostripain (Worthington, I. U. B. 4.4.22.8, Freehold, New Jersey). Products of the digestion were applied to mice exactly as in Experiments I, II, and III. Aliquots of the digestive agent dissolved in appropriate buffer and/or activation solutions were used as controls for these tests. MM BGE extracts were not used for this study because the hydrophobic nature of the solvent interfered with the enzymatic action of trypsin and clostripain.

Trypsin and clostripain are proteolytic enzymes which cleave peptide bonds between the amino acids in proteins and peptides. Trypsin cleaves at the carboxy (C) terminal of the amino acids lysine and arginine (as well as some oxidized derivatives of cysteine). Clostripain cleaves only at the C-terminal of arginine. CNBr is a nonenzymatic chemical which cleaves only at the C-terminal of methionine and causes the methionine to form chemical derivatives diagnostic of CNBr cleavage -- methionine sulfone and sulfoxide.

Experiment V employed BGEs which did not cause death in test mice in Experiment I. In this experiment saliva from selected lorises was diluted ten-fold with physiological saline. One hundred µl aliquots of the diluted saliva were added to FA extracts of BGE and allowed to react at ambient temperature (20-28° C) for 10 to 20 min.

Incubated BGEs were either injected into CD mice as in Experiments I through IV or used for Experiment VI.

In Experiment VI whole BGEs, lethal fractions from Experiment II or BGE incubated with saliva from *N. coucang* were presented to a variety of carnivores at Carnivore Preservation Trust, Pittsboro, North Carolina. Observations of the carnivores' reactions were recorded by an observer unaware of whether the cotton swab contained a sham sample, a BGE extract or a BGE-saliva incubate. Carnivores to which presentations were made included species known to be sympatric with *N. coucang*.

In Experiment VII cotton swabs were rubbed across the dorsal fur of an adult male *Peridicticus potto*, an adult female *P. potto* and her one month old female infant. Cotton swabs were presented to two-spotted African palm civets, (*Nandinia binotata)* known to be predators of pottos (Charles-Dominique, 1977) as well as four other carnivore species (Table 1). Reactions of the carnivores to the swabs were recorded by an observer familiar with carnivore behavior but unaware of the identity of sham (saline) and test swabs.

Experiment VIII was designed to test the possibility that the toothcomb found in lorisids (and other prosimians) promoted capillary transfer of saliva from the mouth to the "occlusal" surface of the lower incisors and might, therefore, serve to inject solvated toxins into an animal being bitten. In this Experiment II mandibles of *N. coucang* and two of comparably sized *Aotus trivirgatus* (Fig. 3A) were placed on a flat surface. Mandibles were placed vertically resting on a plane defined by the angle of the jaw and the posterior surface of the head of the mandible (Fig 3B). Two 4.5" x 8.5" Kimwipes (Kimberly-Clark, Roswell, Georgia) were folded to form a surface of approximately two square inches. The absorbent Kimwipe was gently held against the distal end of the mandibular incisors. A Pasteur pipette was filled with a solution of water colored with Coomassie Blue to facilitate visualization. A quantity of the dye solution was applied to the intersection of the groove between the two medial incisors and the alveolus until a meniscus was visible. The meniscus dissipated, and liquid was wicked to the "occlusal" surface through grooves between the incisors where it was absorbed by the folded Kimwipe. As soon as a meniscus disappeared it was replenished. Throughout the course of the experiment the Kimwipe was removed and weighed on an analytical balance, and the weights were recorded. Liquid reaching the folded Kimwipe had flowed against gravity.

A B

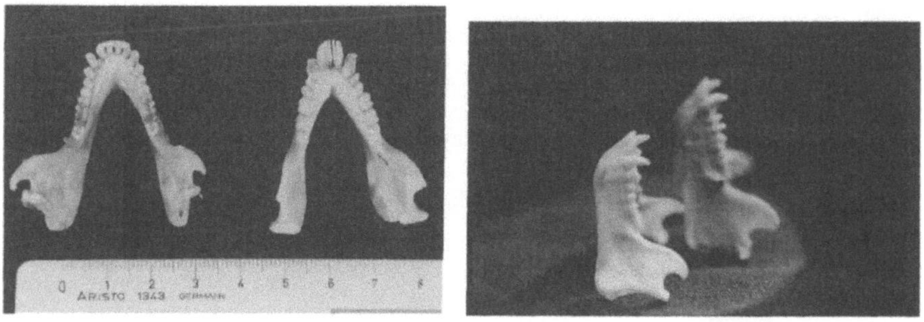

Figure 3. A. Size comparison of mandibles of *Aotus trivirgatus* (left) and *Nycticebus coucang* (right) used in Experiment VIII. B. Mandibles shown in the "vertical" position used to test comparative rates of wicking.

RESULTS

In Experiment I, FA extracts from 2 of 10 slow lorises were lethal to CD mice (n =3 per extract). MM extracts from four of seven lorises were lethal (n =3 per extract).

Results of Experiment II are shown in Fig. 4 and Fig. 5. Homologous fractions from the eight FA BGE extractions in Experiment I which were not previously lethal yielded materials which were lethal upon administration to CD mice.

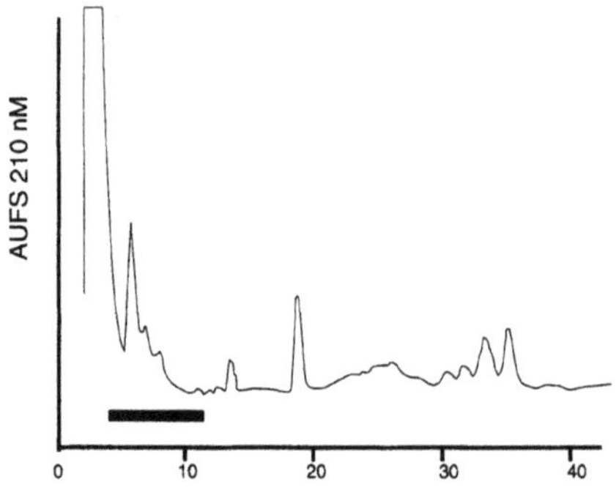

Figure 4. Chromatogram of FA BGE (Experiment II). Shaded bar indicates fractions which were lethal.

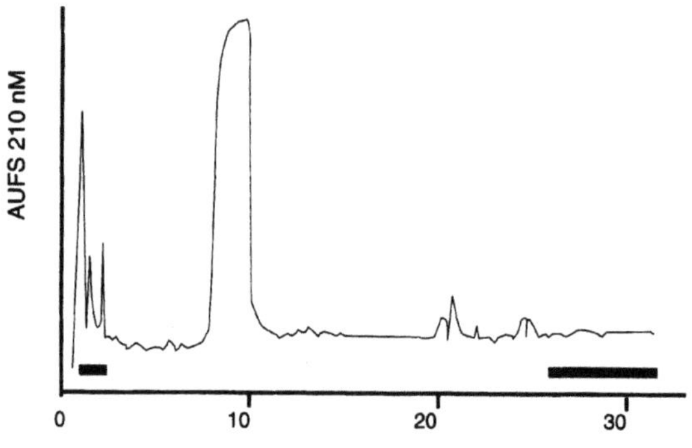

Figure 5. Chromatogram of MM BGE (Experiment II). Shaded bars indicate fractions which were lethal.

Lethal fractions from FA extracts in Experiment II were combined with a late eluting fraction corresponding to fractions seven and eight, (Fig.6) from the same chromatographic development. In all eight cases the administration of mixtures of previously lethal fractions with the late eluting fractions produced no lethality.

In Experiment IV, lethal fractions were incubated with trypsin, clostripain, or CNBr and subsequently chromatographed as in previous experiments. Fractions from the HPLC were administered to CD mice and assayed for lethality. The results are shown in

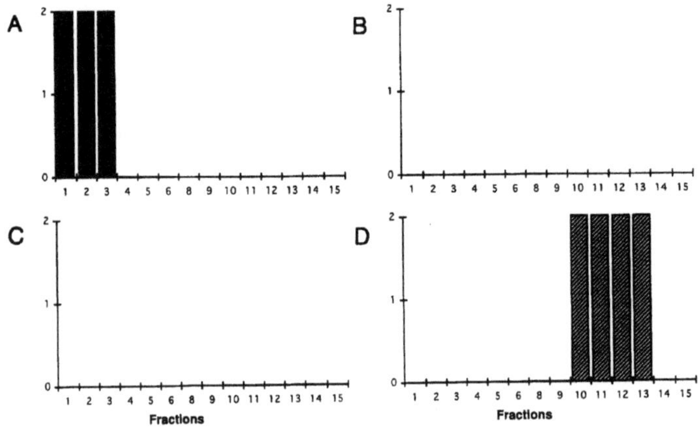

Figure 6. Toxicity tests of FA BGE and chemical and enzymatic digests of FA BGE. Black bars indicate intact BGE. Cross-hatched bars indicate products of digestion. (See Experiment IV.) A. Intact BGE. B. Tryptic digest. C. CNBr digest. D. Clostripain digest.

Fig. 6. Results are presented as one of three alternatives (for consistency with other publications): no bar, no effect; full bar, lethal within the 10 min trial period; half-bar, subject animal was alive at the end of the 10 min trial but either showed marked motor impairment or died immediately after the trial period.

In Experiment V unfractionated FA BGE from the eight lorises whose BGE was not lethal in Experiment I were incubated with diluted *N. coucang* saliva. The incubated mixtures administered to CD mice were lethal in all cases.

In Experiment VI whole BGE, FA extracts of BGE, or BGE incubated with saliva from slow lorises were presented to selected carnivores known to be sympatric with *N. coucang*. Results of these behavioral trials are summarized in Table 1. Likewise, observations of the reaction of three *Nandinia binotata* to swabs taken from the adult and infant *Perodicticus potto* are compiled in this table.

Table 1. Reactions of carnivores to selected swabs from *Nycticebus coucang* and *Perodicticus potto.*

Donor	Recipient[1]	Sample Type[2]	*n*	Average Persistence[3] (5 sec intervals)	Remarks
N. coucang	***Felis tigris***	BGE	2	2.5	Lip curl, tongue protruding
	Felis tigris	BGI		0.5	
	Panthera pardis[4]	BGE	2	2.0	Licked sample, drooled
	Panthera pardis	BGI		1.0	Avoidance
	Arctictis binturong	BGE	5	2.4	
	Arctictis binturong	BGI		1.0	Avoidance
	Neofelis nebulosa	BGE	2	2.0	
	Neofelis nebulosa	BGI		1.0	Salivation, lip curl, flehmen
	Helarctos malayanus	BGE	2	2.5	
	Helarctos malayanus	BGI		1.0	Rapid retreat
	Paradoxurus hemaphroditus (Sumatra)	BGE	1	2.0	
	Paradoxurus hemaphroditus (Sumatra)	BGI		1.0	
	Paradoxurus hemaphroditus (Philippines)	BGE	2	2.0	
	Paradoxurus hemaphroditus (Philippines)	BGI		2.0	
P. potto	***Nandinia binotata***	FD	3	2.5	Feeding response
	Nandinia binotata	MD		3.0	Feeding response
	Nandinia binotata	IFD		1.0	Vocalize, avoidance
	Panthera pardis[5]	FD	2	2.5	
	Panthera pardis	MD		3.0	
	Panthera pardis	IFD		1.0	
	Paradoxurus hemaphroditus (Sumatra)	FD	1	2.0	
	Paradoxurus hemaphroditus (Sumatra)	MD		See remarks	Salivate, ate swab
	Paradoxurus hemaphroditus (Sumatra)	IFD		See remarks	Salivate, ate swab
	Felis caracal[6]	FD	2	1.0	
	Felis caracal	MD		1.0	
	Felis caracal	IFD		3.0	
	Arctictis binturong	FD	1	2.0	
	Arctictis binturong	MD		2.5	
	Arctictis binturong	IFD		2.8	
	Paradoxurus hemaphroditus (Philippines)	FD	2	1.0	
	Paradoxurus hemaphroditus (Philippines)	MD		2.0	
	Paradoxurus hemaphroditus (Philippines)	IFD		1.0	

[1] Bold type indicates that donor and recipient species are partially or totally sympatric.
[2] Sample type: BGE, brachial gland exudate; BGI, BGE incubated with saliva from *N. coucang*; FD, swab rubbed over dorsal pelage of adult female *P. potto*; MD, swab rubbed dorsal pelage of adult male *P. potto*; IFD, swab rubbed over dorsal pelage of infant female *P. potto*.
[3] Number of contiguous 5sec intervals in which recipient species remained within 30 cm of the sample presented. Timing began after first approach by recipient (≤ 5 cm): minimum value for persistence of BGE trials is 1.0.
[4] *Panthera pardis* of Asian origin.
[5] *Panthera pardis* of African origin.
[6] *Felis caracal* may no longer be sympatric with *Perodicticus potto*.

Results of Experiment VIII are shown in Fig. 7. The results in this figure are the averages of three separate trials from each of two mandibles tested of both *N.coucang* and *A. trivirgatus*.

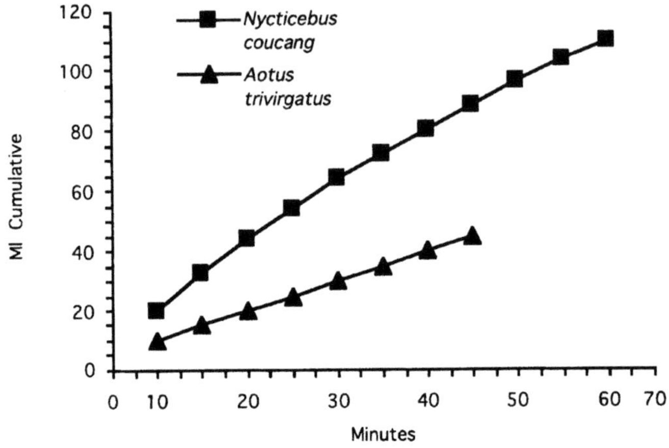

Figure 7. Capillary uptake of vertical mandibles. (See Experiment VIII.)

DISCUSSION

The results of Experiment I indicate that BGE from *N. coucang* is not uniformly toxic to the test animals. Only 20% of the FA test animals died within the rather short time-course of this experiment. Differing lethalities of FA and MM (over half) extracts suggest there are two distinct toxic materials -- one aqueous soluble and the other organic soluble. These differing solubilities are consistent with the administration of toxins through the systemic circulatory system or through release from adipose tissue. These two suggested routes of entry into an animal bitten by a loris in turn suggest a fast-acting toxin and a toxin which is released into the circulation more slowly.

The notion of two distinct toxins is reinforced by the results of Experiment II. In this experiment FA BGE yielded a single contiguous group of toxic fractions (Fig. 4) which eluted near the void volume (V_o) of the HPLC column (approximately 2.5 ml). The fractions are hereafter referred to as FA toxin. Chromatography of MM BGE yielded two distinct regions of toxic fractions (Fig. 5). Because the MM solvent was incompatible with reverse-phase chromatography in low concentrations of ACN, MM extracts were separated under initial conditions of higher ACN than in the case of FA extracts. When the earlier eluting MM toxin (Fig. 5) is rechromatographed under conditions identical to those used for FA extracts it coelutes (data not shown). I assume, therefore, that this toxin is identical to the FA toxin and that FA toxin is somewhat soluble in MM. The later eluting group of toxic fractions is clearly different from FA toxin. Its elution in high concentration ACN indicates that is is lipophilic, and it is subsequently referred to as the MM toxin. *A priori* it is this toxin which I would expect to be the more slow acting.

Experiment III addresses the question of why so few, 20%, of the FA BGE extracts were toxic under the test conditions whereas all of these BGEs were shown to have FA toxins which could be purified by reverse-phase HPLC (see below) or activated by incubation with saliva (Experiment V). FA toxins were prepared from each of the eight *N. coucang* whose FA BGEs had not proven lethal in Experiment I. In all cases, fractions homologous with those of the FA toxin were lethal to test animals. These FA toxins were combined with other fractions which eluted from the same HPLC column development. In all cases FA toxin when combined with fractions seven and eight (Fig. 6) was no longer lethal. These data suggest that there is an FA toxin inhibitor in the BGE of *N. coucang* .

The existence of an aqueous soluble toxin which is reversibly inhibited suggests that the FA toxin is enzymatic and that it is, therefore, a protein or peptide. Because FA toxin eluted just after the V_O of an HPLC column having a nominal pore size of 120Å it is possible to venture that the toxin has a Stokes radius consistent with that of a globular protein of greater than 2500 daltons. The chromatographic separations illustrated in Fig.4 and Fig. 5 are unlikely to have completely purified either toxin, and characterization must await the preparation of more homogeneous compounds. However, it was possible to test the hypothesis that FA toxin is proteinaceous.

Inactivation of FA toxin with proteolytic enzymes or CNBr provides information that either the toxin is proteinaceous or that its action depends upon the presence of a second protein (or peptide) presumably orienting an active site on the toxin or maintaining its conformation. Results from Experiment IV (Fig. 6B) indicate that FA toxin is inactivated by trypsin. Trypsin cleaves peptides at either lysine or arginine. Clostripain also inactivates FA toxin. Whereas trypsin cleaves either lysine or arginine it is tempting to conclude that both trypsin and clostripain had cleaved the same arginine residue. Further inspection of Fig. 6D reveals that this is not the case.

When FA toxin is cleaved by trypsin its toxicity is eliminated. When FA toxin is cleaved by clostripain, the toxicity of the original compound is eliminated, but a later eluting compound is observed to be lethal. Clostripain, therefore, cleaves FA toxin at one or more arginines producing at least two fragments. At least one of these remains toxic. Because tryptic cleavage prevents lethality in all fractions both lysine and arginine must be present in either FA toxin or a putative chaperone protein. Similarly, CNBr inactivation of FA toxin suggests that the toxin contains a methionine.

The FA toxin is therefore likely to be a polypeptide containing the amino acids lysine, arginine and methionine. Cleavage of FA toxin by clostripain into a later eluting toxic compound clearly suggests that BGE may contain FA toxin in the form of a pre- or pro-toxin whose activity depends on post-translational modification by an exogenous compound.

An obvious choice for the source of an exogenous compound or enzyme to accomplish post-translational modification is saliva. It was, after all, this paper's point of departure that slow loris bites had been reported to be toxic and that previous investigations had been unable to identify a toxic compound in the saliva of *N. coucang*. In Experiment VI, BGE adsorbed on cotton swabs was presented to a variety of carnivores known to be sympatric with *N. coucang* . Subject animals were allowed to sniff or lick the swabs as they chose. Identical swabs incubated with *N. coucang* saliva were subsequently presented. In three species (*Neofelis nebulosa,* clouded leopard; *Arctictis binturong,* binturong; *Helarctos malayanus,* sun bear) remarkable reactions occurred. BGE-coated swabs "interested" the carnivores but seemed to provoke little response other than curiosity. BGE-coated swabs incubated with saliva, however, provoked dramatic reactions. Among sun bears (n=2) immediate retreat to the furthest perimeter of the enclosure was noted. Clouded leopards (n=2) exhibited flehmen reactions. Binturongs (n=5) previously curious, retreated and climbed vertical supports in their cages. In these cases, (but not in the case of closely related carnivores allopatric with *N. coucang*) BGE seemed to provoke no remarkable behavior whereas BGE incubated with saliva initiated a noticeable change in behavior.

To what were the carnivores reacting? My brief observations do not allow an answer to this question at this time, however, a hint may be gained from an observation of chromatograms generated during the course of previous experiments. These chromatograms are depicted in Fig. 8. An FA BGE sample incubated with *N. coucang* saliva was chromatographed under the same conditions as those described for Experiment II. A group of at least four chromatographic peaks (seen as two prominent merged peaks in Fig. 8) absent in other chromatograms of FA BGE is apparent. A second aliquot of FA BGE was subsequently incubated with saliva and exposed to a stream of nitrogen for 5 min prior to injection on the HPLC. The chromatographic peaks apparent in Fig. 8A are absent after exposure to nitrogen. Because nitrogen is largely inert under ambient conditions, I conclude that the compounds observed upon incubation were volatile and that the incubation of *N. coucang* saliva with FA BGE (and by inference whole BGE) generates volatiles. Our observations that clouded leopards, binturongs and sun bears reacted to BGE-saliva incubates without licking the swabs upon which they were presented supports the presence of volatile compounds which evoked the observed behaviors.

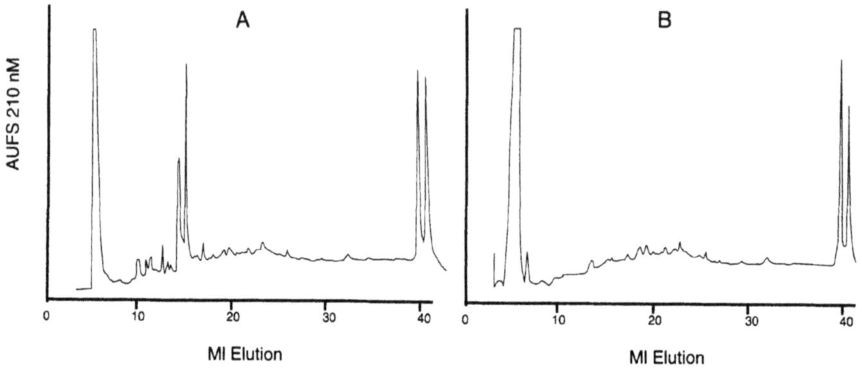

Figure 8. Chromatograms of FA BGE after incubation with saliva from *Nycticebus coucang*. A. FA BGE and saliva chromatographed one minute after mixing. B. FA BGE and saliva chromatographed after a one minute incubation and subsequent five minute exposure to a stream of nitrogen.

Previous reports (Alterman and Hale, 1991) indicated that the BGE of *N. pygmaeus* was also toxic. Because glandular exudates of slow lorises mixed with saliva appear to produce volatile compounds causing aversion in mammals which are potential predators, I wished to examine the possibility that this mechanism was not unique to *Nycticebus*. Although brachial glands are apparent on the arms of *Loris tardigradus*, it was not possible to obtain exudates from individuals in the DUPC colony. The only other lorisine available to me during the course of this work was *Perodicticus potto* which has no apparent brachial glands but is known to produce exudates from genital glands (Manly, 1978). Swabs of an adult male and an adult female were prepared by rubbing the swabs against their dorsal fur. Swabs wiped over the dorsal fur of the female's infant were covered with copious, viscous liquid presumably from grooming by its mother. Results of behavioral trials (Table 1) indicate that whereas the swabs taken from the dorsum of adults evoked aggression from *Nandinia binotata,* swabs taken from the dorsum of the female infant evoked aversive behavior and vocalizations audible for several hundred meters.

It appears then, that a liquid--presumably a salivary mixture--applied to the infant's fur caused a known mammalian predator to retreat and vocalize. This suggests that the saliva applied to the young potto was also generating volatiles. In the cases of both lorisines--*N. coucang* and *P. potto*--there appears to be a chemical role played by some constituent of the saliva in generating volatiles which are potentially defensive. The application of saliva by both species may be of value in promoting survival, but the source of substrate differs.

The infant potto may be likened to a sponge in which substrate of undetermined identity is reacting with salivary compounds (enzymes?) to form noxious volatiles protecting it from attack. In this scenario the mother's toothcomb may act as an applicator which spreads the reactants over a large surface area and ensures that adequate quantities of volatiles are generated while an infant is "parked" otherwise defenseless during its mother's solitary nocturnal foraging.

I predict that the onset of parking will be found to occur when a critical amount of surface area is presented by the infant's developing pelage and that cryptic coloration of the infant is of diminished importance. The white color of some infant pottos suggests, to the contrary, that aposematic coloration may be operative.

The slow loris, however, may deal with the application of saliva by two distinct mechanisms. Recent observations of captive (and therefore habituated) slow lorises by a television camera crew suggest that mothers confronted by large video cameras begin licking their brachial glands and subsequently grooming their infants with their toothcombs (B. Walton, pers. comm.). The observed behavior may be preparation for parking of the infant followed by the mother's retreat. A second route of application for BGE is in salivary solution through bites.

If saliva is a vehicle for the transport of toxins as well as a reactant then the amount of salivary toxin applied to a victim may determine the effectiveness of the bite. Animal dealers who routinely trade in lorises keep a vessel full of water at their side while handling them (pers. observ.). The rationale for this procedure is that coucangs will not readily release a bite. The dealers immerse the limb being bitten with the coucang into the filled vessel as a proven method for getting the animal to abandon its hold. It appears that the loris benefits by prolonged contact of the anterior teeth with its victim.

A potential benefit of maintaining its bite is increased transfer of toxin. One common manner in which toxins in salivary solution are transferred is capillary action. Animals lacking specialized structures for applying venoms (like snakes' fangs or the male platypus' spurs) may employ grooves or channels in their dentition. For example the capillary conduits in some colubrid species are grooves on the posterior dentition (Minton, 1990) while the channel between lower central incisors provides the capillary in *Solendon* species (Pournelle, 1968). I propose that a similar mechanism operates in *Nycticebus* and that the capillaries are channels between the incisors of the toothcomb.

The grooves between lower incisors of *N. coucang* functioned more effectively in capillary transfer than do the spaces between incisors of comparably sized *A. trivirgatus* in Experiment VIII (Fig.7). The transfer of liquid in this experiment was approximately double that for the *N. coucang* specimens examined.

The toothcomb characteristic of living strepsirhines has been viewed as a device to promote grooming (Rosenberger and Strasser, 1985). The results of Experiments III, V, VI, VII and VIII suggest that the toothcomb may function as a chemical tool for the application of defensive volatile-generating compounds which protect infants and as a capillary applicator for the transfer of toxins to allospecifics. If the toothcomb is viewed as a primitive adaptation of early lorisines, then notions of defense by crypsis must be modified to take into account systems which enable small slow-moving primates to avoid predators sensitive to the noxious chemicals generated and/or of the toxic effects they signal. Chemical defenses may therefore be a basis for the adaptation of nocturnal but not inconspicuous primates which are otherwise unable to escape arboreally-specialized predators.

This scenario, however, must be examined in far greater detail than the studies outlined here. The chemical, possibly olfactory, signals I describe probably do not explain the survival of small arboreal animals which are primarily the prey of raptors or reptiles. These predators' use of olfactory and/or vomerolfactory pathways is likely to be different from those of mammals (Weldon and Ferguson, 1993). Because the origins of Primates predate the origins of carnivores used as models in behavioral experiments described here, it is unlikely that chemical defenses employed by modern primates precisely mirror those of their distant ancestors. Chemical defenses used by prey animals and resistance to these defenses by predators constitute rapidly coevolving arms wars. Examples of these coevolutionary spirals are well known among amphibia, reptiles and fishes (Barthalmus and Zielinski, 1988; Zielinski and Barthalmus, 1989).

As adaptive changes in size, locomotion, habitat utilization and social organization have occurred, originally toxic or defensive compounds used against allospecifics may have become socially useful signals among conspecifics and members of closely related taxa. This may to be the case with *Eulemur catta* and *Hapalemur* species which generate exudates used in "stink fights" from carpal glands found on the wrist (Simons and Rumpler, 1988). Histological studies are needed to determine if there is a common origin of these antebrachial glands and their associated brachial glands with the brachial glands of *Nycticebus.* Similarly, we must view phenomena like scent marking from a different perspective asking how, for example, a scent mark "intended" for conspecifics does not habitually become an aid to potential predators or competitors. Can it be that a scent mark like the BGE is only a chemical substrate waiting for an appropriate overmark to give up its informational content and that crypsis in this view is really the possession of a specific activating enzyme not made by a potential enemy?

ACKNOWLEDGEMENTS

The work outlined above could not have been done without the cooperation of E. Simons of the DUPC and his staff. This cooperation continued under the direction of K. Glander. The HPLC and fraction collectors were made available by Gilson Medical Electronics and its representative Thomson Instrument Company. A number of individuals also donated their services. One, in fact, became an example of the toxicity of slow loris bites. These individuals include M. Hale, C. Tilden, J. Perry, R. Ange, D. Brewer, V. Jones, M. Handler and K. Weisenseel. My coeditor K. Izard first suggested that I might be interested in Wilde's 1972 report.

The rare opportunity of observing the reactions of carnivores sympatric with coucangs was afforded me by M. Bleyman and his staff at the Carnivore Preservation Trust. They have literally put their hands into the lion's (actually tiger's and leopard's) mouth to obtain saliva samples.

I am deeply thankful to each of these individuals, organizations and to my colleagues whose "but..." questions continue to generate fruitful areas for further research.

LITERATURE CITED

Alterman,L. & Hale, M. E. (1991) Comparison of toxins from brachial gland exudates of *Nycticebus coucang* and *N. pygmaeus. American Journal of Physical Anthropology, Sup. 12*, 43.

Barrett, E. (1981) The present distribution and status of the slow loris in peninsular Malaysia. *Malaysia Applied Biology, 10*, 205-211.

Barthalamus, G. T. & Zielinski, W. J. (1988) *Xenopus* skin mucus induces oral dyskinesias that promote escape from snakes. *Pharmacology, Biochemistry & Behavior, 30*, 957-959.

Cartmill, M. (1972) Arboreal adaptations and the origin of the order primates. In R. Tuttle (Ed.) *The Functional and Evolutionary Biology of Primates* (97-122), Chicago: Aldine Atherton Inc.

Charles-Dominique, P. (1977) *Ecology & Behaviour of Nocturnal Primates.* New York, Columbia University Press.

Griffiths, M. (1978) *The biology of the monotremes.* New York: Academic Press.

Manley, G. H. (1974) Functions of the external genital glands of *Perodicticus* and *Arctocebus.* In G. A., Doyle, R. D. Martin & A. C. Walker (Eds.) *Prosimian Biology* (313-329) London: Duckworth Press.

Michelsen, A. (1978) Sound reception in different environments. In M. A. Ali (Ed.) *Sensory ecology: Review and perspectives* (345-373) New York: Plenum Press.

Minton, S. A. (1990) Venomous bites by nonvenomous snakes: an annotated bibliography of colubrid envenomation. *Journal of Wilderness Medicine, 1*, 119-127.

Pearson, O. P. (1956) A toxic substance from the salivary glands of a mammal (short-tailed shrew). In E. E. Buckley & N. Porges (Eds.) *Venoms* (55-58), Washington, American Association for the Advancement of Science.

Pournelle, G. H. (1968) Classification, biology and description of the venom apparatus of insectivores of the genera *Solendon, Neomys,* and *Blarina.* In W. Bücherl, E. E. Buckley, & V. Deuloffeu (Eds.) *Venomous animals and their venoms* (31-42) New York: Academic Press.

Pucek, M. (1968) Chemistry and pharmacology of insectivore venoms. In W. Bücherl, E. E. Buckley, & V. Deuloffeu (Eds.) *Venomous animals and their venoms* (43-50) New York: Academic Press.

Rosenberger, A. L. & Strasser, E. (1985) Toothcomb origins: support for the grooming hypothesis. *Primates, 26*, 73-84.

Simons, E. L. & Rumpler, Y. (1988) Eulemur: New generic name for species of Lemur other than Lemur catta. (sic) *Comptes Rendues Academie Scientifique Paris 307*, 547-551.

Tidswell, F. (1906) *Australian Venoms* Sydney: Department of Public Health New South Wales.

Weldon, P. J. & Ferguson, M. W. J. (1993) Chemoreception in crocodilians: anatomy, natural history, and empirical results. *Brain, Behavior and Evolution 41*, 239-245.

Wilde, H. (1972) Anaphylactic shock following bite by a slow loris: *Nycticebus cougang. The American Journal of Tropical Medicine and Hygiene. 21*, 592-594.

Zielinski, W. J. & Barthalmus, G. T. (1989) African clawed frog skin compounds: Antipredatory effects on African and North American water snakes. *Animal Behaviour, 38*, 1083-1085.

Zimmermann, E. (1985) Vocalizations and associated behaviours in adult slow loris (*Nycticebus coucang*). *Folia Primatologica 44*, 239-245.

FOOD DISTRIBUTION AND REPRODUCTIVE CONSTRAINTS IN THE EVOLUTION OF SOCIAL STRUCTURE: NOCTURNAL PRIMATES AND OTHER MAMMALS

Pierre Charles-Dominique

URA1183 CNRS
Muséum National d'Histoire Naturelle
Laboratoire d'Ecologie Générale
4 avenue du Petit Château
91800 Brunoy, France

INTRODUCTION

The enormous amount of research carried out on the sociology of nonhuman primates clearly demonstrates the interest in this group in which we find our roots and a rational explanation for the basis of human behavior. It would thus be logical that the prosimians whose anatomy and physiology conceals a large number of ancestral characteristics (Vallois, 1955; Piveteau, 1958; Le Gros Clark, 1961; Luckett, 1969; Charles-Dominique and Martin, 1970; Martin, 1975, 1979), would attract the attention of primatologists in so far as their study might bring clarity to the most remote origins of the social behavior of our ancestors. The social structures of prosimians have been studied since the 1960s' demonstrating the importance of olfaction in communication and, in nocturnal species, a social organization based on territorial arrangements and the subtle exchange of chemical signals contained in their urinary markings (Sauer and Sauer, 1963; Charles-Dominique, 1972, 1977a, b; Martin, 1972; Niemitz, 1974, 1984; Doyle, 1975; Bearder and Martin, 1980; Schilling, 1979, 1980; Bearder, 1987). Research on species belonging to different orders of placental mammals reveals a great deal of similarity to nocturnal primates, which suggests a common base in mammalian social organization (Eisenberg and Gould, 1970; Eisenberg and Kleiman, 1972; Kleiman, 1974; Emmons, 1975, 1978; Charles-Dominique, 1978a, b; Feer, 1989).

In this article, a tentative model of mammalian social structure is put forward, taking into particular consideration those species adapted to arboreal conditions. The principle is also offered that social characteristics are adaptive responses which represent a compromise between environmental constraints and constraints inherent in the biology of the species. The position of nocturnal primates is then dealt with and discussed in the light of this model.

Creatures of the Dark: The Nocturnal Prosimians
Edited by L. Alterman *et al.*, Plenum Press, New York, 1995

THE GENERAL CONTEXT OF THE DEVELOPMENT OF SOCIAL STRUCTURES IN MAMMALS

Biological Constraints

Each species invests the greater part of its time and energy in two major categories of behavioral activity; sexual behavior in the interests of maintaining genetic identity and recombination in the population, and feeding behavior, necessary to obtain energy used in the course of locomotor, metabolic and reproductive activity. All other behaviors (social, exploratory, defense against predators, nest-building, etc.) are entirely dependent on the first two categories.

Reproduction. The reproductive investment of males and of females is, by nature, radically different. For females, the production of young, their number and survival rate are determining factors, and the chances of transmitting the maternal genome to the generations which follow will depend essentially on the reproductive success of the mother. Metabolic effort (gestation, suckling) necessary for success is directly dependent on the nutritional potential of the environment, each female trying to find ways of gaining access to the necessary nutritional resources, which results in female-female competition.

For males the number of females successfully impregnated is the factor determining transmission of the paternal genome to succeeding generations. Females may thus be considered as a "resource" for males. Male-male competition for females is therefore more important than that associated with access to nutritional resources.

Social structures may therefore be interpreted as a compromise between access to nutritional resources and the appropriation of females by the male. On the one hand the vital domain of females concerns nutritional resources necessary for reproduction and, on the other hand, the vital domain of males concerns imposing themselves on females (at least temporarily).

Nutrition. The quality and quantity of potential nourishment varies considerably with the seasons and also in the way it is distributed spatially.

In a given ecosystem, each species of animal, as a function of its particular specializations, will only exploit a small part of the various kinds of food available (Figure 1a). In general, there is a relationship between the abundance and diversity of these resources, and the density of the population which exploits them. The abundance of each available food category varies with the season, from minimum to maximum, but it may equally well vary from one year to the next (Figure 1b). These variations are more or less pronounced according to the type of food, and also according to the local climate. The environment is alway characterized by a heterogeneity all the more pronounced by a large biological diversity resulting in a patchy distribution of resources (Figure 1c).

If one considers a particular nutritional category (for example, the leaves or flowers of a particular species, its seeds, the pulp of its fruit, etc.), not only is the spatial distribution heterogenous, but the quantity produced per unit is extremely variable. Usually one finds 1) some uncommon and very dispersed individual plants producing a large amount, 2) a number of more common individual plants producing an average amount and, finally, 3) a large number of common individuals producing poorly. This type of distribution, which depends partly on the population structure of plant species, is quite general. It expresses itself in an L-shaped size/frequency distribution curve (Figure 2), which exponentially decreases (Rollet, 1969). In addition, such spatial heterogenity is more or less accentuated depending on the nature of the resource and the type of environment under consideration.

For example, in a tropical forest, this may not be uniform due to the mosaic structure of the plant communities (Oldeman, 1990).

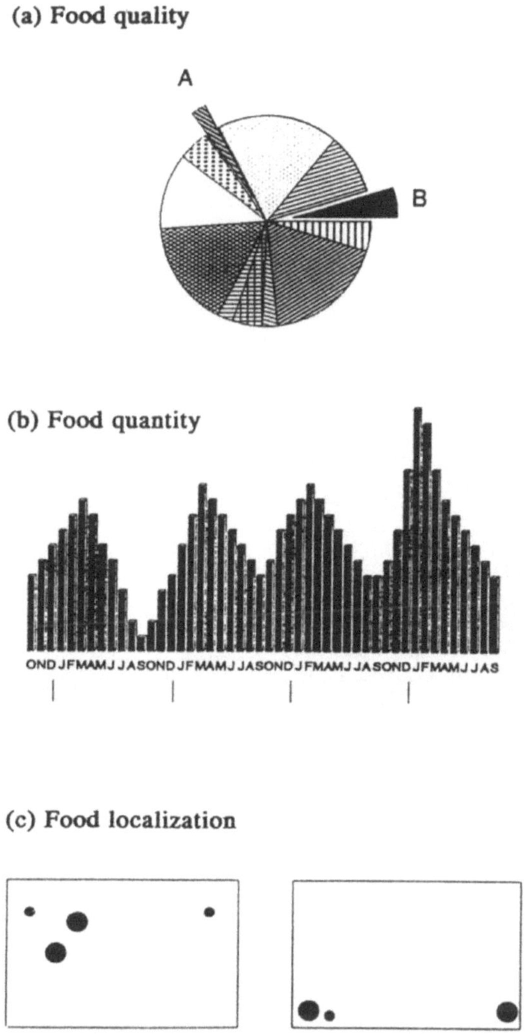

Figure 1. Schematic representation of the nutritional categories available in an ecosystem as a function of their quality, their relative abundance and their spatial distribution.(a) the available food sources (A & B) for an animal species equipped with the necessary specialization; (b) seasonal and year to year quantitative variations of a food category; (c) spatial localization of a food category during two successive seasons.

Strategies for Exploiting Nutrional Resources as a Function of their Local Abundance

One may postulate that the exploitation of resources, corresponding to two extremes in terms of the way in which nutrition is distributed, requires different morphological and sensory adaptations. The use of distribution of a very large amount of food over a broad grid involving great distances from one part to another, is very different from the use of poor production over a compact grid. While "average" production should be accessible to all, some species are adapted for exploiting food resources from a large area, and some for exploiting food resources in a small area.

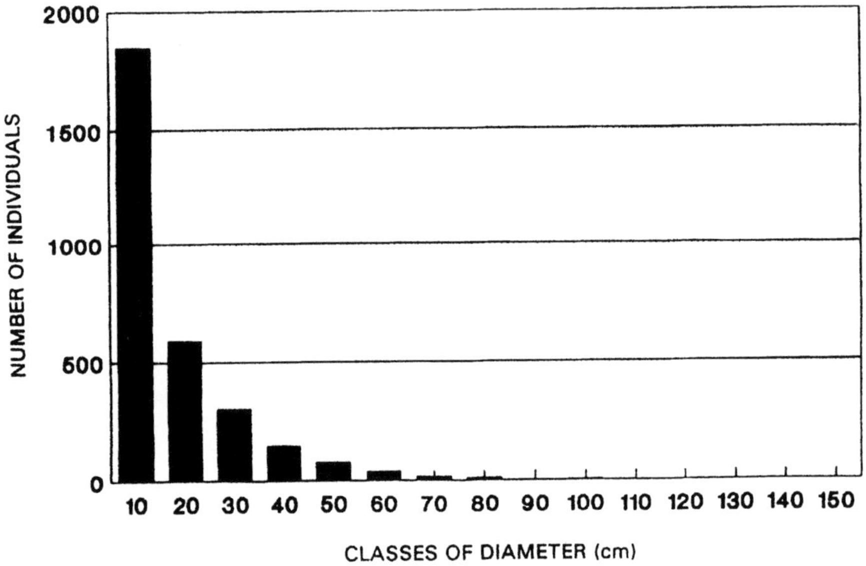

Figure 2. An example, taken from a tropical forest in French Guiana, showing the distribution curve of tree species classed by diameter (from 10 cm) - (by courtesy of B. Riéra - unpublished).

Large production distributed over a broad grid. The exploitation of such resources requires sensory capacity for their detection from a distance, and an efficient mode of locomotion to reach these resources. Local resource abundance allows animals of large size, and thus with major nutritional requirements, to feed themselves adequately, enabling them to invest in the energetic effort required for covering long distances. Moreover this local abundance allows many individuals to feed at the same time (Figure 3a).

Diurnal life, an indispensible condition for the visual detection of small inconspicuous food resources over great distances, will favor this type of exploitation and, for the same reasons, it will favor the development of a social system based on gregariousness. The majority of the examples of gregariousness are found amongst the diurnal mammals: primates, antilopes, carnivores, marsupials, etc. (Charles-Dominique 1975). Nevertheless, there are examples of gregariousness in nocturnal species. For instance, some frugivorous bats are able to exploit clumps of trees producing fruit in large quantities as a result of their ability to fly above the forest and cover long distances at relatively little energetic cost (Thomas, 1975). One should note that these are the largest species of bat (amongst these being Old World fruit bats and New World leaf-nosed bats; Brosset, 1966; Heithaus *et al.*, 1975; Bradbury, 1977; Morrisson, 1978; Fleming and Heithaus, 1981; Thomas, 1982). The use of olfactory cues in air currents circulating above the forest plays a determining role in detecting these resources, which may be spaced great distances apart (Fleming and Heithaus 1986; Laska 1990).

(a)

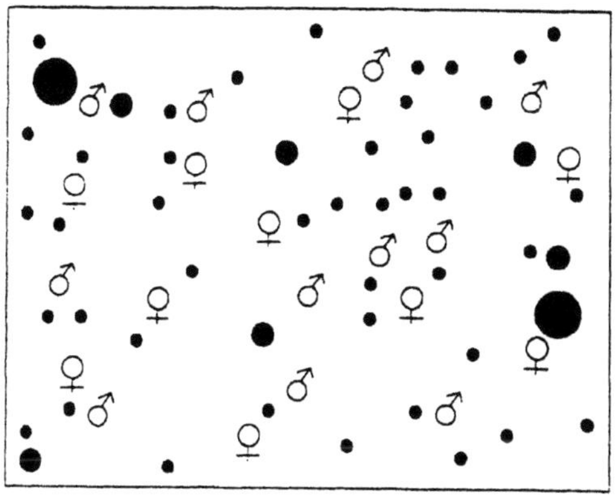

(b)

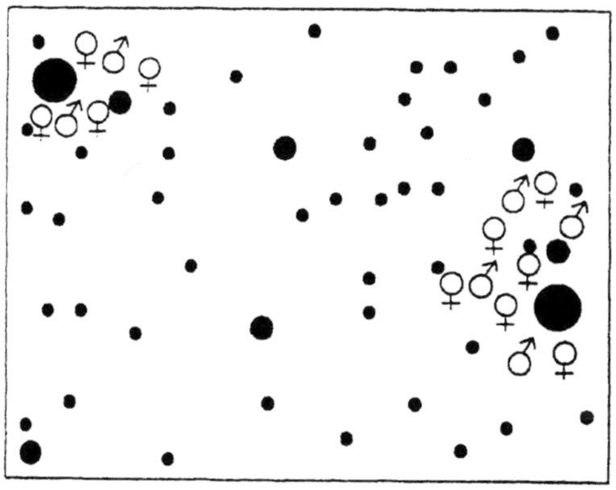

Figure 3. Schematic representation of the distribution of individuals of a solitary nocturnal species, exploiting weak resources distributed over a small grid (a); and the distribution of individuals of a gregarious diurnal species, exploiting abundant resources distributed over a broad grid (b). Full circles represent the level of local food abundance (high, medium and low, according to their size).

Increase in body size, which characterizes all species exploiting resources distributed over a large grid, is made possible as a result of local food abundance. It is also probably a question of an adaptive response which gives individual animals of greater size an advantage in the face of competition from other species using the same resources, and also in the face of intraspecific group competition.

Weak production distributed over a small grid. Due to the proximity of resource areas, a sensory system restricting detection of resources, to short distances is compatible with the exploitation of weak resources in small grids. Nevertheless, the fact that only small quantities are available at each place will favor smaller species whose nutritional needs are low. Animals of greater size, and thus with greater energetic requirements, will not be able to exploit this type of resource. Such resources would require covering very long distances, costing more in energy than the benefits they would receive in covering these distances. For the same reasons this type of resource will not suit gregarious animals (Figure 3b).

The factors common to those species exploiting resources distributed in this way are 1) small body size, and 2) a more solitary way of life. Many examples can be found, particularly among nocturnal species (primates, insectivores, rodents, ungulates, carnivores, etc.), as well as among diurnal species (squirrels, tree shrews). However, nocturnal species constitute the major part of this category, traced back probably to the nocturnality of the first mammals (Charles-Dominique 1975).

Strategies for Exploiting Nutritional Resources as a Function of Seasonal Variation and Distribution

Apart from nutritional resources consistently available for long periods (for example sap exploited by *Cebuella*, Terborgh, 1983), changes in the localization of feeding areas, as a function of phenological plant cyclicity, is the norm. Three types of social strategy may be recognized, corresponding to the first of these variations in the distribution of a resource and subject to the biological constraints suggested above such as seasonal opportunism and individual and group territoriality.

Seasonal opportunism. In the case of seasonal opportunism, there is a single concentration of individuals in the proximity of a food resource (Figure 4a). Animals must move about to find new food items when this resource comes to an end. Due to the irregularity and heterogeneity of production, there will be competition for food based on direct confrontation which will regulate both the distribution and the density of the animal population. In the same way, males will have to compete directly with one another in order to have access to reproductive females, a phenomenon rare among mammals, although it has been reported in the opossum, the most primitive of the marsupials (Atramentowicz, 1982; Charles-Dominique, 1983; Julien-Laferrière and Atramentowicz, 1990).

Individual territoriality. In this case, an individual occupies and defends an exclusive food resource area which is required throughout the entire year, or at least for the greater part of the year, whatever the seasonal fluctuations in abundance and distribution. Competition between females is not related to nutritional resources at any given time, but rather to space which includes potential nutritional resources, providing at the appropriate period of the year sufficient food for her and her young. Once the progeny become independent they disperse and the newly adult females establish themselves in other territories.

In order to reproduce, males must defend a space occupied by females, at least temporarily. Just as a female will have to defend a territory against other females, the male will have to defend his territory against other males to gain access to a female or females. In this case, male territoriality is based on the appropriation of females or, more correctly, on the space which they occupy. Whatever the stage of the female's ovarian cycle, males defend their space against other males. Thus, the moment a female comes into estrus, the resident male will have priority of access. Because the yardstick of male's reproductive success is the number of offspring, i. e. number of females impregnated, the tendency will be

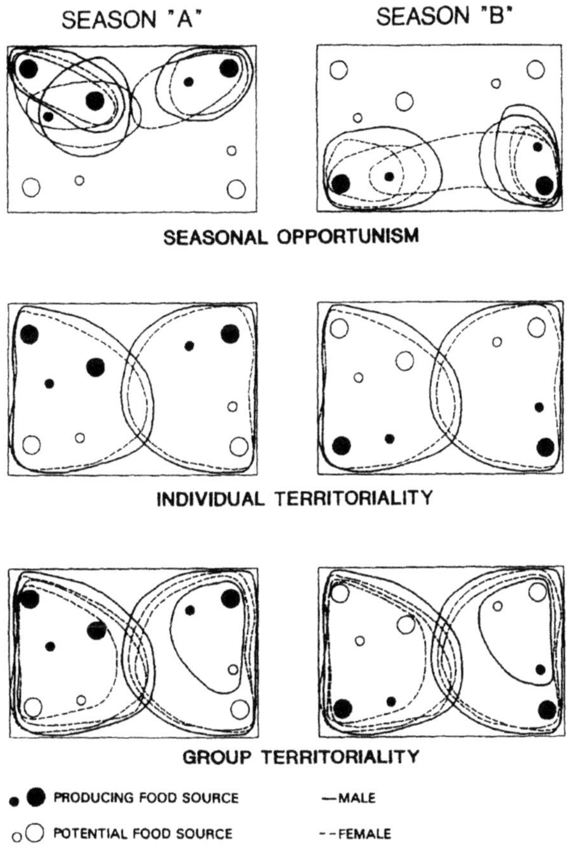

Figure 4. Schematic representation of the spatial distribution of animal belonging to a species practicing (a) seasonal opportunism; (b) individual territoriality; and (c) group territoriality. Two periods of the year are illustrated (left and right), during which the available resources are not found at the same places.

towards polygamy, and the appropriation of the largest number of female territories. Most studies show that it is the presence of females which induces competition between males, not the occupation of an area which has no females (Charles-Dominique, 1978b). A consequence of polygamy will be a decrease in food intake on the part of males, which is to the advantage of females. Supernumery males (celibates) may exploit less profitable areas abandoned by females.

To take advantage of such a system means appropriating a segment where resources have a strong probability of occurrence. For females, this means optimizing access to nutritional resources when they become available and, for males, optimizing access to females when they come into estrus. For this system to function it must be associated with a means of territorial signaling which must include the ability to recognize particular social signals. However, when an adult female dies or when a young individual has to establish a new territory, the establishment of new boundaries may lead to group disruption, the squandering of energy, and increased risk of predation.This type of social structure is found in a number of mammalian lines, e.g. spiny rats (Emmons, 1982), Old World rats and mice (Duplantier, 1988), tree shrews (Martin, 1968; von Holst, 1974), pangolins (Pagès, 1975); sloths (Montgomery and Sunquist, 1975), palm civets (Charles-Dominique, 1978a), elephant shrews (Rathbun, 1976), and water shrews (Berthoud, 1978).

Group territoriality. The third social strategy can be derived from the preceding one with many affiliated females sharing the same territory, rather than an individual female occupying a territory. Young females are sedentary and benefit from their mothers' experience (or that of other associated females) for the best information regarding an area suitable for supplying sufficient food for the entire annual cycle. There is thus a continuity in territorial occupation even when a female dies. Outbreeding is assured by males emigrating at puberty. Control of the space occupied by a group of associated females (a matriarchy) is the focus of intense competition between males, a situation identical to that of individual territoriality described above.

Like individual territoriality, group territoriality is based on signals indicating the presence of a group of females occupying a particular area. Moreover, it is necessary on the one hand, that animals are capable of recognizing individual conspecifics and, on the other hand, that there is a degree of behavioral plasticity enabling a female to modify her attitude towards other females and function according to the position she occupies in society as a whole (her hierarchical position in the same group or in relation to a neighboring group). This requires the establishment of permanent social bonds regularly maintained and reinforced among females.

Group territoriality, found in all primate species, and also in other mammalian lines, may be found in gregarious animals in which all locomotor activities are coordinated and synchronized and in animals which move about more often alone, but which share the same territory with other conspecifics.

In the first case (diurnal, gregarious animals), the fact that partners maintain a constant but small distance from one another favors the development and use of visual and vocal messages for intra-group communication rather than olfactory messages (Roeder, 1985). The fact that behavioral responses to social signals are nearly always immediate, favors group co-ordination, the quick settling of conflicts and regulation of social interactions in general. Territorial group communication (group to group) remains based, however, more often on vocal communication (Gautier and Gautier-Hion, 1977).

In the second case, a solitary lifestyle doesn't mean the social life is simple. In effect, individuals sharing the same territory may remain almost constantly in communication made possible by olfactory signals contained in their urine, or in glandular secretions deposited in the form of markings. However, the deposition of a message and the information it contains may be separated by many hours, even many days, so that behavioral responses to a social signal are often delayed in time (deferred communication). There is thus no synchrony of movement about the territory, but individuals sharing the same territory may meet one another for some of their activities. In this situation, long-range vocal communication may be used for rapid location of conspecifics. Territorial signals (between neighboring territories) are made as often by chemical signals (marking) as by vocal signals. (Jolly, 1966, 1985; Richard, 1974; Roeder and Andersson, 1990).

THE PARTICULAR CASE OF NOCTURNAL PRIMATES

With the exception of the night monkey, *Aotus* (Cebidae) and the Tarsidae (primitive family of the Haplorhinni), the nocturnal primates are all members of a primitive group, the Strepsirhinni. They include the Lorisidae of Africa and Asia, the Cheirogaleidae, the Lepilemuridae, the genus *Avahi* (Indriidae) and the genus *Daubentonia* of Madagascar. The genus *Ptilocercus* is not included in this list since the inclusion of the Tupaidae in the primate order is strongly doubted (Martin, 1968). An abundant literature on the socio-structural traits of nocturnal primates reveals many similarities shared by a large number of species and which may be considered as constituting a basic social model. However, some nocturnal primates diverge to some extent from this basic model, and it is interesting to analyze these exceptions in relation to their respective ecological peculiarities.

The Basic Model

The genera *Galago*, *Otolemur*, *Microcebus*, *Mirza* and *Tarsius* present a cluster of common characteristics which may be schematized as follows; 1) a territory common to many associated females (a matriarchy) overlapping slightly with the territories of neighboring groups. These overlapping areas are often visited by females from two (and sometimes more) neighboring groups. A radio-telemetry study of *Galago alleni* has shown a 10 fold increase in urine marking of these small meeting places visited alternately by non-affiliated females who apparently avoid one another (Charles-Dominique, 1977). These small communal areas, where an individual is less in her own home and more in that of another, may serve to maintain a balance and to regulate territorial competition between neighboring matriarchal groups, 2) males who control one or several females have territories partly overlapping the territories of other neighboring male counterparts. These areas common to males may have a function analogous to those of overlapping females zones, 3) males not actively reproductive (young adults, those of small size, subordinate males) may, depending on the situation, be excluded or tolerated by dominant males. Their exclusion may lead to very strong direct competition (fights), the outcome of which may be an unbalanced sex ratio in favor of females (e.g. *Galago alleni*). In other cases, groups of celibate males may establish themselves at the extremities of female territories (e.g. *Galagoides demidovii*). There may be a degree of tolerance towards promiscuity on the part of dominant males (*G. demidovii*, *Microcebus murinus*; Charles-Dominique, 1972, Martin, 1972; Pagès, 1988). A hierarchy is established, regulated by pheromones contained in urine marks, as has been demonstrated in *M. murinus* (Perret, 1992; Perret and Schilling, 1987a), and 4) social and territorial communication is based on a combination of vocalizations and chemical messages contained in urine marks.

Particular cases

Lorinae. In *Perodicticus potto,* territorial organization is almost identical to that of the basic model, but adult females do not share their territories with their maturing daughters. Moreover, social communication is based almost exclusively on urine marking, as vocal communication is reduced and reserved for some exchanges (mother-young) at close quarters (Charles-Dominique, 1977a). These characteristics, also present in other members of the sub-family (Bearder 1987), may be considered as acquired secondary modifications, which may have appeared at the same time as the sub-family Lorinae developed its own characteristic mode of locomotion based on slow and discrete movement. This camouflage strategy would be incompatible both with very marked and visually noticeable movements and with powerful vocalizations easily detectable by predators.

Lepilemuridae. The major characteristic of this family is that it has adapted itself to the exploitation of leaves and flowers. This has been as a result of the enlargement of the cecum which allows bacterial flora to break down cellulose, and the development of a particular form of caecotrophy which enables the lepilemur, after a twofold digestion of food, to absorb fatty acids coming from cecal fermentation (Charles-Dominique and Hladik, 1971; Hladik *et al.*, 1971; Hladik and Charles-Dominique, 1974). It is this adaptation for exploiting a nutritional resource that is common, abundant, and available in all small territories which influences the social life of *Lepilemur*. Because of the small area exploited (on average 0.1 ha for *L. leucopus* compared to about 4.0 ha for *G. alleni*), an individual may visually control almost all of its territory. Neighboring animals keep an eye on one another from a few meters away, exchanging visual signals (movements of the head) and vocal duets. There are no overlapping zones between neighboring territories, vocal communication predominates, and urine is not used for territorial marking (Charles-Dominique and Hladik, 1971). Other social components corresponding to the basic model are given above. The social organization of *Lepilemur* may be regarded as stemming from the basic model, modified secondarily by ecological specializations peculiar to this family.

Phaner furcifer. The basic source of nourishment of *Phaner* is gum exudates produced by some tree species as a result of parasitic infestation. Gums exudates are a highly reliable source of food, available regularly from year to year. Each female's territory depends on a number of these sources (about 12) producing sufficient amounts of gum (Charles-Dominique and Petter, 1980). As is the case in other nocturnal primates (complying with the basic model), male territories are superimposed on those of females, but there is a clear dominance of females over males. A male, coming to a gum exudate being exploited by a female, waits some meters away until she has finished eating before approaching. If the male is eating when a female arrives he moves away quickly, yielding his place to her. Male and female(s) remain paired and in contact aided by vocalizations as powerful as the distance separating them requires. Calls are emitted spontaneously when one individual starts to move about, or in response to the call of a partner who is moving away. *Phaner* does not urine mark, but the male has a cutaneous gland on the throat which he rubs on the female in the course of allogrooming sessions (Charles-Dominique and Petter, 1980).

Neighboring territories involve overlapping areas as do those of many other species. It should be noted that these zones are established in neutral parts of the forest, parts which contain no gum-producing trees. Neighboring individuals gather together several times a night, emitting powerful vocalizations in unison (vocal chorus). If one of them is placed experimentally in the middle of his territory in a cage, his neighbors will make some forays into his territory despite the number of calls made in response to their calls. If he is removed, neighboring individuals will intrude completely into this territory within an hour (Charles-Dominique and Petter, 1980).

These details must be evaluated in relation to the biology of *Phaner furcifer*. The distances between the small number of places producing gum and some overlapping territorial zones are more or less fixed. As is the case with *Lepilemur*, which also occupies a small territory, a vocal call permits immediate control of certain strategic points. The defense of nutritional resources spread over larger territories, in contrast, requires a combination of vocal signals and urine marks. This type of system is the rule for most nocturnal primates.

DISCUSSION AND CONCLUSIONS

Despite some differences which may be explained by specific ecological adaptations, in most of the nocturnal primates, a homogeneity of social structure can be seen which fits the requirements of reproduction and foraging. A factor common to almost all nocturnal primate species, and also to a number of other mammalian species, is the use of urine as a means of spreading social and territorial signals. This must be regarded as an ancestral mode of communication, and an attribute of macrosmatic animals equipped with a vomeronasal organs (Schilling, 1987). The major use of this type of message rests on a connection between the identity of the individual and the urine markings which it distributes around its territory. In effect, urine contains the metabolic products of an animal's sex hormones and its diet. To a conspecific equipped with the necessary sensory organs, these urine marks may form an uneven olfactory picture of the individual emitting them (identity, hormonal and metabolic activity diet). In some cases of very simple social structure, such a composite picture may serve only in recognition by the male of the female's estrous condition and for the mutual stimulation of sexual physiology (the case of the Didelphid marsupial opossum, Perret and Atramentowicz, 1989). In a number of cases where social structure is more complex, as in nocturnal primates, it appears that the urinary signal may transmit many other messages: These may include individual recognition, the time lapse between deposit and receipt of the signal (Schilling, 1980), pheromonal signals inhibiting sexual activity in some individuals (Perret and Schilling, 1987a,b; Izard 1990), and several messages biasing sex-ratio at conception (Perret, 1990, Perret, this volume).

In mammals, and more particularly in the primates, the change to a diurnal way of life represented an innovation requiring new adaptations in the area of vision, in the method of locating and discriminating certain nutritional resources situated at great distances, not possible at night (Charles-Dominique, 1975). At the same time, access to new types of nutritional resources, locally abundant but distributed over a large area, would have favored the evolution of gregariousness. It is the coming together of gregariousness with a diurnal way of life which may have led to the predominance of vision at the expense of olfaction, thus leading to the establishment of a system of communication more rapid and flexible, necessary for life in a group.

Paleontological studies show that the first mammals were small-sized (Simpson, 1945; Szalay, 1975) and macrosmatic (olfaction predominating). They therefore had great difficulty detecting nutritional resources over long distances, which most probably led them to orient themselves towards small nutritional resources within a more constrained area (at least for non-flying, forest-dwelling species). One may also suppose that they were nocturnal, and that their way of life would have been more conservative with regard to ancestral characteristics. In effect, most nocturnal primates, with some specific exceptions, much like nocturnal species belonging to other mammalian orders, retain a number of primitive characteristics and rely largely on olfaction for communication.

ACKNOWLEDGEMENTS

I thank G.A. Doyle for translation of this manuscript from French and constructive discussion in revising the manuscript, and K. Izard and L. Alterman for their thorough review of the manuscript.

REFERENCES

Atramentowicz, M. (1982). Influence du milieu sur l'activité locomotrice et la reproduction de *Caluromys philander* (L.), *Revue d'Ecologie (Terre et Vie)*, *36*, 376-395.

Bearder, S.K. (1987). Lorises, bushbabies and tarsiers: diverse societies in solitary foragers. In B.B. Smuts, D.L. Cheney, R.M. Seyfarth, R.W. Wrangham, & T.T. Struhsaker (Eds.), *Primate Societies* (pp.11-24). Chicago: University Press.

Bearder, S.K. and Martin, R.D. (1980). Acacia gum and its use by bushbabies *Galago senegalensis* (Primates: Lorisidae), *International Journal of Primatology*, *1*, 103-128.

Berthoud, G. (1978). The movements of the european hedgehog: a preliminary study, *La Terre et la Vie*, *22*, 73-32.

Bradbury, J.W. (1977). Social organization and communication. In W.A. Wimsatt, (Ed.), *Biology of Bats*, vol. II, (pp. 1-72). New York: Academic Press,

Brosset, A. (1966). Les chiroptères du Haut Ivindo (Gabon), *Biologia Gabonica*, *2*, 47-86.

Charles-Dominique, P. (1972). Ecologie et vie sociale de *Galago demidovii* (Fischer 1808; Prosimii) *Zeitschrift für Tierpsychology. Suppl. 9*, 7-41.

Charles-Dominique, P. (1975). Nocturnality and diurnality: an ecological interpretation of these two modes of life by an analysis of the higher vertebrate fauna in tropical forest ecosystems. In W.P. Luckett & F.S. Szalay (Eds.), *Phylogeny of the Primates: a Multidisciplinary Approach*, (pp. 69-88). New York: Plenum Press

Charles-Dominique, P. (1977a). *Ecology and Behaviour of Nocturnal Primates.* London: Duckworth.

Charles-Dominique, P. (1977b). Urine marking and territoriality in *Galago alleni* (Waterhouse, 1837 - Lorisoidea, Primates - a field study by radio-telemetry, *Zeitschrift für Tierpsychology*, *43*, 113-138.

Charles-Dominique, P. (1978a). Ecologie et vie sociale de *Nandinia binotata* (Carnivores, Viverridés): comparaison avec les prosimiens sympatriques du Gabon, *La Terre et la Vie*, *32*, 477-528.

Charles-Dominique, P. (1978b). Ecological position of the Lorisidae family compared to other mammalian families, *Annals of Carnegie Museum of Natural History*, *6*, 26-30.

Charles-Dominique, P. (1983). Ecology and social adaptations in Didelphid marsupials: comparison with Eutherians of similar ecology, In J.F. Eisenberg, & D.G. Kleiman (Eds.), *Advances in the Study of Mammalian Behavior*, (pp.395-422).Washington: Special Publication of American Society of Mammalogy.

Charles-Dominique, P., & Martin, R.D. (1970). Evolution of lorises and lemurs, *Nature*, *227*, 257-260.

Charles-Dominique, P., & Hladik, C.M. (1971). Le lépilémur du Sud de Madagascar - écologie, alimentation et vie sociale, *La Terre et la Vie*, *25*, 3-66.

Charles-Dominique, P., & Petter, J.J. (1980). Ecology and social life of *Phaner furcifer*, In P. Charles-Dominique, H.M. Cooper, A. Hladik, C.M. Hladik, G.F. Pariente, E. Pagès, J.J. Petter, A. Petter-Rousseaux & A. Schilling (Eds.), *Nocturnal Malagasy Primates: Ecology, Physiology and Behavior* (pp.41-73). London, New York: Academic Press.

Doyle, G.A. (1975). Some aspects of urine-washing in four species of prosimian primate under semi-natural laboratory conditions In S. Kondo, H. Kawai and A. Ehara (Eds) *Contemporary Primatology* (pp.232-237). Basel: Karger

Duplantier, J.M. (1988). Biologie évolutive de populations du genre *Mastomys* (Rongeur, Muridé) au Sénégal, Thèse de Doctorat d'Etat, USTL.

Eisenberg, J.F., & Gould, E. (1970). The tenrecs: a study in mammalian behavior and evolution, *Smithsonian Contribution to Zoology*, *27*, 1-137.

Eisenberg, J.F., & Kleiman, D.G. (1972). Olfactory communication in mammals, *Annual Review of Ecology and Systematics*, *3*, 1-31.

Emmons, L.H. (1975). Ecology and behavior of african rainforest squirrels, PhD Thesis, Cornell University, Ithaca, New York.

Emmons, L.H. (1978). Sound communication among african rainforest squirrels, *Zeitschrift für Tierpsychology*, *47*, 1-49.

Emmons, L.H. (1982). Ecology of *Proechymis* (Rodentia, Echimyidae) in south-eastern Peru, *Tropical Ecology*, *23*, 280-290.

Feer, F. (1989). Occupation de l'espace par deux bovidés sympatriques de la forêt dense africaine *Cephalophus callipygus* et *C. dorsalis*, *Revue d'Ecologie (Terre et Vie)*, *44*, 225-248.

Fleming, T.H., & Heithaus, E.R. (1981). Frugivorous bats, seed shadows, and the structure of tropical forests, *Biotropica*, *13*, 45-53.

Fleming, T.H., & Heithaus, E.R. (1986). Seasonal foraging behavior of the frugivorous bat *Carollia perspicillata*. *Journal of Mammalogy*, *67*, 660-671.

Gautier, J.P., & Gautier-Hion, A. (1977). Communication in Old World monkeys. In T.E. Seabock, (Ed.), *How Animals Communicate* (pp.890-964). Bloomington: Indiana University Press.

Heithaus, E.R., Fleming, T.H., & Opler, P. (1975). Foraging patterns and resource utilization in seven species of bats in a seasonal tropical forest, *Ecology*, *56*, 841-854.

Hladik, C.M., & Charles-Dominique, P. (1974). The behaviour and ecology of the sportive lemur (*Lepilemur mustelinus*) in relation to its dietary peculiarities. In G.A. Doyle, & A.C Walker, (Eds), *Prosimian Behaviour*. (pp.23-37,). London. Duckworth.

Hladik, C.M., Charles-Dominique, P., Valdebouze, P. Delort-Laval, J., & Flanzy, J. (1971). La caecotrophie chez un primate phytophage du genre *Lepilemur* et ses corrélations avec les particularités de son appareil digestif, *Comptes Rendus de l'Académie des Sciences Paris*, *272*, 3191-3194.

Izard, M.K (1990). Social influences on the reproductive success and reproductive endocrinology of prosimians primates. In *Socioendocrinology of primate reproduction*. (pp.159-186). Wiley-Liss, Inc.

Jolly, A. (1966). *Lemur behavior: a Madagascar field study*. Chicago University Press.

Jolly, A. (1985). *The evolution of priate behavior, 2nd edition*. New York, Macmillan.

Laska, M. (1990). Olfactory sensitivity to food odor components in the short-tailed fruit bat, *Carollia perspicillata* (Phyllostomatidae), *Journal of Comparative Physiology*, *166*, 395-399.

Julien-Laferrière, D., & Atramentowicz, M. (1990). Feeding and reproduction: the example of three Didelpid marsupials in two neotropical forests (French Guiana), *Biotropica*, *22*, 404-144.

Kleiman, D.G. (1974). Patterns of behaviour in hystrichomorph rodents, *Symposium of Zoological Society of London*, *34*, 171-209.

Le Gros Clark, W.E. (1961). *The Antecedents of Man*, Edinburgh: University Press.

Luckett, P. (1969). Evidence for the phylogenetic relationships of tree shrews (Family Tupaiidae) based on the placenta and foetal membranes, *Journal of Reproduction and Fertility*, *Suppl. 6*, 419-433.

Martin, R.D. (1968). Reproduction and ontogeny in tree-shrews (*Tupaia belangeri*), with reference to their general behaviour and taxonomic relationships, *Zeitschrift für Tierpsychology*, *25*, 409-532.

Martin, R.D. (1972). A preliminary field study of the lesser mouse lemur (*Microcebus murinus* J.F. Miller 1777). *Zeitschrift für Tierpsychology*, Suppl. 9, 43-89.

Martin, R.D. (1975). The bearing of reproduction behavior and ontogeny on Strepsirhine phylogeny. In W.P. Luckett & F.S. Szalay (Eds.), *Phylogeny of the Primates: a Multidisciplinary Approach*, (pp. 265-297). New York: Plenum Press.

Martin, R.D. (1979). Phylogenetic aspects of Prosimian behavior. In G.A. Doyle, & R.D. Martin, (Eds.), *The Study of Prosimian Behavior* (pp.45-77). New York: Academic Press.

Montgomery, G.G., & Sunquist, M.E. (1975). Impact of sloth on neotropical forest energy flow and nutrient cycling. In F.B. Golley, & E. Medina, (Eds.), *Trends in Tropical Ecology* (pp.69-98). Berlin: springer.

Morrisson, D.W. (1978). Foraging ecology and energetics of the frugivorous bat *Artibeus jamaicensis*, *Ecology*, *59*, 716-723.

Niemitz, C. (1974). Outline of the Behavior of *Tarsius bancanus*, In G.A. Doyle, & R.D. Martin, (Eds.) *The Study of Prosimian Behavior*, (pp.631-660). London & New York: Academic Press.

Niemitz, C. (1984). Synecological relationships and feeding behavior of the genus *Tarsius*, In C. Niemitz, (Ed.), *Biology of Tarsiers* (pp.59-75). Stuttgart & New York: G. Fisher,

Oldeman, R.A.A. (1990). *Forests: Elements of Silvology*. Berlin: Springer-Verlag.

Pages, E. (1975). Etude éco-éthologique du Pangolin *Manis tricuspis* par radio-tracking, *Mammalia*, 39, 613-641.

Pagès, E. (1988). Modalités de l'occupation de l'espace et relations interindividuelles chez un prosimien nocturne malgache (*Microcebus murinus*), *Folia Primatologica*, *50*, 204-220.

Perret, M. (1990). Influence of social factors on sex ratio at birth, maternal investment and young survival in a prosimian primate, *Behavioral Ecology and Sociobiology*, *27*, 447-454.

Perret, M. (1992). Environmental and social determinants of sexual function in the male lesser mouse lemur (*Microcebus murinus*), *Folia Primatologica*, *50*, 1-25.

Perret, M., & Atramentowicz, M. (1989). Plasma concentrations of progesterone and testosterone in captive woolly opossums (*Caluromys philander*), *Journal of Reproduction and Fertility*, *85*, 31-41.

Perret, M., & Schilling, A. (1987a). Role of prolactin in a pheromine-like sexual inhibition in the male lesser mouse lemur, *Journal of Endocrinology*, *4*, 279-287.

Perret, M., & Schilling, A. (1987b). Intermale sexual effect elicited by volatile urinary ether extract in *Microcebus murinus* (Prosimian Primates), *Journal of Chemical Ecology*, *133*, 495-507.

Piveteau, J. (1958). *Traité de Paléontologie, tome VI, Origine des Mammifères et leur évolution*, Paris: Masson.

Rathbun, G.B. (1976). Evolution of the rump region in the folden-rumped elephant-shrew, *Annals of Carnegie Museum of Natural History*, *6*, 11-19.

Richard, A. (1974). Patterns of mating in *Propithecus verreauxi verreauxi*. In G.A. Doyle, R.D. Martin, & A.C. Walker (Eds.) *Prosimian Biology*. (pp.49-74). London: Duckworth.

Roeder, J.J. (1985). Evolution des systèmes de communication chez les carnivores et les primates: organisation sociale et modalités de communication, *Comportement*, *2*, 1-80.

Roeder, J.J. and Andersson, J.R. (1990). *Primates: Recherches actuelles.* Paris: Masson.
Rollet, B. (1969). Etudes quantitatives d'une forêt dense humide sempervirente de la Guyane vénézuélienne, Thèse de Doctorat, Université de Toulouse.
Sauer, E.G.F., & Sauer, E.M. (1963). The South West african bushbaby of the *Galago senegalensis* group, *Journal of SouthWest Africa Scientific Society, 16*, 5-36.
Schilling, A. (1979). Olfactory communication in Prosimians. In G.A. Doyle, & R.D. Martin (Eds.), *The Study of Prosimian Behavior* (pp.461-542). New York: Academic Press.
Schilling, A. (1980). The possible role of urine in territoriality of some nocturnal primates. In D.M. Stoddart (Ed.), *Symposium of the Zoological Society of London*, Academic Press, *45*, 165-193.
Schilling, A., 1987, L'organe vomeronasal des mammifères, *Journal de Psychologie. 3/4*, 221-278.
Simpson, G.G. (1945). The principles of classification and a classification of mammals, *Bulletin of American Museum Natural History, 85*, 1-350.
Szalay, F.S. (1975). Phylogeny of primate higher taxa: the basicranial evidence. In W.P. Luckett, and F.S. Szalay, (Eds.) *Phylogeny of the Primates, a multidisciplinary approach*, (pp.91-125), New York: Plenum Press.
Terborgh, J. (1983). *Five New World primates: a study in comparative ecology*, Princeton: University Press.
Thomas, D.W. (1982). The ecology of an african savanna fruit bat community: resource partitioning and role in seed dispersal, PhD Thesis, University of Aberdeen.
Thomas, S.P. (1975). Metabolism during flight in two species of bats, *Phyllostomus hastatus* and *Pteropus gouldii*. *Journal of Experimental Biology, 63*, 273-293.
Vallois, H. (1955). Ordre des primates. In P.P. Grassé (Ed.) *Traité de Zoologie* (pp.1854-2206). Paris: Masson.
Von Holst, D. (1974). Social stress in the tree-shrew: its causes and physiological and ethological consequences. In G.A. Doyle, and A.C. Walker (Eds.), *Prosimian Biology* (pp.389-411). London: Duckworth.

SOCIAL ORGANIZATION IN THE AYE-AYE *(DAUBENTONIA MADAGASCARIENSIS)* AND THE PERCEIVED DISTINCTIVENESS OF NOCTURNAL PRIMATES

Eleanor J. Sterling [1,2]
Alison F. Richard[1]

[1]Department of Anthropology
Yale University
51 Hillhouse Ave
New Haven, CT 06511 USA

[2]Deutches Primatenzentrum
Kellnerweg 4
37077 Göttingen Germany

INTRODUCTION

Nocturnal species have typically received cursory attention in comparative analyses of primate social organization, and it has been assumed that their social organization is both less complex and more homogeneous than that of diurnal forms (e.g. Charles-Dominique and Martin, 1970; Martin, 1972; Charles-Dominique, 1977; Eisenberg, 1981; Fleagle, 1988). This assumption seems to have more to do with the difficulty of seeing the interactions of nocturnal species than with empirical evidence of qualitative differences in complexity. The difficulty arises partly from observation conditions at night, and partly from the fact that the interactants are often displaced from each other in space and time.

The acuteness of this difficulty became evident during efforts to characterize aye-aye social organization based on a two-year field study by E.J.S. on the island of Nosy Mangabe (see also Sterling, 1993a,b). Female aye-ayes maintained exclusive ranges and had no affiliative relationships with one another, whereas males had extensively overlapping ranges and showed a variety of affiliative and agonistic relationships. Males and females had overlapping ranges and interacted in individually distinctive ways. When a female was in estrus, many males gathered and mated with her, and males mated with more than one female. The particular configuration of spatial and social relationships found among male and female aye-ayes is unique among nocturnal species studied to date. Other patterns of interaction were either not present or, as we suspect, masked by long intervals separating the behavior of interactants.

As in many studies of primates in the wild, our knowledge of the aye-ayes of Nosy Mangabe is further limited by our ignorance of the age and genealogical relationships of the study animals. Basic parameters of aye-aye social organization, such as dispersal by maturing animals, remain undocumented, and the existence or boundaries of communities have yet to be established. However, our data do enable us to refute the early suggestion that aye-ayes have a "primitive social structure similar to that of *Microcebus murinus*" (Petter, 1977).

In this paper, we first present data on the social, mating and spacing systems of aye-ayes in support of the overview given above. We then argue that the difficulty of comparing the social organizations of nocturnal and diurnal primates characterized by displaced interactions on the one hand, and by synchronic interactions on the other, is an example of the broad problem of establishing homology among behaviors, i.e. establishing that behaviors are similar because they are retentions from a common ancestor. Comparative studies of closely related nocturnal and diurnal primates could provide an important testing ground for hypotheses about the determinants of social organization if, and only if, a sound basis is found for identifying behavioral homologies.

METHODS AND STUDY SITE

The study took place on Nosy Mangabe Special Reserve, a 520 ha island located 5 km from the east coast of Madagascar in the Bay of Antongil. Nine aye-ayes were moved to this forested island from the mainland in the 1960s as part of a conservation effort (Petter and Peyrieras, 1970). E.J.S. radiocollared eight individuals (six males: Sof, Man, Nam, Tak, FV, Sil) and two females: Nir, Von), and identified another seven individuals (two females and five of unknown sex). With the help of two guides, she followed some combination of these animals at night over a two-year period between 1989 and 1991. The quantitative analyses presented in this paper come from data collected during partial (a minimum of four consecutive hours) or all-night follows of two females (Nir, Von) and four males (Sof, Man, Nam, FV), all of adult size (n = 1,014 h). Radio collars on the other two males (Sil and Tak) worked for less than one month apiece, *ad hoc* observations on these individuals and on the other seven identified but uncollared individuals are presented descriptively, to help clarify inter-individual relationships.

Observations were undertaken on single individuals on any given night. Distance to and identity of conspecifics and activity of the focal animal were recorded at 5 minute intervals. Distance to other animals was divided into five categories: contact, 0.1-4.9 m, 5.0-9.9 m, 10.0-14.9 m, and 15.0-19.9 m. Twenty m was set as the maximum distance for recording interactions, because it was difficult to detect animals without vocalizations or to estimate their location when they were vocalizing at distances greater than 20 m. The total time conspecifics were observed to be within 20 m of the focal animal was noted to the nearest minute. Rare events, such as scent marking and avoidance behavior were recorded *ad libitum*. In some of the analyses presented here, data were subdivided into three seasons: (i) Hot, Wet (HW) = mid-December through mid-May, (ii) Cold, Wet (CW) = mid-May through mid-September, and (iii) Hot, Dry (HD) = mid-September through mid-December.

Three aspects of social organization are examined in this paper: social interactions, mating and spacing. Students of diurnal primates commonly view spacing as a socially or ecologically dependent variable (e.g. Oates, 1987), and we agree that it is difficult to consider in isolation. However, in this study as in many others involving nocturnal prosimians (see Bearder, 1987), more was discovered about spacing than about social behavior, making it easy for evidence concerning spacing to outweigh or swamp sparser evidence of social interactions in our analysis. Separating the analysis of spatial and temporal dimensions of social interactions from examination of the identity of interactants and the content of their interactions helps clarify the nature of social relationships of nongregarious, nocturnal species and thereby facilitates comparison with gregarious, diurnal species.

Several kinds of evidence were used to characterize these elements of aye-aye social organization: vocalizations, observed interactions, inferred interactions, and nest and range use. Calls were identified by sound, and classified according to the context in which they were used (Table 1). Other researchers have described aye-aye vocalizations (Petter and Petter-Rousseaux, 1967; Petter and Charles-Dominique, 1979; Stanger and Macedonia, 1994.), but in the absence of tape recording and sonograms from this study it is difficult to match calls described here with those reported elsewhere.

Table 1. Types of vocalizations given. Aff=Affiliative, Agon=Agonistic, Dist=Distress. Categories follow Stanger and Macedonia 1994.

Category	Vocalization	Context
Aff	Screech Eep-short	Contact call used when non-antagonistic animals fed close together; often used prior to locomotion
Aff	Screech Eep - long	Used over greater distances than short call, context similar
Agon	Screech Aack - short	Directed at conspecifics, context usually aggressive. Used to maintain distance
Agon	Screech Aack-long,loud	Communicated over longer distances when other animals were approaching - a "do not come near" call
Agon	Screech Aack-long,soft	Given commonly during breeding periods when many animals were close by
Mating/Dist	Whimper	Repetitive call given by females during copulation; by young individuals trying to forage in a tree already occupied by an adult; used during fights over food; and in direct encounters between males, and between males and females
Dist/Agon	Scream	Given during capture; or preceding aggressive interactions
Dist	Plea	Given during capture or when surprised by another individual
Dist	Sneeze	Two-part alarm given to humans, other aye-ayes, *Varecia*.
Dist	Snort	Used when animals were agitated, "pacing"
Dist	Hai-hai	Used when animals were roused from nest during capture
Other	Groan	Given when coming out of anaesthesia
Other	Whirring	Sounds like fishing reel running out; frequent during feeding, meaning unclear; also given during recovery from anaesthesia

Data on social relationships come from observed interactions, defined as those in which two or more animals interacted in the presence of the observer, and from inferred interactions, defined as those in which only one of the animals was in sight of the researcher, but it was possible to infer the context (Table 2). Social interactions involving estrous females are considered in the mating section and are excluded from the analysis of social interactions.

Table 2. Types of interactions between individual aye-ayes, Nosy Mangabe 1989-1991

Type	Activity	Description
		Observed interactions
Affiliative	Foraging in tandem	Two or three individuals move in concert from one foraging site to another; vocal contact and coordinated movement
Agonistic	Fighting and chasing	Two individuals meet, try to bite one another, or two individuals come into proximity and one flees and the other pursues
Agonistic	Displacement	Individual moves towards a resource, calls, and individual using resource leaves
Sexual	Mating	Male mounts female for extended period, presumed intromission
		Inferred interactions
Agonistic	Avoidance	Individual moves along an often traveled path between feeding sites, then deviates from path when another individual is in the feeding site ahead

Patterns of nest use and home range overlap were also used to make inferences about social organization. Although animals did not sleep in the same nest at the same time during this study, some animals used the same nest serially. Information on nest use was collected by tracking animals to their nests during daylight hours. On a given day we sometimes located several individuals' nesting sites; each occupied nest was logged separately as one "nest day". Data on home range size and overlap were collected during all-night follows and superimposed on a map of Nosy Mangabe (Sterling, 1993 a,b). Pathfinder 1.5 (M. Winslett, unpublished manuscript) was then used to calculate minimum convex polygons from night range paths.

RESULTS

Social Relationships

Aye-ayes on Nosy Mangabe seemed to communicate primarily by means of calls and scent-marks. Three kinds of scent-marking were observed - ano-genital rubbing, head and chest rubbing, and overmarking - but they are not considered further in this paper. Visual signals were infrequently observed, though animals regularly exhibited a stereotyped pacing when other individuals were nearby.

Study animals were rarely observed interacting outside the mating period, and each spent no more than - and usually much less than - 21% of its time within 20 m of another animal during a season (Table 3). This maximum was undoubtedly influenced by the fact that one female, ♀Nir, was nearing estrus for almost a third of the observations.

Table 3. Percentage of five minute samples in which animals were found within 20 m of another animal. HD = hot, dry season. HW = hot, wet season. CW = cold, wet season. Numbers in parentheses represent number of five-minute samples in which the animals were observed per season.

Animal	HD	HW	CW
♂ Nam	14% (595)	8% (1958)	2% (895)
♂ Man	13% (575)	No data	No data
♂ Sof	8% (102)	No data	No data
♂ FV	15% (701)	12% (206)	8% (812)
♀ Nir	21% (1105)	No data	7% (1063)
♀ Von	5% (906)	12% (1356)	No data

Although interactions were infrequent, certain patterns were evident between individuals and between classes of individuals. Females interacted rarely and were uniformly aggressive toward one another; all three observed encounters resulted in a fight or chase (Table 4). Unlike males and certain male-female pairs, we have no evidence of active avoidance; it seemed simply that females rarely came into proximity. Vocalizations between females were, likewise, rare and aggressive (Table 4).

Table 4. Interactions and vocalizations between males, between males and females, and between females. (Data are from collared females Nir and Von, and collared males Sof, Nam, Man and FV; n=1,014 h.)

	No. of encounters	No. of 5 minute samples			
		Affiliative	Agonistic	Other	Total
Interactions					
Male-male	40	114	171	285	570
Male-female	21	140	70	258	468
Female-female	3	0	18	11	29
Vocalizations					
Male-male	As above	11	7	5	23
Male-female	"	18	6	21	45
Female-female	"	0	3	0	3

Twenty-one encounters were seen between males and females and, in contrast to those between females, their content was quite variable (Table 4). Encounters lasted from 1-153 minutes (mean 63 ± 59 min). Tandem foraging accounted for 30% of time spent interacting, and fighting for 15%. Only twice were males seen to avoid females. Agonistic vocalizations were rare, and nearly half the vocalizations given during encounters were classified as affiliative (Table 4). One pair, ♂FV and ♀Nir, seemed to have a particularly close relationship (by aye-aye standards, anyway). They were seen within 20 m of one another for varying lengths of time on 15 separate occasions, and they were observed foraging in tandem on eight occasions outside the mating season. ♂FV mated with ♀Nir on the second day of her estrus (see below).

Males were observed interacting with one another more frequently than they interacted with females and, certainly, much more frequently than females interacted with one another. Encounters between males varied in content (Table 4), and tended to be briefer (range = 1-79 min, mean = 26±25 min) than those between males and females. For 20% of the five-minute samples during these encounters, males foraged in tandem, for 30% they exhibited agonistic behavior, and the remainder was spent in displacements (5%) or foraging in the same tree without evident interaction (45%). Certain males consistently avoided others (n=12 occurrences). Calls given by males to each other likewise indicated varying levels of tolerance (Table 4).

Although the overall frequency of interaction between any two males was low, the following accounts illustrate the differing quality of relationships evident between particular pairs. The two males Man and Nam met on 14 occasions; five times they foraged in tandem (12 5-min samples), seven times they foraged in the same tree with no apparent interaction (76 5-min samples), and twice they vocalized aggressively over a resource and one animal left (3 5-min samples). ♂FV and ♂Nam encountered one another on eight occasions, all of which resulted in ♂FV leaving. On five other occasions, ♂FV appeared to take active steps to avoid ♂Nam -- as he did with other males, too (♂Sof, n=3; ♂Man, n=2). ♂FV was not significantly smaller than the other adult males, either in weight or limb length (Sterling 1993a).

An additional clue to the nature of relationships between males is provided by observations made during the capture process. Success in capturing animals was greatest during mating periods when many males clustered around the female. Males roused from their nests by efforts to capture them eventually went in search of other nests. Often these other nests were occupied. Reactions by the occupants to attempted entry varied from rebuff to prolonged altercations to grooming sessions. ♂Man and ♂Nam, who were seen foraging in tandem (n=5 encounters) outside mating periods, groomed one another when ♂Nam tried to enter ♂Man's nest. ♂Nam and ♂Sof both rebuffed ♂FV, who also consistently fled from altercations with these males during the night. ♂FV became involved in a prolonged altercation with ♂Sil when ♂FV tried to enter ♂Sil's nest, and ♂Sil eventually left. During the only observed interaction between ♂Sil and ♂FV outside the mating periods, ♂Sil avoided ♂FV.

Outside mating periods, larger numbers of animals (3-4) were seen foraging and traveling together, and seemed to be using calls to coordinate activities (n=6 occasions, 24 5-minute samples). These rare observations are difficult to interpret, but the possibility should not be excluded that there are definable communities within the aye-aye population on Nosy Mangabe. It should be noted that aggregations of several aye-ayes have also occasionally been observed foraging together in forests on the Madagascar mainland (Ancrenaz, 1991; pers. obs. EJS).

Mating

♀Nir and ♀Von were observed mating in October 1990 and February 1991, respectively. Both females exhibited genital swelling and color change for three days, which we take to be physical indicators of estrus (Sterling 1993a). During this time, they exhibited a repetitive pattern of traveling swiftly over 500-1000 meters, then sitting still for an hour, giving long calls that drew males from all directions. Three of the males that followed each estrous female were study animals. At least two others were unknown. During the first two nights of estrus, the females rebuffed all males. Even when a male succeeded in instigating copulation during this time, the female struggled with him throughout. Males chased each other in circles around the female and tried to dislodge whichever male was copulating with the female. Copulation lasted for approximately one hour (mean = 62±12 min, n=6). Afterwards, the female traveled quickly to another location and the pattern was repeated.

During the first two nights of estrus, each female mated with ♂FV. ♂Sof and ♂Sil were observed in pursuit of both females, but were never seen to mate with either. At the end of the second night of ♀Nir's estrus, an unidentified male slept in the open outside her nest. On the third night, while it was still early evening and light, ♀Nir approached this male just after leaving the nest. He fled, she chased him, and eventually they copulated. He was the only male with whom she was seen to copulate that night.

Spacing

The two females studied had non-contiguous home ranges separated by 250-300 m., and no other females were encountered within the study females' home ranges. Uncollared females were occasionally sighted outside the study area regularly traversed by E.J.S., and an uncollared female was frequently encountered between and at the range boundaries of ♀Nir and ♀Von. Home ranges of males overlapped and were larger than those of the females. There was considerable overlap between male home ranges, varying from 40 to 75% (Sterling, 1993a,b).

Data on time spent in proximity to conspecifics were analyzed separately for males and females (Figure 1a and b). Females were never seen touching conspecifics outside the context of mating. They spent little time in proximity to other animals. What time they did spend in proximity was for the most part with males, and their only close contact with other females occurred during rare agonistic encounters. Males touched other males during fights, but most of their time in proximity to other males was spent between 10 and 15 m. Males were less often found in proximity to females but, when they were, they were commonly at closer distances than to other males.

Nests were located in the early morning on 582 nest days. The eight study animals used a total of 116 nests. The number of nests used by both males and females increased as a function of observation time, although females used fewer nests per unit of observation time than males (Fig. 2). Twenty four of these nests were used on different days by two or more individuals. Twenty one nests were serially occupied by two individuals, one was used by three and two were occupied on different nights by four individuals (Table 5). While ♂FV and ♂Nam shared nests sequentially at one time or another with most animals in the area, ♂FV and ♀Nir preferentially shared nests with one another, and ♀Nir rarely shared nests with animals other than ♂FV. It should be emphasized that the context within which sequential sharing occurred is unclear and findings may reflect differences in social relationships or in resource-related ranging behavior.

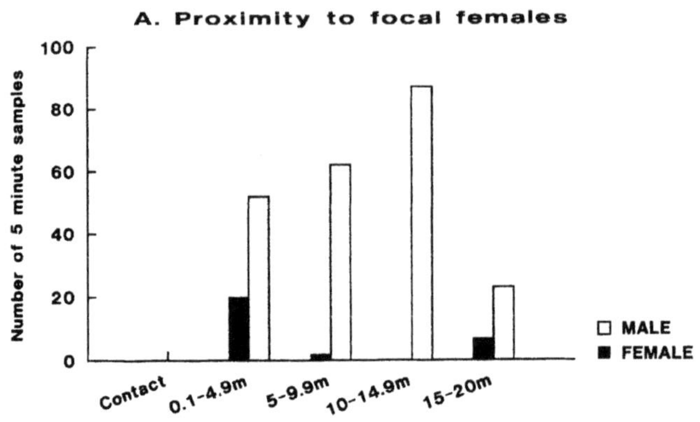

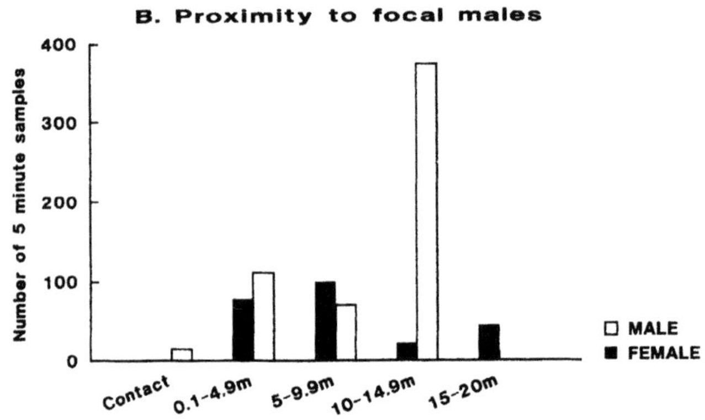

Figure 1 A.. Number of 5-min samples in which females and males were found within five distance categories from two focal females (Females: n=29, males: n=223). B. Number of 5-min samples in which females and males were found within five distance categories from three focal males (Females: n=245, males: n=570).

Table 5. Number of nests shared by animals on separate nights.

Animal	♂FV	♂Nam	♀Nir	♂Man	♂Sof	♀Von	♂Tak	♂Sil
♂ FV	-	5	8	2	4	2	0	0
♂Nam	5	-	2	4	3	0	1	0
♀Nir	8	2	-	0	2	0	0	0
♂ Man	2	4	0	-	2	1	0	0
♂Sof	4	3	2	2	-	0	0	0
♀Von	2	0	0	1	0	-	0	1
♂Tak	0	1	0	0	0	0	-	0
♂Sil	0	0	0	0	0	1	0	-
Total nests	40	35	20	16	16	10	4	2
Total days observed	162	169	110	27	33	75	4	2

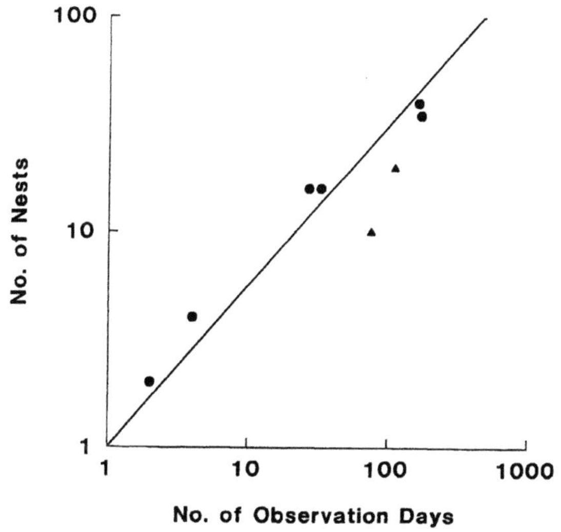

Figure 2. Plot of number of nests against number of observation days. Nests of males are indicated by circles, those of females by triangles. The line is fitted to the male nests only. For males, $\log_{10}$ (nests) = $0.75*\log_{10}$ (days); $p<.001$, $r^2 = .98$.

DISCUSSION

Aye-ayes resemble other nocturnal species in several elements of their social organization in addition to being predominantly non-gregarious. For example, overlap in sleeping site use has also been found in some galagos and in *Lepilemur mustelinus* (Bearder and Doyle, 1974; Hladik and Charles-Dominique, 1974; Charles-Dominique, 1977). When mating, aye-ayes resemble some galagos and *Nycticebus coucang* in that estrous females called in males, possibly from other home ranges (Elliot and Elliot, 1967; Bearder and Martin, 1979). With more data, further similarities to other nocturnal species may emerge. For example, one interpretation of the results of this study is that there are central and peripheral aye-aye males, much as males are differentiated in most galago species (Charles-Dominique, 1977; Bearder and Martin, 1979, Harcourt, 1984; Bearder and Doyle, 1974; Bearder, 1987).

Some elements of aye-aye social organization recall features found in diurnal species. For example, there is striking similarity between observations of mating in aye-ayes and sifaka (*Propithecus verreauxi*). Like estrous aye-aye females, sifaka females in estrus gave loud calls to which males responded by converging on the female. Among sifaka, these males included residents of the same group and visitors from other groups. Like aye-aye males, sifaka males chased and fought with one another, and the female repelled some mating attempts and accepted others (Richard, 1974; 1985; 1992).

In short, similarities between certain behaviors are readily evident among nocturnal primates and between nocturnal and diurnal primates, but the biological and evolutionary significance of this mosaic of similarities remains unclear in the absence of robust criteria by which to establish behavioral homologies. Our use of such broad terms as "agonistic" or "affiliative" to characterize aye-aye behavior masks underlying uncertainty about whether the constituent behaviors that make up these classes are homologous with behaviors so labeled in other primates. Comparisons between nocturnal species are equally problematic. For example, while *Daubentonia* and *Lepilemur* both display shared sleeping site use, the former constructs nests whereas the latter seeks out holes or other pre-existing refuges. It is unclear whether the presence of site-specific sleeping preferences in these two species represents a shared, evolutionarily primitive pattern or a case of convergent evolution.

The idea that differences in primate social organization are due to environmental pressures is deeply rooted in primate studies (*e.g.*, DeVore, 1963; Hall, 1965), although some have long argued that social organization is also strongly determined by phylogenetic constraints (*e.g.*, Chalmers and Rowell, 1971; Struhsaker, 1969). The presence of closely related nocturnal and diurnal primates on Madagascar provides an excellent opportunity to assess the relative importance of certain environmental and phylogenetic effects as well as to explore the concept of behavioral homology more fully. In particular, if differences between the social organizations of nocturnal and diurnal lemurs are greater than those between diurnal lemurs and other diurnal primates, it would suggest that nocturnality is indeed important in shaping social organization. On the other hand, if differences between nocturnal and diurnal lemurs are less than those among lemurs, regardless of their activity pattern, and other diurnal primates, then phylogenetic status becomes a more salient explanation of observed patterns.

Given the potential interest of such comparisons, why have nocturnal primates so often been excluded from comparative studies entirely, or included as a single datum point? The difficulty of identifying behavioral homologies is clearly not the answer to this question, for the same issue is present in comparing diurnal species and yet has been largely overlooked (e.g. Wrangham, 1987). We believe the exclusion of nocturnal primates from consideration is due more to a lack of detailed information about nocturnal species. With little information available, the social organization of nocturnal primates could be, and was, conveniently represented as the "primitive condition" from which evolved the (more complex) social organizations of diurnal species. With nocturnal primates viewed as a primitive and homogenous starting point, accounting for the shifting selective pressure that gave rise to diurnal social systems has taken precedence over examination of the social organization of nocturnal primates per se.

Today, there is mounting evidence for social diversity among nocturnal strepsirhines (e.g. Clark, 1985; Harcourt and Nash, 1986; Bearder, 1987; Crompton and Andau, 1987), but this does not of itself establish if their social organizations are distinctive and, if so, what determines this distinctiveness. Pariente (1979), Clark (1985), and Bearder (1987) have considered this issue, and in the remainder of this paper we briefly examine three propositions emerging from their work: (1) the complexity of communication between members of nocturnal species is constrained by low light levels, and is associated with correspondingly less complex social organization; (2) the lack of cohesive groups in nocturnal prosimians is a consequence of predation pressure, and (3) spatial patterns of nocturnal prosimians are broadly concordant with the hypothesis that the distribution of females depends primarily on the size, quality, and patterning of food patches whereas males try to hold the ranges of females. Linking all three propositions is the idea that social organization is strongly determined by environmental factors.

(1)The solitary habits of most nocturnal species are attributable to differences in their communication abilities (Bearder, 1987). Testing this proposition requires integrated comparative analysis of the content and complexity of vocal, visual and olfactory signals in nocturnal and diurnal primates. At present, the data are not available for such analyses. Morphological and physiological specializations for olfactory communication in nocturnal primates support the idea that compared to diurnal primates, nocturnal primates rely more on olfactory channels and less on visual signals, but not necessarily that they communicate less complex information (Charles-Dominique, 1977; Schilling, 1974; 1979; Bearder, 1987). Yet even the presumed differences in channels of communication are questionable: certain New World monkeys also have morphological specializations and make frequent use of olfactory signals (Epple and Moulton, 1978; Hershkowitz 1977).

The few comparative studies of single channels of communication have failed to show consistent differences between nocturnal and diurnal primate species. In preliminary studies on visual perception, Pariente (1979) found greater differences between Strepsirhines and Haplorhines than between nocturnal and diurnal lemurs. Zimmerman (1985; this volume) reported a vocal repertoire of 18-20 calls for nocturnal bushbabies, numbers similar to those reported by Macedonia (1990) for diurnal lemurs. In contrast, Daubentonia has a very small repertoire of call types compared to the best-studied diurnal primates, but these call types encompass a wide range of structural variation, the possible meanings of which are unknown (Stanger and Macedonia, 1994).

(2) Predators hunt differently by day and by night, and diurnal primates are best protected from diurnal predators by staying together in cohesive groups whereas the crypsis of nocturnal primates helps them elude predators that use auditory signals and stealth to stalk their prey. Clark (1985) pointed out that the largest of the nocturnal primates should be most gregarious, because they are presumably least subject to predation: released from the constraints of predator pressure, animals experience benefits from social bonds and interactions. Clark's (1985) observations of *Galago crassicaudatus* supported this hypothesis, but the evidence from this study does not. Aye-ayes are by far the largest nocturnal primates (2700-3000 g compared to 1600 g for the next largest species), yet in contrast to *G.crassicaudatus* animals aggregated and interacted rarely.

(3) The distribution of females in nocturnal species is primarily determined by the distribution of resources (Clark, 1985). The ranging patterns of aye-aye females on Nosy Mangabe provides no support for this proposition. Seeds of *Canarium* fruit constitute one of the aye-ayes' three most commonly eaten foods; animals spent between 16% and 77% of their time feeding on Canarium by season (Sterling 1994; Sterling et al. 1994). Canarium are large-crowned and abundant on Nosy Mangabe, often occurring in clusters of between 5-20 trees (Carlson, unpubl., Sterling 1993a). Females never congregated in these trees, despite the presence of such a dietarily important, large and highly clumped resources.

Perhaps the most striking point to emerge from consideration of these three propositions is how little we in fact know about nocturnal primates, and about aye-ayes in particular. Based on the sparse evidence available, however, there is little or no support for any of the three. With respect to (2) and (3), aye-ayes may constitute an exceptional case, our observations may be incomplete and misleading, and/or the theoretical formulation of these propositions may itself be flawed by its exclusive focus on environmental determinants.

In summary, we believe that the full integration of research on nocturnal primates into comparative analyses of primate social organization could provide abundant new insights into the mechanisms whereby complex social systems evolve. To achieve such an integration, however, students of nocturnal species will have to gather more detailed information and, in particular, rely less on evidence of spatial patterns in making inferences about social relationships and more on the content of interactions. Renewed attention must also be given to the long-neglected issue of behavioral homology (but see Martin, 1972; Wrangham, 1987). Until a sound basis is developed for comparing the diversity of social organization starting to emerge for nocturnal primates with that long recognized among diurnal species, research on nocturnal species will remain out in the cold - if not in the dark.

ACKNOWLEDGMENTS

The field portion of this study was supported by the following grants to E.J.S: a National Science Foundation Doctoral Dissertation Grant 8913021 (with A.F.R), a Fulbright-Hays Fellowship, the Wenner-Gren Foundation for Anthropological Research, the National Geographic Society, a John F. Enders Fellowship and the Williams Fund. A NATO post-doctoral fellowship provided partial support for data analysis. The Government of Madagascar gave E.J.S. permission to work in Madagascar, and the School of Agronomy of the University of Antananarivo, in particular M. Rakotomanga Pothin, graciously sponsored her research. We thank R.E. Dewar for ideas and assistance with the analysis of nest occupancy rates and for help in meeting Plenum's formatting requirements. We are also grateful to K. Stanger and E. Zimmerman for their comments on an earlier draft of this paper.

REFERENCES

Ancrenaz, M., 1991, "Contribution à l'étude eco-ethologique du aye-aye *(Daubentonia madagascariensis)*", DVM Thesis, Ecole Nationale Veterinaire d'Alfort.

Bearder, S.K., 1987, Lorises, bushbabies, and tarsiers: diverse societies in solitary foragers, in: "Primate Societies," B.B. Smuts, D.L. Cheney, R.M. Seyfarth, R.W. Wrangham, and T.T. Struhsaker, eds., University of Chicago Press, Chicago.

Bearder, S.K. and Doyle, G.A., 1974, Ecology of bushbabies *Galago senegalensis* and *Galago crassicaudatus*, with some notes on their behaviour in the field, in: "Prosimian Biology," R.D. Martin, G.A. Doyle, and A.C. Walker, Duckworth, London.

Bearder S.K. and Martin R.D., 1979, The social organization of a nocturnal primate revealed by radio tracking, in: "A Handbook on Biotelemetry and Radio Tracking," C.J. Amlaner, and D.W. Macdonald, eds., Pergamon Press, Oxford.

Charles-Dominique, P., 1977 "Ecology and Behaviour of Nocturnal Prosimians," Columbia University Press, New York.

Charles-Dominique, P and Martin, R.D., 1970, Evolution of lorises and lemurs, *Nature* (Lond) 227:257.

Chalmers, N.R. , and Rowell, T.E., 1971, Behaviour and female reproductive cycles in a captive group of mangabeys, *Folia Primat.* 14:1.

Clark, A.B., 1985, Sociality in a nocturnal "solitary" prosimian: *Galago crassicaudatus, Int. J. Primatol.* 6:581.

Crompton, R.H. and Andau, P.A., 1987, Ranging, activity rhythms, and sociality in free-ranging *Tarsius bancanus*: A preliminary report, *Int. J. Primatol.* 8:43.

DeVore, I., 1963, A comparison of the ecology and behavior of monkeys and apes, in "Classification and Human Evolution", S.L. Washburn, ed. Aldine, Chicago.

Eisenberg, J.F., 1981, "The Mammalian Radiations," University of Chicago Press, Chicago.

Elliot, O. and Elliot, M., 1967, Field notes on the slow loris in Malaya, *J. Mammal.* 48:497.

Epple, G. and Moulton, D. 1978, Structural organization and communicatory functions of olfaction in nonhuman primates, in: "Sensory Systems of Primates," C.R. Noback, ed., Plenum Press, New York.

Fleagle, J.G., 1988, "Primate Adaptation and Evolution," Academic Press, New York.

Hall, K.R.L., 1965, Social organization of the Old World monkeys and apes, *Symp. Zool. Soc. Lond.* 14:265.

Harcourt, C.S., 1984, The Behavior and Ecology of Galagos in Kenyan Coastal Forest, Ph.D. Dissertation, Cambridge University, Cambridge.

Harcourt, C.S. and Nash, L.T., 1986, Social organization of galagos in Kenyan coastal forests: 1: *Galago zanzibaricus, J. Primatol.* 10:339.

Hershkovitz, P, 1977, "Living New World Monkeys (Platyrrhini), Vol. 1", University of Chicago Press, Chicago.

Hladik, C.M. and Charles-Dominique, P., 1974, The behaviour and ecology of the sportive lemur *(Lepilemur mustelinus)* in relation to its dietary peculiarities, in: "Prosimian Biology," R.D. Martin, G.A. Doyle, and A.C. Walker, eds., Duckworth, London.

Macedonia, J.M., 1990, Vocal Communication and Antipredator Behavior in the Ringtailed Lemur, Lemur catta. Ph.D. Dissertation, Duke University, Durham, NC.

Martin, R.D., 1972, Adaptive radiation and behaviour of the Malagasy lemurs, *Phil. Trans. Roy. Soc. Lond., Series B* 264:295.

Oates, J.F., 1987, Food distribution and foraging behavior, in: "Primate Societies", B.B. Smuts, D.L. Cheney, R.M. Seyfarth, R.W. Wrangham, T.T. Struhsaker, eds, University of Chicago Press, Chicago.

Pariente, G.F., 1979, The role of vision in prosimian behavior, in: "The Study of Prosimian Behavior", G.A. Doyle and R.D. Martin, eds, Academic Press, New York.

Petter, J-J., 1977, The aye-aye, in: "Primate Conservation," Prince Rainier III and G.H. Bourne, eds., Academic Press, New York

Petter, J-J. and Charles-Dominique, P., 1979, Vocal communication in prosimians, in: "The Study of Prosimian Behavior," G.A. Doyle and R.D. Martin, eds, Academic Press, London.

Petter, J-J. and Petter-Rousseaux, A., 1967, The aye-aye of Madagascar, in: "Social Communication Among Primates," S.A. Altmann, ed. University of Chicago Press, Chicago.

Petter, J-J. and Peyrieras, A., 1970, Nouvelle contribution à l'étude d'un lémurien malgache, le aye-aye (*Daubentonia madagascariensis* E Geoffroy), *Mammalia* 34:167.

Richard, A.F., 1974, Patterns of mating in Propithecus verreauxi, in: "Prosimian Biology," R.D. Martin, A.C. Walker, and G. Doyle, eds., Duckworth, London.

Richard, A.F., 1985, Social boundaries in a Malagasy prosimian, the sifaka (*Propithecus verreauxi*), *Int. J. Primatol.* 6:553.

Richard, A.F., 1992, Aggressive competition between males, female-controlled polygyny and sexual monomorphism in a Malagasy primate, *Propithecus verreauxi*, *J. Human Evol.* 22:395.

Schilling, A. 1974. A study of marking behavior in Lemur catta, in: "Prosimian Biology," R.D. Martin, A.C. Walker, and G. Doyle, eds., Duckworth, London.

Schilling, A., 1979. Olfactory communication in prosimians, in: "The Study of Prosimian Behavior," G.A. Doyle and R.D. Martin, eds, Academic Press, London.

Simons Morland, H., 1991, "Social Organization and Ecology of Black and White Ruffed Lemurs (*Varecia variegata variegata*) in Lowland Rainforest, Nosy Mangabe, Madagascar," Ph.D. Dissertation, Yale University, New Haven.

Stanger, K.F. and Macedonia, J.M., 1994, Vocalizations of aye-ayes (*Daubentonia madagascariensis*) in captivity, *Folia Primatol.* 62:160.

Sterling, E.J. 1993a, "Behavioral Ecology of the Aye-aye (*Daubentonia madagascariensis*) on Nosy Mangabe in Madagascar," Ph.D. Dissertation, Yale University, New Haven.

Sterling, E.J., 1993b, The social organization of the aye-aye (*Daubentonia madagascariensis*) on Nosy Mangabe, Madagascar, in: "Lemur Social Systems and their Ecological Basis, " J. Ganzhorn and P. Kappeler, eds, Plenum Press, NY.

Sterling, E.J. 1994. Aye-ayes: Specialists on Structurally Defended Resources. *Folia Primatol.* 62:142.

Sterling, E.J., Dierenfeld, E.S., Ashbourne, C.J. and Feistner, A.T.C., 1994. Dietary intake, food composition and nutrient intake in wild and captive populations of *Daubentonia madagascariensis. Folia Primatol.* 62:115.

Struhsaker, T.T., 1969, Correlates of ecology and social organization among African cercopithecines, *Folia Primatol.* 11:80.

Wrangham, R.W., 1987. The significance of African apes for reconstructing human social evolution, in: "The Evolution of Human Behavior: Primate Models," W.G. Kinzey, ed., SUNY Press, New York.

Zimmerman, E., 1985, The vocal repertoire of the adult senegal bushbaby (*Galago senegalensis senegalensis*), *Behaviour* 94:212.

Zimmerman, E., 1995, Acoustic Communication in Nocturnal Prosimians, this volume.

LOCOMOTION IN NOCTURNAL PROSIMIANS

Holger Preuschoft,[1] Hartmut Witte,[1] and Martin Fischer[2]

[1]Abtlg. Funktionelle Morphologie
Institut für Anatomie
Ruhr-Universität Bochum
[2]Institut für spezielle Zoologie
und Evolutionsbiologie
Friedrich Schiller-Universität Jena

INTRODUCTION

Napier and Napier (1967), whose categorization has often been criticized and sometimes modified, but never replaced, divided the prosimians into the following locomotor categories: slow climbing (and bridging), branch running and walking, vertical clinging and leaping.

All of these categories are represented among nocturnal primates, remarkably so, because leaping and rapid running on branches requires optical orientation. In spite of the low level of light at night, tarsiers and Galagidae (leapers), and Cheirogalinae and Lepilemurinae (branch runners) are largely nocturnal. Nocturnal activity evidently is no problem for slow climbers like the lorises. Where some Lemurinae, *Avahi, and Phaner* are to be placed in terms of locomotor categories was, and still is, a matter of controversy. In addition, some of these forms are not wholly nocturnal.

More important than a discussion of the locomotor categories and the attribution of species to these categories are the questions:

- What are the physical requirements of a given motor or postural behavior?
- By which morphological and physiological characteristics does an animal become "adapted" to one or the other locomotor or postural behavior?

These questions have been the focus of our research for the last 10 years, and here we present some results together with new theoretical considerations.

All primates are characterized by their autopodia, or their prehensile hands and feet. The simplest way of using hands and feet is by placing the soles on the substrate, so that (partial) body weight in the form of compressive force is carried by the soles (Figure 1). Elongated fingers and toes clasp around the branch and allow the transmission of not only compression, but also tensile force and torques (Figure 1). The force between palms or soles and substrate is a fraction of weight plus pressure exerted by the fingers or toes. In suspension, the compressive force between fingers or toes and substrate is body weight

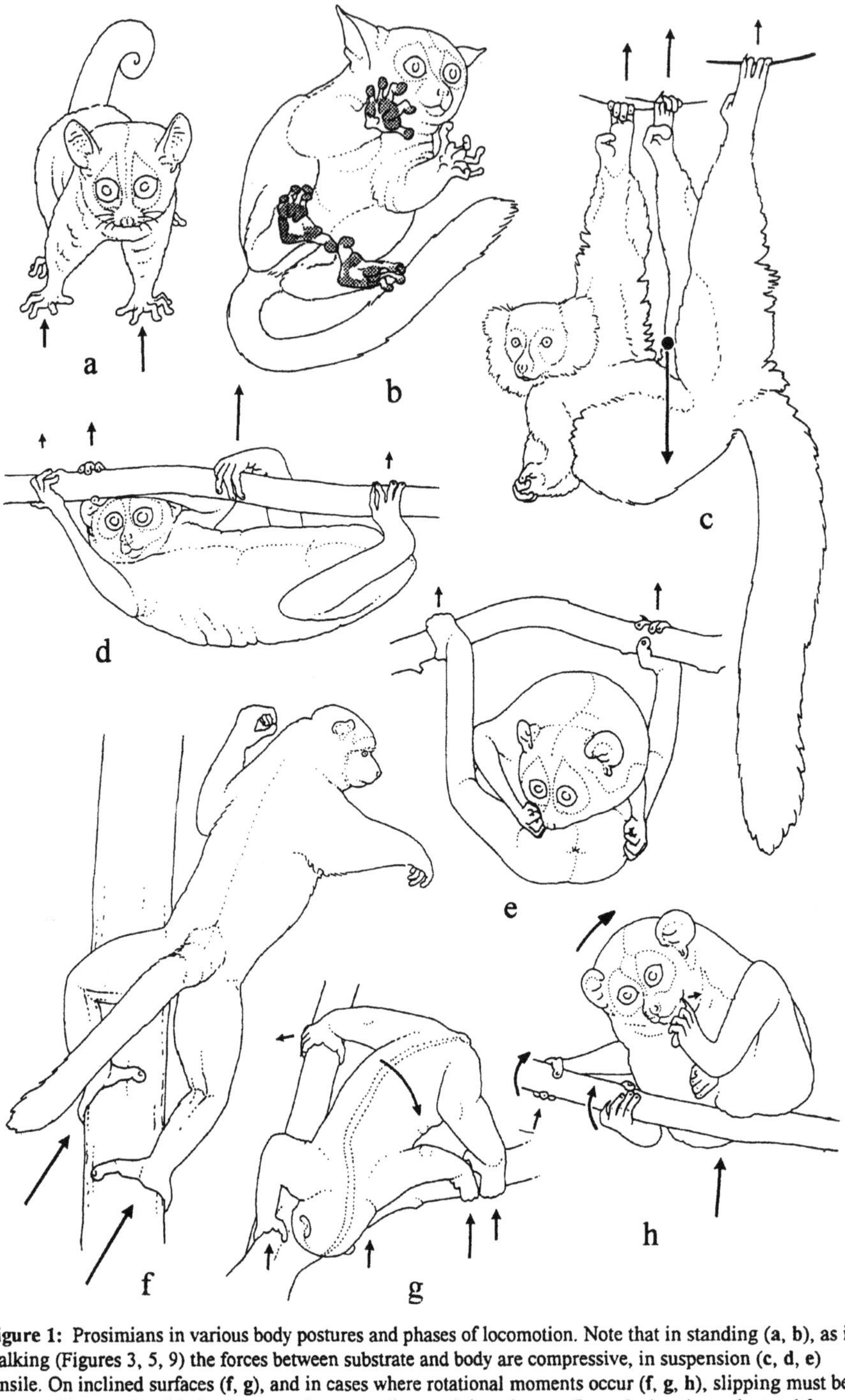

Figure 1: Prosimians in various body postures and phases of locomotion. Note that in standing (**a**, **b**), as in walking (Figures 3, 5, 9) the forces between substrate and body are compressive, in suspension (**c**, **d**, **e**) tensile. On inclined surfaces (**f**, **g**), and in cases where rotational moments occur (**f**, **g**, **h**), slipping must be prevented by creating friction between the autopodium and the substrate. In such situations, the total forces between hands or feet and substrate become greater than body weight, and the limbs transmit moments. Modified from Preuschoft (1989).

alone. The longer the fingers and toes, as separate elements, the better the grip can be adapted to a thick, uneven, or composite substrate and the surer the adhesion (Figure 1, and see below).

RESULTS AND DISCUSSION

The transmission of moments in variable directions becomes possible by splitting the hand or foot into digits I versus II - V, or I and II versus III-V on opposite sides of a branch (Figures 1, 2 and 8). Thus the primate autopodium can itself produce forces which press the fingers and the palm or the toes and the sole against the substrate to create friction (Figure 2). This grasping instrument now has the same qualities of transmitting compressive forces or tensile forces roughly in line with the extremity, as well as the additional potential to transmit moments tending to rotate the hand or foot about the seized object. The soft and mobile pulps can be squeezed into minor holes and excavations on the surface of a substrate. The hands and feet are equipped with well-identified friction skin. The dermatoglyphic lines themselves are a means for increasing friction resistance (Cartmill, 1974, 1979, 1985; Buck and Bär, 1993), and, as Schulz (1994) has shown recently, are arranged according to the strain patterns which occur in the digital pulps under compression and shear.

The necessarily powerful toe flexors lift the heel from the ground, because of their high plantarflexing torque about the ankle joint (Preuschoft, 1970). The carpus is not influenced in the same way, perhaps because of the less acute angle in the carpal joint compared to the ankle. In both extremities, the prehensility of the autopodia necessitate a strong and heavy musculature in the forearms and lower legs - a fact which has consequences for many primates and their locomotion (Preuschoft and Günther, 1994; and below).

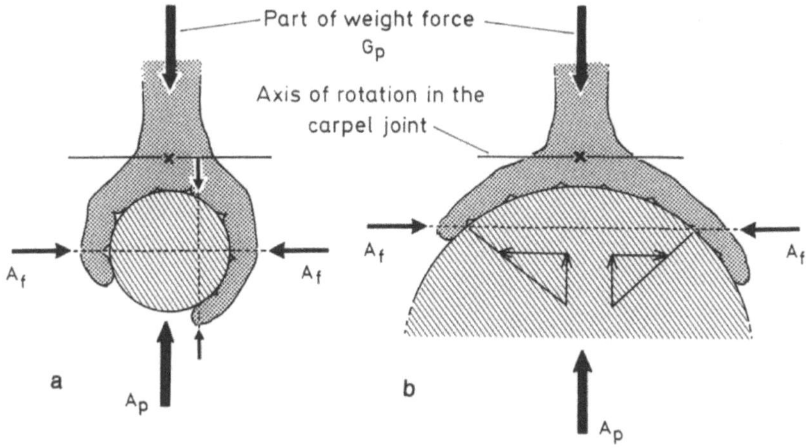

Figure 2: Friction resistence depends on the compressive force between substrate and fingers or toes, and palms or soles. Aside from short phases of rapid movement, a total compressive force higher than body weight, or a compressive force deviating from the vertical, can only be produced if the animal exerts a force, usually with the aid of fingers or toes on one side, and a counterforce of the same size, normally by means of the thumb or the big toe. **a)** thin twig ; **b)** thick branch.
Reproduced from Preuschoft *et al.*, (1993).

The prefered mode of locomotion in the notoriously slow climbing Lorisidae appears to be walking (Figure 3) on top of and along the branches (Nieschalk, 1991). An efficient way to do so is by using the physics of the inverted pendulum (Figure 3). The longer the pendulum chord, the higher the speed of walking. The mechanics of walking are given in Mochon and McMahon (1980, 1981), and their consequences for morphology are detailed in Preuschoft and Demes (1984), Witte *et al.* (1991), Preuschoft and Witte (1991), Preuschoft *et al.* (1992), Preuschoft and Günther (1994)m, Günther and Preuschoft (1992), Christian (1993). The high speed of a long pendulum explains the long limbs in the slender loris, but not their shortness in the slow loris and in the potto. The best explanation offered for their stubby proportions is that shortness of the limbs, as a general rule, makes life safer and saves force during movement (see Preuschoft, 1990). If walking speed is not an essential factor (or not highly selective) in the life of an animal, its limbs will tend to be short. The pivot of rotation in the limb of a primate which walks along a thin branch is not a finger or metacarpo-phalangeal joint, nor the ground foot contact, as in cursorial mammals, but the joint proximal to the "forceps". Consequently, the limb elongation in primates is concentrated on the stylopodia and the zygopodia - which makes primates different from cursorial mammals, in which the metapodia or even the digits are elongated, while the stylopodia are (partly) integrated into the body wall. In the hindlimb, the femur and the tibia

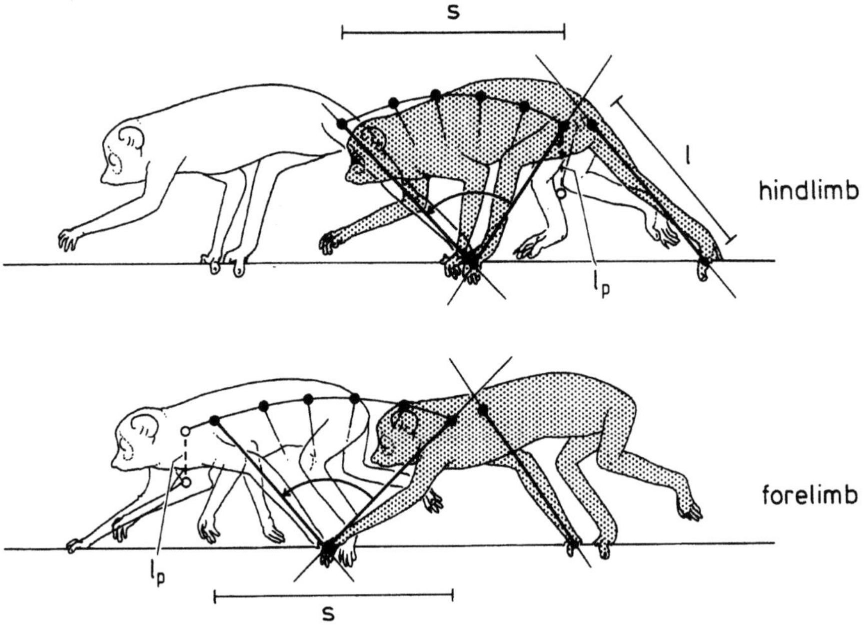

Figure 3: Walking *Loris tardigradus*. During the stance phase, each limb swings like an inverted pendulum about the joint proximal to the "forceps". Walking speed is

$$V = \frac{s}{T}$$

where s is the horizontal distance covered by a part of the body mass, and T is the pendulum period,

$$T = 2 \cdot \pi \cdot \sqrt{\frac{l}{g}}$$

Walking speed, therefore, is proportional to the square root of pendulum length l. In the swing phase, the limbs form "physical" pendula, in which l_p depends on the mass of the limb segments, and the distances of their centers of mass from the shoulder, or hip joints, and the flexion of the limbs. The time needed for the foreswing is given by 1/2 pendulum period, and acceleration as well as deceleration is resisted by the mass moment of inertia: $J = l_i^2 \cdot m_i$

plus fibula tend to be long as do humerus and radius plus ulna. The carpus and metacarpus are usually short in prosimians and arboreal monkeys; the fingers, however, are long (Jouffroy, 1993). The tarsus and metatarsus are short only in the case of the Lorisinae which are highly specialized walkers and use their feet only for procuring a firm hold during locomotion. Other walking primates may possess feet with elongated tarsus or metatarsus which allow them to fulfil other tasks (Schultz, 1963). All arboreal forms possess long fingers and toes in the interest of safe adhesion.

Because of their prefered direction of walking, the substrate reaction force travels along the branch and along digit II (Figure 4). The "rolling off" of the hand or foot along digit II leads to a force concentration on this ray and to a phase of instability when the grip of the lateral digits against thumb or big toe is losened. Both can be minimized by a "functional" or morphological shortening of the metatarsal and the phalanges of ray II (Preuschoft et al., 1993).

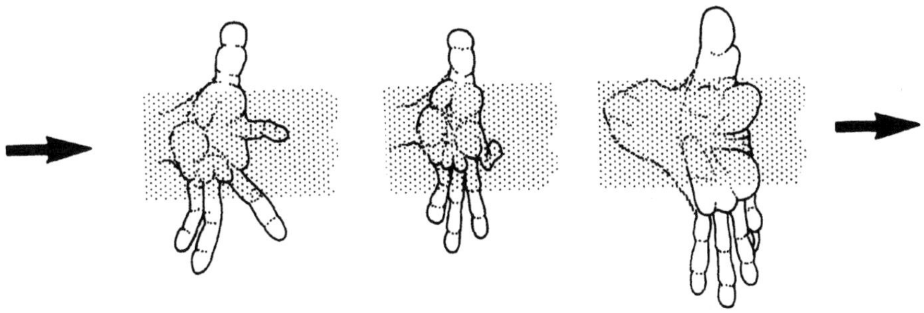

Figure 4: In a walking lorisid, the weight force exerted by body mass times earth acceleration travels usually within the stippled band from the carpus or tarsus to a point between digits I and IV in the ectaxomic auto-podia (Nieschalk and Demes, 1993); in the mesaxonic autopodium of simians, it would be between digits I and III, Rawlins (1993). Concentration of bending stress in the digital ray II is minimized by functional or morphological shortening (see Nieschalk and Demes, 1993).

A tail is mechanically superfluous for walking and slow climbing, but does not confer a mechanical disadvantage. If present, it may be used to maintain balance on very narrow supports.

Specialized walkers usually have front- and hindlimbs of the same length, but efficient quadrupedal walking is also not a big problem for animals which have hindlimbs that are longer than their forelimbs (Figure 5). To keep the swings of the pendula in phase, the animals may either reduce functional hindlimb length by flexion and by placing the sole flat on the substrate, or increase functional forelimb length by moving the shoulder blade fore and aft, so that the pivot of the pendulous swing is moved to the same height as the hip joint. The latter has been documented by Fischer (1993) for several small mammals with the aid of X-ray films on a treadmill and seems to be a general technique used in all "small" mammals (below 2-3 kg as an approximation). There is little doubt, that Cheirogalinae, lemurs, or large *Otolemur* species make use of the same technique. (See also "branch running").

Discontinuities in the three-dimensional meshwork of branches and twigs are overcome by "bridging" (Figure 6). According to our observations, this is mostly done by stretching both arms or one arm out to reach a new support, while the hindlimbs maintain support. This requires an ability to transmit moments from the feet to the substrate. Prehensility and great strength of the feet are clearly of great adaptive value. Inside the animal's body, however, bridging evokes very great internal forces (Figure 6). At the knee and hip joints, the resulting torques can be more difficult to compensate than inside the

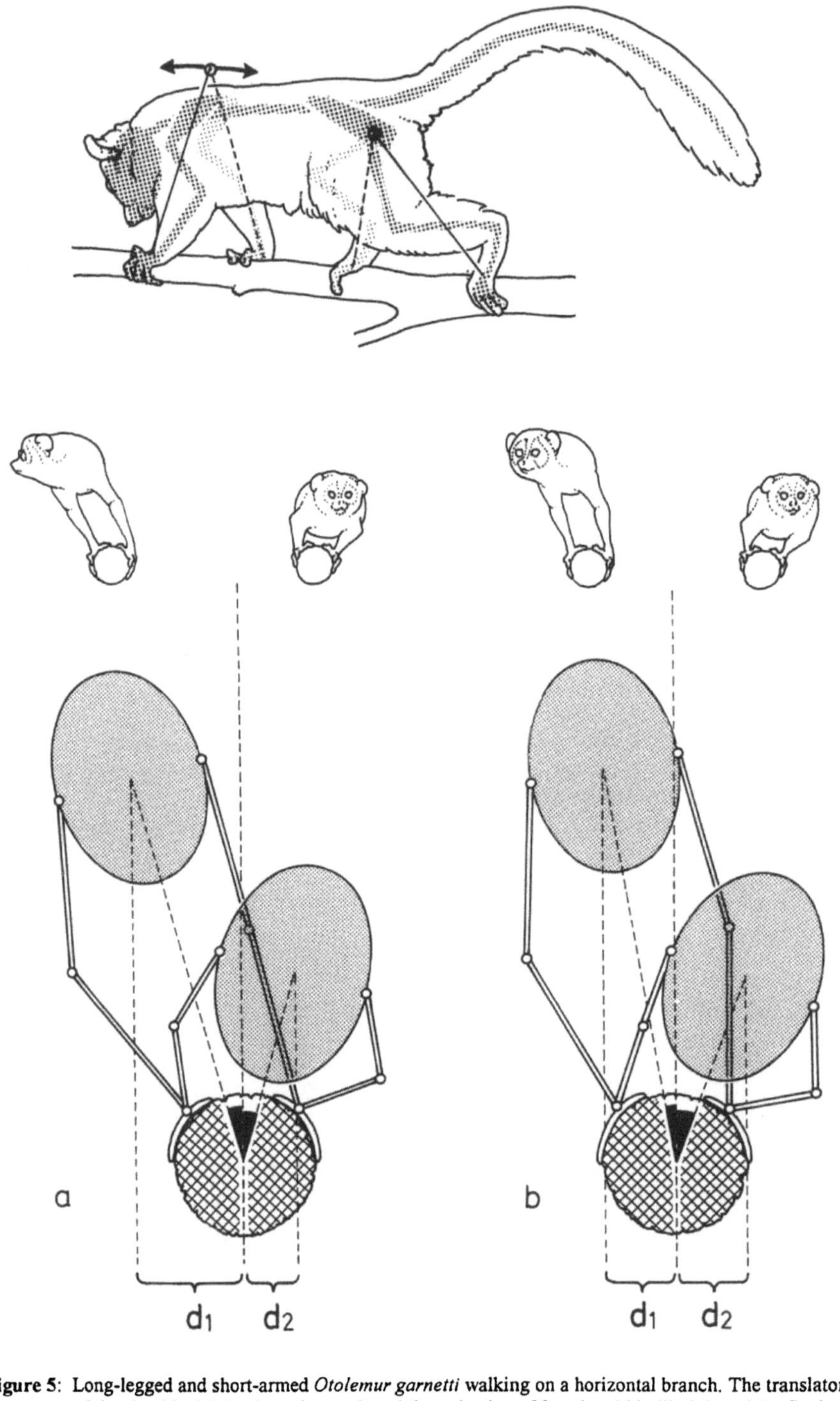

Figure 5: Long-legged and short-armed *Otolemur garnetti* walking on a horizontal branch. The translatory movement of the shoulder blade along the trunk and the reduction of functional hindlimb length by flexion of the joints are shown schematically. The pivots about which the limbs swing in the stance phase are nearly at the same height, in spite of the different lengths of the extremities.

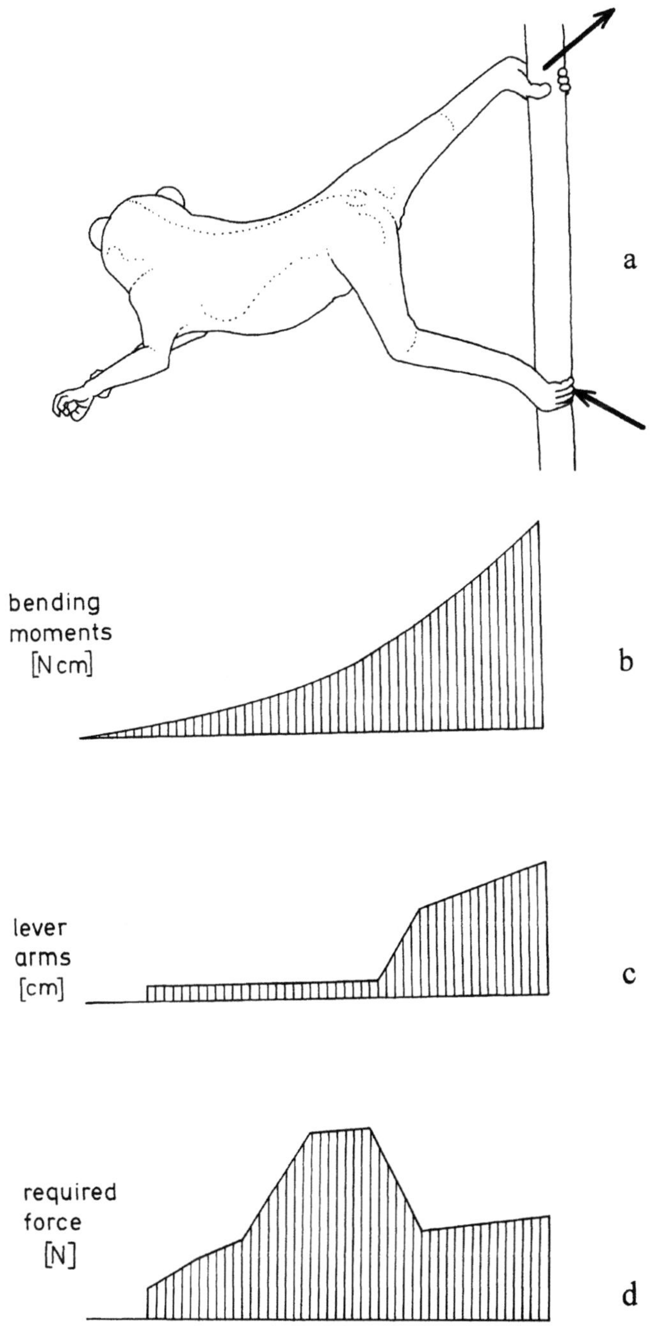

Figure 6: *a Loris* bridging. The trunk itself is under the dorsally convex bending moments shown in the diagram (**b**). These are sustained by the muscles of the trunk wall and the vertebral column. From the hip joint on to the right, the bending moments are resisted by the upper (right) hindlimb under tension, the lower (left) hindlimb under compression. The approximate direction of the external resultant forces is shown in **a**, as is the approximate length of the lever arms (**c**) of the tensile forces against the compression-resistant structures (vertebral column or left leg). If the bending moment at a given point is divided by the lever arm at the same point, the relative compressive or tensile force is the result (**d**). Arbitrary units.

(thicker) trunk. Our calculations indicate that this favors shortness of the limbs and, for the sake of reaching over a distance, a long trunk, especially marked in lorises (which possess up to nine lumbar vertebrae according to Schultz, 1961).

There was recently and still is some discussion about the footfall sequence in primates (Figure 7). In spite of Vilensky's (1989; Vilensky and Moore, 1992) observation that lateral and diagonal footfall sequences can occur in monkeys, it seems that the diagonal sequence prevails among Lorisidae. It is evident that diagonal footfall secures, during longer time intervals, two-limb supports located on either side of a narrow substrate so that no rotational moments occur. Another advantage of the diagonal footfall sequence becomes evident when climbing vertical substrates (Figure 7). To make sufficient friction for sustaining full body weight, the pair of forces exerted by thumb and fingers may be insufficient, in which case the adductors of the arms and legs may be used to increase compression of the support, provided that a counterforce exists to prevent a giving way of the substrate. This counterforce can be exerted by the contralateral hand or foot.

Primates which move rapidly in short bouts are difficult to classify according to their locomotion, and are placed, for want of a more specific category, into the category "branch running". It includes the Cheirogalinae, the Callithricidae,and, to some extent, the lemurs

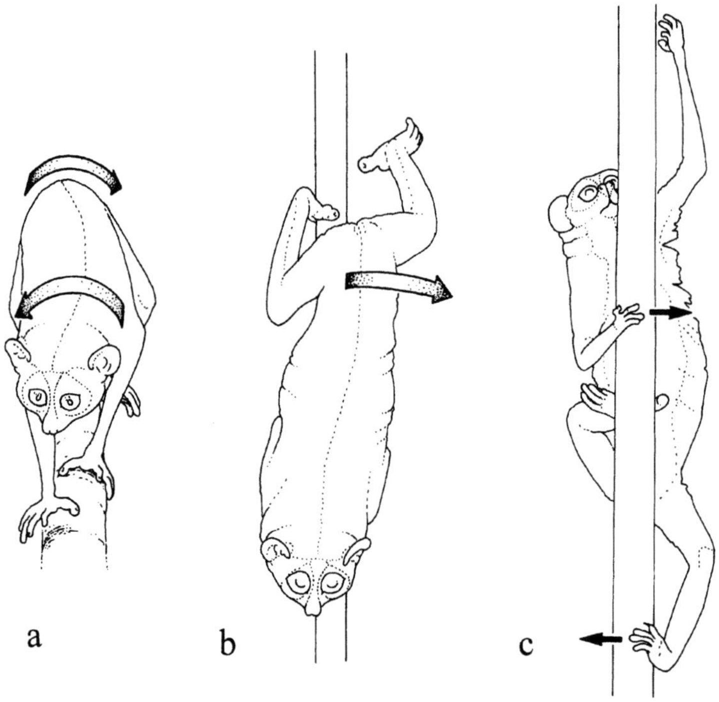

Figure 7: *Loris tardigradus* marching along a branch seen from on top (**a** and **b**). In **a** the tendency of the trunk to rotate to the animal's right side about its left hand is shown; in **b** the tendency to rotate to the animal's left side about its right foot. These rotating moments are opposed in **a** by the torque exerted by a part of trunk weight around the right foot, which rotates to the animal's left. In **b** it is not clear whether rotation is prevented by a counterbalancing torque exerted by the anterior half of the trunk, or by friction. **c** shows an upward climbing *Loris* from below. Note the opposing influences exerted by adduction of the limbs on the substrate, because of the diagonal footfall sequence. In a two-limb support on ipsilateral extremities (which occurs in lateral footfall patterns, but not in lorises) the branch has more time to give way under the combined forces exerted by a hind- and a forelimb.

and the larger galagos. Our own observations provide grounds for assuming that this type of locomotion may be ancestral for mammals, and by no means bound to an arboreal lifestyle. It is, however, obviously favorable for animals of small body size and of sufficient versatility to inhabit arboreal as well as terrestrial environments. Here we should not, however, lose sight of the fact that an arm-thick branch for a *Microcebus* is analogous to a race track for a horse. In combination with prehensile hands, "small animal locomotion" can be used very efficiently in arboreal habitats, a fact largely due to the low center of gravity (Figures 8 and 9, as already pointed out by Jenkins, 1974 and by Cartmill, 1974).

Among the characteristics of small mammal locomotion are permanently flexed joints. This flexion requires a great deal of force for balancing the joints, but it offers three advantages: First, the ability to accelerate very quickly (for rapid escape or sudden change of direction), without the delay of first flexing the joints in order to get acceleration distance; second, the ability to accommodate locomotion to depressions in the substrate without the necessity for lowering the body mass into a minor depression and hoisting it out again on any substrate on which the animal is moving; third, flexion of the limbs implies a center of gravity close to the substrate. This makes the animal and its locomotion insensitive to inclinations of the substrates; that is, it favors the ability to climb (Jenkins, 1974; Cartmill, 1974; see also Figure 9).The disadvantage of high force requirements is compensated for by the allometric surplus of force in small animals (Demes, 1989).

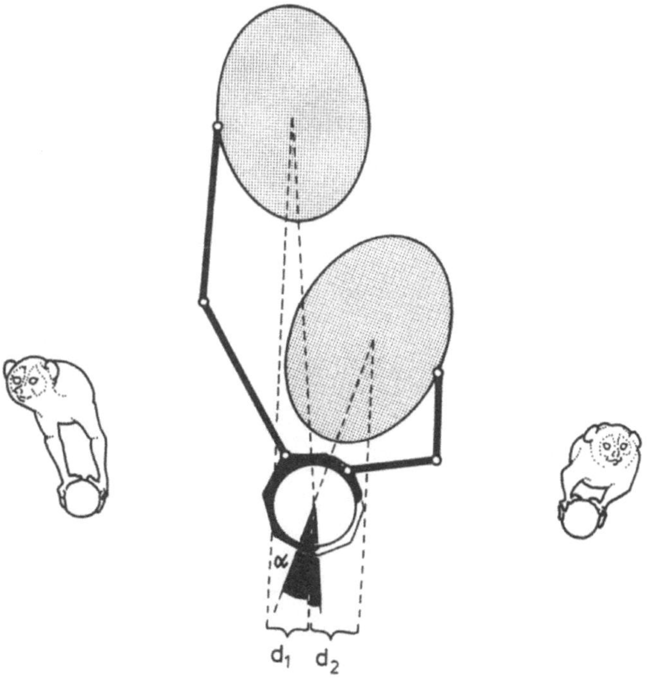

Figure 8: The schematic drawing shows the bodies of prosimians in two positions excentric to the support, so that rotating moments occur between the animal and its support, as well as inside the latter. This is only possible in animals equipped with prehensile autopodia. Note that the long-legged animal (left) exerts no greater torque about the support than the short-legged animal (leaning to the right), provided that the excursion d (not the angle alpha marked by the black triangles) of the center of mass in both animals is the same. (See also Preuschoft *et al.*, 1992; *contra* Cartmill, 1974).

Walking and trotting can be used. In both cases the back of the animal is parallel to the substrate, with hip joint and the vertebral margins of the scapula being at the same height (Fischer, 1993; see also Figures 9 and 10). This means that both extremities have the same pendulum length and thus easily swing in the same period. A more rapid locomotion is "galloping" or, in the terminology of the experts, the "asymmetric" or "in phase" gaits. They are used often by many lemurids, in particular by the Cheirogalinae (Figure 11).

A characteristic of these gaits is the flexion of the trunk. In hyraxes (which have been extensively analysed by Fischer (1993), trunk flexion accounts for 40% of the stride length of the hindlimbs. A similar contribution to stride length seems to be made by trunk flexion in the Cheirogalinae as well (Figure 11). This means of lengthening the stride is confined to rather small mammals. In horses, for instance, flexion of the vertebral column in the canter is not more than 20^0 (Loitsch, 1994). This flexion adds little to stride length, although it has great importance as a spring mechanism (Alexander, 1990) and for acceleration of the swing limbs (Witte *et al.*, 1995).

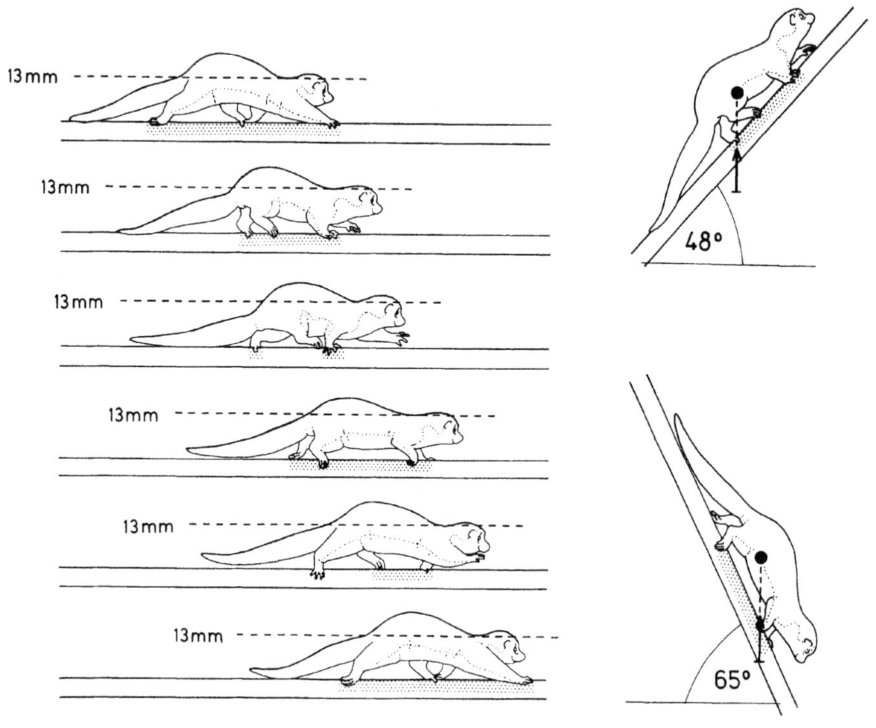

Figure 9: *Cheirogaleus medius* walking along a horizontal bar. The low position of the center of gravity allows locomotion on steeply inclined supports without losing external balance (see also Cartmill, 1974), and without the necessity to transmit, aside from compression, friction from the animal on the substrate. Even on steeply inclined substrates, the center of mass is above and the gravity vertical within the area of support. Drawn from a video film.

The speed of locomotion depends on stride length times stride frequency. Frequency is restricted by the time needed for the recovery swings of the limbs. Stride length can be increased by a phase of suspension between footfalls. This is realized in the trot (Figure 10). Its suspension phase, however, is limited by the angle of excursion and by limb length. When the suspension phase is elongated, the forces on each leg become very great.

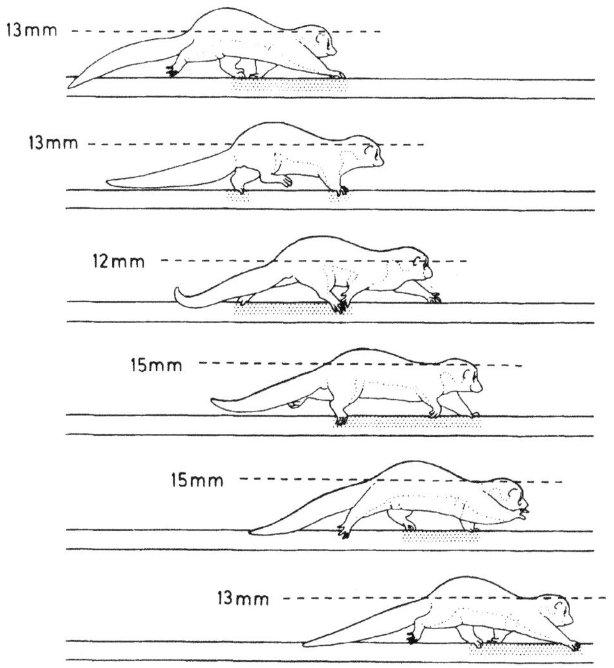

Figure 10: *Cheirogaleus medius* trotting. Drawn from a video-film. The translatory movement of the shoulder blade can be derived from the different positions of the elbow and forearm in anteverted and retroverted postures. Translations of the shoulder were documented for various small mammals on X-ray films by Fischer (1993). To be compared with Figure 9. Note that the trunk is nearly horizontal or parallel to the substrate, and does not change its inclination. This is in contrast to the situation in bounding (Figure 11).

Gait change toward an asymmetric gait

a) further increases stride length of both fore- or hindlimbs at one time, because the suspension phase becomes longer.
b) enables the ground reaction to be distributed on two limbs (leading and trailing) so that one disadvantage of trotting (high loads on the limbs) is avoided.
c) enables forelimbs and hindlimbs to swing about pivots clearly dorsal to the trunk (Figure 12), so that the virtual pendulum length is great.
d) allows the angles of excursion to remain small, which makes muscular activity less costly (Figure 12 in contrast to Figure 13).
e) leads to the necessary muscular force being produced partly by the trunk muscles, not by the extremities (Figure 12). The shorter, and slower contractions necessary for the trunk muscles save energy (Kram and Taylor, 1990).

A problem may be the necessity for moving large parts of trunk mass against each other. Fischer's X-ray films show clearly that, in the trotting of small mammals, at times either the forelimbs gain ground against the hindlimbs, or the reverse occurs. In the former case, the trunk becomes stretched and in the latter, it is shortened by pronounced flexion. In both cases, a correction is made by intercalation of a "bound" or "half bound" sequence. This either brings both hindlimbs simultaneously towards the forelimbs, or it increases the distance between the fore- and hindlimbs by an extended swing phase of both forelimbs. In either case, the force for the movement comes from the trunk musculature.

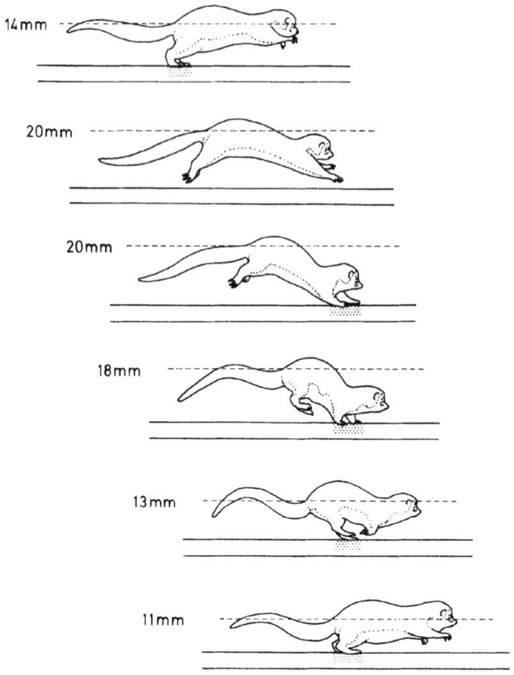

Figure 11: *Cheirogaleus* bounding. Sequence drawn from a video-film. Note the flexion and extension of the trunk, the swinging of the limbs about pivots dorsal to the back, and the variable inclination and height of the trunk in relation to the substrate. To be compared with Figure 12. Like Figures 9 and 10 modified from Nieschalk (1991).

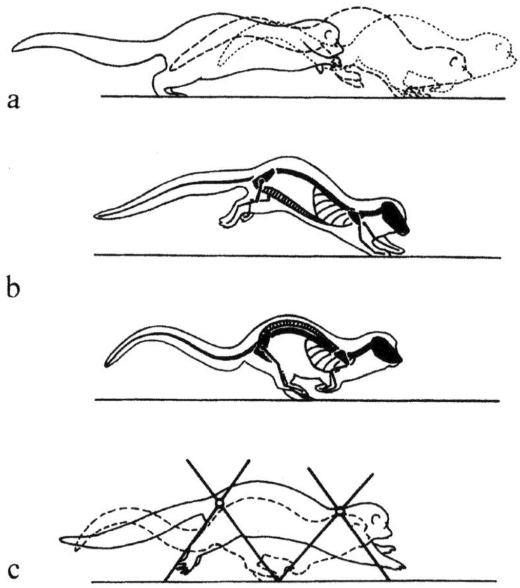

Figure 12: Bounding *Cheirogaleus* as in Figure 11. **a** emphasizes the extension and flexion of the back; **b** indicates the trunk flexors and the trunk extensors which induce trunk movements, but not the static function of the muscles. **c** shows the wide excursions of the limbs (where the phase of suspension is not shown in its full length for the hindlimbs), and the pivot of excursion above the most proximal joints of the limbs in the standing animal. Modified from Nieschalk (1991).

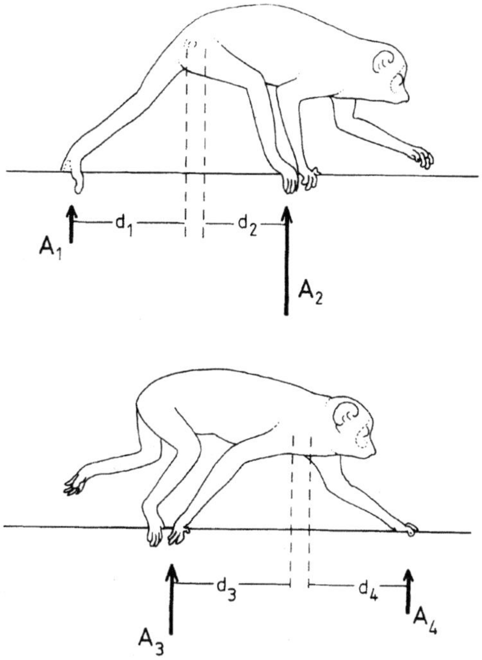

Figure 13: Two phases from the walking cycle in *Loris*, to show the dangerously long lever arms of the substrate reaction which occur due to the great angles of excursion in slow motion.

Independent of the gait, there is a physical rule to be observed in any locomotion on limbs: the sum of the integrals of the ground-foot-forces over time are equal to a continuous action of body weight (Fig. 14). If the hindlimbs are longer, they can rest longer on the ground during each stride cycle. In combination with the caudal shift of the body's center of mass, towards the now stronger and heavier hindlimbs, this leads to the result that their share of the body support increases (Christian 1993; Christian *et al.*, 1994). This effect is emphasized at higher speeds, where the hindlimbs are placed far forward and thus are placed below the center of body mass, (CM), at least during a part of the cycle (Figures 12, 14). The carrying of an increasing part of body weight by the hindlimbs leads directly to bipedality, a tendency inherent in many primates (as well as in many reptiles, see Christian *et al.*,1994). The elongation of the hindlimbs, mentioned repeatedly in this argument, has much to do with the alternative way of overcoming discontinuities in the three-dimensional meshwork of arboreal habitats, namely leaping.

Leaping in an arboreal environment requires a take-off from one spot and the ability to land at another, thereby eliminating the full impulse contained in the flying body within fractions of a second. Faults of coordination are fatal (Peters and Preuschoft, 1984). Such heavy selective pressures do not exist in animals who leap on and from flat, two-dimensional substrates (dog: Alexander, 1972; horse: Preuschoft *et al.*, 1988; Merkens, 1987; macaque: Günther, 1989; Günther *et al.*, 1992). These animals can and do separate horizontal from vertical accelerations and decelerations.

The basic physical necessities for leaping were the topic of Peters and Preuschoft (1984). Because force = mass · acceleration, leaping favors elongation of the hindlimbs to increase the distance available for acceleration. This either increases the impulse or reduces the force necessary for speed. In any event a very strong musculature is necessary. The trajectories of travel through the air can be varied in order to avoid obstacles. According to

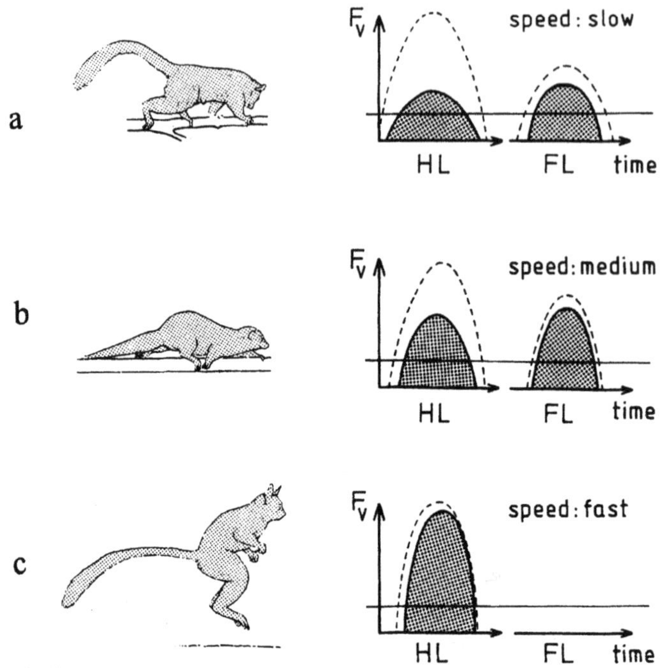

Figure 14: Ground-foot forces over time in animals moving with increasing speed on hindlimbs with increasing relative lengths. **a** Slow walking *Otolemur garnettii*; **b** trotting *Cheirogaleus medius*; **c** hopping *Galago moholi*. The share of "impulse" (force · time, stippled areas) becomes greater for the hindlimb. The dashed lines outline the hypothetical impulses possible under consideration of limb length and stance phase duration. Since the more heavily loaded hindlimbs become stronger and heavier, this effect is enforced by the shift of the CM towards the hindlimbs. If the feet are placed below the CM, bipedality becomes possible and efficient. F_V: vertical force, HL: hindlimb, FL: forelimb. Modified from Christian (1993).

Crompton and Sellers (1994), many prosimians choose a lower trajectory than the energetically optimal 45° take-off angle. These authors have shown that the lower and less curved trajectory allows the animal to reach its aim faster than on a more curved path, which makes sense for escaping as well as for chasing.

What is advantageous for the take off will also be favorable in landing, where higher forces than that of take-off come into action. This has been observed repeatedly (e.g. Preuschoft, 1985, 1988; Günther 1989). The rationale was worked out by Ruder and Gruber (1987) and Gruber *et al.* (1982): The speed of the take-off is limited by the time necessary for muscle contraction, whereas in the landing, at least the hard parts of the skeleton receive a high braking force at the time of initial body contact. This force is transmitted from one skeletal element to the other and changes their state of movement. The greater masses of the soft parts (muscles, trunk) follow with some delay, producing another, negative impulse containing lower force, but distributed over longer time.

These high forces can again be handled best by long and muscular hindlimbs. The problem, however, is that the hindlimbs, which have launched the animal into a leap, are trailing behind, despite being necessary for the landing in front of the body. Thus, a rotation during the airborne phase is necessary. To rotate a part of body mass forward in this phase, another partial mass must be rotated in the opposite direction, or some other external force must be recruited. This gives rise to three questions:

1. How can rotation be facilitated ?
2. Which partial mass rotates against the inertia of the hindlimbs ?
3. From where is the required external force obtained?

1. Rotation is facilitated by assuming a body shape in which mass moments of inertia are minimal. Tarsiers assume after take-off a cylindrical form to allow rapid rotation about a longitudinal axis, which is initiated by the propelling hindlimbs. Later in flight the animal assumes a ball shape which offers least resistance against being rotated about a transverse axis. Another important and efficient means of facilitating rotation while airborne is an appropriate distribution of body mass; as much as possible close to the center of gravity, and as little as possible in the distal segments leads to reduction of the mass moments of inertia of the hindlimbs. Even more efficient, of course, is a mass reduction on the most distal segments, allowing the musculature to be concentrated proximally (Peters and Preuschoft, 1984). The requirements of grasping, however, which lead to heavily muscled zygopodia and autopodia, are in conflict with this tendency.

The rotation of the free-floating body can be either stopped by a tail movement (as Niemitz, 1984, and Peters and Preuschoft, 1984 have observed in tarsiers; Günther, 1989, in galagos), or the whole flight can be stabilized by a spreading out of the extremities, as can often be observed in lemurs or in monkeys. Dunbar (1994) has argued that high weight of hands and feet also increases the mass moments of inertia of the extremities, making them a very efficient tool for stabilization of the flight phase. This is undoubtedly true, although we believe that the reasons for having heavy hands and feet are different!

2. The partial mass, which can initiate the rotating of the hindlimbs forward, can be rather small, if far away from the body's center of gravity. This holds true for a slender and long tail (Peters and Preuschoft, 1984). Growing body mass and growing linear dimensions, however (see Demes 1989), with its high moment of inertia, would necessitate an enormous tail. As a consequence, the tail loses its importance (and length) in large leapers (*Propithecus diadema, Indri*, tail-less macaque species, *Homo*) and the rotation of the body is gradually taken over by the increasingly heavier forelimbs (Demes and Günther, 1989a and b).

The mass concentration on proximal segments is also of importance for other modes of locomotion and is, therefore, widespread (Günther, 1989; Witte *et al.*, 1991; Preuschoft and Witte, 1991; Preuschoft *et al.*, 1992; Günther *et al.*, 1992). This necessity is in conflict with other conditions of primate life, namely forceful gripping (Preuschoft, 1992; Preuschoft and Günther, 1994), which requires a strong musculature and, therefore, much weight on the forearms and hands or lower legs and feet, respectively.

3. During the phase of free flight, only one external force can be applied to the body, namely air resistence. Peters and Preuschoft (1984) have checked this factor experimentally in the case of the short-haired tail of *Tarsius bancanus*. The effect was negligible, *T. bancanus* only use the mass inertia of their tails. It may be, however, that bristle-equipped or bushy-haired tails catch more air-resistance.

On the other hand, the spreading out of the arms and legs also increases the ventral surface of a leaper, and, as a consequence, air resistance. After Günther (1989) had observed that galagos often jump farther than they theoretically should do, on the basis of their take-off force and their take-off angle, Demes *et al.* (1991) provided sound reasons for their making use of air resistance for gliding.

Vertical clinging, which is connected with leaping in the locomotor category, should be treated separately from leaping from a biomechanical point of view and rather be combined with vertical climbing. As Preuschoft & Witte (1991) and Preuschoft *et al.* (1992) have emphasized, there are at least two widely divergent ways to cling vertically; either suspended by the arms (which is the prefered method in apes (Kortlandt, 1976, personal comm.; Preuschoft and Witte, 1991; Preuschoft *et al.*, 1992), or "standing" on the prehensile feet (Figure 15), which is often favored by lemurs. In most primates, intergradations can be observed, although long arms and short hindlimbs make the first alternative less energy-consuming, while long hindlimbs combined with weak (and short) forelimbs are "preadaptive" for the second. Essential in both cases is the adhesive gripping of hands (first case) or feet (second case).

The locomotor behavior in which most "physical work" is spent, next to leaping, is climbing upward. In climbing downward, less force must be invested (although the same muscles may be recruited). In climbing upward, there are again two possibilities, as in vertical clinging: either the animal can hoist itself to a higher place using the arms, or it can push itself upward using the legs. Critical in hoisting is a pull component against the stem, which is difficult to control. During the change of handholds, the hindlimbs have to take over the weight force. Critical in the pushing technique is also the change of the foothold (where the arms have to take over the responsibility). On the other hand, the push is advantageous because it is parallel to or slightly away from the stem and therefore can be controlled easily with little expenditure of force by the arms. The theoretical interdependencies explained here are fully in line with the experimental results obtained by Hirasaki *et al.* (1993)

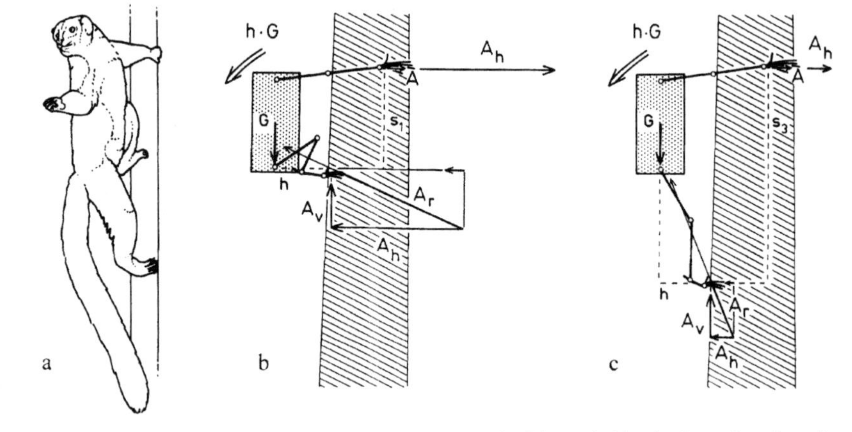

Figure 15: **a** *Lemur fulvus* clinging vertically. **b** Body weight G is carried by the feet, where it evokes a reaction force (A_v). The arms exert a horizontal pull (A_h) to compensate the rotational moment G · h of weight force about the foothold. The greater the distance s_1, the smaller A_h. This horizontal force component must be countered by a force of the same size acting against the feet. The latter (A_h) combines with the reaction force A_v to form a resultant A_r. In **b** the situation is shown for the flexed left hindlimb, in **c** for the extended right hindlimb. The reaction forces A_v are of the same size in both cases. Extension of the hindlimb also reduces load arm length, so that less muscle force is needed. The comparison can be generalized: the longer and more extended the hindlimbs, the smaller the resultant force A_r which must be sustained by the hindlimbs: length of the hindlimbs in this situation saves energy!
Drawing modified from Preuschoft and Witte (1991).

As a matter of fact, the hindlimbs in most primates are much heavier, and that means stronger than the forelimbs. Günther (1992) recently compared the segment-weights in 35 species of primates and 22 species of non-primates. He found that in only seven species of mammals is the weight of the forelimb more than 75% of hindlimb weight, and in only three primate species, and in sloths, do the arms reach the mass of the hindlimbs. The functional hindlimb dominance of primate locomotion, even on flat ground, was first documented by Kimura *et al.* (1979), and recently confirmed by Demes *et al.* (1994).

We suspect, that climbing upward is an important functional reason for the development of the hindlimb dominance in primates. This suggestion is not at variance with the views of Fleagle *et al.* (1981), who also emphasized upward climbing as essential for this characteristic.

SUMMARY

The morphology and the kinematics of various nocturnal prosimians are reviewed in order to identify adaptations of these animals to the requirements of posture and locomotion.

From a mechanical point of view "slow climbing" is to a large extent identical to walking. Walking speed depends on the lengths of the extremities. Limb elongation, however, is in conflict with the advantage of safety and force capacity contained in reduced limb length. Shortness of the limbs also reduces the mass moments of inertia of a limb in the swing phase, and low mass moments of inertia permit higher stride frequencies. The actual limb length of a primate is a compromise of conflicting needs, which can only be found on the basis of a detailed knowledge of the animal's mode of living.

Special emphasis is placed on "branch running". We found that this arboreal locomotion does not deviate too much from the locomotor behavior which is widespread among "small" mammals. X-ray films of living mammals taken by one of us (M.F.) allow the identification of the mechanical principles involved in "small mammal"-locomotion in contrast to the locomotion of "large" mammals. The limbs of small mammals are short, so that the CM is at low height above the substrate, which allows (unchanged) locomotion on steeply inclined surfaces. The combination of relatively long hindlimbs (which also permit rapid acceleration) with relatively short forelimbs allows walking because of the flexed positions of the former and the translatory movements of the shoulder along the body axis. This increases virtual pendulum length of the short forelimb. Extensive trunk flexion and extension in asymmetrical gaits makes possible the recruitment of the trunk musculature for locomotion and allows step lengths increased beyond the values dictated by the lengths of the limbs.

Leaping requires elongation of the hindlimbs only in small animals, large ones possess sufficient acceleration distance even without making full use of their limb's flexibility. The mechanical needs posed by vertical clinging are not *per se* the same as in leaping, although adaptations to leaping can be used to make vertical clinging less energy-consuming: The elongated hindlimbs can also be and are used to minimize the expenditure of force (= energy) in vertical clinging, provided that the feet are prehensile.

ACKNOWLEDGEMENTS

This paper is the product of joint efforts, to which not only the authors, but also other members of the Functional Morphology in Bochum, of the Dept. of Anatomy in Liverpool and of the Institute of Special Zoology in Tübingen have contributed. Some of them show up in the reference list, others not. We would like to mention in particular D. Krebs, W. Maier, R.Crompton, St. Recknagel, U. Nieschalk, A. Christian, M. Schulz, D. Koberg, M. Günther, C. Lesch, and H. Schulze. H. Schulze drew the illustrations and and acted as a publisher, by formating the whole article. We also thank the members of the IWF in Göttingen. Prof. G. Doyle read the paper very carefully and transformed it into reasonable English.

REFERENCES

Alexander, McN. (1974). The mechanics of jumping in a dog (*Canis familiaris*). *Journal of Zoology, London, 173*, 549-573.

Alexander, McN. (1990). Elastic mechanisms in the locomotion of vertebrates. *Netherlands Journal of Zoology, 40*, 93-105.

Alexander, McN. & Bennett-Clark, H.C. (1977). Storage of elastic strain energy in muscle and other tissues. *Nature*, 265, 114-117.

Buck, C. & Bär, H. (1991). Investigations on the biomechanical significance of dermatoglyphic ridges. In H. Preuschoft & D.J. Chivers (Eds.), *Hands of Primates* (pp. 285-306). Wien, New York: Springer-Verlag.

Cartmill, M. (1974). Pads and claws in arboreal locomotion. In F.A. Jenkins (Ed.), *Primate Locomotion* (pp. 45-83). New York: Academic Press.

Cartmill, M. (1979). The volar skin of primates: Its frictional characteristics and their functional significance. *American* Journal of *Physical Anthropology*, 50, 497-510.

Cartmill, M. (1985). Climbing. In M. Hildebrand, D.M. Bramble, K.F. Liem & D.B. Wake (Eds.), *Functional Vertebrate Morphology* (pp. 73-88). Cambridge, Mass. and London: The Belknap Press of Harvard University Press.

Christian, A. (1993). Zur Biomechanik der Lokomotion vierfüßiger Reptilien (besonders der Sqamata). Dissertation, Bochum. Will appear printed in *Courier Forschungsinstitut Senckenberg*, Frankfurt a.M.

Christian, A., Horn, H.-G. & Preuschoft, H. (1994a). Bipedie bei rezenten Reptilien. *Natur und Museum*, 124, 45-57.

Christian A., Horn, H.-G. & Preuschoft, H. (1994b). Biomechanical reasons for bipedalism in reptiles. *Amphibia - Reptilia, 15*, 275-284.

Demes, B. (1989). Biomechanische Allometrie: Wie die Körpergröße Fortbewegung und Körperform von Primaten bestimmt. *Courier Forschungsinstitut Senckenberg*, Frankfurt a.M., *141*, 1-83.

Demes, B. & Günther, M.M. (1989a). Biomechanics and allometric scaling in primate locomotion and morphology. *Folia Primatol*ogica, *53*, 125-141.

Demes, B. & Günther, M.M. (1989b). Wie die Körpermasse den Springstil von Halbaffen und deren Proportionen bestimmt. *Zeitschrift für Morphologie und Anthropologie* 77, 209-225.

Demes, B., Forchap, E. & Herwig, H. (1991). They seem to glide. Are there aerodynamic effects in leaping prosimian primates? *Zeitschrift für Morphologie und Anthrop*ologie, *78*, 373-385.

Demes, B., Larson, S.G., Stern, J.T. & Jungers, W.L. (1994). The hindlimb drive of primates - theoretical reconsideration and empirical examination of a widely held concept. *Journal of Human Evolution, 26*, 353-374.

Demes, B., Jungers W.L. jr. & Nieschalk, U. (1990). Size and speed related aspects of quadrupedal walking in slender and slow lorises. In F.K. Jouffroy, M.H.Stack & C.Niemitz (Eds.), *Gravity, Posture and Locomotion in Primates* (pp. 175-197). Firenze: Il Sedicesimo.

Dunbar, D.C. (1994). The influence of segmental movements and design on whole-body rotations during the airborne phase of primate leaps. *Zeitschrift für Morphologie und Anthropologie, 80*, 1 (in press).

Fischer, M.S. (1993). Die Lokomotion von *Procavia capensis* (Mammalia: Hyracoidea). Ein Beitrag zur Evolution des Bewegungssystems der Säugetiere. Habilitationsschrift, Tübingen.

Fleagle, J.G., Stern, J.T., Jungers, W.L., Susman, R.L., Vangor, A.K. & Wells, J.P. (1981). Climbing: A biomechanical link with brachiation and with bipedalism. *Symposia of the Zoological Society of London, 48*, 359-375.

Günther, M.M. (1989). Funktionsmorphologische Untersuchungen zum Sprungverhalten an mehreren Halbaffenarten (*Galago moholi, Galago* (*Otolemur*) *garnetti, Lemur catta*). Dissertation, Berlin.

Günther, M.M., Preuschoft, H., Ishida. J. & Nakano, Y. (1992). Can prosimian-like leaping be considered a preadaptation to bipedal walking in hominids? In S. Matano, R.H. Tuttle, H. Ishida & M. Goodman (Eds.), *Topics in Primatology* Vol. 3 (pp. 153-165). Tokyo: University of Tokyo Press.

Gruber, K. (1987). Entwicklung eines Modelles zur Berechnung der Kräfte im Knie- und Hüftgelenk bei sportlichen Bewegungsabläufen mit hohen Beschleunigungen. Dissertation, Tübingen.

Gruber, K., Legal, H. & Ruder, H. (1982). Biomechanische Analyse der Bewegungsabläufe an der unteren Extremität. I. Beschleunigungen. *Zeitschrift für Orthop*ädie, *120*, 806-813.

Hirasaki, E., Kumakura, H. & Matano, S. (1993). Kinesiological characteristics of vertical climbing in *Ateles geoffroyi* and *Macaca fuscata*. *Folia Primatologica, 61*, 148-156.

Jenkins, F.A. jr. (1974). Tree shrew locomotion and the origins of primate arborealism. In F.A. Jenkins jr. (Ed.), *Primate Locomotion* (pp. 85-115). New York: Academic Press.

Jouffroy, F.K., Godinot, M. & Nakano, Y. (1993). Biometrical characteristics of primate hands. In H. Preuschoft & D.J. Chivers (Eds.), *Hands of Primates* (pp. 133-171). Wien, New York: Springer-Verlag.
Jungers, W.L. & Stern, J.T. (1984). Kinesiological aspects of brachiation in lar gibbons. In H. Preuschoft, D.J. Chivers, W.Y. Brockelman & N. Creel (Eds.), *The Lesser Apes, Evolutionary and Behavioural Biology* (pp. 119-134). Edinburgh: Edinburgh University Press.
Kimura, T., Okada, M. & Ishida, H. (1979). Kinesiologic characteristics of primate walking: its significance in human walking. In M.E. Morbeck, H. Preuschoft & N. Gomberg (Eds.), *Environment, Behavior and Morphology: Dynamic Interactions in Primates* (pp. 297-311). New York: G. Fischer.
Kram, R. & Taylor, C.R. (1990). Energetics for running: a new perspective. *Nature, 346*, 265-267.
Loitsch, C. (1994). Kinematische Untersuchung über den Galopp von Pferden (*Equus caballus)*. Dissertation, Bochum.
Merkens, H. (1987). Quantitative evaluation of equine locomotion using force plate data. PhD thesis, Utrecht.
Mochon, S. & McMahon, T.A. (1980). Ballistic walking. *Journal of Biomech*echanics, *13*, 49-57.
Mochon, S. & McMahon, T.A. (1981). Ballistic walking: An improved model. *Mathematical Biosc*iences, 241-260.
Napier, J.R. & Napier, P.H. (1967). *A Handbook of Living Primates*. New York: Academic Press.
Niemitz, C. (1984). Activity rhythms and use of space in semi-wild Bornean tarsiers, with remarks on wild spectral tarsiers. In C. Niemitz (Ed.), *Biology of Tarsiers* (pp. 85-115). Stuttgart: G. Fischer.
Nieschalk, U. (1991). Fortbewegung und Funktionsmorphologie von *Loris tardigradus* und anderen kleinen quadrupeden Halbaffen in Anpassung an unterschiedliche Habitate. Dissertation, Bochum.
Nieschalk, U. & Demes, B. (1992). Biomechanical determinants of reduction of the secondary ray in Lorisinae. In H. Preuschoft & D.J. Chivers (Eds.), *Hands of Primates* (pp. 225-234). Wien, New York: Springer-Verlag.
Peters, A. & Preuschoft, H. (1984). External biomechanics in *Tarsius* and its morphological and kinematic consequences. In C. Niemitz (Ed.), *Biology of Tarsiers* (pp. 227-255). Stuttgart: G. Fischer.
Preuschoft, H. (1970). Functional anatomy of the lower extremity. In G.H. Bourne (ed.) *The Chimpanzee*, Vol. 3 (pp. 221-294). Basel: Karger-Verlag.
Preuschoft, H. (1970). Statische Untersuchungen am Fuß der Primaten. II. Statik des ganzen Fußes. *Zeitschrift für Anatomie und Entwicklungsgeschichte, 131*, 156-192.
Preuschoft, H. (1985). On the quality and magnitude of mechanical stresses in the locomotor system during rapid movements. *Zeitschrift für Morphologie und Anthropologie, 75*, 245-262.
Preuschoft, H. (1988). The external forces and internal stresses occuring in the distal limb segments of dressage and jumping horses. *Zeitschrift für Säugetierkunde, 54*, 172-190.
Preuschoft, H. (1989). Body shape and differences between species. *Human Evolution*, 4, 145-156
Preuschoft, H. (1990). Gravity in primates and its relation to body shape and locomotion. In F.K. Jouffroy, M.H. Stack & C., Niemitz (Eds.), *Gravity, Posture and Locomotion in Primates* (pp.107 - 127). Firenze: Il Sedicesimo.
Preuschoft, H. (1992). Biomechanics and functional morphology. *Karger Gazette, 54*, 7.
Preuschoft, H., Godinot, M., Beard, C., Nieschalk, U. & Jouffroy, F.K. (1993). Biomechanical considerations to explain important morphological characters of primate hands. In H. Preuschoft & D.J. Chivers (Eds.), *Hands of Primates* (pp. 245-256). Wien, New York: Springer-Verlag.
Preuschoft H. & Demes, B. (1984). Biomechanics of brachiation. In H. Preuschoft, W.Y. Brockelman, D.J. Chivers & N. Creel (Eds.), *The Lesser Apes. Evolutionary and Behavioral Biology* (pp. 96-118). Edinburgh: Edinburgh University Press.
Preuschoft, H. & Günther, M.M. (1994). Biomechanics and body shape in primates compared with horses. *Zeitschrift für Morphologie und Anthropologie, 80*, 1 (in press).
Preuschoft, H. & Witte, H. (1991). Biomechanical reasons for the evolution of hominoid body shape. In B. Senut & M. Pickford (Eds.), Origine(s) de la Bipédie chez les Hominidés (pp. 59-77). Paris: CNRS.
Preuschoft, H., Witte, H. & Demes, B. (1992). Biomechanical factors that influence overall body shape of large apes and humans. In S. Matano, R.H. Tuttle, H. Ishida & M.Goodman (Eds.), *Topics in Primatology*, Vol. 3 (pp. 259-289). Tokyo: Tokyo University Press.
Rawlins, R.G. (1993). Locomotive and manipulative use of the hand in the Cayo Santiago macaques (*Macaca mulatta)*. In H. Preuschoft & D.J. Chivers (Eds.) *Hands of Primates* (pp. 21-30). Wien, New York: Springer-Verlag.
Sellers, W.I. & Crompton, R.H. (1994). A system for 2- and 3D kinematic and kinetic analysis of locomotion, and its appplication to analysis of the energetic efficiency of jumping locomotion. *Zeitschrift für Morphologie und Anthropologie, 80*, 1, (in press).

Schultz, A.H. (1961). Curvature of the vertebral column. In H. Hofer, A.H. Schultz & D. Starck (Eds.), *Primatologia* Vol. IV-5, Basel: Karger Verlag.

Schultz, A.H. (1963a). The relative lengths of the foot skeleton and its main parts in primates. In J.R. Napier & N.A. Barnicot (Eds.), *The Primates* (pp. 199-206). London: Zoological Society of London.

Schultz, A.H. (1963b). Relations between the length of the main parts of the foot skeleton in primates. *Folia Primatol*ogica, *1*, 150-171.

Schulz, M.H.H. (1994). Die biomechanische Bedeutung der Papillarleistenmuster auf den Fingerbeeren der Primaten. *Zeitschrift für Morphologie und Anthropologie, 80,* (in press).

Vilensky, J.A. (1989). Primate quadrupedalism: How and why does it differ from that of typical quadrupeds? *Brain, Behavior and Evolution, 34,* 357-364.

Vilensky, J.A. & More, A.M. (1992). Utilization of lateral- and diagonal-sequence gaits at identical speeds by individual vervet monkeys. In S. Matano, R.H. Tuttle, H. Ishida & M. Goodman (Eds.), *Topics in Primatology* Vol. 3 (pp. 129-137). Tokyo: University of Tokyo Press,

Witte, H., Preuschoft, H. & Recknagel, St. (1991). Human body proportions explained on the basis of biomechanical principles. *Zeitschrift für Morphologie und Anthropologie, 78,* 407-423.

Witte, H., Lesch, C., Preuschoft, H. & Loitsch, C. (1995). Die Gangarten der Pferde. *Pferdeheilkunde* (in press).

FUNCTIONAL MORPHOLOGY OF LEAPING BEHAVIORS IN GALAGIDS: ASSOCIATIONS BETWEEN LANDING LIMB USE AND DIAPHYSEAL GEOMETRY

Carl J. Terranova

Department of Anthropology*
Northwestern University
Evanston IL 60208

INTRODUCTION

An animals' limb long bones must support its weight during postural and locomotor behaviors. Locomotor behaviors have been shown to result in bending and compressive loads of relatively large magnitude on the long bones. Of the several different types of locomotion so far studied with force plates and/or strain gauges (such as slow and fast quadrupedalism, leaping, suspension), leaping has been shown to result in the largest forces (e.g. Biewener *et al.*, 1983; Biewener, 1990). Of the two phases of the leap where an animal is in contact with a support, take-off and landing, it is landing that results in the largest reaction forces: forces up to 10-15 times body weight have been shown to act on the limb skeleton at landing (Gunther, 1985; Gunther *et al.*, 1991; Preuschoft, 1985). Bending is the most common and dangerous type of loading experienced by the limb skeleton (Biewener *et al.*, 1983; Biewener and Taylor, 1986; Bertram and Biewener, 1988); bone is stronger in compression than it is in bending. The shape of long bone diaphyses dictate the behavior of the bone under load (Lanyon and Rubin, 1985; see also Cowin *et al.*, 1984). The amount (area) and distribution (second moment of area) of diaphyseal cortical bone determine its compressive and bending strength, respectively (e.g. Wainwright *et al.*, 1979). Recent research into the functional morphology of primate locomotion has demonstrated associations between diaphyseal cross-sectional geometry and locomotor patterns in galagids (Burr *et al.*, 1982), lorises (Demes and Jungers, 1989), indriids (Jungers and Minns, 1979; Demes *et al.*, 1991), macaques and other anthropoids (Ruff, 1987, 1988; Burr *et al.*, 1989; see the review of Ruff and Runestad, 1992).

Previous studies of the relationships between diaphyseal morphology and behavior have relied upon contrasts of species divided into relatively gross categories of locomotor behavior (i.e. arboreal quadruped, terrestrial quadruped, leaper, slow climber, e.g. Schaffler *et al.*, 1985). Identification of form-function complexes is complicated by phylogenetic and size variables that can not be controlled in such contrasts. More recently, research has focused on the patterns of limb morphology in closely related species that locomote in broadly similar ways. For example, in an examination of the scaling of limb long bone cross-sectional properties in indriids, Demes and coworkers have shown that the smaller species have slightly increased femoral cortical thickness relative to the larger species (Demes *et al.*, 1991). They concluded "that either changes in locomotor performance must compensate for the weight-related increase in forces and moments or that the larger bodied species operate appreciably closer to the limits of their safety margins" (Demes *et al.*, 1991:537).

It has been demonstrated that large and small bodied terrestrial mammals differ in their locomotion, larger bodied species move more slowly, keeping their limbs in contact with the ground for longer periods of time, and move on straighter, shorter limbs (e.g. Biewener, 1983; 1989; 1990). A change in duty factor, coupled with decreased speed and smaller bony moment arms effectively lessens the peak compressive and bending loads that are engendered during locomotion. Interspecific differences in locomotion, and especially in locomotor behaviors that subject the limbs to large forces, are important to examine if we are to better understand the functional associations between behavior and morphology.

Variability in limb use during landing has previously been discussed by several researchers (Table 1). However, the variation has never been quantified. In determining the relative frequencies with which different limb strike patterns are used it becomes possible to order species along a limb use continuum, from hindlimb only landers (e.g. *G. moholi*) through quadrupedal landers to forelimb only landers (perhaps *G. demidoff*). Most species exhibit patterns that include different styles (e.g. *O. crassicaudatus* and *O. garnettii*; Table 1). The purpose of arranging species according to limb use at landing is to direct investigations relating limb function to limb structure. The main purpose of this paper is to investigate the relationships between limb long bone geometry and leaping performance variation in galagids. Data is presented on limb use at landing in several galagid species. Mechanically important aspects of the shape of humeral and femoral diaphyses are examined within the context of leaping performance variation.

MATERIALS

Behavioral observations of *Galago moholi* (N=7), *Otolemur garnettii* (N=5), and *O. crassicaudatus* (N=7) were made at the Duke University Primate Center (Durham NC). Identification of species was based on DUPC records.

Galagid long bones were measured and radiographed at the American Museum of Natural History (New York), Cleveland Museum of Natural History (Ohio), British Museum (Natural History) (London), Museum National d'Histoire Naturelle (Paris), Laboratoire d'Anatomie Comparee (Paris), Rijksmuseum van Natuurlijke Historie (Leiden), Anthropologisches Institut und Museum (Zurich), and Musee Royal de l'Afrique Centrale (Terverun). Identification of species was based on museum tag designation and locality of capture. If taxon assignment was questionable (e.g. *G. moholi* or *G. zanzabaricus*) notes included with each specimen were relied upon; specimens of questionable taxonomic affiliation are not included in the present study. Specimen identification numbers for both museum and DUPC specimens are available upon request.

TABLE 1. Landing Limb Strike Patterns. Hindlimb, Forelimb, Quadrupedal denote the set of limbs initially contacting the landing support.

Hindlimb	**Forelimb**	**Quadrupedal**
GAMO[1,5]	EUEL[4]	EUEL[1,2]
GASE[1,4,5]	GAZA[1]	GAZA[1]
GAAL[2]	GAAL[1]	OTCR[5]
OTGR[1,5]	GADE[1,4]	OTGR[5]
	OTCR[3,4,5]	

ABBREVIATIONS		REFERENCES
GADE	*Galago demidoff*	1: Nash *et al.*, 1989
GAMO	*Galago moholi*	2: Charles-Dominique, 1977
GASE	*Galago senegalensis*	3: Charles-Dominique& Bearder, 1979
EUEL	*Euoticus elegantulus*	4:Oxnard *et al.,* 1990
GAAL	*Galago alleni*	5: this paper
OTGR	*Otolemur garnettii*	
OTCR	*Otolemur crassicaudatus*	

METHODS

Behavior

Animals were videotaped in a research room in the nocturnal building where a support network was constructed of wooden dowels. The network consisted of equal lengths of vertical, horizontal, and oblique supports (Fig. 1) and was constructed so that the majority of movements occurred in a plane perpendicular to the camera. The greatest distance between supports was 2.5 m and poles extended 2.5 m up from the floor. Single or paired animals of one species were placed on the supports and observed for a period of 8 hrs; two 4 hr periods of observation occurred each day. Taping occurred opportunistically throughout the session, depending on the activity level of the animals. All animals were presented with the same support network.

A Panasonic PV330 camcorder was used for taping locomotor behaviors. The camera was mounted on a tripod and positioned approximately 4 m from the network. The entire support network could be seen without panning the camera. Occasionally, the camera was hand-held to permit close-up views of small animals (e.g. *G. moholi*). Videotaping was done using standard room light.

A Panasonic AG6300 playback deck was used in stop-frame examination of the tapes. Tapes were extensively edited for analysis because of the many non-leaping locomotor, postural, and feeding behaviors captured. In few cases did the tape speed, usually a field every 60th of a second, limit visualization of landing limb strike sequence.

Videotapes were examined to determine the orientation of take-off and landing supports, distance between supports, and the pattern of limb strike at landing.

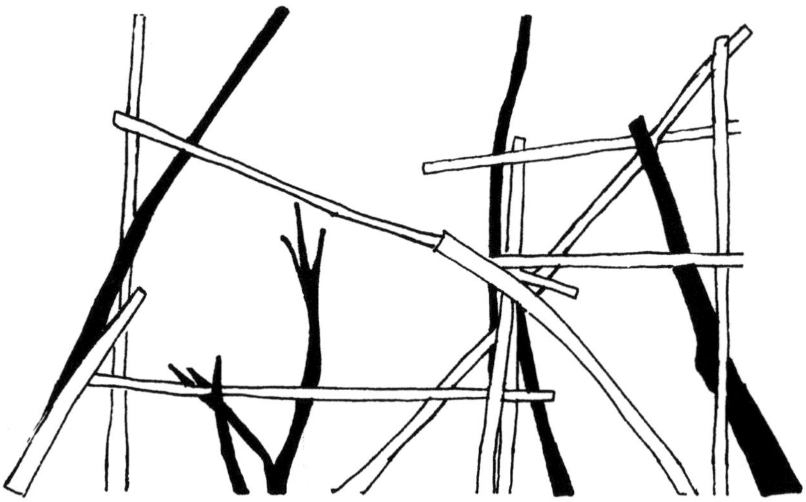

FIGURE 1. Support network used for studying landing behaviors. The network is composed of equal lengths of vertical, horizontal, and oblique supports. Blackened supports are branches from the Duke Forest, clear supports are wooden dowels. All animals studied were presented with this network in order to control for variation in support structure.

Support orientations are defined as follows:

HORIZONTAL:parallel to a horizontal plane, between 0 and 20 degrees;

VERTICAL:perpendicular to a horizontal plane, between 70 and 90 degrees;

OBLIQUE:oriented between 20 and 70 degrees from a horizontal plane.

The categories used to describe limb use at landing are:

FORELIMB: those in which the forelimbs make initial contact with the landing support, hindlimbs trailing (Fig. 2);

HINDLIMB: those in which the hindlimbs make initial contact with the landing support, forelimbs trailing (Fig. 2);

QUADRUPEDAL: those in which all four limbs contact the landing support simultaneously (Fig. 2);

Limb use at landing was tallied for each species (Table 2). Fisher's exact test was used to test if landing style is independent of support orientation for each species. Exact probability values were computed using Statxact (Cytel Software) and are based on permutation distributions, rather than chi-square distributions, allowing analysis of contingency tables with overall small sample size and some 0 cell values. Landing style variation within and between species was tested against a hypothetical distribution of equal use of all landing styles using a G test (Sokal and Rohlf, 1989). The interspecific tests of landing limb use directly address the question of limb use variation among galagids, while the tests for association between landing style and support orientation are a first step toward examining the relationship between support structure and limb use in each species. Additionally, frequencies of different limb strike patterns on the three support types were used in classifying species in Table 1.

Morphology

Humeral and femoral length between articular surfaces, midshaft diameter, thickness of anterior, posterior, lateral and medial cortices, and the medullary cavity were measured from anteroposterior (AP) and mediolateral (ML) radiographs with a needle-point caliper and a 5X three dimensional loupe. Geometric properties at midshaft locations are estimated using a hollow ellipse model; no significant differences between cortex thickness within a plane are apparent in the sample (see Runestad *et al.*, 1993). Cortical bone area, second moment of area (I) and the section modulus (Z) were estimated as follows:

$$\text{Cortical Area} = pi*((Dy*Dx)-(dy*dx))\ mm^2 \quad (1)$$

$$Iap = (pi/4)*((Dy^3*Dx)-(dy^3*dx))\ mm^4 \quad (2)$$

$$Iml = (pi/4)*((Dx^3*Dy)-(dx^3*dy))\ mm^4 \quad (3)$$

$$Zap = Iap/Dy\ mm^3 \quad (4)$$

$$Zml = Iml/Dx\ mm^3 \quad (5)$$

FIGURE 2. Representative landing postures traced from slides taken of a video screen.
TOP LEFT: *Otolemur garnettii* landing fore limbs first. Landing support below take- off support.
TOP RIGHT: *Galago moholi* landing hindlimbs first onto vertical (top) and horizontal (bottom) supports.
BOTTOM LEFT: *Otolemur garnettii* landing quadrupedally onto a vertical support.
BOTTOM RIGHT: *Otolemur crassicuadatus* landing quadrupedally (left) and fore limbs first (right).

where Dy is external AP diameter, dy is internal AP diameter, DX is external ML diameter, and dx is internal ML diameter.

Femoral midshaft is determined as 50% of bone length between articular surfaces, measured proximally from the distal articular surface. At this section location no crests, ridges, or prominences exist. The humeral section is located at 45% of the length, more proximal or distal locations include bony crests and ridges that alter the cylindrical shape. Cortical area is proportional to axial compressive strength, I and Z track eccentric strength or bending rigidity (Roark and Young, 1976; see also Demes and Jungers, 1989; Demes *et al.*, 1991). The section modulus, Z, is sensitive to the amount of cortical bone located away from the neutral axis of the bone and as such is a better measure of the rigidity of the section than is the second moment of area (I) (see also Demes and Jungers, 1993). Both are reported here for comparative purposes.

Indices are used to examine diaphyseal shape (Index 1), cortical bone distribution (Index 2, Index 3), robustness (Index 4) and to estimate bone strength in compression (Index 5). The thickness of cortical bone relative to the total thickness of the bone at midshaft is estimated as the ratio of the medullary cavity thickness to the total diameter (K, Currey and Alexander, 1985).

$$\text{Index 1} = Dy/Dx \tag{6}$$

$$\text{Index 2} = Iap/Iml \tag{7}$$

$$\text{Index 3} = Zap/Zml \tag{8}$$

$$\text{Index 4} = \frac{(Dy*Dx)*100}{\text{bone length}} \tag{9}$$

$$\text{Index 5} = \text{area/body weight} \tag{10}$$

$$Kap = Dy/dy;\ Kml = Dx/Dy \tag{11}$$

For the denominator in equation 10 species mean areas and species mean body weights were used. Body weights were taken from the literature (Table 3; see Dagosto and Terranova, 1992 for references).

Differences of thickness, I, and Z between planes within a bone and between bones in the same plane are tested with T-tests. Because of the number of tests made the critical value used for acceptance of a significant difference is $P<.01$. All larger P-values are reported.

Recent scaling analyses have shown variable patterns within strepsirhines. Demes and Jungers (1993) have shown, in a broad interspecific sample of strepsirhines, femoral midshaft geometric properties scale to weight with slight positive allometry. However, different patterns were found to characterize indriids (isometry) and lorisids (positive allometry) (Demes and Jungers, 1989; Demes *et al.*, 1991). Biewener (1990) hypothesized slight positive allometry between mechanical properties and mass in relatively small sized mammals; positive allometry would result in relatively robust limb bones in larger sized animals. Because static forces are hypothesized to increase with mass in a positively allometric fashion the slight positive allometry of diaphyseal cortical properties would ensure comparable bone strength and rigidity in large and small species. However, reaction force magnitudes have been found to be size independent in a variety of mammals (e.g. Biewener, 1991), with behavioral changes hypothesized to effectively decrease peak forces in large bodied taxa. Examination of scaling patterns in galagids, with their known behavioral diversity, will be used to assess the roles of behavioral and morphological variation relative to body weight. The scaling of cross-sectional properties to body weight is examined with linear regression analyses. Species mean natural logarithm transformed data are used to examine the correlations and scaling coefficients between geometric properties and mass. Both Model I and Model II slopes are reported.

RESULTS

Landing Styles

Patterns of landing behavior are variable between species (Table 2, Fig. 2). A test of the similarity of landing patterns within and among the three galagids, with all support orientations combined, reveals that the three species do not land with similar limb strike patterns (heterogeneity G value= 221.78; with 4 degrees of freedom, 95% critical value=9.49). Intraspecific patterns differ significantly from equal use of all three landing categories (G values for *G. moholi*, *O. garnettii*, and *O. crassicaudatus* are 268.93, 33.36, 12.07, respectively; with 2 degrees of freedom 95% critical value= 5.99). Tests of association between landing limb strike pattern and support orientation reveal that *G. moholi* lands hindlimbs first on all support types (Fisher exact P=.268), while *O. garnettii* and *O. crassicaudatus* use different limb strikes on different support orientations (Fisher exact P=.0003 and .0004, respectively). *O. garnettii* lands quadrupedally on vertical or oblique supports, hindlimbs first on obliques, and least frequently, forelimb first on vertical supports. *O. crassicaudatus*, on the other hand, was observed to land primarily on oblique supports, using quadrupedal and forelimb first landings nearly equally. Further descriptions of support use and landing style for each species are provided below.

Galago moholi This species of galago leaped most frequently (N=141). Take-off support orientation was primarily oblique (48%); vertical and horizontal supports were used 30% and 21%, respectively. Landings occurred on vertical 46% of the time, oblique 39% and horizontal 15%. (Table 2). The landing style of this species was overwhelmingly with the hindlimbs making initial contact with the support (Fig. 2); 97% of all leaps were terminated with hindlimb first landing. Hindlimb landings were seen on vertical and oblique supports approximately equally (45% and 38%, respectively). Forelimb or quadrupedal landings were seen less than 1% of the time on these supports. Landing on horizontal supports was also primarily hindlimb-first, 1% of horizontal landings were accomplished with the forelimbs making initial contact.

Otolemur garnettii A total of 93 leaps are included in the analysis (Table 2). Leaps were begun and ended on vertical supports 58% and 56% of the time, respectively. Just over 30% of take-offs and landings occurred on oblique supports; approximately 10% of all leaps were from or onto horizontal supports. Limb use at landing is more variable than for *G. moholi*. Quadrupedal landings occurred 62% of the time, primarily on vertical but also on horizontal and oblique supports (Fig. 2). Hindlimb and forelimb first landings occurred 22% and 16%, respectively. Most of the hindlimb landings occurred on oblique supports; forelimb landings were more frequent on vertical supports.

Quadrupedal landings occurred 80% of the time with a vertical target support; forelimb first landings occurred just over 10% of the time that verticals are used. The majority of forelimb landings occurred after short (<.5 m) leaps (Fig. 2). Quadrupedal and hindlimb-first landings were approximately evenly used when landing on oblique supports. Eight of ten landings on horizontal supports were quadrupedal.

Otolemur crassicaudatus A total of 35 leaps were recorded. Oblique supports were most frequently used for both take-off and landing, 68% and 70%, respectively (Table 2). Take-off and landing upon vertical and horizontal supports occurred between 10% and 22% of all leaps. Quadrupedal and forelimb-first landings were used equally (46%); hindlimb landings were used 9%, exclusively on horizontal supports. The majority of leaps occurred on oblique supports, terminated with either quadrupedal or forelimb-first landings (Fig. 2). Forelimb landings occurred in long (>1 m) and short (<.5 m) leaps.

TABLE 2. Landing limb use by support type.
Cells in the format percent / number of episodes; percents do not always equal 100 due to rounding.

	Oblique	Vertical	Horizontal	Total
Galago moholi				
Quad	- / -	**1** / 1	- / -	**1** / 1
Fore	**1** / 2	- / -	**1** / 1	**2** / 3
Hind	**38** /53	**45** /64	**14** /20	**97** /137
Total	**39** /55	**46** /65	**15** /21	/141
Otolemur garnettii				
Quad	**17** / 16	**38** / 35	**8** / 7	**62** / 58
Fore	**3** / 3	**13** / 12	- / -	**16** / 15
Hind	**16** / 15	**3** / 3	**2** / 2	**21** / 20
Total	**36** / 34	**54** / 50	**10** / 9	/ 93
Otolemur crassicaudatus				
Quad	**40** / 14	**3** / 1	**3** / 1	**46** / 16
Fore	**34** / 12	**11** / 4	- / -	**46** / 16
Hind	- / -	- / -	**9** / 3	**9** / 3
Total	**74** / 26	**14** / 5	**12** / 4	/ 35

G-Tests:

	Landing style by Support Orientation[1] G Value	prob.	Equal limb use[2] G value	prob.	Homogeneity[3] G Value	prob.
GAMO	5.33	NS	271.76	<.001	230.38	<.001
OTGR	14.45	<.05	35.23	<.001		
OTCR	19.45	<.001	12.77	<.01		

1: test of independence of landing style by support orientation
Critical Values: $X^2(.05,6) = 12.59$; $X^2(.001,6)= 22.46$
2: test of equal use of all landing styles, independent of
support orientation. Critical Values: $X^2(.001,2)=13.82$; $X^2(.01, 2)=10.60$
3: Sample homogeneity test
Critical Values as in 1.

Morphology

Diaphyseal Shape Humeral and femoral midshaft diaphyseal shape is variable among galagids (Table 3). *O. crassicaudatus* and *Galago alleni* have similarly shaped anteroposterior (AP) and mediolateral (ML) humeral diameters, while all other species have greater AP than ML diameters (paired T-test, P<.02) except *G. moholi,* whose ML diameter is greater than AP diameter (P=.02). Femoral diaphyseal shape follows a similar pattern with *O. crassicaudatus* and *G. alleni* exhibiting similarly shaped AP and ML diameters (P>.02), while all other species have significantly greater AP diameters (P≤.002). The femoral dimensions are significantly greater than the humeral dimensions in both planes for all species (P≤ .01).

Diaphyses that have similarly shaped AP and ML dimensions are equally resistant to forces applied from either the sagittal or coronal planes. The large bodied galagids exhibit this pattern in both the humerus and femur. Smaller galagids have one plane that is preferentially reinforced and therefore more resistant to bending forces acting on the humerus and femur. Bending forces that occur in the sagittal plane, likely engendered during flexion/extension movements of the limbs during landing, are accommodated by the increased amount of bone in the AP plane in all smaller bodied galagids except the humerus of *Galago moholi* where the ML plane is enlarged.

Cortical Area For all galagids, the femur is capable of withstanding greater compressive loads than is the humerus (Table 3). A ratio of humeral to femoral compressive strength (Table 5) shows that *G. moholi* (a devoted hindlimb lander) has femora nearly twice as strong as humeri whereas *O. crassicaudatus* (a quadrupedal or forelimb-first lander) has femora that are approximately 25% stronger than humeri. Comparisons of the index between cheirogaleids, lorids, galagids, and *Avahi* indicates that the quadrupedal species (lorises, pottos, and cheirogaleids, except *Mirza*) have very similar values for the humerus and femur. *O. crassicaudatus* has values that fit nicely with this pattern. *Mirza* exhibits quite strong femora, comparable to *G. alleni* and *E. elegantulus*. On the other hand, the species customarily regarded as highly hindlimb-dominated leapers (e.g. *Avahi*, and the Senegal and Mohol galagos) have femora that are approximately twice as strong as humeri. The galagids that tend to leap less frequently and land with a greater contribution from the fore limb have intermediate values.

Second Moment of Area and Section Moduli As a gauge of the distribution of cortical bone in the section, second moments of area (I) and section moduli (Z) were computed (Tables 3,4). These variables reflect the bending rigidity of the section. Judging by both I's and Z's the femur is significantly more resistant to bending in the AP plane than in the ML plane (P<.01). This is also true for the humerus of *G. senegalensis* and *Euoticus*, but there is no significant difference in *O. crassicaudatus* and *G. alleni*. The humerus of *G. moholi* is more resistant to ML bending (P=.01). Indices of AP to ML second moments of area and section moduli (Index 2 and 3, respectively; Table 4) summarize the differences in bending strength between the two planes (see below).

Two species, *G. alleni* and *E. elegantulus*, are of similar enough size for a direct comparison of the values. These species differ in their limb use patterns. The former is a more frequent leaper that lands using either hind or fore limb strike, while the latter is a more quadrupedal species that lands after a leap either quadrupedally or fore limbs first. A comparison of AP Z values of the humerus and femur indicates that the more quadrupedal species, in terms of overall locomotor behavior and landing style, has more similar bending rigidities of the two bones. *Euoticus* has stronger humeri in both bending and compression; this may be related to the postures the animal adopts when feeding on gums head downward on large diameter supports. *Euoticus* is more robustly built than *G. alleni*, in both the humerus and femur. The pattern is difficult to understand if it is true that *G. alleni* is a frequent ricochetal leaper (as summarized in Oxnard *et al.*, 1990).

TABLE 3. Tests of differences in diaphyseal measures of the humerus (HUM) and femur (FEM). N= number of humeri/number of femora. An asterisk indicates a significant difference (P<.02) between anteroposterior (ap) and mediolateral (ml) dimensions; 1 indicates ap > ml, 2 indicates ml > ap. For humeral vs. femoral cortical area contrasts all significant differences indicate that femoral area is greater than humeral area. NS: differences are not significant. See text for derivation of variables.

	N	Hum Diameter AP vs. ML	Fem Diameter AP vs. ML	Hum vs. Fem Cortical Area	Hum Second Moment of Area AP vs. ML	Fem Second Moment of Area AP vs. ML	Hum Section Modulus AP vs. ML	Fem Section Modulus AP vs. ML
GALAGIDS								
Otolemur crassicaudatus 1150 grams	14/17	NS	NS	*	NS	$*^{1}$	NS	$*^{1}$
Galago alleni 295 grams	7/ 7	NS	NS	*	NS	$*^{1}$	NS	$*^{1}$
Euoticus elegantulus 275 grams	17/18	$*^{1}$	$*^{1}$	*	$*^{1}$	$*^{1}$	$*^{1}$	$*^{1}$
Galago senegalensis 215 grams	19/21	$*^{1}$	$*^{1}$	*	$*^{1}$	$*^{1}$	$*^{1}$	$*^{1}$
Galago moholi 160 grams	5/5	$*^{2}$	$*^{1}$	*	$*^{2}$	$*^{1}$	$*^{2}$	$*^{1}$
CHEIROGALEIDS								
Phaner furcifer 440 grams	1/1	NS	NS	NS	NS	NS	NS	NS
Cheirogaleus major 450 grams	3/3	NS	NS	NS	NS	*1	NS	NS

TABLE 3. continued

	N	Hum Diameter AP vs. ML	Fem	Hum vs. Fem Cortical Area	Hum Second Moment of Area AP vs. ML	Fem	Hum Section Modulus AP vs. ML	Fem
Mirza coquereli 330 grams	1/1	NS	NS	NS	NS	NS	NS	NS
Cheirogaleus medius 233 grams	7/7	NS	NS	NS	NS	NS	NS	NS

K-Values Humeral and femoral K-values (Table 6) indicate the thickness of the cortical walls relative to the external diameter of the section locations. The smaller K-values are associated with thicker cortical walls. The femoral K values in the AP plane are significantly less than those of the ML plane for all species, except *E. elegantulus* (P=.09). Humeral K-values show that the AP plane has thicker cortices than the ML plane for all species except *E. elegantulus* and *G. senegalensis* (P>.03). In fact, AP cortices tend to be thicker than ML cortices in both bones (see also Index 1, Table 4). The pattern is statistically significant for all species except *O. crassicaudatus* and *G. alleni*. The femoral AP K-values are significantly less than those for the humerus for all species except *O. crassicaudatus* and *G. alleni* (P=.40 for both species).

Robusticity Robusticity (Index 4, Table 5) was calculated as the product of the external diameters over bone length. For all species the humerus is more robust than the femur. This pattern is driven by the great length of the femora and the relative shortness of the humeri. More instructive in terms of the mechanical capabilities of a long bone is an examination of compressive strength. Compressive strength was determined as cortical area relative to body weight (Index 5, Table 5).

TABLE 4. Species mean values (±1 standard deviation) for indices describing the shape of humeral and femoral diaphyses. Index 1=ap external diameter/ml external diameter; Index 2=Iap/Iml; Index 3=Zap/Zml.

	INDEX 1 HUM	INDEX 1 FEM	INDEX 2 HUM	INDEX 2 FEM	INDEX 3 HUM	INDEX 3 FEM
Otolemur crassicaudatus	1.02 ±.06	1.04 ±.07	1.06 ±.12	1.13 ±.13	1.03 ±.06	1.08 ±.06
Galago alleni	.97 ±.07	1.07 ±.06	.95 ±.14	1.18 ±.12	.98 ±.08	1.11 ±.06
Euoticus elegantulus	1.10 ±.10	1.09 ±.05	1.20 ±.20	1.24 ±.14	1.08 ±.09	1.14 ±.07
Galago senegalensis	1.08 ±.13	1.11 ±.09	1.19 ±.28	1.30 .21	1.08 ±.12	1.17 ±.10
Galago moholi	.89 ±.07	1.21 ±.06	.76 ±.12	1.54 ±.17	.85 ±.07	1.28 ±.08
Phaner furcifer	.99	1.06	.95	1.12	.96	1.06

TABLE 4. continued

	INDEX 1		INDEX 2		INDEX 3	
	HUM	FEM	HUM	FEM	HUM	FEM
Cheirogaleus major	.95 ±.07	1.14 ±.05	.91 ±.15	1.28 ±.11	.95 ±.09	1.13 ±.05
Mirza coquereli	1.01	1.10	1.02	1.18	1.01	1.07
Cheirogaleus medius	1.04 ±.05	1.05 ±.12	1.06 ±.10	1.10 ±.23	1.01 ±.05	1.04 ±.11

TABLE 5. Robusticity; Index 4=((ap diameter*ml diameter)*100)/length) and Compressive Strength; Index 5=midshaft cortical area/body weight for the humerus and femur. See text for discussion of the ratio of humeral to femoral compressive strength (last column). Data for lorises and *Avahi* from Demes and Jungers, 1989 and Demes *et al.*, 1991, respectively.

	INDEX 4		INDEX 5		INDEX 5 RATIO (H/F)
	HUM	FEM	HUM	FEM	
Otolemur crassicaudatus	15.18 ± .93	12.07 ± .53	10.66	14.24	.75
Galago alleni	13.98 ± .75	10.72 ± .47	13.26	22.04	.60
Euoticus elegantulus	16.22 ±1.53	11.03 ± .49	17.34	26.44	.66
Galago senegalensis	17.10 ±1.53	11.16 ± .41	20.20	34.86	.58
Galago moholi	14.71 ± .83	11.24 ± .37	18.52	38.11	.49
Phaner furcifer	17.08	12.56	13.27	16.64	.80
Cheirogaleus major	19.66 ±1.17	13.68 ± .40	16.18	16.83	.96
Mirza coquereli	17.47	14.03	19.07	25.18	.76
Cheirogaleus medius	18.37 ±.90	13.57 ±.56	12.42	14.04	.89

TABLE 5. continued

	INDEX 4		INDEX 5		INDEX 5 RATIO (H/F)
	HUM	FEM	HUM	FEM	
Loris tardigradus					.91
Nycticebus coucang					.96
Avahi laniger					.48

TABLE 6. Galagid K-values: the ratio of internal to external diameter. Midshaft sections of femur and humerus, species means ± 1 standard deviation. Low values indicate thick walls, see text for further discussion.

	HUMERUS		FEMUR	
	Kap	Kml	Kap	Kml
Otolemur crassicaudatus	.56 $\pm$.04	.60 $\pm$.05	.56 $\pm$.04	.65 $\pm$.04
Galago alleni	.60 $\pm$.07	.63 $\pm$.04	.57 $\pm$.05	.64 $\pm$.07
Euoticus elegantulus	.68 $\pm$.03	.64 $\pm$.05	.56 $\pm$.06	.66 $\pm$.04
Galago senegalensis	.59 $\pm$.04	.56 $\pm$.10	.51 $\pm$.06	.65 $\pm$.04
Galago moholi	.67 $\pm$.03	.56 $\pm$.04	.53 $\pm$.03	.65 $\pm$.03

TABLE 7. Scaling of lengths and geometric properties to mass in galagids. Natural logarithm transformed species means. Ordinary least squares (OLS) and Reduced major axis (RMA) slopes are reported. Sample size for all is 5. SEE=standard error of the estimate. Isometry for linear measures is .33, areas: .67, section moduli (Z): 1.00, second moments of area (I): 1.33.

	r	Y intercept	OLS slope (se)	RMA slope	SEE
HUMERUS:					
LENGTH	.986	1.57	.368 (.036)	.373	.055
AP DIAMETER	.921	-1.00	.371 (.091)	.403	.138
ML DIAMETER	.988	- .91	.355 (.032)	.359	.049

TABLE 7. continued

	r	Y intercept	OLS slope (se)	RMA slope	SEE
AREA	.967	-2.73	.745 (.114)	.770	.173
Iap	.943	-5.91	1.513 (.309)	1.607	.469
Iml	.974	-5.52	1.443 (.193)	1.482	.293
Zap	.952	-5.56	1.129 (.210)	1.186	.319
Zml	.972	-5.27	1.079 (.151)	1.110	.229
FEMUR:					
LENGTH	.988	3.06	.210 (.019)	.213	.029
AP DIAMETER	.942	.13	.226 (.047)	.240	.071
ML DIAMETER	.983	-.30	.285 (.031)	.290	.047
AREA	.956	-.81	.505 (.090)	.528	.136
Iap	.951	-1.74	.982 (.184)	1.033	.280
Iml	.973	-2.64	1.098 (.150)	1.128	.228
Zap	.954	-2.49	.741 (.134)	.777	.204
Zml	.971	-3.00	.806 (.114)	.830	.173

Body Size All morphological variables are highly correlated with weight (Table 7). The humerus and femur scale differently with weight among galagids. Humeral length, area, second moment of area and section modulus are isometric with mass; femoral length and cross-sectional properties are negatively allometric (see also Jungers, 1985). This suggests that the femora of the larger species, which include *O. crassicaudatus, G. alleni*, and *E. elegantulus*, are not only relatively shorter but also less resistant to compressive and bending loads. The humeri of the large species, however, are relatively more robustly built than those of the small species. In terms of limb use during high impact behaviors it is apparent that the larger species are better equipped to dissipate forces with the use of all four limbs while smaller species can use hind limb contact to dissipate forces.

The scaling patterns found here differ from those found in a study of long bone diaphyseal geometry in species representative of all strepsirhines (Demes and Jungers, 1993). They found that for the entire sample of strepsirhines these variables, bone lengths, diameters, areas, second moments of area and section moduli, scale in a positive allometric manner. However, Demes *et al.* (1991) have shown that the femur in indriids is isometric (ML diameter, area, ML second moment of area) or negatively allometric (AP diameter, AP second moment of area, total length) with weight while Demes and Jungers (1989) have shown the lorid femur to scale with positive allometry. The differences found among subsets of strepsirhines suggests that size alone does not dictate limb long bone design; phylogenetic affiliation and the mechanical demands of locomotor and postural behaviors also affect scaling patterns. In a study currently underway I examine the scaling of lemurid diaphyseal geometry to body weight, allowing for more complete discussion of the effects of locomotion, size, and phylogenetic affinity on strepsirhine limb long bone design.

DISCUSSION

The heterogeneity of landing patterns within the galagids indicates that the inclusion of all species in one locomotor category obscures mechanically important, and variable, aspects of the included species' locomotion. Table 1 contains a categorization of galagids based on limb strike patterns at landing. The data presented above support the notion that patterns of limb use are variable between species. Importantly, limb strike patterns are also somewhat variable within species. Only two species consistently land hindlimbs first (*G. senegalensis* and *G. moholi*): *G. demidoff* may be unique in landing forelimbs first (Nash *et al.*, 1989; Oxnard *et al.*, 1990), while all other species land with some combination of quadrupedal and fore or hind limb strikes (see Table 1). It will be important to quantify the frequency with which different limb postures are used by the species not included in the behavioral component of this study. At this time, however, it appears reasonable to use this categorization as a first attempt to examine patterns of diaphyseal shape relative to landing style.

Based on the data presented here and compiled from literature sources (Table 1) it is possible to make several generalizations about the landing behaviors of galagos. The most hindlimb-dominated species are the Mohol and Senegal galagos, Allen's galago has been noted to land with either fore or hind limbs but not quadrupedally, and Garnett's galago will use hind limbs or all limbs for long distance leaps but fore limbs for leaps of less than .5 m. The thick-tailed and Demidoff's galago have been suggested to not be able to land hind limbs first, with the former landing quadrupedally or on its fore limbs and the latter primarily on its fore limbs. The needle-clawed galago appears similar to the thick-tailed galago. This diversity of limb use patterns allows for contrasts to be made between species that rely on one set of limbs and those that tend to be more quadrupedal.

The three species examined here demonstrate that the concept of hindlimb dominated locomotion describes only a portion of the kinds of movements used by a species. The important aspect of this variation is that it likely results in differences in forces experienced by a limb during locomotion. Although not directly tested here, it appears reasonable to assert that differences in the number of limbs used to decelerate after a leap results in variable stress magnitudes (and perhaps altered directions) experienced by the limbs used to contact the landing support. The aspect of limb bone design that is responsible for accommodating compressive loads is the area of cortical bone. Galagos that rely on all four limbs to dissipate landing impact forces have humeri and femora that are of approximately equal strength. In contrast, species that rely on only the hindlimbs to take up the forces of landing have femora that are nearly twice as strong as their humeri.

Comparable behavioral data exists for several other groups of strepsirhines. With this data it is possible to compare the limb design of these species with galagids to determine the pervasiveness of particular morphological patterns in species that customarily impact load their hindlimbs but not the fore limbs, land quadrupedally loading all four limbs simultaneously, or land with initial contact of the forelimbs. For example, landing behavior in indriids appears rather stereotyped; landings occur with the outstretched hind limbs initially contacting the landing support (Oxnard *et al.*, 1990). Indriid limb cross-sectional geometry has been described by Demes *et al.* (1991). Locomotor and morphological data also exist for the slow climbing lorids (Demes & Jungers, 1989). Cheirogaleid locomotion has been described as primarily quadrupedal by Walker (1979), and Oxnard has figured slightly greater frequency of leaping in *Mirza* versus *Phaner* and *Cheirogaleus* (Oxnard *et al.*, 1990). Cheirogaleid limb morphology data is presented in Table 3 (Terranova, 1994). Contrasting indriids, lorids, and cheirogaleids with galagids allows an examination of the strepsirhine locomotor continuum, spanning slow quadrupedalism to ricochetal leaping with hindlimb landing.

Among the cheirogaleids, no significant differences are found in diameter, second moment of area or section moduli between planes in the humerus and femur or in contrasts of the same plane in the two bones; a similar pattern is found in *O. crassicaudatus*, *G. alleni*, and lorids (Demes and Jungers, 1989). This is in contrast to diaphyseal design in indriids and some galagids

(*G. moholi*, *G. senegalensis*, *E. elegantulus*) where the AP plane is more resistant to bending. These leaping forms may experience greater bending forces in this plane as a result of muscular contraction producing acceleration via rapid extension of hip, knee, and ankle joints at take-off as well as when landing on partially flexed limbs. Inspection of humeral and femoral compressive strengths (Index 5, Table 5) shows that the cheirogaleids tend to have fore and hind limb bones that are quite similar in strength (except *Mirza*), again, a pattern similar to that seen in *O. crassicaudatus*, lorids, and to a lesser extent *E. elegantulus*. The value for *Avahi* shows that, in contrast, the hindlimb-dominated leapers (i.e. those that land with initial contact of the hind limbs including *G. senegalensis* and *G. moholi*) have femora that are nearly twice as strong as humeri.

Differences in bending rigidity are also apparent between frequently quadrupedal species (including cheirogaleids, except *Mirza*, and *O. crassicaudatus* and perhaps *Euoticus*) and the more saltatory galagids (Table 3). Femoral second moments of area (I), as well as section moduli (Z), tend to be almost twice as large for the femur as for the humerus in leaping forms, whereas in the quadrupedal forms the differences in bending rigidity between the humerus and femur is much less pronounced.

Ruff and Runestad (1992) have recently shown similar differences between leaping and non-leaping prosimians. Specifically, they show that lorids have humeri that are stronger (resist compressive forces) than they are rigid (resist bending forces) whereas leapers exhibit the opposite pattern (their Fig. 7, pg. 420). A similar pattern can be discerned in contrasts of galagid and cheirogaleid humeri (Table 3). It appears that species which use forelimbs in landing (i.e. *O. crassicaudatus*, *G. alleni*, *Euoticus*) have humeri that are designed to accommodate bending forces, whereas the hindlimb landers (especially *G. senegalensis* and *G. moholi*) have humeri that are better able to accommodate axial compressive forces, perhaps engendered by muscular contraction during relatively low-impact quadrupedal scurrying and climbing.

The relationship between K values and either size or locomotion is unclear. Comparing the data obtained here with that for lorids and indriids (Demes and Jungers, 1989; Demes *et al.*, 1991) suggests that thicker bones tend to characterize leapers of small size, small quadrupedal species have somewhat thinner bones. The situation is different in larger species (those around one kilogram in weight). However, *Nycticebus* has low femoral K values, lower than for either *O. crassicaudatus* or *Avahi laniger*, both leapers of comparable size. The robustness of this slow climbing lorids' limbs is difficult to explain.

Scaling patterns of cross-sectional parameters and body weight in indriid and galagid humeri are similar; lengths, diameters, areas and distributional variables are essentially isometric with weight. The patterns for the femora are quite different, however. In both groups femoral length exhibits slight (indriid) to strong (galagid) negative allometry. Femoral midshaft diaphyseal cortical area in indriids is isometric with mass, in galagids this measure is negatively allometric. Section moduli and second moments of area also follow this trend. An implication of these scaling patterns is that large bodied galagids have relatively underdesigned femora (or small bodied galagids have overdesigned femora). The landing patterns of these larger bodied galagids (e.g. *O. crassicaudatus*, *O. garnettii*) demonstrate a greater reliance on quadrupedal dissipation of landing impacts. In fact, the differences between mechanical properties of their humeri and femora are very similar to the primarily quadrupedal, infrequently leaping cheirogaleids and the non-leaping lorids (see also Ruff and Runestad, 1992). While large bodied indriids have slightly less accommodating femora than isometry would predict, their hind limbs are likely strong enough to dissipate landing impacts without recruitment of the fore limbs. Certainly these two groups, formerly included in the category vertical clinger and leaper by Napier and Walker (1967), display far too much behavioral and morphological variation to be included in a single locomotor category.

CONCLUSIONS

Patterns of limb long bone strength track aspects of landing behavior in galagids and other strepsirhines. The functional differentiation of limb bones is presumably underlain by differences in the forces experienced during high impact behaviors, such as when landing after a leap. Those species that preferentially use hindlimbs in landing, for example *G. moholi, G. senegalensis, Avahi laniger*, are characterized by having divergent compressive strengths of the humerus and femur, with the femur being almost twice as strong than the humerus. Quadrupedal landers and slow climbing quadrupeds, all species that load their fore limbs during locomotor behaviors more frequently than the species mentioned above, for example *O. crassicaudatus, Cheirogaleus*, and the lorids, have comparable amounts of cortical bone in the humerus and femur. Further, species that specialize in using hindlimbs to land after a leap exhibit significantly more rigid bones in the AP plane, the more quadrupedal species appear to have bones that are equally resistant to bending in AP and ML planes.

Functional analysis of diaphyseal geometry in closely-related species can result in a better understanding of the interplay between behavior and morphology (Demes *et al.*, 1991; Ruff and Runestad, 1992). It is important to also explicitly consider mechanically important aspects of the locomotion of the group being investigated. The behavioral data presented here demonstrate variation in just such a mechanically important behavior. Assuming that all galagids engage in `similar' leaping behaviors would have obscured an important contribution to limb design. The variable patterns that have been uncovered in the different groups of strepsirhines (e.g. lorids vs galagids vs indriids) implies that limb bone geometry is affected by a variety of factors including locomotor and postural biomechanics, body size, and phylogenetic affiliation. A comprehensive examination of strepsirhine long bone geometry relative to limb use during leaping behaviors (Terranova, 1994) can help to clarify the differences noted here between galagids and indriids in terms of both behavior and morphology. Inclusion of lemurids will allow an examination of all strepsirhines, permitting an assessment of the relative effects of phylogenetic affiliation, size, and limb use during locomotion on limb design. The data derived from such study can be applied to the reconstruction of the locomotor repertoires of extinct forms, such as the adapids and omomyids, allowing a reconstruction of the acquisition and radiation of modern primate locomotor repertoires.

Acknowledgments K. Izard, L. Alterman, and G. Doyle are gratefully acknowledged for organizing the Creatures of the Dark Conference. I sincerely thank the curators and staffs of the museums where I studied and radiographed strepsirhine limb bones. I am indebted to K. Glander and the staff of the Duke University Primate Center (DUPC) for facilitating the behavioral research. B.T. Shea and D. Martin allowed me to use video recording and playback equipment. M. Dagosto provided comments on the manuscript and has clarified and improved the research upon which this report is based. A.C. Walker allowed me to study his films of lemurs and galagos. Special thanks go to B.S. Coffman for orchestrating my use of animals at the DUPC and drawing the figures in the text. Constructive comments were offered by the editors and reviewers. This research was supported, in part, by Sigma Xi, Northwestern University Graduate School, and NSF DBS 9211930. This is publication #586 of the DUPC.

REFERENCES

Biewener, AA., 1983, Allometry of quadrupedal locomotion:the scaling of duty factor,bone curvature and limb orientation to body size, *Journal of experimental Biology* ,*105*,147-171.

1989 Scaling body support in mammals: limb posture and muscle mechanics, *Science, 245,*45-48.

1990 Biomechanics of terrestrial locomotion, *Science, 250,*1097-1103.

Biewener, AA and R. Taylor, 1986, Bone strain: a determinant of gait and speed? *Journal of experimental Biology, 123,*383-400.

Biewener, AA; JJ Thomason; LE Lanyon, 1983, Mechanics of locomotion and jumping in the forelimb of the horse (Equus):*in vivo* stress developed in the radius and metacarpus, *Journal of Zoology, London, 201,* 67-82.

Bertram, JEA and AA Biewener, 1988, Bone curvature:sacrificing strength for load predictability? *Journal of theoretical Biology,131,* 75-92.

Burr, DB; G Piotrowski; RB Martin; P Nong Cook, 1982, Femoral mechanics in the lesser bushbaby (*Galago senegalensis*): structural adaptations to leaping in primates, *Anatomical Record, 202,* 419-429.

Burr, DB; CB Ruff; C Johnson, 1989, Structural adaptations of the femur and humerus to arboreal and terrestrial environments in three species of Macaque, *American Journal of Physical Anthropology, 79,* 357-367.

Charles-Dominique, P,1977, *Ecology and Behavior of Nocturnal Primates*, New York: Columbia University Press.

Cowin, SC, LE Lanyon, and G Rodan, 1984, The Kroc Foundation Conference on functional adaptation in bone tissue, *Calcified Tissue International* 36:S1-S6.

Currey, JD and R McN Alexander, 1985, The thickness of the walls of tubular bones, *Journal of Zoology, London,* (A)*206,* 453-468.

Dagosto, M and CJ Terranova, 1992, Estimating the body size of Eocene primates: a comparison of results from dental and postcranial variables, *International Journal of Primatology, (13)3,* 307-344.

Demes, B and MM Gunther,1989, Biomechanics and allometric scaling in primate locomotion and morphology, *Folia Primatologia, 53,* 125-141.

Demes, B and WL Jungers, 1989, Functional differentiation of long bones in lorises, *Folia Primatologia, 52,* 58-69.

Demes, B. and WL Jungers,1993, Long bone cross-sectional dimensions, locomotor adaptations and body size in prosimian primates, *Journal of human Evolution, 25,* 57-74.

Demes, B; WL Jungers; K Selpien, 1991, Body size, locomotion, and long bone cross-sectional geometry in indriid primates, *American Journal of Physical Anthropology, 86,* 537-547.

Gunther, MM
1985 Biomechanische voraussetzungen beim Absprung des Senegalagos, *Zeitschrift fur Morphologie und Anthropologie, 75,* 287-306.

Gunther, MM, H Ishida, H Kumakura and Nakano, 1991, The jump as a fast mode of locomotion in arboreal and terrestrial biotopes, *Zeitschrift fur Morphologie und Anthropologie, 78(3),* 341-372.

Jungers, WL, 1985, Body size and scaling of limb proportions in primates. In WL Jungers, (Ed.), *Size and Scaling in Primate Biology,* (pp.345-381) New York: Plenum Press.

Jungers, WL and RJ Minns, 1979, Computed tomography and biomechanical analysis of fossil longbones, *American Journal of Physical Anthropology, 50,* 285-290.

Lanyon, LE and CT Rubin, 1985, Functional adaptation in skeletal structures. IN: M. Hildebrand, DM Bramble, KF Liem, and DB Wake, (Eds.), *Functional Vertebrate Morphology,* (pp.1-25). The Belknap Press, Cambridge, Massachusettes..

Nash, LT, SK Bearder and TR Olson,1989, Synopsis of *Galago* species characteristics, *International Journal of Primatology, (10)1,* 57-80.

Napier, J and AC Walker, 1967, Vertical clinging and leaping: a newly recognized category of locomotor behaviour of primates, *Folia Primatologia, 6,* 204-219.

Oxnard, CE, RH Crompton and SS Lieberman,1990, *Animal Lifestyles and Anatomies,* University of Washington Press.

Preuschoft, H,1985, On the quality and magnitude of mechanical stresses in the locomotor system during rapid movements, *Zeitschrift fur Morphologie und Anthropologie, 75,* 245-262.

Roark,RJ and WC Young. 1976, *Formulas for Stress and Strain,* McGraw-Hill.

Ruff, CB 1987 Structural allometry of the femur and tibia in Hominoidea and *Macaca*, *Folia Primatologia, 48,* 9-49.

1988 Hindlimb articular surface allometry in Hominoidea and *Macaca*, with comparisons to diaphyseal scaling, *Journal of Human Evolution, 17,* 687-714.

Ruff, CB and JA Runestad, 1992, Primate limb bone structural adaptations, *Annual Reviews of Anthropology 21,* 407-433.

Runestad, JA, CB Ruff,JC Nieh, RW Thorington,JR and MF Teaford, 1993, Radiographic estimation of long-bone cross-sectional geometric properties, *American Journal of Physical Anthropology, 90,* 207-213.

Schaffler, MB; DB Burr; WL Jungers; CB Ruff,1985, Structural and mechanical indicators of limb specialization in primates, *Folia Primatologia, 45,* 61-75.

Sokal, RR and FJ Rohlf,1989, *Biometry*, San Francisco: Freeman and Company.

Terranova, CJ, 1994, Leaping behaviors and the cross-sectional geometry of strepsirhine primate long bones. Ph.D. Thesis, Northwestern University.

Wainwright, SA, WD Biggs, JD Currey, and JM Gosline, 1976, *Mechanical Design in Organisms*, New Jersey: Princeton University Press.

Walker, AC, 1979, Prosimian locomotor behavior. In: G Doyle and RD Martin, (Eds), *The Study of Primate Behavior*, London: Duckworth.

LOCOMOTOR ADAPTATIONS OF EOCENE PRIMATES: ADAPTIVE DIVERSITY AMONG THE EARLIEST PROSIMIANS

Herbert H. Covert
Department of Anthropology
University of Colorado, Boulder
Boulder, Colorado 80309-0233

INTRODUCTION

Since the initial discovery of Eocene primates well over a century ago it has been widely recognized that these animals more closely resemble the prosimians than any other group of living creatures. This fact, along with the realization by evolutionary biologists that form - function - adaptive role relationships among extant creatures provide the basis for making inferences about the adaptations of extinct creatures, indicates that to understand the adaptations of the Eocene primates we must have a firm understanding of the adaptations of extant prosimians. For much of this century our ability to make suggestions about the adaptations of Eocene primates was hampered by the lack of skeletal material for the vast majority of these taxa and the lack of field observations on extant prosimian primates. This has changed dramatically during the past 35 years. Samples of post-cranial remains of Eocene primates have improved significantly and there is now a tremendous amount of information on the ecology of the extant prosimians so that we are now able to offer a range of informed suggestions about the behavior of these earliest primates. Extensive reviews of the ecology and behavior of the extant prosimians can be found in Martin *et al.* (1974), Charles-Dominique *et al.* (1980), Fleagle (1988), and this volume. Recent reviews of some of this fossil material can be found in Covert (1986), Fleagle (1988), Gebo (1988), Gebo *et al.* (1991), Conroy (1990), and Dagosto (1993).

The Adapids and Omomyids

The vast majority of fossil primates dating to the Eocene are placed into the families Adapidae or Omomyidae. These are the earliest occurring animals to show a wide suite of primate shared-derived features including flattened nails, an opposable hallux, a postorbital bar, and a petrosal bulla. Each of these families experienced diverse and successful adaptive radiations. These radiations are best known from samples collected in the western interior of the United States and western Europe with much smaller samples known from Asia and Africa. Adapids have been traditionally referred to as "lemur-like" and omomyids as "tarsier-like" with these terms being indicative of both phylogenetic relations and behavioral reconstructions. As outlined below, and as discussed by Covert and Williams (1994), these terms have little utility today as our understanding of the biology of both the extant and extinct prosimians has grown.

Locomotor Diversity Among Extant Prosimians

Extant prosimians show a wide range of locomotor behaviors. Field research on these creatures during the past 35 years has clearly demonstrated that these animals represent a series of successful adaptive radiations and that locomotion is one of the ways in which they often significantly diverged from one another. The anatomy of these creatures has also been studied in great detail during this time period. One of the primary goals of this research has been to gain an understanding of the relationship among specific anatomical features and locomotor behaviors. The combination of field and laboratory research has yielded a good understanding of the wide range of locomotor adaptations shown by these creatures. This information also serves as baseline data for research on locomotor adaptations of the Eocene primates. I briefly review some of the distinctive locomotor patterns shown by extant prosimians to remind the reader of the impressive diversity of these creatures.

The following information is drawn primarily from Walker (1974, 1979), MacArdle (1981), Tattersall (1982), Gebo (1987a, 1987b, 1989), Gebo and Dagosto (1989) and Fleagle (1988).

Lemuridae. *Lemur* -- lemurs [diurnal, cathemeral]. These animals are all fairly large and active arboreal quadrupeds. Running is the predominant mode of locomotion but leaping and climbing are also fairly common for each of these species. The ring-tailed lemur is unusual among these creatures in spending a significant amount of time on the ground.

Varecia -- ruffed lemurs [diurnal]. Ruffed lemurs are also arboreal quadrupeds. They appear to leap slightly less frequently and climb slightly more frequently than do the other lemurs.

Hapalemur -- bamboo lemurs [diurnal, crepuscular]. These animals are more specialized in locomotor and positional behavior than are the other members of this family. While they are arboreal quadrupeds, they also spend a great deal of time clinging to vertical substrates. They are frequent and powerful leapers.

Indriidae. *Avahi* -- wooly lemurs [nocturnal]. *Propithecus* -- sifakas [diurnal]. *Indrii* -- indriis [diurnal]. Although there is a substantial body size range among these creatures (wooly lemurs weigh less than a kilogram whereas indriis weigh ten times as much) each of these arboreal creatures are acrobatic and powerful leapers that often feed and rest in upright postures. They are also more frequent climbers on vertical substrates than are most other prosimians.

Daubentoniidae. *Daubentonia* -- aye-aye [nocturnal]. The primary locomotor mode utilized by the aye-aye is fairly deliberate arboreal quadrupedalism.

Lepilemuridae. *Lepilemur* -- sportive lemurs [nocturnal]. The arboreal sportive lemur appears to closely resemble indriids in locomotor and postural adaptations. They are acrobatic leapers and often cling to vertical substrates.

Cheirogaleidae. *Microcebus* -- mouse lemurs [nocturnal]. These creatures are arboreal quadrupeds that walk, run, and leap through the small substrate found in the undercanopy of the Malagasy forests. *Mirza* and *Cheirogaleus* -- dwarf lemurs [nocturnal]. These creatures are also arboreal quadrupeds that walk, run, and leap. It appears that *Mirza* and the mouse lemurs are more acrobatic in their locomotor behavior than is *Cheirogaleus*. *Phaner* -- fork-marked lemurs [nocturnal]. These creatures are specialized gum foragers. They are arboreal quadrupeds that have keeled and claw-like nails allowing them to move on relatively large vertical substrates.

Lorisinae. All members of the subfamily Lorisinae are highly specialized nocturnal slow moving arboreal quadrupeds. They do, however, vary in the types of substrate and canopy level they prefer to inhabit.

Loris -- slender loris and *Arctocebus* -- angwantibo. These creatures appear to prefer to move on small substrates such as the vines and small branches of the forest's understory.

Nycticebus -- slow loris and *Perodicticus* -- potto. These animals appear to prefer the primary canopy and spend most of their time on fairly large substrates.

Galaginae. All members of this nocturnal subfamily show specializations of the hindlimb and particularly the foot that are associated with leaping. While all are capable leapers there is a good deal of variation among them in the frequency in which they do leap and in their habitat preference.

Galago -- Senegal bushbabies. This is an acrobatic leaper who prefers to move about in the small substrates of the lower forest levels.

Otolemur -- thick-tailed bushbabies. They walk, run, and leap in the lower forest levels. They tend to move on larger substrates than the senegal bushbaby.

Euoticus -- needle-clawed bushbabies. These creatures resemble the fork-marked lemur in having keeled nails that are used to clinging to large substrates. These bushbabies leap and run through a range of forest levels.

Galagoides -- dwarf bushbabies. These tiny creatures prefer the dense foliage of various forest levels. They are active runners and leapers.

Tarsiidae. *Tarsius* -- tarsiers [nocturnal]. This family includes a single extant genus with four species. Each of these creatures are highly specialized acrobatic leapers. They also frequently cling to vertical substrates. In fact, it is likely that tarsiers are the most specialized extant primate when locomotion, diet, neurological system, and social system are considered.

Important Anatomical Correlates of Locomotion

Arboreal Features. The majority of extant primates are arboreal and possess a wide range of features associated with this habitat. Long and dexterous digits are adaptations to an arboreal environment and function particularly well for locomotion on fairly small substrates. Terrestrial primates tend to show some reduction in digit length. Arboreal primates have shoulder, elbow, wrist, hip, and ankle joints that are more mobile than those of most nonprimate mammals. This mobility allows primates to move with ease across a range of uneven substrates that characterize the arboreal environment. Terrestrial primates often reduce the mobility in at least some of these joints. Primates also have a mobile transverse tarsal joint and a large opposable hallux. These allow them to invert or evert the plantar surface of the foot and to grasp onto the uneven substrates they constantly encounter. Terrestrial primates also frequently reduce the size of the hallux.

These anatomical features associated with arboreality among extant primates provide paleontologists important data for interpreting locomotor adaptations among the prosimians of the Eocene. Specifically, fossil primate remains that illustrate these arboreal correlates are assumed to have been arboreal.

Features that Separate Frequent Leapers from less Frequent Leapers. Leaping is a common adaptation among extant prosimian primates. Frequent and powerful leaping and associated anatomical features have evolved on a number of occasions within this group.

In both general and specific comparisons leapers tend to have longer hindlimbs than non-leapers (and frequent leapers tend to have even longer hindlimbs than occasional leapers). Thus the frequency of leaping is inversely related to the intermembral index value (Napier and Walker, 1967; Fleagle, 1977, 1988; MacArdle, 1981; Anemone, 1988, 1990, 1993).

Femoral Morphology. Leaping primates consistently differ from their close relatives that leap less frequently in a number of femoral features. Frequent leapers have a relatively long femur, proximally located trochanters, a short femoral neck, and deep femoral condyles that project ventral to the shaft. The functional relationship of each of these features to leaping has been discussed in detail (e.g. Fleagle, 1988; Anemone, 1990, 1993).

Tibial Features. There are also a number of features of the tibia that tend to be associated with leaping among primates. A mediolaterally narrow tibia is common for leapers and is probably an adaptation to counteract anteroposterior bending. The distal tibiofibular facet is often relatively long among leapers, subsequently decreasing rotation of the fibula around the tibia (Fleagle, 1988).

Calcaneal Proportions. Leaping primates often have a calcaneus that is lengthened anteriorly. This serves to increase the lever arm for leaping. Leapers also tend to have a straight plantar calcaneal surface, and a shorter, more convex posterior calcaneal facet (Gebo, 1988).

Additional Tarsal Features. Gebo (1988) has recently provided a detailed review of the foot morphology of prosimians in relation to locomotor variation. He noted that the talus of leaping prosimians is often distinctive in a number of features. It tends to have a long straight neck, a high talar body, a deeply grooved talar trochlear, and parallel trochlear rims. He also noted that the navicular is often lengthened among leaping prosimians.

This long series of anatomical features often associated with leaping among extant prosimian primates (and other primate groups) provide paleontologists important data for interpreting locomotor adaptations among Eocene primates.

Features associated with specialized careful and slow quadrupedalism. Dagosto (1983) and Gebo (1988) have identified a number of features shared by the lorisines that appear to be adaptations for slow and careful quadrupedal climbing. Dagosto generated an extensive list of such features, but only a few will be mentioned here. Lorisines tend to exhibit increased mobility in their joints, particularly for nonsagittal movements. Their hindlimbs are not elongated compared to their forelimbs. The humeral bicipital groove is shallow and the deltopectoral crest is thick. The femoral head is oriented proximomedially and the greater trochanter is equal in height to the femoral head. The distal femur is as wide if not wider than it is deep.

Gebo's (1988) review included a list of tarsal features that are related to climbing including a calcaneus that tends to have an elevated posterior calcaneal facet, a plantar surface that is strongly curved anteroposteriorly, a short anterior region, and a long midcalcaneal region. Gebo characterized the talus of these creatures as having a dished trochlea; a dorsoplantarly narrow, mediolateral wide, and plantarflexed head; and low body. He also noted that these creatures have a short navicular.

As with the features associated with leaping, these associated with climbing provide paleontologists a rich data base for interpreting the locomotor adaptations of Eocene primates.

Although some prosimians are highly specialized leapers (i.e. tarsiers, senegal bushbabies, and indriids) and some are highly specialized slow moving arboreal quadrupeds (i.e. lorisines), most utilize a wide range of locomotor behaviors. As noted by Gebo (1988), the morphology of less specialized prosimians is often intermediate between the more specialized leapers and slow movers. This is presumably related to the wide range of locomotor behaviors utilized. Thus, as we analyze the locomotor adaptations of extinct creatures we need to be cognizant of the possibility that they were fairly generalized in adaptation.

MATERIALS AND METHODS

During the past 15 years I have examined collections of Eocene mammals housed in museums in both North America and Europe. Since 1987 I have also collected an extensive sample of these creatures from the Washakie Basin of Wyoming for the University of Colorado Museum of Natural History. The following discussion focuses on analyses of primate postcrania that occur in these collections (Table 1).

The Comparative Method

I believe that the best tool for understanding the adaptations of extinct animals is the comparative method. If we have a sound understanding of the biology of extant animals and if there are interpretable patterns of anatomical traits and adaptive roles among these creatures, we can use this information to make suggestions about the biology of extinct creatures. Such research is not without problems, many of which are discussed in some detail by Kay and Cartmill (1977), Kay and Covert (1984), and Anthony and Kay (1993). However, the comparative method is an important tool for understanding the behavior of extinct animals if utilized in the following way. One must carefully consider the functional relationship between a specific morphological feature and the adaptive role of interest. If one finds that there is a consistent or nearly consistent association of the adaptive role with the specific morphological feature (but not necessarily the reverse) this adaptive role can be attributed to extinct animals that are characterized by this specific morphological trait. An understanding of these relations with a consideration of the body size of the animals of interest allows researchers to make informed and plausible suggestions about the adaptations of extinct creatures.

DISCUSSION

Locomotion is only one part of the ecology of a primate. I find it useful to consider it in combination with other aspects of the ecology of a creature such as body size, activity pattern, and diet when possible. I briefly mention these attributes below so that a better understanding of the adaptive profile of these Eocene primates is gained.

It has been noted by many researchers during the past few years that various body parts significantly scale with body size for various mammalian groups including primates (e.g. Kay and Simons, 1980; Gingerich *et al.*, 1982). Thus, body size estimates can be generated for extinct primates by assuming they scaled in a way similar to extant primates. Here I use adapid body size estimates published by Fleagle (1988). Estimation of body size for omomyids appears to be more challenging than for adapids. As noted by Gingerich (1981), omomyids may have had teeth that were relatively large for their body size. Dagosto and Gebo (pers. comm.) have also noted this and analysis of omomyid post cranial remains in our collections support this position. Some of the body size estimates used here are from Dagosto and Terranova (1992), others are generated from Gingerich's (1981) equation.

Suggestions of the activity patterns of Eocene primates are based on the size of their orbits. As noted by Cartmill (1970) and Kay and Cartmill (1977), among modern primates, nocturnal forms have relative large orbits whereas diurnal forms have relatively small ones. If we assume that this relationship held in the past we can plot the orbit size of extinct primates with data on extant forms. If the orbit of the extinct primate is relatively large given its skull size I suggest it was nocturnal and if the orbit is relatively small I suggest it was diurnal.

Hypotheses about the diets of these creatures are based on consideration of shearing capabilities of their molars in relation to their estimated body size. Among extant prosimians (and other primates) insectivorous and folivorous creatures tend to have better developed shear on their molars than do frugivores. Furthermore, insectivorous primates are small creatures, usually less than 300 g in size, whereas folivores are moderate to large creatures with a body mass of more than 800 g. By estimating the amount of shear development on the molars of an extinct primate (generating a shear quotient -- see Kay and Covert, 1984 and Covert, 1986) one can determine if a creature had relatively more or less shear than expected given its molar size. If it had less shear than expected a frugivorous diet can be attributed to it. If the animal had more shear than expected and a small body size (less than 300 g) it was probably insectivorous. If the animal had more shear than expected and appears to have been a moderate to large-sized creature (greater than 800 g) it was probably folivorous.

By combining these suggestions about body mass, locomotor adaptation, dietary specialization and activity patterns we can generate a series of adaptive profiles for these earliest primates. What we see in the following discussion is that the omomyids and adapids of the Eocene experienced a broad adaptive radiation in both North America and Europe (Table 2 and 3).

TABLE 1. Known skeletal material of Eocene primates.

Adapidae

Cantius ralstoni talus, calcaneus, hallucial metatarsal, phalanges Gebo *et al.* (1991)
Cantius mckennai distal tibia, calcaneus, talus, entocuneiform, hallucial metatarsal, cuboid, navicular, phalanges Covert (1985, 1988) Gebo *et al.* (1991)
Cantius trigonodus Humerus, radius, ulna, pelvis, femur, tibia, calcaneus, talus, navicular, cuboid, entocuneiform, hallucial metatarsal, Matthew (1915) Rose and Walker (1985) Gebo *et al.* (1991)
Cantius abditus Femur, tibia, talus, calcaneus Covert (1983, 1985) Gebo *et al.* (1991)
Cantius frugivorus Humerus, pelvis, femur, tibia, calcaneus, talus, entocuneiform, first metatarsal Covert (1983, 1986) Gebo *et al.* (1991)
Cantius venticolus Capitate, calcaneus, talus, cuboid, entocuneiform Hamrick and Covert (1991), Gebo *et al.* (1991)
Copelemur australotutus calcaneus, talus Covert (1993)
Notharctus tenebrosus Skeleton Gregory (1920)
Notharctus robustior Most skeletal elements Gregory (1920)
Smilodectes gracilis Skeleton Covert (1985)
Adapis parisiensis Humerus, femur, tibia, calcaneus, talus, metatcarpals, metatarsals, phalanges carpals Decker and Szalay (1974), Dagosto (1983), Beard and Godinot (1988)
Leptadapis magnus Humerus, femur, tibia, calcaneus, talus, cuboid, metacarpals, metatarsals, phalanges Decker and Szalay (1974), Dagosto (1983)
Adapidae Three partial skeletons Franzen and Koenigswald (1988)
Europolemur koenigswaldi forelimb Franzen and Koenigswald (1988)
Pronycticebus neglectus partial skeleton Thalmann *et al.* (1989)

Omomyidae

Teilhardina belgica talus, calcaneus Szalay (1976)
Tetonoides n. sp. talus, calcaneus Covert and Williams (1994)
Tetonoides pearcei talus Covert and Williams (1994)
Arapahovius gazini femur, tibia, talus, calcaneus, navicular, cuboid Savage and Waters (1978) Williams *et al.* (1991)
Absarokius abbotti humerus, tibia, talus, calcaneus, navicular Covert and Hamrick (1993)
Hemiacodon gracilis humerus, vertebrae, pelvis, femur, tibia, talus, calcaneus, cuboid, entocuneiform, first metatarsal Simpson (1940), Dagosto (1993)
Omomys cf. *carteri* Rosenberger and Dagosto (1992), Covert and Murphey (1994)
Necrolemur zitteli calcaneus Schmid (1979)
Necrolemur sp. humerus, femur, tibiofibula, calcaneus Hurzeler (1948), Szalay and Dagosto (1980)
Microchoerus erinaceus humerus, femur, calcaneus Schmid (1979), Szalay and Dagosto (1980), Dagosto (1993)
Nannopithex raabi partial hindlimb Weigelt (1933)

Adaptive Diversity Among Omomyidae

Skeletal elements are known for only a handful of the more than 70 recognized species of Omomyidae (Table 1). Recent collecting, however, has begun to yield more substantial samples of such material. The following discussion outlines the adaptive profiles of five of the best known members of this family.

Absarokius. While this is usually the most commonly occurring omomyid in middle and late Wasatchian collections in North America, nothing was known about its locomotor adaptations until this decade. During the past few field seasons the University of Colorado research teams have collected a handful of specimens that are assigned to this genus with a good deal of confidence. Covert and Hamrick (1993) described the distal humerus, tibia, talus, and calcaneus of this creature. We also now have a navicular and an additional calcaneus that are best assigned to *Absarokius*. As described by Covert and Hamrick, the humeral fragment of this creature has a cylindrical humeral trochlea separated from a spherical capitulum by a distinct groove, as do most early primates. These features are associated with a mobile elbow that was capable of a wide range of flexion, extension, supination, and pronation. This range of mobility is characteristic of arboreal creatures. The tibia of *Absarokius* is quite interesting in showing an extensive distal tibiofibular articulation. Among extant primates only *Tarsius* shows such an extensive contact between these elements (*Tarsius* has a fused distal tibiofibula). As described by Fleagle and Simons (1983), such an extensive tibiofibular articulation is associated with leaping among extant primates. Covert and Hamrick concluded that this morphology indicated that *Absarokius* was a frequent leaper. They described the talus of this creature as having a fairly long, straight neck and high body with parallel trochlear rims. Each of these features are associated with leaping among extant prosimians (Gebo, 1988). Covert and Hamrick described the *Absarokius* calcaneus as showing a moderate amount of anterior elongation, most closely resembling that of *Microcebus murinus*. The calcaneus also has a short and strongly convex posterior calcaneal facet. These are features that Gebo (1988) has found to be associated with leaping among prosimians. On the other hand, however, the plantar surface of this calcaneus is moderately curved. In this feature, this element is more similar to that of climbing prosimians than to that of leapers. The navicular of *Absarokius* is distinctive in being fairly long, another feature that indicates this creature was a frequent leaper. Finally, the articulations among the calcaneus, talus, and navicular indicate that *Absarokius* had a mobile transverse tarsal joint characteristic of most arboreal primates.

In sum, the skeletal elements of *Absarokius* indicate this was an arboreal creature that was probably a frequent leaper and runner. Of the extant prosimians it most closely resembles the mouse lemur in overall morphology. It differs only in minor details that may indicate slightly more frequent leaping by this early Eocene primate. Based on the size of its dentition this creature had a body mass of approximately 200 g; its relatively large orbits indicate it was probably nocturnal, and and its low crowned molars suggest it was probably primarily frugivorous (Table 2).

Arapahovius gazini. This middle Wasatchian omomyid is known only from the Washakie Basin of Wyoming. The fossil sample includes well preserved dentitions and a handful of skeletal elements including the proximal femur, distal tibia, talus, calcaneus, cuboid, and navicular. The femur is of particular interest because it represents the earliest known example of this element for this family. It resembles that of middle Eocene omomyids, *Hemiacodon* and *Omomys* (see below) in having a cylindrical head placed on a short neck that is oriented perpendicular to the femoral shaft. This morphology is associated with frequent leaping among extant primates. The tarsal elements of this taxon are each somewhat elongated and resemble those of cheirogaleines in overall morphology. They also resemble those of *Absarokius*, possibly differing in showing slightly less elongation. It is likely that *Arapahovius* was a frequent runner and leaper. Based on the size of its dentition it had a body mass of approximately 200 g and its low crowned molars indicate that it was probably frugivorous (Table 2).

Hemiacodon gracilis. Simpson (1940) described a fragmentary skeleton of the middle Eocene *Hemiacodon gracilis* that included a few vertebrae, most of the left side of the pelvis, proximal and distal femora and tibiae, a calcaneus, navicular, cuboid, entocuneiform, and hallucial metatarsal. We also have a talus that can be attributed to this taxon in the University of Colorado Museum's collections. Like the omomyids described above, the tarsals of *Hemiacodon* show some anterior elongation. The femoral fragments exhibit a cylindrical femoral head on a short neck oriented perpendicular to the shaft and an anteroposteriorly deep femoral condyle with a narrow patellar trochlea raised anteriorly off of the femoral shaft. As Dagosto (1993) stated, these femoral features resemble those of both *Galago* and *Tarsius*. She further notes that the femoral head of *Hemiacodon* is slightly more similar to that of *Tarsius* than that of *Galago* in not being as flattened on its dorsal and posterior aspects. The distal tibial fragments are noteworthy because they preserve an extensive distal tibiofibular articular facet. The surface of this facet is rough and indicative of a syndesmotic joint (Dagosto, 1993) and one in which the fibula could not rotate on the tibia. In most aspects of its preserved elements *Hemiacodon* expresses morphology characteristic of arboreal running and some leaping (resembling the locomotion of the cheirogaleines). The morphology of its femur and distal tibia, however, is indicative of more frequent leaping than that shown by the cheirogaleines. This creature probably had a body mass of between 250 and 400 g (based on tarsal size, Dagosto and Terranova, 1992), a frontal of this taxon indicates that it had relatively large orbits and was probably nocturnal. Its molars have fairly well developed shear and are indicative of a diet that included substantial amounts of insects and fruit (Table 2).

Table 2. Adaptive profiles for various omomyids.

Taxon	*Absarokius*	*Arapahovius*	*Hemiacodon*
Body size	200 g	200 g	250-400 g
Diet	Frugivorous	Frugivorous	Frugivorous - Insectivous
Locomotion	Active arboreal runner and leaper	Active arboreal runner and leaper	Active arboreal runner and leaper
Activity Pattern	Nocturnal	?	Nocturnal
Analog	*Microcebus/Cheirogaleus*	*Microcebus/Cheirogaleus*	*Cheirogaleus*

Taxon	*Omomys*	*Necrolemur*
Body size	200-350 g	200-350 g
Diet	Frugivorous - Insectivorous	Frugivorous
Locomotion	Active arboreal runner and leaper	Active arboreal leaper
Activity Pattern	Nocturnal	Nocturnal
Analog	*Cheirogaleus*	*Galagoides alleni*

***Omomys* cf. *carteri*.** During the summer of 1993, Dr. Peter Robinson made the largest collection of skeletal elements known for a single omomyid species to date. This material represents fragmentary skeletons of six or so individuals of the middle Eocene genus *Omomys* and includes multiple fragments of the scapula, humerus, ulna, radius, pelvis, femur, tibia, fibula, calcaneus, talus, cuboid, navicular, entocuneiform, and hallucial metatarsal. Although analysis of this material is in preliminary stages the following observations are noteworthy. The humeral head is nearly spherical and tapers slightly distally as do those of *Microchoerus* and *Necrolemur* (see Dagosto, 1993). The humeral head also projects above the humeral tubercles indicating that this creature had a shoulder that was capable of a wide range of motion. The distal humerus is similar to that of other omomyids in overall morphology and is indicative of a wide range of pronation and supination at the elbow. The hindlimb has a number of features commonly associated with leaping including a cylindrical femoral head, an anteroposteriorly deep distal femur, a mediolaterally compressed tibial shaft, a fairly extensive distal tibiofibular articulation, and somewhat elongated tarsals. This combined morphology indicates that *Omomys* was an active arboreal runner, leaper, and climber. In tarsal morphology *Omomys* most closely resembles the cheirogaleines whereas in the femoral and tibial morphology it more closely resembles the acrobatic leaping galagos and tarsiers. The tarsal and limb bones of this creature indicate that it had a body mass in the 200 - 350 g range. In addition, maxillary fragments indicate that this creature had relatively enormous orbits indicating a nocturnal activity pattern and fairly well developed shear on its molars indicative of a diet with substantial amounts of insects and fruit (Table 2).

Necrolemur. This omomyid is a fairly common element of the late Eocene faunas of Europe. It is known from various hindlimb elements including the tibiofibula, calcaneus, and talus (Schmid, 1979; Godinot and Dagosto, 1983; and Dagosto, 1985). The morphology of the tibiofibula and calcaneus is quite interesting because it appears to show evidence of a highly specialized leaping adaptation for this genus. The tibia and fibula are distally fused into a single element resembling that of tarsiers. The calcaneus shows significant distal elongation resembling galagos and tarsiers. Both of these features are thought to be important functional correlates of leaping and thus indicate that *Necrolemur* was a specialized leaper. *Necrolemur* had a body mass in the 200 - 350 g range, its orbits are relatively large indicating a nocturnal activity pattern and poorly developed shear on its molars is indicative of a frugivorous diet (Table 2).

Adaptive Diversity among the Adapidae

The skeleton of adapid primates is better known than that of omomyids and as with omomyids, recent discoveries have greatly expanded our knowledge about the range of locomotion shown by the adapids during the Eocene (Table 1).

Cantius mckennai. This is a commonly occurring adapid from early Eocene deposits of North America. While the vast majority of specimens are dental remains, the sample does include a distal tibia, calcanei, tali, cuboid, naviculares, entocuneiform, and first metatarsal (Covert, 1988). The morphology of these elements indicates that this creature had a mobile ankle, a transverse tarsal joint that allowed a great deal of inversion, an opposable first digit, and grasping digits. This, combination, suggests that *C. mckennai* was arboreal. The preserved features are not indicative of either acrobatic leaping or specialized slow movement. *Cantius mckennai* appears to have had a body mass of approximately 1,500 g. The molars had relatively poorly developed shearing features that are indicative of a frugivorous diet (Covert, 1986).

Cantius trigonodus. Rose and Walker (1985) described a partial skeleton of this early Eocene North American primate. These authors note that its hindlimb was substantially longer than its forelimb. Humeral, radial, and ulnar fragments are indicative of powerful forelimb musculature. Joints of the hip and the foot show the enhanced mobility characteristic of arboreal creatures. In overall morphology, the skeleton of *Cantius trigonodus* is quite similar to those of the middle Eocene notharctines *Notharctus* and *Smilodectes* (see below) and Rose and Walker concluded that it was an active arboreal quadruped with a propensity for leaping. I agree with these conclusions in general but note that the distal femur of *C. trigonodus* does not show the characteristic morphology of a frequent leaper to the extent of its middle Eocene relatives. Members of this species had a body mass of approximately 1,700 g. The molars had poorly developed shearing capabilities indicating that it was primarily frugivorous (Covert, 1986; Table 3).

Notharctus tenebrosus. This creature was a common member of the middle Eocene mammalian fauna of North America and is known from a number of partial skeletons collected in the Bridger Basin of Wyoming. Its anatomy was described in detail by Gregory (1920) and his observations remain the authorative source on this material. Both forelimb and hindlimb morphology of this creature closely resembles the larger extant Malagasy primates such as the indriids and *Lemur.* Most joints indicate a wide range of motion characteristic of arboreal creatures. In addition, *Notharctus* shows a handful of features that are often associated with leaping including a hindlimb that is much longer than the forelimb (intermembral index = 59.6) and a narrow patellar trochlear floor markedly raised above the femoral shaft (Covert, 1985). In a very general way the forelimb is similar to that of the extant large-bodied Malagasy primates however, it is distinctive in a handful of features. The muscles crossing the shoulder from the thorax to the humerus insert relatively distally and the origin of brachioradialis is situated relatively proximally on the shaft of the humerus. These features indicate that the arm of this creature was heavily muscled. This information, in combination with a humeral head that projects cranially relatively to the tubercles suggest that these animals may have been doing some forelimb suspension or powerful climbing that including hoisting-up the body with the forelimbs. Furthermore, the ulnar styloid process appears to be relatively short for these animals and may indicate that its wrist was slightly more mobile than that of the extant Malagasy primates. In sum, *Notharctus tenebrosus* was probably a frequent and powerful leaper that was also a capable climber. This creature had a body mass of approximately 3,100 g, it had relatively small orbits indicative of a diurnal activity pattern, and its molars had well developed shear crests that suggest a folivorous diet (Covert, 1986; Table 3).

Table 3. Adaptive profiles for various adapids.

Taxon	*Cantius trigonodus*	*Cantius venticolus*	*Notharctus tenebrosus*
Body Size	1,700 g	3,500 g	3,100 g
Diet	frugivorous	frugivorous	folivorous
Locomotion	Active arboreal runner and leaper	Active arboreal runner and leaper	Active arboreal leaper and runner
Activity Pattern	?	?	diurnal
Analog	*Lemur fulvus*	*Varecia*	*Propithecus*
Taxon	*Smilodectes gracilis*	*Adapis parisiensis*	*Pronycticebus neglectus*
Body Size	1,950 g	1,500 g	800 - 900 g
Diet	folivorous	folivorous	folivorous
Locomotion	Active arboreal leaper and runner	Slow, deliberate arboreal quadruped	Active arboreal runner and leaper
Activity cycle	diurnal	diurnal	nocturnal
Analog	*Propithecus*	?	?*Lepilemur*?

Smilodectes gracilis. This is another North American middle Eocene adapid known from nearly complete skeletons collected in the Bridger Basin. In overall morphology this creature closely resembles the large extant Malagasy primates such as the indriids and *Lemur*. In two features related to leaping (as with *Notharctus tenebrosus*, see above) *Smilodectes* particularly resembles the indriids. *Smilodectes* has an extremely low intermembral index (60.8) and a narrow patellar trochlear floor markedly raised above the femoral shaft (Covert, 1985). These features, along with the short femoral neck, proximally located femoral trochanters, and mediolaterally compressed tibia provide clear evidence that *Smilodectes* was a frequent and powerful leaper. As with *Notharctus tenebrosus*, *Smilodectes* had a heavily muscled arm, a somewhat cranially positioned humeral head, and a reduced ulnar styloid process. These features probably are indicative of frequent climbing and some forelimb suspensory behavior. *Smilodectes gracilis* was approximately 1,950 g in body mass, the orbits of this creature were not particularly large indicating that it was probably diurnal, and its molars had extremely well developed shear crests indicating a folivorous diet (Covert, 1986). In overall morphology, *Smilodectes* is quite similar to *Notharctus tenebrosus*, differing primarily in being a bit more gracile in limb structure and having better developed shearing molars. These differences probably indicate that *Smilodectes* was slightly more acrobatic in locomotor behavior and more folivorous in dietary adaptation than its close relative (Table 3).

Adapis parisiensis. This late Eocene European creature is known from a very large sample of dental remains, some beautifully preserved crania, and a substantial number of primarily isolated skeletal remains. Dagosto's (1983) description of the skeletal elements of this creature stands as an authorative statement on the locomotor behavior of this creature. Its hindlimbs and forelimbs were approximately of equal length and likely short relative to trunk length. In addition, it appears to parallel lorisines in a long series of skeletal features. The humeral bicipital groove is shallow and the deltopectoral crest is thick. The femoral head is oriented proximomedially and the greater trochanter is equal in height to the femoral head. The distal femur is as wide as it is deep. The talus is dorsoplantarly narrow, and mediolaterally wide. In each of these features (and others, see Dagosto, 1983) *Adapis parisiensis* appears to be adapted to careful and slow quadrupedalism. *Adapis parisiensis* was probably around 1,500 g in body mass, its orbits are relatively small indicating a diurnal activity pattern, and its molars have extremely well developed shear crests that document a folivorous diet (Covert, 1986; Table 3).

Pronycticebus neglectus. Thalmann *et al.* (1989) reported on a nearly complete specimen of the middle Eocene adapid primate from the Geiseltal brown coal of Germany. The skeleton of this animal appears to more closely resemble those of the North American adapids than that of *Adapis*. Specifically, it appears to lack any specialization for slow quadrupedalism. Further, in the anatomy of both fore- and hindlimb joints this animal appears to have been arboreal. In addition, Thalmann *et al.* (1989) estimate that the intermembral index of this creature would have been less than 73 and thus consistent with fairly frequent leaping behavior. It lacks the specializations of the extant slow quadrupeds (and those shown by *Adapis parisiensis*) or those of the extremely active leapers and thus appears to have been an active arboreal quadruped. These authors suggested that *P. neglectus* had a body mass of approximately 800 to 850 g. Its orbits appear to be relatively large (as with *P. gaudryi*) and was probably nocturnal. The molars of this specimen (the only specimen of this species that I am familiar with) are heavily worn but those of *P. gaudryi* have well developed shear crests indicative of a folivorous diet (Covert, 1986; Table 3).

Partial skeletons from the Messel brown coal. As described by Franzen and Koenigswald (1988) four partial adapid primate skeletons are known from the middle Eocene Messel brown coal deposits and are believed to be approximately the same age as *Pronycticebus neglectus*. Two preserve the hindlimb anatomy and two the forelimb. Only one preserves sufficient craniodental information to identify its taxonomy at a level more specific than the family Adapidae and this specimen is the type and only known sample of *Europolemur koenigswaldi*. The other three specimens are probably too large to belong to this species but all may belong to a single adapid species. The anatomy of these specimens most closely resemble that of *Cantius*. The hindlimb material appears to be fairly long and gracile. Neither fore- or hindlimb elements show specializations for slow quadrupedalism or extremely active leaping and thus these creatures were probably active arboreal quadrupeds. Franzen and Koenigswald (1988) reported that the orbit preserved for *E. koenigwaldi* was not particularly large and would be consistent with a diurnal activity pattern. The molars of this creature show well developed shearing, indicating that it was probably a folivore.

CONCLUSIONS

As noted in the introduction the adapids have long been labeled lemur-like and the omomyids as tarsier-like. Our present understanding of the ecology of extant prosimians allows us to test the implications of these labels. As is evident from the preceding reviews these labels are not accurate and thus are not appropriate for describing the ecological niches of the Eocene primates. Instead, both the omomyids and the adapids appear to resemble the modern prosimians in having experienced a dramatic and broad adaptive radiation. The omomyids ranged in size from approximately 35 to 800 g; some of these creatures appear to have been nocturnal whereas others were diurnal (*Rooneyia*, see Kay and Covert, 1984). While most of these creatures were primarily frugivorous, others appear to have been insectivorous, and while our information about the locomotor adaptations of these creatures is quite limited, the majority known from postcranial elements appear to have most closely resembled mouse and dwarf lemurs in locomotor adaptations (e.g. *Omomys* and *Absarokius*), with at least one genus (*Necrolemur*) being a frequent and acrobatic leaper similar to the tarsiers or smaller galagos (Table 2). The adapids ranged in size from approximately 100 to 6,000 g; most of these creatures appear to have been diurnal whereas some were nocturnal; while most of these creatures were primarily frugivorous, many appear to have been folivorous and a few insectivorous. As with the Omomyidae our information about the locomotor adaptations of these creatures is limited. Some of these creatures apparently resembled common lemurs in locomotor adaptations (e.g. *Cantius* and some of the Messel adapids) whereas others more closely resembled the indriids in locomotor adaptations (e.g. *Notharctus* and *Smilodectes*), and one species (*Adapis parisiensis*) most closely resembled the lorisids in locomotor adaptations (Table 3).

It has also been stated in this review that the omomyids and adapids most closely resemble the modern prosimians in their adaptations. This does not mean, however, that specific extinct species are nearly identical to the modern analogs in morphological detail or adaptation. Further, these similarities need not indicate anything specifically about phylogenetic relationships. For example, while *Omomys* most closely resembles *Mirza* in overall morphology and thus adaptive profile, its hip, knee, and ankle more closely resemble those of *Tarsius* and the more acrobatic members of *Galago*. *Omomys* is not a cheirogaleid and is quite unlikely that it is in or near the ancestry of this family. Likewise, while *Smilodectes* most closely resembles *Propithecus* in overall morphology and thus adaptive profile, its dentition is distinctively that of a notharctine and its forelimb is more robust than any indriid. *Smilodectes* is not an indriid and does not share a specific phylogenetic relationship with this family. Rather, *Omomys* and *Smilodectes* represent the adaptive radiations of the omomyids and adapids respectively that overlap broadly with the adaptive radiation exhibited by the extant prosimians.

ACKNOWLEDGMENTS:

I thank L. Alterman and K. Izard for inviting me to participate in "Creatures of the Dark". Special thanks go to D. Ayers-Darling, L. A. Jolley, and B. Wright for comments on an earlier version of this manuscript. I also thank D. Gebo, M. Dagosto, B. Williams, and M. Hamrick for informative discussions and comments on the morphology of early primates. I thank the following persons and institutions for access to skeletal material of extant and extinct primates in their collections: the American Museum of Natural History, G. Musser and R.H. Tedford; the National Museum of Natural History (Smithsonian Institution), R. Thorington and R. Emry; Yale Peabody

Museum, J.H. Ostrom; the Museum of Paleontology at the University of Michigan, P.D. Gingerich; the Museum of Comparative Zoology at Harvard University, F.A. Jenkins, Jr. and M. Rutzmoser; the U.S. Geological Survey (Denver) T.M. Bown; the University of Wyoming Geological Museum, J.A. Lillegraven; the Duke University Primate Center, E.L. Simons and P. Chatrath; the Museum National d'Histoire Naturelle, Paris, D.E. Russell; Naturhistorisches Museum, Basel, B. Engesser; and the Laboratoire de Paleontologie, University of Montpellier, M. Godinot. I am particularly grateful for the generosity of P. Robinson of the University of Colorado Museum for access to the *Omomys* skeletal material. Finally, I thank the Wyoming State Office of the Bureau of Land Management for permission to work in the Washakie Basin.

LITERATURE CITED

Anemone, R.L. (1988). The functional morphology of the prosimian hindlimb: Some correlates between anatomy and positional behavior. Ph.D. Dissertation, University of Washington, Univ. Microfilms, Ann Arbor, Michigan.

Anemone, R.L. (1990). The VCL hypothesis revisited: patterns of femoral morphology among quadrupedal and saltatorial prosimian primates. *American Journal of Physical Anthropology, 83*,373-393.

Anemone, R.L. (1993). The functional anatomy of the hip and thigh in primates. In D.L. Gebo (Ed.), *Postcranial Adaptation in Nonhuman Primates* (pp. 150-174). Dekalb: Northern Illinois University Press.

Anthony, M.R.L. and Kay, R.F. (1993). Tooth form and diet in ateline and alouattine primates: reflections on the comparative method. *American Journal of Science, 293-A*, 356-382.

Cartmill, M. (1970). The orbits of arboreal mammals: a reassessment of the arboreal theory of primate evolution. Ph.D. Dissertation, The University of Chicago, Univ. Microfilms, Ann Arbor, Michigan.

Charles-Dominique, P., Cooper, H.M., Hladik, A. Hladik, C.M., Pages, E., Pariente, G.F., Petter-Rousseaux, and Schilling, A. (1980). (Eds.), *Nocturnal Malagasy Primates*. New York: Academic Press.

Conroy, G.C. (1990). *Primate Evolution*. New York: Norton Press.

Covert, H. H. (1983). Newly discovered post-cranial remains of *Cantius*. *American Journal of Physical Anthropology, 60*,185-86.

Covert, H.H. (1985) Adaptations and evolutionary relationships of the Eocene primate family Notharctidae. Ph.D. Dissertation, Duke University, Univ. Microfilms Ann Arbor, Michigan.

Covert, H.H. (1986) Biology of early Cenozoic primates. In D.W. Swindler and J. Erwin (Eds.), *Comparative Primate Biology, Volume 1: Systematics, Evolution, and Anatomy* (pp. 335-359). New York: Alan R. Liss.

Covert, H.H. (1988). Ankle and foot morphology of *Cantius mckennai*: adaptations and phylogenetic implications. *Journal of Human Evolution, 17*,57-70.

Covert, H.H. and Hamrick, M.W. (1993). Description of new skeletal material of the early Eocene anaptomorphine primate *Absarokius* (Omomyidae) and a discussion about its adaptive profile. *Journal of Human Evolution 25*,351-362.

Covert, H.H. and Murphey, P.C. (1994). New middle Eocene omomyine skeletal material. *American Journal of Physical Anthropology*, Supl. *18*,72.

Covert, H.H. and Williams, B.A. (1994). Recently recovered specimens of North American Eocene omomyids and adapids and their bearing on debates about anthropoid origins. In J.G. Fleagle and R.F. Kay (Eds.), *Anthropoid Origins* (pp.29-54). New York: Plenum Press.

Dagosto, M. (1983). Postcranium of *Adapis parisiensis* and *Leptadapis magnus* (Adapiformes, Primates): adaptations and phylogenetic significance. *Folia Primatologica, 41*,49-101.

Dagosto, M. (1985). The distal tibia of primates with special reference to the Omomyidae. *International Journal of Primatology, 6*,45-75.

Dagosto, M. (1993). Postcranial anatomy and locomotor behavior in Eocene primates. In D.L. Gebo, (Ed.), *Postcranial Adaptation in Nonhuman Primates*, pp. 199-219. Dekalb: Northern Illinois University Press.

Dagosto, M. and Terranova, C.J. (1992). Estimating the body size of Eocene primates: a comparison of results from dental and postcranial variables. *International Journal of Primatology, 13*,307-344.
Decker, R.L. and Szalay, F.S. (1974). Origins and function of the pes in the Eocene Adapidae (Lemuriformes, Primates). In F.A. Jenkins, Jr. (Ed.), *Primate Locomotion*, pp. 261-291. New York: Academic Press.
Fleagle, J.G. (1977). Locomotor behavior and skeletal anatomy of sympatric Malaysian leaf-monkeys. *Yearbook of Physical Anthropology, 20*,440-453.
Fleagle, J.G. (1988). *Primate Adaptation and Evolution*. San Diego: Academic Press.
Fleagle, J.G. and Simons, E.L. (1983). The tibio-fibular articulation in *Apidium phiomense*, an Oligocene anthropoid. *Nature, 301*,238-239.
Gebo, D.L. (1987a). Functional anatomy of the tarsier foot. *American Journal of Physical Anthropology, 73*,9-31.
Gebo, G.L. (1987b). Locomotor diversity in prosimian primates. *American Journal of Primatology, 13*,271-281.
Gebo, D.L. (1987c). Humeral morphology of *Cantius*, an early Eocene adapid. *Folia Primatologica, 49*,52-56.
Gebo, D.L. (1988). Foot morphology and locomotor adaptations in Eocene primates. *Folia Primatologica, 50*,3-41.
Gebo, D.L. (1989). Postcranial adaptation and evolution in Lorisidae. *Primates, 30(3)*,347-67.
Gebo, D.L. and Dagosto, M. (1989). Foot anatomy, climbing, and the origin of the Indriidae. *Journal of Human Evolution, 17*,135-154.
Gebo, D.L., Dagosto, M., and Rose, K.D. (1991). Foot morphology and evolution in early Eocene *Cantius*. *American Journal of Physical Anthropology, 86*,51-73.
Gingerich, P.D. (1981). Early Cenozoic Omomyidae and the evolutionary history of tarsiiform primates. *Journal of Human Evolution, 10*,345-374.
Gingerich, P.D., Smith, B.H., and Rosenberg, K. (1982). Allometric scaling in the dentition of primates and prediction of body weight from tooth size in fossils. *American Journal of Physical Anthropology, 58*,81-100.
Godinot, M. and Dagosto, M. (1983). The astragalus of *Necrolemur* (Primates, Microchoerinae). *Journal of Paleontology, 57*,1321-24.
Gregory, W.K. (1920). On the structure and relations of *Notharctus*, an American Eocene primate. *Memoirs of the American Museum of Natural History*, new series, *3(2)*,49-243.
Hamrick, M.W. and Covert, H.H. (1991). Late Wasatchian notharctine fossils from the Washakie Basin, Wyoming and a comment on omomyid diversity. *American Journal of Physical Anthropology, 12*,86.
Hurzeler, J. (1948). Zur Stammesgeschichte der Necrolemuriden. *Schweizerishche Palaontologishe Abhandlungen, 66*,1-46.
Kay, R.F. and Cartmill, M. (1977). Cranial morphology and adaptation of *Palaechthon nacimienti* and other Paromomyidae, with a description of a new genus and species. *Journal of Human Evolution, 6*,19-53.
Kay, R.F. and Covert, H.H. (1984). Anatomy and the behavior of extinct primates. In D.J. Chivers, B.A. Wood, and A. Bilsborough (Eds.), *Food Acquisition and Processing in Primates*. pp. 467-508. New York: Plenum Press.
Kay, R.F. and Simons, E.L. (1980). The ecology of Oligocene African Anthropoidea. *International Journal of Primatology, 1*,21-37.
MacArdle, J.E. (1981). Functional morphology of the hip and thigh of the lorisiformes. *Contributions to Primatology*, vol. *7*, F.S. Szalay (Ed.). Basil: S. Karger.
Martin, R.D., Doyle, G.A., and Walker, A.C. 1974 (Eds.), *Prosimian Biology*. New York: Academic Press.
Matthew, W.D. (1915). A revision of the lower Eocene Wasatch and Wind River faunas. Part IV. Entelonychia, Primates, Insectivora (part). *Bulletin of the American Museum of Natural History, 34*,429-483.
Napier, J.R. and Walker, A.C. (1967). Vertical clinging and leaping - a newly recognized category of locomotion behavior of primates. *Folia Primatologica, 6*,204-219.
Rose, K.D. and Walker, A.C. (1985). The skeleton of early *Cantius*, oldest lemuriform primate. *American Journal of Physical Anthropology, 66*,73-89.

Rosenberger, A.L. and Dagosto, M. (1992). New craniodental and postcranial evidence of fossil tarsiiforms. In S. Matano, R.H. Tuttle, H. Ishida, and M. Goodman (Eds.), *Topics in Primatology*, Vol. *3* (pp.37-51). Kyoto: University of Kyoto Press.

Savage, D.E. and Waters, B.T. (1978). A new omomyid primate from the Wasatch formation of Southern Wyoming. *Folia Primatologica, 30*,1-29.

Schmid, P. (1979). Evidence of microchoerine evolution from Dielsdorf (Zurich region, Switzerland) - a preliminary report. *Folia Primatologica, 31*,301-311.

Simons, E.L. (1972). *Primate Evolution: an Introduction to Man's Place in Nature.* New York: MacMillan.

Simpson, G.G. (1940). Studies of the earliest primates. *Bulletin of the American Museum of Natural History, 77*,185-212.

Szalay, F.S. (1976). Systematics of the Omomyidae (Tarsiiformes, Primates): taxonomy, phylogeny, and adaptations. *Bulletin of the American Museum of Natural History 156*,157-450.

Szalay, F.S. and Dagosto, M. (1980). Locomotor adaptations as reflected in the humerus of Paleogene primates. *Folia Primatologica, 34*,1-45.

Szalay, F.S. and Delson, E. (1979). *Evolutionary History of the Primates.* New York: Academic Press.

Tattersall, I. (1982). *The Primates of Madagascar.* Columbia Press: New York.

Thalmann, U. Haubold, H. and Martin, R.D. (1989). *Pronycticebus neglectus* - an almost complete adapid primate specimen from the Geiseltal (GDR). *Palaeovertebrata, 19*,115-130.

von Koenigswald, W. (1979). Ein Lemurenrest aus dem eozanen Olschiefer der Grube Messel bei Darmstadt. *Paleontologische Zeitschrift, 53*,63-76.

Walker, A.C. (1974). Locomotor adaptations in past and present prosimian primates. In F.A. Jenkins, Jr., (Ed.), *Primate Locomotion* (pp. 349-381). New York: Academic Press.

Walker, A.C. (1979). Prosimian locomotor behavior. In G.A. Doyle & R.D. Martin (Eds.), *The Study of Prosimian Behavior* (pp.543-565). New York: Academic Press.

Ward, S.C. and Sussman, R.W. (1979). Correlates between locomotor anatomy and behavior in two sympatric species of *Lemur. American Journal of Physical Anthropology, 50*,575-590.

Weigelt, J. (1933). Neue Primaten aus der mitteleozanen (oberlutetischen) Braunkohle des Geiseltals. *Nova Acta Leopoldina, 1*,97-156.

Williams, B.A., Ayers-Darling, D., and Covert, H.H. (1991). New specimens of *Arapahovius gazini* from early Eocene deposits of the Washakie Basin, Wyoming. *American Journal of Physical Anthropology* supl. *12*,86.

CAPTIVE CONSERVATION AND THE ROLE OF THE AZA PROSIMIAN ADVISORY GROUP

Ingrid Porton

St. Louis Zoological Park
Forest Park
St. Louis, MO 63110

The reduction of wildlife populations all over the world is perhaps most dramatically illustrated by the number of tropical forest species that are endangered. The rapid rate of deforestation in the tropics has resulted in the reduction and fragmentation of populations which directly impacts species vulnerability to extinction. As a taxonomic group found almost exclusively in the tropics, primates are particularly hard hit. At a time when we still have numerous questions concerning the biology of such species as the nocturnal prosimians, many are endangered.

The conservation of endangered species requires a multipronged, interorganizational, and interdisciplinary approach. Habitat preservation and/or restoration, a comprehensive in-situ management program, basic and applied research, public education, and captive breeding programs are all facets of a conservation strategy. To that end, the American Association of Zoos and Aquariums (AZA) declared wildlife conservation as the Association's highest priority (Hutchins & Wiese, 1991).

Over the past ten years the management of captive breeding programs has changed dramatically. Rather than each zoo independently planning a propagation program for the groups of animals in their collection, zoos now strive to manage the many disparate populations of a particular species as one.

Collective management requires information on all the individuals contained within the population. A studbook, which is a genealogical record of a species or subspecies in captivity and includes information on the sex, birthdate, sire, dam, location(s), and death date of each individual, serves this function. Studbooks can be international or regional in scope. For example, there is an international studbook for *Varecia variegata*, *Eulemur macaco*, and *E. mongoz*, while North American regional studbooks exist for the Asian prosimians (including five taxa: *Nycticebus coucang*, *N. pygmaeus*, *Loris tardigradus*, *Tarsius syrichta*, and *T. bancanus*) and the *Cheirogaleids* (including three taxa: *Cheirogaleus medius*, *Microcebus murinus*, and *Mirza coquereli*. Studbook information is updated annually to accurately reflect the status of the captive population.

In North America the scientifically based cooperative breeding program is the Species Survival Plan (SSP). All zoological institutions that display SSP taxa and want to participate in the program sign a Memorandum of Participation which indicates they will abide by the management recommendations set forth by the SSP. The goal of an SSP is to develop a self-sustaining population for a predetermined period of time, often set at 100 to 200 years. Should it become possible to reestablish a species in the wild or reinforce a small fragmented population, individuals from the SSP would be available.

Creatures of the Dark: The Nocturnal Prosimians
Edited by L. Alterman *et al.*, Plenum Press, New York, 1995

Basically, the genetic goal of an SSP is to retain a high percent of the founding population's genetic diversity which is most successfully accomplished by equalizing family size for all founders. Genetic analysis of the studbook provides data on the actual reproductive success of the founder lineages. SSP breeding recommendations formulated to decrease discrepancies in founder representation can then be made. These recommendations may require certain individuals to cease reproducing and others to be moved from one institution to another to form genetically appropriate pairs.

SSP breeding recommendations have to be made within the constraint of limited captive space. Restricting an SSP to the minimum number of individuals necessary to maintain a self-sustaining population allows inclusion of more endangered species into propagation programs. To illustrate the problem, consider that the World Conservation Union (IUCN) recognizes 512 primate taxa, of which approximately 215 are maintained in captivity. The Primate CAMP (Stevenson, *et al.*, 1992) has recommended breeding programs for 158 taxa, 89 of which are currently found in captivity (although some populations number fewer than ten individuals). That translates into the need to bring in 69 taxa not now in captive breeding programs. Conversely, the remaining 126 taxa found in captivity are not of conservation concern and are recommended to be carefully managed as small populations or phased out.

Developing a regional collection plan to identify which taxa should be the target of North American captive conservation programs is the responsibility of the AZA's Taxon Advisory Group TAG. There are four primate TAGs: The Old World Primate TAG, New World Primate TAG, Great Ape TAG, and Prosimian TAG, formed in 1991. Decisions concerning which species to manage in SSPs are based on the current and predicted future space available to the taxonomic group in question (obtained through housing surveys), the size, genetic, and demographic characteristics of the existing captive populations, and information on the status of wild populations.

Similar taxonomically broad captive management programs are being developed in other regions of the world. Indeed, a European Prosimian TAG has been formed, and several other regions have SSP-like breeding programs for selected prosimian species. Coordinating regional programs, either through the global management of a taxa or by regions cooperatively selecting different taxa on which to concentrate their efforts, will result in the most efficient use of limited resources.

The Prosimian TAG is also working to open up communication between the scientific and zoo community to achieve a meaningful exchange of information. Perhaps even more important is expediting the potential to pool resources and collaborate on seemingly unrelated projects that together contribute to the preservation of this group of primates.

REFERENCES

Hutchins, M. & Wiese, R. (1991). Beyond genetic and demographic management: the future of the species survival plan and related conservation efforts, *Zoo Biology*, *10*, 285-292.

Stevenson, M., Baker, A. & Foose, T. (1992). Conservation assessment and management plan for primates, in "CAMP Summary Report," IUCN.

THE ROLE OF STUDBOOKS IN CONSERVATION AND CAPTIVE MANAGEMENT OF NOCTURNAL PROSIMIANS

Helena Fitch-Snyder

Center for Reproduction of Endangered Species
Zoological Society of San Diego
P.O. Box 551
San Diego, CA 92112

Studbooks can be a valuable tool in the management and conservation of captive populations. Besides tracking individual animals, these data provide an effective mean for measuring the success of a captive population and a basis for making species management decisions.

A studbook is a historical inventory of all known animals in the captive population. Each animal is assigned a permanent studbook number based on its acquisition date. This can provide a Taxon pedigree that links animals through past generations to wild-born ancestors.

All animals, including stillbirths and abortions, are included in this data base. The data contain all successive imports, births, deaths and transactions, along with information such as: place of birth, sex, parent identification, and individual identification numbers. In some cases, medical history and necropsy data may be incorporated. Additional information could also include identification of all possible sires and dams, karyotyping, and any other individual characteristics.

Data from a studbook can be utilized in many ways. At a management level, it can be used to locate and place surplus animals and prevent inbreeding. Reproductive information, including generation lengths, reproductive success of individuals, birth season, litter size, fecundity and inter-birth interval can be determined through analysis of studbook data. Inbreeding coefficients, age structure, and demography information can also be available in studbooks.

With this information, management questions can be answered, such as: Which animal should reproduce? Is this a population we want to grow? Should further imports be considered?

Besides the actual listing of living and historical populations, studbooks include additional facts such as description, systematics, status of wild populations, current and former distribution, and ecology of wild populations. Studbooks also include a comprehensive bibliography for the species.

A studbook could cover only one subspecies or it could include an entire Taxon. It may be kept on a regional level only, such as North American, or include the entire international captive population. Regional studbooks focus on a specific geographic area, but also include all ancestors regardless of location. Regional studbooks can later expand to an international level if there is a need, especially if animals are frequently transferred from region to region. If there are both international and regional studbooks, keepers will usually share information and may also use the same set of studbook numbers.

A studbook keeper is usually affiliated with a zoological institution or university, and is responsible for collecting all the data and updating the publication on an annual basis. A proposal to start a new studbook must be approved by American Association of Zoological Parks and Aquariums along with relevant captive management committees.

Creatures of the Dark: The Nocturnal Prosimians
Edited by L. Alterman *et al.*, Plenum Press, New York, 1995

Most studbooks today are started with the help of the International Species Information System. This database provides a record of all the animals that have been previously reported, but this information needs to be carefully verified and checked for discrepancies. This is accomplished by comparing with the institutional records, published articles, import records, and through personal communication with former owners. The official computer program used for most studbooks is called SPARKS.

Studbooks are distributed to all institutions and persons holding the species and are also made available to other interested parties for the development of management plans. Many specialized libraries also keep limited sets of studbooks.

Table 1. Current List of Studbooks of Nocturnal Prosimian Species.

Species	Studbook Keeper	Scope
ASIA		
Nycticebus coucang	H. Fitch-Snyder	N. America
N. coucang	J. Buchan	United Kingdom
N. pygmaeus	H. Fitch-Snyder	N. America
N. pygmaeus[1]	R. Ratajszczak	International
Loris tardigradus	H. Fitch-Snyder	N. America
L. t. nordicus	H. Schulze	International
Tarsius syrichta	H. Fitch-Snyder	N. America
T. bancanus	H. Fitch-Snyder	N. America
AFRICA		
Galago alleni[1]		
G. moholi[1]	M. Edwards	N. America
G. senegalensis[1]	M. Edwards	N. America
G. inustus[1]	M. Edwards	N. America
G. demidovii[1]	M. Edwards	N. America
G. elengantulus[1]	M. Edwards	N. America
G. zanzibaricus[1]	M. Edwards	N. America
G. inustus[1]	M. Edwards	N. America
O. crassicaudatus[1]	M. Edwards	N. America
Perodicticus potto[1]	C. Fredricks	International
Arcticebus calabarensis[1]	C. Fredricks	International
MADAGASCAR		
Cheirogaleus medius	B. Coffman	N. America
Microcebus murinus	B. Coffman	N. America
M. coquereli	B. Coffman	N. America
Daubentonia madagascariensis[1]	J. Bryan Carroll	International

[1]Newly approved or in process of being published

THE MADAGASCAR FAUNA GROUP

Andrea S. Katz

The Madagascar Fauna Group
Duke University Primate Center
3705 Erwin Road
Durham, North Carolina 27705

INTRODUCTION

In 1987, representatives of the government of Madagascar and of American and European zoos and universities attended a conference at St. Catherine's Island in Georgia to discuss and develop multiple strategies to support the long-term survival of Malagasy lemurs. There the Malagasy and international participants signed a formal agreement called the St. Catherine's Convention for Collaboration. The Convention established a framework for future cooperative activities to benefit the endangered Malagasy lemurs. The goals of the St. Catherine's Convention were to:

1) protect natural habitats in Madagascar
2) conserve endangered lemur species through captive breeding programs both in and outside of Madagascar
3) conduct basic research on Malagasy lemurs
4) provide relevant training for Malagasy students, technicians and scientists

The Madagascar Fauna Group was formed in 1988 to carry out the goals of the St. Catherine's Convention. The group is a consortium of 15 American and European zoological parks and institutions including the Duke University Primate Center, the Brookfield Zoo, the Los Angeles Zoo, the New York Zoological Society, the Oklahoma City Zoo, the San Antonio Zoo, the San Francisco Zoo, the St. Louis Zoo, the Jersey Wildlife Preservation Trust, the Marwell Preservation Trust, the Mulhouse Zoo, the Columbus Zoo, the Denver Zoo, the Atlanta Zoo and the Cincinnati Zoo.

The stated purpose of the Madagascar Fauna Group (MFG) is to help all classes of Malagasy fauna, particularly those whose populations and numbers appear to be falling below the minimums considered viable by conservation biologists and national authorities (Anderson, 1990). The MFG provides funds, materials, expertise and training to the Malagasy authorities, to promote collaborative captive breeding, education and reintroduction programs. In Madagascar, the MFG works in close collaboration with the

Creatures of the Dark: The Nocturnal Prosimians
Edited by L. Alterman *et al.*, Plenum Press, New York, 1995

Ministry of Water and Forests and the Ministry of Universities. To date the MFG's projects have focused on capital improvements, staff training and captive breeding programs for Malagasy fauna at Madagascar's two zoos- the Botanical and Zoological Park of Tsimbazaza in Antananarivo and the Zoological Park of Ivoloina near Tamatave.

ACTIVITIES AT IVOLOINA

Since 1987, Andrea Katz and Charles Welch have worked at Ivoloina as Technical Advisors on behalf of both the MFG and the Duke University Primate Center. Their objective there has been to assist Madagascar's Water and Forests Department with the rehabilitation of Ivoloina, an old holding facility for lemurs and tortoises which were confiscated by the Malagasy authorities (Katz and Welch, 1992). The goal was to build, in phases, a regional conservation center and zoo with sustainable programs of captive breeding of endangered eastern rainforest species, conservation education for zoo visitors and school groups, training in endangered species husbandry and management, and research to help develop sound reintroduction projects for captive-bred and confiscated lemurs.

At Ivoloina, the MFG has supported the construction of new animal enclosures and a small veterinary hospital. Together with Wildlife Preservation Trust International and the consortium of Strasbourg University, the Saarbrucken Zoo, the Cologne Zoo and the Mulhouse Zoo, the MFG has helped create a 2.5 hectare peninsula for free-ranging groups of lemurs. The peninsula serves as a site for studies on the adaptability of captive lemurs to secondary and primary forests in Madagascar.

Mouse lemurs (*Microcebus rufus*), dwarf lemurs (*Cheirogaleus major*), and aye-aye (*Daubentonia madagascariensis*) are presently housed at Ivoloina. However only the aye-aye has been targeted as a captive breeding priority for the zoo. One pair of aye-aye are housed at Ivoloina and two females and one male are housed at Parc Tsimbazaza in Antananarivo, bringing the total number of aye-aye in captivity worldwide to 20. Concerning other nocturnal lemur species such as *Lepilemur* and *Avahi*, Ivoloina will only hold individuals that might be confiscated, at which time some studies could be conducted on their captive husbandry.

ACTIVITIES AT PARC TSIMBAZAZA

The MFG has also provided funds for caging improvements and technical assistance to Parc Tsimbazaza, Madagascar's national zoo. Both Francine Woods of the Los Angeles Zoo and Hilary Morland, the past Project Coordinator for the MFG, have worked closely with Tsimbazaza's staff to improve animal husbandry and management practices. They have also assisted with the development of sound administrative and animal collection policies. Parc Tsimbazaza now has a system for charging entrance fees, providing the zoo with an independent and sustainable source of income. New gift items are now for sale at the zoo, and profits go directly back to zoo operations. The MFG also sponsored the head of Tsimbazaza's education department to attend a training course for zoo educators at the Jersey Zoo.

ACTIVITIES IN THE WILD

The integration of captive breeding programs for endangered lemurs with efforts to increase the viability of wild populations is a primary goal of the MFG. To this end, the MFG is involved in the preliminary phase of a re-stocking project for the black and white ruffed lemur (*Varecia variegata variegata*) into the Natural Reserve of Betampona, located near Ivoloina in eastern Madagascar (Welch and Katz, 1992). Nocturnal lemurs also found in Betampona include *D. madagascariensis*, *C. major*, *M. rufus*, *Lepilemur mustelinus*, *Avahi laniger* and perhaps *Phaner furcifer* and *Allocebus trichotis*. If the re-stocking program progresses successfully, greater protection of these nocturnal species and other fauna in Betampona will be a secondary, though no less significant, result.

CONCLUSION

The Madagascar Fauna Group is actively working for the animals and habitat of Madagascar. Although the aye-aye is now the only nocturnal prosimian included in the MFG's captive conservation program, perhaps the organization can serve as a model for other "Creatures of the Dark".

REFERENCES

Anderson, D.E. (1990). Formation of the Madagascar Fauna Captive Propagation Group, *American Association of Zoological Parks and Aquariums Annual Conference Proceedings*, pp. 37-40.

Katz, A.S. & Welch, C.R. (1992). Project Ivoloina: report of activities, March 1991-February 1992, Unpublished report to the Madagascar Fauna Group, Wildlife Preservation Trust International, and the Duke University Primate Center.

Welch, C.R. & Katz, A.S. (1992). Survey and census work on lemurs in the natural reserve of Betampona in eastern Madagascar with a view to re-stocking, *Dodo, the Journal of the Wildlife Preservation Trusts*, *28*,45-58.

AYE-AYE CONSERVATION: THE ROLE OF THE JERSEY WILDLIFE PRESERVATION TRUST

Anna T.C. Feistner and J. Bryan Carroll

Jersey Wildlife Preservation Trust
Les Augrès Manor, Trinity
Jersey JE3 5BF
Channel Islands, British Isles

INTRODUCTION

Gerald Durrell founded his Zoo for the breeding of endangered species in 1959, on the island of Jersey in the British Channel Islands. Four years later, in 1963, the Jersey Wildlife Preservation Trust (JWPT) was founded. In 1973 Wildlife Preservation Trust International (WPTI), based in the USA, was formed, and in 1985 Wildlife Preservation Trust Canada (WPTC). These three organisations, collectively known as the Wildlife Preservation Trusts (WPTs), work to help preserve species threatened with extinction.

This is achieved through support of the **propagation** of rare and endangered species in captivity as an aid to their survival; the **reintroduction** of rare and endangered species to their native habitats, and the **restoration** of these habitats to preserve the species in the wild; **research** on these species in captivity and in the wild, to achieve a better understanding of their biology, ecology and behavior, and to develop plans for their sustained management; the **education** of all persons about the value of wildlife, particularly those from developing countries living in close association with rare and endangered species; the **professional training** of zoologists and conservation biologists, particularly those from the developing countries of the world; and the formulation of **strategies** and **policy activities** for the conservation of endangered species and their habitats.

The Jersey Wildlife Preservation Trust became involved with the conservation of nocturnal prosimians in 1990, when six aye-ayes (*Daubentonia madagascariensis*) were imported from Madagascar under the Accord with the Malagasy Government (Durrell, 1991) to found a captive breeding program. All the aye-ayes outside Madagascar and their progeny belong to the Malagasy authorities.

MAINTENANCE AND CAPTIVE BREEDING

The JWPT currently holds eight aye-ayes (2.6): two adult males, three adult females, a female subadult, a female juvenile, and a female infant (Table 1). Six individuals (2.4) were imported in late 1990, an immature female died in 1992, a captive-bred infant female was born in 1992, another in 1993, and an adult female from the Parc Zoologique de Paris was imported on breeding loan in 1993.

Table 1. Aye-ayes held at JWPT.

Name	Sex	Age class	Origin
Alain	male	adult	wild caught
Patrice	male	adult	wild caught
Mina	female	adult	wild caught
Faharoa	female	infant	captive bred
Juliet	female	adult	wild caught
Zazakaly	female	subadult	wild caught
Sambany	female	juvenile	captive bred
Mamy	female	adult	wild caught[1]

[1] import on loan from Parc Zoologique de Paris

Housing

Details of husbandry and housing are given in Carroll and Beattie (1993) and the information presented here is summarized from their description. The aye-ayes are housed in either custom-built or specially modified enclosures. Each animal is housed separately as they are largely solitary in the wild (E.Sterling, pers. comm.), with the exception of the youngsters Sambany and Faharoa, who are still housed with their mothers. Enclosure sizes vary: there are six with floor areas of 14-25 m^2 and heights of 5 m; and three smaller areas measuring 4-9 m^2 and 3 m high (see Table 2). Each enclosure is linked to at least two other enclosures, either by tunnels or slides, which facilitates movement and mixing of animals. All the enclosures are extensively furnished with branches, ropes and logs. The floor substrate is bark chippings. There are humidifiers and heaters in each cage which keep the areas at an average temperature of 22-23 °C and an average humidity of 65-70 %. The heaters and humidifiers are controlled by thermostats and hygrostats.

Lighting

The aye-ayes are kept on a reverse lighting schedule which mimics Malagasy photoperiod changes, but within a northern hemisphere framework. They are provided with "true light" strip lights during their day and dimmed infrared spots for their night. Extra lights are on 30 min prior to light change and remain on 30 min after the lights change. These are "dawn" and "twilight" lights which help lessen the abruptness of the light changes. All lighting is operated automatically using digital timers.

Table 2. Aye-aye enclosure sizes at JWPT

Cage no.	Floor surface area	Maximum height (m)
1	17.5 m^2	5.0
2	17.0 m^2	5.0
3	14.0 m^2	5.0
4	13.7 m^2	4.9
5	15.8 m^2	5.0
6	25.5 m^2	5.7
7	4.0 m^2	3.0
8	7.5 m^2	3.0
9	9.0 m^2	3.0

Nesting Facilities

Every animal has access to two nest-boxes, and pregnant females have three. These are made of hardwood to withstand the aye-aye's formidable teeth and measure 35 x 35 x 35 cm with an entrance hole measuring approximately 23 cm in diameter. The nest-boxes have hooks on the back so they can be suspended on vertical mesh panels fixed to the wall. The aye-ayes also make use of tunnel entrances and hollow logs as sleeping sites. They are regularly provided with nesting material (pampas grass, woodwool, leafy branches, etc.) which is used inside the nest-boxes, but have never been seen to build nests in the open. As well as using the tunnels, the aye-ayes can be moved non-invasively using the nest-boxes. The aye-aye is kept in the box by covering the entrance while it is asleep, and it can then be transferred without being handled. This method can also be used to weigh the animal.

Feeding

The aye-ayes are fed once a day shortly after their active period starts. Water is changed daily. The food is spread around the cage to promote foraging behavior and the animals eat on and off during their entire active period. An average daily feed per animal consists of 2-3 nuts, insect larvae, 4-5 pieces of fruit, 1-2 pieces of vegetable, muesli, and an extra item such as coconut, sugar cane, or egg, etc. The aye-ayes are quite selective, with each individual having its own likes and dislikes, which are catered for. Experience shows that varying the food as much as possible is important, since the aye-ayes often stop eating a certain item if it is presented too often. Ripeness is also essential, as unripe fruits such as avocados, mangos, and papayas are rejected.

Management

Aye-ayes are a remarkably clean species. They do not use latrines and their feces are small and well formed. Cleaning is therefore minimal. Surfaces are cleaned as necessary, which is usually every few weeks. The bark chippings are changed every 3-4 months or more frequently if necessary. All left-over food is picked up daily before the lights change over, while the animals are asleep.

Breeding

Daily records of the adult females' vulval condition are maintained. Male and female are mixed when the female is considered to be at the peak of estrus as judged by the changes in the vulva (Carroll and Beattie, 1993). This has proved to be a successful breeding strategy resulting in successful births in August 1992 and August 1993 (Beattie et al., 1992; Feistner and Carroll, 1993a). All other information is recorded on computer using ARKS (Animal Records Keeping System) software. The JWPT also holds the International Studbook for the species.

Environmental and Behavioral Enrichment

All of the aye-ayes are provided with leafy branches, twigs, and other material (see above) for nesting material as often as possible. They appear to prefer larger leaves and nest building occupies a lot of their time. Rotten logs, sometimes infested with beetle larvae, are provided when available. These promote natural extractive foraging behavior, as does the daily provision of bamboo stems filled with insect larvae. Items such as eggs, nuts, and coconuts are presented whole. Perching is replaced when branches are available, to provide new arboreal pathways. Additionally, stumps and odd pieces of wood are often presented. Aye-ayes are extremely curious and exploratory and anything new stimulates tapping behavior. It is also important to provide them with enough things to gnaw so that their constantly growing incisors do not become too long.

INTERNATIONAL COOPERATION

The International Studbook

The International Studbook for the aye-aye was started in 1992, and currently lists all aye-ayes, living or dead, that have been held in captivity since 1986. It is hoped that a retrospective addition of historical records can be made in the future.

The current captive population comprises 23 animals, three of which were born in captivity. They are held at five institutions, two within Madagascar, Parc Botanique et Zoologique de Tsimbazaza and Zoo Ivoloina, and three outside Madagascar, Duke University Primate Center (DUPC), Jersey Wildlife Preservation Trust (JWPT), and Parc Zoologique de Paris.

The first captive birth was recorded at DUPC in April 1992 to a female that was already pregnant when captured. The first captive breeding occurred in August 1992 at JWPT with the birth of an infant conceived in captivity after a gestation period of 158 days (Beattie et al., 1992). Captive-conceived infants were born at DUPC in October and December 1992, but regrettably the latter did not survive. Another was born at JWPT in August 1993.

While there is clearly some way to go before a thriving captive breeding program is achieved, the aye-aye births that occurred in 1992 constitute a significant advance towards such a goal, and represent an important contribution towards the maintenance of Madagascar's biodiversity (Feistner and Carroll, 1993b).

In the future, the studbook will be an essential database on which long-term population management strategies can be based. The genetic management of small populations is of vital importance in minimising the loss of variability within captive breeding programs. Strategies for implementing such genetic management rely on sound demographic data for the captive population. These data can only be reliably provided through an accurate studbook. By starting the studbook now while the population is still small, we will reduce the likelihood of error brought about by retrospective compilation of data.

The Madagascar Fauna Group

The Madagascar Fauna Group (MFG) is an international consortium of 15 zoological institutions concerned with the conservation and protection of Malagasy wildlife (Katz, Chapter). JWPT was a founder member of the MFG and is a member of the Steering Committee.

RESEARCH

JWPT has undertaken an extensive research program on the captive aye-ayes since their arrival in November 1990 (Feistner and Price, 1992; Feistner, in prep.). Research areas include work on husbandry and maintenance methods (Carroll and Beattie, 1993), study of substrate use, locomotion and posture (Curtis, 1992; Curtis and Feistner, this volume, in press) and diet and nutrition (Ashbourne, in prep.). Additionally, long-term behavioral observations on four mother-offspring pairs focus on interactions and infant development (Feistner and Ashbourne, this volume, in press; Feistner and Steyn, in prep.; Feistner and Sterling, in prep.). Studies have also been conducted on adult social interactions, reproductive behavior and breeding (Beattie et al., 1992, Feistner et al., in prep.). Other research includes work on scent discrimination (Price and Feistner, in press) and handedness (Feistner et al., submitted). All research is undertaken with minimal disturbance to the animals. Purpose-built research facilities allow observation of the animals in both public display and off-exhibit areas. The WPTs have also contributed to field studies through WPTI funding of surveys of Betampona Natural Reserve in Madagascar (Welsh and Katz, 1992).

EDUCATION

The aye-ayes at JWPT are on display to the general public. They can thus act as a 'flagship' species for raising awareness about the environmental problems Madagascar faces. In the 'Twilight World of the Aye-aye' exhibit at JWPT there are color panels describing aye-aye biology and behavior, with information on characteristics such as teeth, hand and ear morphology; on diet, feeding and foraging methods; and on activity, posture, vocalizations, and habitat. Nearby, color panels illustrate the problems of environmental degradation, forest destruction, and erosion in Madagascar. The quality of this exhibit was recently recognized by a Zoo Animal Welfare Award from the Universities Federation for Animal Welfare.

TRAINING

The International Training Centre (ITC) for the Captive Breeding and Conservation of Endangered Species was opened at the JWPT in 1984. It runs 12-16 week courses in practical animal management (Waugh, 1983; 1985). These courses involve practical work on zoo sections, lectures, workshops, discussions, and a research project. Further study involving the preparation of a dissertation may lead to a Diploma in Endangered Species Management ratified by the University of Kent at Canterbury. These courses are particularly aimed at training people from developing countries (Waugh and Wemmer, 1994). The ITC also runs a three week Summer School on Breeding and Conservation of Endangered Species, attended by 24 students each year. The Education Department has run two seven-week International Zoo Educator courses (Coffey, 1992), aimed at enhancing the environmental and conservation education potential of zoological institutions in developing countries, through training of zoo personnel. Through these courses JWPT has trained 12 Malagasy nationals, all directly involved with endangered species (see Table 3).

Table 3. Malagasy personnel trained at Jersey Wildlife Preservation Trust.

RAKOTOARISOA, Gilbert	1983/84
RANDRIANAIVORAVELONA, Joseph	1983/84
RANDRIANASOLO, Bodo	1987
RANDRIANASOLO, Voara	1987
RAKOTONDRAPARANY, Felix	1989
RAKOTOBEARISON, Germain	1989
RANDRIANASY, Jeannot	1989
RAKOTOARISON, Nasolo	1992
RANDRIAMAHAZO, Herilala	1992
MORA, Roger	1992
RAMANAMPAMONJY, Julien	1993
RAVELOMANANA RAVOKATRA, Paulette	1993

The majority of personnel were from the Parc Botanique et Zoologique de Tsimbazaza and from Parc Ivoloina, six of whom are directly involved with aye-aye husbandry and education.

SUMMARY

Although widespread and occurring in several protected areas, the long-term future of aye-ayes in the wild is still far from secure, as Madagascar continues to struggle against environmental degradation and economic decline. This bizarre creature of the dark, the only living representative of a whole primate family, constitutes a flagship species for conservation - not only for nocturnal prosimians but for the biodiversity of Madagascar. Cooperative management of the species in captivity, integrated with scientific study of captive and wild aye-ayes, and continued collaboration with the Malagasy Government will all be necessary to ensure the continued survival of the aye-aye in the forests of Madagascar.

ACKNOWLEDGEMENTS

ATCF's attendance at the Creatures of the Dark conference was supported by the Center for International Studies at Duke University, the Wenner-Gren Foundation via a support grant to K. Izard and L. Alterman, and the Jersey Wildlife Preservation Trust. Thanks to the conference organisers for their invitation to participate in the meeting.

REFERENCES

Beattie, J.B., Feistner, A.T.C., Adams, N.M.O., Barker, P., & Carroll, J.B. (1992). First captive breeding of the aye-aye *Daubentonia madagascariensis*. *Dodo, Journal of the Wildlife Preservation Trusts, 28*,23-30.

Carroll, J.B. & Beattie, J.C. (1993). The maintenance and husbandry of aye-ayes *Daubentonia madagascariensis* at the Jersey Wildlife Preservation Trust. *Dodo, Journal of the Wildlife Preservation Trusts, 29*,45-54.

Coffey, P. (1992). Report of the International Zoo Educator Training Course held at the Jersey Wildlife Preservation Trust in 1991. *Dodo, Journal of the Wildlife Preservation Trusts, 28*,14-22.

Curtis, D.J. (1992). Substrate use in captive aye-ayes *Daubentonia madagascariensis*. *Dodo, Journal of the Wildlife Preservation Trusts, 28*,30-44.

Curtis, D.J. & Feistner, A.T.C. (1993). Positional behaviour in captive aye-ayes *Daubentonia madagascariensis*, *Creatures of the Dark, The Nocturnal Prosimians, Duke University*. Abstract.

Curtis, D.J. & Feistner, A.T.C. (in press). Positional behaviour in captive aye-ayes *Daubentonia madagascariensis*, *Folia Primatologica*.

Durrell, L. (1991). Notes on the Durrell Expedition to Madagascar September-December 1990. *Dodo, Journal of the Jersey Wildlife Preservation Trust, 27*,9-18.

Feistner, A.T.C. & Ashbourne, C. (1993). Infant development in a captive-bred aye-aye *Daubentonia madagascariensis* over the first 24 weeks of life, *Creatures of the Dark, The Nocturnal Prosimians, Duke University*. Abstract.

Feistner, A.T.C. & Ashbourne, C. (in press). Infant development in a captive-bred aye-aye *Daubentonia madagascariensis* over the first year of life. *Folia Primatologica*.

Feistner, A.T.C. & Carroll, J.B. (1993a). Captive breeding of the aye-aye, *Daubentonia madagascariensis*. *Primate Eye* 49:16-17.

Feistner, A.T.C. & Carroll, J.B. (1993b). Breeding aye-ayes: an aid to preserving biodiversity. *Biodiversity and Conservation*, 2,283-389.

Feistner, A.T.C., Price, E.C. & Milliken, G. (submitted). Hand preference patterns for tapping, digit feeding, and food holding in captive aye-ayes *Daubentonia madagascariensis*. *Folia Primatologica*.

Feistner, A.T.C., & Price, E.C. (1992). Behavioural research on captive aye-ayes, *Daubentonia madagascariensis*, at the Jersey Wildlife Preservation Trust. *Abstracts of the XIVth Congress of the International Primatological Society*: 203. Abstract.

Price, E.C. & Feistner, A.T.C. (in press) Responses of captive aye-ayes (*Daubentonia madagascariensis*) to the scent of conspecifics: A preliminary investigation. *Folia Primatologica*.

Waugh, D.R. (1983). The Wildlife Preservation Trust Training Programme. *Dodo, Journal of the Jersey Wildlife Preservation Trust, 20*,12-16.

Waugh, D.R. (1985). The Training Programme: Assistance, assessment and reconnaissance, *Dodo, Journal of the Jersey Wildlife Preservation Trust, 22*,12-18.

Waugh, D.R. & Wemmer, C. (1994). Training in zoo biology: Two approaches to enhance the conservation role of zoos in the tropics. In P.J. Olney, G.M. Mace & A.T.C. Feistner (Eds.), *Creative Conservation: Interactive Management of Wild and Captive Animals* (pp. 207-240). London: Chapman and Hall.

Welsh, C. & Katz, A. (1992). Survey and census work on lemurs in the Natural reserve of Betampona in eastern Madagascar with a view to restocking. *Dodo, Journal of the Wildlife Preservation Trusts, 28*,45-58.

THE NOCTURNAL PROSIMIAN COLONY AT THE DUKE UNIVERSITY PRIMATE CENTER

Barbara S. Coffman

Colony Manager (Currently Research Associate)
Duke University Primate Center
3705 Erwin Road
Durham NC 27705

INTRODUCTION

The Duke University Primate Center (DUPC) in Durham, North Carolina, houses the world's largest and most diverse collection of nocturnal prosimian primates. The collection includes 12 species which are maintained for the purposes of research, conservation, and education. These species are:

aye-aye (*Daubentonia madagascariensis*)
lesser mouse lemur (*Microcebus murinus*)
fat-tailed dwarf lemur (*Cheirogaleus medius*)
greater dwarf lemur (*Cheirogaleus major*)
Coquerel's dwarf lemur (*Mirza coquereli*)
slow loris (*Nycticebus coucang*)
pygmy loris (*Nycticebus pygmaeus*)
slender loris (*Loris tardigradus*)
potto (*Perodicticus potto*)
Garnett's galago (*Otolemur garnettii*)
Mohol galago (*Galago moholi*)
Philippine tarsier (*Tarsius syrichta*)

The current colony is comprised mostly of individuals born at the DUPC. Those that were captured in the wild were imported under agreements with the governments of the countries of origin, and under CITES permits issued by the U.S. Fish and Wildlife Service. Animals are exchanged with zoos and other research facilities to maintain a genetically and demographically healthy colony.

Nocturnal prosimians at the DUPC are housed indoors under a controlled light cycle to mimic the North Carolina photoperiod. This is essential for the Malagasy prosimians which are seasonal breeders. Animals are housed as pairs, large family groups (in the case of *Microcebus*, *Mirza*, and *Cheirogaleus*), or in mixed species groups. One pair of *O.*

Creatures of the Dark: The Nocturnal Prosimians
Edited by L. Alterman *et al.*, Plenum Press, New York, 1995

garnetii and its one offspring have been housed outdoors year-round with a pair of red-bellied lemurs (*Eulemur rubriventer*). Rooms or cages are decorated with natural branches and vines, and supplied with polyvinyl chloride tubing or wooden boxes for sleeping and nesting. Izard (1989) provides details on caging, staff, and the history of the DUPC animal colony.

Reproduction is carefully monitored through visual inspection of estrus, or vaginal smears. Population growth is regulated by separating males and females which are not designated to breed, although it is hoped that eventually contraceptive methods currently in use for diurnal lemurs (see Porton, et. al., 1990) can be used with the nocturnal species as well. Contraception will allow large family groups of nocturnals to be housed together, without the problem of unchecked population growth. Bachelor groups and mixed species groups of nocturnals are currently used as a means to provide social stimulation without unwanted reproduction.

RESEARCH

Research projects involving nocturnal prosimians at the DUPC have included studies of social behavior (e.g., Hoffman and Foerg, 1982), reproduction (e.g., Foerg, 1982; Izard and Rasmussen, 1985), morphology (e.g., Glander, 1994), phylogeny (e.g., Cartmill, 1975), locomotor patterns (e.g., Terranova, this volume), and vocalizations (e.g., Stanger and Macedonia, 1994). In order to accommodate a variety of research interests, populations of at least 15 animals are maintained for most species. This allows scientists to study large samples of a variety of nocturnal prosimians. The recent addition of large research rooms (4.5 m wide X 5.5 m deep) provides opportunities for behavioral studies as well as any projects that require the isolation of individual or groups of animals (see Terranova, this volume).

In addition to live animals, the DUPC maintains a collection of frozen cadavers and deep-frozen tissues of animals which died of natural causes (as permitted by Endangered Species permit PRT-679043). Tissues are generally used to acquire DNA for studies of phylogeny, while cadavers are available for anatomical studies or museum collections.

CONSERVATION

The conservation goals of the DUPC are the breeding of endangered prosimians in captivity for the purpose of preserving genetic diversity and for eventual reintroduction, and to learn as much as possible about captive propagation and husbandry in order to facilitate this process. The DUPC maintains detailed pedigrees, and participates in the American Zoo and Aquarium Association's (AZA) Species Survival Plans (SSP). Several of DUPC's staff are members of or advisors to the AZA's Prosimian Taxon Advisory Group (TAG) and have helped to formulate plans for the future management of the U.S. captive populations. In 1993 the DUPC published the AZA approved North American Regional Studbook for the Cheirogaleidae (Coffman, 1993), and is currently supporting the publication of studbooks for brown lemurs (*Eulemur fulvus*) and sifakas (*Propithecus*) and an SSP for the mongoose lemur (*Eulemur mongoz*).

The most notable recent events contributing to conservation efforts are the successful births of two aye-aye infants at the DUPC. Much has been learned of the husbandry and management of the aye-aye since the importation of 3.5 individuals between 1987 and 1991 (see Coffman, et. al., 1993). Together with the recent aye-aye births at the Jersey Wildlife Preservation Trust (see Feistner and Carroll, this volume), the DUPC births represent a major first step toward a successful captive breeding program for this unique and highly endangered primate. Pedigree information on the aye-ayes at the DUPC is supplied to the international studbook maintained at Jersey; it is hoped that in the future animals will be exchanged between these institutions for breeding, with the goal of preserving genetic diversity in the captive population.

EDUCATION

The DUPC is open to the public for guided tours. Tours are designed to educate the public about the plight of these endangered and threatened species in the wild, and the role of captive breeding and research programs in contributing to conservation efforts. Volunteers present slide shows at local schools where teachers are encouraged to participate. An adopt-a-lemur program has been initiated, through which individuals or classes of school children may help fund the maintenance of a chosen animal, and in the process learn more about their adopted species.

REFERENCES

Cartmill, M. (1975). Strepshirhine basicranial structures and the affinities of the Cheirogaleidae. In Luckett, W.P. and Szalay, F.S (Eds.) *Phylogeny of the Primates: A Multidisciplinary Approach*. Pp. 313-354.

Coffman, B.S. (1993). Cheirogaleid North American regional studbook. *Duke University Primate Center Occasional Papers* No. 1, 126pp.

Coffman, B.S., Hess, W.R., Glander, K.E., Feeser, P.E., and Simons, E.L. (1993). Management of a breeding colony of aye-ayes (*Daubentonia madagascariensis*) at the Duke University Primate Center. *AAZPA 1993 Regional Proceedings*, Lake Monroe, FL. Pp. 161-167.

Feistner, A.T.C. and Carroll, J.B. (1994). Aye-aye conservation: The role of the Jersey Wildlife Preservation Trust. In L. Alterman, G. Doyle, & M.K. Izard (Eds.) *Creatures of the Dark* (Pp. 519-525). New York: Plenum Press.

Foerg, R. (1982). Reproduction in *Cheirogaleus medius*, *Folia Primatologica* 39:49-62.

Glander, K.E. (1994). Morphometrics and growth in captive aye-ayes (*Daubentonia madagascariensis*). *Folia Primatologica* 62:108-144.

Hoffman, R. and Foerg, R. (1982). Development of locomotion and social behavior in infants of the fat-tailed dwarf lemur (*Cheirogaleus medius*). *Sonderdruck aus Zeitschrift fur Saugetierkunde Bd.* 48. H.3, S. 129-136.

Izard, M.K. (1989). Duke University primate center, *Journal of Medical Primatology*, 18:85-98.

Izard, M.K., and Rasmussen, D.T. (1985). Reproduction in the slender loris (*Loris tardegradus malabaricus*), *American Journal of Primatology* 8:153-165.

Porton, I., Asa, C., and Baker, A. (1990). Survey results on the use of birth control methods in primates and carnivores in North AMerican zoos. *AAZPA 1990 Annual Conference Proceedings*, Indianapolis, IN. Pp. 489-497.

Stanger, K.F. and Macedonia, J.M. (1994). Vocalizations of aye-ayes (*Daubentonia madagascariensis*) in captivity. *Folia Primatologica* 62:160-169..

Terranova, C.J (1994). Functional morphology of leaping behaviors in galagids: Associations between landing limb use and diaphyseal geometry. In, L. Alterman, G. Doyle, & M.K. Izard (Eds.) *Creatures of the Dark* (Pp. 473-493). New York: Plenum Press.

A NEW ROLE FOR THE MILITARY: ENVIRONMENTAL PROCECTION AND RESTORATION

Major General E. D'Souza, PVSM

Indian Army Retired

The very concept of the Military being used for the protection and restoration of the rapidly degenerating environment may seem incongruous to many. It is a concept often ridiculed by some, as exemplified at an environmental seminar in India. There a group of women activists demanded the expulsion of the speaker for daring to make such a proposal as, in their opinion, the Military are often described as wanton destructors of wildlife and despoilers of nature.

At a recent international meeting of retired Generals and Admirals convened in London in April 1993, attended by over 20 participants from Africa, Asia, the Americas, Europe, the Gulf States and Russia, the new concept of security was discussed and redefined. No longer can security be defined in purely politico-military terms. It now covers a much wider perspective which includes social security, a quality of life for all and last but by no means the least, environmental harmony. In the past, the Military's role was twofold: ensurance of the integrity of national borders and internal peace. After World War II, two more roles were added viz: disaster relief and international peacekeeping so much in evidence today. In the last decade, a fifth dimension has been introduced based on the fact that with the diminishment of the nuclear threat and international confrontation leading to war, the greatest threat to world peace is enviornmental degradation exemplified by rampant deforestation and habitat destruction, piercing of the ozone layer, the greenhouse effect and species destruction. The reduction in tension has resulted in frontline troops in Europe and elsewhere being withdrawn to peacetime locations. Thus, the question confronting most Governments is how best to use this valuable resource of trained manpower on productive tasks. Those skeptics who feel the Military have no role to play any longer are living in a fools' paradise, as no nation today can afford the luxury of doing away with the Armed Forces. Yet, the inevitable question that now arises is how best to use this resource on productive tasks, international peacekeeping and disaster relief notwithstanding. Based on the role model of the Indian Army with which this author has been associated since 1942, it has been conclusively proved that the Military has and can play a role in the protection and restoration of the environment in a positive manner **without blunting the cutting edge of the sword.**

QUALIFICATIONS OF THE MILITARY TO DO SO

The Military is well qualified to perform this projected role by virtue of its infrastructure, organization into formations, units and subunits, enabling it to be deployed in large or small self-contained groups, in its leadership, motivation, discipline, training,

RETIRED GENERALS AND ADMIRALS FOR PEACE: THE NEW THRUST

The novel idea of employing retired generals and admirals for environmental protection has been the subject of presentations in various forums pertaining to the environment. It received an impetus when the President of WWF, Prince Philip of England, invited the author to address an international gathering in Divonne on how the Indian Army has been involved in this productive role. Since then, there has been a chain reaction with presentations being made in Brasilia, the UK, Bulgaria, Austria, Bordeaux at the International Press Institute's Annual Congress, at the National Parks and Protected Areas World Congress in Caracas, Oxford, London, the House of Lords, The British Ministry of Defense and in India. The outcome from the Caracas meeting has been most encouraging, in that a cellule d'action has been introduced by Stevens of the University of Hawaii to which 26 countries have pledged their support. The University of Durham, UK has established a chair to carry out a global study of the use of the Military for environmental protection, charged with the responsibility of producing a global compendium to be made available to all interested countries. The brochure "What is proper soldiering", produced by Brigadier Harbottle of the Centre for Peacemaking, London, includes one chapter on this new role for the Military. An impressive start has been made.

THE INDIAN ARMY'S ACHIEVEMENTS

In precise terms what exactly has the Indian Army achieved in the last decade? The work done covers: afforestation on defense lands and degraded areas; reduction in use of non-renewable sources of energy through the introduction of renewable sources such as geothermal, thermal, wind, solar and biogas; pollution control in Military cantonments through regular maintenance of vehicles; strict sanitary and hygiene measures; water management; afforestation of hillsides to control landslides and soil erosion; population control by being exemplars; and creating awareness. To ensure coordination with the Army, a cell has been set up at Army Headquarters under the Director General of Military Training. To date, two annual meetings have been held, attended by representatives from all six Army Commands, the Border Roads, the Territorial Army, the National Cadet Corps, and the Navy and Air Force. In addition, senior retired Army Officers who are environmental activists, and there are many, are also invited. An important decision made at the annual meeting in November, 1992, was to include in brochures prepared for senior visiting officers a paragraph on environmental tasks undertaken and problems faced. In addition, in covering ecologically rich and remote areas, observations made of endangered and threatened species and status of habitats should be included in patrol reports. In doing so, it must be understood that this task is not specific but subsidiary to the patrol's main military tasks. In saying so it must be realized that the environmental observations become part of the main task.

LINKING THESE POSSIBLITIES WITH PROSIMIANS

How should these recommendations be linked with the creatures of the dark, nocturnal prosimians endemic to India? In other words, how can the Indian Military contribute to species and habitat protection of *Loris tardigradus* found in South India, and *Nycticebus coucang* in Eastern India? It is an exercise in futility to expect Indian troops operating in these areas to undertake this role without providing them with some information on these two species, their habitat and the threat to them.

In making such information available it is pointless giving troops scientific data. The information provided should be such that it is within the ambit of the Indian infantryman who is normally required to operate in such habitats in counter-insurgency operations. For instance the following information on *Loris tardigradus* would be within his ken:

Nycticebus coucang has a far more extensive range and its habitat is the densest parts of the tropical rain forests of Eastern India, where troops are constantly on the move. It is larger than the slow loris. Other characteristics of this species are: a vestigial tail; pelage white turning to buff and gray with a dark midline on the neck and back; found in the primary and secondary forests of Assam and the numerous bamboo groves; slow and deliberate climber with its powerful and pronounced fingers and toes, which are claw-like in appearance; feeds on large mollusks, small birds and small mammals; strikes with unusual speed; the male marks its territory with urine; when aroused it lets of a low buzzing hiss or a pure whistle.

Both these species are threatened. The slow loris is becoming increasingly popular as a house pet, though present government orders prohibit its capture. The author saw such a house pet in Gokak, South India and was therefore able to do a close study of the external characteristics. It had been trapped by tribals inhabiting the Western Ghats of South India. Its aggressiveness was evident when teased.

The author was also able to see a close-up flashlight photograph of *Loris tardigradus* taken recently in its natural habitat in the Western Ghats of South India.

Thus, the Military, briefed on the easily identifiable characteristics and the habitats of these two nocturnal creatures, should be in a position to report on them during the course of their movement in the habitats described, in South and East India, thus adding to the sketchy knowledge available. The Indian Military, by their very presence in these areas, would help in habitat protection so essential for the future of these species. Perhaps the Military in other countries where these Creatures of the Dark are found, such as Madagascar, parts of Africa and Asia could be equally motivated and their in-built infrastructure used for this purpose.

CONCLUSION

In conclusion, there can be little doubt that in the current scenario, the Military the world over has a very definite part to play in the protection of nature and natural resources and its regeneration or restoration. By virtue of its deployment, more often than not in ecologically rich and remote areas, its infrastructure and its availability, the Armed Forces can carry out this mission, based on the role model of the Indian Military. One caveat remains: "To do so without blunting the cutting edge of the sword."

PROSIMIANS: FROM OBSCURITY TO EXTINCTION?

Robert D. Martin

Anthropologisches Institut und Museum
Universität Zürich-Irchel
Winterthurerstrasse 190
8057 Zürich, Switzerland

INTRODUCTION

This concluding presentation provides a general background for selected issues raised by other contributions to the symposium Creatures of the Dark rather than a general synthesis. Four interconnected issues are raised: (1) phylogenetic relationships of modern prosimians; (2) evolution of Malagasy lemurs; (3) speciation in lemurs; (4) lemur conservation.

Evolutionary relationships of prosimian primates are briefly reviewed, with special emphasis on nocturnal habits. Relationships among primates of modern aspect are discussed first with respect to the proposed phylogenetic division between strepsirhines and haplorhines. Relationships among strepsirhines are then considered, concentrating on the hypothesis that Cheirogaleidae and Lorisiformes are sister groups.

Evolution of the lemurs of Madagascar represents a spectacular example of adaptive radiation among primates. The highly diverse modern lemur fauna, which until recently included numerous species now known only as subfossils, apparently originated largely or exclusively within Madagascar. Such extensive speciation and specialization within a single island demands special explanation. An explanatory model based on physical subdivision of Madagascar into semi-discrete zones with different climatic characteristics, initially proposed by the author in 1972, is reviewed in the light of new evidence. The evolution of the aye-aye is then considered as an extreme case of progressive specialization that can be explained on the basis of this model.

Evolutionary diversification begins with speciation within individual lemur genera. The process is illustrated with two examples. In the first case, very good evidence for sportive lemurs (Lepilemur species) indicates that chromosomal change was associated with the evolution of barriers between species, if not with the process of speciation itself. In the second example, there is convincing quantitative evidence for a valid distinction between two species of lesser mouse lemurs (Microcebus murinus and M. rufus) but as yet no evidence of chromosomal change involved in speciation. This latter example reveals that diverse data sources should be exploited in analysis of speciation patterns in natural populations of primates.

There has been increasing recognition of the urgent need for effective conservation measures for various prosimian species, notably for several larger-bodied lemur species. Ironically, several prosimian species have been discovered or rediscovered

Creatures of the Dark: The Nocturnal Prosimians
Edited by L. Alterman *et al.*, Plenum Press, New York, 1995

in recent years, yet virtually all of these face a particularly severe threat of extinction. Furthermore, the lemurs of Madagascar constitute a special case among primates because their original geographical ranges are extremely small. The menace arising from progressive deforestation and habitat fragmentation is correspondingly greater and the need for action all the more urgent.

PHYLOGENETIC RELATIONSHIPS OF PROSIMIANS

Nocturnal Habits and Primate Phylogeny

Reconstruction of phylogenetic relationships among prosimians continues to be an active focus of interest. A general review of evidence from living and fossil forms conducted by the author (Martin, 1990) was recently extended by consideration of new fossil evidence (Martin, 1993). Most authorities now agree that the six main groups of living primates (lemurs, lorises, tarsiers, New World monkeys, Old World monkeys, apes and humans) together constitute a monophyletic group to the exclusion of Plesiadapiformes. For some authors, the living primates together with their undisputed fossil relatives ("primates of modern aspect") constitute a monophyletic group labelled euprimates (Szalay & Delson, 1979) representing the sister-group of "archaic primates" (Plesiadapiformes). Other authors (e.g. Cartmill, 1972; 1974b; Martin, 1990; 1993) regard the proposed link with Plesiadapiformes as unconvincing and therefore favor restriction of the order Primates to euprimates.

A central issue concerning the evolutionary tree of primates of modern aspect is the inferred division between "wet-nosed" strepsirhines (lemurs and the loris group) and "dry-nosed" haplorhines (tarsiers and simians) indicated by numerous features of the living representatives (Martin, 1990). Although this division is questioned by some authors, there seems to be a growing consensus that it reflects a genuine early bifurcation in the evolutionary tree of primates. Comparison of living primates has yielded extensive evidence that the strepsirhine primates form a monophyletic group characterized, among other things, by the possession of a tooth-scraper or some secondary dental condition derived from this. Although monophyly of the strepsirhines is widely accepted, however, doubts have been expressed about the proposed link between tarsiers and simians. The position of the tarsiers continues to be interpreted in different ways, although the present author is convinced that the haplorhine primates also constitute a monophyleticgroup. The proposition that there is a link between tarsiers and simian primates has been further strengthened by biochemical comparisons of the d- and b-globin genes (Koop et al., 1989) and of the involucrin gene (Djihan & Green, 1991).

Unfortunately, it has not yet been possible to clarify the phylogenetic position of tarsiers using chromosomal evidence, although this approach has proved to be of considerable value for investigating relationships in other parts of the primate evolutionary tree (see Rumpler & Dutrillaux, this volume). Indeed, there are directly conflicting reports in the literature with respect to the chromosomal affinities of tarsiers. An abstract published by Poorman et al. (1985) originally stated that 67% of the chromosomes of Tarsius bancanus had direct counterparts in Galago crassicuadatus. This implies that tarsiers are linked to strepsirhines rather than to simians. Yet Dutrillaux and Rumpler (1988) subsequently reached the incompatible conclusion that the chromosomes of Tarsius syrichta have undergone such extensive rearrangement that it is impossible to reconstruct their relationships to other primates on the basis of chromosomal similarities. Although it is just possible that these markedly different conclusions reflect a real difference between the two tarsier species investigated (T. bancanus and T. syrichta), such extreme divergence between species of the same genus is unlikely. Coleman Healy (this volume) has now presented impressive new data from in situ hybridization of chromosomes that go a long way toward resolving the chromosomal affinities of Tarsius. Firstly, she has shown that T. bancanus and T. syrichta actually have very similar karyotypes, differing only by minor rearrangements confined to two chromosomes. Secondly, she found that most of the homologies between Tarsius and Galago inferred by Poorman et al. were artefacts, leaving only two apparently similar chromosomes as possible exceptions to the conclusions reached by Dutrillaux and Rumpler. Tarsiers do, indeed, have a very derived karyotype, as these latter authors proposed, and extensive rearrangement of their chromosomes rules out any simple reconstruction of their relationships on the basis of chomosomal banding patterns. More precise information of the kind provided by Coleman Healy is now needed to assess their affinities to other primates on the basis of chromosomal evidence. For present purposes, the division between strepsirhine and haplorhine primates is of particular interest because the former are predominantly nocturnal (only a minority of surviving lemur species being diurnal), whereas the latter (with the exceptions of Tarsius and Aotus) are characteristically diurnal. It is informative to examine the distribution of nocturnal and diurnal habits among living primates in relation to their inferred phylogenetic relationships (Fig. 1). The occurrence of nocturnal habits (indicated by the black zone in the figure) is a common feature of living strepsirhine primates and it is reasonable to conclude that they are derived from a nocturnal ancestor. This inference is supported by the observation that strepsirhine primates (including most of the diurnal forms) typically possess a reflecting tapetum lucidum behind the retina, which may be interpreted as an adaptation for vision in poor light conditions. Present evidence indicates that riboflavin crystals constitute the active principle of the strepsirhine tapetum and that this is unique among mammals (Martin, 1990). It is therefore reasonable to postulate the existence of a strepsirhine ancestor characterized by the possession of such a tapetum. Strepsirhine primates (especially the nocturnal forms) are also characterized by retention of several primitive features of the olfactory system that have been lost by all haplorhines. The retention of a rhinarium by strepsirhines, contrasting with the diagnostic loss of this feature by haplorhines, is connected with the general retention of a well-developed olfactory system in lemurs and lorises (see Martin, 1990 for a review).

It should be noted, incidentally, that some lemurs of the family Lemuridae (e.g. Eulemur mongoz) do not fit the usual clear demarcation between nocturnal and diurnal habits. These lemurs can be active both by day and by night. For this reason, Tattersall (1988) introduced the term "cathemeral" to label this unusual activity pattern. Subsequently, it was suggested by Engqvist & Richard (1992) that the cathemeral activity of certain lemur species may represent an adaptation for spreading the ingestion of slowly digested items (e.g. fibrous plant foods) across a wider time-span. This possibility now needs testing with appropriate field data. Implications of cathemeral activity for social organization have been discussed by Overdorff and Rasmussen (this volume).

In contrast to strepsirhines, the occurrence of diurnal habits (indicated by the white zone in Fig. 1) is almost universal among haplorhine primates. Apart from the rare exceptions of tarsiers and owl monkeys, haplorhines are all strictly diurnal, and it is therefore reasonable to conclude that they are derived from a diurnal haplorhine ancestor. It follows that the common ancestor of the simian primates was similarly diurnal. These inferences are supported by a number of indications that Tarsius and Aotus are derived from a diurnal ancestor and have secondarily redeveloped nocturnal habits. The inferred early switch to diurnal habits in haplorhine primates can be linked, among other things, to a reduction in the olfactory apparatus, increased emphasis on vision and a more rapid evolutionary increase in the relative size of the brain in comparison to strepsirhine primates (Martin, 1990).

The hypothetical reconstruction in Fig. 1 indicates that those lemurs which now show diurnal habits diverged from nocturnal ancestors quite independently of the

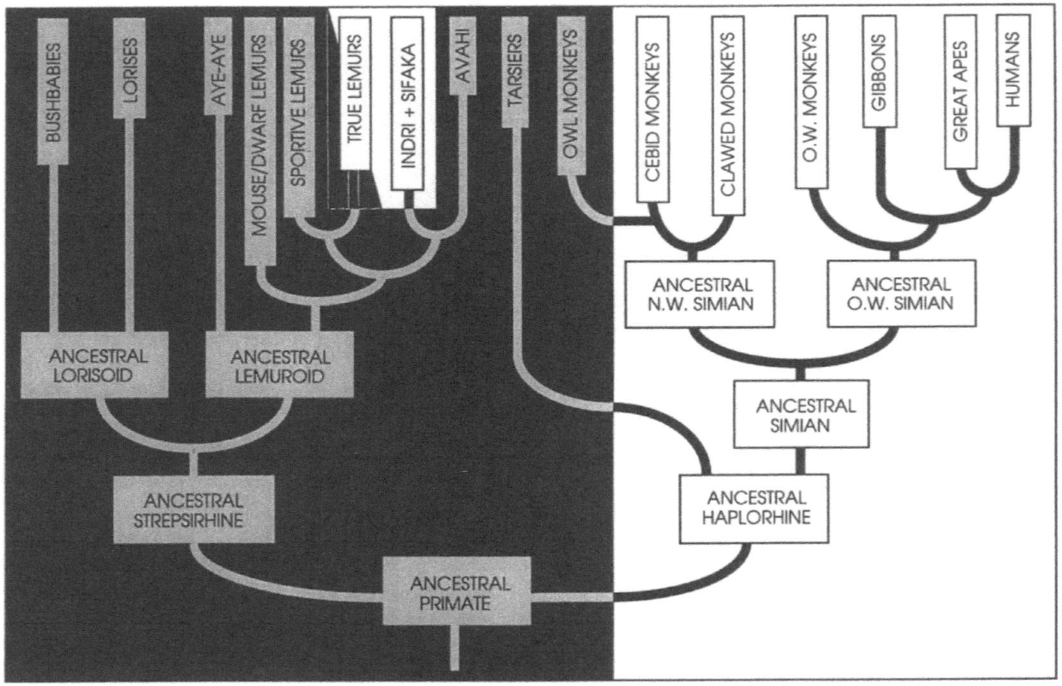

Figure 1: Phylogenetic tree illustrating inferred relationships between living primates and a basic division between nocturnal forms (black zone) and diurnal forms (white zone). Note that some true lemurs (Lemuridae) show an unusual blend of nocturnal and diurnal habits (cathemeral pattern). Fossil forms have not been considered; conditions in hypothetical ancestral stocks have been inferred exclusively from comparisons of living forms. (Adapted from Martin, 1979; 1990.)

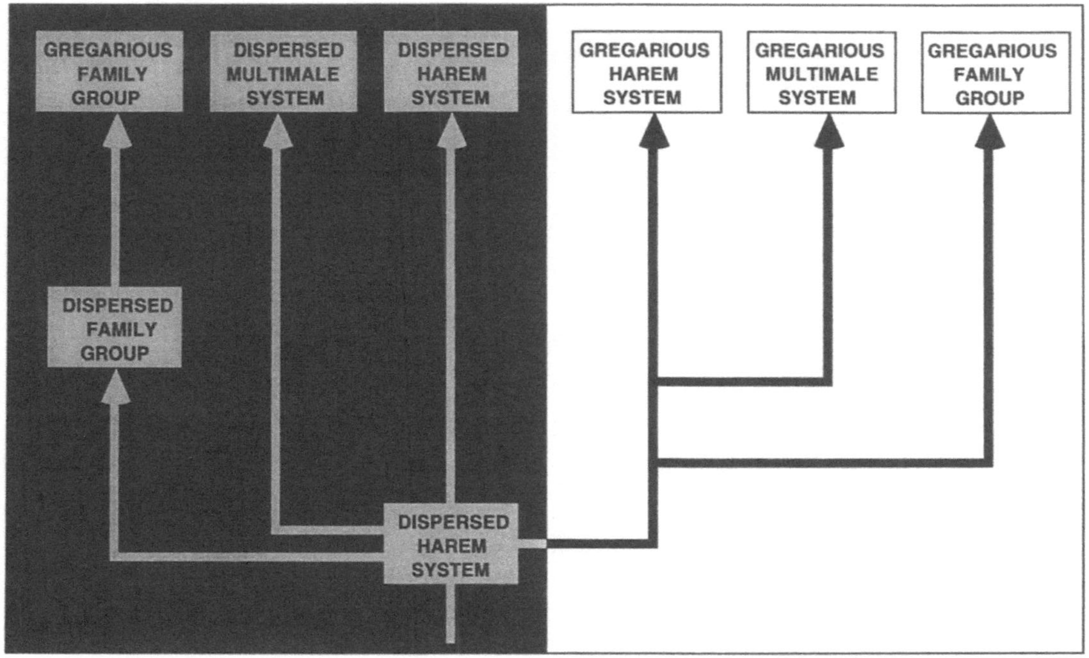

Figure 2: Diagram indicating possible evolution of patterns of social organization among primates, emphasizing the influence of nocturnal versus diurnal habits. Nocturnal primates tend to be solitary foragers with dispersed social systems. The hypothetical ancestral pattern is a dispersed harem, subsequently retained by various living nocturnal prosimian species. Dispersed multimale systems and dispersed family groups probably represent secondary developments among nocturnal prosimians. A few nocturnal primate species live in gregarious family groups. Gregarious habits are otherwise restricted to diurnal primates, which can similarly exhibit harem systems, multimale systems or family groups.

emergence of the diurnal ancestral haplorhine. In this connection, it is interesting to note that Allman (this volume) states that the diurnal Propithecus has a nocturnal type of lateral geniculate nucleus. The diurnal lemurs accordingly constitute an interesting test case. On the one hand, the independent evolution of diurnal forms among the Madagascar lemurs permits testing of associations between diurnal habits and other features inferred for diurnal haplorhines (e.g. for sensory organs, brain morphology, body size and social organization). With respect to morphological features, for example, this expectation is fulfilled. Like haplorhines, diurnal lemurs do indeed show some reduction of the olfactory apparatus and some improvements in visual capacities (e.g. with respect to color vision) relative to nocturnal strepsirhines (Martin, 1990). For this reason, diurnal lemurs can provide an informative model of the condition likely to have been present in ancestral simians (Tattersall, 1993).

On the other hand, the independent switch from nocturnal to diurnal habits in certain lemurs and in ancestral haplorhine primates is likely to be reflected by individual differences of detail. It is, for instance, to be expected that adaptations for color vision in diurnal lemurs and diurnal haplorhines should show different trends. Shared features of modern simians indicate that the ancestral simian had a dichromatic system of color vision with short-wave and long-wave cones having absorbance maxima in the region of 430 nm and 565 nm, respectively. Following their divergence from this ancestor, Old World simians developed a trichromatic system by adding a third pigment (probably with an absorbance maximum in the region of 535 nm), whereas New World simians developed a peculiar system involving polymorphism of the gene locus for the long-wave pigment on the X-chromosome. The result is that all males are dichromats whereas about two thirds of females are trichromats (Bowmaker, 1991). By contrast, apparent behavioral evidence of trichromatism in true lemur species (L. catta and E. fulvus) is probably attributable to a synergistic mechanism involving two cone pigments (with peak sensitivities at about 437 nm and 545 nm in both species) in combination with rod signals. In these lemurs, there is no trichromatism as in Old World simians and no polymorphism as in New World simians (Jacobs and Deegan, 1993). Further, it seems quite possible that the long-wave cones originated separately in lemurs and in simians.

The probable independent evolution of diurnal habits in certain lemurs is also of particular interest with respect to the relationship between activity period and social organization (Fig. 2). It has been common practice to refer to all or most nocturnal prosimians as "solitary" because individuals are commonly seen alone at night. However, contrary to the impression given by some syntheses (e.g. Crook and Gartlan, 1966), "solitary" is not the opposite of "social" (Charles-Dominque, 1978). Several field studies have documented well-defined social networks in a wide variety of nocturnal prosimian species (e.g. Bearder, 1987; Charles-Dominique, 1977; Charles-Dominique et al., 1980; see also chapters by Bearder, Charles-Dominique and Sterling & Richard, this volume). Bearder used the term "solitary forager" for nocturnal prosimians, to underline the contrast with diurnal primates which typically form recognizable feeding groups that range as a more-or-less cohesive unit. The correct contrast is accordingly between "solitary" nocturnal primates and "gregarious" diurnal primates; all primates have social networks of some kind. A further point arising from this distinction is that nocturnal primates, like diurnal primates, show diverse patterns of social organization. Depending on the species, social patterns of nocturnal solitary foragers show parallels to harem systems, multimale systems and family groups of diurnal primates.

Having recognized the diversity of social systems found in nocturnal primates, one can proceed to a hypothetical reconstruction of the evolution of social organization among primates (Fig. 2). As a starting-point, it can be assumed that the common ancestor of living primates was a small-bodied nocturnal form (see Fig. 1). Nocturnal primates are typically solitary foragers and their social systems hence usually show a dispersed pattern. It seems likely that the ancestral social system was of the dispersed harem type, with some males possessing home ranges overlapping those of several females and with surplus males peripheralized in some way as a result of relative intolerance between adult males. This pattern of social organization has been retained by a number of living nocturnal prosimian species. Subsequent evolution of social systems among nocturnal primates led to the emergence of new patterns. Increasing tolerance between males permitted the origin of dispersed multimale systems, while restriction of each male to a single female led to the development of dispersed family groups. A few nocturnal primate species such as the avahi live in relatively cohesive family groups that move and forage together, and this constitutes one of the rare tendencies towards gregarious habits to be found among nocturnal primates (referred to as "pre-gregarious" behavior by Charles-Dominque, 1978). Apart from such isolated exceptions, gregarious habits are restricted to diurnal primates. In this case, too, the independent development of diurnal habits in certain lemur species provides an instructive parallel. Like diurnal simians, diurnal lemurs are gregarious and form cohesive foraging groups. Among diurnal lemurs, there are clear-cut examples of multimale groups (e.g. Lemur catta; Propithecus verreauxi) and of family groups (e.g. Indri), although definite harem groups have yet to be reported. In any event, the consistent occurrence of gregarious habits among diurnal lemurs confirms the interpretation that nocturnal habits in some way favor solitary foraging whereas diurnal habits favor the formation of cohesive foraging groups. Once again, however, convergent evolution of the same feature (gregarious habits) in diurnal lemurs and diurnal haplorhines is betrayed by differences of detail (as noted by Charles-Dominque, 1978). It has, for example, been reported for several diurnal lemur species that female dominance is characteristic (see Kubzdela et al., 1992 for a recent review). Further, in contrast to many simian primates, sexual dimorphism in body weight and canine teeth is absent among diurnal lemurs (see Masters, this volume) and their social groups do not show female-biased sex ratios.

Adaptive Radiation of the Strepsirhine Primates

Although it is widely accepted that the Malagasy lemurs (Lemuriformes) and the loris group (Lorisiformes) together constitute a monophyletic assemblage of strepsirhine primates, opinions continue to differ with respect to the divergence of lineages from the common ancestor. Two main alternative hypotheses have been proposed (Fig. 3). The first is that the lemuriforms and lorisiforms constitute separate monophyletic groups, in which case the present clear geographical separation between the lemurs of Madagascar and the Afro-Asian lorisiforms might reflect an ancient division between land masses. According to the second hypothesis, however, the lemuriforms are not monophyletic. Numerous authors have suggested that the phylogenetic ties of the mouse and dwarf lemurs (family Cheirogaleidae) lie with the lorisiforms rather than with other lemurs. The evidence for this suggestion is primarily derived from the basicranial region of the skull and there are, indeed, intriguing similarities between cheirogelids and lorisiforms in the configuration of the internal carotid system. The issue has recently been re-analyzed by Yoder (1992), who linked new ontogenetic information for Microcebus with pre-existing data from Galago. As with the adult systems, the ontogenetic patterns are closely similar in Microcebus and Galago. In adults, the promontory and stapedial arteries are greatly reduced and replaced by an ascending pharyngeal artery that enters the skull through the foramen lacerum. Adult lorisiforms have a carotid rete mirabile, whereas cheirogaleids do not. But a new character identified by Yoder, also interpreted as linking cheirogaleids to lorisids, is a vascular plexus ventral to the otic capsule. In lorisiforms, this is later involved in the arterial rete mirabile; in cheirogaleids it is subsequently involved in the venous

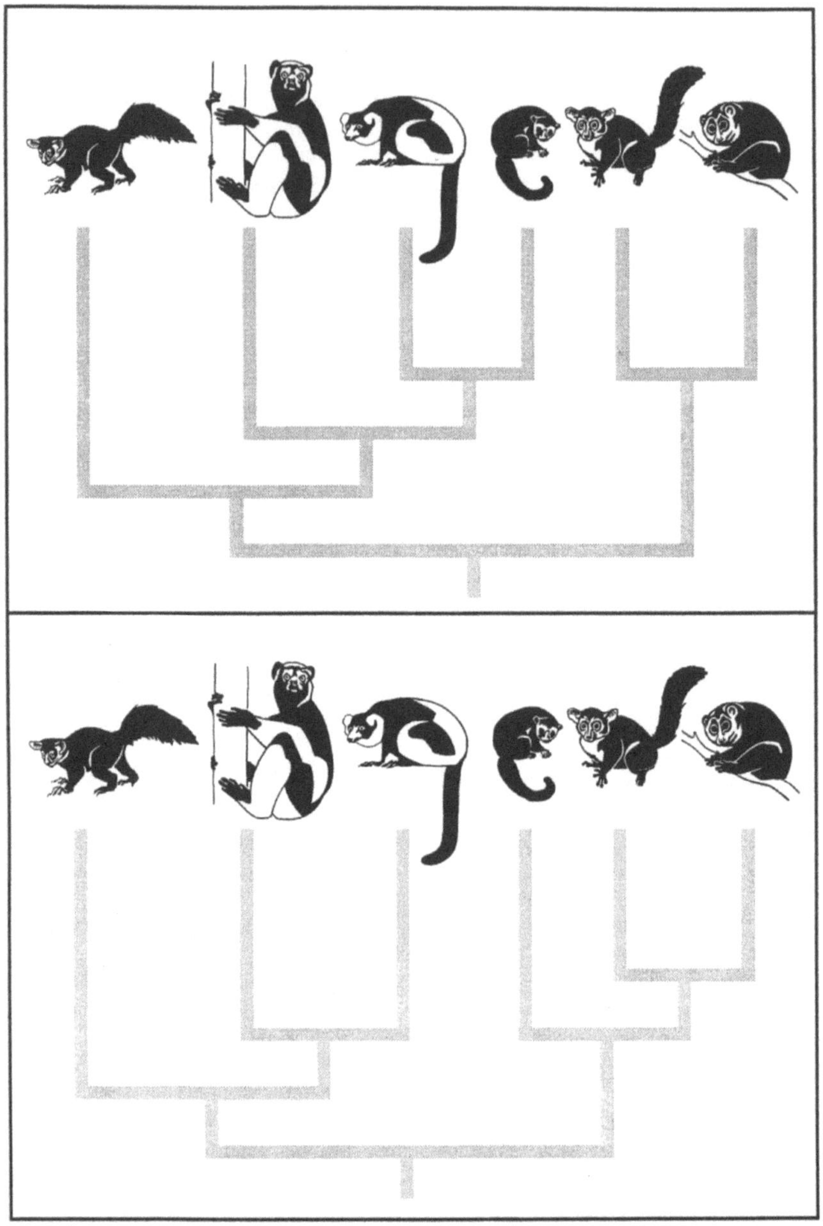

Figure 3: Alternative intepretations of the phylogenetic relationships among the major groups of strepsirhine primates. One possibility (above) is that lemurs and the loris group represent separate monophyletic units. An alternative interpretation (below) is that the correct division is between one monophyletic group containing the loris group and the mouse lemur family (Cheirogaleidae) and a second monophyletic group containing by the remaining lemurs. This interpetation is supported by basicranial evidence.The primates portrayed are representatives of individual families (from left to right) as follows: an aye-aye (family Daubentoniidae), an indri (family Indriidae), a variegated lemur (family Lemuridae), a dwarf lemur (family Cheirogaleidae), a bush-baby (family Galagidae) and a slow loris (family Lorisidae). (Primate illustrations by Lucrezia Bieler-Beerli.)

system instead. Yoder concludes that outgroup comparisons and the principle of geological precedence (i.e. fossil evidence) both support the hypothesis that the condition observed in cheirogaleids and lorisiforms is derived. Yet she also concludes that her own ontogenetic evidence neither supports nor contradicts this hypothesis. However, Yoder does not discuss the ontogeny of the internal carotid system in lemurs other than cheirogaleids, so there remains a possibility that some similarities between Microcebus and Galago may be primitive for strepsirhines generally. Nevertheless, there do seem to be striking similarities between cheirogaleids and lorises in the basicranial region that are unlikely to be primitive retentions. These similarities may therefore represent either shared derived features, indicating the prior existence of a specific common ancestor of cheirogaleids and lorisiforms, or convergent features. Several primate morphologists have concluded that they represent shared derived features (e.g. Groves, 1974; Schwartz et al., 1978; Szalay & Katz, 1973; Tattersall & Schwartz, 1974; 1975).

Over the past two decades, an increasing contribution has been made by chromosomal and biochemical evidence to our understanding of phylogenetic relationships between prosimians. An update on chromosomal evidence provided by Rumpler & Dutrillaux (this volume) confirms previous indications from this source that Lemuriformes and Lorisiformes are separate monophyletic groups, thus dampening current enthusiasm for linking the cheirogaleids to the lorisiforms. Given that a wide array of biochemical evidence similarly fails to indicate a specific link between cheirogaleids and lorisiforms (see Martin, 1990), there are very good reasons to suspect that any derived similarities in the basicranial region shared by these two groups are, in fact, a result of convergent evolution. This provides a prime example of a direct conflict between phylogenetic inferences based on two kinds of data (morphological versus chromosomal/biochemical). At present, there is still a general tendency for morphologists to ignore the conflicting chromosomal and biochemical reconstructions, and several authors have proposed new classifications incorporating the uncorroborated hypothesis that cheirogaleids and lorisiforms are sister-groups. This is not a constructive move. The most cautious interpretation would be that the question of the relationships of cheirogaleids is unresolved and that no change in classification is justifiable. In a previous review (Martin, 1990), it was concluded that the balance of the evidence in fact suggests that the lemuriforms and the lorisiforms are monophyletic sister-groups.

This latter interpretation has recently received supported from a study conducted by Yoder (1994). In the first truly comprehensive approach to anlysis of the evidence, she conducted a combined phylogenetic analysis of morphological data (osteological characters) and new molecular data from PCR-amplified DNA sequences (the first 720 bases of the mitochondrial cytochrome b gene). She obtained molecular data from 8 lemur species and 3 lorisid species, with additional data from Saimiri and Tupaia as outgroups, and conducted phylogenetic reconstructions using a computer-based approach (PAUP programme). Analysis of the morphological data alone provided weak support for monophyly of the lemurs, while analysis of the molecular data alone yielded strong support. Analysis of the two data sets combined produced a result convincingly favoring monophyly of the lemurs.

It should also be noted that the chromosomal data reported by Rumpler and Dutrillaux clearly indicate that the aye-aye branched off first among lemurs. This conflicts with inferences based on morphology apparently indicating a link between Daubentonia and the indri family. In fact, somewhat surprisingly, there is only weak chromosomal evidence for monophyly of the entire strepsirhine assemblage, despite the fact that this is one point on which there is general agreement among morphologists. The failure of chromosomal reconstructions to provide unequivocal confirmation of strepsirhine monophyly may well be attributable to the fact that their resolving power declines as one passes back in time. Because of superimposition of successive rearrangements, a certain amount of doubt attaches to reconstructions of very early stages.

EVOLUTIONARY BIOLOGY OF LEMURS

The question of the adaptive radiation of the lemurs in Madagascar is, of course intimately connected with the issue of the relationship between lemuriforms and lorisiforms. It is widely accepted that the ancestral lemurs probably invaded Madagascar from the African mainland, although other less likely possibilities have been suggested, such as invasion from Asia or back-migration to Africa (see Masters this volume). It is also widely accepted that some kind of filter operated, limiting the early mammalian invasions of Madagascar to carnivores, insectivores, primates and rodents. Hence, there is a reasonable likelihood that mammalian invaders of Madagascar arrived by rafting from Africa, as suggested by Millot (1952). The timing of these invasions is, however, very difficult to establish. Given that Madagascar separated from Africa as part of the general process of continental drift, invasion of the island by land-living mammals doubtlessly became increasingly unlikely during the Tertiary. Today, Madagascar is separated from Africa by a formidable barrier represented by the Mozambique Channel, 350 km wide at its narrowest point and some 1.5 km in depth. It is therefore probable that the lemurs (and most other terrestrial mammals) reached Madagascar at a relatively early stage. Tattersall (1993) follows a common trend in inferring that lemurs colonized Madagascar during the Eocene. An early Tertiary date was also originally suggested by the present author (Martin, 1972a). On this basis, one might date the lemur invasion of Madagascar at about 50 million years ago (mya).

There is, however, a major problem in that separation between Madagascar and the African mainland was apparently initiated before the end of the Jurassic. Masters (this volume) accordingly times the isolation of Madagascar at between 160 and 120 mya. Even the youngest date of 120 mya in this range is 70 million years earlier than the commonly inferred Eocene date for the invasion of Madagascar by ancestral lemurs, and it seems likely that a substantial sea-water barrier between Madagascar and Africa was already present by mid-Eocene times. It is highly improbable that primates existed in early Cretaceous times (120 mya), so the suggested timing of the initial separation between Madagascar and Africa reinforces the inference that the ancestral lemurs would have had to cross some kind of water barrier to reach Madagascar. Nevertheless, it is exceedingly unlikely that these ancestral lemurs (along with ancestral carnivores, insectivores and rodents) would have been able to cross a sea channel perhaps half as wide as the modern Mozambique Channel in mid-Eocene times. A potential solution to this problem is provided by the proposal that dates of divergence in the evolutionary tree of primates (and of other mammals) are considerably earlier than commonly inferred by palaeontologists. It has been common practice to date the origin of any given group from its first known fossil representative, but calculations based on the small proportion of fossil primate species so far discovered indicate that real dates of origin are likely to be considerably earlier (Martin, 1993). On this basis, the likely time of origin of primates may be pushed back to about 90 mya (see also the tree in Novacek and Wyss, 1988), and the lemurs may therefore have originated more than 80 mya, during the later Cretaceous. Obviously, the earlier the date that is set for an invasion of Madagascar by ancestral lemurs (and ancestral forms of other groups of terrestrial mammals), the more compatible this will be with current reconstructions of continental positions during the Cretaceous.

Whatever the timing of the invasion of Madagascar by ancestral lemurs, it seems to have been an unlikely event that subsequently became most improbable. This being the case, it is difficult to accept the proposition that the ancestors of cheirogaleids reached

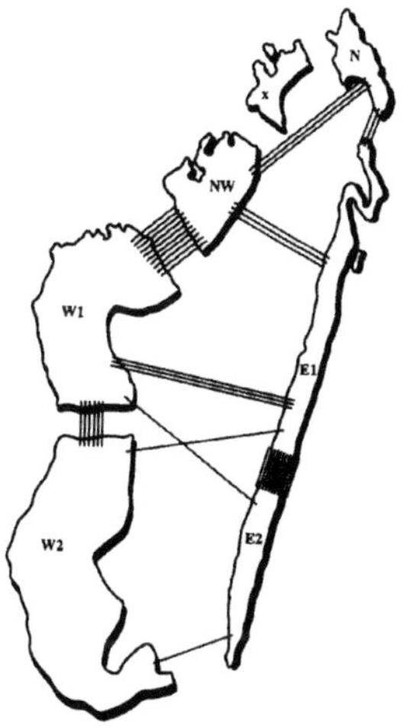

Figure 4: Exploded map of Madagascar showing 6 main coastal subzones with distinctive climatic features separated by geographical barriers (adapted from Martin, 1972): northern and southern sections of the east coast rainforest (E1, E2); deciduous forest of the west coast (W1); dry forest of the west coast (W2); deciduous forest of the northwest coast (NW); deciduous forest of the north coast (N). The additional small subzone in the northwest (x) is unusual in that its main biological affinities seem to be with subzone E1 rather than with either of the adjacent subzones (NW; N). There is, however, little information on lemurs occurring in subzone x so it has been omitted from the analysis. Grey bars indicate numbers of lemur species shared between the six main subzones. Species sharing between subzones shows the following main features: (1) a basic chain around the island; (2) stronger links in the east and west, respectively; (3) a hiatus between subzones E2 and W2 in the south.

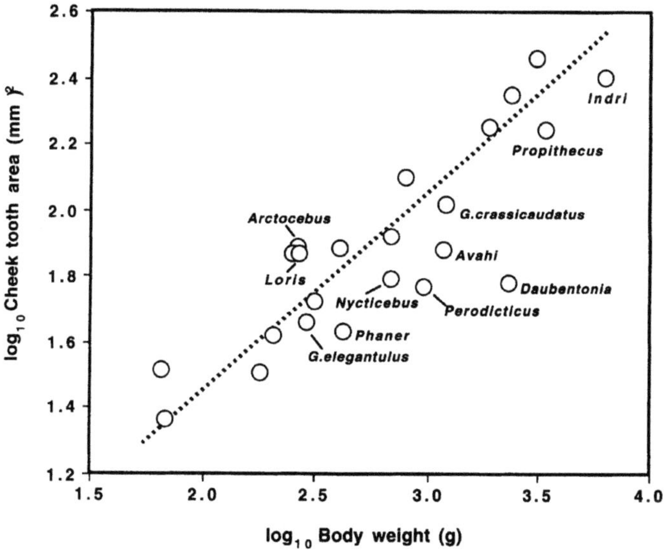

Figure 5: Double logarithmic plot of cheek tooth area (maximum length x maximum width of the upper row of premolars and molars in the adult dentition) against body weight for strepsirhine primates. The best-fit line is the reduced major axis. Three species that feed extensively on gums (Galago elegantulus; Perodicticus potto; Phaner furcifer) all lie below the line, as do all three members of the family Indriidae (Avahi laniger, Indri indri, Propithecus verreauxi). The point for Nycticebus coucang lies quite close to that for P. potto, suggesting that Nycticebus is likely to be a gum-feeder as well. In contrast, Arctocebus calabarensis and Loris tardigradus both lie above the line. As Arctocebus is known not to be a gum-feeder, the same can be inferred for Loris. Daubentonia has by far the smallest relative cheek tooth size among strepsirhines.

Madagascar later than the ancestor of other lemurs (or, alternatively, that a lorsiform ancestor migrated back to Africa following a divergence between cheirogaleids and other lemurs within Madagascar). As was noted by Cartmill (1975), this factor weighs any comprehensive analysis in favor of lemur monophyly.

Regardless of whether the modern lemurs can be traced back to one common ancestor or two, the known diversity of this group of primates is spectacular. This, in itself, requires special consideration. Masters (this volume) has referred to the "explosive speciation" of lemurs in Madagascar, noting that in the modern fauna there are 44 lemur species (including recently extinct subfossil forms) as opposed to only 16 lorisid species in Africa, despite the fact that the land area of Madagascar is only 2.5% of that of Africa. In terms of number of species relative to land area, the adaptive radiation of the lemurs is undoubtedly outstanding among primates. Although it seems highly likely that the number of lorisiform species in Africa has generally been markedly underestimated (see contributions by Bearder and Scheffrahn, this volume), the fact remains that there are far more lemur species per unit area in Madagascar. Masters notes that Darwin emphasized the importance of available land area as well as geographical isolation for evolution and one must therefore ask how such diversity could arise among the lemurs within the confines of an island the size of Madagascar.

Numerous authors have commented on the rich climatic subzonation of Madagascar (e.g. see Tattersall, 1982). The present author (Martin, 1972a) proposed a basic model in which major rivers and areas of high ground combine to act as barriers separating individual regions with distinctive climatic, and hence botanical, characteristics. It was suggested that these barriers are sufficiently impermeable to favor local speciation but that occasional migration across them could be followed by competition between quite closely related species with resulting divergent specialization. Repeated speciation and subsequent specialization could eventually generate a great diversity of species.

The island of Madagascar can, in fact, be conceptualized as a network of smaller habitat islands isolated by barriers that are that are sufficently pronounced to permit speciation but permeable enough to permit occasional invasion leading to enhanced interspecific competition. After considering the distribution patterns of modern lemurs in the light of geographical, climatic and botanical evidence, the author (Martin, 1972a) recognized six main coastal subzones separated by geographical barriers. (Although the elevated central plateau of Madagascar should, ideally, also be considered, little information is available for lemur distributions there, primarily because of massive deforestation. Given the generally cooler climate on the central plateau, it is possible that it acted as a barrier to migration at least for some lemur species, but reliable data are lacking.)

Using available information on lemur distribution patterns (Petter et al. 1977; Tattersall, 1982; Harcourt and Thornback, 1990), it is possible to make a conservative estimate of the number of species present in each coastal subzone. Using this information, the sharing of species between the subzones can be illustrated with an exploded map of Madagascar (Fig. 4). It can be seen that the subzones are basically linked in a chain running around the island, with a marked hiatus between subzones E2 and W2 in the south. (These two subzones present a stark contrast in climatic conditions, with the xerophytic vegetation of subzone W2 differing radically from the rainforest of subzone E2.) Only the single species Eulemur fulvus occurs in both E2 and W2, and the ranges of the two subspecies concerned (E. fulvus rufus and E. fulvus collaris, respectively) are not in contact. Indeed, these two forms of the brown lemur are regarded as separate species by some authorities. There is also a fairly clear subdivision between the eastern and western subzones. The two sections of the east coast rainforest (E1, E2) share numerous species and the same applies to the dryer forests of the west coast (W1, W2, NW). A relatively complex situation exists in the north of the island. The forest of the north coast subzone (N) apparently has closer affinities with subzone NW than with the intervening subzone x, which was included by Martin (1972a) with subzone E1. Unfortunately, there is little information on the lemur species present in subzone x, so it had to be omitted from the analysis, but it seems likely that lemur distribution patterns show an unusual pattern of crossing over (NW + N; E1 + x) in the north of the island. Overall, the present pattern of lemur distribution could easily have arisen from multiple sequences of isolation, speciation and subsequent expansion among subzones, as proposed in the original model.

It should be noted, incidentally, that the subzones shown in Fig. 4 are by no means fixed or exclusive. It has, for instance been shown by Thalmann & Rakotoarison (in press) that the distribution of lemur species within subzone W1 is quite complex, so further subdivisions are needed. The model presented here is no more than an approximation, based on the major geographical subdivisions between modern lemur species. Further refinements will undoubtedly be necessary. On the other hand, it should not be forgotten that the proposed dynamic process of repeated speciation and competition requires that individual species should spread into adjacent subzones from time to time.

An extreme example of specialization among lemurs is provided by the aye-aye (Daubentonia madagascariensis). It is possible to explain the evolution of this species as a result of cascading competition in line with the model proposed above. The modern aye-aye has a highly derived, rodent-like incisors and a very thin middle finger that is used as a probe. Although it is now well established that aye-ayes also include various plant items in their diet, it is clear that the rodent-like incisors and the digital probe are particularly used in conjunction for feeding on wood-boring insect larvae. As has now been shown by Erickson (this volume), the aye-aye can use its thin middle finger to tap trunk surfaces to locate hidden cavities with great precision ("percussive foraging"). In one of the most memorable papers published in the field of prosimian biology, Cartmill (1974b) argued that these adaptations occurred because the aye-aye came to occupy an ecological niche that is filled by woodpeckers or their analogues in other parts of the world. (A woodpecker analogy was, in fact, first suggested by Sandwith, 1858.) Cartmill also noted that a parallel development occurred in Australasia, in the form of the striped possum Dactylopsila. This nocturnal marsupial also has a rodent-like dentition and a finger adapted as a probe. This hypothesis might therefore explain why the aye-aye eventually developed such marked specializations, but the way in which such extreme modifications could arise through evolution within Madagascar remains to be explained.

It seems likely that the common ancestor of the lemurs was a small-bodied nocturnal species with a typical tooth-scraper containing six teeth in the lower jaw. The original function of this tooth-scraper has been the subject of much debate. Many authors have concluded that its primary function lies in grooming of the fur, but an alternative interpretation (Martin, 1972; 1979; 1990) is that the tooth-scraper was developed primarily in association with feeding on plant exudates (e.g. gums) and that the grooming function is secondary. The tooth-scraper is particularly well developed in two nocturnal prosimians known to be specialized gum-feeders (Euoticus elegantulus and Phaner furcifer) and field observations of Galago moholi have confirmed that the tooth-scraper is indeed used during gum-feeding. It is therefore quite possible that a small-bodied ancestral lemur equipped with a tooth-scraper similarly used it to feed on plant exudates as one of the components of its diet. There is a significant connection here with the major dietary specialization of the aye-aye in that the acacia gum consumed by G. moholi is produced as a reaction to wood-boring insect larvae (Bearder & Martin, 1980). It is therefore conceivable that the aye-aye's special adaptations for feeding on such larvae developed through a number of intermediate stages from the ancestral condition, progressively shifting from the harvesting of plant exudates to predation on the insect larvae responsible for the exudate flow. The problem, however, is that none of these hypothetical intermediate stages has been preserved in the modern lemur fauna.

At this point, Cartmill's identification of a parallel between Daubentonia and Dactylopsila is once again relevant. Dactylopsila belongs to the marsupial family Phalangeridae and an intermediate stage in the transition from feeding on plant exudates to predation on wood-boring insect larvae is provided by the glider-possum, Petaurus australis. Long-term field observations of these marsupials (Wakefield, 1970) have revealed that, although they feed heavily on nectar, they also occasionally remove bark from various eucalypt trees in order to feed on their sugary sap. Indeed, they use their procumbent lower incisors to make large V-shaped incisions in the trunk, such that the sap is channelled to a single point where it can be regularly harvested. Indirect evidence indicates that P. australis also removes patches of bark to feed on insect larvae, although no excavation of deep cavities containing wood-boring larvae seems to occur. The glider-possum therefore seems to combine sap-feeding with predation on insect larvae living directly beneath bark. This provides some support for a hypothetical gradual transition from a gum-feeding ancestral lemur to the modern aye-aye, involving repeated speciation and competition-driven specialization according to the model illustrated in Fig. 4.

An additional link between exudate feeding and predation on wood-boring larvae is indicated by an analysis of the relative size of the cheek teeth (premolars and molars). The aye-aye is an extreme outlier among living prosimians in having a very small total cheek tooth area. It emerges that prosimians tthat feed on exudates also have relatively small cheek teeth (Fig. 5), so an extension of this trend could theoretically lead to the condition in the aye-aye. It must be noted, however, that indriids also have relatively small cheek teeth, so an explanation of this is also required for sound interpretation of the extreme condition found in the aye-aye.

PROCESSES OF SPECIATION AMONG LEMURS

Sportive Lemurs

One of the most striking and informative studies of speciation within a lemur genus has been furnished by a recent theoretical analysis of chromosomal differences between different populations of sportive lemurs (Lepilemur). In standard reference works (e.g. Napier and Napier, 1967), the sportive lemurs have traditionally been allocated to the single species Lepilemur mustelinus, although the existence of several fairly distinct populations representing at least subspecies has recently been widely accepted (e.g. Martin, 1990; Tattersall, 1982; Wolfheim, 1983). However, initial study of the karyotypes of different populations of sportive lemurs at once revealed major distinctions.This was taken as an indication that there could be as many as seven different species (Rumpler, 1975) The number of species subsequently identified was six (Rumpler & Albignac, 1978). As a result, explicit recognition of several different sportive lemur species has since been favored by a number of authors (e.g. Petter et al., 1977; Petter & Petter-Rousseaux, 1979).

A more penetrating investigation of chromosomal differences among these different Lepilemur populations has now yielded very convincing evidence that they are indeed separate species (Ishak et al., 1992). At the same time, it has been shown how these chromosomal differences could generate effective reproductive barriers between the various species that may now be recognized. Because of the fact that sportive lemurs are very difficult to maintain (let alone breed) in captivity, a direct test of reproductive barriers has not so far been possible. Analysis of chromosomal features of different Lepilemur species indicates, however, that hybrids between them would suffer from reduced male fertility or complete sterility. This conclusion is based on meiotic configurations predicted to occur with the mixed karyotypes of hybrids (Ishak et al., 1992). The predominant form of chromosomal change inferred for the evolution of modern sportive lemur species from a single parental species is Robertsonian translocation (= "centric fusion"). The net effect of this process is that two different one-armed acrocentric chromosomes become united to form one two-armed metacentric chromosome. This change initially occurs as a polymorphism in a population, but it may then become fixed such that two pairs of acrocentric chromosomes in the original parent population are eventually replaced by one pair of metacentric chromosomes. In itself, either during this polymorphic stage or following later hybridization with individuals retaining the karyotype of the parental population, this change need not cause any problem at meiotic division. This is indicated by the fact that natural populations of primates may exhibit balanced polymorphisms involving such chromosomal rearrangements. At the pachytene stage of meiosis, two acrocentric chromosomes can line up with the two homologous arms of the derived metacentric chromosome, thus forming a trivalent instead of the normal two bivalents. A problem arises only when two populations diverge and in each of them a given acrocentric chromosome becomes "fused" with a second acrocentric chromosome that differs between the two populations (Fig. 6). The situation can become even more complex if additional different translocations become fixed in the two populations. At the pachytene stage of meiosis, because of the chromosomal differences between the two species resulting from these translocations, hybrids between them would show complex multivalents involving four or more chromosomes (in addition to normal bivalents and perhaps one or more trivalents). Such multivalent chains are (for understandable reasons) known to be associated with a marked reduction in fertility in various mammals, so it can be concluded that successful hybridization is unlikely wherever they occur.

Ishak et al. (1992) examined the banded karyotypes of six different sportive lemur populations and predicted the meiotic configurations that would occur in hybrids between them. In every case, the predicted hybrid condition would be associated with at least one complex multivalent chain at meiosis. It can therefore be inferred with some confidence that interbreeding among the different Lepilemur species in Madagascar would lead to the production of hybrids with markedly reduced fertility or even sterility.

On the basis of the chromosomal changes thought to have taken place, it is also possible to infer phylogenetic relationships among the different Lepilemur species and hence to infer the likely sequence of speciation events. On this basis, Ishak et al. determined the following overall pattern, starting from an initial common ancestor: (1) separation between a western species and an eastern species (the latter being L. mustelinus); (2) isolation between the western species population and a new species in the northernmost region of Madagascar (subzone N in Fig. 4: L. septentrionalis); (3) isolation from the residual western species population of a new species in the midwestern region of Madagascar (northern part of subzone W2 in Fig. 4: L. ruficaudatus); (4) isolation from the residual western species population of a new species in the southwestern region of Madagascar (remaining southern part of subzone W2 in Fig. 4: L. leucopus); (5) isolation from the residual western species population of a new species in the northwestern region of Madagascar (subzone x in Fig. 4: L. dorsalis); (6) the remaining residual population in the northern part of the western region constitutes the sixth species (subzone W1 in Fig. 4: L. edwardsi).

This reconstruction matches quite closely the geographical zonation reviewed above as a basis for successive waves of speciation in Madagascar (Fig. 4). The following common features deserve emphasis: Firstly, there is a primary division between the

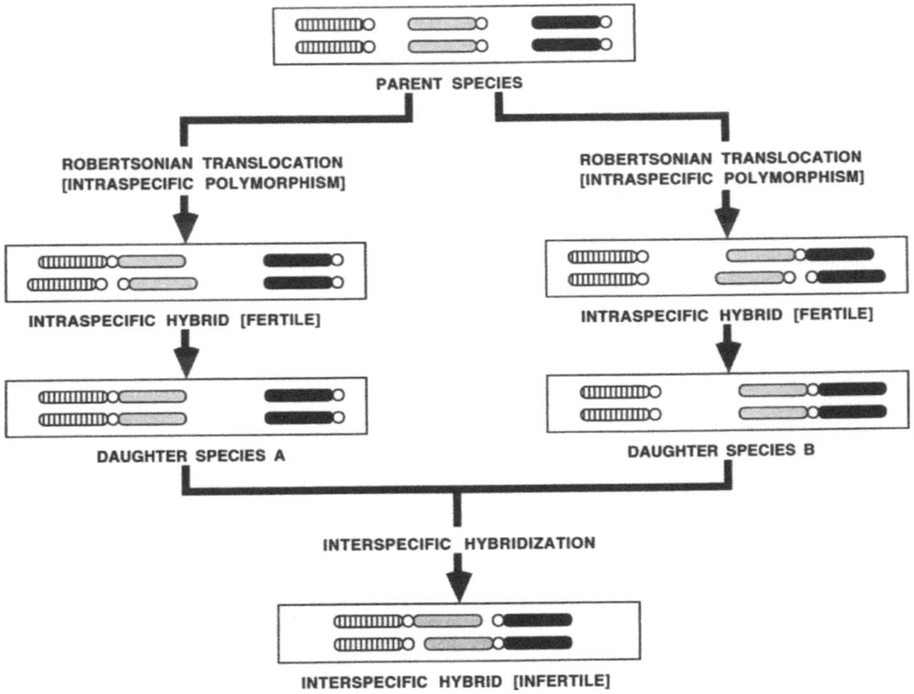

Figure 6: Schematic illustration of divergent chromosomal change likely to lead to a reproductive barrier between species. (For simplicity, only three chromosome pairs are shown for the parent species.) Robertsonian translocation ("fusion") generating a two-armed metacentric chromosome from two one-armed acrocentrics has been a common feature in the chromosomal evolution of Lepilemur. Initially, such translocations occur as a polymorphism in any given population, but may then become fixed as the typical condition. Two different Robertsonian translocations involving the same pair of original acrocentric chromosomes (the grey chromosomes of the parent species) may become fixed in two separate populations. If this occurs, any hybridization between the two populations will lead to formation of a multivalent chain when the chromosomes are paired at meiosis. In this schematic illustration, a chain of four chromosomes would be formed in the hybrid, probably leading to infertility.

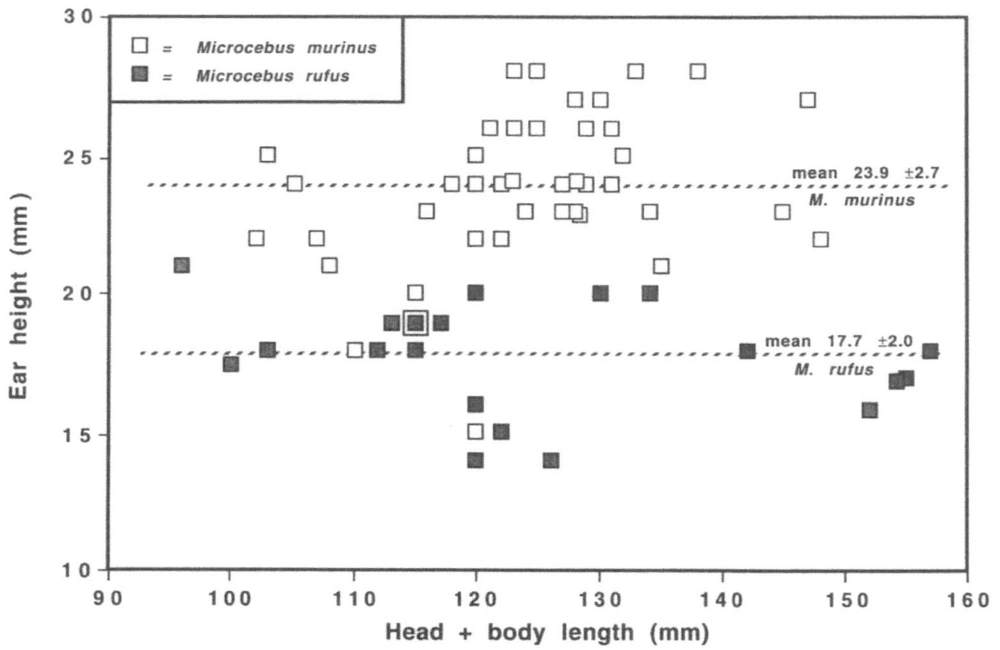

Figure 7: Plot of the ear pinna height against head-and-body length for M. murinus and M. rufus (mostly based on original data transcribed from collector's labels in museum collections). On average, the ears are 35% higher in M. murinus and the two distributions barely overlap. The average size of the ear does not change with increasing head-and-body length, so this distinction between the two species appears to be independent of age. The outlined point indicates a brown mouse lemur found living sympatrically with grey mouse lemurs near Morondava, west Madagascar. (Specimen collected by J.-J. Petter in 1971.)

eastern and western regions of Madagascar, which provided the basis for the first inferred division between Lepilemur species. Secondly, zonation is more marked, and speciation is hence more active, in the western region than in the eastern region. (In the case of Lepilemur, all inferred speciation events following the initial division into two species are confined to the western region.) Thirdly, the proposed geographical subdivisions between species in the western region are broadly similar.

This relatively close correspondence between the reconstructed sequence of speciation among sportive lemurs and the theoretical inference of speciation patterns based on geographical zonation in Madagascar (see Fig. 4) permits two more general conclusions. In itself, the correspondence between the specific model of Lepilemur speciation and the more general model generates valuable reciprocal confirmation of the inferred patterns. In addition, however, the close agreement between the apparent geographical ranges of the individual Lepilemur species and distinct climatic zones separated by major obstacles (e.g. rivers) suggests that speciation in this case may not be a result of chromosomal changes alone. Although it is in principle possible that chance fixation of different Robertsonian translocations within a single population could lead to reproductive isolation of subpopulations, it seems more likely that some combination of geographical, climatic and chromosomal factors led to the development of species barriers in this case. One should therefore be wary of taking the neatly analyzed case of Lepilemur as providing support for any concept of "chromosomal speciation" operating in isolation.

Lesser Mouse Lemurs

Another instructive practical illustration of speciation among lemurs is provided by the lesser mouse lemurs. Following the classical paper by Schwarz (1931), all lesser mouse lemurs were allocated to the single species Microcebus murinus in many subsequent texts (e.g. Petter, 1962; Napier and Napier, 1967; Wolfheim, 1983). It was, however, recognized from the outset that two fairly distinct forms can be separated, at least at the subspecific level: a grey long-eared form typically inhabiting dry forest areas (M. m. murinus) and a brown short-eared form typically inhabiting rainforest areas (M. m. rufus). It was briefly noted by Martin (1972) that new information on geographical distribution and morphology indicated that these forms "may well be separate species", and there has since been an increasing tendency to recognize them as such. Petter et al. (1977) were perhaps the first authors thereafter to advocate formal recognition of the separate species M. murinus and M. rufus and this distinction was consolidated in a subsequent paper on prosimian classification by Petter and Petter-Rousseaux (1979). That was followed first by Tattersall (1982), who cited personal communications from the present author in recognizing the species distinction, and then by Harcourt and Thornback (1990). Tattersall jusifiably described the nomenclature of the Microcebus species as "horribly entangled" but concluded that the names M. murinus and M. rufus are, indeed, appropriate for the main populations of the grey and brown mouse lemurs, respectively.

In addition to the main populations of western grey and eastern brown lesser mouse lemurs, Petter (1962) noted that a rufous lesser mouse lemur occurs sympatrically with the western grey form in certain areas. This relatively rare western rufous form generally resembles the eastern brown mouse lemur, but Petter stated that the ears of the western rufous form are larger. Originally, Petter (1962) specifically mentioned only Ankarafantsika with respect to the occurrence of a brown lesser mouse lemur in the west, but subsequently Petter et al. (1977) also reported its presence in Morondava. Tattersall (1982) also noted that a brown lesser mouse lemur is found in some areas of western Madagascar, living sympatrically with the grey lesser mouse lemur. In fact, his map (Fig

3.33) shows the brown form as occurring only in the region of Morondava in the west. He reserved judgement on the taxonomic status of this western brown form. Harcourt and Thornback (1990) recognised the occurrence of western brown mouse lemurs at Ankarafantsika and near Morondava. Quite recently, there has been a report of the likely co-existence of two forms of the lesser mouse lemur in a western forest area at Bemaraha, to the north of Morondava, although only one form was actually sighted (Thalmann, 1992; Rakotoarison et al., 1993). A photograph taken by T. Mutschler (pers. comm.) of one lesser mouse lemur in the region can be identified on grounds of relative height of the ear as being similar to M. rufus rather than to M. murinus. Such evidence is insufficient to assess the degree of difference between the eastern and western brown lesser mouse lemurs, but it seems clear that both are quite distinct from the western grey lesser mouse lemur. As a minimum, therefore, one can recognise the existence of a grey species (M. murinus) in the west and a brown species (M. rufus) occurring mainly in the east but also present at certain localities in the west.

Although the basic distinction between western grey and eastern brown lesser mouse lemurs has been widely accepted and incorporated into several published distribution maps (e.g. Petter et al., 1977; Tattersall, 1982), few quantitative data on the distinctions between them have been published. Field measurements cited by Tattersall give average ear heights of 24.8 mm for M. murinus (n = 29) and of 17.5 mm for M. rufus (n = 8). These figures closely match figures of 23.9 ±2.7 mm (n = 42) for M. murinus and of 17.7 ±2.0 mm (n = 20) for M. rufus compiled from various collections by the author. As can be seen from Fig. 7, there is relatively little overlap between measurements for the two species and their different values for average ear height remain relatively constant regardless of body size (as indicated by head-and-body length). The ear height for one western brown mouse lemur, collected by Petter near Morondava, was found to be 19 mm, very close to the average for eastern brown mouse lemurs and considerably less than the average for grey mouse lemurs. Apart from the pelage color, however, this difference in ear height is the only obvious external feature that separates the two species. There are no significant differences in head-and-body length or hindfoot length, although brown lesser mouse lemurs tend to have a shorter tail (not significant) and a somewhat lower body weight.

The author's attention was first drawn to possible differences in skull morphology by an early publication by Forsyth Major (1894) dealing with cheirogaleids, including Microcebus. In particular, the figures in that paper suggested that the distance between the temporal lines on the braincase may differ markedly between M. murinus and M. rufus (Fig. 8). These lines indicate the dorsal boundary of the jaw muscles attached to the braincase and thus provide an indication of the degree of development of the temporomandibular musculature.The distance between the lines is reduced with increased muscle development. This potential distinction was examined by averaging three measurements of the distance between the temporal lines (anterior: just behind the orbits; central; posterior: at the junction between the parietal bones and the occipital region) for a series of skulls. It is to be expected that the relative degree of development of the jaw musculature might increase with increasing body size, thus leading to a progressive decrease in the distance between the temporal lines. The average distance between the temporal lines was therefore plotted against skull length for M. murinus and M. rufus and revealed a marked separation between the two species (Fig. 9, A). The average distance between the temporal lines is clearly smaller in M. murinus at any given skull size, indicating greater development of the temporomandibular musculature in this species. When the trend toward decrease in the distance between the temporal lines with increasing skull size is taken into account, virtually no overlap is found between the two distributions. Two skulls of brown lesser mouse lemurs from Ankarafantsika in

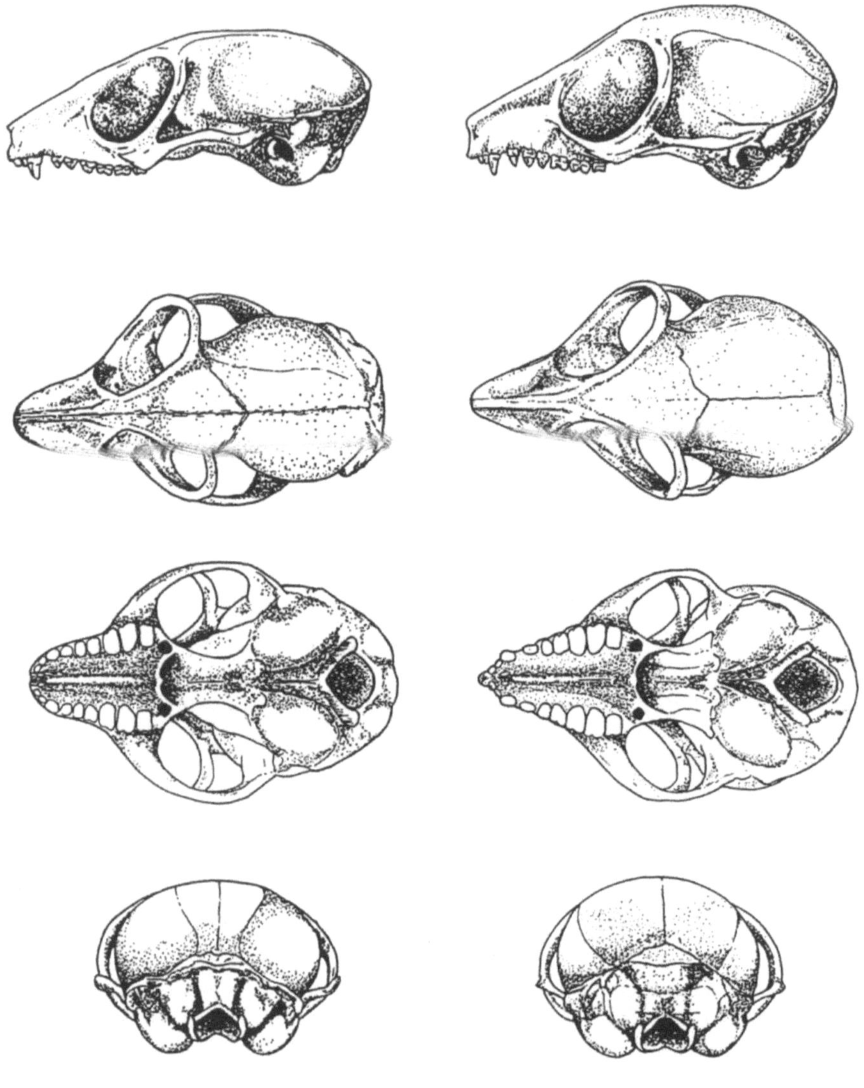

Figure 8: Line drawings of the skulls of M. murinus (left) and M. rufus (right). In lateral view, the braincase of M. murinus can be seen to be lower and less rounded than that of M. rufus. In dorsal view, the temporal lines on the braincase (indicating the upper boundary of the temporomandibular musculature) are much farther apart in M. rufus than in M. murinus. There is also a subtle difference in the contours of the orbits. In ventral view, it can be seen that the palate is markedly longer in M. rufus than in M. murinus. M. rufus also has a small rostrum projecting in front of the upper incisors, whereas M. murinus does not. In rear view, the difference in shape of the braincase and in the distance between the temporal lines is also clearly apparent. (Drawings by Anne-Elise Martin based on skulls collected by J-J. Petter.)

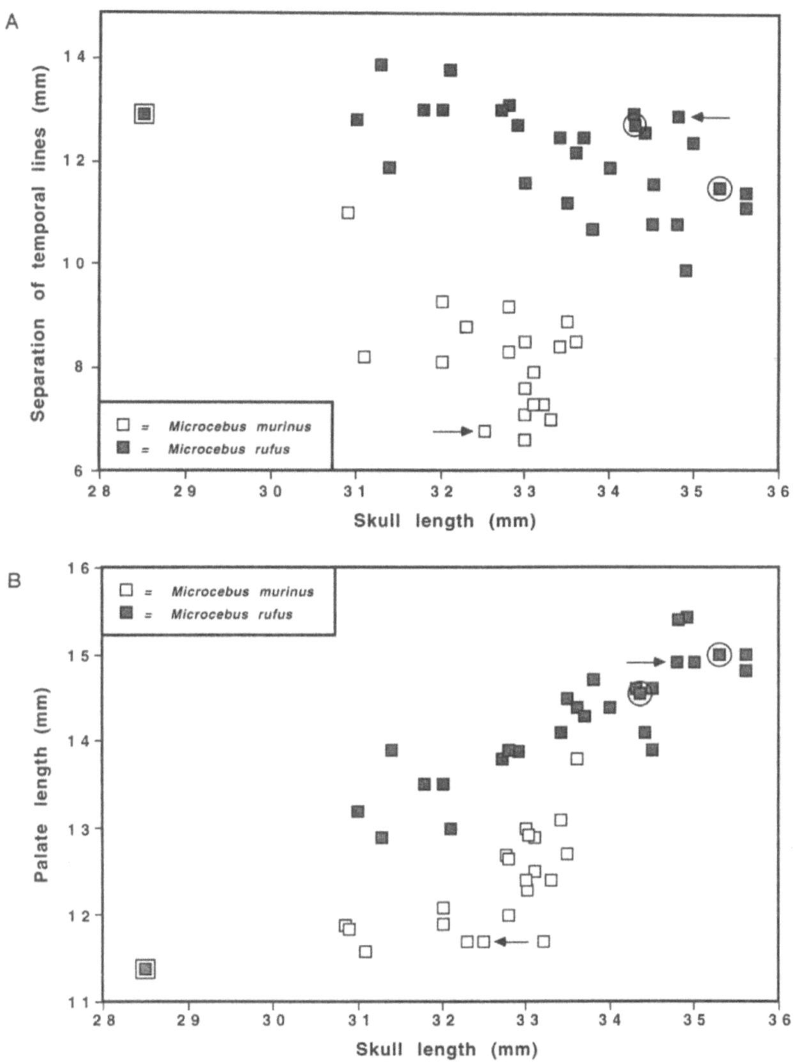

Figure 9: Plots of the average distance between the temporal lines on the skull (A) and of palate length (B) against skull length for M. murinus and M. rufus (data collected by the author from skulls in various collections). In both cases, arrows indicate the type specimens of M. murinus and M. rufus and outlined points indicate brown mouse lemurs found sympatrically with grey mouse lemurs in western Madagascar (square outline = Morondava; circular outline = Ankarafantsika). In both species, the degree of separation between the temporal lines decreases with increasing skull length, reflecting increasing development of the temporal musculature, whereas the length of the palate simply increases with increasing skull length as a reflection of overall growth of the skull. When these trends are taken into account, there is very little overlap between the two species.

northwestern Madagascar (collected by Petter) clearly fall within the distribution of points for M. rufus. The skull of a brown lesser mouse lemur from the Morondava region in western Madagascar (also collected by Petter) is very small, so it is difficult to dtermine whether it fits the size-related trend for eastern M. rufus. However, the distance between the temporal lines is very large and therefore most like that of typical M. rufus.

Comparison of skulls of the two main species also revealed that there is a marked difference in palate length (Fig. 8). In this case, the expectation is that palate length should increase in a consistent fashion with skull length. A plot of palate length against skull length for M. murinus and M. rufus (Fig. 9, B) again reveals a marked difference between the two species, with little overlap when the scaling trend is taken into account. In the case of the palate, too, the two skulls of brown lesser mouse lemurs from Ankarafantsika clearly fall within the distribution of points for M. rufus. As before, the very small skull of the brown lesser mouse lemur from the Morondava region in western Madagascar seems to be most like that of typical M. rufus. Interestingly, the palate is longer in M. rufus than in M. murinus. A plot of palate length against skull length for insectivores, tree-shrews and primates (Martin, 1990) indicates that a relatively long palate seems to be typical of species that feed extensively on insects. In view of this, it seems reasonable to conclude that M. rufus (with its relatively long palate) is adapted for greater insectivory, whereas M. murinus (with its relatively powerful jaw musculature) is adapted for greater herbivory.

There are also more subtle morphological differences between the skulls of M. murinus and M. rufus (Fig. 8). For instance, the snout of M. rufus has a small rostrum projecting in front of the upper incisors, a feature lacking from M. murinus. The cheek teeth of M. rufus tend to have relatively high, sharp cusps, whereas the cusps on the cheek teeth of M. murinus tend to be relatively low and rounded. This dental difference, which is difficult to quantitify, further suggests that M. rufus is adapted for greater insectivory while M. murinus is adapted for greater herbivory. The profile of the braincase also differs between the two species, perhaps in connection with the differential degree of development of the temporomandibular musculature. In M. murinus the braincase is flatter and less rounded than in M. rufus.

All of these features taken together document a fairly sharp distinction between the grey and brown lesser mouse lemurs, and the brown specimens occasionally found in the west generally resemble the eastern form. The sympatric occurrence of brown and grey forms in western Madagascar, combined with the close proximity between the two forms in the Fort Dauphin region of southern Madagascar (Martin, 1972b) provides further confirmation that at least two species should be recognized. Using a combination of pelage coloration, ear height and skull dimensions, it is possible to identify individual specimens unequivocally as either M. murinus or M. rufus, regardless of their geographical origin. Specimens identified by the author in this way have been summarized in a distribution map (Fig. 10). These observations broadly confirm previously published distribution maps, but also clearly indicate that grey and brown mouse lemurs occur sympatrically (or at least in close proximity) in a minimum of three regions of western Madagascar: near Ankarafantsika, near Morondava and at Namoroka (south of Soalala). Two specimens in the Paris collection that were collected at Namoroka (Nos. 1932.3370 and 1932.3373) are undoubtedly of the rufus type, contrasting with a typical murinus specimen in the London collection that was also collected near Soalala but closer to the coast. This increases the documented occurrence of brown lesser mouse lemurs in western Madagascar. Overall, the distribution map in Fig. 10 once again emphasizes a primary division between the western and eastern domains, accompanied by greater complexity in the western region.

A recent field study of lesser mouse lemurs in a forest area to the north-east of Morondava in western Madagascar has yielded important new information on the taxonomic status of the western brown mouse lemur (Schmid & Kappeler, in press). In that area, two different types of mouse lemur (one rufous and one grey) were found living in direct sympatry and a series of measurements was conducted on a large sample of trapped individuals. The body weights of the grey mouse lemurs were found to be significantly greater than those of the rufous mouse lemurs (average values: 58g and 31g, respectively). The rufous form was found to have shorter ears (average values: 17.0 mm and 22.4 mm, respectively) and a relatively longer tail. The average ear size given for the rufous form is quite close to the overall average of 17.7 mm found by the author for M. rufus while the average ear size given for the grey form is quite close to the overall average of 23.9 mm found for M. murinus . On the basis of their new evidence, Schmid & Kappeler identify the western brown mouse lemur as a third species, Microcebus myoxinus.

So far, no chromosomal differences have been found between the two main species M. murinus and M. rufus. Indeed, in terms of gross morphology and banding patterns of the chromosomes, it is not even possible to separate the genera Microcebus and Cheirogaleus. Recent studies of highly repetitive DNA have now revealed distinctions between Cheirogaleus major, M. murinus and Microcebus (Mirza) coquereli, but it remains to be tested whether M. murinus and M.rufus can be differentiated on this basis. In any case, the absence of gross chromosomal differences between the lesser mouse lemurs shows that such differentiation does not always accompany speciation.

Comparison of these two cases of speciation, sportive lemurs and lesser mouse lemurs, yields an additional point that is worth noting. In the case of the sportive lemurs, there is only limited morphological evidence on which to base any distinction between species. The main impetus towards modern recognition of several species came from chromosomal studies which yielded a strong indication of separation among distinct populations. Perhaps for this reason, morphologists have generally been slow to accept the proposed distinctions between species. With lesser mouse lemurs, by contrast, chromosomal evidence supporting the proposed distinction between species is completely lacking, and it is morphological evidence that strongly indicates a separation between species in this case. Here, it is karyologists who have been reluctant to accept a species distinction based purely on morphological evidence. The lesson to be learned is that the study of species differences should be a multidisciplinary undertaking. Those concerned with taxonomy, be they morphologists, ethologists, karyologists or molecular biologists, should make a determined effort to understand and apply evidence from other disciplines. It is, for example, clear that vocalizations can provide an extremely useful tool for identifying distinctions between species (see chapters by Bearder, Zimmermann, and Stanger, this volume).

CONSERVATION BIOLOGY OF LEMURS

As is true of primates generally, there are acute conservation issues concerning prosimians, particularly the Malagasy lemurs. On the positive side, the rediscovery of Allocebus trichotis (Meier and Albignac, 1991) and several discoveries of previously unrecognized lemur species have been reported in recent years. Propithecus tattersalli (Simons, 1988) and Hapalemur aureus (Meier et al., 1987; Rumpler et al., 1988) have been added as new species, and a population of Avahi of unknown taxonomic status has been discovered outside the recognized range of distribution in western Madagascar (Thalmann, 1992; Thalmann et al., 1992; Rakotoarison et al., 1993). It has also become increasingly obvious that several additional Galago species must now be recognised (e.g.

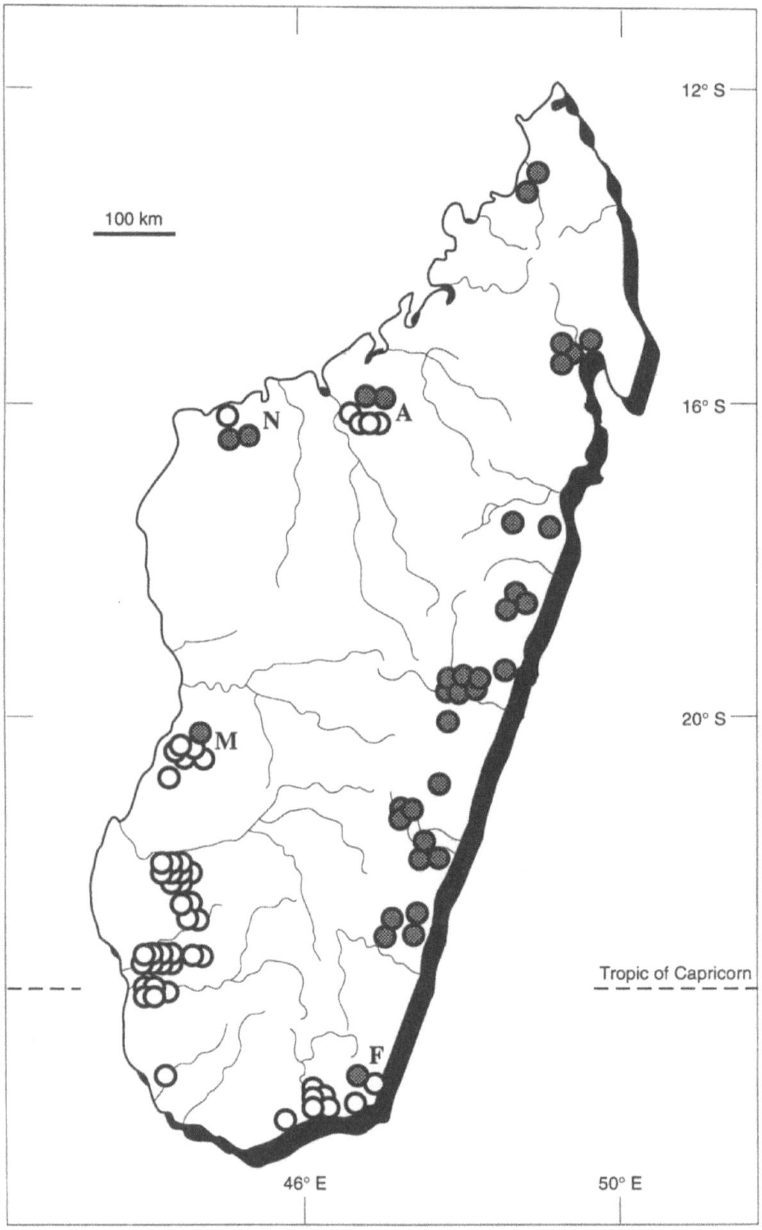

Figure 10: Map of Madagascar showing the geographical distributions of M. murinus (white circles) and M. rufus (shaded circles), based on museum specimens examined by the author. Specimens of M. rufus are mainly confined to the east, whereas specimens of M. murinus are restricted to the northwest, the west and the south. A limited number of specimens resembling east coast brown lemurs have been found in the west (M = Morondava) and in the northwest (A = Ankarafantsika; N = Namoroka). At Mandena, just north of Fort Dauphin (F), brown and grey mouse lemurs are found in close proximity without evidence of interbreeding.

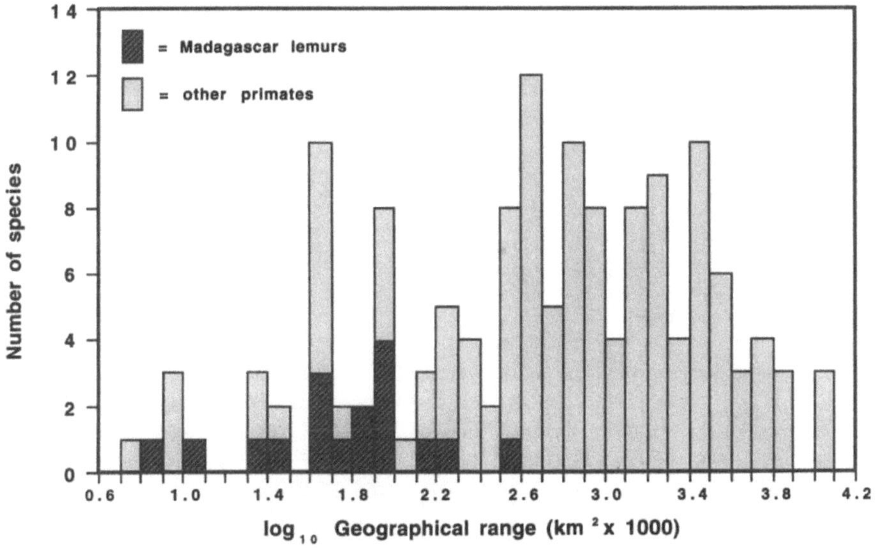

Figure 11: Histogram showing sizes of geographical ranges for individual primate species (n = 144; data from Wolfheim, 1983). Logarithmic values for the geographical ranges have been used simply to compress the distribution. It can be seen that range sizes for all of the Malagasy lemur species fall at the lower end of the distribution, along with some of the most vulnerable primate species from other groups.

see chapters by Bearder and Scheffrahn, this volume). Finally, a fifth tarsier species has now been identified:Tarsius dianae (Niemitz et al., 1991). Ironically, however, these new or rediscovered species are in many cases among those most threatened by extinction simply because of the small population sizes that delayed their discovery.

The lemurs of Madagascar are particularly vulnerable to extinction. Their populations have declined because of a combination of natural climatic change and habitat loss due to human activities (Tattersall, 1993), but this is true of many primate species. In the case of the lemurs, the threat is particularly severe because their natural geographical ranges are so small in comparison to those of most other primates (Fig. 11). As a side-effect of the explosive speciation of lemurs on Madagascar, small populations of species are commonly confined to the individual subzones shown in Fig. 4 and this explains why they are especially threatened by extinction. It is therefore an objective fact that conservation of the lemurs must be regarded as a priority for urgent action.

The lemurs hence provide a prime example of the r™le to be played by the emerging discipline of conservation biology. This synthetic discipline depends not only on field undertakings ranging from general surveys through studies of behavioral ecology and on to local education programs, but also on captive breeding and on the integration of baseline information extending from broad evolutionary principles down to the practicalities of genetic monitoring. In this context, it is very encouraging that a Taxonomic Advisory Group has now been established specifically for prosimian primates (Porton, this volume). Conservation biologists will hopefully help us to prevent the prosimians from passing from the obscurity of their original nocturnal habits to the final obscurity of extinction.

ACKNOWLEGMENTS

Thanks are due to the following people for their assistance in providing access to preserved specimens of lesser mouse lemurs: Jean-Jacques Petter (then at the Muséum National d'Histoire Naturelle, Brunoy), Francis Petter (Mammalogie, Muséum National d'Histoire Naturelle, Paris), Pru Napier (then at the British Museum of Natural History, London). I am grateful to Urs Thalmann for his help in providing a computerized baseline map of Madagascar and for giving advice on the estimation of geographical areas within Madagascar. Thanks are also due to Lucrezia Bieler-Beerli for preparing the outline illustrations of living primates in Fig. 3 and to Anne-Elise Martin for the drawings of mouse lemur skulls in Fig. 8. The following people provided helpful advice and discussion at various stages: Simon Bearder, Jean-Jacques Petter, Yves Rumpler, Ian Tattersall, Urs Thalmann. Last but not least, I would like to thank the co-organizers of a most rewarding conference, K. Izard and L. Alterman, for inviting me to take part. Their invitation stimulated me to pick up loose threads spun during my early visits to Madagascar in 1968 and 1972 and to interweave them with certain recent developments.

REFERENCES

Bearder, S.K. (1987). Lorises, bushbabies, and tarsiers: Diverse societies in solitary foragers. In B.B. Smuts, D. Cheney, R.M. Seyfarth, R. Wrangham, & T. Struhsaker (Eds.), *Primate Societies* (pp. 11-24). Chicago: Chicago University Press.

Bearder, S.K. & Martin, R.D. (1980). *Acacia* gum and its use by bushbabies, *Galago senegalensis* (Primates, Lorisidae). *International Journal of Primatology, 1*, 103-128.

Bowmaker, J.K. (1991). Visual pigments and colour vision in primates. In A. Valberg & B.B. Lee (Eds.), *From Pigments to Perception* (pp. 1-9). New York: Plenum Press.

Cartmill, M. (1972). Arboreal adaptations and the origin of the order Primates. In R.H. Tuttle (Ed.). *The Functional and Evolutionary Biology of Primates* (pp. 97-122). Chicago: Aldine-Atherton.

Cartmill, M. (1974a). Rethinking primate origins. *Science*, 184, 436-443.

Cartmill, M. (1974b). *Daubentonia, Dactylopsila* and klinorhynchy.In R.D. Martin, G.A. Doyle, & A.C. Walker (Eds.), *Prosimian Biology* (pp. 655-670). London: Duckworth.

Cartmill, M. (1975). Strepsirhine basicranial structures and the affinities of the Cheirogaleidae. In W.P. Luckett & F.S. Szalay (Eds.), *Phylogeny of the Primates* (pp. 313-354). New York: Plenum Press.

Charles-Dominique, P. (1977). *Ecology and Behaviour of Nocturnal Primates.* (Translated by R.D. Martin), London: Duckworth.

Charles-Dominique, P., Cooper, H.M., Hladik, A., Hladik, C.M., Pagès, E., Pariente, G.F., Petter-Rousseaux, A., Petter, J.-J., & Schilling, A. (1980). *Nocturnal Malagasy Primates.* New York: Academic Press.

Charles-Dominique, P. (1978). Solitary and gregarious prosimians: evolution of social structure in prosimians. In D.J. Chivers & K.A. Joysey (Eds.), *Recent Advances in Primatology. vol. 3: Evolution* (pp. 139-149). London: Academic Press.

Crook, J.H. & Gartlan, J.S. (1966). On the evolution of primate societies. *Nature, London, 210*, 1200-1203.

Djihan, P. & Green, H. (1991). Involucrin gene of tarsioids and other primates: Alternatives in evolution of the segment of repeats. *Proceedings of the National Academy of Sciences, USA, 88*, 5321-5325.

Dutrillaux, B. & Rumpler, Y. (1988). Absence of chromosomal similarities between tarsiers (*Tarsius syrichta*) and other primates. *Folia Primatologica*, 50, 130-133.

Engqvist, A. & Richard, A.F. (1992). Diet as a possible determinant of cathemeral activity in primates. *Folia Primatologica, 57*, 169-172.

Forsyth Major, C.I. (1894). Über die malagassischen Lemuriden-Gattungen *Microcebus, Opolemur*, und *Chirogale. Novitates Zoologiae, 1*, 1-39.

Groves, C.P. (1974) .Taxonomy and phylogeny of prosimians. In R.D. Martin, G.A. Doyle, & A.C. Walker (Eds.), *Prosimian Biology* (pp. 449-473). London: Duckworth.

Harcourt, C. & Thornback, J. (1990). *Lemurs of Madagascar and the Comoros. The IUCN Red Data Book.* Gland, Switzerland: IUCN.

Ishak, B., Warter, S., Dutrillaux, B., & Rumpler, Y. (1992). Chromosomal rearrangements and speciation of sportive lemurs *(Lepilemur* species). *Folia Primatologica, 58*, 121-130.

Jacobs, G.H. & Deegan, J.F. (1993). Photopigments underlying color vision in ringtail lemurs (*Lemur catta*) and brown lemurs (*Eulemur fulvus*). *American Journal of Primatology, 30*, 243-256.

Koop, B.F., Siemieniak, D., Slightom, J.L., Goodman, M., Dunbar, J., Wright, P.C., & Simons, E.L. (1989). Tarsius δ- and β-globin genes: conversions, evolution, and systematic implications. *Journal of Biological Chemistry, 264*, 68-79.

Kubzdela, K.S., Richard, A.F., & Pereira, M.E. (1992). Social relations in semi-free-ranging sifakas (*Propithecus verreauxi coquereli*) and the question of female dominance. *American Journal of Primatology, 28*, 139-145.

Martin, R.D. (1972a). Adaptive radiation and behaviour of the Malagasy lemurs. *Philosophical Transactions of the Royal Society (London) B, 264*, 295-352.

Martin, R.D. (1972b). A preliminary field-study of the Lesser Mouse Lemur *(Microcebus murinus*, J.F. Miller 1777). *Zeitschrift für Tierpsychologie, Beiheft, 9*, 43-89.

Martin, R.D. (1979). Phylogenetic aspects of prosimian behavior. In G.A. Doyle & R.D. Martin (Eds.), *The Study of Prosimian Behavior* (pp. 45-77). New York: Academic Press.

Martin, R.D. (1990). *Primate Origins and Evolution: A Phylogenetic Reconstruction.* London/New Jersey: Chapman Hall/Princeton University Press.

Martin, R.D. (1993). Primate origins: plugging the gaps. *Nature, London, 363*, 223-234.

Meier, B. & Albignac, R. (1991). Rediscovery of *Allocebus trichotis* Günther 1875 (Primates) in Northeast Madagascar. *Folia Primatologica, 56*, 57-63.

Meier, B., Albignac, R., Peyriéras, A., Rumpler, Y. & Wright, P. (1987). A new species of *Hapalemur* (Primates) from South East Madagascar. *Folia Primatologica, 48*, 211-215.

Millot, J. (1952). La faune malgache et le mythe Gondwanien. *Mémoires de l'Institut Scientifique de Madagascar (série* A), 7, 1-36.

Napier, J.R. & Napier, P.H. (1967). *A Handbook of Living Primates.* London: Academic Press.

Niemitz, C., Nietsch, A., Warter, S., & Rumpler, Y. (1991). *Tarsius dianae*: A new primate species from Central Sulawesi (Indonesia). *Folia Primatologica, 56,* 105-116.

Novacek, M.J. & Wyss, A.R. (1988). Higher-level relationships of the recent eutherian orders: Morphological evidence. *Cladistics, 2,* 257-287.

Petter, J.-J. (1962). Recherches sur l'écologie et l'éthologie des lémuriens malgaches. *Mémoires du Muséum National d'Histoire Naturelle, série A (Zoologie), 27,* 1-146.

Petter, J.-J., Albignac, R., & Rumpler, Y. (1977). Mammifères lémuriens (Primates prosimiens). *Faune de Madagascar, 44,* 1-513.

Petter, J.J. & Petter-Rousseaux, A. (1979). Classification of the prosimians. In G.A. Doyle and R.D. Martin (Eds.), *The Study of Prosimian Behavior* (pp. 1-44.) New York: Academic Press.

Poorman, P.A., Cartmill, M., & MacPhee, R.D.E. (1985). The G-banded karyotype of *Tarsius bancanus* and its implications for primate phylogeny. *American Journal of physical Anthropology, 66,* 215.

Rakotoarison, N., Mutschler, T., & Thalmann, U. (1993). Lemurs in Bemaraha (World Heritage Landscape, Western Madagascar). *Oryx, 27,* 35-40.

Rumpler, Y. (1975). The significance of chromosomal studies in the systematics of the Malagasy lemurs. In I. Tattersall and R.W. Sussman (Eds.), *Lemur Biology* (pp. 25-40). New York: Plenum Press.

Rumpler, Y. & Albignac, R. (1978). Chromosome studies of the lepilemur, an endemic Malagasy genus of lemurs: Contribution of the cytogenetics to their taxonomy. *Journal of Human Evolution, 7,* 191-196.

Rumpler, Y., Meier, B., & Rakotosamimanana, B. (1988). La découverte d'un nouveau primate. *La Recherche,* 19, 1092-1094.

Sandwith, M.H. (1858). Sur les moeurs de l'aye-aye de Madagascar. *Annales des Sciences naturelles, 4ème série, 10,* 377-381.

Schmid, J. & Kappeler, P.M. (in press). Sympatric mouse lemurs (*Microcebus* species) in western Madagascar. *Folia Primatologica.*

Schwartz, J.H., Tattersall, I. & Eldredge, N. (1978). Phylogeny and classification of the primates revisited. *Yearbook of Physical Anthropology, 21,* 95-133.

Schwarz, E. (1931). A revision of the genera and species of Madagascar Lemuridae. *Proceedings of the Zoological Society of London, 1931,* 399-428.

Simons, E.L. (1988). A new species of *Propithecus* (Primates) from Northeast Madagascar. *Folia Primatologica, 50,* 143-151.

Szalay, F.S. & Delson, E. (1979). *Evolutionary History of the Primates.* New York: Academic Press.

Szalay, F.S. & Katz, C.C. (1973). Phylogeny of lemurs, galagos and lorises. *Folia Primatologica, 19,* 88-103.

Tattersall, I. (1982). *The Primates of Madagascar.* New York: Columbia University Press.

Tattersall, I. (1988). Cathemeral activity in primates: A definition. *Folia Primatologica, 49,* 200-202.

Tattersall, I. (1993). Madagascar's lemurs. *Scientific American, 268* (1), 110-117.

Tattersall, I. & Schwartz, J.H. (1974). Craniodental morphology and the systematics of the Malagasy lemurs (Primates, Prosimii). *Anthropological Papers of the American Museum of Natural History, 52,* 139-192.

Tattersall, I. & Schwartz, J.H. (1975). Relationships among the Malagasy lemurs: The craniodental evidence. In W.P.Luckett & F.S. Szalay (Eds.) *Phylogeny of the Primates* (pp. 299-312). New York: Plenum Press.

Thalmann, U. (1992). Lemur survey in Bemaraha, Western Madagascar. *Madagascar Environment Newsletter,* 2, 2.

Thalmann, U. & Rakotoarison,N. (in press). Distribution of lemurs in central western Madagascar, with a regional distribution hypothesis. *Folia Primatologica.*

Thalmann, U., Geissmann, T., Mutschler, T., & Rakotoarison, N. (1992). Field observations on a recently discovered population of *Avahi* in western Madagascar. In J. Ruffié, J.-J. Petter, Y. Rumpler, & N. Herrenschmidt (Eds.), *Abstracts of the XIVth Congress of the International Primatological Society* (pp. 240). Strasbourg: International Primatological Society.

Wakefield, N.A. (1970). Notes on the glider-possum, *Petaurus australis* (Phalangeridae, Marsupialia). *Victorian Naturalist, 87,* 221-236.

Wolfheim, J.H. (1983). *Primates of the World: Distribution, Abundance, and Conservation.* Seattle: University of Washington Press.

Yoder, A.D. (1992). The applications and limitations of ontogenetic comparisons for phylogeny reconstruction: the case of the strepsirhine internal carotid artery. *Journal of Human Evolution, 23,* 183-195.

Yoder, A.D. (1994). Relative position of the Cheirogaleidae in strepsirhine phylogeny: A comparison of morphological and molecular methods and results. *American Journal of physical Anthropology, 94,* 25-46.

Participants of an International Conference on

CREATURES OF THE DARK: THE NOCTURNAL PROSIMIANS
held
June 9 to 12, 1993
in
Durham, North Carolina

INDEX

FSC
www.fsc.org
MIX
Papier aus verantwortungsvollen Quellen
Paper from responsible sources
FSC® C105338

If you have any concerns about our products,
you can contact us on
ProductSafety@springernature.com

In case Publisher is established outside the EU,
the EU authorized representative is:
Springer Nature Customer Service Center GmbH
Europaplatz 3, 69115 Heidelberg, Germany

Printed by Libri Plureos GmbH
in Hamburg, Germany